W9-DHM-743

Backbone of the Americas: Shallow Subduction, Plateau Uplift, and Ridge and Terrane Collision

edited by

Suzanne Mahlburg Kay
Department of Earth and Atmospheric Sciences
Snee Hall
Cornell University
Ithaca, New York 14853
USA

Víctor A. Ramos
Laboratorio de Tectónica Andina
Facultad de Ciencias Exactas y Naturales
Universidad de Buenos Aires–
Consejo Nacional de Investigaciones Científicas y Técnicas
Ciudad Universitaria Pabellón 2
Buenos Aires
Argentina

William R. Dickinson
Department of Geosciences
University of Arizona
Tucson, Arizona 85721
USA

Memoir 204

3300 Penrose Place, P.O. Box 9140 ▪ Boulder, Colorado 80301-9140, USA

2009

Published by The Geological Society of America, Inc.
3300 Penrose Place, P.O. Box 9140, Boulder, Colorado 80301-9140, USA
www.geosociety.org

Printed in U.S.A.

GSA Books Science Editor: Marion E. Bickford and Donald I. Siegel

Library of Congress Cataloging-in-Publication Data

Backbone of the Americas : shallow subduction, plateau uplift, and ridge and terrane collision / edited by Suzanne Mahlburg Kay, Víctor A. Ramos, William R. Dickinson.
p. cm. — (Memoir ; 204)
Includes bibliographical references.
ISBN 978-0-8137-1204-8 (cloth)
1. Orogeny—America. 2. Geology, Structural—America. I. Kay, Suzanne Mahlburg, 1947–. II. Ramos, Víctor A. III. Dickinson, William R.

QE621.5.A45B33 2009
551.8′2091812—dc22

2009009035

Cover: SRTM (Satellite Radar Topographic Mission) image showing the central and southern part of the Central Andean Puna-Altiplano Plateau. From left to right are the Pacific Ocean, the forearc valleys and ranges, the high plateau (white and gray tones) with high volcanic peaks and giant ignimbrite calderas (largely red and purple tones), and the foreland thrust belts (green and brown tones). See Chapter 11 ("Shallowing and steepening subduction zones, continental lithospheric loss, magmatism, and crustal flow under the Central Andean Altiplano-Puna Plateau") for discussion of these features.

This SRTM image was processed by Jay Hart of Trumansburg, New York. Elevation data were compiled into a single graphic from Costa Rica to the Falklands, with shadows thrown from the east and a palette designed to isolate local variation. This excerpt from *Cordillera* is used with permission from Hart (http://www.earthpattern.com).

10 9 8 7 6 5 4 3 2 1

Contents

Preface

This volume has its genesis in the "Backbone of the Americas: Patagonia to Alaska" conference that took place in the city of Mendoza, Argentina, from 2 to 7 April 2006. The conference was convened by the Asociación Geológica Argentina and the Geological Society of America in collaboration with the Sociedad Geológica de Chile and was the first meeting on this scale to be jointly convened by North and South American geologic societies. Some 400 participants from 20 countries along the western margins of the Americas and elsewhere participated. The purpose of the meeting was to explore common scientific themes affecting the evolution of the Cordilleras of the Americas from a multidisciplinary international perspective. Major themes included collision of active and aseismic oceanic ridges, shallowing and steepening subduction zones, plateau and orogenic uplift, and terrane collision. This volume presents a selection of papers by the invited plenary speakers and other represenative presentations of the symposium.

The volume begins with overview papers on the Precambrian to Holocene evolution of the North, Central, and South American Cordilleras and a review of marine tectonic processes along the margin. These are followed by a series of papers highlighting distinctive processes and tectonic regions along the margin. The discussions begin in North America, proceed through Central America, and progress into South America. The presentations here are complemented by overviews in five field trip guides prepared in conjunction with the meeting to highlight the meeting themes in South America. These guides appear in a separate volume edited by Kay and Ramos, which is entitled *Field Trip Guides to the Backbone of the Americas in the Southern and Central Andes: Ridge Collision, Shallow Subduction, and Plateau Uplift*. The volume was published in 2008 as FLD013 in the Geological Society of America Field Guide series.

The first chapter here, by William R. Dickinson (University of Arizona), sets the stage in North America by presenting an overview of the evolution, setting, and global context of the 5000-km-long North American orogen. Unlike the Central and South American and Alaskan cordilleras, much of the margin of the North American cordillera is currently characterized by dextral transform faults. The history of this margin initiated with the breakup of the Rodinian supercontinent, which created a Neoproterozoic to early Paleozoic passive margin that was later modified by Permian-Triassic transform faults from California into Mexico, late Paleozoic to Mesozoic accretion of oceanic island arcs and subduction complexes, and subduction after the Triassic. Mesozoic to Cenozoic terrane accretion occurred as far south as northern South America, and subduction produced retroarc basins and Mesozoic to Cenozoic batholiths. The subduction era largely ended with the shallow subduction episode that produced the Laramide breakup of the Cordilleran foreland. The history ends with extension in the Basin and Range, and margin-parallel dextral strike slip that displaced elongate segments of the coastal region to the north.

The second chapter, by Víctor A. Ramos (Universidad de Buenos Aires, Argentina), presents a complementary overview of the evolution of the western South American margin, which is the longest continental margin currently dominated by subduction. The review shows that subduction began shortly after the Late Proterozoic breakup of the Rodinia supercontinent with terrane amalgamation occurring throughout the Neoproterozoic and Paleozoic all along the margin. Episodes of rifting, detachment, and re-accretion led to a complex system of exotic to para-autochthonous Neoproterozoic and Paleozoic terranes in the Central Andes. Generalized rifting associated with the Mesozoic breakup of the Pangean supercontinent and opening of the South Atlantic Ocean followed this period. The subsequent northeastward absolute motion of Gondwana created negative trench rollback leading to a subduction regime characterized by backarc extension that lasted until the late Early Cretaceous. The Early Cretaceous to middle Miocene history in the north is distinctive in that island-arc and oceanic plateau collisions along the Venezuelan to Ecuadorian margin occurred in association with Caribbean plate motion. The Cenozoic is marked by episodes of shallowing and steepening of the subducting plate leading to repeated broadening and narrowing of the volcanic arc interspersed with times of magmatic quiescence, development of foreland fold-thrust systems, and intermittent delamination of continental lithosphere and crust.

The third chapter, by Roland Von Huene (U.S. Geological Survey, University of California–Davis) and Cesar Ranero (Instituciô Catalana de Recerca i Estudis Avançats, Barcelona, Spain), highlights Neogene collision and deformation along the eastern Pacific convergent margin. All along the margin, features with high-standing relief on the subducting oceanic plate are shown to be colliding with continental slopes. Common processes include the uplift of the continental margin, accelerated seafloor and basal subduction-erosion, and shallow subduction zones. At a smaller scale, the subduction of seamounts and lesser ridges attached to the subducted plate produce temporary surface uplift. The collision and subduction of the Yakutat terrane in North America, the Cocos Ridge in Central America, the Nazca and Juan Férnandez Ridges in central South America, and the south Chile triple junction in southern South America are considered.

The fourth chapter, by Eugene Humphreys (University of Oregon), discusses the relation of flat subduction to magmatism and deformation in the western United States from a geophysical point of view. Humphreys argues that flat subduction of the Farallon plate during the Laramide orogeny was due to a combination of subduction of an oceanic plateau and suction in the mantle wedge as the shallowing slab approached the North American cratonic root. The shallow slab caused the lithosphere to dehydrate, and removal of the flat slab put the asthenosphere in contact with the basal lithosphere, leading to magmatism that was particularly intense where the basal lithosphere was fertile, below the present Basin and Range Province. Heating of the weakened lithosphere in that region led to convective instability. The strongest part of the lithosphere remained a plateau, and the weak park to the west collapsed to form the Basin and Range. The development of a transform plate boundary on the western margin led to the weakened westernmost zone being entrained within the Pacific plate.

The fifth chapter, by George Davis (University of Arizona) and Alex Bump (BP Exploration and Production Technology, Houston), analyzes the tectono-structural evolution of the Colorado Plateau. These authors point to the role of the Neoproterozoic, Paleozoic, and Mesozoic sedimentary rocks and the heterogeneous latest Paleoproterozoic and Mesoproterozoic crystalline basement, which they mechanically overlie in producing the dominant Late Cretaceous to early Tertiary Laramide basement-cored uplifts and associated monoclines. They argue that the geometries of these uplifts reflect "trishear" fault-propagation, that the uplifts largely formed as a response to tectonic inversion of Neoproterozoic normal-displacement shear zones, and that the principal NE-SW– and NW-SE–directed compressive stresses generating the uplifts were transmitted through the basement in response to plate-generated stresses. The deformation is attributed to the Colorado Plateau being caught in a bi-directional tectonic vise with the northwest side subjected to SE-directed compressive stress, and the base of the lithosphere subjected to viscous NE-directed undershearing produced by the shallowly subducting Farallon slab. The plateau is argued to have systematically deformed along the weakest deep basement links with variable reactivation of old shear zones leading to a disparate array of uplifts.

The sixth chapter, by Eric Erslev and Nicole Koenig (Colorado State University), focuses on the three-dimensional kinematics of Laramide Rocky Mountain deformation with an emphasis on contractional, basement-involved foreland deformation. Their perspective comes from the analyses of minor faults and geographic information system (GIS)–enhanced structural maps. They discuss current kinematic hypotheses for the Laramide orogeny, which include single-stage NE- to E-directed shortening, sequential multidirectional shortening, and transpressive deformation partitioned between NW-striking thrust and N-striking strike-slip faults. Proposed driving forces range from external stresses paralleling plate convergence to internal stresses related to gravitational collapse of the Cordilleran thrust belt. Laramide deformation is argued to show a primary external influence of ENE-directed shortening parallel to North American–Farallon plate convergence parameters, complexities related to localized preexisting weaknesses, and impingement by the adjoining Cordilleran thrust belt. Obliquities between convergence directions and the northern and southeastern boundaries of the Laramide province are argued to have created transpressive arrays of en echelon folds and arches without major through-going strike-slip faults.

The seventh chapter, by Michelangelo Martini (Universidad Nacional Autónoma de México [UNAM], Querétaro, Mexico), Luca Ferrari (UNAM, Querétaro), Margarita López-Martínez (Ensenada, Mexico), Mariano Cerca-Martínez (UNAM, Querétaro), Victor Valencia (University of Arizona), and Lina Serrano-Durán (UNAM, Querétaro), focuses on the causes of Cretaceous to Eocene magmatism and Laramide deformation in southwestern Mexico. They use new mapping and structural studies along with new $^{40}Ar/^{39}Ar$ and U-Pb zircon dates from a broad region in the central-eastern part of the widely discussed Guerrero terrane to argue against models that attribute the Laramide-age deformation in this region to accretion of allochthonous terranes. They favor a model in which most of the Guerrero terrane is composed of autochthonous or parautochthonous units that formed on the thinned continental margin of the North American plate, and in which the Mesozoic magmatic and sedimentary rocks record a long-lasting migrating, west-facing arc and related extensional backarc and forearc basins. The driving force to generate the Late Cretaceous–early Tertiary shortening and shearing of the southern margin of the North American plate remains poorly understood.

In the eighth chapter, Gerhard Wörner (Universität Göttingen, Germany), Russell Harmon (U.S. Army Research Laboratory), and Wencke Wegner (Universität Göttingen) use magmatic rocks from the, until recently, poorly studied Cordillera de Panama to put the evolution and closure of the Central American land bridge into a tectonic framework. They argue that the Isthmus of Panama reflects intermittent magmatism and oceanic plate interactions over the last 100 m.y., and they trace its early evolution using the chemistry of the magmatic rocks in the Caribbean large igneous province. Evidence for the initiation of arc activity comes from the 66–45 Ma Chagres igneous complex, which could constitute much of the upper crust in the region. The chemistry of this complex shows that by 66 Ma, the mantle under the region was more depleted and variable in composition than the Galapagos hotspot mantle that gave rise to older magmas. The chemistry of the subsequent 20–5 Ma andesites shows that any plume influence was gone by the Miocene. The end of this arc stage at about 5 Ma coincided with the collision of a series of aseismic ridges. The chemistry of the younger lavas (<2 Ma) is argued to reflect the melting of an oceanic ridge subsequent to collision rather than the competing hypothesis, which states that their chemistry reflects addition of continental lithosphere into the mantle through forearc subduction-erosion.

Moving into northern South America, the ninth chapter, by Cristian Vallejo (ETH [Eidgenössische Technische Hochschule], Zürich), Wilfried Winkler (ETH Zürich), Richard A. Spikings (University of Geneva), Leonard Luzieux, (ETH Zürich), Friedrich Heller (ETH Zürich), and François Bussy (Université de Lausanne, Switzerland), presents an updated analysis on the mode and timing of the collision of oceanic terranes against the continental margin of Ecuador. This paper presents new field data on the volcanic rocks of the Western Cordillera, new paleomagnetic constraints, and a new stratigraphic framework for oceanic plateau postcollisional volcanic rocks based on an extensive set of new U-Pb and ^{40}Ar-^{39}Ar geochronologic ages. The authors propose a new model for the geodynamic evolution of the accreted terranes and the subduction geometry of the colliding plateau in the latest Cretaceous.

The tenth chapter, by François Michaud (Nice, France), Cesar Witt (Escuela Politécnica, Ecuador), and Jean-Yves Royer (Brest, France), deals with fact and fiction over the influence of the subduction of the Carnegie volcanic ridge on Ecuadorian geology, particularly with respect to the timing of the initiation of ridge subduction and processes associated with collision against the continental margin. They conclude that there is no clear link between deformation along the continental margin and coastal uplift with the subduction of the Carnegie Ridge. They further argue for relatively recent initiation of ridge subduction based on seismic evidence against a shallowly subducting slab beneath Ecuador, alternatives to partial melting of the subducting ridge to explain the adakitic signatures in magmatic rocks, and insignificant vertical uplift related to ridge subduction. Finally, they use global positioning system (GPS) results to argue that the northward migration of the North Andean block can explain inhibition of ridge-induced vertical strain.

The eleventh chapter, by Suzanne Mahlburg Kay (Cornell University) and Beatriz Coira (Consejo Nacional de Investigaciones Científicas y Técnicas, Argentina), presents an overview of the Neogene evolution and uplift of the Central Andean Altiplano-Puna Plateau, which is the world's highest and most extensive plateau formed in the absence of continental collision. Magmatic, geophysical, and structural data are used to construct lithospheric-scale models of three distinctive transects in the southern Altiplano, northern Puna, and southern Puna regions that reveal variable roles for shallowing and steepening of the subducting plate, crustal shortening, delamination of thickened lower crust and lithosphere, mantle and crustal melting, and deep crustal flow. Temporal similarities in processes in these transects are correlated with changes in the rate of westward drift of South America and slab rollback, and along-strike temporal differences are correlated with variations in Nazca plate geometry that reflect the southward progression of the subduction of the aseismic Juan Férnandez Ridge. The large concentration of giant ignimbrites in the northern Puna region is attributed to the greatest change in slab geometry above a region where an early Miocene amagmatic flat slab was followed by steepening slab angles, leading to a massive ignimbrite flare-up.

The twelfth chapter, by Patricia Alvarado (Universidad Nacional de San Juan, Argentina), Mario Pardo (Universidad de Chile), Hersh Gilbert (Purdue University), Silvia Miranda (Universidad Nacional de San Juan, Argentina), Megan Anderson (Colorado College), Mauro Saez (Universidad Nacional de San Juan), and Susan Beck (University of Arizona), presents a discussion of the seismic characteristics of the Chilean-Pampean flat-slab region, with a particular emphasis on the region under the Sierras Pampeanas in central Argentina. The paper reviews and presents new results from regional seismic networks on the seismicity and shape of the subducted Nazca plate, the lithospheric structure above the flat-slab region, and the character of the overlying crust. The seismic character of the crust is explored in terms of inheritance of crustal properties from the pre-Mesozoic terranes that are thought to characterize the crust in the region and a model of partial eclogitization of a thickened crust in the western region. This chapter complements the geological discussion of the Chilean-Pampean flat-slab region in the field guides (see below).

The overviews in volume 13 of the Geological Society of America Field Guide series round out the discussion of processes in central and southern South America. The two chapters by Ramos feature the Paleozoic through Neogene evolution of the modern amagmatic Pampean flat-slab region over the shallowly subducting Nazca plate. The first emphasizes the Argentine Frontal and Main Andean Cordilleras near the southern boundary of the flat-slab segment, and the second features a transect from the Sierras Pampeanas across the Precordillera fold-and-thrust belt through the main Andean Range to Chile. There is a parallelism here with western North American geology as the evolution of the Pampean flat slab region is the most quoted analogue in analyses of the North American Laramide orogeny. The chapter by Zapata, Zamora Valcarce, Folguera, and Yagupsky highlights the Andean Cordillera and retroarc of the Neuquén Andes in Argentina (~38.5°S to 37°S) in a region where both contractional and extensional deformation have occurred. The text discusses the multiple Cretaceous to Holocene deformational events, which are interpreted with respect to changes in the configuration of the convergent plate boundary and the dip of the subducting slab.

The chapter by Kay, Coira, and Mpodozis highlights the Neogene evolution of the southern Puna plateau between 23°S and 27.5°S latitude and the southernmost Central Andean volcanic zone arc in Chile at 26.5°S to 27.5°S latitude. The text emphasizes the magmatic record, particularly, the giant Cerro Galán ignimbrites and distinctive mafic flows of the southern Puna region, the late Miocene to Pliocene migration of the frontal magmatic arc from the Maricunga belt to the Central volcanic zone, and the implications for forearc subduction-erosion.

Finally, the chapter by Gorring highlights the region in southern Patagonia, east of the Chile triple junction region, near where ridge-trench collision is occurring at 46.5°S latitude. The text features the southern Patagonian Cordillera, where distinctive backarc deformational and magmatic features reflect the northward-propagation of a series of slab windows related to ridge-trench collisional events beginning at ca. 14 Ma.

Last, the editors would like to sincerely thank all of the reviewers for their constructive comments on the papers. In alphabetical order, the reviewers were: Christopher Andronicos (Cornell University), Larry D. Brown (Cornell University), Elena Centeno Garcia (Universidad Nacional Autónoma de México, Mexico), Peter Clift (University of Aberdeen, UK), Darrel S. Cowan (University of Washington), Jelle de Boer (Wesleyan University, USA), Donna Eberhart-Phillips (University of California, Davis), Jaillard Etienne (IRD–LGCA [Institut de Recherche pour le Développment–Laboratoire de Géodynamique des Chaînes Alpines], Maison des Géosciences, France), Andreas Folguera (Universidad de Buenos Aires, Argentina), John Geissman (University of New Mexico), Estanislao Godoy (Servicio Nacional de Geología y Minería, Chile), Stephen Grand (University of Texas, Austin), Jeffrey Hedenquist (consultant, Canada), Teresa Jordan (Cornell University), Robert Kay (Cornell University), David Lageson (Montana State University), Stephen Marshak (University of Illinois), René Maury (Université de Bretagne Occidentale, France), Tony Monfret (Nice University, France), Constantino Mpodozis (Antofagasta Minerals, Chile), Jairo Alonso Osorio Naranjo (Colombian Geological Survey, Bogotá), Warren Pratt (consultant, Great Britain), Jose A. Perello (Antofagasta Minerals SA, Santiago, Chile), Augusto Rapalini (Universidad de Buenos Aires, Argentina), Jeremy Richards (University of Alberta, Canada), Pablo Samaniengo (Escuela Politécnica Nacional, Ecuador), Eric Sandvol (University of Missouri), David Scholl (U.S. Geological Survey, Menlo Park), Walter Snyder (Boise State

University), Robert Trumbull (GeoForschungsZentrum Potsdam, Germany), Gene Yogodzinski (University of South Carolina), and Dante Morán Zenteno (Universidad Nacional Autónoma de México, Mexico).

REFERENCE CITED

Kay, S.M., and Ramos, V.A., eds., 2008, Field Trip Guides to the Backbone of the Americas in the Southern and Central Andes: Ridge Collision, Shallow Subduction, and Plateau Uplift: Boulder, Colorado, Geological Society of America Field Guide 13, 181 p.

The Geological Society of America
Memoir 204
2009

Anatomy and global context of the North American Cordillera

William R. Dickinson
Department of Geosciences, University of Arizona, Tucson, Arizona 85721, USA

ABSTRACT

The Cordillera of western North America occupies the central 5000 km of the circum-Pacific orogenic belt, which extends for 25,000 km along a great-circle path from Taiwan to the Antarctic Peninsula. The North American Cordillera is anomalous because dextral transform faults along its western flank have supplanted subduction zones, the hallmark of circum-Pacific tectonism, along much of the Cordilleran continental margin since mid-Cenozoic time. The linear continuity of the Cordilleran orogen terminates on the north in the Arctic region and on the south in the Mesoamerican region at sinistral transform faults of Mesozoic and Cenozoic age, respectively.

The Cordilleran margin of Laurentia was formed initially by rift breakup of the supercontinent Rodinia followed by development of the Neoproterozoic to early Paleozoic Cordilleran miogeocline along a passive continental margin, but it was modified in California and Mexico by Permian to Triassic transform truncation of Paleozoic tectonic trends. Late Paleozoic and Mesozoic accretion of oceanic island arcs and subduction complexes expanded the width of the Cordilleran orogen both before and after Triassic initiation of ancestral circum-Pacific subduction beneath the Cordilleran margin. Mesozoic to Cenozoic extensions and counterparts of Cordilleran accreted terranes extend southward into the Caribbean Antilles and northern South America.

The development of successive forearc and retroforeland basins accompanied the progress of Cordilleran orogenesis over time, and coeval Mesozoic to Cenozoic batholith belts reflect continuing plate consumption at subduction zones along the continental margin. The assembly of subduction complexes along the Cordilleran continental margin continued into Cenozoic time, but dextral strike slip along the Pacific flank of the Cordilleran orogen displaced elongate coastal segments of the orogen northward during Cenozoic time. In the United States and Mexico, Laramide breakup of the Cordilleran foreland during shallow slab subduction and crustal extension within the Basin and Range taphrogen also expanded the width of the Cordilleran orogen during Cenozoic time.

INTRODUCTION

The North American Cordillera occupies the central 5000 km of the circum-Pacific orogenic belt, which extends for 25,000 km along a great circle of the globe from Taiwan to the Antarctic Peninsula (Fig. 1). The North American segment of the orogenic belt is atypical because the modern Queen Charlotte and San Andreas transform systems along the Cordilleran continental margin have largely supplanted the hallmark of circum-Pacific tectonism, subduction of oceanic plates. The consequent lack of active arc volcanism, except for the Cascades volcanic chain of the Pacific Northwest (Fig. 1), obscures the continuity of the North American Cordillera with the remainder of the circum-Pacific orogenic belt, where trenches and volcanic arcs are characteristic. On the

Dickinson, W.R., 2009, Anatomy and global context of the North American Cordillera, *in* Kay, S.M., Ramos, V.A., and Dickinson, W.R., eds., Backbone of the Americas: Shallow Subduction, Plateau Uplift, and Ridge and Terrane Collision: Geological Society of America Memoir 204, p. 1–29, doi: 10.1130/2009.1204(01).

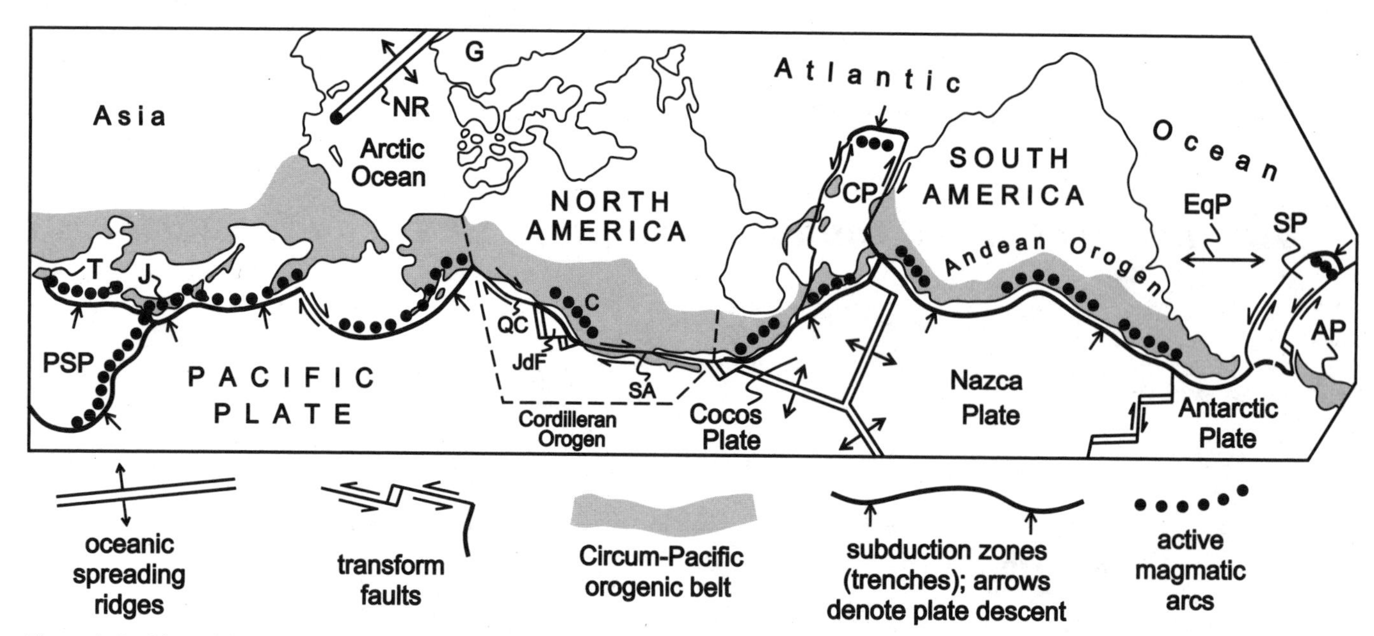

Figure 1. Position of the North American Cordillera as the central segment of the circum-Pacific orogenic belt aligned along the equatorial great circle of a Mercator projection with its pole at 25°N, 15°E after Dickinson et al. (1986b). Abbreviations: AP—Antarctic Peninsula; C—Cascades volcanic chain; CP—Caribbean plate; EqP—equatorial plane of Mercator projection; G—Greenland; J—Japan; JdF—Juan de Fuca plate; NR—Nansen Ridge (northern extremity of modern Atlantic spreading system); PSP—Philippine Sea plate; QC—Queen Charlotte fault; SA—San Andreas fault; SP—Scotia plate; T—Taiwan.

other hand, deep erosion of ancient subduction zones and extinct arc assemblages has exposed crustal levels in the North American Cordillera that are hidden beneath seawater or surficial volcanic cover along much of the Pacific rim. The atypical nature of the North American Cordillera thus affords geologic insights into the rock masses of arc-trench systems at depth that are more difficult to gain elsewhere along the circum-Pacific orogenic belt.

The North American Cordillera extends as an approximately linear and generally coherent entity from the Gulf of Alaska and the Yukon Province of Canada on the north to the mouth of the Gulf of California in southern Mexico on the south (Fig. 1). Beyond those limits, geotectonic trends bend and diverge into the Arctic and Caribbean realms. The method of this paper is to examine the evolution of the North American Cordillera in space and time using paleotectonic maps that illustrate salient aspects of its internal configuration, its relationship to Arctic and Caribbean tectonic elements, and key stages in its geotectonic evolution. The only palinspastic reconstruction on the paleotectonic maps is closure of the young (<6 Ma) Gulf of California for all pre-Neogene maps, but the restoration of Baja California against mainland Mexico is not as close as postulated by a recent reconstruction placing the tip of the Baja California peninsula as far south as Cabo Corrientes, south of the Tres Marias Islands (Wilson et al., 2005).

My aims are to highlight the global context of the North American Cordillera and to show how its internal anatomy relates to circum-Pacific tectonics, but discussion of its geologic history in detail is beyond the scope of the paper. Moreover, the broad time frames chosen to show overall geotectonic relations in compact fashion do not allow some aspects of diachronous behavior to be illustrated. To avoid repeating previous arguments and reproducing voluminous citations to past literature supporting them, reference is made wherever appropriate to my own published papers.

A subsidiary goal is to reconcile the different interpretations of Cordilleran plate tectonics through time that have been developed by geologists working in Canada, the United States, and Mexico. Pre-Cretaceous tectonic connections across the Canadian border are obscured by Cretaceous batholiths and Cenozoic lava fields of the Pacific Northwest, and pre-Cretaceous tectonic ties from the United States into Mexico were broken by the late Mesozoic Border rift belt (Dickinson and Lawton, 2001b), which penetrated across the continental block near the U.S.–Mexico border from the northwestern corner of the opening Gulf of Mexico to the rear flank of the Cordilleran batholith belt.

INITIATION OF CORDILLERAN MARGIN

The Cordilleran margin of North America was first established by Neoproterozoic rifting of Laurentia, the Precambrian core of the continent, from another continental block or blocks within the supercontinent Rodinia, which had been assembled during Mesoproterozoic Grenville orogenic events. As there is no current consensus on the identity of the conjugate continental margin or margins that rifted away from the ancestral Cordilleran margin of Laurentia, Figure 2 shows schematically the alternate suggested configurations of Rodinia prior to continental

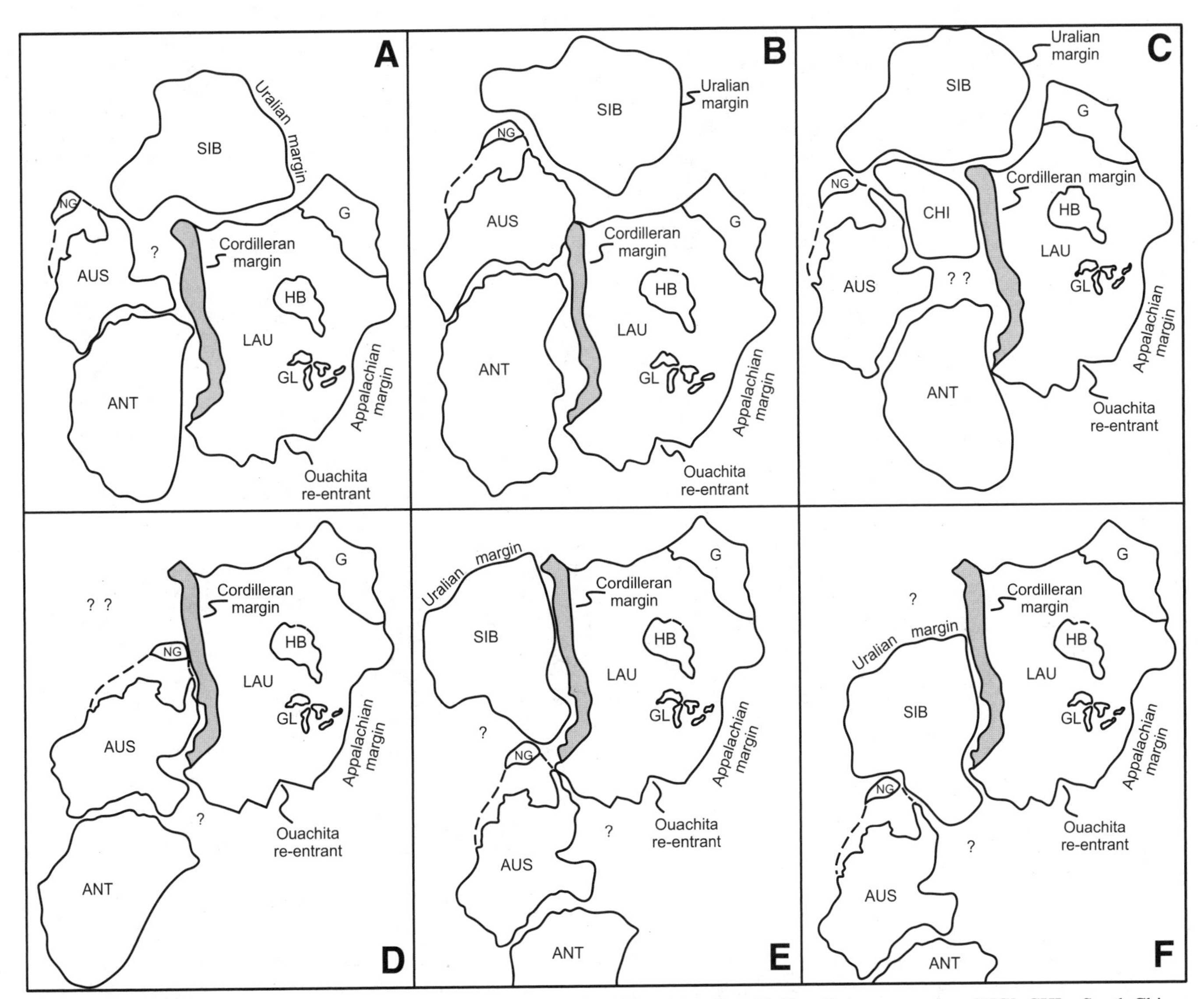

Figure 2. Configurations of continental blocks (ANT—East Antarctica; AUS—Australia with New Guinea appendage [NG]; CHI—South China; SIB—Siberia) adjacent to the Cordilleran margin (gray) of Laurentia (LAU) for alternate reconstructions of Rodinia: (A) SWEAT (southwest U.S.–East Antarctica) hypothesis (Hoffman, 1991; Moores, 1991; Dalziel, 1991); (B) SWEAT adjusted for best paleomagnetic fit (Weil et al., 1998) with Siberia expanded; (C) SWEAT with intervening South China block (Li, 1999); (D) AUSWUS (Australia–southwest U.S.) hypothesis (Karlstrom et al., 1999, 2001); (E) AUSMEX (Australia–Mexico) hypothesis (Wingate et al., 2002) with Siberia added here; (F) preferred Siberian connection (Sears and Price, 1978, 2000, 2003). Laurentian geographic features: G—Greenland; GL—Great Lakes; HB—Hudson Bay.

separation. East Gondwanan continents once affiliated with the Appalachian margin of Laurentia are not shown.

Figure 2A is the classic SWEAT (Southwest U.S.–East Antarctica) hypothesis of 1991, Figure 2B is the SWEAT configuration as adjusted later for best paleomagnetic fit, and Figure 2C is the SWEAT configuration with the South China block intervening between Laurentia and Australia. The SWEAT and modified SWEAT ties all postulate close connections within Rodinia between southwest Laurentia and Antarctica, with Siberia lying north of Laurentia within a paleo-Arctic realm having nothing directly to do with the Cordilleran margin (Pisarevsky and Natapov, 2003). Difficulties with documenting Laurentia–Antarctica connections of Precambrian-age belts led to the AUSWUS (Australia–western U.S.) hypothesis (Fig. 2D), and further difficulties with rift timing led to the AUSMEX (Australia–Mexico) hypothesis (Fig. 2E). In both these postulated configurations of Rodinia, Australia rather than Antarctica lay adjacent to southwest Laurentia (either the United States or Mexico, respectively). Siberia has been added gratuitously to the AUSMEX configuration of Figure 2E to show its potentially close similarity to the Siberian connection (Fig. 2F), which was advanced well before SWEAT, and which has been recently developed in detail as the configuration

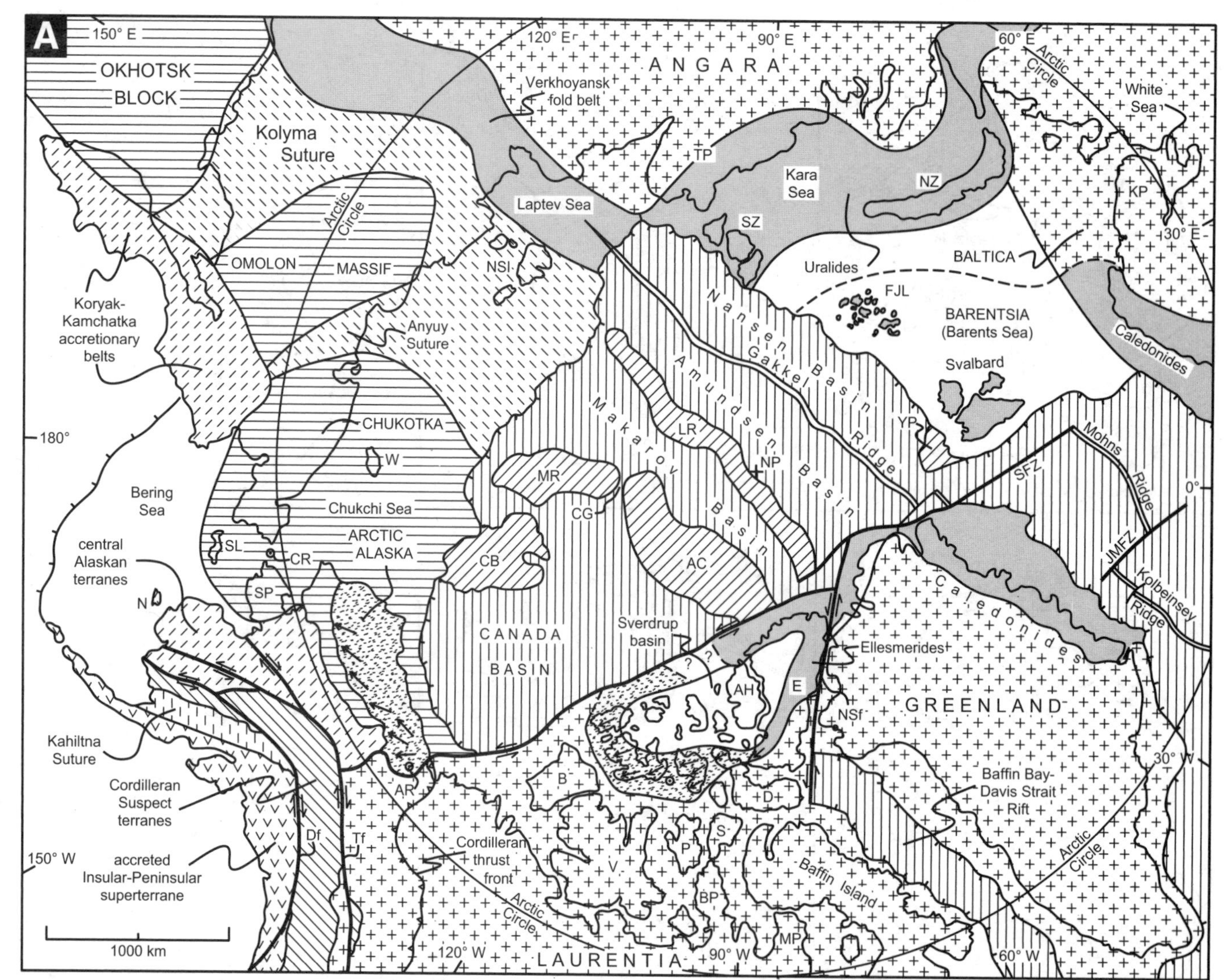

Figure 3. Tectonic elements of the Arctic region (polar projection): (A) present configuration modified after Grantz et al. (1990b) and Jakobsson et al. (2003); (B) inferred mid-Jurassic configuration. Eurasian orogens (Verkhoyansk, Uralides, dismembered Caledonides) are shaded, crosses denote continental cratons (Angara, Baltica, Laurentia), including linked off-shelf assemblages along the Cordilleran margin of Laurentia, and horizontal rules denote other coherent Siberian-Alaskan crustal blocks (except offshore Barentsia and Bering Sea blank because of uncertain affinity). Modern shelf breaks are marked as lines with ticks, ocean basins with vertical rules (segments of modern Atlantic seafloor spreading system shown as double lines), aseismic intraoceanic ridges and off-shelf plateaus with diagonal rules, and accreted Cordilleran-Siberian suspect terranes and suture belts by varied ornamentation distinguishing principal crustal increments. Sources of data (counterclockwise from upper left): northeast Siberia modified after Dickinson (1978), Fujita and Newberry (1982, 1983), Natal'in et al. (1999), Sokolov et al. (2002), and Miller et al. (2002); Bering Sea after Worrall (1991); interior Alaska adapted after Jones et al. (1987) and Plafker and Berg (1994); suspect terranes and thrust front of Cordilleran orogen after Figures 4, 6, and 9; Devonian (-Mississippian) clastic wedge of Arctic Alaska and the Canadian Arctic islands (stippled with arrows denoting paleocurrent trends) after Donovan and Tailleur (1975), Embry and Klovan (1976), Nilsen (1981), Nilsen and Moore (1982), Embry (1988), and Handschy (1998); Upper Paleozoic to Lower Mesozoic Sverdrup Basin (masking older assemblages) after Sweeney (1976) and Balkwill (1978); Nares Strait transform fault after McWhae (1986); Ellesmerides (Franklinian geosyncline or Pearya) and Greenland Caledonides modified after Surlyk and Hurst (1984), Miall (1986), and Trettin (1987, 1991); Gakkel Ridge after Michael et al. (2003); Arctic Uralides adapted after Hamilton (1970) and Inger et al. (1999); front of Caledonide orogen east of Franz Joseph Land and Svalbard in offshore Barentsia after Lorenz et al. (2007). The alternate (rotated) reconstruction of Arctic Alaska follows Grantz et al. (1990a) and Embry (1990). Rotation poles: CR—for Chukotka counterclockwise back-rotation of 60° relative to Arctic Alaska (for preferred reconstruction); AR—for Arctic Alaska clockwise back-rotation of 66° relative to Laurentia (for alternate reconstruction). Barbed half-arrows in A denote sense of strike slip on selected faults (Df—Denali; NSf—Nares Strait transform; Tf—Tintina), with restoration of net slip indicated by balls on fault traces in B (net Denali slip = 370 km after Lowey, 1998; net Tintina slip = 485 km after Dickinson, 2004). Mid-ocean-ridge fracture zones: JMFZ—Jan Mayen; SFZ—Spitzbergen. Arctic aseismic ridges (CG is Coronation Gap): AC—Alpha Cordillera; LR—Lomonosov Ridge; MR—Mindeleev Ridge. Submerged offshore tablelands: CB—Chukchi Borderland; VP—Vöring Plateau; YP—Yermak Plateau. Selected Islands: AH—Axel Heiberg; B—Banks; D—Devon; E—Ellesmere; FJL—Franz Joseph Land; NSI—New Siberian Islands; N—Nunivak; NZ—Novaya Zemlya; P—Prince of Wales; S—Somerset; SL—St. Lawrence; SZ—Severnaya Zemlya; V—Victoria; W—Wrangell. Prominent peninsulas: BP—Boothia; KP—Kola; MP—Melville; SP—Seward; TP—Taimyr. NP—North Pole. (*Continued on following page.*)

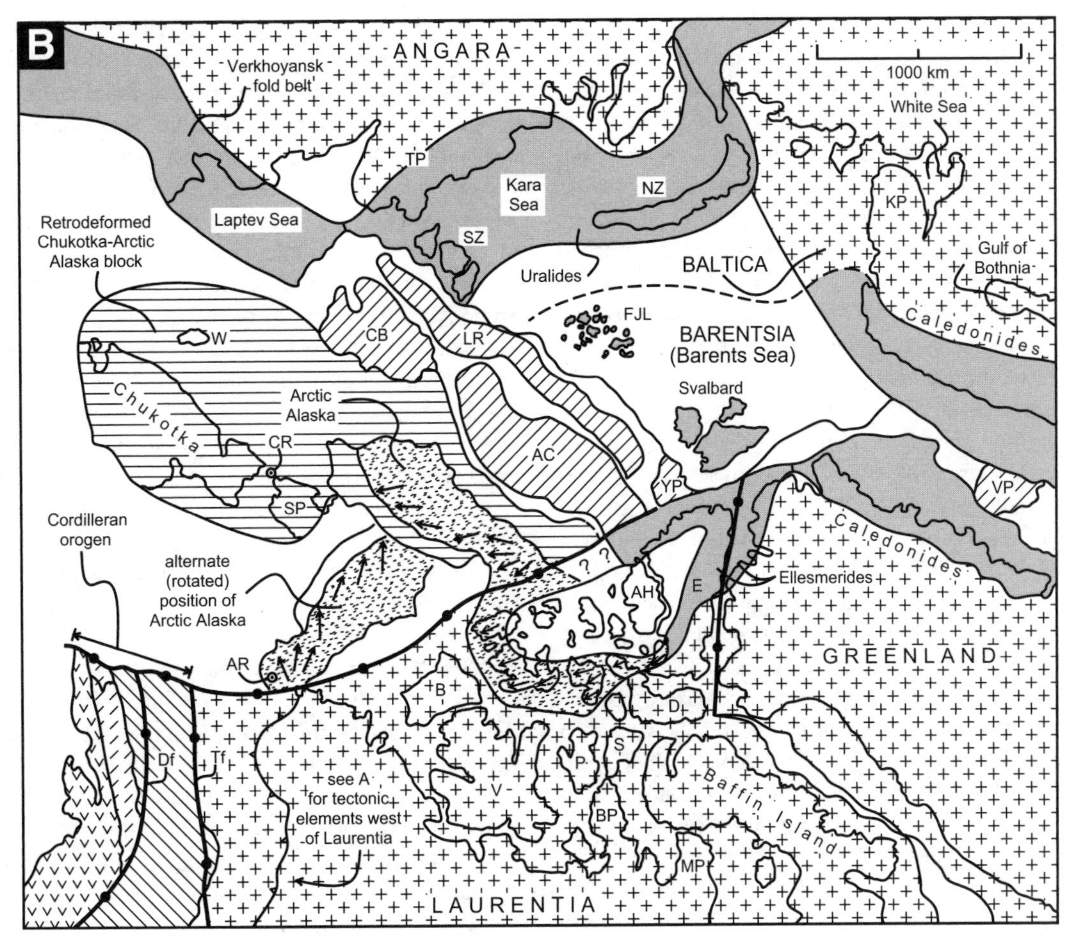

Figure 3 (*continued*).

of Rodinia preferred here. A close match of Precambrian strata in the Death Valley region of southwestern Laurentia and the Sette-Daban Range of eastern Siberia seemingly confirms the Siberian connection (Sears et al., 2005), for which the northern margin of the modern Siberian or Angara craton is the principal continental margin conjugate to the rifted Cordilleran margin of Laurentia.

The correct configuration of Rodinia remains unclear, however, and there is continuing support for AUSWUS (Burrett and Berry, 2000; Fioretti et al., 2005), as well as the Siberian connection (or AUSMEX). Hints that two separate rifting events may have been involved in the development of the Cordilleran continental margin (see later discussion) may ultimately provide the conceptual leverage needed to solve the Rodinian puzzle.

CORDILLERAN–ARCTIC RELATIONS

An understanding of interactions among tectonic elements of the Cordilleran and Arctic realms is important because the Yukon is no more distant from the Verkhoyansk and Uralian fold belts of Eurasia than it is from the Gulf of California (Fig. 1).

For Arctic plate reconstructions, an appreciation of the nature of several aseismic ridges that cross the Arctic seafloor is critical. There is general agreement that the Lomonosov Ridge (Fig. 3A) is a ribbon continent that was rifted away from the Barents Shelf edge north of Scandinavia by seafloor spreading along the Gakkel Ridge (Weber and Sweeney, 1990), which is the Arctic extension of the Mid-Atlantic spreading system. Restoration of the Lomonosov Ridge against Barentsia (Fig. 3B) closes the Nansen and Amundsen Basins (Fig. 3A), which jointly form the Eurasian Basin between Greenland and Siberia. Coordinate shift of the Canadian Arctic islands relative to Greenland by restoring slip across the Nares Strait transform packs Ellesmere Island against the Yermak Plateau (Fig. 3A) at the northwest tip of Barentsia (Fig. 3B).

Retrospective closure of the Makarov and Canada Basins, jointly forming the Amerasian basin south of the Lomonosov Ridge, is more contentious because obscure patterns of seafloor magnetic anomalies do not define spreading directions during Arctic opening (Vogt et al., 1982). Standard reconstructions (Lawver et al., 2002; Golonka et al., 2003) posit clockwise back-rotation of Arctic Alaska to restore the north flank of the Brooks Range against the continental margin of northern Canada (alternate reconstruction of Fig. 3B). There is, however, no robust paleomagnetic evidence for this postulated rotational motion (Lane, 1997), which involves a number of questionable assumptions.

For rotational closure of the Canada Basin, the Alpha Cordillera (Fig. 3A) is treated as a hotspot track made of overthickened oceanic crust formed in the wake of rotating Chukotka (Fig. 3A), which forms a western extension of Arctic Alaska. The magnitude of Alpha Ridge aeromagnetic anomalies is, however, 2.5 times the magnitude of the largest aeromagnetic anomalies observed over intraoceanic features like the Iceland plateau, and this suggests that the Alpha Cordillera is composed of stretched continental crust (Coles and Taylor, 1990). Moreover, the postulated sphenochasmic origin of the Makarov and Canada Basins implies from the geometry of required plate motions that the southern margin of the Lomonosov Ridge is a paleotransform (Jackson and Gunnarsson, 1990), whereas its jagged morphology resembles a rifted margin. By contrast, the abrupt and rather straight continental margin of Arctic Canada resembles a paleotransform, but it is required by the rotational model for the Canada Basin to be an orthogonally rifted margin. In addition, back-rotation of Arctic Alaska and Chukotka to swing them to initial positions against the continental margin of Canada would produce crustal overlap of the Chukchi continental borderland (CB of Fig. 3A) with the Canadian Arctic islands.

More satisfactory closure of the Canada and Makarov Basins is achieved by treating the continental margin off the Canadian Arctic islands as a paleotransform along which the Alpha Cordillera and Arctic Alaska slid differentially westward with respect to Canada (Fig. 3B). By this translational hypothesis, continental Alpha Cordillera and Alaska–Chukotka crustal elements rifted sequentially away from the Lomonosov Ridge before seafloor spreading along the Gakkel Ridge separated the Lomonosov Ridge from Barentsia. The timing of rifting and translation is not well constrained, but the geologic history of the Canadian continental margin implies that motion began in Late Jurassic to Early Cretaceous time before 125 Ma, and a postdeformational breakup unconformity was created across the newly formed passive continental margin by 100 Ma in mid-Cretaceous time (Embry and Dixon, 1990). A retrodeformation procedure for the Alaskan and Bering oroclines (Fig. 3A), to straighten conjoined Alaska–Chukotka, allows restoration of Arctic Alaska and Chukotka against an ancestral Eurasian continental margin that spanned between Greenland and Siberia (Fig. 3B). Preserving the notion of rotating Arctic Alaska away from the Canadian Arctic requires postulating undocumented strike slip across the Bering Strait between Alaska and Chukotka, and significant internal deformation within Chukotka (Miller et al., 2006). These ancillary modes of deformation are unnecessary if rotation of Arctic Alaska is abandoned in favor of translation. Recovery of oroclinal bending in the Alaska–Bering region (Fig. 3A) is required, however, to avoid overlap of restored Chukotka with the core of Siberia (Fig. 3B).

The most powerful argument against rotation of Arctic Alaska is the geometry of a clastic wedge of Devonian to Mississippian strata that was once contiguous along depositional strike from the Canadian Arctic islands into the Brooks Range of Arctic Alaska (Fig. 3B). The strata of Middle to Upper Devonian age in Canada and Upper Devonian to Lower Mississippian age in Alaska grade in both areas from fluvial-deltaic on the east-northeast to marine on the west-southwest (Nilsen, 1981). As noted by Nilsen and Moore (1982), back-rotation of Arctic Alaska would place paleocurrent trends and facies patterns for the two segments of the clastic wedge in Canada and Alaska into near opposition, as well as leaving the closely related strata separated paleogeographically (see alternate reconstruction of Fig. 3B). By contrast, the translational paleotransform option for the continental margin of Arctic Canada restores the Brooks Range of Arctic Alaska adjacent to the Canadian Arctic islands, with the Alpha Cordillera restored against the Lomonosov Ridge, and paleocurrent trends in the Canadian and Alaskan segments of the clastic wedge are then subparallel, with facies patterns that are coordinate (Fig. 3B). Southeast-vergent Paleozoic thrusts in easternmost Arctic Alaska (Oldow et al., 1987) are compatible with southeast-vergent structures of the Ellesmerides in the Innuitian orogen of the Canadian Arctic islands if they are translated laterally for the preferred reconstruction of Figure 3B, but their vergence would be anomalous if the structures were rotated 66° for the alternate reconstruction of Figure 3B.

Derivation of detritus in the Canadian segment of the Devonian clastic wedge from the Franklinian-Caledonian orogen of Ellesmere Island (and eastward) is implied by detrital zircon populations from the Canadian Arctic islands (McNicoll et al., 1995). The Alaskan segment of the clastic wedge was probably derived from analogous sources along tectonic strike in the Alpha Cordillera or Lomonosov Ridge, then adjacent to Barentsia (Fig. 3B). Detrital zircons in Paleozoic strata of Chukotka imply derivation of clastic sediment from the Verkhoyansk or Taimyr regions of Siberia (Fig. 3B), rather than from Laurentia (Miller et al., 2006). Late Proterozoic granitic magmatism on the Seward Peninsula of western Alaska also favors Barentsian rather than Laurentian affinity for Arctic Alaska and Chukotka (Patrick and McClelland, 1995).

The tectonic complexity of interior Alaska (Gottschalk et al., 1998; Johnston, 2001; Bradley et al., 2007) and eastern Siberia (Miller et al., 2002) defies facile interpretation, but it does not preclude the translational model for origin of the Canada Basin. Restoration of post–mid-Cretaceous dextral strike slip on the Denali and Tintina fault systems (Fig. 3B) back-slips southern Alaskan terranes south of the projection of the paleotransform of the Canadian Arctic, along which Arctic Alaska was translated westward in pre–mid-Cretaceous time. Jurassic-Cretaceous accretionary episodes in central Alaska and eastern Siberia were broadly coeval with translation of Arctic Alaska and Chukotka westward, and were probably related along tectonic strike (Fig. 3A), but the intervening Bering Sea masks presumed tectonic continuity between the two regions.

PRE–MIDDLE TRIASSIC CORDILLERA

Contrasts in the configurations of Paleozoic geotectonic elements in northern (Fig. 4) and southern (Fig. 5) segments of the North American Cordillera stem largely from differences in the timing of premiogeoclinal Neoproterozoic rifting, Mesozoic

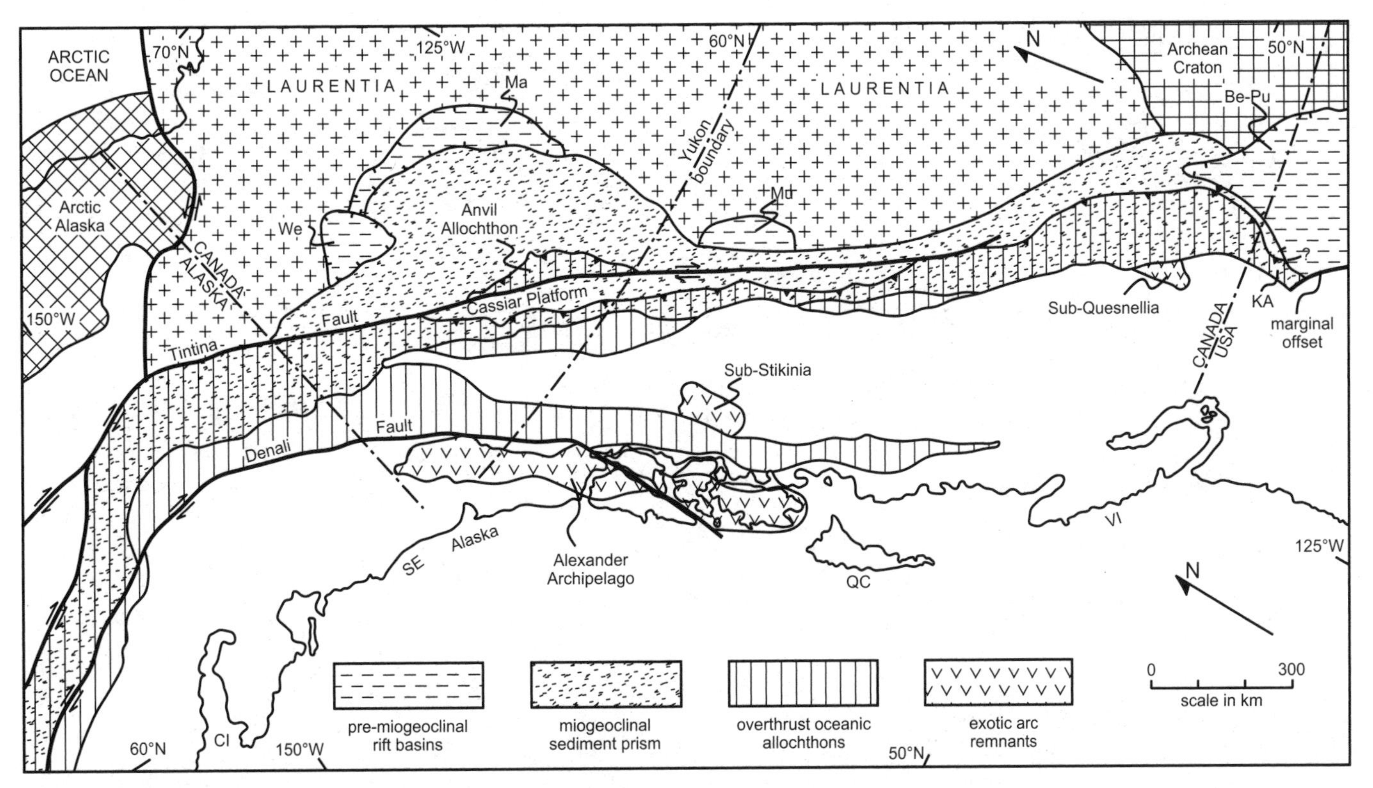

Figure 4. Pre–Middle Triassic tectonic elements of the northern Cordillera (Alaska–Canada) showing relation of accreted terranes to miogeoclinal belt. Premiogeoclinal rift assemblages (pre-Rodinian and intra-Rodinian): Be-Pu—Belt-Purcell (1470–1370 Ma); Ma—Mackenzie Mountains (975–775 Ma); Mu—Muskwa (1760–1660 Ma); We—Wernecke (1820–1710 Ma). Map is modified after Plafker and Berg (1994) and Dickinson (2004). Pacific coast geographic features: CI—Cook Inlet; QC—Queen Charlotte Islands; VI—Vancouver Island. Kootenay structural arc (KA) spans the U.S.–Canada border north of the trans-Idaho discontinuity (marginal offset).

oroclinal bending that redoubled Paleozoic lithic belts of the northern Cordillera, and interaction of the southern Cordillera with the Pennsylvanian-Permian Ouachita(-Marathon) suture belt between Laurentia and Gondwana.

Miogeoclinal Sedimentation

Neoproterozoic continental separation by rifting initiated deposition of a westward-thickening miogeoclinal prism of shelfal and off-shelf slope strata that persisted until mid–Late Devonian time along a passive Cordilleran continental margin that formed the rifted flank of Laurentia facing the paleo–Pacific Ocean (Gabrielse and Campbell, 1992; Fritz et al., 1991; Geldsetzer and Morrow, 1991; Poole et al., 1992). Differences in rift timing north and south of a marginal offset in the rifted Cordilleran margin just south of the U.S.–Canada border (Figs. 4 and 5) suggest that an unidentified continental block beyond the Uralian margin of Siberia (Fig. 2F) lay west of Canada to the north of the marginal offset. The elongate trend of the miogeoclinal belt is offset >250 km across the ancestral structure (Figs. 4 and 5), which was termed the trans-Idaho discontinuity by Yates (1968).

At intervals along the length of the northern Cordillera (Fig. 4), thick sedimentary successions were deposited locally within intracontinental pre-Rodinian rift basins, which were later transected by the Neoproterozoic rift trend that delineated the Cordilleran miogeocline. These include the Paleoproterozoic Wernecke Supergroup (1820–1710 Ma) and Muskwa assemblage (1760–1660 Ma) in northern Canada, and the Mesoproterozoic Belt-Purcell Supergroup (1470–1370 Ma) spanning the U.S.–Canada border (Evans et al., 2000; Ross et al., 2001; Luepke and Lyons, 2001). Later deposition of the Mackenzie Mountains Supergroup (975–775 Ma), exposed along an elongate belt subparallel to the miogeocline in the Yukon (Rainbird et al., 1996), more closely preceded miogeoclinal sedimentation and may reflect a precursor phase of the rifting that led to continental separation. Overlying strata of the Windermere Supergroup form the lower part of the miogeoclinal sediment prism (Fig. 4) of marine strata that is continuous along the length of the northern Cordillera (Ross, 1991). Basaltic rocks interbedded with glaciomarine diamictite in basal horizons of the Windermere Supergroup have been dated isotopically at 770–735 Ma (Devlin et al., 1988; Colpron et al., 2002), which can be taken as the age span of premiogeoclinal rifting north of the trans-Idaho discontinuity.

In the southern Cordillera (Fig. 5), coeval nonmarine strata of the Uinta Mountain Group (770–740 Ma), which fill an

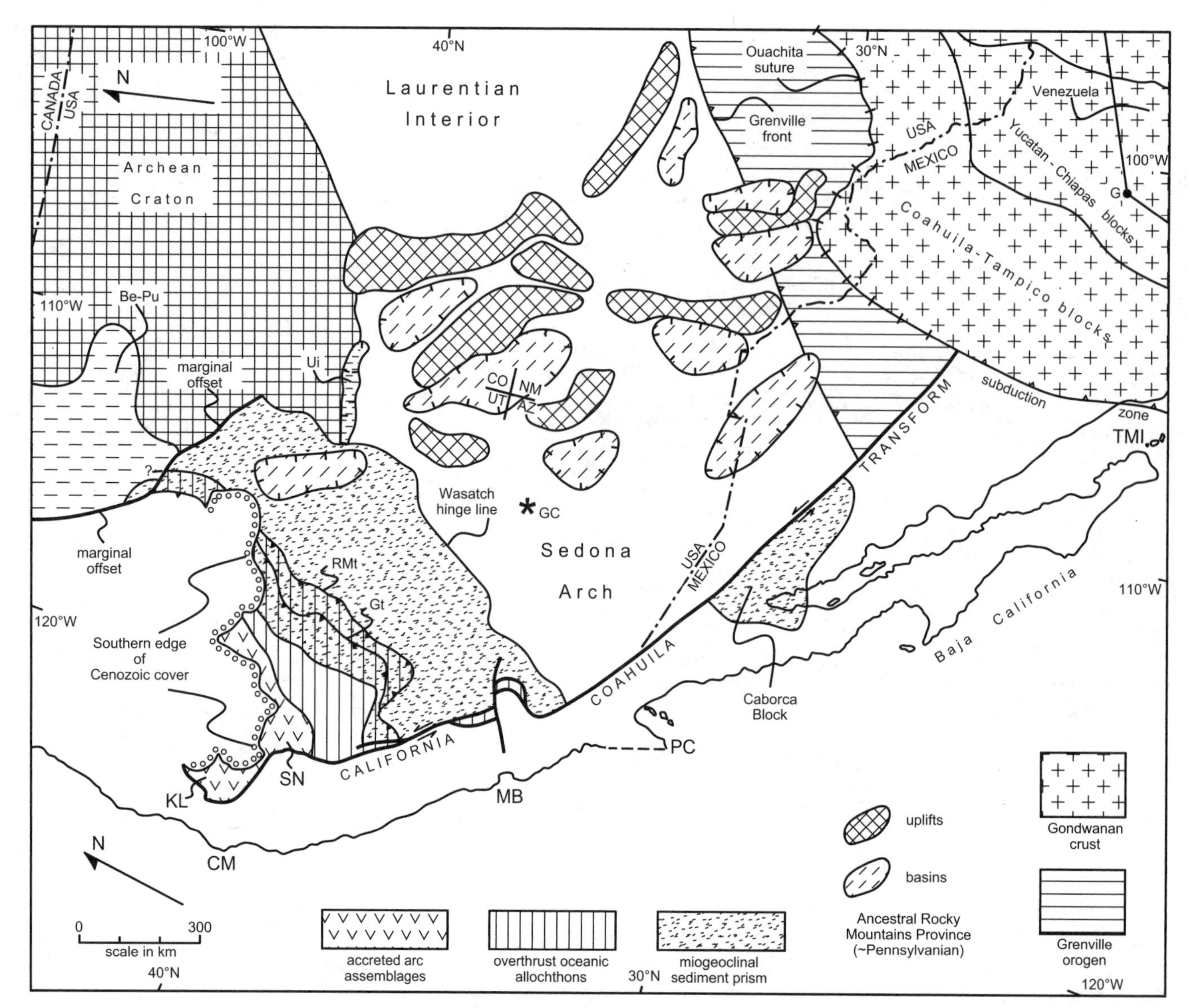

Figure 5. Pre–Middle Triassic tectonic elements of the southern Cordillera (U.S.–Mexico) showing relation of accreted terranes to miogeoclinal belt (Be-Pu—premiogeoclinal Belt-Purcell rift assemblage; Ui—Uinta aulacogen). Key faults: Gt—Golconda thrust; RMt—Roberts Mountains thrust. Gulf of California and Gulf of Mexico are closed (G—restored predrift position of Guajira Peninsula of Venezuela). Map is modified after Dickinson (2000, 2004), Dickinson and Lawton (2001a, 2003), and Walker et al. (2002). Pacific coast geographic features: CM—Cape Mendocino; MB—Monterey Bay; PC—Point Conception; TMI—Tres Marias Islands. Interior cross (CO-NM-UT-AZ) denotes the Four Corners state junction, GC indicates the location of the Grand Canyon, and the marginal offset near the U.S.–Canada border follows the trans-Idaho discontinuity. Accreted arc assemblages include those of the eastern Klamath Mountains (KL) and the northern Sierra Nevada (SN).

aulacogen oriented at a high angle to the Cordilleran margin, and of the Chuar Group (775–735 Ma) in the Grand Canyon were deposited within intra-Rodinian rift troughs developed well before continental separation occurred south of the trans-Idaho discontinuity (Timmons et al., 2001; Dehler et al., 2005). The distinctly older Unkar Group (1255–1105 Ma) of the Grand Canyon had been deposited in the Laurentian foreland north of the Grenville thrust front (Fig. 5) during protracted orogenesis that assembled Rodinia (Timmons et al., 2005).

Backstripping of miogeoclinal Paleozoic strata exposed in the Great Basin of the western United States, and in southernmost Canada near the marginal offset in the miogeoclinal trend, indicates that postrift thermotectonic subsidence was delayed until Early Cambrian time, during the interval 525–515 Ma (Dickinson, 2004, 2006). By analogy with the modern Atlantic continental margin of Laurentia, where ~55 m.y. elapsed between the initiation of Triassic synrift troughs and the emplacement of Jurassic oceanic crust offshore (Manspeizer and Cousminer, 1988), premiogeoclinal rifting along the continental margin south of the trans-Idaho discontinuity can be estimated at ca. 575 Ma. That inference is compatible with an age of 570 ± 5 Ma for synrift volcanic rocks (Colpron et al., 2002) exposed in southern

Canada approximately along strike from the marginal offset in the continental margin.

The date of ca. 575 Ma inferred for initiation of miogeoclinal rifting in the southern Cordillera is much later than pre-Windermere miogeoclinal rifting at ca. 750 Ma in the northern Cordillera. This diachroneity implies that the Cordilleran miogeocline was the product of two distinct Neoproterozoic rifting events that tore two disparate continental blocks off the western flank of Laurentia to the north and to the south, respectively, of the marginal offset in the passive continental margin along the trans-Idaho discontinuity. Volcanic rocks of intermediate age (715–685 Ma) at the base of the miogeoclinal succession near the marginal offset (Link et al., 2005) may reflect prolonged Neoproterozoic deformation along that structure.

Accretionary Episodes

In late Paleozoic time, internally disrupted allochthons (Figs. 4 and 5) composed of oceanic facies were thrust for 100+ km over the Cordilleran miogeocline from the paleo-Pacific realm to the west. Lithic constituents are predominantly seafloor chert, argillite, and quartzose turbidites, but they also include pillow basalts of oceanic crust or seamounts, and peridotite or serpentinite derived from subjacent oceanic mantle. In the northern Cordillera (Fig. 4), Paleozoic thrusts are largely overprinted by Mesozoic thrusting and other deformation, but they are preserved as the Roberts Mountains and Golconda thrusts (Fig. 5) in the Great Basin of the intermountain region in the southern Cordillera.

The Roberts Mountains and Golconda allochthons were emplaced during two discrete episodes of incipient continental subduction termed the Antler and Sonoma orogenies, respectively (Dickinson, 2004). The two events were spaced ~110 m.y. apart during brief intervals of ~25 m.y. each in latest Devonian to earliest Mississippian and Late Permian to earliest Triassic time, respectively (Dickinson, 2006). The two allochthons were the subduction complexes of an intraoceanic arc complex now exposed as the volcanic assemblages of Devonian-Mississippian and Permian frontal and remnant arcs that were accreted to Laurentia in the Klamath–Sierran region (Fig. 5), when successive episodes of slab rollback drew the arc complex toward the miogeoclinal margin (Dickinson, 2000). The Antler and Sonoma orogenies reflected arc-continent collision between the west-facing miogeocline and the east-facing arc complex, which first subducted offshore paleo-Pacific seafloor and then the continental margin downward to the northwest.

To the north, lack of evidence for the presence of either allochthon along the trend of the marginal offset marked by the trans-Idaho discontinuity may stem from engulfment by younger batholiths, widespread erosion, burial beneath volcanic cover, or anomalous tectonic behavior along that atypical segment of the Cordilleran miogeoclinal margin. Allochthons of both Antler and Sonoma age have been identified, however, in the Kootenay structural arc (Fig. 4), which spans the U.S.–Canada border north of the trans-Idaho discontinuity (Dickinson, 2004).

In the northern Cordillera (Fig. 4), post-Triassic tectonic transport of Paleozoic allochthons during Mesozoic arc-continent collision and retroarc thrusting has largely obscured their original character and positions (Dickinson, 2004; Colpron et al., 2006). The Sylvester allochthon emplaced above the Cassiar platform (Fig. 4) is composed exclusively of post-Devonian rocks (Nelson, 1993), and it apparently represents only the younger allochthon of Sonoma age. Nearby, however, blueschists of both Early Mississippian (ca. 345 Ma) and Late Permian (270–260 Ma) age are present within the allochthonous Yukon–Tanana terrane (Erdmer et al., 1998) lying structurally above miogeoclinal strata both west of the Cassiar platform and within the isolated Anvil allochthon (Fig. 4), which has been offset from farther northwest by dextral strike slip along the Tintina fault. Early Mississippian (360–350 Ma) and Late Permian (ca. 260 Ma) peaks in felsic magmatism (Colpron et al., 2006), closely matching the blueschist ages, reflect pulses of arc magmatism associated with late Paleozoic subduction. In detail, however, the metamorphosed Yukon–Tanana terrane includes parautochthonous miogeoclinal strata as well as overthrust allochthons in both the Yukon and adjacent Alaska (Fig. 4).

Remnants of both Devonian and Permian arc assemblages (sub-Stikinia and sub-Quesnellia of Fig. 4), which locally underlie Mesozoic Stikinia (Gunning et al., 2006) and Quesnellia (Beatty et al., 2006) arc assemblages in the northern Cordillera, are regarded here as counterparts of the Klamath–Sierran arc assemblages accreted to the southern Cordillera (Fig. 5) during Antler-Sonoma orogenesis. Deformed allochthonous assemblages exposed both east and west of the Cassiar platform (Fig. 4) also include Devonian (365–340 Ma) and Permian (ca. 260 Ma) granitic plutons of arc affinity (Mortensen, 1992). Moreover, Permian arc as well as seafloor volcanics are present in the Sylvester allochthon, which structurally overlies the Cassiar platform (Nelson, 1993), and in analogous correlative assemblages exposed farther south (Ferri, 1997). All these telescoped Paleozoic arc remnants (the Finlayson and Klinkit arcs of Colpron et al., 2006) are regarded here as remnants of intraoceanic arcs accreted to the northern Cordillera in late Paleozoic time. By contrast, the Paleozoic arc assemblage of the Alexander Archipelago (Fig. 4) was not accreted along the coastal fringe of the northern Cordillera until mid-Mesozoic time, following oroclinal deformation of the allochthonous Paleozoic assemblages lying immediately to the east (see following discussion).

Others have inferred that the Paleozoic arc assemblages, interpreted here as remnants of an east-facing intraoceanic arc complex that collided with the Cordilleran margin, formed instead as a west-facing offshore fringing arc separated from the Cordilleran margin by an elongate backarc basin of uncertain width (Miller et al., 1992; Colpron et al., 2006). Still others have suggested an initially west-facing arc complex that was associated with backarc spreading to open a marginal sea between the arc system and Laurentia, followed by reversal of arc polarity to draw the fringing arc complex, then east-facing, back against the Cordilleran margin during a short-lived episode of west-dipping subduction (Dusel-Bacon et al., 2006; Piercey et al., 2006; Nelson et al., 2006; Colpron et al., 2007). The existence of a west-facing fringing arc only separated from the Cordilleran margin of Laurentia by backarc spreading during arc evolution is discounted here from the lack of Paleozoic magmatism affecting known Laurentian basement. Moreover, in the southern Cordillera, where structural telescoping of Paleozoic arcs and subduction complexes is minimal, there is no evidence for a west-facing arc at any time during the development of the frontal and remnant arcs that form the Klamath–Sierran accreted arc complex (Dickinson, 2000). The distance that the intraoceanic arc system initially stood off Laurentia is uncertain, but some arc plutons intrude deformed strata composed of detritus that was transported westward from the continental block (Nelson and Gehrels, 2007).

Antler-Sonoma Foreland

Tectonic loads of overthrust Antler-Sonoma allochthons downflexed the western flank of Laurentia to form asymmetric proforeland basins that extended across the miogeoclinal belt onto the fringe of the interior craton (Savoy and Mountjoy, 1995; Dickinson, 2006). An apron of clastic sediment composed of quartzolithic detritus (Dickinson et al., 1983; Gordey, 1992) derived from Antler-age allochthons was shed eastward into the foreland toward carbonate platforms of the continental interior (Smith et al., 1993), but Sonoma-age downflexure of the craton is indicated only by the inland limit of Triassic marine transgression (Dickinson, 2006). Intrabasinal extension recorded by syndepositional normal faulting in the northern Cordillera (Gordey et al., 1987) was induced by flexural deformation of the basin floor in response to thrust loads imposed from the west (Smith et al., 1993).

During the time interval between Antler and Sonoma thrusting along the Cordilleran margin, a broad tract of the foreland region in the southern Cordillera was disrupted into local uplifts and yoked basins of the Ancestral Rocky Mountains province (Fig. 5). The intracontinental deformation was a response to torsional effects produced by suturing of Gondwana to the southern flank of Laurentia along the Pennsylvanian-Permian Ouachita orogen, which formed a western extension of the Hercynian orogenic belt during the assembly of Pangea (Dickinson and Lawton, 2003). Deformation was centered on Pennsylvanian time, though precursor effects occurred in latest Mississippian time on the east and residual effects lingered into Early Permian time farther west. Interior strain within the continental block developed as Laurentia was drawn progressively into the subduction zone along the northern margin of Gondwana by sequential intercontinental suturing that proceeded from east to west.

Continental Truncation

Once the northern margin of Gondwana had lodged against Laurentia along the Ouachita suture, continued subduction beneath the western flank of Gondwanan crust, newly incorporated into Pangea, produced truncation of the Cordilleran margin along the California-Coahuila transform (Fig. 5), which offset the miogeoclinal Caborca block from California into Mexico (Dickinson and Lawton, 2001a). Continuing subduction beneath Gondwanan crust (Fig. 5) generated the Permian-Triassic (284–232 Ma) East Mexico continental-margin arc along the western edge of Pangea after Gondwana was conjoined with Laurentia (Torres et al., 1999). The age span of the East Mexico arc dates movement on the California-Coahuila transform as mainly Permian to mid-Triassic (Dickinson, 2000), although incipient slip may have begun in Pennsylvanian time (Stevens et al., 2005).

The truncation of Antler-Sonoma orogenic trends oriented northeast-southwest in the southern Cordillera set the stage for Mesozoic circum-Pacific subduction along a nascent continental margin that extended northwest-southeast (Dickinson, 2000; Stevens et al., 2005), in line with the flank of the northern Cordillera. Continuations of pre-truncation Antler and syntruncation Sonoma orogens across the California-Coahuila transform remain uncertain (Stewart, 2005), although counterparts may be present as oceanic facies thrust over parts of the Caborca block (Stewart et al., 1990). Close tectonic connections postulated (Poole et al., 2005) between overthrust assemblages of the Caborca block and the Ouachita suture belt seem unlikely, however, because the Caborca block was still in place as an integral southern continuation of the Cordilleran miogeoclinal belt along the western flank of Laurentia at the same time that Ouachita orogenesis was deforming the southern flank of Laurentia far to the east.

MIDDLE MESOZOIC CORDILLERA

Following Antler-Sonoma accretionary expansion of the Cordilleran margin of Laurentia, lodging of the southern flank of Laurentia against Gondwana along the Ouachita suture, and modification of the Cordilleran margin by truncation along the California-Coahuila transform, a magmatic arc developed along the whole length of the Cordillera in response to subduction of oceanic lithosphere beneath the continental margin (Figs. 6 and 7). Inception of the native early Mesozoic arc-trench system of the Cordillera marked the onset of integrated circum-Pacific subduction along the paleo-Pacific margin of Pangea. The oldest widespread segments of the native Cordilleran magmatic arc are Late Triassic in age (Dickinson, 2004, 2008), but precursor phases of arc plutonism are as old as Early Triassic in southern California (Barth and Wooden, 2006). These older plutons in the southern Cordillera suggest that Cordilleran arc magmatism may initially have propagated northward from the Permian-Triassic East Mexico arc as new patterns of plate motion evolved along the Laurentian continental margin.

Parallel to the California-Coahuila transform (Fig. 5), native Triassic-Jurassic arc assemblages transect miogeoclinal and Laurentian cratonal crust (Schweickert and Lahren, 1993), which had been placed in a position close to the continental margin by transform truncation of the continental block. The arc trend extended

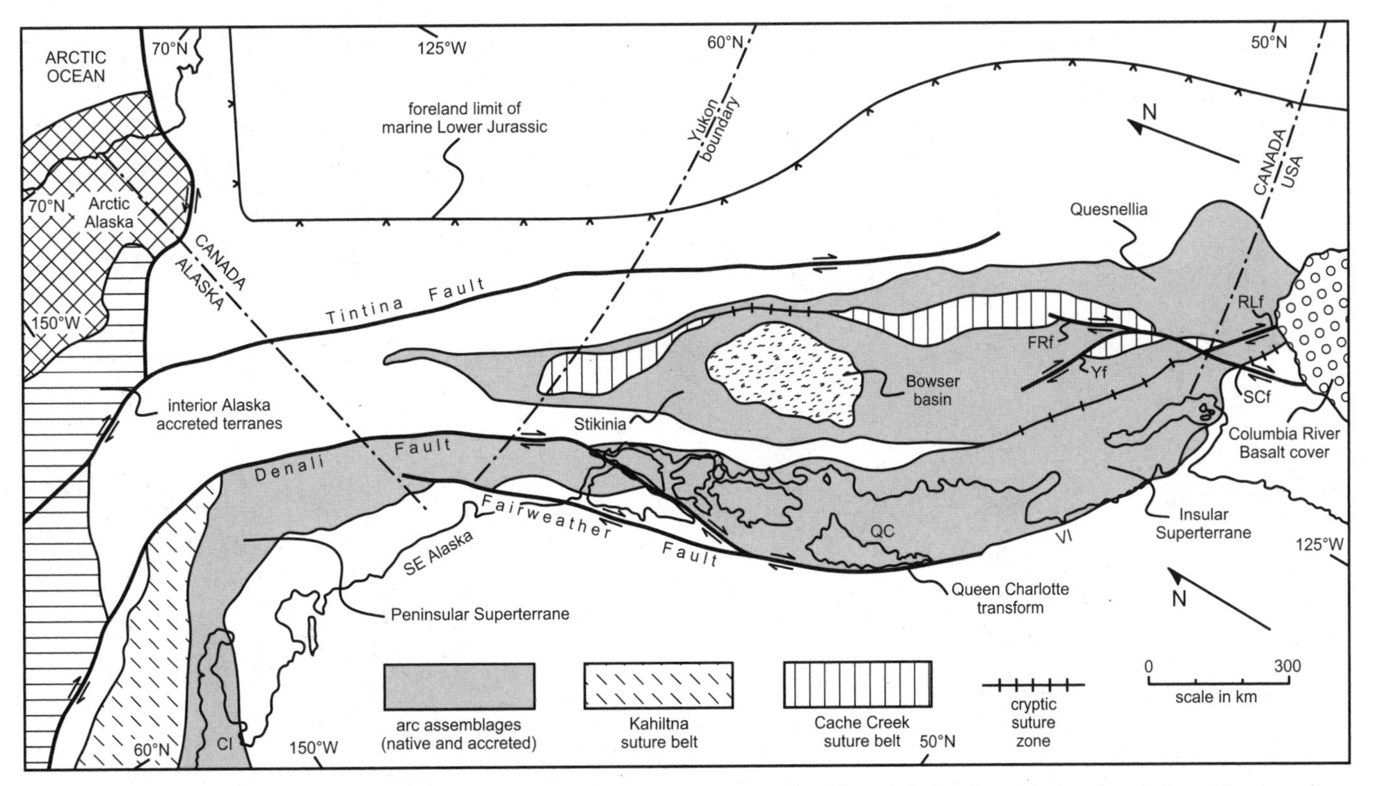

Figure 6. Middle Triassic to Early Cretaceous tectonic elements of the northern Cordillera (Alaska–Canada) showing native and accreted arc assemblages. Faults: FRf—Fraser River; SCf—Straight Creek; RLf—Ross Lake; Yf—Yalakom. Map is modified after Plafker and Berg (1994) and Dickinson (2004). Pacific coast geographic features: CI—Cook Inlet; QC—Queen Charlotte Islands; VI—Vancouver Island.

southward (Fig. 7), however, across the Ouachita suture into Gondwanan crust of eastern Mexico (Dickinson and Lawton, 2001a), and northward (Fig. 6) to merge with the Quesnellia arc built in the Canadian Cordillera on Antler-Sonoma assemblages thrust over the continental margin during Paleozoic time (Dickinson, 2004). Past speculation that Quesnellia was a free-standing intraoceanic arc accreted later to North America by collapse of an intervening marginal ocean basin is countered by recent isotopic studies (Unterschutz et al., 2002; Erdmer et al., 2002; Petersen et al., 2004; Nelson and Friedman, 2004). The backarc region in the northern Cordillera was flooded by marine waters (Fig. 6), but it was occupied behind much of the southern Cordillera by desert ergs and associated fluvial assemblages (Fig. 7). Accommodation space for both kinds of sedimentary successions was provided by westward tilt of the flank of the interior craton as the result of geodynamic subsidence under the influence of a subducted slab in the mantle beneath (Mitrovica et al., 1989).

West of the native Triassic-Jurassic arc assemblage, there is a paired subduction complex composed of mélanges and variably deformed thrust panels of oceanic strata termed the Cache Creek belt (Mortimer, 1986), from its designation in the northern Cordillera. The Cache Creek belt (Figs. 6 and 7) forms a suture zone trapped between the Triassic-Jurassic continental margin and varied segments of exotic or displaced island arcs of Triassic to Jurassic age that were accreted to the continental margin from the paleo-Pacific realm to the west as the result of collisions between the west-facing continental-margin arc and east-facing intraoceanic arcs (Moores, 1998; Godfrey and Dilek, 2000; Ingersoll, 2000; Dickinson, 2004, 2008). Blueschists of the Cache Creek belt have yielded Late Triassic isotopic ages of 230–210 Ma in both the northern and southern Cordillera (Erdmer et al., 1998), and fossiliferous stratal components of the Cache Creek belt range in age locally from Carboniferous to Early or Middle Jurassic (Dickinson, 2004, 2008). The Cache Creek oceanic realm had closed by Late Jurassic time as far south as central California, but the analogous Arperos-Mezcalera suture zone along strike in Mexico (Fig. 7) was not trapped between the native continental-margin arc and the exotic offshore arc complex of the Guerrero superterrane until Early Cretaceous time (Dickinson and Lawton, 2001a; Dickinson, 2004).

Northern Cordillera

Midway up the length of the northern Cordillera (Fig. 6), the northernmost extremity of the Cache Creek belt is enclosed between the native Quesnellia arc to the east and the Stikinia arc, which is interpreted as an oroclinally deformed extension of the Quesnellia arc now lying west of the Cache Creek belt (Mihalynuk et al., 1994). The western flank of Stikinia is underlain by Paleozoic assemblages (Fig. 4) that were apparently

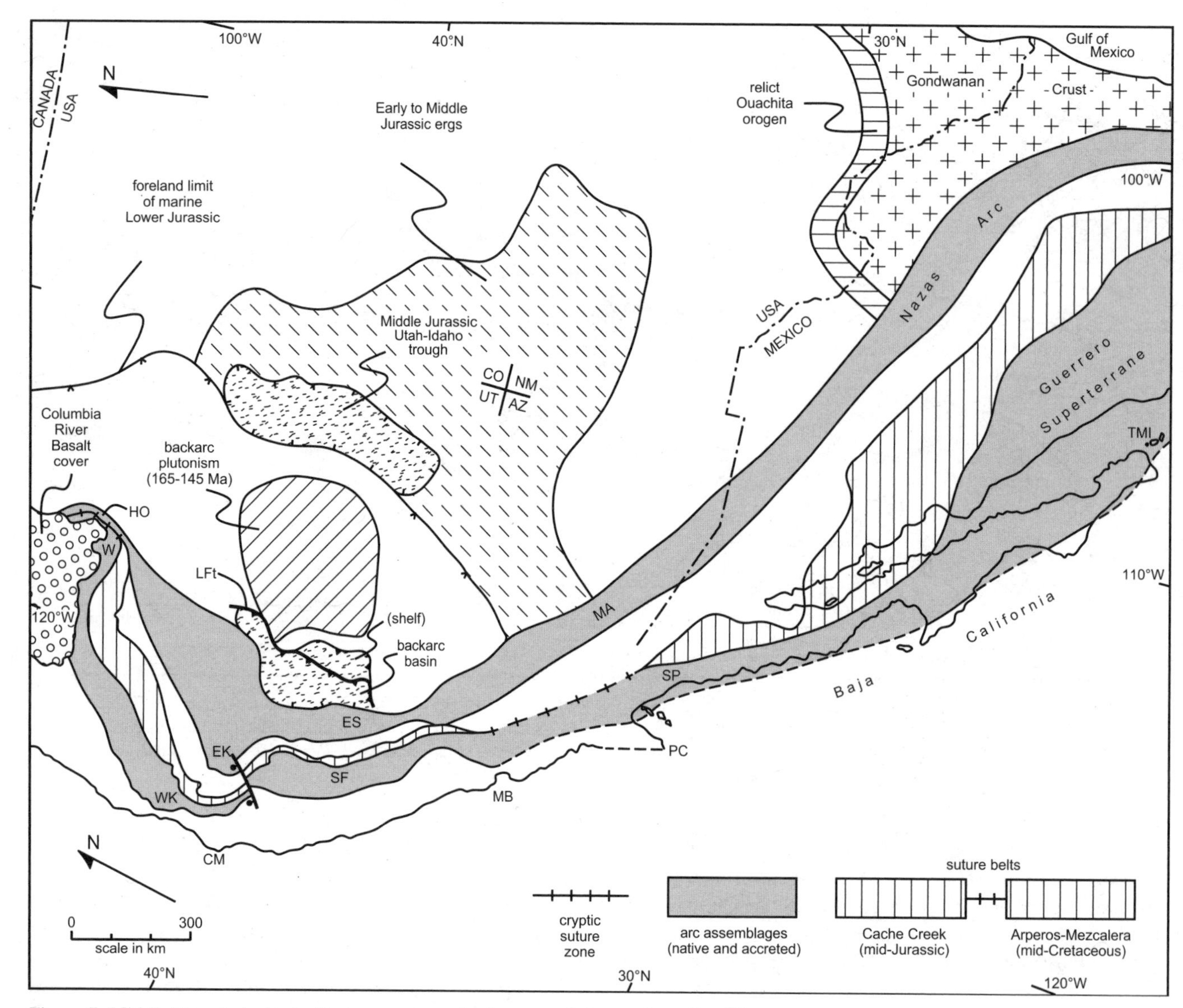

Figure 7. Middle Triassic to Early Cretaceous tectonic elements of the southern Cordillera (U.S.–Mexico) showing native and accreted arc assemblages. Gulf of California is closed. LFt—Luning-Fencemaker retroarc thrust system (175–165 Ma). Key U.S. arc segments: EK—eastern Klamath; ES—eastern Sierran; HO—Huntington–Olds Ferry; MA—Mojave-Arizona; SF—Sierran foothills (–Great Valley subsurface); SP—Santiago Peak; W—Wallowa (Blue Mountains); WK—western Klamath. Map is modified after Dickinson and Lawton (2001a) and Dickinson (2004, 2006, 2008). Pacific coast geographic features: CM—Cape Mendocino; MB—Monterey Bay; PC—Point Conception; TMI—Tres Marias Islands. Interior cross (CO-NM-UT-AZ) denotes the Four Corners state junction.

peeled away from the continental margin north of Quesnellia by backarc spreading during oroclinal rotation of Stikinia. Analysis of tectonic relations along the eastern flank of Stikinia shows that the Stikinia arc was east-facing at the time of its reattachment to the continental block, and a forearc basin was located between the arc assemblage and the Cache Creek belt, across which Stikinia was amalgamated with Quesnellia by arc-arc collision, completed early in Middle Jurassic time (Dickinson, 2004; English and Johnston, 2005). Longitudinal variations in the ages of the youngest stratal components of the Cache Creek belt entrapped between Quesnellia and Stikinia suggest progressive southward closure of the Cache Creek suture from the oroclinal hinge point on the north where Quesnellia bends around into Stikinia (Dickinson, 2004). The Quesnellia–Stikinia relationship is the only instance of such pronounced oroclinal deformation of an arc assemblage known within the North American Cordillera.

The Late Jurassic to Early Cretaceous Bowser Basin (Fig. 6) developed atop Stikinia as a successor basin that postdated Stikinia–Quesnellia arc-arc collision, and much of the contact between Stikinia and the Insular-Peninsular superterrane (Fig. 6) to the west is masked by correlative Upper Jurassic to Lower Cretaceous strata of the elongate Gravina Basin (McClelland et al.,

1992). Mid-Cretaceous thrusting later carried rocks east of the contact zone over the Gravina Basin and the eastern flank of the Insular-Peninsular superterrane. Locally, however, sub-Gravina metavolcanic rocks that overlap the contact zone document initial Middle Jurassic (ca. 175 Ma) accretion of the Insular-Peninsular superterrane to the western flank of Stikinia (Gehrels, 2001). As Stikinia was then an east-facing island arc, there was no subduction along its western flank to draw the Insular-Peninsular superterrane toward the continental margin. Accordingly, Early to Middle Jurassic arc magmatism (190–165 Ma) displayed on Vancouver Island, the Queen Charlotte Islands, and peninsular Alaska is inferred to reflect activation of subduction along the eastern or inboard flank of the Insular-Peninsular superterrane (Dickinson, 2004). Consumption of intervening paleo-Pacific seafloor west of the continental margin thereby allowed the Insular-Peninsular superterrane to accrete to the backarc flank of Stikinia. To the north, the Kahiltna suture belt (Fig. 6) marks the zone along which the Insular-Peninsular superterrane lodged against interior Alaska, but to the south, near the U.S.–Canada border, the suture zone is cryptic where it has been largely obliterated by the intrusion of Late Cretaceous batholiths (Fig. 6). The Early Jurassic Talkeetna igneous assemblage at the northern end of the Insular-Peninsular superterrane has the isotopic signature of an intraoceanic island arc (Rioux et al., 2007).

The Insular-Peninsular superterrane includes the Paleozoic arc assemblage of the Alexander Archipelago (Fig. 4), composed of largely pre-Devonian rocks overlain by less deformed Devonian to Permian strata (Butler et al., 1997), and the Wrangellia terrane, composed largely of Permian arc volcanics overlain by Triassic basalt capped by Upper Triassic limestone (Jones et al., 1977). These two major components of the superterrane were amalgamated together by Carboniferous time (Gardner et al., 1988). Before its accretion to North America along the backside of Stikinia during Jurassic time, the Insular-Peninsular superterrane apparently drifted within the paleo-Pacific realm as a compound intraoceanic arc structure between the paleolatitudes of the Arctic and the Pacific Northwest (Butler et al., 1997; Colpron et al., 2007). The complex Jurassic plate motions required to achieve oroclinal deformation and tight face-to-face juxtaposition of the Quesnellia and Stikinia arc assemblages, coupled with accretion of the Insular-Peninsular superterrane to the Cordilleran margin within the same time frame, are indeterminate with present information.

Southern Cordillera

In the southern Cordillera, south of Miocene Columbia River Basalt cover (Fig. 7), exotic Triassic-Jurassic intraoceanic island arcs that were accreted west of the Cache Creek suture belt in mid-Jurassic time as far south as central California include variously named local terranes of the Blue Mountains, western Klamath Mountains, and Sierra Nevada foothills (Dickinson, 2004, 2008). The accreted arc assemblages are everywhere separated from coeval native Cordilleran arc assemblages to the east by mélanges and associated deformed and metamorphosed strata forming the southern extension of the Cache Creek suture belt, and incorporating oceanic facies that have yielded fossils as young as Early to Middle Jurassic in age.

A segment of the suture zone between native and accreted arc assemblages in southern California has been overprinted by intrusion of Late Cretaceous batholiths (Todd et al., 2003) to form the cryptic suture of Figure 7. Farther south, the accreted arc assemblage forms the Guerrero superterrane of coastal Mexico and Baja California (Dickinson and Lawton, 2001a), with extensions as far north as the Santiago Peak volcanics (Fig. 7), which were displaced northward into present-day coastal California by Cenozoic strike slip along the San Andreas transform system. The Guerrero superterrane is composed dominantly of Upper Jurassic to Lower Cretaceous volcanogenic assemblages, but, in Baja California, it includes Upper Triassic to Middle Jurassic components coeval in age with native Cordilleran arc assemblages of the southwestern United States and northeastern Mexico (Fig. 7). In southern Mexico, the composite Guerrero superterrane includes a Jurassic-Cretaceous assemblage of conjoined intraoceanic arcs (Talavera-Mendoza et al., 2007). The Arperos-Mezcalera suture of central Mexico (Fig. 7) was the southern counterpart of the Cache Creek suture belt farther north, but it did not close between the Guerrero superterrane and the pre-Mesozoic continental framework of eastern Mexico until ca. 120 Ma, in Early Cretaceous time (Dickinson and Lawton, 2001a).

The Mesozoic arc assemblages lying offshore from the Cache Creek and Arperos-Mezcalera suture belts in the southern Cordillera are interpreted as segments of an east-facing intraoceanic arc complex that was not accreted to the continental margin until an intervening oceanic plate, termed the Mezcalera plate (Dickinson and Lawton, 2001a), was consumed. The name Mezcalera is taken from a suture assemblage in central Mexico, but the northern extremity of the Mezcalera plate was apparently the oroclinal hinge where the Quesnellia and Stikinia arcs are conjoined longitudinally around the northern limit of the Cache Creek suture belt in the northern Cordillera (Fig. 6). The deepest structural levels of the accreted arc structures are ophiolitic in character, with no hint of continental affinity (Dickinson, 2004, 2008). Subduction of the Mezcalera plate westward beneath the intraoceanic arc and eastward beneath the native Cordilleran arc eventually produced arc-continent collisions to accrete exotic intraoceanic arc segments to the Cordilleran continental margin. Accretion of the intraoceanic arc complex expanded the Pacific margin of the Cordillera and triggered initial subduction of the Farallon plate, lying beyond the accreting arc system that was built on its eastern edge.

An alternate interpretation, that the Mezcalera oceanic plate occupied an elongate backarc basin behind a west-facing fringing arc that stood not far offshore from the southern Cordillera, encounters the difficulty that longitudinal counterparts in both the northern Cordillera and the Caribbean region farther south (see following discussions) were east-facing intraoceanic features. Moreover, there is a fundamental impediment to the generation of simultaneous arc magmatism along both an offshore intraoceanic arc and the native Cordilleran continental-margin arc for a prolonged interval of time if the intervening oceanic basin is assumed to have been a narrow backarc basin. At typical trench convergence rates of 60–90 mm/yr (Jarrard, 1986), the continuation of native Cordilleran arc activity throughout the 50–100 m.y. history of active arc magmatism within the offshore intraoceanic arc complex requires an intervening ocean basin that is 3000–9000 km wide between the continental margin and the intraoceanic arc system offshore.

The following tectonic features in the southern Cordillera east of the Sierra Nevada can be related to consumption of the Mezcalera plate and accretion of the intraoceanic arc on the leading edge of the Farallon plate (Dickinson, 2006): (1) Middle Jurassic inversion of a Late Triassic to Middle Jurassic backarc basin by thrusting of basinal facies over shelf facies along the Luning-Fencemaker thrust system, (2) formation of the Middle Jurassic Utah-Idaho rift trough farther inland, and (3) a broad domain of Middle to Late Jurassic backarc plutonism (165–145 Ma) between the two sedimentary basins (Fig. 7).

Subsidence of the backarc basin immediately behind the Cordilleran arc was fostered by backarc extension that ended at the same approximate time in the Middle Jurassic as the accretion of the intraoceanic arc that marked final consumption of the Mezcalera plate. Inversion of the basin by thrusting stemmed from crustal shortening across the arc during arc-continent collision. The intracontinental rifting that formed the Utah-Idaho trough was produced by slab rollback of the Mezcalera plate as subduction slowed along the continental margin (Lawton and McMillan, 1999). The pulse of backarc plutonism, locally overprinting the Luning-Fencemaker thrust system, was triggered by thermal effects produced by subterranean slab breakoff of the subducted Mezcalera plate in the mantle beneath the southern Cordillera after closure of the crustal suture zone to the west (Cloos et al., 2005). Slab breakoff is inferred to have fostered upwelling and partial melting of asthenospheric mantle to trigger inland magmatism not directly linked to arc activity. Delayed arrival beneath the Cordillera of the leading edge of the Farallon plate, which was subducted beneath the continental margin only after arc accretion was complete, provided a time window for the episode of backarc magmatism induced by Mezcalera slab breakoff.

Pacific Northwest Relations

The longitudinal correlation of pre–mid-Cretaceous tectonic elements across the Pacific Northwest from Canada into the United States is obscured by widespread Neogene volcanic cover and overprinting of tectonic trends by complex structural deformation near the U.S.–Canada border (Fig. 8A). A satisfactory tectonic reconstruction (Fig. 8B) is achieved, however, by reversing 105–110 km of Eocene dextral slip on the Fraser River–Straight Creek fault zone and 110–115 km of previous Cretaceous to Eocene dextral slip on the intersecting Yalakom–Ross Lake fault system (Dickinson, 2004), coupled with back-rotation of Siletzia and the Blue Mountains to recover clockwise tectonic rotations during Eocene time (Heller et al., 1987). The amounts of coordinated clockwise rotation inferred for Figure 8 (see figure caption) improve the reconstruction of Dickinson (2004), which was based on postulated rotations that were less well constrained. Further recovery of Cenozoic extensional deformation recorded by core complexes and the Kishenehn-Flathead rift basins near the U.S.–Canada border (Fig. 8) might achieve a slightly closer fit of pre-Mesozoic tectonic elements across the Pacific Northwest, but it was not attempted for the reconstruction of Figure 8 because the amount of net Cenozoic extension remains uncertain.

From the reconstruction (Fig. 8B), native arc assemblages of Triassic-Jurassic age in the Eastern Klamath Mountains and Sierra Nevada are shown to connect through a coeval arc assemblage of the eastern Blue Mountains with the continental-margin Quesnellia arc of the northern Cordillera. Fault-dismembered segments of the Jurassic-Cretaceous Tyaughton-Methow trough, spanning the U.S.–Canada border, formed a forearc basin along the flank of Quesnellia that was converted to a sutural basin by accretion of Stikinia, which the basin onlaps westward (Dickinson, 2004). Immediately to the west, the Cache Creek suture belt of the northern Cordillera is contiguous southward into central mélange belts of the Blue Mountains and Klamath Mountains.

Farther west, because critical relations are buried under Cenozoic volcanic cover, there is uncertainty whether the accreted arcs of the western Klamath Mountains and Sierra Nevada foothills are southern counterparts of the Stikinia arc or of the Insular-Peninsular superterrane in the northern Cordillera. Correlation with the latter is favored because the accreted Wallowa terrane west of the Cache Creek suture belt in the Blue Mountains (Fig. 8A) has an internal stratigraphy that closely matches the stratigraphy of Wrangellia within the Insular-Peninsular superterrane (Jones et al., 1977). Characteristic Wrangellia stratigraphy is not present south of the Blue Mountains, but the accreted Klamath-Sierran arc assemblages resemble counterparts in the Queen Charlotte Islands within the Insular-Peninsular superterrane. Beyond the Shuksan thrust system, which bounds the Insular-Peninsular superterrane on the south near the U.S.–Canada border (Figs. 8A and 8B), an underthrust assemblage of Upper Jurassic to Lower Cretaceous blueschists and clastic strata resembles the late Mesozoic Franciscan subduction complex thrust beneath the western Klamath Mountains (Dickinson, 2004, 2008). Along the Pacific Northwest coast, Siletzia is an oceanic Paleocene-Eocene seamount chain accreted to the continental margin early in Eocene time (Dickinson, 2008) and buried beneath a postaccretion Eocene and younger forearc basin before tectonic rotation (Heller and Ryberg, 1983).

POST–MID-CRETACEOUS CORDILLERA

By Late Cretaceous time, a batholith belt marking the trend of the Cordilleran magmatic arc was emplaced along the length of the Cordillera inland from coastal subduction complexes that

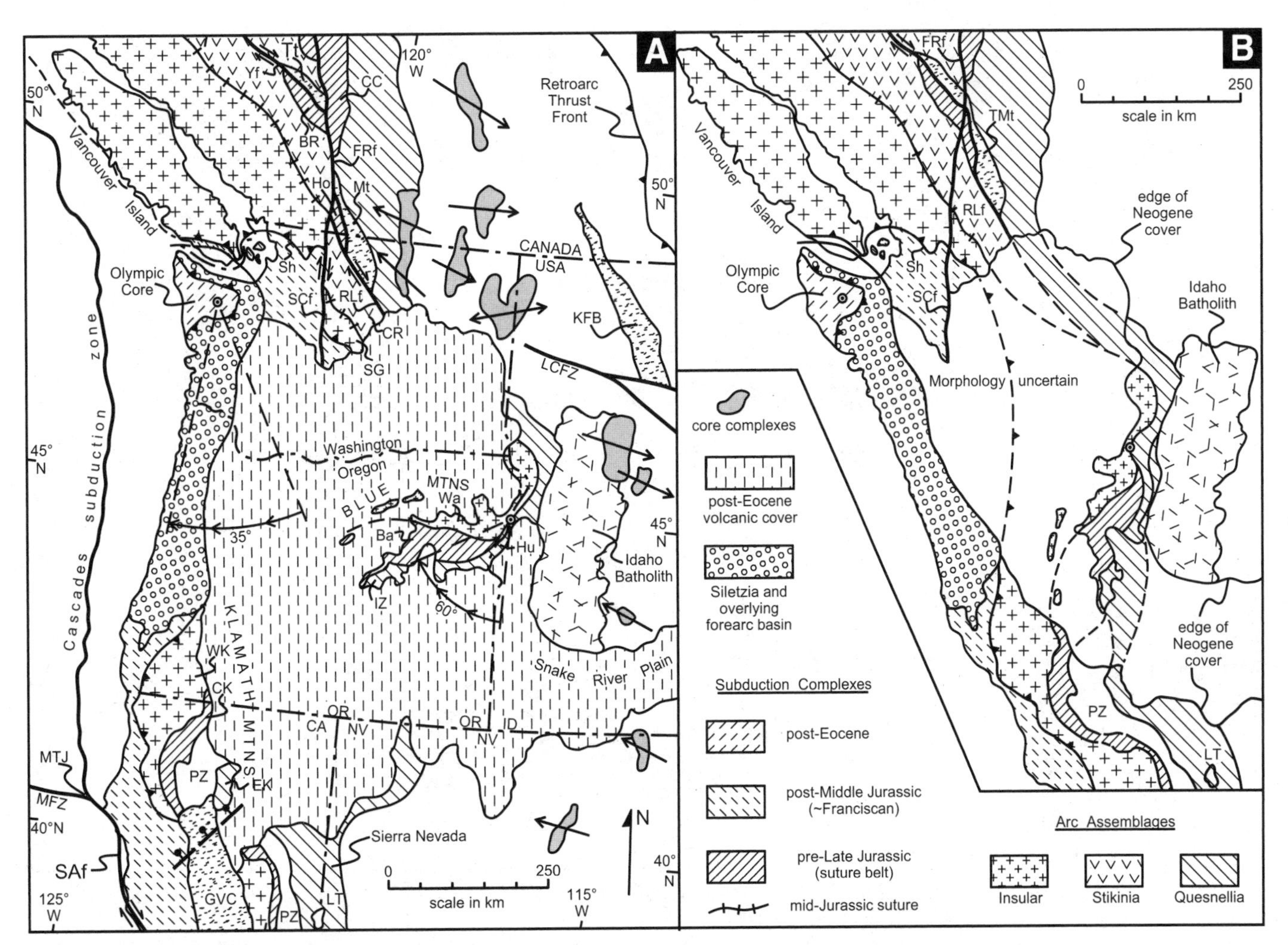

Figure 8. Correlations of pre–mid-Cretaceous tectonic elements spanning the Pacific Northwest: (A) present; and (B) Cretaceous-Paleogene (restored). Pivot points (dotted circles) and circular arc segments with arrows denote bulk Paleogene rotations of (1) the Oregon-Washington Coast Range estimated as 35° by subtracting Neogene continuum distortion of ~15° (England and Wells, 1991) from net paleomagnetic rotation of ~50° (Wells and Heller, 1988), and (2) the Blue Mountains province of eastern Oregon by 60° (Wilson and Cox, 1980). Basin fill (stippled): GVC—Cretaceous-Cenozoic Great Valley of California; KFB—Cenozoic (48–20 Ma) Kishenehn-Flathead rift basins; TMt—Jurassic-Cretaceous Tyaughton-Methow forearc-sutural trough (Tt-Mt—offset segments). Arc assemblages (west to east including tectonic correlatives along strike): (1) exotic Insular superterrane (SG—Swakane Gneiss; Wa—Wallowa-Seven Devils terrane; WK—western Klamath Mountains terranes); (2) accreted Stikinia (-Cadwallader) terrane (CR—Cascade River–Holden belt); (3) native Quesnellia arc terrane (EK—eastern Klamath Mountains; IZ—Triassic-Jurassic Izee forearc-sutural basin; Hu—Huntington arc or Olds Ferry terrane). Pre–Late Jurassic mélange terranes (suture belt): Ba—Baker; BR—Bridge River; CC—Cache Creek; CK—central Klamath Mountains; Ho—Hozameen. Faults: FRf—Fraser River; LCFZ—Lewis and Clark zone; MFZ—Mendocino fracture zone (MTJ—Mendocino triple junction); SAf—San Andreas; RLf—Ross Lake; SCf—Straight Creek; Sh—Shuksan thrust system (schematic); Yf—Yalakom. Other features: LT—Lake Tahoe (approximate in B); PZ—accreted Paleozoic terranes of the eastern Klamath Mountains and northern Sierra Nevada. Arrows on core complexes denote vergence (direction of movement of upper plate). Maps are modified after Dickinson (1979, 2004), Constenius (1996), and Reed et al. (2005).

were accreted incrementally to the expanding continental margin during late Mesozoic and early Cenozoic time (Figs. 9 and 10). The batholiths were largely intruded into previously accreted tectonic elements, except along their eastern fringe in the United States, where they were injected into native Laurentian crust. Stitching plutons that represent early phases of batholith development tied disparate terranes together in Canada and the United States by Middle to Late Jurassic time (van der Heyden, 1992; Friedman and Armstrong, 1995; Schweickert et al., 1999; Dickinson, 2008), but were not emplaced farther south in Mexico until Early Cretaceous time (Dickinson and Lawton, 2001a).

By Late Jurassic time in Canada (Cant and Stockmal, 1989) and by Early Cretaceous time in the United States (Dickinson, 2006), crustal shortening across the Cordilleran orogen gave rise to a retroarc thrust belt along the flank of a retroforeland basin (DeCelles, 2004), which was downflexed by the load of thrust sheets emplaced from the west (Figs. 9 and 10). To the north (Fig. 9), the thrust belt linked Arctic Alaska to the Cordilleran

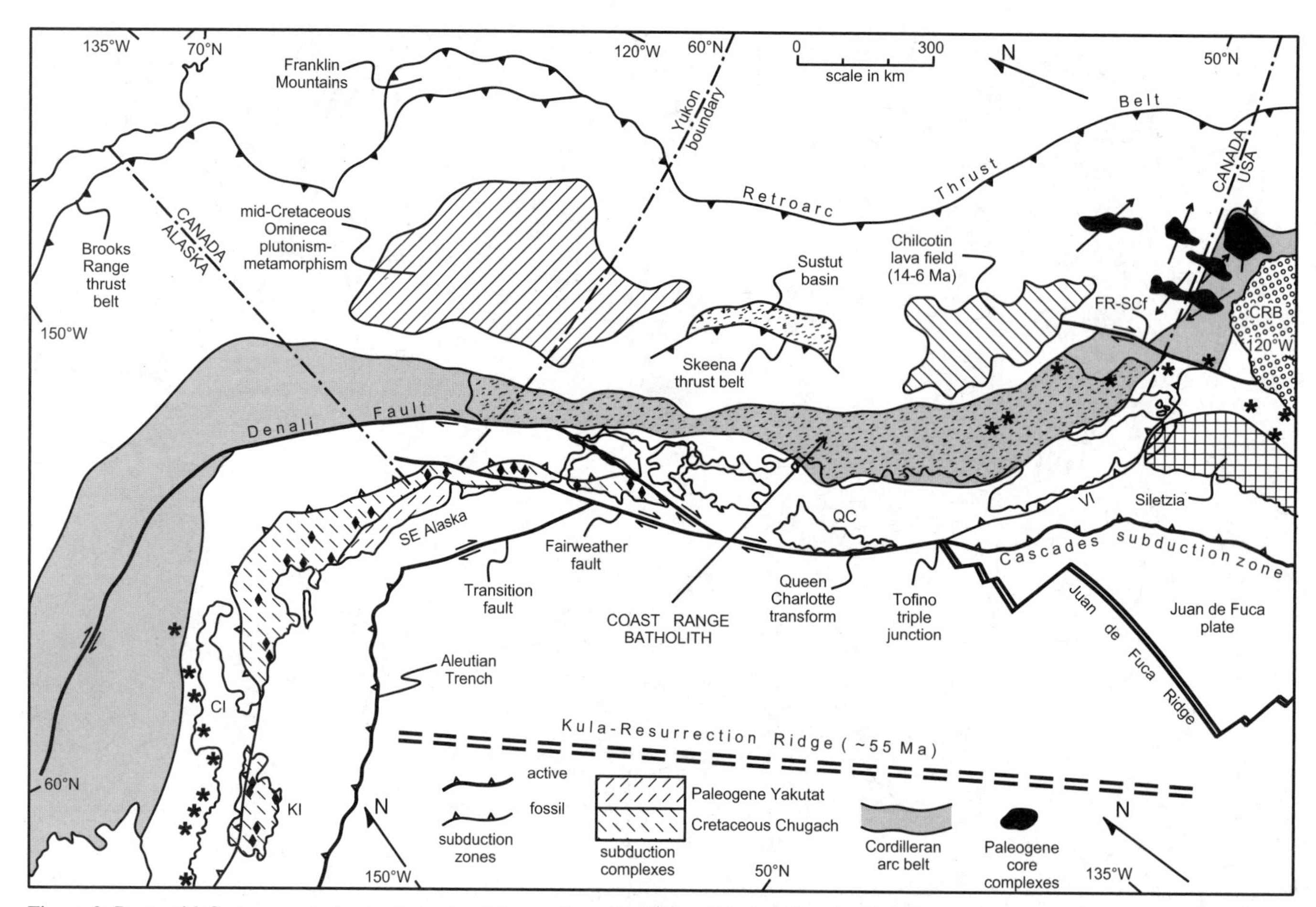

Figure 9. Post–mid-Cretaceous tectonic elements of the northern Cordillera (Alaska–Canada) showing the major batholith belt (magmatic arc) and associated tectonic features. Asterisks denote modern arc volcanoes linked to subduction at the Aleutian Trench or Cascades subduction zone. Diamonds denote near-trench (forearc) Paleocene-Eocene plutons (61 Ma on west; 50 Ma on east) linked to migration of the triple junction where the Kula–Resurrection seafloor spreading ridge (depicted at 55 Ma) intersected the ancestral Cordilleran subduction zone. Arrows for Cordilleran core complexes (58–42 Ma) denote the transport directions of the upper plates. CRB—Columbia River basalt cover; Siletzia—accreted province of Paleogene (62–49 Ma) seamounts. FR-SCf—Fraser River–Straight Creek fault. Map is modified after Plafker and Berg (1994), Haeussler et al. (2003), Dickinson (1997, 2002, 2004), and Reed et al. (2005). Pacific coast geographic features: CI—Cook Inlet; KI—Kodiak Island; QC—Queen Charlotte Islands; VI—Vancouver Island.

orogen for the first time. A high-standing intramontane plateau (Dilek and Moores, 1999) that bridged between the Cordilleran batholith belt and retroarc thrust highlands resembled the Altiplano of the modern Andes, and, by analogy, it has been termed the Nevadaplano in the United States (DeCelles, 2004). Intraorogen deformation well to the west of the leading edge of the retroarc thrust system generated the Skeena thrust belt, with its associated Sustut basin, in the northern Cordillera (Fig. 9) and the Eureka thrust belt in the southern Cordillera (Fig. 10).

During early Cenozoic time, basement-cored uplifts and intervening sedimentary basins of the Laramide province grew within the Cordilleran foreland of the United States (Fig. 10), but they did not extend into Canada or Mexico. During late Cenozoic time, the southern Cordillera in both the United States and Mexico was broadened by crustal stretching within the extensional Basin and Range taphrogen (Dickinson, 2002), a postorogenic feature that did not extend into Canada (Fig. 11). Coastal slivers of the Cordilleran orogen were displaced northward, however, by dextral Cenozoic transform faults that supplanted subduction zones along much of the full length of the continental margin, except where subduction has continued west of the Cascades volcanic arc (Figs. 9 and 11).

Northern Cordillera

In the northern Cordillera (Fig. 9), plutonism along the Coast Range batholith continued from Cretaceous into Paleogene time, and the youngest mid-Eocene plutons were intruded along the eastern fringe of the batholith (Armstrong and Ward, 1993). Arc magmatism was arrested along most of the Canadian continental margin when the oceanic spreading ridge between the Kula and Resurrection plates was drawn against the continental

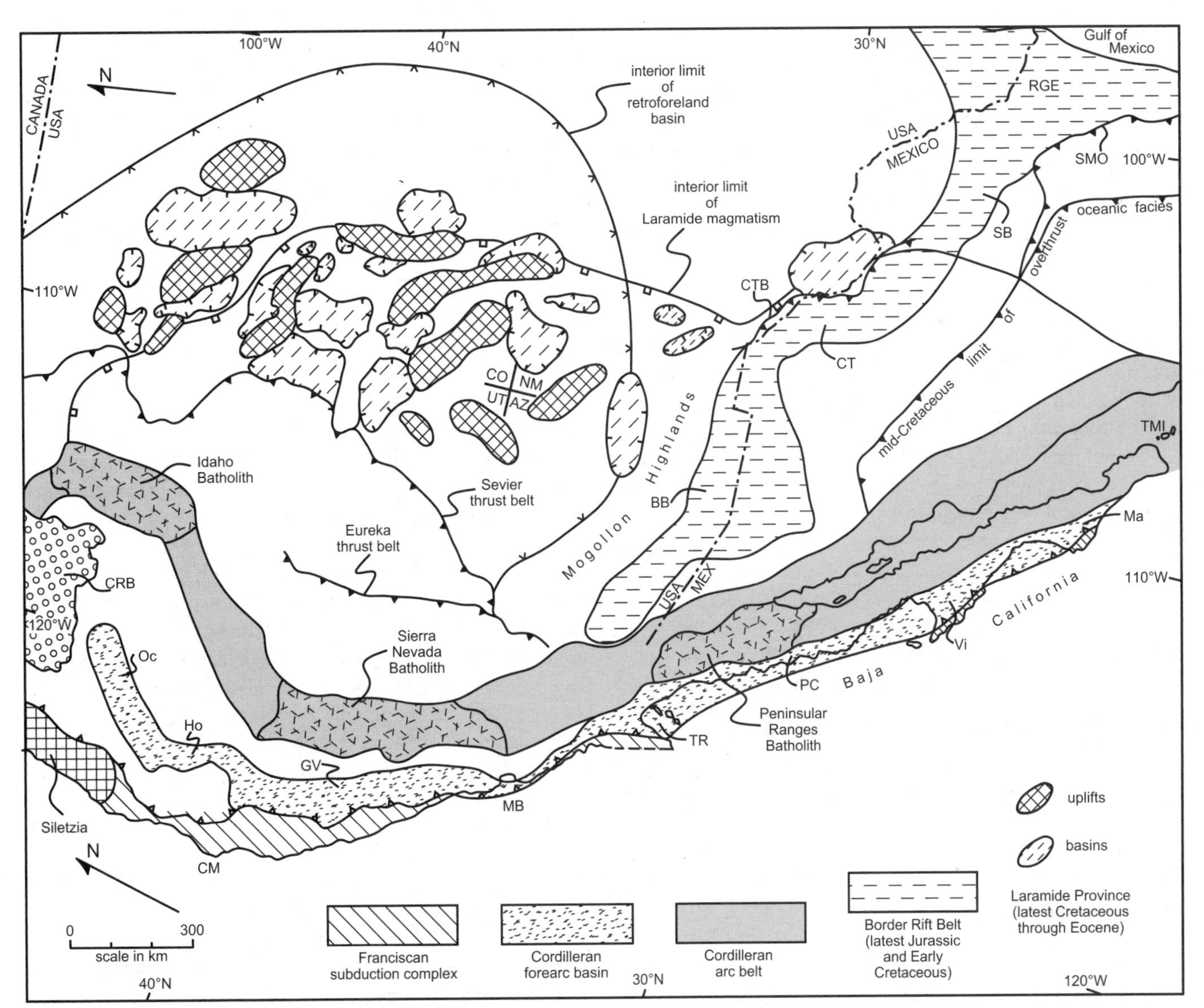

Figure 10. Mid-Cretaceous to mid-Cenozoic tectonic elements of the southern Cordillera (U.S.–Mexico) showing the major batholith belt (magmatic arc) and associated tectonic features. Gulf of California is closed, and California Transverse Ranges are back-rotated. Key segments of Cordilleran forearc basin: GV—Great Valley; Ho—Hornbrook; Ma—Magdalena; Oc—Ochoco; PC—peninsular coast; TR—Transverse Ranges; Vi—Vizcaino. Key segments of Border rift belt: BB—Bisbee Basin; CT—Chihuahua trough; SB—Sabinas Basin. Other geologic features: CRB—Columbia River basalt cover; CTB—Chihuahua tectonic belt (basin inversion); RGE—Rio Grande embayment; SMO—Sierra Madre Oriental thrust front; Siletzia—accreted Paleogene (62–49 Ma) seamounts. Map is modified after Dickinson et al. (1988), Dickinson and Lawton (2001a, 2001b), and Dickinson (1996, 2004). Pacific coast geographic features: CM—Cape Mendocino; MB—Monterey Bay; TMI—Tres Marias Islands. Interior cross (CO-NM-UT-AZ) denotes the Four Corners state junction.

margin by progressive Paleogene subduction of the Resurrection plate (Haeussler et al., 2003). The final consumption of the Resurrection plate at ca. 50 Ma initiated dextral transform slip at the continental margin along the Queen Charlotte fault system, still active between North America and the Pacific plate, which amalgamated with the Kula plate not long after the demise of the Resurrection plate. On the north, the past migratory intersection of the Kula–Resurrection ridge with the Aleutian Trench is marked by a diachronous belt of forearc plutons (Bradley et al., 2003), which were intruded into the accretionary prism of the Alaskan segment of the Cordilleran arc during the interval from 61 Ma on the west to 50 Ma on the east (Fig. 9). Cenozoic translation along the Queen Charlotte transform system disrupted the late Mesozoic forearc belt along the Canadian continental margin (Fig. 9) and transported Cretaceous-Paleogene Chugach–Yakutat subduction complexes northward into coastal Alaska (Sample and Reid, 2003).

Past speculation (Monger, 1993; Cowan et al., 1997), based solely on paleomagnetic data, that the western Canadian Cordillera, including a large segment of the Cretaceous batholith belt, was transported northward by Cretaceous-Paleogene coastwise transform slip from an origin along the continental margin of the southwestern United States or Mexico, encounters an apparently insuperable difficulty. No segment of the continental-margin arc-trench system is missing from the southwestern United States or Mexico (Fig. 10). The Franciscan subduction complex and a parallel trend of linked forearc basins are continuous seaward from the batholith belt along the entire continental margin of the southern Cordillera, as is the batholith belt itself. There is no missing segment of the U.S.–Mexico batholith belt that could have been seized by an oceanic plate and carried northward to Canada along a coastwise transform fault. The aberrant paleomagnetic data from coastal Canada can be attributed instead to pluton tilt and compaction shallowing of paleomagnetic vectors (Butler et al., 2001).

East of the batholith belt in the northern Cordillera, backarc mid-Cretaceous plutons in the Omineca region (Fig. 9) were not an integral facet of arc magmatism, but they were instead probably derived from sources within continental crust underthrust from the east along the retroarc thrust belt. Cenozoic backarc magmatism in the northern Cordillera, including the Chilcotin lava field (Fig. 9), coeval with eruption of the Columbia River Basalt farther south, was probably related to the generation of subterranean slab windows associated with subduction of the Resurrection plate and evolution of the Queen Charlotte transform (Madsen et al., 2006). The development of large core complexes near the U.S.–Canada border (Fig. 9) reflected backarc superextension that was kinematically similar to early phases of extension within the Basin and Range taphrogen farther south (see following discussion).

Southern Cordillera

In addition to the Basin and Range taphrogen, there are two major tectonic features in the backarc region of the southern Cordillera (Fig. 10) that are not present in the northern Cordillera. The older of the two is the Jurassic-Cretaceous Border rift belt (Dickinson and Lawton, 2001b; Lawton, 2004), and the younger is the domain of basement uplifts and associated basins formed by Cretaceous-Paleogene Laramide deformation involving both basement and cover over a broad region inland from the Sevier retroarc thrust belt (Dickinson and Snyder, 1978). Thick-skinned Laramide deformation overlapped thin-skinned Sevier thrusting in time, but it persisted later and disrupted the retroforeland basin formed in front of the Sevier thrust system.

The Border rift belt is an elongate intracontinental arm of the Gulf of Mexico, where oceanic crust was first generated in mid-Jurassic time, and it represents an extension of Mesoamerican tectonic patterns into the southern Cordillera (Dickinson et al., 1986a). Late Jurassic rifting was associated with slab rollback of the Mezcalera plate beneath the native Cordilleran arc belt of the U.S.–Mexico border region (Lawton and McMillan, 1999), and postrift thermotectonic subsidence continued through Early Cretaceous time. Late Cretaceous Sevier thrusting did not extend southward beyond the western limit of the Border rift belt, and the Sevier retroforeland basin was bounded on the south by the Mogollon Highlands, which formed as an uplifted rift shoulder along the northern flank of the Border rift belt (Fig. 10). Although somewhat diachronous in timing, backarc rifting and backarc thrusting thus developed in contrasting geodynamic realms of the Cordilleran backarc region.

Laramide deformation began near ca. 70 Ma in latest Cretaceous time throughout the Laramide Rocky Mountains, but its termination was diachronous from mid-Eocene time on the north, where the end of Laramide deformation was coincident with termination of Sevier thrusting (DeCelles, 2004), to near the end of Eocene time on the south (Dickinson et al., 1988; Cather, 2004). Movements on thin-skinned retroarc thrusts of the Sierra Madre Oriental (Fig. 10) in Mexico were coeval with Laramide deformation farther north, but they had different tectonic style. Thick-skinned Laramide deformation that broke up the continental block is a tectonic episode thought to have been controlled by shallow slab descent induced by subduction of a buoyant oceanic plateau embedded in the subducting Farallon plate. Laramide deformation was accompanied by migration of arc magmatism inland (Fig. 10), and by a marked diminution of arc activity to produce a Paleogene magmatic null over the region between Laramide uplifts and the coast (Dickinson, 2006). On the south, Laramide deformation impinged on volcanic fields representing the arc magmatism that had migrated inland during the Laramide episode of shallow subduction (Seager, 2004).

Basin and Range Province

Following Laramide deformation in the southern Cordillera (Fig. 10), the Basin and Range taphrogen (Fig. 11) was superimposed upon the Cordilleran orogen beginning in mid-Cenozoic time. The Basin and Range Province is a compound taphrogen (Dickinson, 2002), within which early phases of extension were associated with slab rollback of the subducted Farallon oceanic plate. Later phases of classic basin-range block faulting were related to torsion of the continental block under the influence of transform shear between the North American plate and the Pacific plate across the San Andreas transform system at the continental margin to the west (Dickinson and Wernicke, 1997). Late Oligocene and younger (27–0 Ma) slab-window volcanism in coastal California (Fig. 11) was spatially coordinated in time with the evolution of the San Andreas transform (Dickinson, 1997; Wilson et al., 2005). The controlling slab window, which had an intricate configuration in detail, formed when subduction of oceanic microplates along the California continental margin was progressively terminated as the Mendocino and Rivera triple junctions (Fig. 11), at the ends of the transform plate boundary, migrated northward and southward, respectively.

The relationship of early phases of extensional deformation in the Basin and Range taphrogen to slab rollback of the subducted Farallon plate in the mantle beneath the region is shown by patterns of migratory arc magmatism and the development of core

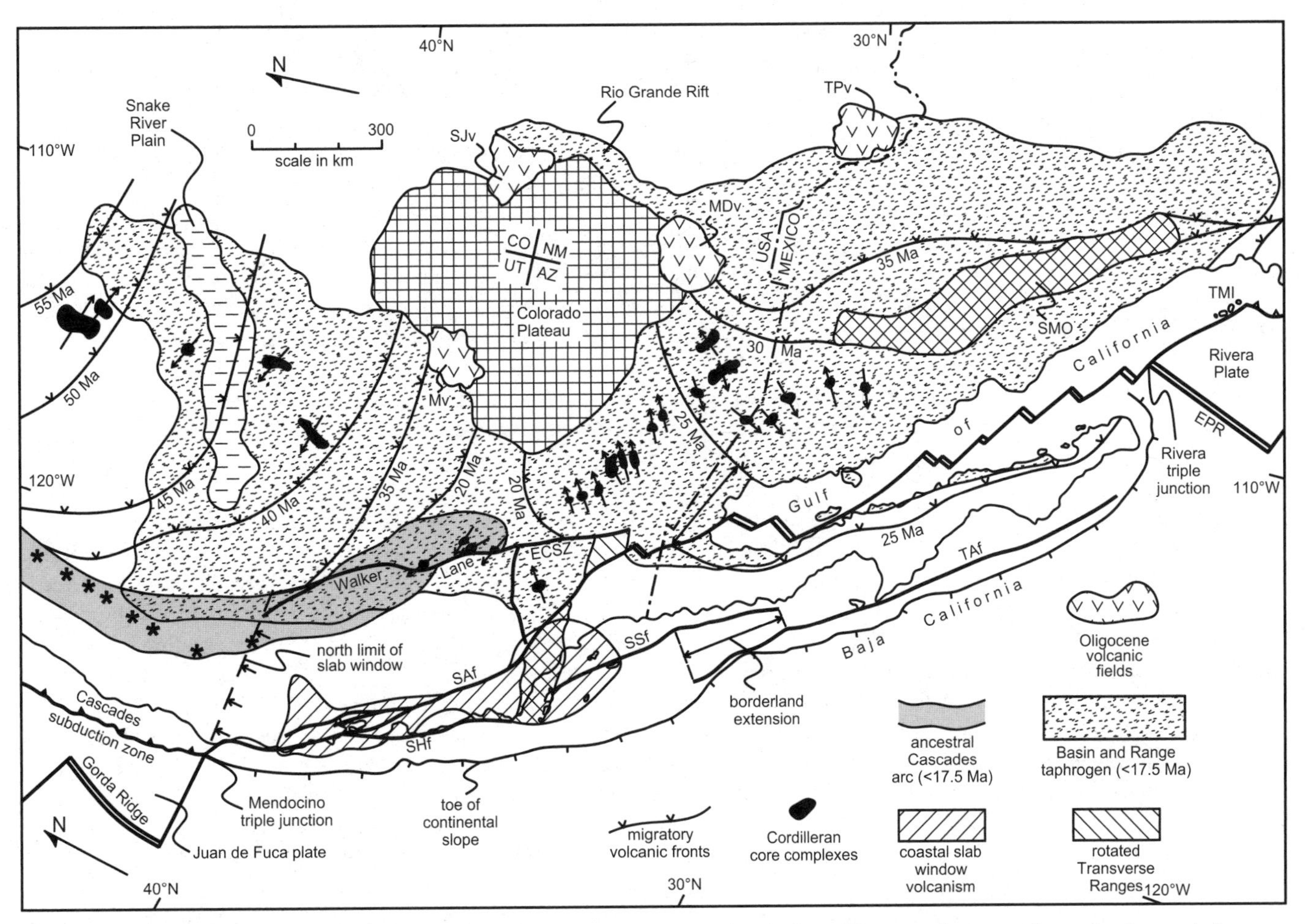

Figure 11. Post–mid-Cenozoic tectonic elements of the southern Cordillera showing Basin and Range taphrogen and associated tectonic features. Asterisks denote active volcanoes of Cascades arc. Arrows for Cordilleran core complexes denote transport directions of upper plates (48–40 Ma north of Snake River Plain, 36–22 Ma south of Snake River Plain, 28–14 Ma southwest of Colorado Plateau, 12–8 Ma west of Colorado Plateau near Walker Lane). SMO—Sierra Madre Occidental (unfaulted enclave). Interior mid-Tertiary volcanic fields: Mv—Marysvale; MDv—Mogollon-Datil; SJv—San Juan; TPv—Trans-Pecos. Principal strands of Neogene San Andreas fault system (dextral strike slip): SAf—San Andreas; SHf—San Gregorio–Hosgri; SSf—San Clemente–San Isidro; TAf—Tosco–Abreojos (ECSZ—eastern California shear zone). Map is modified after Dickinson (1996, 1997, 2002, 2004), Dickinson and Wernicke (1997), Dickinson et al. (2005), Reed et al. (2005), Wilson et al. (2005), and Henry and Faulds (2005). EPR—East Pacific Rise; TMI—Tres Marias Islands; interior cross (CO-NM-UT-AZ) denotes the Four Corners state junction.

complexes by local superextension of continental crust (Fig. 11). By the beginning of Oligocene time, inland migration of Laramide magmatism had carried subduction-related igneous centers to positions near the eastern limit of later Basin and Range extension (Fig. 10). The eruption of post-Laramide Oligocene volcanic fields near the Colorado Plateau, well inland from the continental margin (Fig. 11), was sustained by continued subduction of the Farallon plate along the continental margin to the west.

The pace of post-Laramide slab rollback is indicated by successive positions of migratory volcanic fronts (Fig. 11) as arc magmatism swept back toward the continental margin of the southern Cordillera from early Eocene to early Miocene time (Dickinson, 2002, 2006). By ca. 17.5 Ma, arc magmatism west of the Colorado Plateau was aligned with the ancestral Cascades arc near the continental margin (Fig. 11). Initiation of the back-sweep in magmatism earlier to the north than to the south resulted from earlier termination of Laramide deformation in the north than in the south. As arc magmatism swept progressively back toward the coast in a regional sense, volcanic fronts actually propagated southward to the north and west of the Colorado Plateau, but toward the northwest south of the Colorado Plateau (Fig. 11). This geometric pattern of migratory magmatism can be ascribed to decoupling of subterranean segments of the subducted Farallon plate to the north and to the south of the growing slab window near the coast.

Core complexes in the southern Cordillera (Fig. 11) formed as backarc extensional features when arc magmatism swept past their respective sites. The timing of core-complex superextension varied from 48 to 40 Ma north of the Snake River Plain, to 36–22 Ma south of the Snake River Plain but west of the Colorado

Plateau, and finally to 28–14 Ma southwest of the Colorado Plateau (Dickinson, 2002). The distribution of the core complexes in space and time tracked the migration of magmatism in detail, with only a short time lag (<5 m.y.), but superextension nowhere preceded the passage of migratory volcanic fronts. This relation argues that both the migratory magmatism and the core-complex superextension were related to slab rollback, but that the magmatism was not triggered by the superextension. Extension directions for the core complexes (Fig. 11) were subparallel to the trends of the migratory volcanic fronts, and not parallel to the directions of arc migration. This relation argues that backarc crustal extension was driven by deviatoric intracrustal stresses oriented normal to the axis of crustal thickening along the Cordilleran orogen, and it was not controlled in azimuth by the migration path of the evolving arc axis.

Basin-range block faulting of classic style began at ca. 17.5 Ma in the northern Basin and Range Province, within the Numic subtaphrogen (Dickinson, 2002), which was located to the west and northwest of the Colorado Plateau, but this faulting was delayed until ca. 12.5 Ma in the southern Basin and Range Province forming the Piman subtaphrogen (Dickinson, 2002), located south and southeast of the Colorado Plateau. The differential onset of block faulting in the two segments of the Basin and Range taphrogen was related to the time that subduction of microplates ended along the northern and southern segments of the continental margin lying off the southern Cordillera, and shear interaction could thereby first be transmitted directly from the Pacific plate to the North American plate along the San Andreas transform (Dickinson, 2002). The southern Basin and Range Province is split into two parallel domains, the Meseta Central and the Gulf of California extensional provinces (Fig. 11), by the largely unfaulted enclave of the Sierra Madre Occidental (Clark, 1976).

Multiple strands of the transform plate boundary have subsequently evolved along the coastal fringe of the southern Cordillera to slice the California margin into multiple slivers and to calve Baja California away from mainland Mexico through rifting within the Gulf of California (Fig. 11). The most inland strand of the transform system is the Walker Lane belt (Henry and Faulds, 2005), which links southward through the Eastern California shear zone to the head of the Gulf of California (Fig. 11). Strike slip along the Walker Lane, which is propagating northward as the Mendocino triple junction (Fig. 11) migrates north, is progressively detaching the California fringe of the continental block from the inland part of the southern Cordillera. As the Mendocino triple junction moves north, the southern end of the Cascades subduction zone retreats northward ahead of the triple junction, and subduction at the continental margin is supplanted farther south by transform shear that reaches into the continental block as far as the Walker Lane. Future propagation of the Walker Lane to the northwest will eventually forge a tectonic linkage to the Cascades subduction zone along the coast, and California will then be calved off North America as Baja California has already been calved off farther south (Henry and Faulds, 2005). Superextension at crossover rifts linking strands of the Walker Lane fault belt has produced anomalously young Miocene (12–8 Ma) core complexes (Fig. 11) along the western fringe of the Basin and Range Province (Dickinson, 2002).

CORDILLERAN–CARIBBEAN RELATIONS

Geotectonic patterns in the Caribbean or Mesoamerican region (Fig. 12) reveal the nature of connections between the North and South American Cordilleras. Active modern magmatic arcs (Fig. 12A) include the Trans-Mexico and Central American arcs facing the Pacific Ocean parallel to the Middle America Trench, and the Lesser Antilles island arc facing the Atlantic Ocean. Small ocean basins intervening between North and South America include the Gulf of Mexico, the Yucatan Basin, the Columbian and Venezuelan Basins of the Caribbean Sea separated by the intraoceanic and aseismic Beata Ridge, and the backarc Grenada Basin, which opened in Paleogene time between the Lesser Antilles frontal arc and the Aves Ridge remnant arc. Gondwanan continental crust underlies the Chortis block of nuclear Central America, including the offshore Nicaraguan Rise, the eastern flank of Mexico, including Yucatan and the offshore Campeche Bank, and the southern fringe of the United States, including Florida. The bulk of the Bahama Platform southeast of Florida is an intraoceanic hotspot track that formed when Africa and South America jointly rifted away from North America, and the loop of the Greater and Leeward Antilles linked by the Aves Ridge represents an accreted intraoceanic arc derived from the Pacific realm.

Any paleotectonic reconstruction of the Mesoamerican realm must begin with closure of the rhombochasm forming the elongate Cayman Trough (Fig. 12A) and reversal of the strike slip since middle Eocene time (ca. 50 Ma) along its bounding faults. The flanks of the trough are active transform faults and dormant fracture zones bounding oceanic crust along the trough floor, and they are related to a short spreading center transverse to the length of the trough (Fig. 12A). The oldest seafloor geomagnetic anomalies of the Cayman Trough near the continent-ocean boundaries at the ends of the trough are dated at ca. 49 Ma (Leroy et al., 2000). The total longitudinal extent of oceanic crust along the trough floor is ~1100 km (Rosencrantz and Sclater, 1986; Rosencrantz et al., 1988; Mann, 2007). Equivalent offset of the Chortis block relative to southern Mexico was required along the Motagua-Polochic transform fault (Fig. 12A), transecting Central America in order for the Cayman rhombochasm to open. Recovery of ~1100 km of Motagua-Polochic sinistral strike slip restores the Chortís block to a position south of Mexico (Fig. 12B) across a segment of coast truncated by post–50 Ma strike slip (Dickinson and Lawton, 2001a; Silva-Romo, 2008; Mann, 2007). As the Chortis block slid eastward with respect to Mexico in post–mid-Eocene time, the trench off southern Mexico (Fig. 12A) was incrementally lengthened south of the Trans-Mexico volcanic belt. The change in the azimuth of transform motion from the coastline of southern Mexico to the trend of the Cayman Trough was accommodated by internal deformation of the Caribbean plate

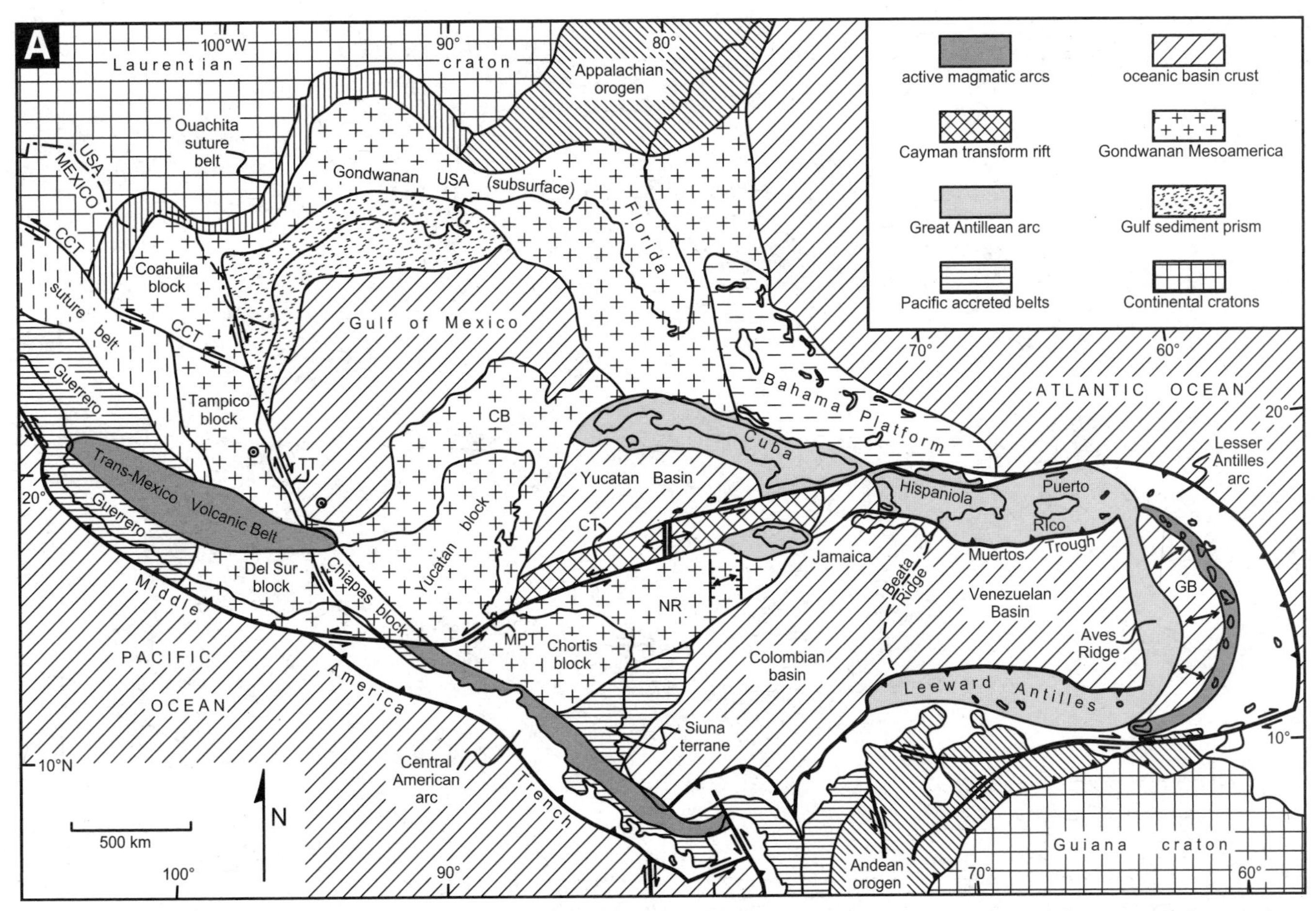

Figure 12. Tectonic elements of the Mesoamerican region south of the North American Cordillera: (A) present; (B) Cretaceous-Paleogene (pre-Mesozoic Chortis block of nuclear Central America is restored westward relative to Mexico before Cenozoic Cayman translation; postaccretion Franciscan subduction and Alisitos magmatism are not shown in Baja California); and (C) mid-Jurassic (before Atlantic and proto-Caribbean seafloor spreading). Barbs denote overriding plates at subduction zones (heavy lines) and Subandean thrust front. Half-arrows denote strike slip on transform faults (CCT—California-Coahuila; CTT—Coahuila-Tamaulipas; MPT—Motagua-Polochic; TT—Tehuantepec; YT—Yucatan). Double-headed arrows denote younger than 50 Ma motion at spreading ridge in Cayman Trough (CT), seafloor spreading in backarc Grenada Basin (GB), and crustal extension (schematic) within eastern Nicaraguan Rise (NR). Blocks of pre-Mesozoic Gondwanan crust in eastern Mexico and Central America: Chiapas now attached to Yucatan, Chortis including offshore Nicaraguan Rise, Coahuila contiguous with southern fringe of Gondwanan United States, Tampico–Del Sur overprinted by Trans-Mexico volcanic belt, Yucatan including offshore Campeche Bank (CB). Map was adapted after Case and Holcombe (1980), Dickinson and Coney (1980), Mann and Burke (1984), Pindell (1985), Speed (1985), Klitgord and Schouten (1986), Viele and Thomas (1989), Pindell and Barrett (1990), Erikson et al. (1990), Mann et al. (1991), Salvador (1991), Hempton and Barros (1993), Marton and Buffler (1994), Schouten and Klitgord (1994), Venable (1994), Pindell et al. (1998, 2005), Mann (1999), Bird et al. (1999), Dickinson and Lawton (2001a), Ostos et al. (2005), Reed et al. (2005), Silva-Romo (2008), Mann (2007), and Rogers et al. (2007a, 2007b). (*Continued on following two pages.*)

south of the transform system (Wadge and Burke, 1983). Greater displacement of Jamaica by ~1300 km from an initial position at the southern edge of the Yucatan block (Fig. 12B) reflects crustal stretching within the Nicaraguan Rise (Fig. 12A) in addition to Cayman strike slip. Eastern extensions of the Cayman transform system beyond the limit of the Cayman Trough disrupted and displaced segments of the Greater Antilles arc in islands lying to the south and east of Cuba (Figs. 12A and 12B).

Restoration of Cayman-related strike slip clarifies the relation of the ancestral Cretaceous-Paleogene Antillean arc system, which includes increments as old as mid-Cretaceous in age (Pindell and Barrett, 1990), to North and South America (Fig. 12B). Along the northern fringe of Cuba, the north-facing Greater Antilles island arc partially subducted the southern edge of Florida–Bahama carbonate platforms in late Eocene time (45–40 Ma). The western end of the colliding arc system was bounded by the Yucatan transform (Fig. 12B), along which the western end of the arc slid northward with respect to North America. The Yucatan basin was formed as a backarc basin that was later trapped north of the Cayman Trough (Fig. 12A) by lateral displacement of the continental Chortis block and Nicaraguan Rise. Westward from the Caribbean realm, the colliding Antillean arc system was

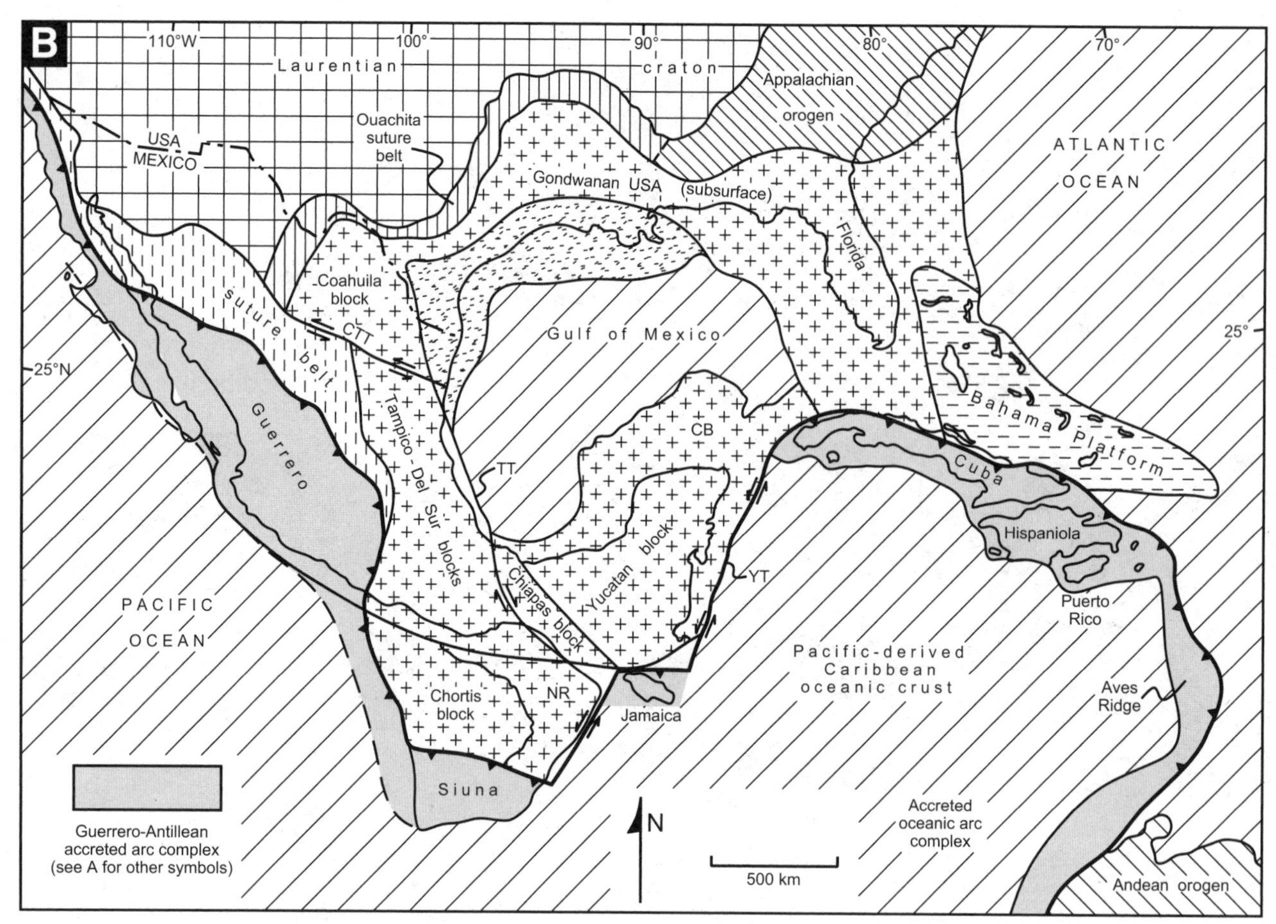

Figure 12 (*continued*).

connected through the Siuna terrane (Venable, 1994) of nuclear Central America with the Guerrero superterrane of western Mexico (Rogers et al., 2007a). This connection shows that the Cenozoic arc-continent collision between Cuba and Florida was a longitudinal continuation of the sequential Mesozoic arc-continent collisions along the Cordilleran continental margin farther north (Figs. 12A and 12B). Late Cretaceous (80–70 Ma) partial subduction of the southern tip of the Yucatan block (Pindell and Barrett, 1990), presumably beneath restored Jamaica (Fig. 12B), constitutes a link in timing between accretion of the Guerrero superterrane to western Mexico in Early Cretaceous time and accretion of the Greater Antilles arc to the Bahama Platform and the shelf south of Florida in Paleogene time (Dickinson and Lawton, 2001a; Rogers et al., 2007b).

Southeast of Paleogene Cuba, the ancestral Antillean arc extended past Puerto Rico down the Aves Ridge to northwestern South America (Fig. 12B). Arc volcanism that continued until mid-Oligocene time (ca. 30 Ma) in Puerto Rico was succeeded by the Oligocene and younger arc activity of the Lesser Antilles (Pindell and Barrett, 1990). The Leeward Antilles off present-day South America (Fig. 12A) represent the continuation of the ancestral Antillean arc, as disrupted and displaced eastward into distributed segments by Neogene deformation associated with transport of the modern Caribbean plate eastward with respect to South America as well as North America. Farther south in coastal Colombia and Ecuador, counterparts of the Antillean intraoceanic arc and associated oceanic crust were accreted to the Andean margin of South America during Cretaceous and Paleogene time (Aspden and McCourt, 1986; Jaillard et al., 1997; Cosma et al., 1998; Reynaud et al., 1999). Collision of east-facing intraoceanic arcs with the Cordilleran system was an integral facet of Mesozoic-Cenozoic geotectonic evolution for both Americas.

The modern Caribbean seafloor (Fig. 12A) was brought into position between North and South America behind the Antillean arc system from origins within the Pacific realm to the west (Fig. 12B), and it supplanted proto-Caribbean seafloor that formed during the Mesozoic separation of North and South America by seafloor spreading tied to the opening of the Atlantic Ocean to the east (Mann, 1999). Proto-Caribbean oceanic crust of Atlantic affinity was entirely subducted beneath the migratory Antillean arc of Pacific affinity. The modern Caribbean plate was separated from the Pacific realm when the Central American arc extended southward along its western edge by Oligocene time; the Lesser Antilles forms its eastern edge (Fig. 12A).

Reconstruction of the Atlantic realm using seafloor magnetic anomalies packs South America tightly against North America

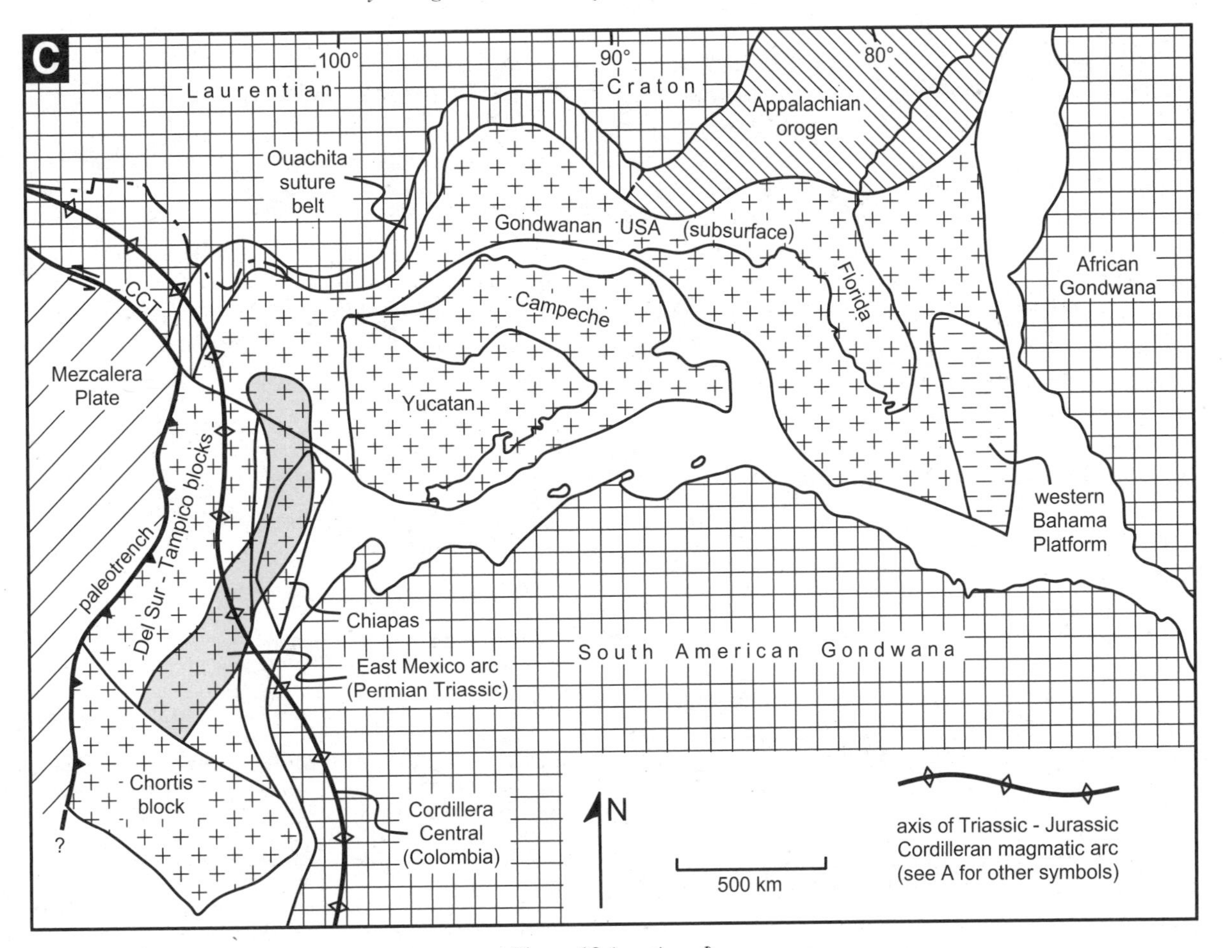

Figure 12 (*continued*).

in Jurassic time across an intervening belt of Gondwanan crustal blocks including Yucatan and the Campeche Bank (Fig. 12C). During separation of North and South America by extensions of the Atlantic spreading system, Yucatan and the Campeche Bank rotated counterclockwise to open the Gulf of Mexico (Pindell, 1985; Marton and Buffler, 1994; Schouten and Klitgord, 1994). Before rifting of South America from North America, the native Triassic-Jurassic magmatic arc along the Cordilleran margin of North America was contiguous with its counterpart along the Andean margin of South America (Fig. 12C), but the course of arc evolution during continental separation by rifting is not yet understood. Another unresolved question, essentially unaddressed to date, is how closing phases of the Gondwanide orogen reaching into southern South America related in time and space to nascent phases of the circum-Pacific orogen in northern South America and the North American Cordillera.

PALEOGEOGRAPHIC EVOLUTION: SUMMARY

Lithic assemblages older than ca. 750 Ma in the northern Cordillera and older than ca. 575 Ma in the southern Cordillera formed within the supercontinent Rodinia or pre-Rodinian Laurentia. Following rift breakup of Rodinia late in Neoproterozoic time, the Cordilleran miogeocline was deposited along a passive continental margin open to the paleo–Pacific Ocean until mid-Late Devonian time. From Late Devonian to Early Triassic time, subduction complexes associated with an east-facing intraoceanic arc complex that accreted to the Cordilleran margin were thrust over the miogeoclinal assemblage. Late in the same time frame, the southern Cordilleran margin was truncated by the California-Coahuila transform linked to the Permian-Triassic East Mexico magmatic arc, which was built along the western edge of Gondwanan Pangea after closure of the Ouachita suture between Gondwana and Laurentia. Torsional strain associated with sequential closure of the Ouachita suture had earlier induced Pennsylvanian intracontinental deformation of the Cordilleran foreland to produce the Ancestral Rocky Mountains.

Circum-Pacific subduction of oceanic lithosphere began by Late Triassic time to initiate the Cordilleran orogen and persisted all along the continental margin until conversion of segments of the offshore trench to transform faults in Cenozoic time by interaction between the subduction zone and oceanic spreading centers. Salient facets of the Cordilleran orogen include a Cretaceous batholith belt intruded along the arc trend and a retroarc thrust belt inland from the arc trend. From mid-Jurassic to mid-Cretaceous time, diachronous arc-continent collisions between east-facing intraoceanic arcs and the west-facing continental-margin arc expanded the Cordilleran margin of Laurentia, and the regional pattern of arc accretion extended into Cenozoic time to the south in the Mesoamerican region.

Jurassic-Cretaceous modifications of the Cordilleran orogen included lateral transport of Arctic Alaska past the northern end of the Cordillera, and rupture of the continental block along the Border rift belt by an arm of the Gulf of Mexico spreading system during separation of South America from North America. Cenozoic modifications of the Cordilleran orogen included breakup of the southern end of the downflexed retroforeland basin lying inland from the retroarc thrust belt by Laramide deformation during an interval of shallow slab subduction, lateral expansion of the southern Cordillera by crustal extension within the Basin and Range taphrogen, and displacement of crustal slivers along the seaward flank of the orogen by strike slip on strands of continental-margin transform fault systems.

ACKNOWLEDGMENTS

A preliminary oral version of this paper was presented at the 2006 Backbone of the Americas Conference in Mendoza, Argentina. I appreciate recent correspondence and discussions about Arctic relations with L.A. Lawver and E.L. Miller; the Siberian connection with R.A. Price; tectonic elements in the Pacific Northwest with R.J. Dorsey, J.M. English, S.T. Johnston, J.W.H. Monger, A.W. Snoke, J.E. Wright, and S.J. Wyld; California tectonics with Mark Cloos, D.S. Cowan, R.A. Schweickert, and R.V. Ingersoll; the Walker Lane with J.E. Faulds; the Greater Southwest with J.E. Spencer; and the Mesoamerican region with T.F. Lawton and J.L. Pindell, but none of those colleagues is responsible for my conclusions about any aspects of Cordilleran geology. Jim Abbott of SciGraphics in Tucson prepared the figures.

REFERENCES CITED

Armstrong, R.L., and Ward, P.L., 1993, Late Triassic to earliest Eocene magmatism in the North American Cordillera: Implications for the western interior basin, *in* Caldwell, G.W.E., and Kauffman, E.G., eds., Evolution of the Western Interior Basin: Geological Association of Canada Special Paper 39, p. 31–47.

Aspden, J.A., and McCourt, W.J., 1986, Mesozoic oceanic crust in the central Andes of Colombia: Geology, v. 14, p. 415–418, doi: 10.1130/0091-7613(1986)14<415:MOTITC>2.0.CO;2.

Balkwill, H.R., 1978, Evolution of the Sverdrup basin, Arctic Canada: American Association of Petroleum Geologists Bulletin, v. 62, p. 1004–1028.

Barth, A.P., and Wooden, J.L., 2006, Timing of magmatism following initial convergence at a passive margin, southwestern U.S. Cordillera, and ages of lower crustal magma sources: The Journal of Geology, v. 114, p. 231–245, doi: 10.1086/499573.

Beatty, T.W., Orchard, M.J., and Mustard, P.S., 2006, Geology and tectonic history of the Quesnel terrane in the area of Kamloops, British Columbia, *in* Colpron, M., and Nelson, J.L., eds., Paleozoic Evolution and Metallogeny of Pericratonic Terranes at the Ancient Pacific Margin of North America, Canadian and Alaskan Cordillera: Geological Association of Canada Special Paper 45, p. 483–504.

Bird, D.E., Hall, S.A., Casey, J.F., and Millegan, P.S., 1999, Tectonic evolution of the Grenada Basin, *in* Mann, P., ed., Caribbean Basins: Amsterdam, Elsevier, p. 389–416.

Bradley, D., Kusky, T., Haeussler, P., Goldfarb, R., Miller, M., Dumoulin, J., Nelson, S.W., and Karl, S., 2003, Geologic signature of early Tertiary ridge subduction in Alaska, *in* Sisson, V.B., Roeske, S.M., and Pavlis, T.L., eds., Geology of a Transpressional Orogen Developed during Ridge-Trench Interaction along the North Pacific Margin: Geological Society of America Special Paper 371, p. 19–49.

Bradley, D.C., McClelland, W.C., Wooden, J.L., Till, A.B., Roeske, S.M., Miller, M.L., Karl, S.M., and Abbott, J.G., 2007, Detrital zircon geochronology of some Neoproterozoic to Triassic rocks in interior Alaska, *in* Ridgway, K.D., Trop, J.M., Glen, J.M.G., and O'Neill, J.M.,, eds., Tectonic Growth of a Collisional Continental Margin: Crustal Evolution of Southern Alaska: Geological Society of America Special Paper 431, p. 155–189, doi: 10.1130/2007.2431(07).

Burrett, C., and Berry, R., 2000, Proterozoic Australia–western United States (AUSWUS) fit between Laurentia and Australia: Geology, v. 28, p. 103–106, doi: 10.1130/0091-7613(2000)28<103:PAUSAF>2.0.CO;2.

Butler, R.F., Gehrels, G.E., and Bazard, D.R., 1997, Paleomagnetism of Paleozoic strata of the Alexander terrane, southeastern Alaska: Geological Society of America Bulletin, v. 109, p. 1372–1388, doi: 10.1130/0016-7606(1997)109<1372:POPSOT>2.3.CO;2.

Butler, R.F., Gehrels, G.E., and Kodama, K.P., 2001, A moderate translation alternative to the Baja British Columbia hypothesis: GSA Today, v. 11, no. 6, p. 4–10, doi: 10.1130/1052-5173(2001)011<0004:AMTATT>2.0.CO;2.

Cant, D.E., and Stockmal, G.S., 1989, The Alberta foreland basin: Relationship between stratigraphy and Cordilleran terrane-accretion events: Canadian Journal of Earth Sciences, v. 26, p. 1964–1975.

Case, J.E., and Holcombe, T.L., 1980, Geologic-Tectonic Map of the Caribbean Region [3 sheets]: U.S. Geological Survey Miscellaneous Investigations Series Map I-1100, scale 1:2,500,000.

Cather, S.M., 2004, Laramide orogeny in central and northern New Mexico and southern Colorado, *in* Mack, G.H., and Giles, K.A., eds., The Geology of New Mexico; a Geologic History: New Mexico Geological Society Special Publication 11, p. 203–248.

Clark, K.F., 1976, Geologic section across Sierra Madre Occidental, Chihuahua to Topolobampo, Mexico, *in* Woodward, L.A., and Northrop, S.A., eds., Tectonics and Mineral Resources of Southwestern North America: New Mexico Geological Society Special Publication 6, p. 26–38.

Cloos, M., Sapiie, B., Quarles van Ufford, A., Weiland, A.J., Warren, P.Q., and McMahon, T.R., 2005, Collisional Delamination in New Guinea: The Geotectonics of Subducting Slab Breakoff: Geological Society of America Special Paper 400, 51 p.

Coles, R.L., and Taylor, P.T., 1990, Magnetic anomalies, *in* Grantz, A., Johnson, L., and Sweeney, J.F., eds., The Arctic Ocean Region: Boulder, Colorado, Geological Society of America, Geology of North America, v. L, p. 119–132.

Colpron, M., Logan, J.M., and Mortensen, J.K., 2002, U-Pb zircon age constraint for late Neoproterozoic rifting and initiation of the lower Paleozoic passive margin of western Laurentia: Canadian Journal of Earth Sciences, v. 39, p. 133–143, doi: 10.1139/e01-069.

Colpron, M., Nelson, J.L., and Murphy, D.C., 2006, A tectonostratigraphic framework for the pericratonic terranes of the northern Canadian Cordillera, *in* Colpron, M., and Nelson, J.L., eds., Paleozoic Evolution and Metallogeny of Pericratonic Terranes at the Ancient Pacific Margin of North America, Canadian and Alaskan Cordillera: Geological Association of Canada Special Paper 45, p. 1–23.

Colpron, M., Nelson, J.L., and Murphy, D.C., 2007, Northern Cordilleran terranes and their interactions through time: GSA Today, v. 17, no. 4/5, p. 4–10, doi: 10.1130/GSAT01704-5A.1.

Constenius, K.N., 1996, Late Paleogene extensional collapse of the Cordilleran foreland fold and thrust belt: Geological Society of America Bulletin, v. 108, p. 20–39, doi: 10.1130/0016-7606(1996)108<0020:LPECOT>2.3.CO;2.

Cosma, L., Lapierre, H., Jaillard, E., Laubacher, G., Bosch, D., Desmet, A., Mamberti, M., and Gabriele, P., 1998, Pétrographie et géochemie des unités magmatiques de la Cordillère occidentale d'Équateur (0°30′S): Implications tectoniques: Société Géologiques de France Bulletin, v. 169, p. 739–751.

Cowan, D.S., Brandon, M.T., and Garver, J.I., 1997, Geologic tests of hypotheses for large coastwise displacements—A critique illustrated by the Baja British Columbia controversy: American Journal of Science, v. 297, p. 117–173.

Dalziel, I.W.D., 1991, Pacific margins of Laurentia and East Antarctica–Australia as a conjugate rift pair: Evidence and implications for an Eocambrian supercontinent: Geology, v. 19, p. 598–601, doi: 10.1130/0091-7613(1991)019<0598:PMOLAE>2.3.CO;2.

DeCelles, P.G., 2004, Late Jurassic to Eocene evolution of the Cordilleran thrust belt and foreland basin system, western U.S.A.: American Journal of Science, v. 304, p. 105–168, doi: 10.2475/ajs.304.2.105.

Dehler, C.M., Sprinkel, D.A., and Porter, S.M., 2005, Neoproterozoic Uinta Mountain Group of northeastern Utah: Pre-Sturtian geographic, tectonic, and biologic evolution, *in* Pedersen, J., and Dehler, C.M., eds., Interior Western United States: Boulder, Colorado, Geological Society of America Field Guide 6, p. 1–25.

Devlin, W.J., Brueckner, H.K., and Bond, G.C., 1988, New isotopic data and a preliminary age for volcanics near the base of the Windermere Supergroup, northeastern Washington, U.S.A.: Canadian Journal of Earth Sciences, v. 25, p. 1906–1911.

Dickinson, W.R., 1978, Plate tectonic evolution of North Pacific rim: Journal of Physics of the Earth, v. 26, supplement, p. S1–S19.

Dickinson, W.R., 1979, Mesozoic forearc basin in central Oregon: Geology, v. 7, p. 166–170, doi: 10.1130/0091-7613(1979)7<166:MFBICO>2.0.CO;2.

Dickinson, W.R., 1996, Kinematics of Transrotational Tectonism in the California Transverse Ranges and Its Contribution to Cumulative Slip along the San Andreas Fault System: Geological Society of America Special Paper 305, 46 p.

Dickinson, W.R., 1997, Tectonic implications of Cenozoic volcanism in California: Geological Society of America Bulletin, v. 109, p. 936–954, doi: 10.1130/0016-7606(1997)109<0936:OTIOCV>2.3.CO;2.

Dickinson, W.R., 2000, Geodynamic interpretation of Paleozoic tectonic trends oriented oblique to the Mesozoic Klamath-Sierran continental margin in California, *in* Soreghan, M.J., and Gehrels, G.E., eds., Paleozoic and Triassic Paleogeography and Tectonics of Western Nevada and Northern California: Geological Society of America Special Paper 347, p. 209–245.

Dickinson, W.R., 2002, The Basin and Range Province as a composite extensional domain: International Geology Review, v. 44, p. 1–38, doi: 10.2747/0020-6814.44.1.1.

Dickinson, W.R., 2004, Evolution of the North American Cordillera: Annual Review of Earth and Planetary Sciences, v. 32, p. 13–45, doi: 10.1146/annurev.earth.32.101802.120257.

Dickinson, W.R., 2006, Geotectonic evolution of the Great Basin: Geosphere, v. 2, p. 353–368, doi: 10.1130/GES00054.1.

Dickinson, W.R., 2008, Accretionary Mesozoic–Cenozoic expansion of the Cordilleran continental margin in California and adjacent Oregon: Geosphere, v. 4, p. 329–353, doi: 10.1130/GES00105.1.

Dickinson, W.R., and Coney, P.J., 1980, Plate tectonic constraints on the origin of the Gulf of Mexico, *in* Pilger, R.H., Jr., ed., The Origin of the Gulf of Mexico and the Early Opening of the Central Atlantic Ocean: Baton Rouge, Louisiana State University, p. 17–36.

Dickinson, W.R., and Lawton, T.F., 2001a, Carboniferous to Cretaceous assembly and fragmentation of Mexico: Geological Society of America Bulletin, v. 113, p. 1142–1160, doi: 10.1130/0016-7606(2001)113<1142:CTCAAF>2.0.CO;2.

Dickinson, W.R., and Lawton, T.F., 2001b, Tectonic setting and sandstone petrofacies of the Bisbee Basin (USA-Mexico): Journal of South American Earth Sciences, v. 14, p. 475–504, doi: 10.1016/S0895-9811(01)00046-3.

Dickinson, W.R., and Lawton, T.F., 2003, Sequential intercontinental suturing as the ultimate control for Pennsylvanian Ancestral Rocky Mountains deformation: Geology, v. 31, p. 609–612, doi: 10.1130/0091-7613(2003)031<0609:SISATU>2.0.CO;2.

Dickinson, W.R., and Snyder, W.S., 1978, Plate tectonics of the Laramide orogeny, *in* Matthews, V., III, ed., Laramide Folding Associated with Basement Block Faulting in the Western United States: Geological Society of America Memoir 151, p. 355–366.

Dickinson, W.R., and Wernicke, B.P., 1997, Reconciliation of San Andreas slip discrepancy by a combination of interior Basin and Range extension and transrotation near the coast: Geology, v. 25, p. 663–665, doi: 10.1130/0091-7613(1997)025<0663:ROSASD>2.3.CO;2.

Dickinson, W.R., Harbaugh, D.W., Saller, A.H., Heller, P.L., and Snyder, W.S., 1983, Detrital modes of upper Paleozoic sandstones derived from Antler orogen in Nevada: Implications for nature of Antler orogeny: American Journal of Science, v. 283, p. 481–509.

Dickinson, W.R., Klute, M.A., and Swift, P.N., 1986a, The Bisbee Basin and its bearing on late Mesozoic paleogeographic and paleotectonic relations between the Cordilleran and Caribbean regions, *in* Abbott, P.L., ed., Cretaceous Stratigraphy, Western North America: Los Angeles, Pacific Section, SEPM (Society for Sedimentary Geology), Book 46, p. 51–62.

Dickinson, W.R., Swift, P.N., and Coney, P.J., 1986b, Tectonic Strip Maps of Alpine-Himalayan and Circum-Pacific Orogenic Belts (Great Circle Projections): Geological Society of America Map and Chart Series MC-58, scale 1:20,000,000.

Dickinson, W.R., Klute, M.A., Hayes, M.J., Janecke, S.U., Lundin, E.R., McKittrick, M.A., and Olivares, M.D., 1988, Paleogeographic and paleotectonic setting of Laramide sedimentary basins in the central Rocky Mountain region: Geological Society of America Bulletin, v. 100, p. 1023–1039.

Dickinson, W.R., Ducea, M., Rosenberg, L.I., Greene, H.G., Graham, S.A., Clark, J.C., Weber, G.E., Kidder, S., Ernst, W.G., and Brabb, E.E., 2005, Net Dextral Slip, Neogene San Gregorio–Hosgri Fault Zone, Coastal California: Geologic Evidence and Tectonic Implications: Geological Society of America Special Paper 391, 43 p.

Dilek, Y., and Moores, E.M., 1999, A Tibetan model for the early Tertiary western United States: Journal of the Geological Society of London, v. 156, p. 929–941, doi: 10.1144/gsjgs.156.5.0929.

Donovan, T.J., and Tailleur, I.L., 1975, Map Showing Paleocurrent and Clast-Size Data from the Devonian-Mississippian Endicott Group, Northern Alaska: U.S. Geological Survey Miscellaneous Field Studies Map MF-692, scale 1:7,500,000.

Dusel-Bacon, C., Hopkins, M.J., Mortensen, J.K., Dashevsky, S.S., Bressler, J.R., and Day, W.C., 2006, Paleozoic tectonic and metallogenic evolution of the pericratonic terranes of east-central Alaska and adjacent Yukon, *in* Colpron, M., and Nelson, J.L., eds., Paleozoic Evolution and Metallogeny of Pericratonic Terranes at the Ancient Pacific Margin of North America, Canadian and Alaskan Cordillera: Geological Association of Canada Special Paper 45, p. 25–74.

Embry, A.F., 1988, Middle-Upper Devonian sedimentation in the Canadian Arctic islands and the Ellesmerian orogeny, *in* McMillan, N.J., Embry, A.F., and Glass, D.J., eds., Devonian of the World: Volume II. Sedimentation: Canadian Society of Petroleum Geologists Memoir 14, p. 15–28.

Embry, A.F., 1990, Geological and geophysical evidence in support of the hypothesis of anticlockwise rotation of northern Alaska: Marine Geology, v. 93, p. 317–329, doi: 10.1016/0025-3227(90)90090-7.

Embry, A.F., and Dixon, J., 1990, The breakup unconformity of the Amerasian Basin, Arctic Ocean: Evidence from Arctic Canada: Geological Society of America Bulletin, v. 102, p. 1526–1534, doi: 10.1130/0016-7606(1990)102<1526:TBUOTA>2.3.CO;2.

Embry, A., and Klovan, J.E., 1976, The Middle-Upper Devonian clastic wedge of the Franklinian geosyncline: Bulletin of Canadian Petroleum Geology, v. 24, p. 485–639.

England, P., and Wells, R.E., 1991, Neogene rotations and quasicontinuous deformation of the Pacific Northwest continental margin: Geology, v. 19, p. 978–981, doi: 10.1130/0091-7613(1991)019<0978:NRAQDO>2.3.CO;2.

English, J.M., and Johnston, S.T., 2005, Collisional orogenesis in the northern Canadian Cordillera: Implications for Cordilleran crustal structure, ophiolite emplacement, continental growth, and the terrane hypothesis: Earth and Planetary Science Letters, v. 232, p. 333–344, doi: 10.1016/j.epsl.2005.01.025.

Erdmer, P., Ghent, E.D., Archibald, D.A., and Stout, M.Z., 1998, Paleozoic and Mesozoic high-pressure metamorphism at the margin of ancestral North America in central Yukon: Geological Society of America Bulletin, v. 110, p. 615–629, doi: 10.1130/0016-7606(1998)110<0615:PAMHPM>2.3.CO;2.

Erdmer, P., Moore, J.M., Heaman, L., Thompson, R.I., Daughtry, K.L., and Creaser, R.A., 2002, Extending the ancient margin outboard in the Canadian Cordillera: Record of Proterozoic crust and Paleocene regional metamorphism in the Nicola horst, British Columbia: Canadian Journal of Earth Sciences, v. 39, p. 1605–1623, doi: 10.1139/e02-072.

Erikson, J.P., Pindell, J.L., and Larue, D.K., 1990, Mid-Eocene–early Oligocene sinistral transcurrent faulting in Puerto Rico associated with formation of the northern Caribbean plate boundary zone: The Journal of Geology, v. 98, p. 365–384.

Evans, K.V., Aleinikoff, J.N., Obradovich, J.D., and Fanning, J.M., 2000, SHRIMP U-Pb geochronology of volcanic rocks, Belt Supergroup, western Montana: Evidence for rapid deposition of sedimentary strata: Canadian Journal of Earth Sciences, v. 37, p. 1287–1300, doi: 10.1139/cjes-37-9-1287.

Ferri, F., 1997, Nina Creek Group and Lay Range assemblage, north-central British Columbia: Remnants of late Paleozoic oceanic and arc terranes: Canadian Journal of Earth Sciences, v. 34, p. 854–874.

Fioretti, A.M., Black, L.P., Foden, J., and Visonà, D., 2005, Grenville-age magmatism at the South Tasman Rise (Australia): A new piercing point for the reconstruction of Rodinia: Geology, v. 33, p. 769–772, doi: 10.1130/G21671.1.

Friedman, R.M., and Armstrong, R.L., 1995, Jurassic and Cretaceous geochronology of the southwestern Coast Belt, British Columbia, 49° to 51°, *in* Miller, D.M., and Busby, C., eds., Jurassic Magmatism and Tectonics of the North American Cordillera: Geological Society of America Special Paper 299, p. 95–139.

Fritz, W.H., Cecile, M.P., Norford, B.S., Morrow, D., and Geldsetzer, H.H.J., 1991, Cambrian to Middle Devonian assemblages, *in* Gabrielse, H., and Yorath, C.J., eds., The Cordilleran Orogen in Canada: Boulder, Geological Society of America, Geology of North America, v. G-2, p. 153–218.

Fujita, K., and Newberry, J.T., 1982, Tectonic evolution of northeastern Siberia and adjacent regions: Tectonophysics, v. 89, p. 337–357, doi: 10.1016/0040-1951(82)90043-9.

Fujita, K., and Newberry, J.T., 1983, Accretionary tectonics and tectonic evolution of northeast Siberia, *in* Hashimoto, M., and Uyeda, S., eds., Accretion Tectonics in the Circum-Pacific Region: Tokyo, Terra Scientific, p. 43–57.

Gabrielse, H., and Campbell, R.B., 1992, Upper Proterozoic assemblages, *in* Gabrielse, H., and Yorath, C.J., eds., The Cordilleran Orogen in Canada: Boulder, Geological Society of America, Geology of North America, v. G-2, p. 127–150.

Gardner, M.C., Bergman, S.C., Cushing, G.W., MacKevett, E.M., Jr., Plafker, G., Campbell, R.B., Dodds, C.J., McClelland, W.C., and Mueller, P.A., 1988, Pennsylvanian stitching of Wrangellia and the Alexander terrane, Wrangell Mountains, Alaska: Geology, v. 16, p. 967–971, doi: 10.1130/0091-7613(1988)016<0967:PPSOWA>2.3.CO;2.

Gehrels, G.E., 2001, Geology of the Chatham Sound region, southeast Alaska and coastal British Columbia: Canadian Journal of Earth Sciences, v. 38, p. 1579–1599, doi: 10.1139/cjes-38-11-1579.

Geldsetzer, H.H.J., and Morrow, D.W., 1991, Upper Devonian carbonate strata of the foreland belt (Rundle assemblage), *in* Gabrielse, H., and Yorath, C.J., eds., The Cordilleran Orogen in Canada: Boulder, Geological Society of America, Geology of North America, v. G-2, p. 127–150.

Godfrey, N.J., and Dilek, Y., 2000, Mesozoic assimilation of oceanic crust and island arcs into the North American continental margin in California and Nevada: Insights from geophysical data, *in* Dilek, Y., Moores, E.M., Elthon, D., and Nicolas, A., eds., Ophiolites and Oceanic Crust: New Insights from Field Studies and the Ocean Drilling Program: Geological Society of America Special Paper 349, p. 365–382.

Golonka, J., Bocharova, N.Y., Ford, D., Edrich, M.E., Bednarczyk, J., and Wildharber, J., 2003, Paleogeographic reconstructions and basins development of the Arctic: Marine and Petroleum Geology, v. 20, p. 211–248, doi: 10.1016/S0264-8172(03)00043-6.

Gordey, S.P., 1992, Devonian-Mississippian clastics of the foreland and Omineca belts, *in* Gabrielse, H., and Yorath, C.J., eds., The Cordilleran Orogen in Canada: Boulder, Geological Society of America, Geology of North America, v. G-2, p. 230–242.

Gordey, S.P., Abbott, J.G., Tempelman-Kluit, D.J., and Gabrielse, H., 1987, "Antler" clastics in the northern Cordillera: Geology, v. 15, p. 103–107, doi: 10.1130/0091-7613(1987)15<103:AEITCC>2.0.CO;2.

Gottschalk, R.R., Oldow, J.S., and Avé Lallemant, H.G., 1998, Geology and Mesozoic structural history of the south-central Brooks Range, Alaska, *in* Oldow, J.S., and Avé Lallemant, H.G., eds., Architecture of the Central Brooks Range Fold and Thrust Belt, Arctic Alaska: Geological Society of America Special Paper 324, p. 1–8.

Grantz, A., May, S.D., Taylor, P.T., and Lawver, L.A., 1990a, Canada basin, *in* Grantz, A., Johnson, L., and Sweeney, J.F., eds., The Arctic Ocean Region: Boulder, Colorado, Geological Society of America, Geology of North America, v. L, p. 379–402.

Grantz, A., Green, A.R., Smith, D.G., Lahr, J.C., and Fujita, K., 1990b, Major Phanerozoic tectonic features of the Arctic Ocean region, *in* Grantz, A., Johnson, L., and Sweeney, J.F., eds., The Arctic Ocean Region: Boulder, Colorado, Geological Society of America, Geology of North America, v. L, Plate 11, scale 1:6,000,000.

Gunning, M.H., Hodder, R.W., and Nelson, J.L., 2006, Contrasting volcanic styles and their tectonic implications for the Paleozoic Stikine assemblage, western Stikine terrane, northwestern British Columbia, *in* Colpron, M., and Nelson, J.L., eds., Paleozoic Evolution and Metallogeny of Pericratonic Terranes at the Ancient Pacific Margin of North America, Canadian and Alaskan Cordillera: Geological Association of Canada Special Paper 45, p. 201–227.

Haeussler, P.J., Bradley, D.C., Wells, R.E., and Miller, M.I., 2003, Life and death of the Resurrection plate: Evidence for its existence and subduction in the northeastern Pacific in Paleocene-Eocene time: Geological Society of America Bulletin, v. 115, p. 867–880, doi: 10.1130/0016-7606(2003)115<0867:LADOTR>2.0.CO;2.

Hamilton, W., 1970, The Uralides and the motion of the Russian and Siberian platforms: Geological Society of America Bulletin, v. 81, p. 2553–2576, doi: 10.1130/0016-7606(1970)81[2553:TUATMO]2.0.CO;2.

Handschy, J.W., 1998, Regional stratigraphy of the Brooks Range and North Slope, Arctic Alaska, *in* Oldow, J.S., and Avé Lallemant, H.G., eds., Architecture of the Central Brooks Range Fold and Thrust Belt, Arctic Alaska: Geological Society of America Special Paper 324, p. 1–8.

Heller, P.L., and Ryberg, P.T., 1983, Sedimentary record of subduction to forearc transition in the rotated Eocene basin of western Oregon: Geology, v. 11, p. 380–383, doi: 10.1130/0091-7613(1983)11<380:SROSTF>2.0.CO;2.

Heller, P.L., Tabor, R.W., and Suczek, C.A., 1987, Paleogeographic evolution of the United States Pacific Northwest during Paleogene time: Canadian Journal of Earth Sciences, v. 24, p. 1652–1667.

Hempton, M.R., and Barros, J.A., 1993, Mesozoic stratigraphy of Cuba: Depositional architecture of a southeast-facing continental margin, *in* Pindell, J.L., and Perkins, B.E., eds., Mesozoic and Early Cenozoic Development of the Gulf of Mexico and Caribbean Region: Houston, Gulf Coast Section, SEPM (Society for Sedimentary Geology), p. 193–209.

Henry, C.D., and Faulds, J.E., 2005, The Walker Lane and Gulf of California: Related expressions of Pacific–North America plate boundary development: Geological Society of America Abstracts with Programs, v. 37, no. 7, p. 274.

Hoffman, P.F., 1991, Did the breakout of Laurentia turn Gondwanaland inside-out?: Science, v. 252, p. 1409–1412, doi: 10.1126/science.252.5011.1409.

Inger, S., Scott, R.A., and Golionko, B.G., 1999, Tectonic evolution of the Taymyr Peninsula, northern Russia: Implications for Arctic continental assembly: Journal of the Geological Society of London, v. 156, p. 1069–1072, doi: 10.1144/gsjgs.156.6.1069.

Ingersoll, R.V., 2000, Models for origin and emplacement of Jurassic ophiolites of northern California, *in* Dilek, Y., Moores, E.M., Elthon, D., and Nicolas, A., eds., Ophiolites and Oceanic Crust: New Insights from Field Studies and the Ocean Drilling Program: Geological Society of America Special Paper 349, p. 365–382.

Jackson, H.R., and Gunnarsson, K., 1990, Reconstruction of the Arctic: Mesozoic to present: Tectonophysics, v. 172, p. 303–322, doi: 10.1016/0040-1951(90)90037-9.

Jaillard, E., Benítez, S., Carlier, J., and Mourier, T., 1997, Les deformations palèogéne de la zone d'avant- arc sud-équatorienne en relation avec l'évolution géodynamique: Société Géologique de France Bulletin, v. 168, p. 403–412.

Jakobsson, M., Grantz, A., Kristoffersen, Y., and Macnab, R., 2003, Physiographic provinces of the Arctic Ocean seafloor: Geological Society of America Bulletin, v. 115, p. 1443–1455, doi: 10.1130/B25216.1.

Jarrard, R.D., 1986, Relations among subduction parameters: Reviews of Geophysics, v. 24, p. 217–284, doi: 10.1029/RG024i002p00217.

Johnston, S.T., 2001, The great Alaskan terrane wreck: Reconciliation of paleomagnetic and geological data in the northern Cordillera: Earth and Planetary Science Letters, v. 193, p. 259–272, doi: 10.1016/S0012-821X(01)00516-7.

Jones, D.L., Silberling, N.J., and Hillhouse, J., 1977, Wrangellia—A displaced terrane in northwestern North America: Canadian Journal of Earth Sciences, v. 14, p. 2565–2577, doi: 10.1139/e77-222.

Jones, D.L., Silberling, N.J., Coney, P.J., and Plafker, G., 1987, Lithotectonic Terrane Map of Alaska (West of the 141st Meridian): U.S. Geological Survey Miscellaneous Field Studies Map MF-1874-A, scale 1:2,500,000.

Karlstrom, K.E., Williams, M.L., McLelland, J., Geissman, J.W., and Ahäll, K.-I., 1999, Refining Rodinia: Geologic evidence for the Australia–western U.S. connection in the Proterozoic: GSA Today, v. 9, no. 10, p. 1–7.

Karlstrom, K.E., Ahäll, K.-I., Harlan, S.S., Williams, M.L., McLelland, J., and Geissman, J.W., 2001, Long-lived (1.8–1.0 Ga) convergent orogen in southern Laurentia, its extensions to Australia and Baltica, and implications for defining Rodinia: Precambrian Research, v. 111, p. 5–30, doi: 10.1016/S0301-9268(01)00154-1.

Klitgord, K.D., and Schouten, H., 1986, Plate kinematics of the central Atlantic, *in* Vogt, P.R., and Tucholke, B.E., eds., The Western North Atlantic Region: Boulder, Colorado, Geological Society of America, Geology of North America, v. M, p. 351–378.

Lane, L.S., 1997, Canada Basin, Arctic Ocean: Evidence against a rotational origin: Tectonics, v. 16, p. 363–387, doi: 10.1029/97TC00432.

Lawton, T.F., 2004, Upper Jurassic and Lower Cretaceous strata of southwestern New Mexico and northern Chihuahua, Mexico, *in* Mack, G.H., and

Giles, K.A., eds., The Geology of New Mexico; a Geologic History: New Mexico Geological Society Special Publication 11, p. 153–168.

Lawton, T.F., and McMillan, N.J., 1999, Arc abandonment as a cause for passive continental rifting: Comparison of the Jurassic borderland rift and the Cenozoic Rio Grande rift: Geology, v. 27, p. 779–782, doi: 10.1130/0091-7613(1999)027<0779:AAAACF>2.3.CO;2.

Lawver, L.A., Grantz, A., and Gahagan, L.M., 2002, Plate kinematic evolution of the present Arctic region since the Ordovician, *in* Miller, E.L., Grantz, A., and Klemperer, S.L., eds., Tectonic Evolution of the Bering Shelf–Chukchi Sea–Arctic Margin and Adjacent Landmasses: Geological Society of America Special Paper 360, p. 333–358.

Leroy, S., Mauffret, A., Patriat, P., and de Lépinay, B.M., 2000, An alternative interpretation of the Cayman trough evolution from a reidentification of magnetic anomalies: Geophysical Journal International, v. 141, p. 539–557, doi: 10.1046/j.1365-246x.2000.00059.x.

Li, X.-H., 1999, U-Pb zircon ages of granites from the southern margin of the Yangtze block: Timing of Neoproterozoic Jinning orogeny in SE China and implications for Rodinia assembly: Precambrian Research, v. 97, p. 43–57, doi: 10.1016/S0301-9268(99)00020-0.

Link, P.K., Corsetti, F.A., and Lorentz, N.J., 2005, Pocatello Formation and overlying strata, southeastern Idaho: Snowball Earth diamictites, cap carbonates, and Neoproterozoic isotopic profiles, *in* Pedersen, J., and Dehler, C.M., eds., Interior Western United States: Boulder, Colorado, Geological Society of America Field Guide 6, p. 251–259.

Lorenz, H., Gee, D.G., and Whitehouse, M.J., 2007, New geochronological data on Palaeozoic igneous activity and deformation in the Severnaya Zemlya Archipelago, Russia, and implications for the development of the Eurasian Arctic margin: Geological Magazine, v. 144, p. 105–125, doi: 10.1017/S001675680600272X.

Lowey, G.W., 1998, A new estimate of the amount of displacement on the Denali fault system based on the occurrence of carbonate megaboulders in the Dezadeash Formation (Jura-Cretaceous), Yukon, and the Nutzotin Mountains sequence (Jura-Cretaceous), Alaska: Bulletin of Canadian Petroleum Geology, v. 46, p. 379–386.

Luepke, J.J., and Lyons, T.W., 2001, Pre-Rodinian (Mesoproterozoic) supercontinental rifting along the western margin of Laurentia: Geochemical evidence from the Belt-Purcell Supergroup: Precambrian Research, v. 111, p. 79–90, doi: 10.1016/S0301-9268(01)00157-7.

Madsen, J.K., Thorkelson, D.J., Friedman, R.M., and Marshall, D.D., 2006, Cenozoic to Recent plate configurations in the Pacific Basin: Ridge subduction and slab window magmatism in western North America: Geosphere, v. 2, p. 11–34, doi: 10.1130/GES00020.1.

Mann, P., 1999, Caribbean sedimentary basins: Classification and tectonic setting from Jurassic to present, *in* Mann, P., ed., Caribbean Basins: Amsterdam, Elsevier, p. 3–31.

Mann, P., 2007, Overview of the tectonic history of northern Central America, *in* Mann, P., ed., Geologic and Tectonic Development of the Caribbean Plate Boundary in Northern Central America: Geological Society of America Special Paper 428, p. 1–19.

Mann, P., and Burke, K., 1984, Cenozoic rift formation in the northern Caribbean: Geology, v. 12, p. 732–736, doi: 10.1130/0091-7613(1984)12<732:CRFITN>2.0.CO;2.

Mann, P., Dengo, G., and Lewis, J.F., 1991, An overview of the geologic and tectonic development of Hispaniola, *in* Mann, P., Draper, G., and Lewis, J.F., eds., Geologic and Tectonic Development of the North America–Caribbean Plate Boundary in Hispaniola: Geological Society of America Special Paper 262, p. 1–28.

Manspeizer, W., and Cousminer, H.L., 1988, Late Triassic–Early Jurassic synrift basins of the U.S. Atlantic margin, *in* Sheridan, R.E., and Grow, J.A., eds., The Atlantic Continental Margin: Boulder, Geological Society of America, Geology of North America, v. I-2, p. 197–216.

Marton, G., and Buffler, R.T., 1994, Jurassic reconstruction of the Gulf of Mexico basin: International Geology Review, v. 36, p. 545–586.

McClelland, W.C., Gehrels, G.E., and Saleeby, J.B., 1992, Upper Jurassic–Lower Cretaceous basinal strata along the Cordilleran margin: Implications for the accretionary history of the of the Alexander- Wrangellia-Peninsular terrane: Tectonics, v. 11, p. 823–835, doi: 10.1029/92TC00241.

McNicoll, V.J., Harrison, J.C., Trettin, H.P., and Thorsteinsson, R., 1995, Provenance of the Devonian clastic wedge of Arctic Canada: Evidence provided by detrital zircon ages, *in* Dorobek, S.L., and Ross, G.M., eds., Stratigraphic Evolution of Foreland Basins: SEPM (Society for Sedimentary Geology) Special Publication 52, p. 77–93.

McWhae, J.R., 1986, Tectonic history of northern Alaska, Canadian Arctic, and Spitzbergen regions since the Early Cretaceous: American Association of Petroleum Geologists Bulletin, v. 70, p. 430–450.

Miall, A.D., 1986, Effects of Caledonian tectonism in Arctic Canada: Geology, v. 14, p. 904–907, doi: 10.1130/0091-7613(1986)14<904:EOCTIA>2.0.CO;2.

Michael, P.J., Langmuir, C.H., Dick, H.J.B., Snow, J.E., Goldstein, S.L., Graham, D.W., Lehnert, K., Kurras, G., Jokat, W., Mühe, R., and Edmonds, H.N., 2003, Magmatic and amagmatic seafloor generation at the ultraslow-spreading Gakkel Ridge, Arctic Ocean: Nature, v. 423, p. 956–961, doi: 10.1038/nature01704.

Mihalynuk, M.G., Nelson, J.L., and Diakow, L.J., 1994, Cache Creek terrane entrapment: Oroclinal paradox within the Canadian Cordillera: Tectonics, v. 13, p. 575–595, doi: 10.1029/93TC03492.

Miller, E.L., Miller, M.M., Stevens, C.H., Wright, J.L., and Madrid, R., 1992, Late Paleozoic paleogeographic and tectonic evolution of the western U.S. Cordillera, *in* Burchfiel, B.C., Lipman, P.W., and Zoback, M.L., eds., The Cordilleran Orogen: Conterminous U.S.: Boulder, Geological Society of America, Geology of North America, v. G-3, p. 57–106.

Miller, E.L., Gelman, M., Parfenov, L., and Hourigan, J., 2002, Tectonic setting of Mesozoic magmatism: A comparison between northeastern Russia and the North American Cordillera, *in* Miller, E.L., Grantz, A., and Klemperer, S.L., eds., Tectonic Evolution of the Bering Shelf–Chukchi Sea–Arctic Margin and Adjacent Landmasses: Geological Society of America Special Paper 360, p. 313–332.

Miller, E.L., Toro, J., Gehrels, G., Amato, J.M., Prokopiev, A., Tuchkova, M.I., Akinin, V.V., Dumitru, T.A., Moore, T.M., and Cecile, M.P., 2006, New insights into Arctic paleogeography and tectonics from U-Pb detrital zircon geochronology: Tectonics, v. 25, TC3013, 19 p.

Mitrovica, J.X., Beaumont, C., and Jarvis, G.T., 1989, Tilting of continental interiors by the dynamical effects of subduction: Tectonics, v. 8, p. 1079–1094, doi: 10.1029/TC008i005p01079.

Monger, J.W.H., 1993, Cretaceous tectonics of the North American Cordillera, *in* Caldwell, G.W.E., and Kauffman, E.G., eds., Evolution of the Western Interior Basin: Geological Association of Canada Special Paper 39, p. 31–47.

Moores, E.M., 1991, Southwest U.S.–East Antarctica (SWEAT) connection: A hypothesis: Geology, v. 19, p. 425–428, doi: 10.1130/0091-7613(1991)019<0425:SUSEAS>2.3.CO;2.

Moores, E.M., 1998, Ophiolites, the Sierra Nevada, "Cordilleria," and orogeny along the Pacific and Caribbean margins of North America: International Geology Review, v. 40, p. 40–54.

Mortensen, J.K., 1992, Pre–mid-Mesozoic tectonic evolution of the Yukon-Tanana terrane, Yukon and Alaska: Tectonics, v. 11, p. 836–853, doi: 10.1029/91TC01169.

Mortimer, N., 1986, Late Triassic, arc-related, potassic igneous rocks in the North American Cordillera: Geology, v. 14, p. 1035–1038, doi: 10.1130/0091-7613(1986)14<1035:LTAPIR>2.0.CO;2.

Natal'in, B.A., Amato, J.M., Toro, J., and Wright, J.E., 1999, Paleozoic rocks of northern Chukotka Peninsula, Russian far east: Implications for tectonics of the Arctic region: Tectonics, v. 18, p. 977–1003, doi: 10.1029/1999TC900044.

Nelson, J.L., 1993, The Sylvester allochthon: Upper Paleozoic marginal-basin and island-arc terranes in northern British Columbia: Canadian Journal of Earth Sciences, v. 30, p. 631–643.

Nelson, J., and Friedman, R., 2004, Superimposed Quesnel (late Paleozoic–Jurassic) and Yukon-Tanana (Devonian-Mississippian) arc assemblages, Cassiar Mountains, northern British Columbia: Field, U-Pb, and igneous petrochemical evidence: Canadian Journal of Earth Sciences, v. 41, p. 1201–1235, doi: 10.1139/e04-028.

Nelson, J., and Gehrels, G., 2007, Detrital zircon geochronology and provenance of the southeastern Yukon–Tanana terrane: Canadian Journal of Earth Sciences, v. 44, p. 297–316, doi: 10.1139/E06-105.

Nelson, J.L., Colpron, M., Piercey, S.J., Dusel-Bacon, C., Murphy, D.C., and Roots, C.F., 2006, Paleozoic tectonic and metallogenetic evolution of pericratonic terranes in Yukon, northern British Columbia, and eastern Alaska, *in* Colpron, M., and Nelson, J.L., eds., Paleozoic Evolution and Metallogeny of Pericratonic Terranes at the ancient Pacific Margin of North America, Canadian and Alaskan Cordillera: Geological Association of Canada Special Paper 45, p. 323–360.

Nilsen, T.H., 1981, Upper Devonian and Lower Mississippian redbeds, Brooks Range, Alaska, *in* Miall, A.D., ed., Sedimentation and Tectonics in Alluvial Basins: Geological Association of Canada Special Paper 23, p. 187–219.

Nilsen, T.H., and Moore, T.E., 1982, Fluvial-facies model for the Upper Devonian and Lower Mississippian(?) Kanayut Conglomerate, Brooks Range,

Alaska, *in* Embry, A.F., and Balkwill, H.R., eds., Arctic Geology and Geophysics: Canadian Society of Petroleum Geologists Memoir 8, p. 1–12.
Oldow, J.S., Avé Lallemant, H.G., Julian, F.G., and Seidensticker, C.M., 1987, Ellesmerian(?) and Brookian deformation in the Franklin Mountains, northeastern Brooks Range, Alaska, and its bearing on the origin of the Canadian Basin: Geology, v. 15, p. 37–41, doi: 10.1130/0091-7613(1987)15<37:EABDIT>2.0.CO;2.
Ostos, M., Yoris, F., and Avé Lallemant, H.G., 2005, Overview of the southeast Caribbean–South American plate boundary zone, *in* Avé Lallemant, H.G., and Sisson, V.B., eds., Caribbean–South American Plate Interactions, Venezuela: Geological Society of America Special Paper 394, p. 53–89.
Patrick, B.E., and McClelland, W.C., 1995, Late Proterozoic granitic magmatism on Seward Peninsula and a Barentsian origin for Arctic Alaska–Chukotka: Geology, v. 23, p. 81–84, doi: 10.1130/0091-7613(1995)023<0081:LPGMOS>2.3.CO;2.
Petersen, N.T., Smith, P.L., Mortensen, J.K., Creaser, R.A., and Tipper, H.W., 2004, Provenance of Jurassic sedimentary rocks of south-central Quesnellia, British Columbia: Implications for paleogeography: Canadian Journal of Earth Sciences, v. 41, p. 103–125, doi: 10.1139/e03-073.
Piercey, S.J., Nelson, J.L., Colpron, M., Dusel-Bacon, C., Simard, R.-L., and Roots, C.F., 2006, Paleozoic magmatism and crustal recycling along the ancient Pacific margin of North America, northern Cordillera, *in* Colpron, M., and Nelson, J.L., eds., Paleozoic Evolution and Metallogeny of Pericratonic Terranes at the Ancient Pacific Margin of North America, Canadian and Alaskan Cordillera: Geological Association of Canada Special Paper 45, p. 281–322.
Pindell, J.L., 1985, Alleghenian reconstruction and subsequent evolution of the Gulf of Mexico, Bahamas, and proto-Caribbean: Tectonics, v. 4, p. 1–39, doi: 10.1029/TC004i001p00001.
Pindell, J.L., and Barrett, S.F., 1990, Geological evolution of the Caribbean region: A plate-tectonic perspective, *in* Dengo, G., and Case, J.E., eds., The Caribbean Region: Boulder, Colorado, Geological Society of America, Geology of North America, v. H, p. 405–432.
Pindell, J.L., Higgs, R., and Dewey, J.F., 1998, Cenozoic palinspastic reconstruction, paleogeographic evolution, and hydrocarbon setting of the northern margin of South America, *in* Pindell, J.L., and Drake, C.L., eds., Paleogeographic Evolution and Non-Glacial Eustasy, Northern South America: SEPM (Society for Sedimentary Geology) Special Publication 58, p. 45–85.
Pindell, J., Kennan, L., Maresch, W.V., Stanke, K.-P., Draper, G., and Higgs, R., 2005, Plate-kinematics and crustal dynamics of circum-Caribbean arc-continent interactions: Tectonic controls on basin development in proto-Caribbean margins, *in* Avé Lallemant, H.G., and Sisson, V.B., eds., Caribbean–South American Plate Interactions, Venezuela: Geological Society of America Special Paper 394, p. 7–52.
Pisarevsky, S.A., and Natapov, L.M., 2003, Siberia and Rodinia: Tectonophysics, v. 375, p. 221–245, doi: 10.1016/j.tecto.2003.06.001.
Plafker, G., and Berg, H.C., 1994, Overview of the geology and tectonic evolution of Alaska, *in* Plafker, G., and Berg, H.C., eds., The Geology of Alaska: Boulder, Colorado, Geological Society of America, Geology of North America, v. G-1, p. 989–1021.
Poole, F.G., Stewart, J.H., Palmer, A.R., Sandberg, C.A., Madrid, R.J., Ross, R.J., Jr., Hintze, L.F., Miller, M.M., and Wrucke, C.T., 1992, Latest Precambrian to latest Devonian time; development of a continental margin, *in* Burchfiel, B.C., Lipman, P.W., and Zoback, M.L., eds., The Cordilleran Orogen: Conterminous U.S.: Boulder, Geological Society of America, Geology of North America, v. G-3, p. 9–56.
Poole, F.G., Perry, W.J., Jr., Madrid, R.J., and Amaya-Martinez, R., 2005, Tectonic synthesis of the Ouachita-Marathon-Sonora orogenic margin of southern Laurentia: Stratigraphic and structural implications for timing of deformational events and plate-tectonic model, *in* Anderson, T.H., Nourse, J.A., McKee, J.W., and Steiner, M.B., eds., The Mojave-Sonora Megashear Hypothesis: Development, Assessment, and Alternatives: Geological Society of America Special Paper 393, p. 543–596.
Rainbird, R.H., Jefferson, C.W., and Young, G.M., 1996, The early Neoproterozoic sedimentary succession B of northwestern Laurentia: Correlations and paleogeographic significance: Geological Society of America Bulletin, v. 108, p. 454–470, doi: 10.1130/0016-7606(1996)108<0454:TENSSB>2.3.CO;2.
Reed, J.C., Jr., Wheeler, J.G., and Tucholke, B.E., 2005, Geologic Map of North America: Boulder, Colorado, Geological Society of America, scale 1:5,000,000.
Reynaud, C., Jaillard, E., Lapierrre, H., Mamberti, M., and Mascle, G.H., 1999, Oceanic plateau and island arcs of southwestern Ecuador: Their place in the geodynamic evolution of northwestern South America: Tectonophysics, v. 307, p. 235–254, doi: 10.1016/S0040-1951(99)00099-2.
Rioux, M., Hacker, B., Mattinson, J., Kelemen, P., Bluzstajn, J., and Gehrels, G., 2007, Magmatic development of an intra-oceanic arc: High-precision U-Pb zircon and whole-rock isotopic analyses from the accreted Talkeetna arc, south-central Alaska: Geological Society of America Bulletin, v. 119, p. 1168–1184, doi: 10.1130/B25964.1.
Rogers, R.D., Mann, P., and Emmet, P.A., 2007a, Tectonic terranes of the Chortis block based on integration of regional aeromagnetic and geologic data, *in* Mann, P., ed., Geologic and Tectonic Development of the Caribbean Plate Boundary in Northern Central America: Geological Society of America Special Paper 428, p. 65–88.
Rogers, R.D., Mann, P., Emmet, P.A., and Venable, M.E., 2007b, Colon fold belt of Honduras: Evidence for Late Cretaceous collision between the continental Chortis block and intra-oceanic Caribbean arc, *in* Mann, P., ed., Geologic and Tectonic Development of the Caribbean Plate Boundary in Northern Central America: Geological Society of America Special Paper 428, p. 129–149.
Rosencrantz, E., and Sclater, J.G., 1986, Depth and age in the Cayman Trough: Earth and Planetary Science Letters, v. 79, p. 133–144, doi: 10.1016/0012-821X(86)90046-4.
Rosencrantz, E., Ross, M.I., and Sclater, J.G., 1988, Age and spreading history of the Cayman Trough as determined from depth, heat flow, and magnetic anomalies: Journal of Geophysical Research, v. 93, p. 2141–2157, doi: 10.1029/JB093iB03p02141.
Ross, G.M., 1991, Tectonic setting of the Windermere Supergroup revisited: Geology, v. 19, p. 1125–1128, doi: 10.1130/0091-7613(1991)019<1125:TSOTWS>2.3.CO;2.
Ross, G.M., Villenueve, M.E., and Theriault, R.J., 2001, Isotopic provenance of the lower Muskwa assemblage (Mesoproterozoic, Rocky Mountains, British Columbia): New clues to correlation and source area: Precambrian Research, v. 111, p. 57–77, doi: 10.1016/S0301-9268(01)00156-5.
Salvador, A., 1991, Origin and development of the Gulf of Mexico basin, *in* Salvador, A., ed., The Gulf of Mexico Basin: Boulder, Colorado, Geological Society of America, Geology of North America, v. J, p. 389–444.
Sample, J.C., and Reid, M.R., 2003, Large-scale, latest Cretaceous uplift along the northeast Pacific rim: Evidence from sediment volume, sandstone petrography, and Nd isotope signatures of the Kodiak Formation, Kodiak Islands, Alaska, *in* Sisson, V.B., Roeske, S.M., and Pavlis, T.L., eds., Geology of a Transpressional Orogen Developed during Ridge-Trench Interaction along the North Pacific Margin: Geological Society of America Special Paper 371, p. 51–70.
Savoy, L.E., and Mountjoy, E.W., 1995, Cratonic-margin and Antler-age foreland basin strata (Middle Devonian to Lower Carboniferous) of the southern Canadian Rocky Mountains and adjacent plains, *in* Dorobek, D.L., and Ross, G.M., eds., Stratigraphic evolution of foreland basins: SEPM (Society for Sedimentary Geology) Special Publication 52, p. 213–231.
Schouten, H., and Klitgord, K.D., 1994, Mechanistic solutions to the opening of the Gulf of Mexico: Geology, v. 22, p. 507–510, doi: 10.1130/0091-7613(1994)022<0507:MSTTOO>2.3.CO;2.
Schweickert, R.A., and Lahren, M.M., 1993, Triassic-Jurassic magmatic arc in eastern California and western Nevada: Arc evolution, cryptic tectonic breaks, and significance of the Mojave–Snow Lake fault, *in* Dunne, G.C., and McDougall, K.A., eds., Mesozoic Paleogeography of the Western United States, Volume II: Los Angeles, Pacific Section, SEPM (Society for Sedimentary Geology), Book 71, p. 227–246.
Schweickert, R.A., Hanson, R.E., and Girty, G.H., 1999, Accretionary tectonics of the western Sierra Nevada metamorphic belt, *in* Wagner, D.L., and Graham, S.A., eds., Geologic Field Trips in Northern California: California Division of Mines and Geology Special Publication 119, p. 33–79.
Seager, W., 2004, Laramide (Late Cretaceous–Eocene) tectonics of southwestern New Mexico, *in* Mack, G.H., and Giles, K.A., eds., The Geology of New Mexico; a Geologic History: New Mexico Geological Society Special Publication 11, p. 183–202.
Sears, J.W., and Price, R.A., 1978, The Siberian connection: A case for the Precambrian separation of the North American and Siberian cratons: Geology, v. 6, p. 267–270, doi: 10.1130/0091-7613(1978)6<267:TSCACF>2.0.CO;2.
Sears, J.W., and Price, T.A., 2000, New look at the Siberian connection: No SWEAT: Geology, v. 28, p. 423–426, doi: 10.1130/0091-7613(2000)28<423:NLATSC>2.0.CO;2.

Sears, J.W., and Price, R.A., 2003, Tightening the Siberian connection in western Laurentia: Geological Society of America Bulletin, v. 115, p. 943–953, doi: 10.1130/B25229.1.

Sears, J.W., Khudoley, A.K., Prokopiev, A.V., Chamberlain, K., and MacLean, J.S., 2005, Lithostratigraphic matches of Meso- and Neoproterozoic strata between SE Siberia and SW Laurentia: Geological Society of America Abstracts with Programs, v. 37, no. 7, p. 41–42.

Silva-Romo, G., 2008, Guayape-Papalutla fault system: A continuous Cretaceous structure from southern Mexico to the Chortís block: Tectonic implications: Geology, v. 36, p. 75–78, doi: 10.1130/G24032A.1.

Smith, M.T., Dickinson, W.R., and Gehrels, G.E., 1993, Contractional nature of Devonian-Mississippian Antler tectonism along the North American continental margin: Geology, v. 21, p. 21–24, doi: 10.1130/0091-7613(1993)021<0021:CNODMA>2.3.CO;2.

Sokolov, S.D., Bondarenko, G.Y., Morozov, O.L., Shekhovtsov, V.A., Glotov, S.P., Ganelin, A.V., and Kravchenko-Berezhnoy, I.R., 2002, South Anyui suture, northeast Arctic Russia: Facts and problems, *in* Miller, E.L., Grantz, A., and Klemperer, S.L., eds., Tectonic Evolution of the Bering Shelf–Chukchi Sea–Arctic Margin and Adjacent Landmasses: Geological Society of America Special Paper 360, p. 209–224.

Speed, R.C., 1985, Cenozoic collision of the Lesser Antilles arc and continental South America and the origin of the El Pilar fault: Tectonics, v. 4, p. 41–69, doi: 10.1029/TC004i001p00041.

Stevens, C.H., Stone, P., and Miller, J.S., 2005, A new reconstruction of the Paleozoic continental margin of southwestern North America: Implications for the nature and timing of continental truncation and the possible role of the Mojave-Sonora megashear, *in* Anderson, T.H., Nourse, J.A., McKee, J.W., and Steiner, M.B., eds., The Mojave-Sonora Megashear Hypothesis: Development, Assessment, and Alternatives: Geological Society of America Special Paper 393, p. 597–618.

Stewart, J.H., 2005, Evidence for Mojave-Sonora megashear—Systematic left-lateral offset of Neoproterozoic to Lower Jurassic strata and facies, western United States and northwestern Mexico, *in* Anderson, T.H., Nourse, J.A., McKee, J.W., and Steiner, M.B., eds., The Mojave-Sonora Megashear Hypothesis: Development, Assessment, and Alternatives: Geological Society of America Special Paper 393, p. 209–231.

Stewart, J.H., Poole, F.G., Ketner, K.B., Madrid, R.J., Roldan-Quitana, J., and Amaya-Martinez, R., 1990, Tectonics and stratigraphy of the Paleozoic and Triassic southern margin of North America, Sonora, Mexico, *in* Gehrels, G.E., and Spencer, J.E., eds., Geologic Excursions through the Sonoran Desert Region, Arizona and Sonora: Arizona Geological Survey Special Paper 7, p. 183–202.

Surlyk, F., and Hurst, J.M., 1984, The evolution of the early Paleozoic deep-water basin of north Greenland: Geological Society of America Bulletin, v. 95, p. 131–154, doi: 10.1130/0016-7606(1984)95<131:TEOTEP>2.0.CO;2.

Sweeney, J.F., 1976, Evolution of the Sverdrup Basin, Arctic Canada: Tectonophysics, v. 36, p. 181–196, doi: 10.1016/0040-1951(76)90015-9.

Talavera-Mendoza, O., Ruiz, J., Gehrels, G.E., Valencia, V.A., and Centeno-Garcia, E., 2007, Detrital zircon U-Pb geochronology of southern Guerrero and western Mixteca arc successions (southern Mexico): New insights for the tectonic evolution of southwestern North America during the late Mesozoic: Geological Society of America Bulletin, v. 119, p. 1052–1065, doi: 10.1130/B26016.1.

Timmons, J.M., Karlstrom, K.E., Dehler, C.M., Geissman, J.W., and Heizler, M.T., 2001, Proterozoic multistage (ca. 1.1 and 0.8 Ga) extension recorded in the Grand Canyon Supergroup and establishment of northwest- and north-trending tectonic grains in the southwestern United States: Geological Society of America Bulletin, v. 113, p. 163–180, doi: 10.1130/0016-7606(2001)113<0163:PMCAGE>2.0.CO;2.

Timmons, J.M., Karlstrom, K.E., Heizler, M.T., Bowring, S.A., Gehrels, G.E., and Crossey, L.J., 2005, Tectonic inferences from the ca. 1255–1100 Ma Unkar Group and Nankoweap Formation, Grand Canyon: Intracratonic deformation and basin formation during protracted Grenville orogenesis: Geological Society of America Bulletin, v. 117, p. 1573–1595, doi: 10.1130/B25538.1.

Todd, V.R., Shaw, S.E., and Hammerstrom, J.E., 2003, Cretaceous plutons of the Peninsular Ranges batholith, San Diego and westernmost Imperial Counties, California: Intrusion across a Late Jurassic continental margin, *in* Johnson, S.E., Paterson, S.R., Fletcher, J.M., Girty, G.H., Kimbrough, D.L., and Martín-Barajas, A., eds., Tectonic Evolution of Northwestern Mexico and the Southwestern USA: Geological Society of America Special Paper 374, p. 185–235.

Torres, R., Ruiz, J., Patchett, P.J., and Grajales, J.M., 1999, Permo-Triassic continental arc in eastern Mexico: Tectonic implications for reconstructions of southern North America, *in* Bartolini, C., Wilson, J.L., and Lawton, T.F., eds., Mesozoic Sedimentary and Tectonic History of North-Central Mexico: Geological Society of America Special Paper 340, p. 191–196.

Trettin, H.P., 1987, Pearya: A composite terrane with Caledonian affinities in northern Ellesmere Island: Canadian Journal of Earth Sciences, v. 24, p. 224–245.

Trettin, H.P., 1991, Tectonic framework, *in* Trettin, H.P., ed., Geology of the Innuitian Orogen and Arctic Platform of Canada and Greenland: Boulder, Colorado, Geological Society of America, Geology of North America, v. E, p. 59–66.

Unterschutz, J.L.E., Creaser, R.A., Erdmer, P., Thompson, R.I., and Daughtry, K.L., 2002, North American origin of Quesnel terrane strata in southern Canadian Cordillera: Inferences from geochemical and Nd isotopic characteristics of Triassic metasedimentary rocks: Geological Society of America Bulletin, v. 114, p. 462–475, doi: 10.1130/0016-7606(2002)114<0462:NAMOOQ>2.0.CO;2.

van der Heyden, P., 1992, A Middle Jurassic to early Tertiary Andean-Sierran model for the Coast Belt of British Columbia: Tectonics, v. 11, p. 82–97, doi: 10.1029/91TC02183.

Venable, M.E., 1994, A Geologic, Tectonic and Metallogenetic Evaluation of the Siuna Terrane [Ph.D. dissertation]: Tucson, University of Arizona, 154 p.

Viele, G.W., and Thomas, W.A., 1989, Tectonic synthesis of the Ouachita orogenic belt, *in* Hatcher, R.D., Jr., Thomas, W.A., and Viele, G.W., eds., The Appalachian-Ouachita Orogen in the United States: Boulder, Colorado, Geological Society of America, Geology of North America, v. F-2, p. 695–728.

Vogt, P.R., Taylor, P.T., Kovacs, L.C., and Johnson, G.I., 1982, The Canada basin: Aeromagnetic constraints on structure and evolution: Tectonophysics, v. 89, p. 295–336, doi: 10.1016/0040-1951(82)90042-7.

Wadge, G., and Burke, K., 1983, Neogene Caribbean plate rotation and associated Central American tectonic evolution: Tectonics, v. 2, p. 633–643, doi: 10.1029/TC002i006p00633.

Walker, J.D., Martin, M.W., and Glazner, A.F., 2002, Late Paleozoic to Mesozoic development of the Mojave Desert and environs, California, *in* Glazner, A.F., Walker, D.W., and Bartley, J.M., eds., Geologic Evolution of the Mojave Desert and Southwestern Basin and Range: Geological Society of America Memoir 195, p. 1–18.

Weber, J.R., and Sweeney, J.F., 1990, Ridges and basins in the central Arctic Ocean, *in* Grantz, A., Johnson, L., and Sweeney, J.F., eds., The Arctic Ocean Region: Boulder, Colorado, Geological Society of America, Geology of North America, v. L, p. 305–336.

Weil, A.B., Van der Voo, R., Mac Niocaill, C., and Meert, J.G., 1998, The Proterozoic supercontinent Rodinia: Paleomagnetically derived reconstructions for 1100 to 800 Ma: Earth and Planetary Science Letters, v. 154, p. 13–24, doi: 10.1016/S0012-821X(97)00127-1.

Wells, R.E., and Heller, P.L., 1988, The relative contributions of accretion, shear, and extension in Cenozoic tectonic rotation in the Pacific Northwest: Geological Society of America Bulletin, v. 100, p. 325–338, doi: 10.1130/0016-7606(1988)100<0325:TRCOAS>2.3.CO;2.

Wilson, D., and Cox, A., 1980, Paleomagnetic evidence for tectonic rotation of Jurassic plutons in Blue Mountains, eastern Oregon: Journal of Geophysical Research, v. 85, p. 3681–3689, doi: 10.1029/JB085iB07p03681.

Wilson, D.S., McCrory, P.A., and Stanley, R.G., 2005, Implications of volcanism in coastal California for the Neogene deformation history of western North America: Tectonics, v. 24, TC3008, doi: 10.1029/2003TC001621, 22 p.

Wingate, M.T.D., Pisarevsky, S.A., and Evans, D.A.D., 2002, Rodinia connections between Australia and Laurentia: No SWEAT, no AUSWUS?: Terra Nova, v. 14, p. 121–128, doi: 10.1046/j.1365-3121.2002.00401.x.

Worrall, D.M., 1991, Tectonic History of the Bering Sea and the Evolution of Tertiary Strike-Slip Basins of the Bering Shelf: Geological Society of America Special Paper 257, 120 p.

Yates, R.G., 1968, The trans-Idaho discontinuity: 23rd International Geological Congress Proceedings, v. 1, p. 117–123.

Manuscript Accepted by the Society 5 December 2008

Printed in the USA

The Geological Society of America
Memoir 204
2009

Anatomy and global context of the Andes: Main geologic features and the Andean orogenic cycle

Víctor A. Ramos
Laboratorio de Tectónica Andina, Facultad de Ciencias Exactas y Naturales, Universidad de Buenos Aires–Consejo Nacional de Investigaciones Científicas y Técnicas, Ciudad Universitaria Pabellón 2, Buenos Aires, Argentina

ABSTRACT

The Andes make up the largest orogenic system developed by subduction of oceanic crust along a continental margin. Subduction began soon after the breakup of Rodinia in Late Proterozoic times, and since that time, it has been intermittently active up to the present. The evolution of the Pacific margin of South America during the Paleozoic occurred in the following stages: (1) initial Proterozoic rifting followed by subduction and final re-amalgamation of the margin in Early Cambrian times, as depicted by the Puncoviscana and Tucavaca Basins and related granitoids in southern Bolivia and northern Argentina; (2) a later phase of rifting in the Middle Cambrian, and subsequent collisions in Middle Ordovician times of parautochthonous terranes derived from Gondwana, such as Paracas, Arequipa, and Antofalla, and exotic terranes originating in Laurentia, such as Cuyania, Chilenia and Chibcha; (3) final Permian collision between South America and North America to form Pangea during the Alleghanides orogeny, leaving behind rifted pieces of Laurentia as the Tahami and Tahuin terranes in the Northern Andes and other poorly known orthogneisses in the Cordillera Real of Ecuador in the Late Permian–Early Triassic; and (4) amalgamation of the Mejillonia and Patagonia terranes in Early Permian times, representing the last convergence episodes recorded in the margin during the Gondwanides orogeny. These rifting episodes and subsequent collisions along the continental margin were the result of changes of the absolute motion of Gondwana related to global plate reorganizations during Proterozoic to Paleozoic times. Generalized rifting during Pangea breakup in the Triassic concentrated extension in the hanging wall of the sutures that amalgamated the Paleozoic terranes. The opening of the Indian Ocean in Early Jurassic times was associated with a new phase of subduction along the continental margin. The northeastward absolute motion of western Gondwana produced a negative trench roll-back velocity that controlled subduction under an extensional regime until late Early Cretaceous times. The Northern Andes of Venezuela, Colombia, and Ecuador record a series of collisions of island arcs and oceanic plateaus from the Early Cretaceous to the middle Miocene as a result of interaction with the Caribbean plate. The remaining Central and Southern Andes record periods of orogenesis and mountain building alternating with periods of quiescence and absence of deformation as recorded in parts of the Oligocene. Based on the generalized occurrence of flat-slab subduction episodes through time, as recorded in most

Ramos, V.A., 2009, Anatomy and global context of the Andes: Main geologic features and the Andean orogenic cycle, *in* Kay, S.M., Ramos, V.A., and Dickinson, W.R., eds., Backbone of the Americas: Shallow Subduction, Plateau Uplift, and Ridge and Terrane Collision: Geological Society of America Memoir 204, p. 31–65, doi: 10.1130/2009.1204(02). For permission to copy, contact editing@geosociety.org.

of the Andean segments in Cenozoic and older times, this paper presents a conceptual orogenic cycle that accounts for the sequence of quiescence, minor arc magmatism, expansion and migration of the volcanic fronts, deformation, subsequent lithospheric and crustal delamination, and final foreland fold-and-thrust development. These episodes are related to shallowing and steepening of the subduction zones through time. This conceptual cycle, similar to the Laramide orogeny in North America, may be recognized wherever a subduction system is or was active in a continental margin.

INTRODUCTION

The Andes are the largest active orogenic system developed by subduction of oceanic crust beneath a continental margin. This continuous and complex mountain belt is the expression of a series of processes associated with subduction that led to the formation of the Andean Cordillera. The Andes developed along more than 8000 km of the Pacific margin of South America, from the Caribbean Sea in the north to the northern Scotia Ridge east of Tierra del Fuego Island in the south. Although the Andes are the type locality of Andean-type orogens as defined by Dewey and Bird (1970), the geologic history involves an intricate record of accretions, collisions, and subduction of different types of oceanic crust. These created a complicated segmentation where tectonic, magmatic, and sedimentation processes changed through time and space and formed the characteristics of the major morphostructural provinces seen in the present Andes (see Fig. 1).

There are several classifications of the Andes Mountains, but the pioneering studies of Gansser (1973) were the first to correlate the observed geological features with plate tectonics. Based on the presence of metamorphic rocks and ophiolitic belts of Andean age, Gansser divided the system into the Northern, Central, and Southern Andes. The Northern and Southern segments are characterized by Jurassic and Cretaceous metamorphic rocks as well as diverse occurrences of oceanic crust obducted to the continental margin during Andean times. On the other hand, the Central Andes lack Mesozoic and Cenozoic metamorphic and ophiolitic rocks, were formed by subduction of oceanic crust, and are the type locality of an Andean-type orogen (Ramos, 1999). This threefold classification is in fact more complex due to other geological processes superimposed on the major plate-tectonic settings.

Barazangi and Isacks (1976, 1979) identified the first two well-documented segments along the Andes without late Cenozoic arc magmatism and ascribed them to flat-slab subduction (Fig. 1). This cold subduction was associated with a subhorizontal Benioff zone observed in the retroarc area and characterized by large and frequent intracrustal earthquakes driven by significant basement shortening. As a result, significant foreland basement uplift events took place in late Cenozoic time, giving rise to a broken foreland with basement uplifts and basins (Jordan et al., 1983a, 1983b). Further detailed seismotectonic studies in the Northern Andes recognized a flat slab segment in the northern Colombian Andes with similar characteristics (Pennington, 1981; Gutscher et al., 2000).

The Andes Mountains have an almost continuous north-trending strike, yet at both the northern and southern ends, they are truncated by east-west–trending continental transform and strike-slip faults. These tectonic features link South America with the Caribbean and Scotia plates, which, regardless of the first-order similarities, share complex geological histories but have a different tectonic evolution.

In order to summarize the main characteristics of the Andes, the initiation of orogenic processes of the present western margin of South America must be combined with the creation of the Pacific Ocean, which initiated the Terra Australis accretionary orogen (Cawood, 2005). This orogen formed the paleo-Andes during Paleozoic time, which led to the formation of Pangea. Subsequent subduction formed the Andean system as we know it today.

ANDEAN-CARIBBEAN INTERACTION

The configuration of the Northern Andes is affected by interaction with the Caribbean plate. The present tectonic setting is the result of a complex tectonic evolution that started soon after the breakup of Pangea and the separation of the North and South American plates (Fig. 2). The interaction can be described as the result of lateral displacement of Mesoamerica toward the east as part of the Caribbean plate. The collision of oceanic rocks and associated sediments of the Caribbean plate against the South American margin produced the development of fold-and-thrust belts with south vergence and foreland basins on stable South America.

Global positioning system (GPS) data show a combination of northward displacement of the Northern Andean block and eastward translation of the Maracaibo block and the northern related areas of Guajira and the Cordillera de la Costa of Venezuela (Fig. 2) (Trenkamp et al., 2002; Pérez et al., 2006). The northward displacement was the result of the oblique collision of the Carnegie Ridge against the southern part of the Northern Andes, which displaced the Northern Andes via a series of active northeast-trending strike-slip faults, like the Guaicáramo and Boconó faults (Audemard and Audemard, 2002). As a consequence of this northeast displacement, the Northern Andes together with the Maracaibo block began to override the Caribbean oceanic plate, and a new subduction system formed in late Neogene time.

The Panamá microplate collision associated with the docking and eastward displacement of the Chocó terrane (Duque Caro, 1990; Taboada et al., 2000) produced the intense strike-slip right-lateral displacements in the Oca and Pilar faults that control the neotectonics of northern Venezuela along the Cordillera de la Costa (see locations in Fig. 2) (Audemard et al., 2006; Pérez et al., 2006).

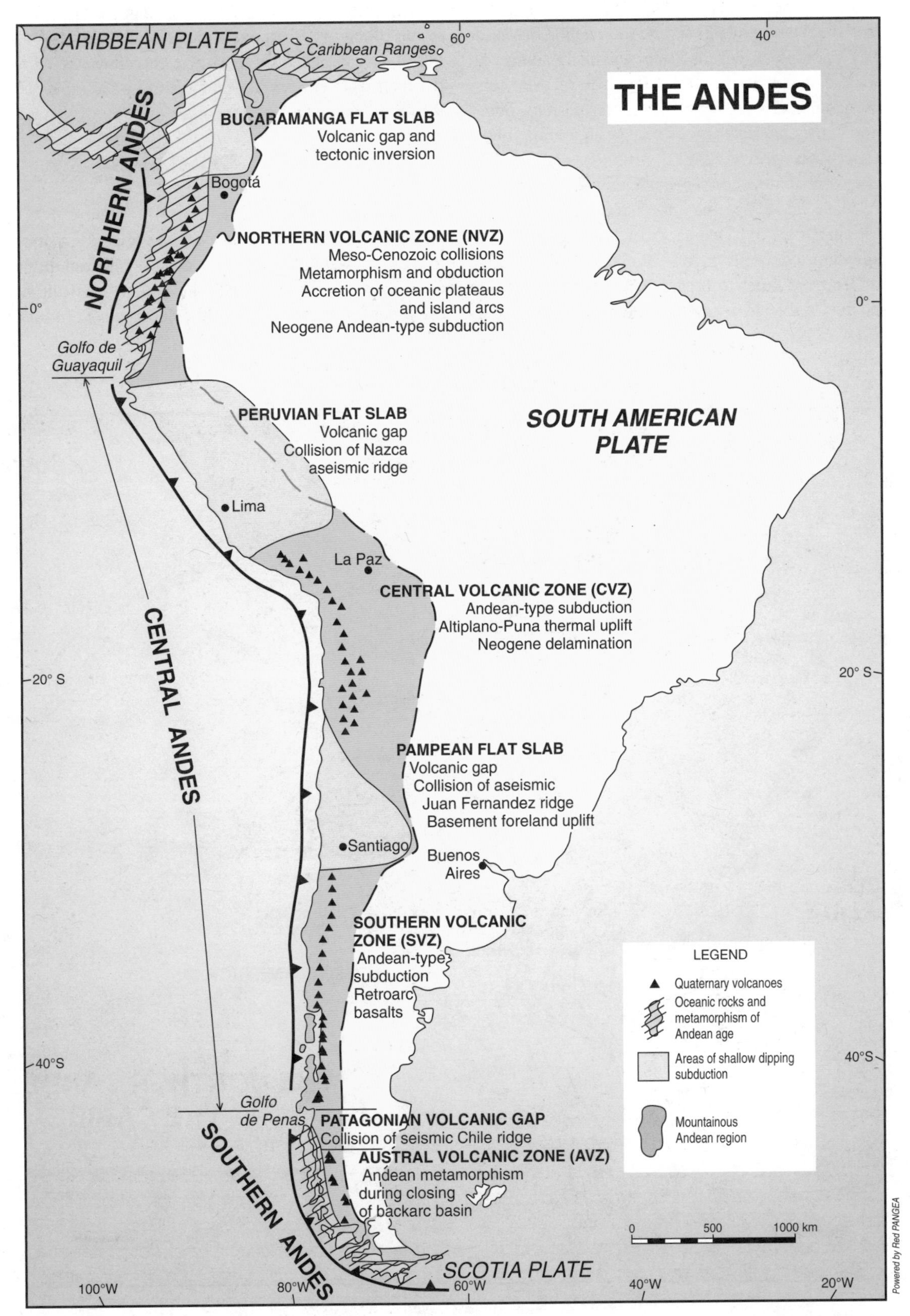

Figure 1. Major segments of the Andes and the main Quaternary volcanic zones and tectonic processes (modified from Ramos, 1999).

South of Panamá, the Coiba microplate (Pennington, 1981), which split from the Nazca plate to release the relative displacement with the Cocos plate, has no deep seismicity along the Colombia trench (Sallarés et al., 2003). The lack of deep seismicity and volcanism in this segment is consistent with the direct transfer of stress to the continent and increasing shortening in the Serranía de Baudo (Acosta et al., 2007).

In summary, the interaction between the Caribbean and South America plates added oceanic crust to the Northern Andes and northern South America. Island arc crust, oceanic rocks, and associated sedimentary deposits have been obducted from Early Cretaceous to Cenozoic times to form the Western and Coastal Cordilleras and the Caribbean ranges (see later discussion on collisional tectonics). This process is still going on as a consequence of the Barbados prism overriding South America (Jácome et al., 2003), as evidenced by the active thrusting east of the Serranía del Interior in eastern Venezuela and by the rapid subsidence of the Orinoco late Quaternary delta.

ANDEAN-SCOTIAN INTERACTION

The southern tip of the Andes is bounded by the Scotia plate, which in general has a geometry and distribution similar to the Caribbean plate, but it also has significant differences. The Caribbean plate collided with the South American plate by adding oceanic material, whereas the Scotia plate, through a series of

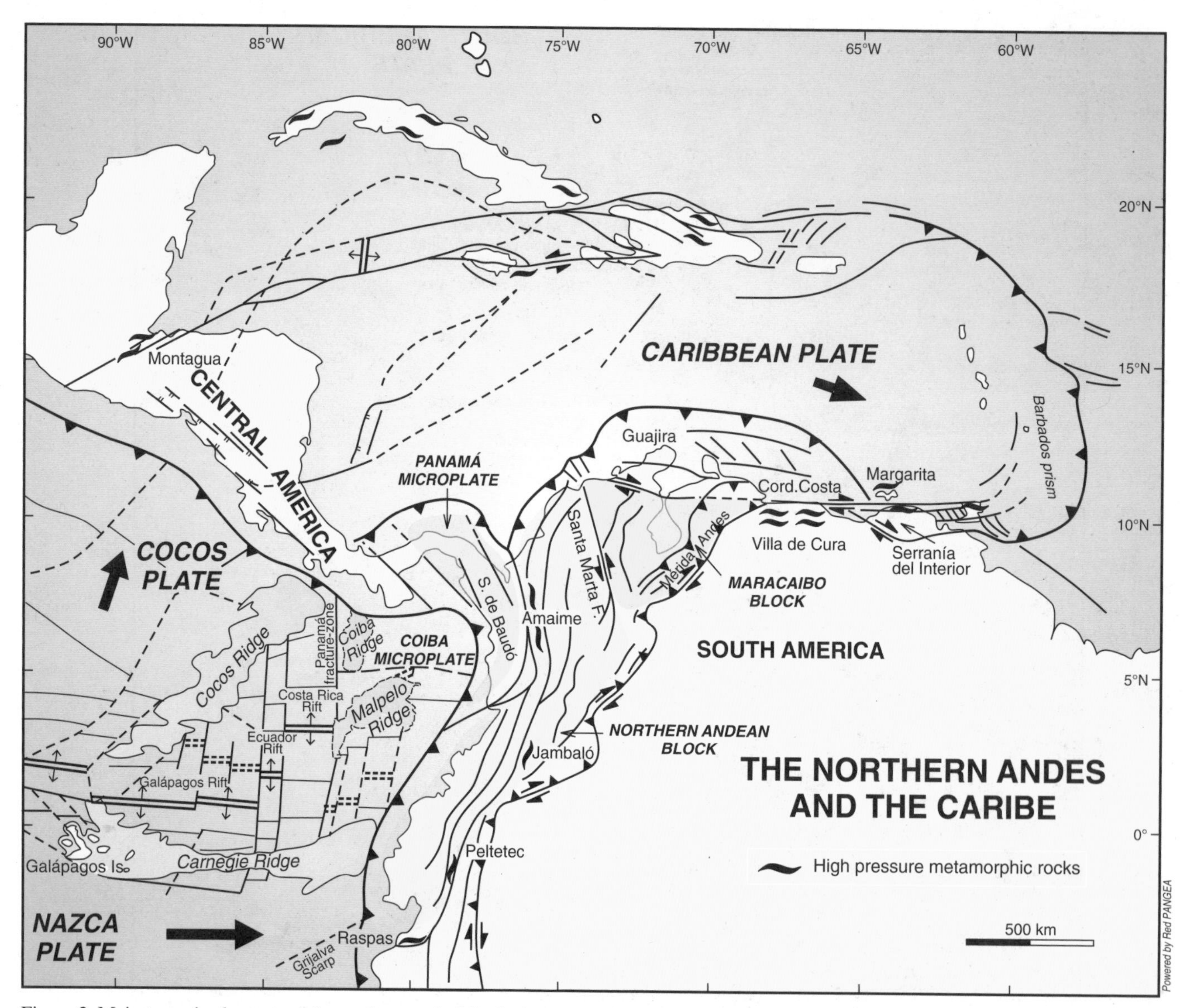

Figure 2. Main tectonic elements of the northern end of the Andes and associated Caribbean Ranges. This sector is characterized by the emplacement of oceanic crust over the South American plate linked to high-pressure metamorphism of Early Cretaceous age (based on Pindell et al., 2006; Costa et al., 2006a; Acosta et al., 2007; other sources discussed in the text).

strike-slip faults, is truncating and tearing asunder pieces of continental crust, which are transported hundreds of kilometers away to form the northern Scotia Ridge, as represented by a series of sialic blocks from the Burwood Bank to the Georgia Island Bank (Klepeis, 1994a; Cunningham et al., 1995; Barker, 2001). These faults are still active and are responsible for major intracrustal earthquakes and neotectonic activity (see Costa et al., 2006b; Smalley et al., 2007).

The basement of Aurora and Shag Banks, as well as South Georgia Island basement (Fig. 3), consists of fragments of Tierra del Fuego Island that have been transported in the last 30 m.y. Interaction between the Scotia plate and South America by means of the Fagnano transform fault has controlled the evolution of the Tierra del Fuego fold-and-thrust belt since the Oligocene (Dalziel et al., 1974; Klepeis, 1994b; Olivero, 1998; Lodolo et al., 2003; Ghiglione and Ramos, 2005).

The formation of the oceanic Scotia plate led to the opening of the Drake Passage in Oligocene times through a complex system of rifting and pull-apart basins, which were recently described by Ghiglione et al. (2008). That opening, which is argued to have changed world climate, produced significant changes in the oceanic current pattern and led to the isolation and cooling of the Antarctic plate (Kennett, 1977; Lawver and Gahagan, 2003; Eagles et al., 2006; Lodolo et al., 2006; Ghiglione and Cristallini, 2007). The evolution of that area, although of great importance to the geology of the southern tip of the Andes, is beyond the scope of this paper (see recent reviews of Barker, 2001; Livermore et al., 2005).

An understanding of the interactions among the Caribbean, Scotia, and South American plates is important for analysis of the tectonic elements of the Northern and Southern Andes, and their associated processes, and will be discussed in the following chapters.

INITIATION OF THE ANDEAN MARGIN

The Pacific margin of South America formed during the breakup of Rodinia. Early proposals of Moores (1991), Dalziel (1991), and Hoffman (1991) led to a wave of geologic and geochronologic studies of Andean basement that demonstrated the importance of a Grenville-age signature in most of the terranes along the entire western continental margin (see Fig. 4). Grenville-age basement terranes are widely distributed in South America (see Fuck et al., 2008), and they form a nearly continuous belt along the western margin from Colombia to Patagonia, in southern Chile and southern Argentina.

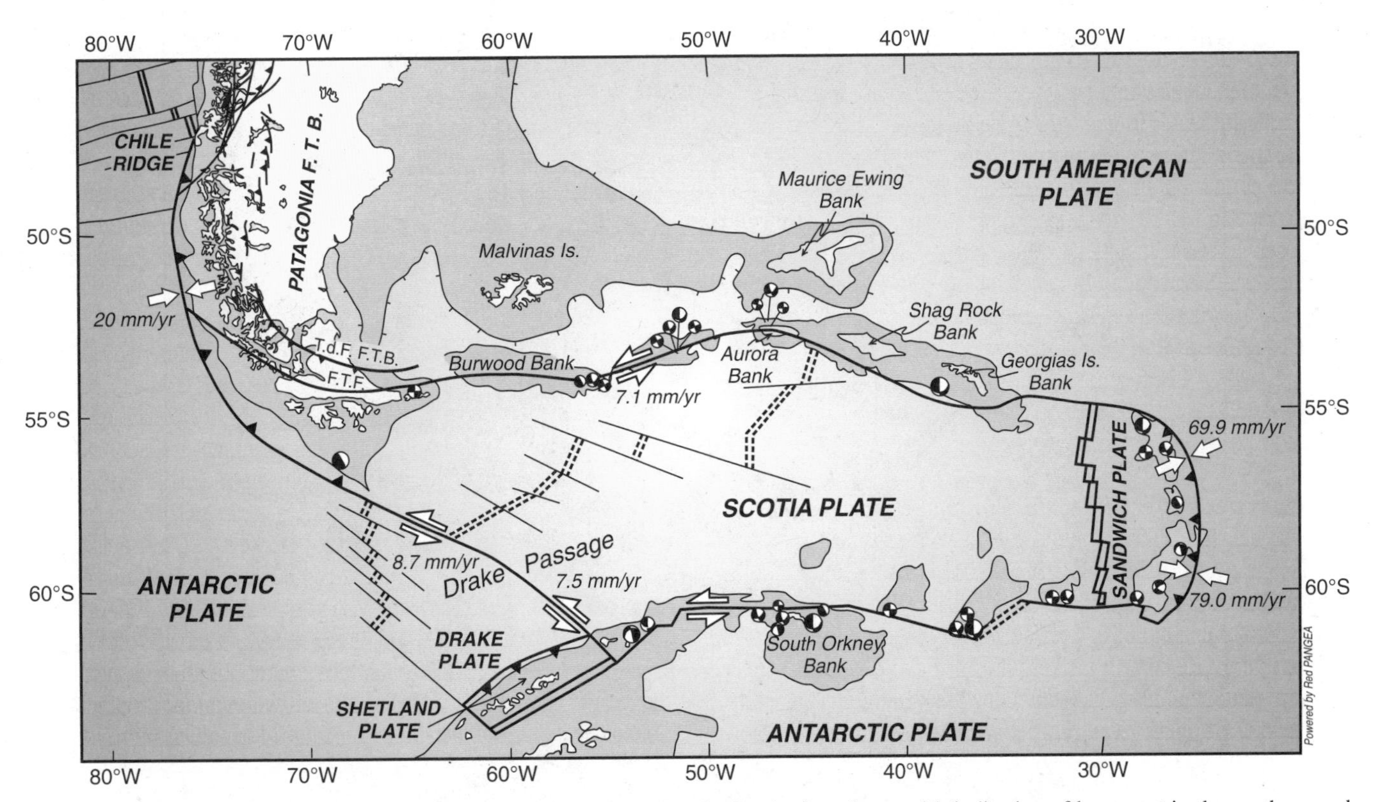

Figure 3. Main tectonic elements and kinematics of the Scotia plate and related microplates, with indication of basement in the northern and southern Scotia Ridges and the continental transform faults that tear Tierra del Fuego Island (based on Barker, 2001; Eagles et al., 2006; Lodolo et al., 2006; other sources discussed in the text). T.d.F. F.T.B.—Tierra del Fuego fold-and-thrust belt; F.T.F.—Fagnano transform fault; F.T.B.—fold-and-thrust belt.

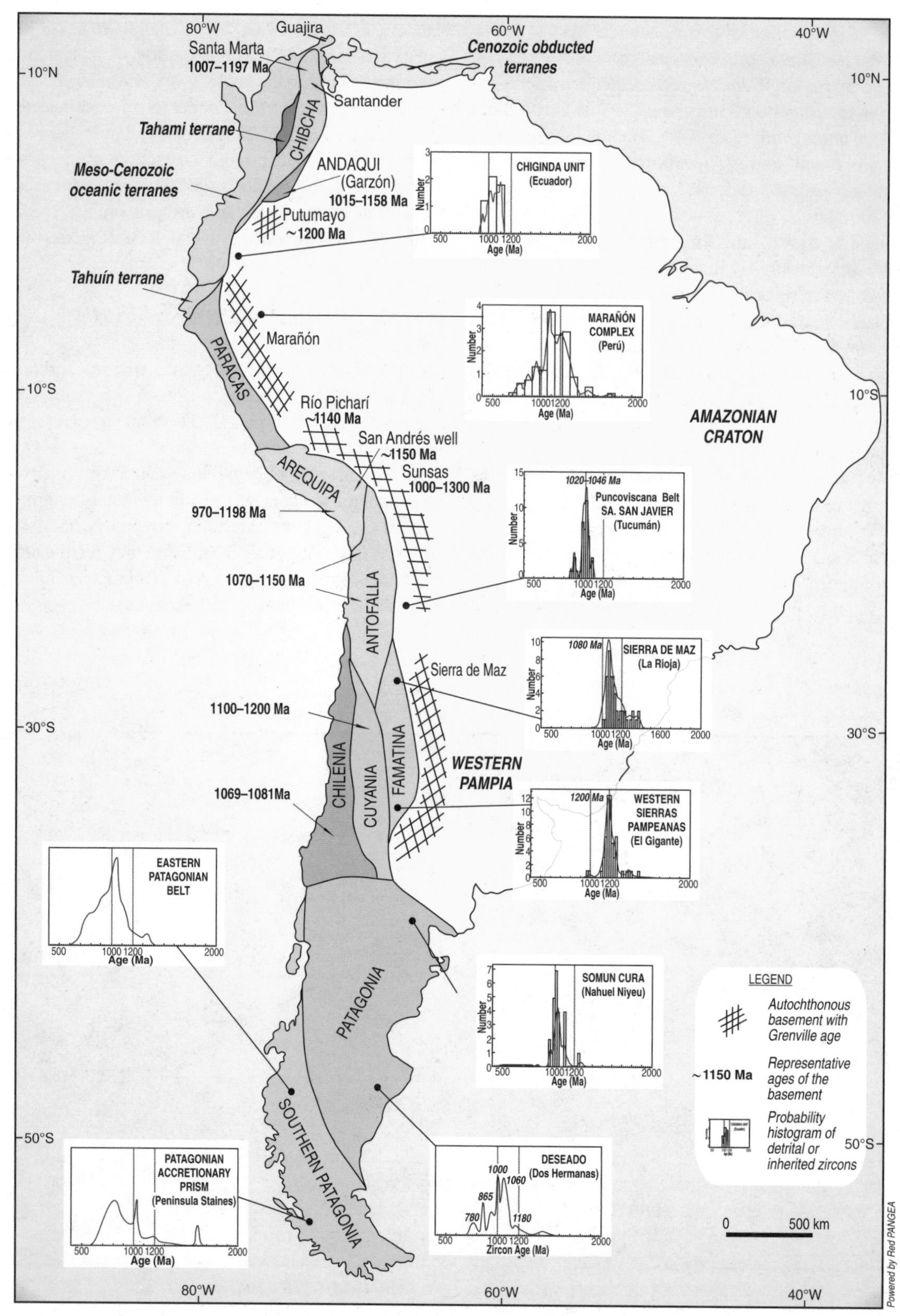

Figure 4. Grenville ages in basement terranes along the western margin of Gondwana based on U-Pb zircon ages from Restrepo Pace et al. (1997); Cordani et al. (2005); Chew et al. (2007); Wasteneys et al. (1995); Kay et al. (1996); Casquet et al. (2008); Ramos (2004, 2008a, 2008b); Rapela et al. (2007); Pankhurst et al. (2003, 2006); and Hervé et al. (2003). The probability age histograms are indicated using detrital zircons from the cover and inherited zircons from igneous rocks (in those areas where the Grenville basement is not exposed).

The Central Colombian terrane, also known as the Chibcha terrane (Restrepo and Toussaint, 1988; Restrepo-Pace, 1993; Alemán and Ramos, 2000), has been palinspastically restored to its pre-Andean position (based on Pindell et al., 1998). The reconstruction of displacement of the Santa Marta, Guajira, and Santander basement blocks forms a single terrane with the Chibcha, outlining the protomargin of Gondwana (Fig. 4). The Andaqui terrane, which was part of autochthonous South America, also has Grenville ages (Cordani et al., 2005), similar to the basement of the Putumayo Basin.

The paleogeography of the Ecuador protomargin is more difficult to establish due to the intense deformation during post-Paleozoic times of the eastern and main orographic feature that corresponds to the Cordillera Real (Litherland et al., 1985). Identification of a Grenville-age source was indirectly based on the studies of Chew et al. (2007). The Cordillera Real basement, which was considered autochthonous by Pratt et al. (2005), has a sedimentary cover of detrital zircons of Grenville age that were derived from a basement adjacent to the Andes. In the Oriente Basin, located east of the Andes, the older platform deposits derived from the basement have a stronger Grenville-age signature than the Andean synorogenic deposits (Martin-Gombojav and Winkler, 2008).

Along the Peruvian protomargin, there are limited exposures of igneous and metamorphic rocks of Mesoproterozoic age (Dalmayrac et al., 1980). Recently, the studies of Chew et al. (2007) and Cardona et al. (2007) have shown an almost constant inheritance of Grenville-age zircons in the igneous and metamorphic Paleozoic rocks north of the Abancay deflection (approximately 13°50 S, at the latitude of the Paracas peninsula). These rocks are aligned with the Sunsás orogen (Fig. 4), a well-known belt of metamorphic and igneous rocks of Grenville age described in eastern Bolivia along the border with Brazil by Litherland et al. (1985, 1989). The Arequipa cratonic block is a Mesoproterozoic terrane interpreted to have originated as the tip of a pre-Grenville Laurentian promontory (comprising Labrador, Greenland, and Scotland) incorporated into the Grenville orogen (Dalziel, 1994; Wasteneys et al., 1995). Pb isotopic compositions seem to contradict this model, indicating instead closer ties with the Amazon craton (Tosdal, 1996; Loewy et al., 2003, 2004). A reconstruction of Grenville orogen remnants in South America (Sadowski and Bettencourt, 1996) indicates that the Peruvian Central Andes basement corresponds to an independent area between the magmatic arc, represented by the Sunsas igneous province in eastern Bolivia and western Brazil, and the Grenville front of southeastern Canada and eastern United States (Ramos, 2008a). This explains the similar trends encountered in the Proterozoic outcrops along the Andes and of the Brazilian Shield (Litherland et al., 1985, 1989). Paleoproterozoic ages indicated by U-Pb zircon geochronology represent the pre-Grenville Laurentian-Amazonian protolith, and Mesoproterozoic ages of granulite-facies metamorphism indicated by U-Pb zircon geochronology represent the main collisional events of the Grenville orogen (Wasteneys et al., 1995; Sadowski and Bettencourt, 1996; Tosdal, 1996). Rifting during breakup of Rodinia in the Neoproterozoic led to separation of Laurentia from Amazonia, leaving behind the parautochthonous Arequipa cratonic block attached to Amazonia (Jaillard et al., 2000; Díaz Martínez et al., 2000; Ramos, 2008a).

Exposures of Grenville-age rocks have been reported in northern Chile and westernmost Bolivia (Damm et al., 1990; Wörner et al., 2000a, 2000b) and in large boulders of coarse Miocene conglomerates in northern Bolivia (Tosdal, 1996).

The late Proterozoic protomargin of Gondwana in the western Pampia terrane of northern and central Argentina shows numerous records of Grenville-age rocks (Vujovich et al., 2004; Escayola et al., 2007; Rapela et al., 2007; Casquet et al., 2008). These rocks show that Pampia was involved in the assemblage of Rodinia, and it was detached during the Rodinia breakup (Baldo et al., 2006). Similar age patterns are observed in the Cuyania (Ramos et al., 1993; Kay et al., 1996; Ramos, 2004) and Chilenia basements (Ramos and Basei, 1997).

The different basement exposures of Patagonia also have an important Grenville-age signature, as indicated by zircons in accretionary prism rocks (Hervé et al., 2003), although some consider these zircons to have been derived from the adjacent Kaapvaal craton of South Africa during Gondwana times. However, recent studies have shown a persistent signature of Grenville ages in igneous and metamorphic rocks that point to a more local origin (see Pankhurst et al., 2003, 2006).

In summary, a Mesoproterozoic age basement ranging in age between 1000 and 1200 Ma is an important component in all terranes along the Pacific margin. The only exception is the Paracas terrane, where the limited basement inliers preserved in northern Perú after the detachment of Oaxaquia terrane have not been dated (see Keppie and Dostal, 2007; Ramos, 2008a). There are similar ages in the autochthonous protomargin of west Gondwana, not only in the well-exposed Sunsás belt of Bolivia, but also in all the other stable platforms bounding the allochthonous and/or parautochthonous terranes, such as in Colombia, Perú, and Argentina.

As a result of the fragmentation of the Rodinia supercontinent, some of the blocks remained attached to present South America, whereas others were disrupted and stayed in Laurentia. The breakup of Rodinia precipitated the opening of the Pacific Ocean and the development of the Terra Australis accretionary orogen along the active western margin of Gondwana during the Paleozoic (Cawood, 2005).

THE PROTO-ANDEAN MARGIN OF WESTERN GONDWANA

The apron of Grenville-age terranes that bounds the present Pacific margin of South America represents either independent blocks from Laurentia trapped in the collision between the two continents that led to the formation of Rodinia (e.g., the Central Colombian terranes), or detached fragments from Laurentia that were left adjacent to the Amazonian craton after the Mesoproterozoic collision (e.g., Arequipa terrane). There are also terranes detached from Laurentia in Paleozoic times that later docked into Gondwana (e.g., Cuyania and Chilenia terranes). All of these transfers between Laurentia and Gondwana show a complex tectonic pattern (see Keppie and Ramos, 1999), but they also indicate the proximity of the two continents after the breakup of Rodinia during the opening of the southern Iapetus Ocean in the early Paleozoic (Dalziel, 1997). Next, I provide a description of the different basement provinces and continental terranes involved in the early evolution of the proto-Andean margin in three segments, the northern, central, and southern Andean segments.

Northern Andean Segment

The basement underlying the early Paleozoic platform of the Los Llanos and Oriente Basins in Colombia and Ecuador, respectively, is poorly known. The Grenville-age basement known in the Putumayo Basin seems to have experienced some extension at the end of the Proterozoic. The Chigüiro-1 and Pato-1 exploration wells in the northern part of Los Llanos Basin intersect low-grade metamorphic sedimentary rocks bearing algal tissue and sphaeromorph acritarchs of Vendian age (Cáceres et al., 2003). These rocks are of similar age and character to those in the Tucavaca and Puncoviscana belts in Bolivia and Argentina, which have been interpreted as extensional basins related to the breakup of Rodinia (Aceñolaza and Aceñolaza, 2005; Ramos, 2008a).

The lower Paleozoic cover is represented by a Middle Cambrian marine transgression that produced the carbonate facies bearing the Celtic trilobite *Paradoxides* (Harrington and Kay, 1951; Bordonaro, 1992). This trilobite is a typical genus from Avalonia, which supports the link proposed by Murphy et al. (2006) between eastern Avalonia and the passive margin of Colombia. Trümpy (1943) described the platform deposits of Serranía de la Macarena (east of the Andaqui terrane), formed of Middle Cambrian carbonates, shales, and sandstones of Tremadoc to Early Ordovician age. There are no fossils from the Late Ordovician. These platform deposits are bordered to the west by turbidites and shales known as the graptolitic facies, which bear Early Ordovician fossils (Ordóñez-Carmona et al., 2006). The Ordovician rocks in the Quetame Massif in the Chibcha terrane are partially metamorphosed into greenschist facies. Mafic and ultramafic rocks are reported from the headwaters of the Ariarí River associated with submarine basalts representing oceanic rocks (Cáceres et al., 2003). These rocks are interpreted as a potential suture formed during the early Paleozoic between the basement of the Quetame Massif in the Chibcha terrane and the Grenville-age autochthonous basement (Fig. 5).

On the eastern side of the Cordillera Central, low-grade schists bearing Early Ordovician graptolites unconformably overlie the El Vapor gneisses of Grenville age, which are part of the Chibcha terrane. West of the suture between the Chibcha and Tahami terranes, most of the suspect Precambrian basement has yielded zircon U-Pb ages no older than Early Permian (Ordóñez-Carmona et al., 2006; Vinasco et al., 2006). This medium- to high-grade basement is formed of orthogneisses and mylonites,

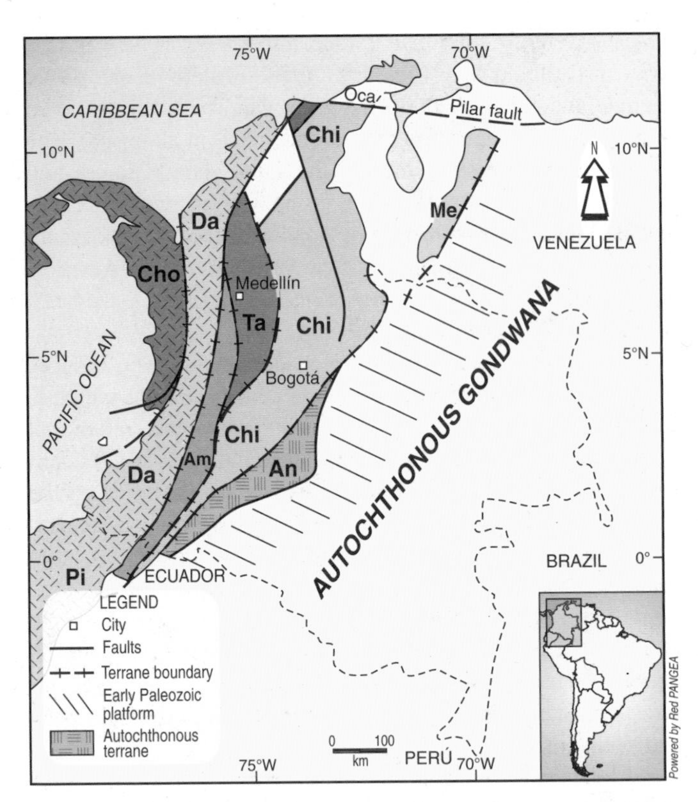

Figure 5. Main allochthonous terranes of the Northern Andes (based on Alemán and Ramos, 2000; Ordoñez-Carmona et al., 2006). Accreted in the Paleozoic: Chi—Chibcha; Me—Mérida; Ta—Tahami. Accreted in the Cretaceous: Am—Amaime, Da—Dagua; Pi—Piñón. Accreted in the Miocene: Cho—Choco terrane. Probable autochthonous terrane: An—Andaqui terrane.

like the Las Palmitas and Abejorral granitic gneisses, which have a calc-alkaline signature and represent crustal melts that probably were produced in an arc setting. U-Pb ages in zircons of ca. 275 Ma are associated with Ar-Ar ages between 240 and 215 Ma related to the breakup of Pangea. These rocks are in tectonic contact with a belt of garnet-bearing amphibolites, peridotites, and stratified dunites east of Medellín, which represent remnants of an ophiolitic suite of possible Permian age (Restrepo, 2003; Martens and Dunlap, 2003), reworked during the Triassic breakup event (Restrepo et al., 2008). Kinematic indicators of the foliated amphibolites show a vergence to the northeast (Pereira et al., 2006).

This multiepisodic evolution is characterized by a series of superposed metamorphic episodes alternating with extensional periods as envisaged by Restrepo and Toussaint (1982, 1988). Evidence for this evolution is preserved in the protomargin of the Northern Andes. A series of tectonic stages is illustrated in Figure 6, including

(1) amalgamation of Rodinia in the Mesoproterozoic;

(2) intracontinental rifting during the breakup of Rodinia in the late Proterozoic;

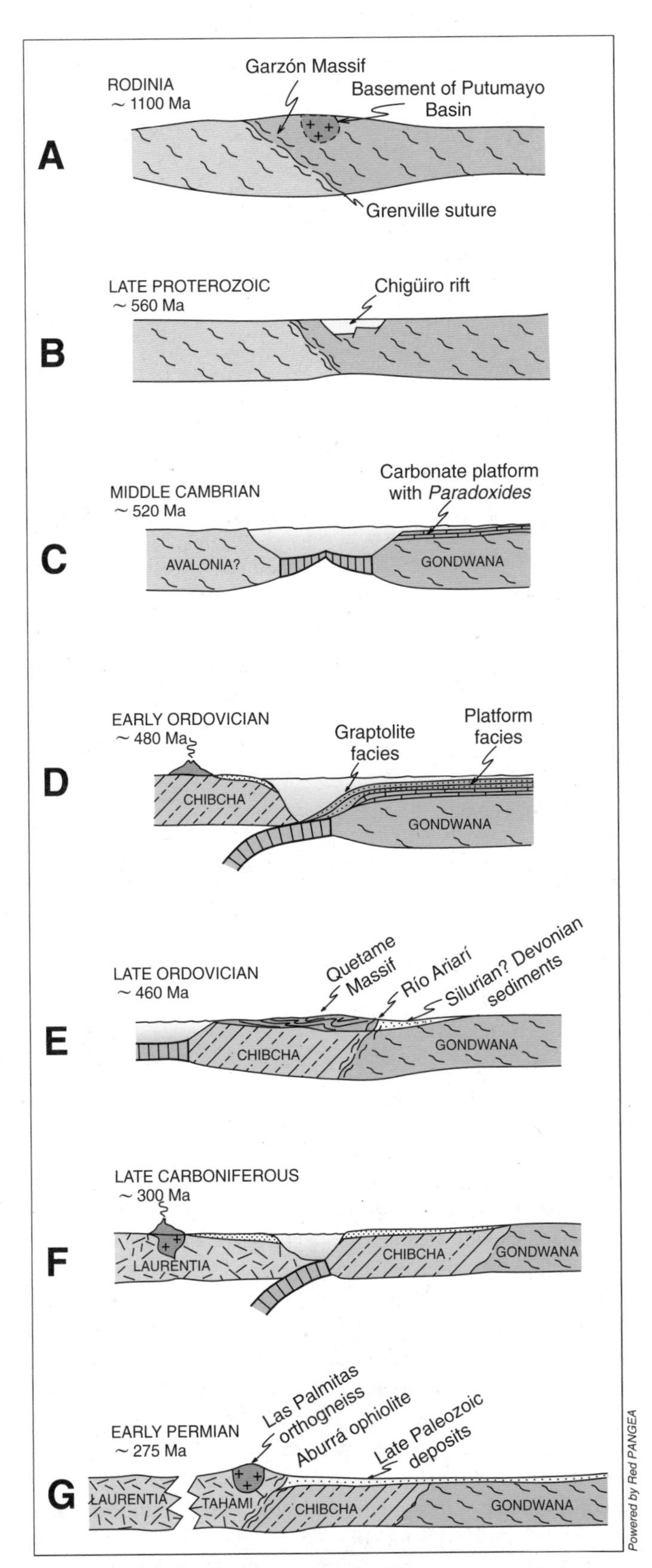

Figure 6. Proposed tectonic evolution of the northern Andean margin (based on Alemán and Ramos, 2000; Ordoñez-Carmona et al., 2006; Pereira et al., 2006; Vinasco et al., 2006).

(3) development of a carbonate passive margin bearing the trilobite *Paradoxides* as an index fossil after Eastern Avalonian terranes drifted away in the Middle Cambrian (Murphy et al., 2006);

(4) thermal subsidence lasting until Middle Ordovician times, which developed into a clastic platform and slope facies associated with a magmatic arc;

(5) Ordovician deformation and low-grade metamorphism related to the docking of the Chibcha terrane to the protomargin and emplacement of the Ariarí River oceanic rocks along the suture;

(6) a magmatic arc, preserved as orthogneisses of late Paleozoic age developed in the western flank of the present Central Cordillera on a Laurentian basement; and

(7) a mafic to ultramafic belt of Permian ophiolites emplaced in the Tahami terrane during the Alleghanides orogeny, which led to the formation of a wide foreland basin that covered the basement of the present Eastern Cordillera and the adjacent Llanos Basin. Subsequent Triassic rifting of Laurentia left behind a piece, known as the Tahami terrane, on the Gondwana side.

Gneisses and granitoids of late Paleozoic to Triassic ages such as the Tres Lagunas and Jubones foliated granitoids, similar to the Tahami terrane, are well known along the eastern margin of the Cordillera Real of Ecuador (Aspden et al., 1992a; Litherland et al., 1994; Sánchez et al., 2006). These gneissic rocks and foliated granitoids are also exposed in the Amotape-Tahuín block along the Huancabamba deflection (Fig. 7), which was interpreted by Feininger (1987) and Mourier et al. (1988) as an allochthonous terrane. These rocks are associated with the Río Piedras amphibolite, which has been described as a mafic complex by Feininger (1978) and Litherland et al. (1994), and which represents possible oceanic rocks of late Paleozoic age. Recent Ar-Ar dating on the orthogneisses and foliated granitoids has yielded ages of 227–211 Ma (Vinasco, 2004; Sánchez et al., 2006). These Triassic ages are interpreted to be associated with crustal melts during the collapse of the Alleghanides orogen and subsequent breakup of Pangea (Alemán and Ramos, 2000; Cediel et al., 2003; Martin-Gombojav and Winkler, 2008). The collision took place during the Permian as established by U/Pb sensitive high-resolution ion microprobe (SHRIMP) and Ar-Ar dating of the metamorphic event that affected the Illescas Massif (Cardona et al., 2008), which represents the southernmost extension of this belt along the northeast coast of Perú. I propose that the Tahuín terrane of Feininger (1987) collided against the Gondwana margin in Early Permian times, possibly as a part of Laurentia, and it was left on the Gondwana side after being detached from Laurentia. Later tectonic reactivations during the Mesozoic produced the rotation detected by paleomagnetism as inferred by Mourier et al. (1988).

Upper Paleozoic deposits have been intersected by several wells in the Oriente Basin of Ecuador (Moreno-Sanchez, 2004), and they are interpreted as evidence of a foreland basin that formed during the amalgamation of Gondwana. No Permian records have been reported for Los Llanos Basin.

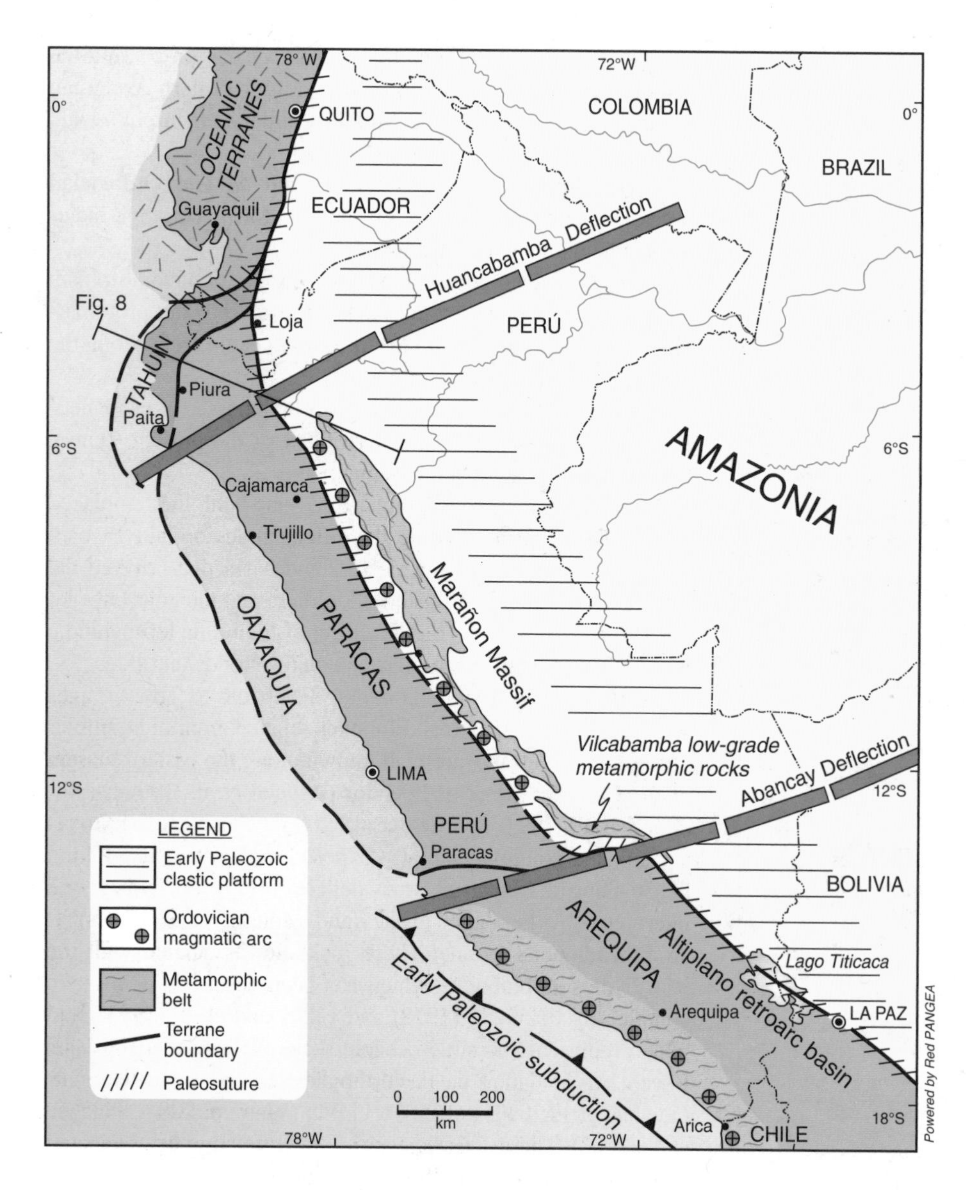

Figure 7. Basement terranes of the northern Central Andes (based on Litherland et al., 1994; Jaillard et al., 2000; Chew et al., 2007; Ramos, 2008a). The eastern suture indicates the protomargin of Gondwana in Ecuador and northern and central segments of Perú. Arequipa has been slightly detached from the margin, but it developed as an early Paleozoic ensialic retroarc basin in the present Altiplano along southern Perú.

Central Andean Segment

The Andean basement of the Central Andes has distinct features from Perú to central Argentina. It is characterized by two belts, an eastern and a western belt.

The eastern belt defines the protomargin of Gondwana, which developed on the western margin of the Amazonia and Pampia cratonic terranes as described by Ramos and Alemán (2000). This protomargin shows minor evidence in Ecuador and Perú of the Brazilian–Pan African orogenic cycle with subduction-related magmatism. This is indicated by the frequent presence of inherited zircon grains in orthogneisses of 0.65–0.45 Ga (Martin-Gombojav and Winkler, 2008; Chew et al., 2007), which show an active Gondwana margin during late Proterozoic–early Paleozoic times. These Neoproterozoic ages may be related to the subsequent reaccretion of the Grenville-age blocks against Gondwana (Fig. 4). The lack of positive evidence for rifting of the Paracas and Arequipa terranes in the northern Perú segment, in comparison with what had happened in the southern Antofalla segment (Fig. 2), raises the question of whether or not both segments were part of a single terrane (see following discussion). The opening of the Puncoviscana and Tucavaca Basins of Bolivia and northern Argentina can be explained by the relative rotation of the Pampia cratonic block, which was not attached at that time to the Río de la Plata craton (see Fig. 9a of Ramos, 2008a).

The eastern belt of Perú is represented by the plutonic, metamorphic, and metasedimentary rocks exposed along the Eastern Cordillera, and by some minor exposures to the north that reach the boundary with Ecuador (Fig. 7).

This basement has been studied in the Cordillera de Marañón by Cardona et al. (2005a, 2005b) and Chew et al. (2007), who recognized an Early to Middle Ordovician magmatic arc in the Eastern Cordillera, which was deformed at ca. 475 Ma, as inferred by the age of the metamorphic belt. This plutonic and metamorphic belt is offset southwestward, and it continues along the western margin of the Arequipa basement (Ramos, 2008a), where again an Ordovician magmatic arc and important metamorphism were documented by Loewy et al. (2004) prior to the intrusion of massive granodiorites at 473 Ma.

These data demonstrate the presence of an Early Ordovician magmatic and metamorphic belt that runs along the western margin of the Arequipa Massif basement, and which is offset northeastward into the Eastern Cordillera (Fig. 7). Chew et al. (2007) suggested that the change in strike of the belt results from the presence of an original embayment on the western Gondwanan margin during the early Paleozoic, whereas Ramos (2008a) favored the hypothesis that a basement block, the Paracas parautochthonous terrane, collided during the late Early Ordovician against the Gondwana margin. A remnant of this continental metamorphic basement block is observed in the continental shelf as the Paracas High (Ramos and Alemán, 2000). The presence of this sialic basement in the offshore platform of central Perú north of the Abancay deflection at ~14°S, between the localities of Paracas and Trujillo, is well established based on gravimetric and refraction data. The data show a high-density (2.7–2.8 g/cm^3) and high-velocity (5.9–6.0 km/s) continuous continental ridge (Thornburg and Kulm, 1981; Atherton and Webb, 1989), which is exposed in the Las Hormigas de Afuera Islands at the latitude of Lima, and which has been intersected in some exploration wells at the latitude of Trujillo (Ramos, 2008a). The detachment of a basement block from the Peruvian margin north of the Abancay deflection, as proposed by Keppie and Ortega Gutiérrez (1995) and Ramos and Alemán (2000), may explain the geological affinities of the Oaxaquia terrane (Fig. 7) with this part of the Gondwana margin. Both terranes share a common high-grade metamorphic basement of Grenville age and similar unique Gondwanan trilobites as described by Moya et al. (1993), which are different from the typical Laurentian fauna. As a result, following the collision of the Paracas block with the margin in the Early Ordovician, the Oaxaquia terrane was detached from Gondwana in Late Ordovician–Silurian times and preserved in central Mexico (Keppie and Dostal, 2007).

South of the Abancay deflection at 14°S latitude, the early Paleozoic history is different (Fig. 7), and two crustal blocks with different origin are present: the Arequipa cratonic block (Dalziel and Forsythe, 1985; Ramos et al., 1986; Jaillard et al., 2000) and the Amazon craton (Teixeira et al., 1989). The Arequipa cratonic block was rifted during the breakup of Rodinia in the Neoproterozoic, and again in the early Paleozoic, to form a backarc sialic basin in the Altiplano (Sempere, 1995). The boundary zone between the Arequipa block and the Gondwanan crust is located beneath the eastern Altiplano and the Eastern Cordillera. It represents a paleosuture and crustal weakness zone inherited from the Mesoproterozoic evolution of the Grenville orogen (Jaillard et al., 2000; Ramos, 2008a). This zone remained active during the Paleozoic, and ever since, characterized by variable behavior depending on the regional state of stresses (Ramos, 1988a; Dorbath et al., 1993; Forsythe et al., 1993).

The southern Peruvian and northern Bolivian segment records a history of subduction along the present margin in Ordovician times, including a coeval backarc basin that developed in a heavily attenuated crust (Ramos, 2008a). The backarc basin was filled with thick platform clastic sequences preserved in the Altiplano and the Eastern Cordillera (Sempere, 1995; Suárez Soruco, 2000). The basin was closed during the Ocloyic deformation in Middle to Late Ordovician times (Ramos, 1986; Bahlburg, 1990).

Variations in the metamorphic conditions of the Lower Paleozoic deposits are remarkable along the eastern belt in the Eastern Cordillera of Perú and Bolivia, as clearly established by Dalmayrac et al. (1980). Orthogneisses and mica-schists in amphibolite facies in northern Perú correspond to the region where the Paracas terrane was accreted in Ordovician times. Southern Perú and most of Bolivia only have sedimentary rocks of that age, heavily deformed along the western Altiplano. Thick Silurian and Devonian clastic facies accumulated during the foreland stage after the Ocloyic deformation in most of the Altiplano and Eastern Cordillera of Bolivia (Sempere, 1995). The transition between the two regions is in the Cordillera de Vilcabamba (Fig. 7), where the pre-Ocloyic deposits are preserved in low-grade greenschist facies (Dalmayrac et al., 1980). The Abancay deflection coincides with a disruption in the strike of the Eastern Cordillera and with a significant change in the metamorphic conditions. There is also an important change in the basement input of Cretaceous magmatism as noticed by Petford et al. (1996), who assert that the Abancay deflection in Perú correlates with a change from the southern segment, which has negative ε_{Nd} isotopic compositions, to more primitive positive values in northern Perú. That change also coincides with attenuated crust in the Paracas terrane after detachment of the Oaxaquia terrane, and a thicker Paleoproterozoic crust in the Arequipa Massif.

The Huancabamba deflection coincides with the southern end of the Tahuín terrane of Feininger (1987), which encompasses the late Paleozoic amphibolites, orthogneisses, and schists exposed in the Amotapes Ranges of southern Ecuador and northern Perú north of Piura (Fig. 7) (Litherland et al., 1994). That block collided against the Gondwanan margin in Early Permian times, after the docking of the Paracas terrane (Fig. 8).

The evolution of northern Chile and the adjacent Bolivian basement again shows some differences (Fig. 9). The high-grade metamorphic basement in the western Puna of northern Argentina near Antofalla (Segerstrom and Turner, 1972) was the first evidence of apparent Precambrian outcrops west of the Gondwana protomargin (Coira et al., 1982). These outcrops were connected with the Arequipa basement inliers, assuming a single Arequipa-Antofalla Massif (Ramos, 1988a). The studies of Baeza and Pichowiak (1988) following the early descriptions of González Bonorino and Aguirre (1970) recognized a metamorphic basement belt in the Precordillera of northern Chile (~18°S to 23°S lat.) along the Belén, Choja, Sierra Moreno, and Limón Verde outcrops. Based on a preliminary U-Pb age (Damm et al., 1986),

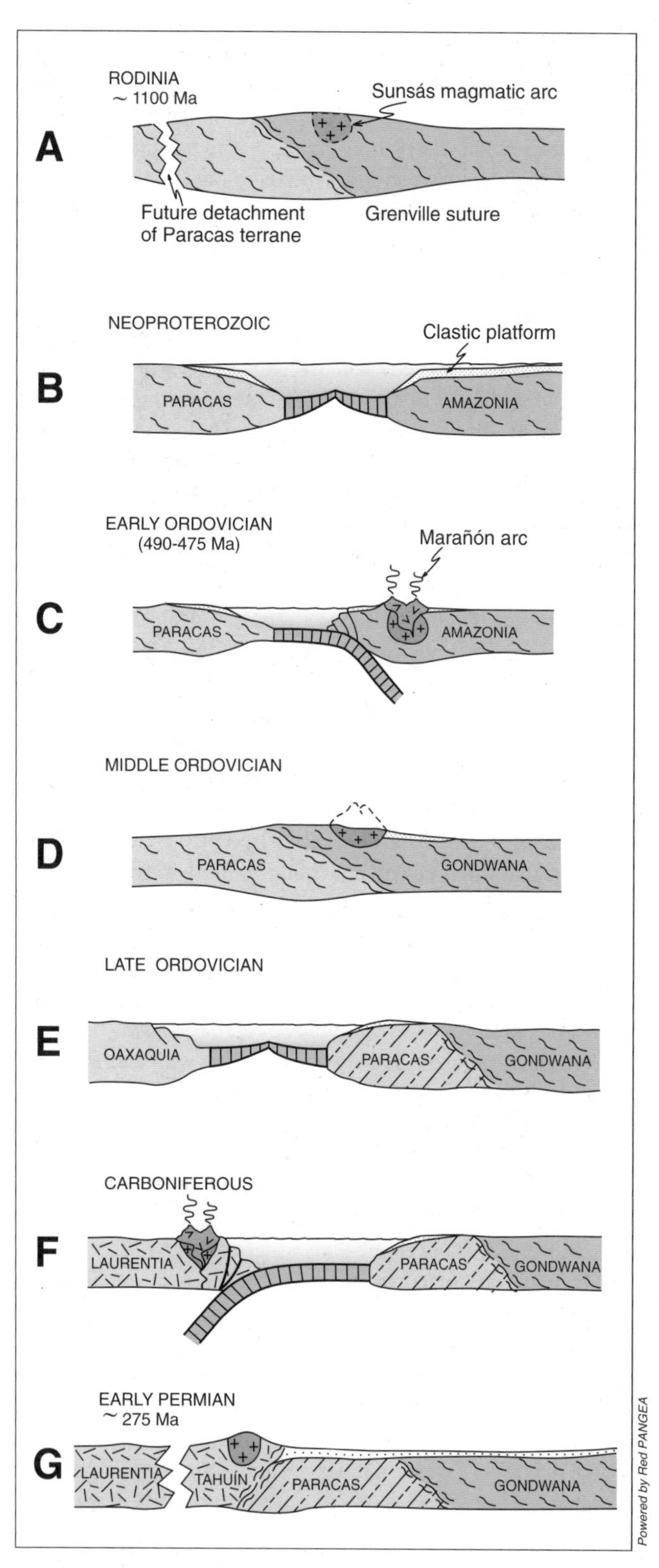

Figure 8. Sequence of basement collisions in the accretionary orogen of the northern sector of Central Andes (based on Feininger, 1987; Litherland et al., 1994; Jaillard et al., 2000; Ramos, 2008a). See Figure 7 for the transect location.

it was assigned to the Mesoproterozoic. However, the Pb isotopic studies of Loewy et al. (2003, 2004) and Wörner et al. (2000a, 2000b) showed that the Antofalla basement is different from that of the Arequipa segment. Although the northern Antofalla basement has evidence of Grenville-age metamorphism, the southern segment appears to have been completely reworked between 0.5 and 0.4 Ga (Franz et al., 2006). The northern segment has a characteristic Pb isotope signature that clearly indicates a relatively unradiogenic basement of Precambrian age, according to Mamaní et al. (2007). The basement of Antofalla in northern Chile and northwestern Argentina has a conspicuous Bouguer gravity anomaly that is characteristic of a distinctive and common sialic basement underlying the thick Andean volcaniclastic cover (Götze et al., 1994).

A belt of low-grade metasediments, grading from slates to schists, is recorded in the Puna and adjacent Eastern Cordillera of northwestern Argentina and southernmost Bolivia (Aceñolaza and Aceñolaza, 2005). This belt extends in a north-south direction from 22°S to 27°S for over 800 km. The several-thousand-meter-thick sedimentary sequence, known as the Puncoviscana Formation, is composed of turbidites, pelagic clays, and minor shallow-water limestones, with local lenses of conglomerates at the base (Ramos and Coira, 2008). The rocks of the Puncoviscana Formation in Bolivia are correlated with the Tucavaca Group (Durand, 1993; Omarini et al., 1999). Detrital zircons and ichnofossils indicate a Neoproterozoic–Cambrian age for the Puncoviscana Basin and the Tucavaca northern extension (Ramos, 2008a). These strata were intruded by tonalites and other granitoids at 530 Ma and intensively deformed during the Pampean deformation in the Early Cambrian.

A series of ophiolitic belts and their associated magmatic arcs indicate activity in a long-lasting accretionary orogen in west-central Argentina (Ramos et al., 1986, 2000; Ramos, 1988a). Initial accretion of the Pampia cratonic block and the Córdoba terrane in latest Proterozoic–earliest Cambrian times characterized the Pampean orogeny in the eastern Sierras Pampeanas (Ramos and Vujovich, 1993; Rapela et al., 1998, 2007; Leal et al., 2004; Escayola et al., 2007).

The U-Pb zircon ages as well as the K-Ar hornblende ages of the upper amphibolite–facies metamorphic rocks indicate significant deformation by the end of the Ordovician related to the Ocloyic collision (Ramos, 1986; Wörner et al., 2000a). Rocks of similar ages, between 490 and 470 Ma, represent the Famatinian cycle in northern Chile and the Argentine Puna. These rocks are well known in the western Faja Eruptiva de la Puna (Palma et al., 1986; Niemeyer, 1989; Coira et al., 1999) and in the eastern Faja Eruptiva de la Puna (Ramos, 1986; Bahlburg and Hervé, 1997; Viramonte et al., 2007). The geochemical and isotopic characteristics of the Puna western belt are typical of a magmatic arc (Coira et al., 1982, 1999). However, the Puna eastern belt, the northern continuation of the Famatina Late Cambrian–Ordovician magmatic arc of Sierras Pampeanas (Ramos, 1988a; Pankhurst and Rapela, 1998; Quenardelle and Ramos, 1999), loses its typical continental arc signature and dies out north of 23°S

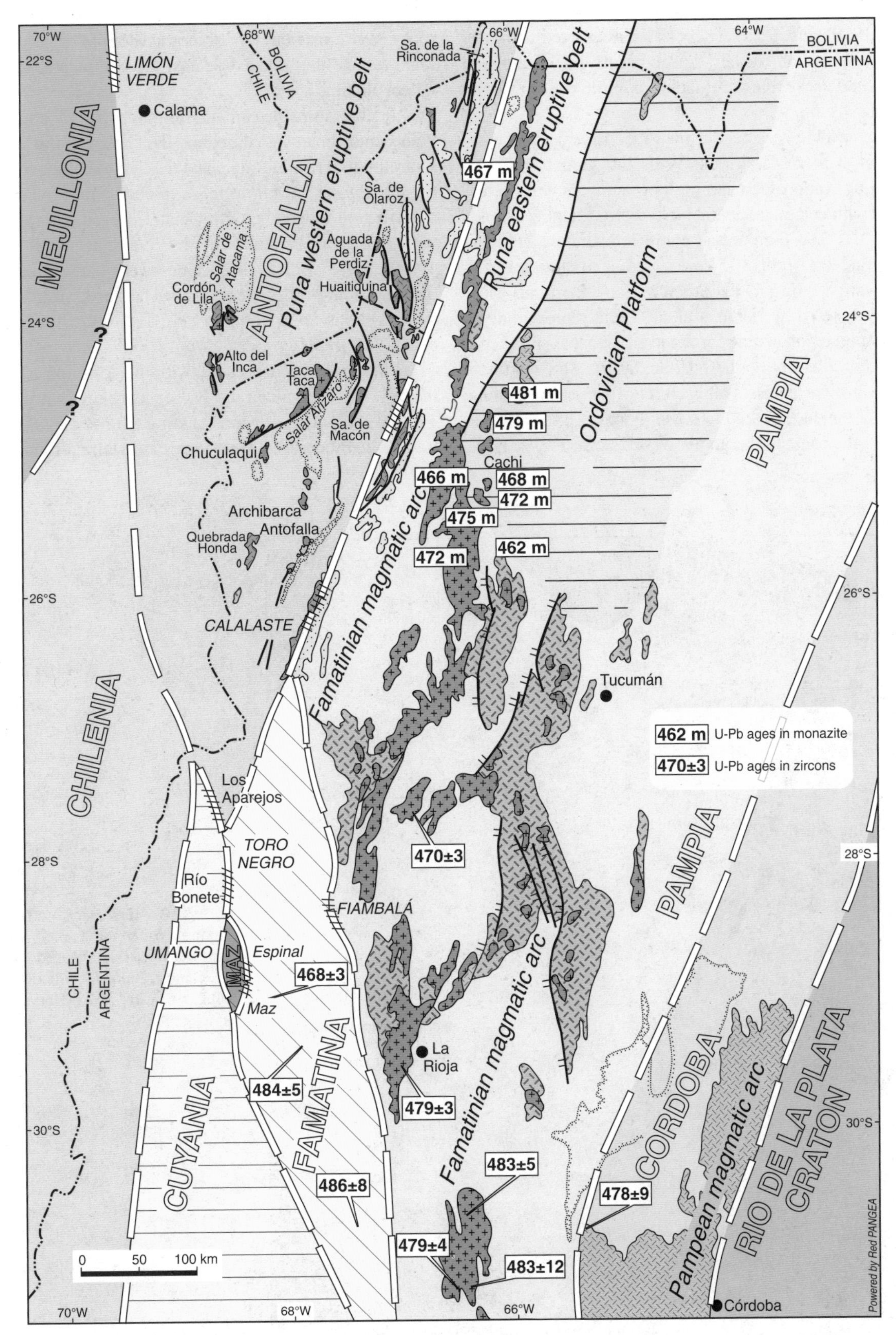

Figure 9. Main tectonic features of the southern Central Andes (based on Ramos, 2000; Martina et al., 2005; Casquet et al., 2008; Ramos and Coira, 2008). Dashed lines are sutures of terranes (names of which are shown in open letters). Ages are in Ma. MAZ—suspect terrane proposed by Casquet et al. (2008).

(Coira et al., 1999) (Fig. 9). These rocks are associated with Late Cambrian to Ordovician backarc clastic basins, which are well dated due to extensive biostratigraphic control (Astini, 2003; Benedetto, 2003).

The metamorphic basement of the Precordillera of northern Chile exposed at Sierra de Limón Verde has garnet amphibolites of tholeiitic composition and high-pressure gneisses in tectonic contact with calc-alkaline gabbros, diorites, and granitoids (Baeza, 1984). These high-pressure–low-temperature (HP-LT) rocks correspond to a depth of 45 km, according to Lucassen et al. (1999), and were dated by U-Pb zircon as Early Permian. These rocks were interpreted by Hervé et al. (1985) as representing an accretionary subduction setting for the late Paleozoic. They crop out in the Precordillera of northern Chile, far from the continental margin east of the gneisses and schists of upper amphibolite facies of the basement exposed in the Mejillones Peninsula and in Río Loa along the coast further to the west (Baeza, 1984). These rocks were considered by Ramos (1988a) to be part of the Mejillonia terrane that docked against the continental margin in late Paleozoic times.

Accretions of the Famatina terrane, a parautochthonous block with Gondwanan signature, and the Cuyania terrane (Fig. 10), a Laurentian-derived terrane, and the first allochthonous terrane in the Gondwana margin at these latitudes, resulted in significant deformation in Middle Ordovician time (ca. 460 Ma) (Astini et al., 1995, 1996; Pankhurst et al., 1998; Quenardelle and Ramos, 1999). These collisional episodes were part of the Famatinian orogeny, a first-order event that can be traced from southern central Argentina up to northern Perú and southern Ecuador (Chew et al., 2007).

The processes associated with the accretion of Cuyania are the best recorded along the Pacific margin (Ramos and Keppie, 1999). Cuyania is a composite terrane formed by the amalgamation of the Precordillera and Pie de Palo terranes in Mesoprotero-

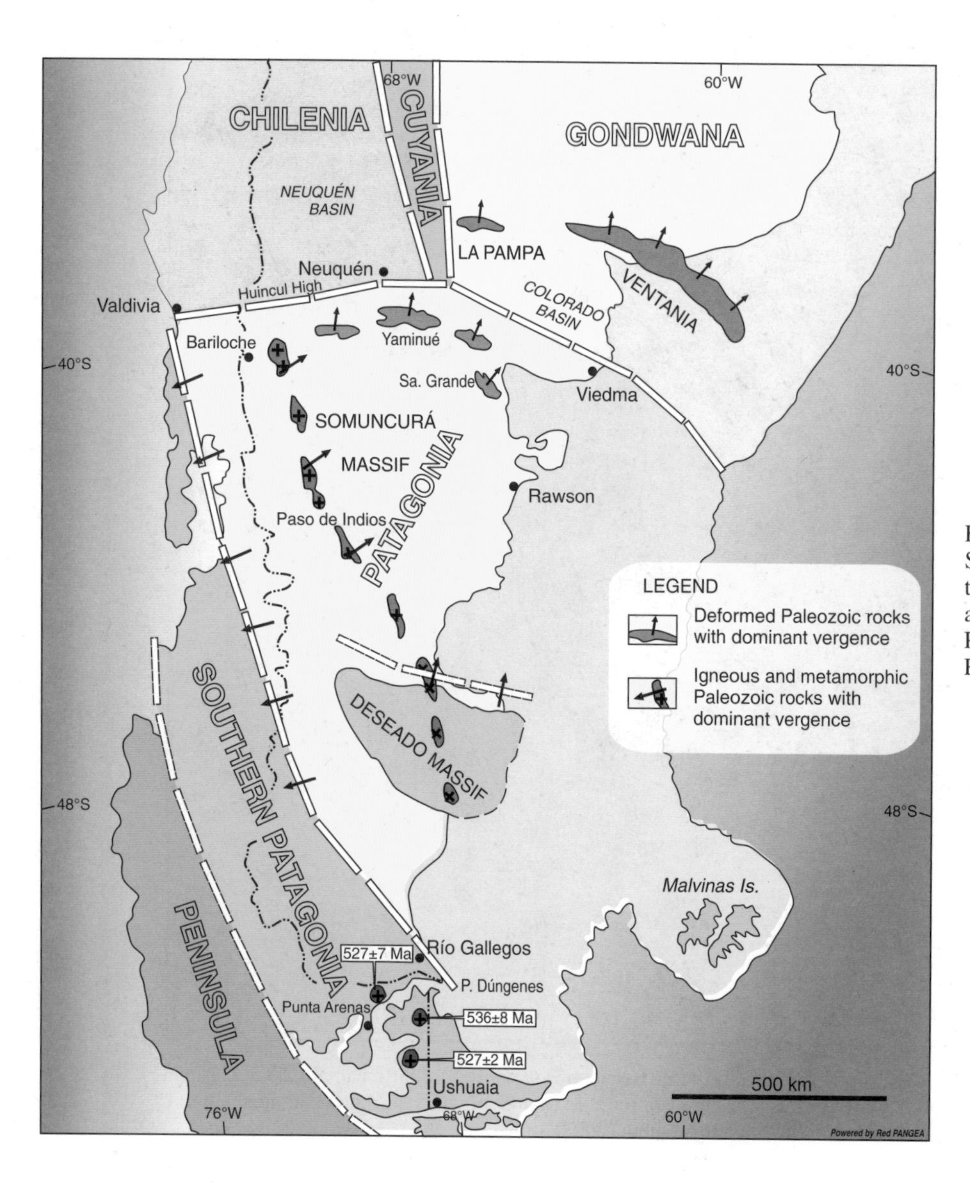

Figure 10. Basement terranes of the Southern Andes and their potential sutures (dashed lines; based on Mosquera and Ramos, 2006; Chernicoff et al., 2008; Pankhurst et al., 2006; Hervé et al., 2008; Ramos, 2008b).

zoic times, which collided with Gondwana in Middle Ordovician times. Biostratigraphic, isotopic, chronological, paleomagnetic, and structural data constrain the origin, transfer, docking, and deformation of the Cuyania terrane, and a general consensus exists on its derivation from Laurentia (see summaries of Thomas and Astini, 2003; Ramos, 2004). However, recent papers still dispute the Laurentian source based on the complexity of the provenance interpreted from detrital zircon data (see Finney, 2007, for other alternatives).

Southern Andean Segment

The southern boundaries of the terranes described in the central Andean segment are covered by late Cenozoic sediments in the extra-Andean foreland (Fig. 10). Therefore, their recognition is based on airborne magnetic maps where a strong positive anomaly is related to mafic and ultramafic rocks emplaced along the sutures (Chernicoff et al., 2008). Seismic interpretation of industry three-dimensional (3-D) data within the Neuquén Basin shows that the NNE-trending basement fabrics north of the Huincul High are truncated by east-west fabric south of the Huincul High. This important change has been related to the deformation produced by the collision of Patagonia (Mosquera and Ramos, 2006).

The Patagonian region shows two different metamorphic and magmatic belts: the northern and the western belts. The northern metamorphic and magmatic belt is preserved parallel to the southern margin of the Neuquén and Colorado Basins along the Río Limay valley from the city of Bariloche in the west up to the Sierra Grande region along the Atlantic coast (Fig. 10) (Ramos, 1984; Varela et al., 1998; Basei et al., 1999). The western metamorphic and magmatic belt crosses central Patagonia with a north-northwestern–south-southeastern trend and continues into the Deseado Massif and further south along the Punta Dúngenes High (Pankhurst et al., 2006; Ramos, 2008b).

The northern metamorphic belt of the Patagonia terrane was studied at Yaminué, where Llambías et al. (2002) described Late Carboniferous orthogneisses metamorphosed in the amphibolite facies and intruded by undeformed Late Permian granitoids. Recent studies based on the magnetic fabric show a ductile deformational fabric consistent with SW-NE compression, which is absent in the Late Permian granites (Rapalini et al., 2008). The second NNW- to NW-trending western belt of magmatic and metamorphic rocks of Patagonia is exposed from Bariloche to Paso de Indios (Fig. 10). The tonalitic gneisses and biotitic granodiorites of Devonian to Carboniferous ages with magmatic arc affinities were interpreted as an oblique late Paleozoic magmatic arc by Ramos (1983), Pankhurst et al. (2006), and Ramos (2008b). These studies failed to recognize a suture between Patagonia and older basement south of the magmatic arc, here called "Southern Patagonia." This Southern Patagonian block has evidence of basement older than Early Cambrian. U-Pb zircon ages from a granodiorite reported by Söllner et al. (2000), as well as new SHRIMP ages obtained from the basement of Tierra del Fuego, indicate ages around 325 Ma (Hervé et al., 2008). These last authors also reported Early Cambrian protolith ages in the Darwin Cordillera.

If it is accepted that the Antarctic Peninsula was detached from Southern Patagonia in Early to Middle Jurassic times (Ghidella et al., 2002), the collision that produced the western metamorphic and igneous rocks in Patagonia can be related to the Peninsula terrane, either by itself or associated with the southern Patagonia block, which could have been a composite terrane together with the Antarctic Peninsula (Fig. 10).

Most of the continental basement of the northern, central, and southern Andes had been amalgamated by the end of the Paleozoic. The late Paleozoic was characterized by significant extension and rifting that followed the formation of the Pangea supercontinent.

BREAKUP OF PANGEA IN THE ANDES

The first evidence of rifting can be seen in the northern Andes with the separation of Laurentia and the opening of the Mexican Gulf, which predated the formation of the Atlantic Ocean. Several rift systems, such as the Uribantes, Machiques, and Espino grabens in Venezuela (Alemán and Ramos, 2000; Cediel et al., 2003), or the various rifts recognized in the subsurface basement of the Eastern Cordillera of Colombia (Sarmiento-Rojas et al., 2006), developed between the latest Triassic and Early Jurassic. At that time, most of the basement accreted against the protomargin of Gondwana was subjected to extension. Rift structures were concentrated in the hanging wall of the previous sutures, as seen in the Pucará rift basins of central and northern Perú and described by Mathalone and Montoya (1995) (Fig. 11). Extension continued along the hanging wall of the suture of the Arequipa terranes in southern Perú, along with deposition of the Late Permian–Early Triassic Mitu Group, a sequence of red beds, evaporates, and alkaline basalts with minor acidic rocks described by Kontak et al. (1990), Carlotto (2002), and Sempere et al. (2002).

These rocks are also associated with some small granitic stocks and syenite intrusions. As pointed out by Kontak (1985) and Sempere (1995), the interface between the Amazonia craton and the orogen was a weak zone that controlled the emplacement of granites, like the Carabaya batholith, and some alkaline volcanic rocks far away from the magmatic arc. Díaz Martínez et al. (2000) and Ramos (2008a) recognized that this old suture between the Amazonia craton and the Arequipa terrane was reactivated during extension, and also by strike-slip reactivation in the early Paleozoic and in late Paleozoic–Early Triassic times. The emplacement of within-plate peraluminous high-K granites and shoshonitic volcanic rocks continued during the late Oligocene and Miocene along the suture in what Jiménez and López-Velázquez (2008) identified as the Huarinas belt.

A large ignimbritic province of rhyolitic composition known as the Choiyoi province (Kay et al., 1989) developed from northern Chile at the latitude of Salar de Atacama (23°30′S lat.) throughout the main Andes to extra-Andean Patagonia (42°S)

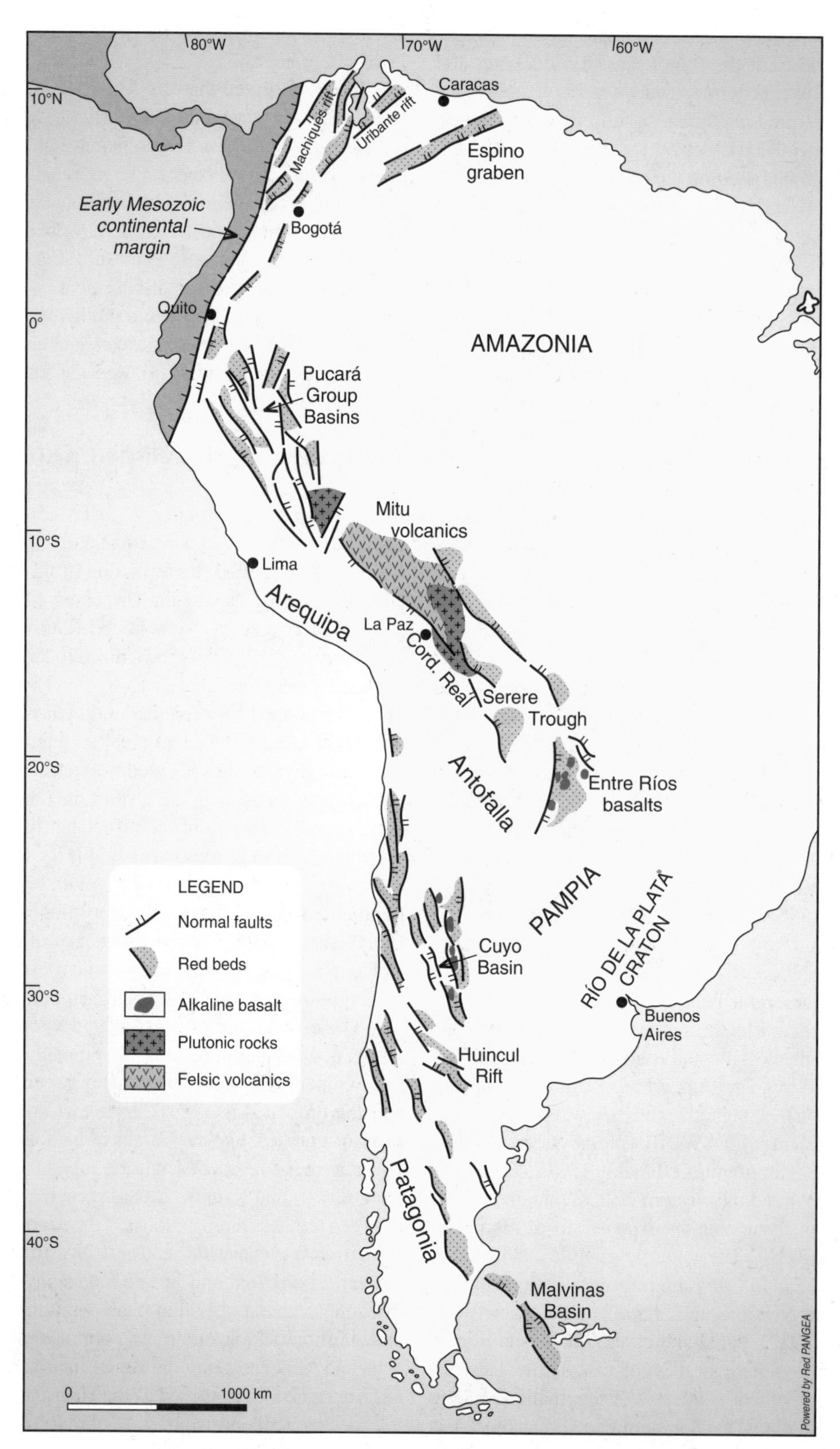

Figure 11. Late Triassic–Early Jurassic generalized extension in the Andean basement. Note the concentration of rifting in the hanging wall of the Paleozoic sutures (based on Uliana and Biddle, 1988; Daly, 1989; Parnaud et al., 1995; Mathalone and Montoya, 1995; Alemán and Ramos, 2000).

over 2000 km of the cordillera. This province is composed of late Paleozoic granitoids and associated acidic volcanic rocks that show a change from a subduction-related magmatic arc setting to widespread extensional volcanism (Fig. 12A) in the Late Permian–Early Triassic (Mpodozis and Ramos, 1989; Mpodozis and Kay, 1992; Llambías and Sato, 1995). This change has been attributed to an extensional collapse after a significant compressive deformation event known as the San Rafael orogenic phase (Kay et al., 1989) or to steepening of the subduction zone after large foreland migration of the arc associated with a flat-slab period (Martínez, 2005; Ramos and Folguera, 2009). This extension propagated to the south and to the east in the Chon Aike province, another rhyolitic ignimbritic province of Early to Middle Jurassic age (Kay et al., 1989) that was deposited in a volcaniclastic rift setting predating the opening of the South Atlantic (Ramos, 2002).

A series of Triassic rift systems with a NW trend in northern Chile and Argentina is well known from the early work of Charrier (1973). The rifts were emplaced in the hanging wall of the previous sutures among the different Paleozoic terranes (Fig. 12B), as established by Uliana and Biddle (1988) and Ramos and Kay (1991).

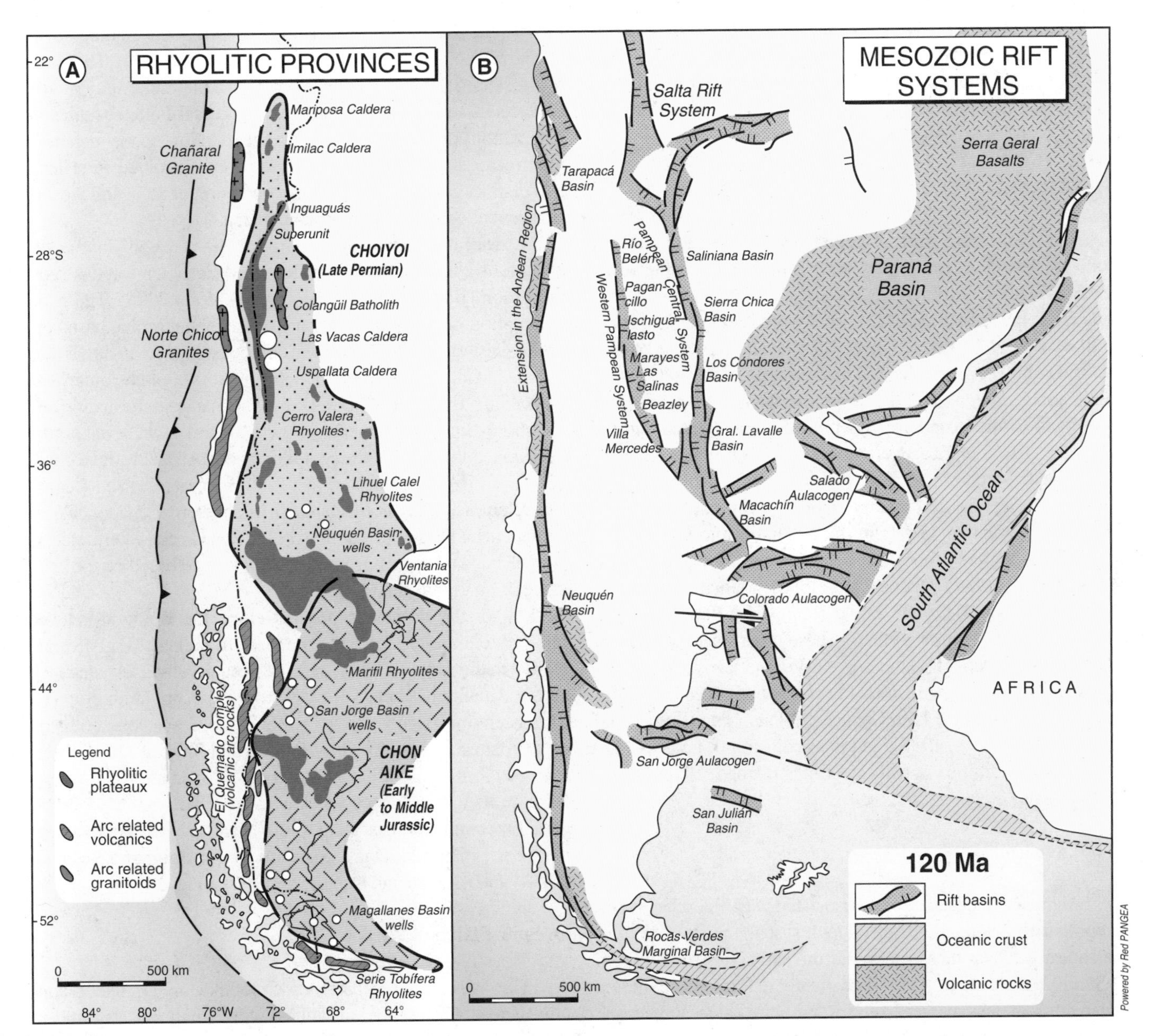

Figure 12. (A) Extensional rhyolitic Choiyoi and Chon Aike provinces deposited in half-grabens developed in southern South America (based on Kay et al., 1989). (B) Rift systems that followed the breakup of Pangea and were related to the opening of the South Atlantic Ocean (based on Uliana and Biddle, 1988; Ramos and Alemán, 2000).

The Triassic–Early Jurassic rift system, which was closely linked to the breakup of Pangea, was followed in southern South America by Middle Jurassic extension, which developed as the Cañadón Asfalto rift basin in central Patagonia (Figari et al., 1996). Extension had developed as far south as the Rocas Verdes Marginal Basin in Late Jurassic times (Dalziel et al., 1974; Hervé et al., 2000). These basins formed as a consequence of the opening of the Weddell Sea between the Antarctic Peninsula and the Patagonian Cordillera at 160 Ma, as denoted by the age of the magnetic anomalies and paleogeographic reconstructions (Ghidella et al., 2007; Ramos and Ghiglione, 2008).

The western margin of central and southern South America underwent major extension in Late Jurassic to Early Cretaceous that resulted in the development of the Salta rift system in northern Argentina and western Bolivia (Salfity and Marquillas, 1994). This extension propagated to the Atlantic margin by means of a series of north-south–trending rifts, which concluded with the opening of the South Atlantic Ocean (Fig. 12B).

COLLISIONAL TECTONICS IN THE ANDES

In his pioneering work, Irving (1971) recognized that most of the western Northern Andes were formed by oceanic rocks west of the Dolores-Romeral fault system (presently known as the Romeral-Peltetec fault; Fig. 13). This work was complemented by the proposal of obduction of this oceanic crust (Restrepo and Toussaint, 1974), which led to the present interpretation of a collisional orogen. This concept was developed by Barrero (1979) and Restrepo and Toussaint (1988) in Colombia, and by Feininger (1987) in Ecuador and northern Perú. Geochemical, petrological, isotopic, and geochronological data, together with evidence for intense deformation, associated in part with blueschist metamorphism, constrain the evolution into three major stages following the proposals of Aspden and McCourt (1986), Mégard (1987, 1989), Duque Caro (1990), Aleman and Ramos (2000), Cediel et al. (2003), López Ramos and Barrero (2003), and Vallejo (2007), among others. These three stages are: initial collision of a series of island arcs during the Early Cretaceous (Aspden et al., 1987); collision of the Caribbean oceanic plateau along the margin in the Late Cretaceous–Paleocene (Dewey and Pindell, 1986; Burke, 1988); and final collision of an island arc of Caribbean affinities in Miocene times (Dengo, 1983; Duque Caro, 1990).

Island Arc Collisions

Fragments of oceanic rocks associated with blueschist metamorphic facies along the Romeral fault system are interpreted as the suture between these rocks and the early Mesozoic margin of South America (Fig. 13). These greatly deformed and altered oceanic fragments have received several names since the terrane proposal of Etayo Serna and Barrero (1983). One of the more detailed studies was done in the Amaime area, where high-pressure lawsonite-glaucophane schists and, locally, eclogites were tectonically emplaced into the Paleozoic metamorphic rocks during the Mesozoic (Aspden and McCourt, 1984). Associated with accretion, there was a major period of dynamic metamorphism throughout the Central Cordillera of Colombia and the Cordillera Real of Ecuador. This episode, known in southern Ecuador as the Peltetec event, occurred around 125–132 Ma, based on the age of the high-pressure metamorphic rocks considered to represent the emplacement age of the blueschists (Feininger, 1982; Aspden et al., 1992b; Litherland et al., 1994). This event reflects the accretion of an Early Cretaceous island arc along the western margin of the Cordillera Real, as recognized by Aspden and Litherland (1992) and Litherland et al. (1994). They recognized the Early Cretaceous suture further south, as well as the emplacement of the Raspas ophiolitic complex and associated blueschists at 132 Ma (Apsden et al., 1995; Reynaud et al., 1999). The study by Bosch et al. (2002) concluded that the latest Jurassic–Early Cretaceous suture observed along the Romeral-Peltetec fault system exhibits high-pressure assemblages and low-grade metamorphic rocks, suggesting that this suture was involved in at least two distinct episodes related to the accretion of an island arc and subsequent collision of an oceanic plateau.

Other authors have interpreted part of these collisions as the closure of a backarc basin that recorded Early Cretaceous sedimentation (Bourgois et al., 1987; Nivia et al., 2006). The main assumption is related to the nature of the metamorphic basement of the western margin of the backarc basin known as the Arquía terrane in Colombia, which for some authors could be older than Paleozoic (Nivia et al., 2006). Due to the intense deformation and the strike-slip displacements of the Romeral fault, it cannot be overlooked that part of the protolith of this Arquía terrane could have been derived from the Paleozoic basement of the Central Cordillera. Similar problems exist in southern Ecuador, where the Chaucha terrane of Feininger (1986) has been interpreted as nonoceanic or different from the Piñón terrane based on the existence of quartzitic sandstones.

All of the oceanic exposures along the Romeral-Peltetec fault system are grouped as a single episode here (Fig. 13) and related to the early accretion of some oceanic island terranes such as the Amaime, Peltetec, Raspas, and Chaucha, among others. This accretion occurred prior to the Late Cretaceous collision of the plateau as proposed by Pindell and Tabbutt (1995), Aleman and Ramos (2000), and Bosch et al. (2002). During Middle Jurassic to Early Cretaceous time, a magmatic arc also developed on continental basement as the island arc was approaching the Central Cordillera, as recognized by Toussaint and Restrepo (1994) and Aleman and Ramos (2000) (see Fig. 14).

Oceanic Plateau Collision

The Western Cordilleras of Colombia and Ecuador contain Upper Cretaceous turbidites with small but significant fault-bounded slivers of basalts and ultramafic rocks that were formerly interpreted as a mid-ocean-ridge basalt (MORB) rock assemblage (Lebras et al., 1987) but later recognized as an oce-

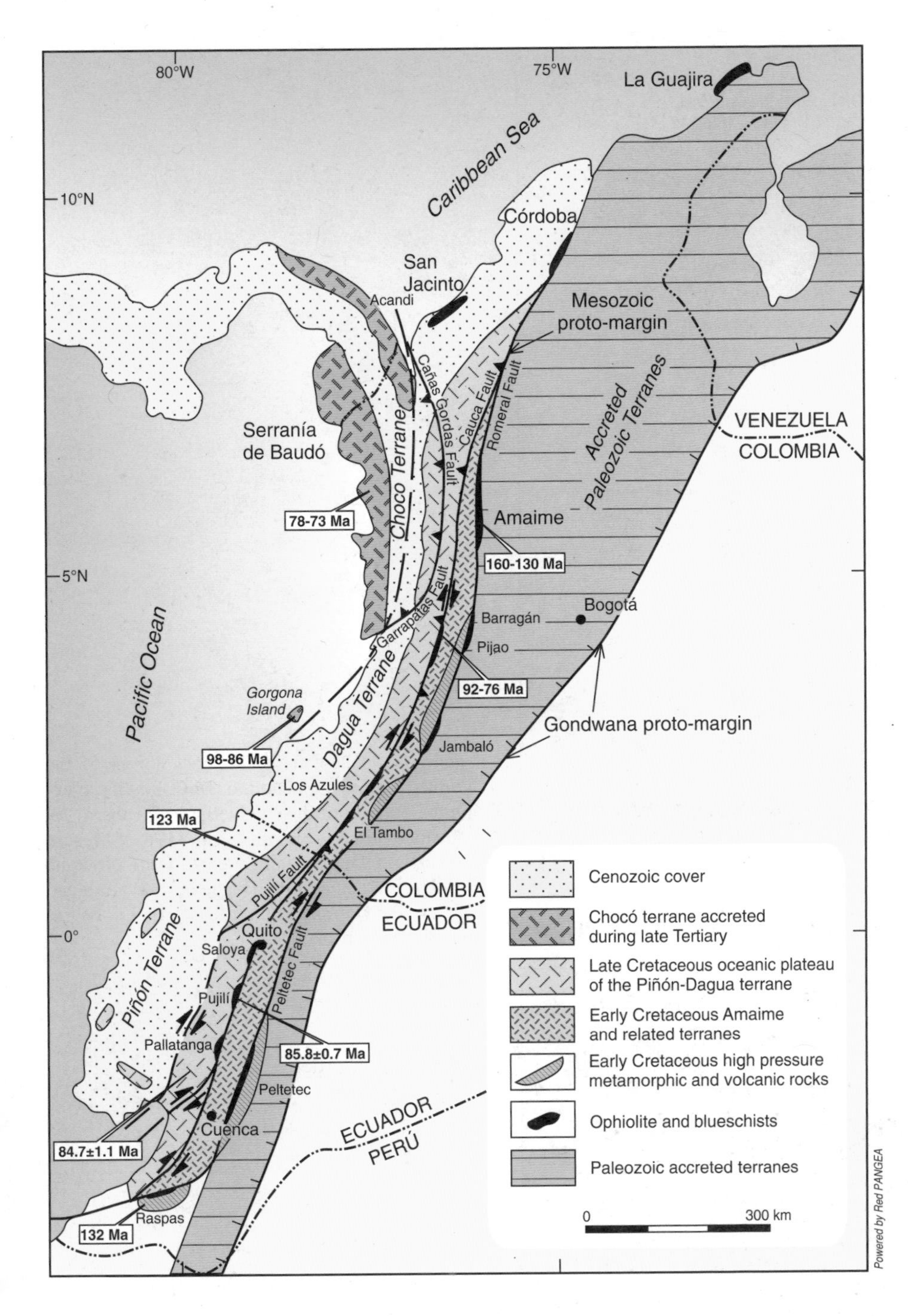

Figure 13. Different oceanic rocks west of the Romeral-Peltetec fault system in western Colombia and Ecuador based on McCourt et al. (1984), Aspden et al. (1992a, 1992b), Reynaud et al. (1999), Aleman and Ramos (2000), Kerr et al. (2003), Cediel et al. (2003), and Vallejo et al. (2006, this volume). Note the accretion in three successive stages: the Amaime terrane and related rocks (Early Cretaceous); the Dagua-Piñón oceanic plateau (latest Cretaceous); and the Chocó terrane (middle Miocene). The collision of the Chocó terrane in the middle Miocene thrusted part of the Dagua terrane to the east along the Cañas Gordas fault, north of Garrapatas fault.

anic plateau sequence (Cosma et al., 1998; Reynaud et al., 1999; Lapierre et al., 2000) and a tectonic mélange.

There have been various interpretations for the characteristic and genesis of this oceanic plateau. The original proposal of Feininger (1986), that pieces of oceanic rocks grouped in a single Piñón terrane were accreted to the continental margin, was challenged by subsequent workers (Kerr et al., 2002; Hughes and Pilatasig, 2002, among others), who proposed a series of collisions between Late Cretaceous and Eocene times. However, recent work based on petrological, geochronologic, and paleomagnetic grounds, has demonstrated that multiple plateaus (or terranes) can be precluded. For example, the paleomagnetic data indicate a single oceanic plateau that fragmented during collision with the South American plate, giving rise to distinct structural blocks (Luzieux et al., 2006). Likewise, new U-Pb SHRIMP zircon data show a limited time span between 87.10 ± 1.66 Ma and 85.5 ± 1.4 Ma for the formation of the oceanic plateau (Vallejo et al., 2006, this volume).

The oceanic plateau continues into Colombia as a part of the Dagua terrane (Etayo Serna and Barrero, 1983; Alemán and Ramos, 2000), although other authors such as Cediel et al.

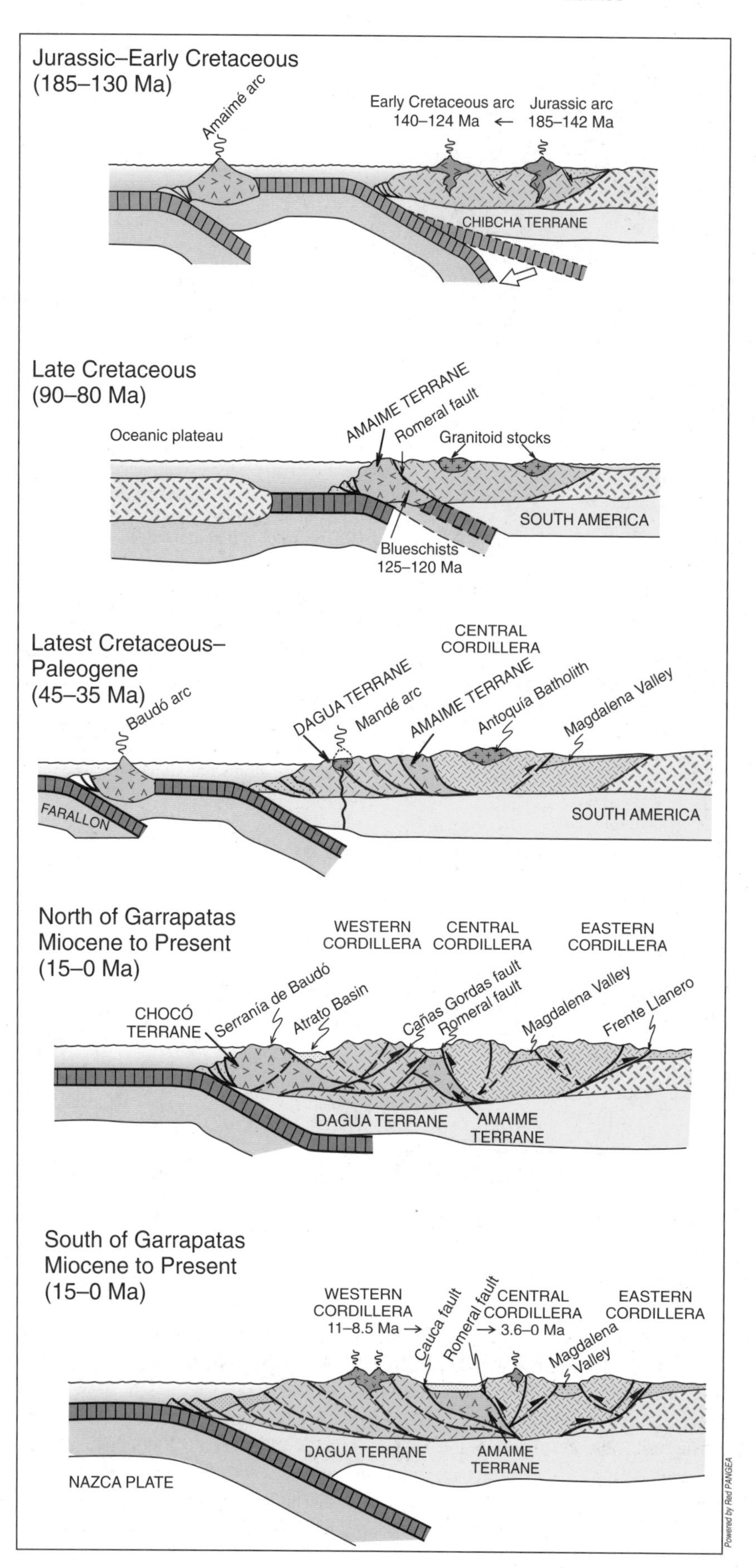

Figure 14. Schematic tectonic evolution of the Northern Andes at the latitude of Colombia showing the polarity of the subduction and subsequent collisions based on Lebras et al. (1987), Mégard (1987), and Aleman and Ramos (2000). Note the different vergence of the Dagua terrane north and south of the Garrapatas fault (based on López Ramos and Barrero, 2003).

(2003) and López Ramos and Barrero (2003) favor multiepisodic collisions. Available geochemical data show the uniform composition of the oceanic rocks, which is also interpreted as the leading edge of the Caribbean plateau (Kerr et al., 1998) that collided with the South American plate at 75–65 Ma (Ramos and Alemán, 2000; Cediel et al., 2003).

Subduction occurred above the oceanic plateau on the Pacific side, as has been demonstrated by volcanic arc assemblages that have been described overlying the plateau (Lebras et al., 1987; Toussaint and Restrepo, 1994; Reynaud et al., 1999; Alemán and Ramos, 2000). At the same time, the approaching plateau generated a magmatic arc on the Central Cordillera of Colombia and the Cordillera Real of Ecuador until the Late Cretaceous. The Paleocene granites of the large Antioquia batholith (59–57 Ma), which partially postdate the collision of the Dagua Plateau, may represent a slab break-off episode that developed in the northern Central Cordillera east of Medellín.

Chocó Collision

The Serranía de Baudó along the Pacific margin of Colombia has been described as an allochthonous oceanic terrane accreted to South America during Cenozoic times (Dengo, 1983; Case et al., 1984; Restrepo and Toussaint, 1988). This area has been interpreted as part of a larger block that includes the Darien of Panamá, the Acandí arc (Fig. 13), and a sliver of the Western Cordillera, named the Chocó terrane by Duque Caro (1990), after Dengo (1983). Based on precise biostratigraphic dating, Duque Caro (1990) constrained the docking of the Chocó terrane at ca. 13 Ma. The basement of this terrane has been interpreted as a different part of the Caribbean plateau based on geochemical and isotopic data (Kerr and Tarney, 2005), although it has an age similar to the Dagua oceanic plateau.

An important sector of the Western Cordillera at the latitudes of the Chocó terrane, north of the Garrapatas fault, is considered to be an independent terrane (Cediel et al., 2003) named the Cañas Gordas terrane following Etayo Serna and Barrero (1983). Based on the same isotopic and geochemical characteristics to the north and south of this fault, Aleman and Ramos (2000) considered the Cañas Gordas block to be part of the Dagua terrane. The unusual tectonic eastward vergence of the Cañas Gordas terrane (Bourgois et al., 1987; López Ramos and Barrero, 2003), when compared with the southern part of the Dagua terrane, can be explained by a transposition of the structure related to the Miocene deformation associated with the docking of the Chocó terrane (Fig. 14).

Isotopic and paleomagnetic data led Kerr and Tarney (2005) to consider the mafic and ultramafic rocks of the Gorgona Island to be part of the Chocó terrane and, at the same time, independent of the Western Cordillera oceanic plateau.

The general tectonic evolution of the oceanic rocks of the western sector of the Northern Andes that formed as part of the Caribbean plateau and formed above the Galápagos hot spot is summarized in Figure 14.

CHANGE IN THE ANDEAN TECTONIC REGIME

The Andean Cordillera as a whole was subjected to an important change in tectonic regime during the mid-Cretaceous. This change from an extensional regime to a compressional regime has been recognized since the early work of Vicente (1970) and Auboin et al. (1973), but the real causes for this change were overlooked. Mpodozis and Ramos (1989) noted the significant variation in the tectonic regime and magmatism from extension and poorly evolved mafic rocks in the Early Cretaceous to compression and andesitic to dacitic rocks in Late Cretaceous to Cenozoic along the Argentine-Chilean Andes. These authors, following Levi and Aguirre (1981) and Levi et al. (1985), assigned this change to the development and closure of an ensialic marginal basin due to stress reorganization controlled by changes in the convergence rate.

The study of Daly (1989) in the Andes of Ecuador was one of the first to relate these changes to the trench roll-back velocity, a criteria followed by Ramos (1999, 2000), who recognized this change in the tectonic regime from Colombia to the southernmost Andes in the mid-Cretaceous. The studies of Somoza (1996) and Somoza and Zaffarana (2008) demonstrated that South American paleopoles and the moving hotspot framework provide a valid kinematic scenario to explain the tectonic regimes. They related the extensional tectonics in the early stages with episodic divergence between the trench and the continental interior. The beginning of contraction was related to the westward acceleration of South America, suggesting that the continent began overriding the trench in the Late Cretaceous. This change in Andean tectonic regime was probably associated with a major plate reorganization at ca. 95 Ma, which resulted from the final breakup of Africa and South America during the opening of the South Atlantic Ocean.

The change in the tectonic regime is clearly shown by the extensional faults that dominate Andean subduction from the Jurassic to the Early Cretaceous, including a period of no or negligible compression in the Aptian-Albian and subsequent compression in the Late Cretaceous associated with the change in the absolute motion of South America (Ramos, 1999) (Fig. 15). Evidence for this change has been described in the northern Andes by Gómez et al. (2005) and in the central Andes by Mpodozis et al. (2005).

This scenario began as a consequence of the breakup of western and eastern Gondwana at ca. 180 Ma and the opening of the Indian Ocean, which coincided with the beginning of subduction along the Pacific margin of South America. The swing between rifting related to the breakup of Pangea and extensional subduction due to the absolute motion of western Gondwana to the northeast is clearly shown by the change in the nature of magmatism. The alkaline mafic within-plate volcanic rocks in the Triassic (Ramos and Kay, 1991; Mpodozis and Cornejo, 1997) gave way to mafic calc-alkaline rocks typical of magmatic arcs (Gana et al., 1994). This extensional setting lasted until the final opening of the South Atlantic Ocean, when compression began (Ramos, 2000).

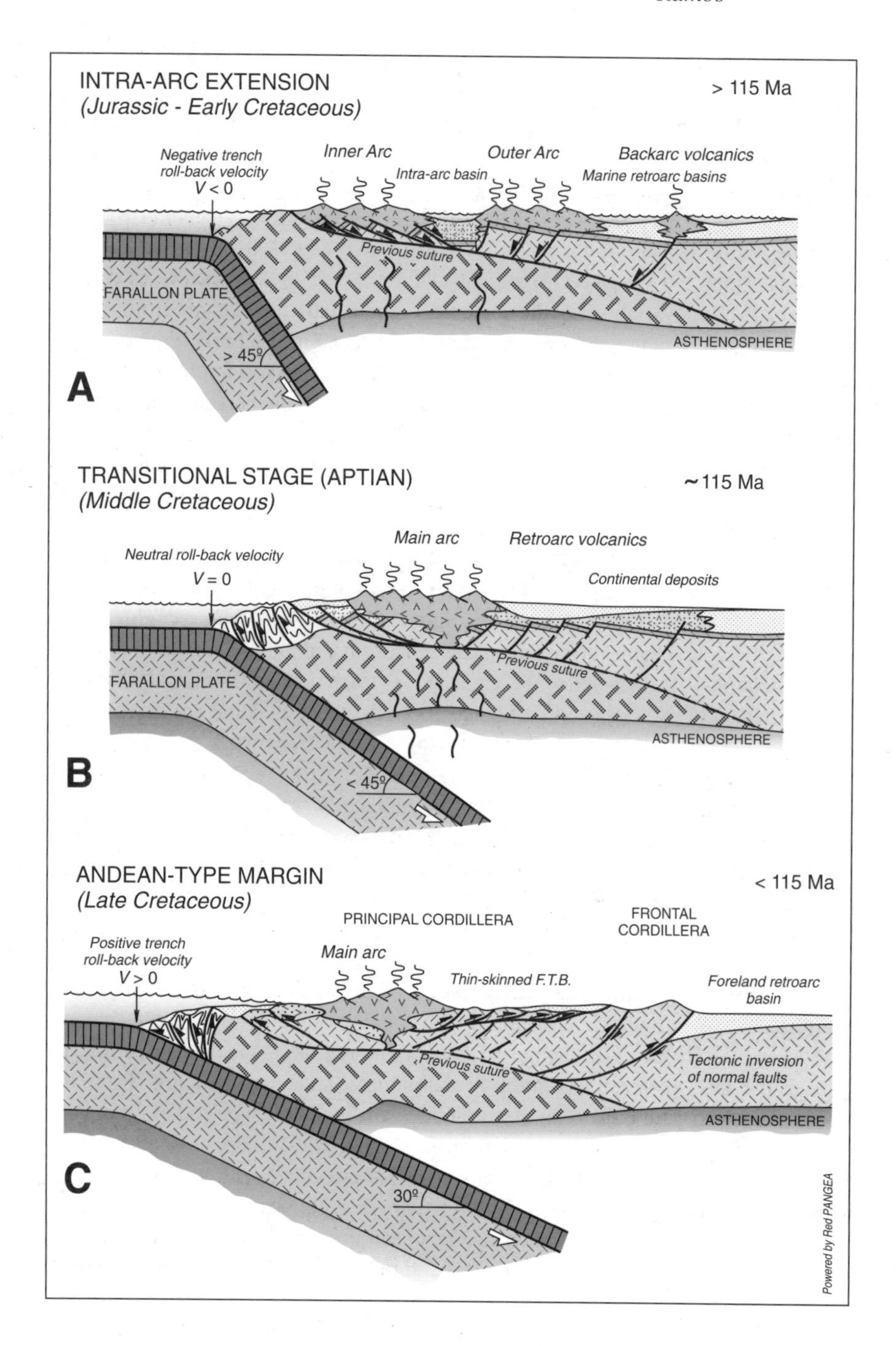

Figure 15. Three distinct stages of subduction during Mesozoic-Cenozoic times (conceptual model based on Daly, 1989; Ramos, 2000). F.T.B.—fold-and-thrust belt.

OROGENIC CYCLE IN AN ANDEAN-TYPE MARGIN

The concept that orogenic processes were intermittent and sporadic in an Andean system has been entrenched in all the hypotheses and theories of mountain building, such as the geosynclinal (Groeber, 1951), or the geoliminar (term coined by Auboin and Borrello, 1966) Andean cycle. With the new approaches derived from the application of plate tectonics, several attempts have been made to explain the paucity and sporadic character of mountain-building episodes through time. The first proposals, like that of Charrier (1973), related changes to relative convergence rates. For Charrier (1973), there was a close correlation between interruptions in spreading and global compressive periods followed by extension, which alternated in a rhythmic pattern. This proposal was improved by Frutos (1981), for whom changes in the directions and rates of sea-floor spreading controlled short compressive orogenic phases alternating with relatively longer periods of extension. He envisaged the beginning of periods of relatively higher rates of plate convergence, such as at 110–85, 76–70, 63–60, 49–45, 35–33, 16–13, and 7 Ma, as responsible for the compression and an orogenic periodicity after episodes of relative quiescence. The discovery that long periods of extension were punctuated by short episodes of compression provided the framework for the Andean model of the 1980s (Aguirre, 1987).

However, the models addressing the variations of the convergence rates developed by Pardo Casas and Molnar (1987) and Somoza (1998) demonstrated that deformation at some times sped up when convergence rates decreased, as in the last 20 m.y. (Oncken et al., 2006). These data also showed that alternation between compression and extension was neither rhythmic nor global, not even along the Andes, as previously accepted. Deformation episodes were not coeval, and diastrophic phases, when precisely dated, did not coincide in the different segments along the Andes (Mpodozis and Ramos, 1989).

Flat-slab subduction periods have been shown to have significance in mountain building, based on the precise correlation between deformational episodes and magmatism in different segments (Jordan et al., 1983a, 1983b; Kay et al., 1987, 1988, 1991). Recent studies have shown similar processes in the Bucaramanga flat slab in northern Colombia as in the present Peruvian and Pampean flat-slab segments (Gutscher et al., 2000; Ramos et al., 2002). Recent work has revealed that the present configuration of flat-slab subduction was not unique, but rather has been a common feature through time. Shallowing and steepening processes affecting the Andean margin has been recognized in different segments, as described by Coira et al. (1993), Sandeman et al. (1995), Kay et al. (1999), and James and Sacks (1999). New evidence allows recognition of an almost complete sequence of flat-slab subduction along the Andean margin (Fig. 16) during late Mesozoic and Cenozoic times, as described by Ramos and Folguera (2009). The proposal that those episodes are not so extraordinary has led to the suggestion that these processes are key factors in the control of mountain building and in the recognition of an orogenic cycle that resembles the processes described long ago by Dickinson and Snyder (1979) in western North America.

Based on this new evidence, a conceptual orogenic cycle is proposed with the following stages. These stages have led to significant deformation, magmatism, and repeated episodes of mountain building in the Andean margin (Fig. 17).

Quiescence, Absence of Deformation, and Incipient Arc Magmatism

Periods of quiescence, where no deformation has occurred, have been recognized at diverse times in separate Andean segments. One of the most remarkable was the beginning of a new period of active subduction after the breakup of the Farallon plate after 27 Ma, marking a new period of arc magmatism. The more orthogonal subduction of the recently formed Nazca plate coincided with an incipient magmatism. The volcanic activity started with mafic, poorly evolved magmatism that subsequently changed to a more andesitic to dacitic composition formed in the volcanic front at ~280–300 km from the trench. There is scarce evidence of compressional deformation at this time. Crustal thickness was normal or slightly attenuated. There was some localized basin subsidence still related to active extension in the Chilean forearc (Jordan et al., 2006, 2007), but in the retroarc, it was mainly related to the thermal sag of previous attenuation. The Oligocene was a period of significant extension through all the Andes related to a significant decrease of the absolute motion of South America (Russo and Silver, 1996).

Expansion of the Magmatic Arc and Related Deformation

Magmatic activity in some cases started expanding toward the foreland, while in others, it stayed constant (Kay et al., 1987, 1988; Mpodozis and Ramos, 1989). This shift in the volcanic front varied from one segment to the next. Some areas may have undergone significant crustal erosion by subduction, as proposed by Ramos (1988b), Stern (1991), and Stern and Mpodozis (1991), while others are stable. In segments where significant crustal erosion has been advanced, a striking correlation among composition, arc shifting, and deformation has been recognized by Haschke et al. (2002) at 21°S–23°S and Kay et al. (2005) at 33°S–36°S latitudes. Migration of magmatism increases heat flux and subsequent development of brittle-ductile transitions in the crust, favoring basement deformation. Thermal weakening of the crust produces a wave of shortening, which is clearly depicted across the Andes (Ramos, 1988b; Mpodozis and Ramos, 1989; Kay and Abbruzzi, 1996), and also confirmed by numerical modeling, which shows that shifting of deformation toward the foreland is associated with thermal heating (Quinteros et al., 2006). Basin formation is controlled by incipient foreland basin sedimentation (Jordan, 1995).

Shallowing of the Subduction Zone and Crustal Thickening

Those areas where shallowing of the Benioff zone has occurred show a clearly marked broadening of the arc, as in the segment between 30°S and 33°S latitude. Associated thermal weakening enhances crustal thickening in those areas where sutures or weak rheologies focus deformation in the basement (Ramos et al., 2002), leading to foreland fragmentation as incipiently seen in the Sierras Pampeanas (Jordan and Allmendinger, 1986).

As a consequence of crustal thickening, the composition of the magmas and their arc products change (Kay et al., 1991; Kay and Mpodozis, 2002). The development of the magmatic arc in a thickened crust causes magmas to stall at the base of the crust in the garnet stability field, producing an eclogitic residue at the crust-mantle interface (Kay et al., 1991).

Tectonic inversions of extensional faults formed in previous settings are common, but deformation is concentrated in the orogenic front where thick- and thin-skinned tectonics are active. Basin formation accompanies the migration of the thrust front (Jordan et al., 1988).

Flat-Slab Subduction, Magmatic Lull, and Deformation

Those areas where flat-slab subduction was developed concentrate important coupling between the upper and lower plates, leading to significant intracrustal deformation. Crustal earthquakes related to basement shortening are seen in the central

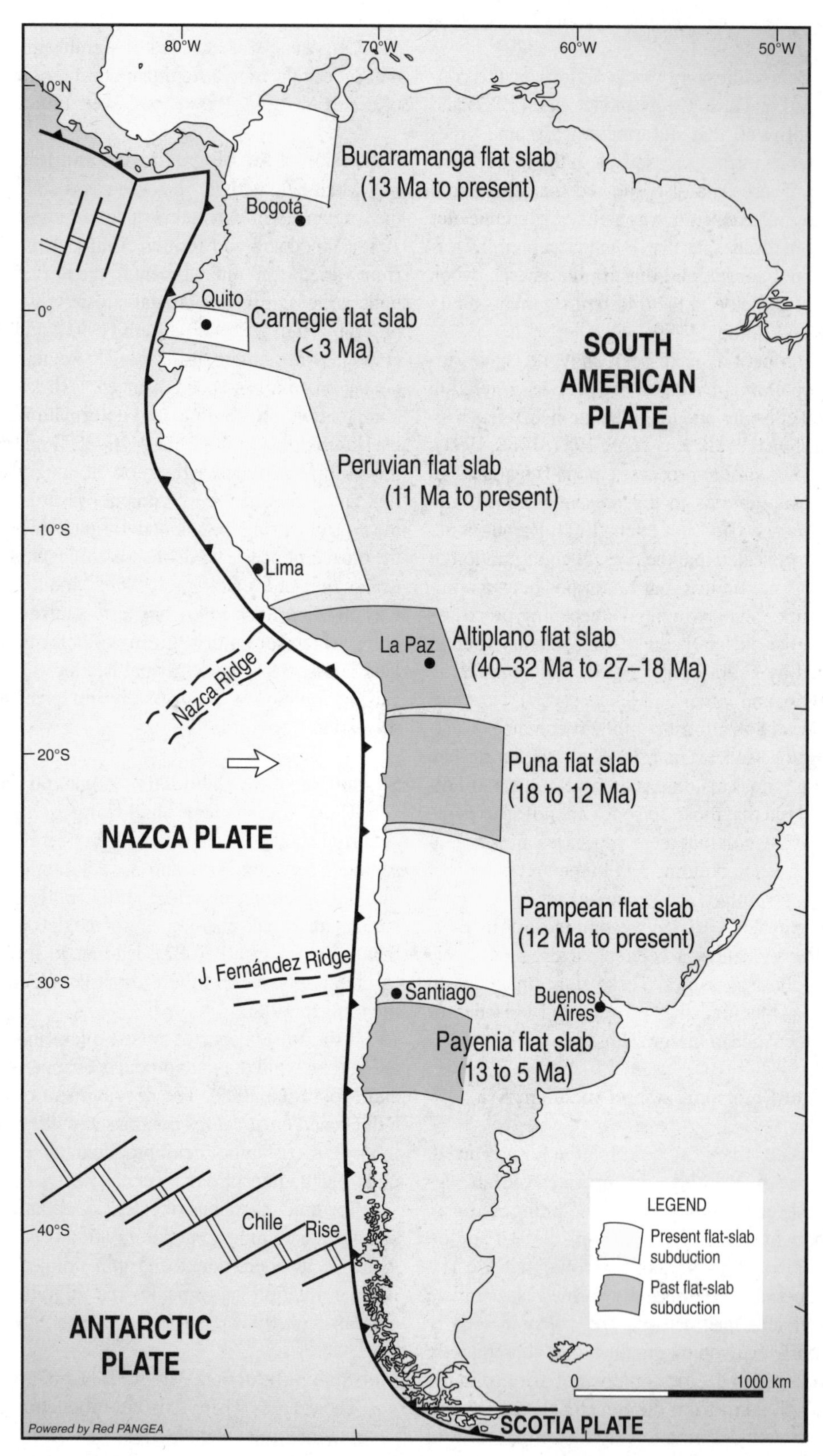

Figure 16. Flat slabs throughout the Andes in Cenozoic times. Modified from Ramos and Folguera (2009).

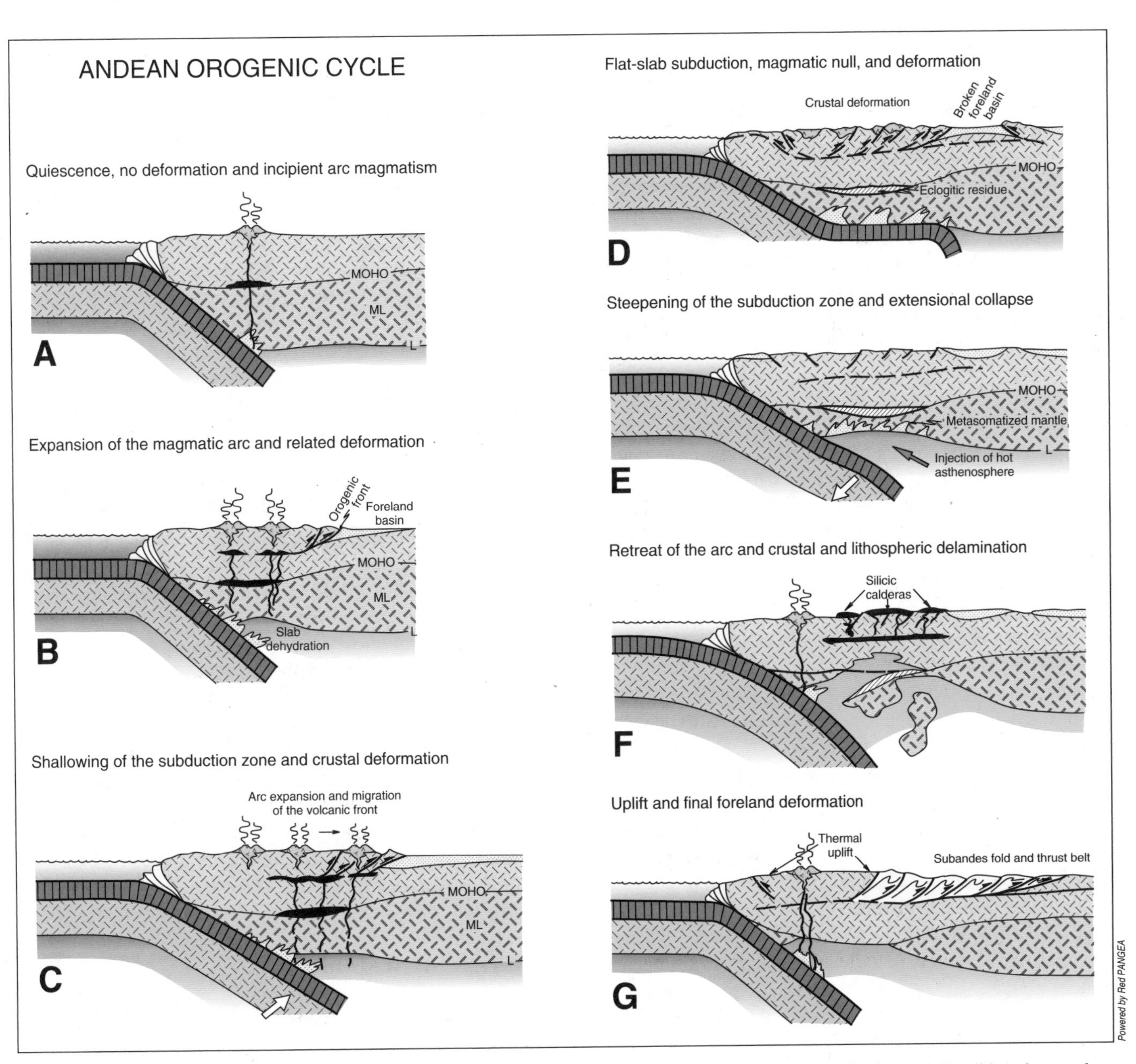

Figure 17. Idealized tectonic evolution of an Andean orogenic cycle (based on many authors; see discussion in the text). L—lithosphere-asthenosphere boundary; ML—lithospheric mantle.

Perú and Pampean segments (Dorbath et al., 1991; Cahill and Isacks, 1992; Smalley et al., 1993). These areas concentrate the maximum seismic energy release, as depicted in Bucaramanga, a concentrated area of energy release, Perú, and the Pampean segments (Gutscher et al., 2000). Dehydration of the oceanic slab does not contribute to the asthenospheric wedge, producing the last ephemeral activity as far as 700 km away from the trench, as seen at 33°S latitude (Kay and Gordillo, 1994). Instead, dehydration contributes to lithospheric mantle metasomatism, as proposed by James and Sacks (1999).

Crustal deformation is at its maximum, and broken foreland basins dominate (Jordan et al., 1983a, 1993). Intermontane basins are frequent, like those in central Perú (Marocco et al., 1995) and in the *bolsones* of the Sierras Pampeanas (Ramos, 2000).

The shallowing of the subduction zone does not always achieve a flat-slab stage, and several examples show that steepening of the subduction may start prior to the horizontalization of the oceanic slab in the retroarc area (Ramos and Folguera, 2005; Ramos and Kay, 2006; Kay et al., 2006; Kay and Coira, this volume).

Steepening of Subduction and Extensional Collapse

After a period of no magmatism, the magmatic arc resumes in the outer eastern areas and starts retreating toward the trench, as described in southern Perú by James and Sacks (1999) and in the northern Puna high plateau by Kay et al. (1999). This period coincides with instability of the stress field, as depicted by Allmendinger et al. (1997) in the Bolivian Altiplano, and localized extensional faults are active together with monogenetic within-plate basic magmatism (Allmendinger et al., 1989; Riller et al., 2001; Matteini et al., 2002).

A contrasting evolution from the previously described thick crust of the Altiplano or Puna, which exceeds 65 km thickness (Götze et al., 1994; Götze and Krause, 2002; Beck and Zandt, 2002), is seen in those areas where steepening is produced in a thin or normal crust. In areas, such as the southern Mendoza segment (35°S to 36°S), steepening is associated with extensive basic ocean-island basalt (OIB) magmatism in the foreland (Ramos and Kay, 2006; Kay et al. 2006) and significant extensional collapse (Folguera et al., 2003, 2007, 2008).

Retreat of Arc Magmatism and Crustal and Lithospheric Delamination

The retreat of a magmatic arc in a thickened crust occurs partially coeval with extensive development of calderas and ignimbritic flows, such as those in the Puna-Altiplano volcanic complex (de Silva, 1989; Kay et al., 1999; Caffe et al., 2002; Zandt et al., 2003). This widespread dacitic magmatism is associated with plateau uplift, as described by Coira et al. (1993), Gubbels et al. (1993), Beck and Zandt (2002), and Garzione et al. (2006). These pervasive dacitic melts are the expression of crustal delamination, postulated by Kay and Mahlburg-Kay (1993) to be a consequence of the crustal instability generated by a denser eclogitic lower crust in contact with a less dense asthenosphere. In recent years, many geophysical experiments have confirmed the abnormal heat flux of this area, the lack of dense mantle lithosphere, the partial sinking of the mantle lithosphere into the asthenosphere, the acidic composition of the lower crust, and several other features produced during steepening of the slab and injection of hot asthenosphere into the metasomatized mantle (see Oncken et al., 2006). For further details on the Puna-Altiplano region and delamination, see Kay and Coira (this volume).

Uplift and Final Foreland Deformation

Thermal uplift, as predicted by Isacks (1988), has been confirmed by a new set of data. The studies of Roeder (1988) and Roeder and Chamberlain (1995) established that palinspastic restorations of structural sections of the upper crust show not only a considerable shortening exceeding 300 km, but also a significant amount of crustal imbalance. These studies showed that part of the lower crust was missing, as well as the corresponding lithospheric mantle. The studies (Beck et al., 2002) in the Bolivian Altiplano show that the old interface between the craton and the accreted Arequipa terrane, a weakness zone that controlled extension and within-plate magmatism through time (Jiménez and López-Velázquez, 2008), is again reactivated with crustal delamination. This was also confirmed by receiver function studies that showed the detachment of lithospheric mantle beneath the Puna (Schurr et al., 2006).

All these processes led to an increase of heat flow of the Altiplano and Puna, which favors a ductile behavior of the lower crust, as envisaged by Isacks (1988), and the subsequent extrusion of the upper crust to form the fold-and-thrust belt of the Subandes. This belt has shortened more than 130 km in the last 7–6 m.y. (Baby et al., 1995; Allmendinger et al., 1997; McQuarrie, 2002; DeCelles and Horton, 2003).

The different stages recognized in the proposed orogenic cycle are summarized in Figure 17. Although this is a conceptual model, it addresses different segments of the Andes during the Cenozoic, and its application is proposed for some segments of the Mesozoic, and even speculatively for Paleozoic times. It is remarkable because it is a testable model that can explain the structural, magmatic, and the basin formation in an Andean-type margin through time.

CONCLUDING REMARKS

The present overview of the anatomy of the Andes described at different times illustrates significant information that can be extrapolated to other systems. The present Andes may be segmented into several discrete areas that underwent a unique evolution and had a complex geological history, but that have in common a series of processes that modeled their tectonic characteristics with different intensities. One of the major controls is the present flat-slab subduction configuration, which not only is responsible for the presence or absence of magmatism, but also the structural evolution of the area. Some other factors, such as the collisions of seismic ridges, have important roles in deformation, magmatic activity, and basin formation. Their analysis is beyond the present overview (for details see Ramos, 2005; Michaud et al., this volume).

The climatic control on the uplift, location, and rate of exhumation, and structural geometries is another first-order factor in the tectonic evolution of the Andes (see, for example, Montgomery et al., 2001; DeCelles and Horton, 2003; Strecker et al., 2007; Blisniuk et al., 2006; Ramos and Ghiglione, 2008). The analysis of climatic control in the formation of the Andes is also beyond the scope of this manuscript.

Concluding remarks can be grouped into the following aspects:

1. The beginning of the Andean subduction system was related to the breakup of Rodinia, which imprinted a strong Grenville signature on all the continental terranes accreted to the margin, and to the protomargin of Gondwana itself. This breakup marked the birth of the Paleozoic Andes. The Neoproterozoic was characterized by opening and closures along most of the margin, but it is still poorly known in the tropical areas because of dense vegetation cover and the poor quality of the outcrops.

2. Significant exchange of continental fragments occurred as a consequence of the accretionary tectonics that dominated the early Paleozoic. Exotic fragments derived from Laurentia alternated with docking of parautochthonous terranes previously detached from the Gondwana protomargin. These extensional and compressive stages in the evolution of the margin were controlled by the absolute motion of Gondwana linked to global plate reorganizations, as proposed in the Terra Australis orogen by Cawood (2005).

3. Laurentian remnants at the Gondwana margin are a testimony in the basement of the Northern Andes to the Laurentia-Gondwana collision during the Early Permian. High-grade metamorphic terranes such as Tahami in Colombia and Tahuin in southern Ecuador and northern Perú were detached from Laurentia and left behind after the collision. Other exposures, such as the Tres Lagunas foliated granitoids, could also have been part of this suite of terranes, but more petrological and geochemical data are still necessary to confirm this assertion.

4. Other late Paleozoic terranes are known in the Central and Southern Andes, such as the Mejillonia and Patagonia terranes, but their origins are not related to a continent to continent collision. They have Gondwanan affinities and are interpreted as parautochthonous terranes.

5. Triassic to Early Jurassic time was associated with the breakup of Pangea. Extension was concentrated along the Pacific margin and was focused on the hanging wall of the previous sutures. Interfaces between cratonic areas and accreted terranes favored within-plate magmatism and extension, as in the Cordillera Real of Bolivia and its continuation to the north and south in Perú and Argentina, respectively.

6. Inception of subduction in the Early Jurassic along the margin was controlled by the breakup of western and eastern Gondwana and the opening of the Indian Ocean. Subduction with negative trench roll-back velocity dominated early Mesozoic subduction, magmatism, and the extensional regime. This extensional regime was controlled by the absolute motion of western Gondwana toward the northeast, in the same direction as the adjacent Farallon plate.

7. A major change in Andean geodynamics occurred at the end of the Early Cretaceous when South America detached from Africa. The beginning of absolute motion to the west and northwest gave rise to the present compressive tectonic setting that controls crustal thickening and Andean uplift. The change also substantially modified the nature and distribution of magmatism along the volcanic arc.

8. The generalized evidence for present and past flat-slab subduction supports a conceptual Andean orogenic cycle that begins after a quiescent period with incipient deformation and magmatism; migration and expansion of the magmatic arc; shifting of the orogenic front and subsequent foreland basin formation during shallowing of the subduction zone; crustal shortening and broken foreland basins at the flat-slab stage; subsequent steepening and extensional collapse; crustal and lithospheric delamination; thermal uplift; and, finally, lower crustal ductile contraction and thin-skinned fold-and-thrust belt formation in the adjacent foreland areas.

This conceptual orogenic cycle could be affected at any stage by changes in the subduction dynamics and, therefore, in the order and intensity of the geological processes. Occasionally, after a short period of shallowing of the subduction zone, an early steepening may inhibit the flat-slab formation, but a series of mild and similar processes can still be detected. On the other hand, the initial conditions of the upper plate give rise to different magmatic and tectonic behaviors. The consequences of the processes are different in initially thick or thin crust. A thick crust may lead to significant crustal delamination and generalized silicic magmatism, and a thin crust may lead to generalized retroarc basaltic magmatism with or without crustal delamination along the axis of the cordillera. If the shallowing is interrupted, minor localized extension and scarce magmatism may be the only features that bear witness to the slight change in the subduction geometry.

The importance of this Andean orogenic model is that it can be tested and it can be applied to the Paleozoic Andes, and to other locations where a subduction system was or is active.

ACKNOWLEDGMENTS

A preliminary oral version of this paper was presented at the 2006 "Backbone of the Americas Conference" in Mendoza, Argentina. I am grateful to Constantino Mpodozis, Suzanne M. Kay, Luis Pilatasig, Jorge Restrepo, Victor Carlotto, Osvaldo Ordoñez, and many others, as well as all my colleagues of the Laboratorio de Tectónica Andina of the Universidad de Buenos Aires, for years of lively and creative discussions that have contributed to our present understanding of the evolution of the Andes. I appreciate the comments and critical reviews of Terry Jordan, Estanislao Godoy, and Jairo Osorio. Special thanks are due to Lidia Lustig for her thoughtful comments. Mario Campaña prepared the illustrations.

REFERENCES CITED

Aceñolaza, F.G., and Aceñolaza, G.F., 2005, La Formación Puncoviscana y unidades estratigráficas vinculadas en el Neoproterozoico–Cámbrico Temprano del Noroeste Argentino: Latin American Journal of Sedimentology and Basin Analysis, v. 12, p. 65–87.

Acosta, J., Velandia, F., Osorio, J., Lonergan, L., and Mora, H., 2007, Strike-slip deformation within the Colombian Andes, *in* Ries, A.C., Butler, R.W.H., and Graham, R.H., eds., Deformation of the Continental Crust: The Legacy of Mike Coward: Geological Society of London Special Publication 272, p. 303–319.

Aguirre, L., 1987, Andean modeling: Geology Today, v. 3, p. 47–48.

Alemán, A., and Ramos, V.A., 2000, The Northern Andes, *in* Cordani, U.J., Milani, E.J., Thomaz Filho, A., and Campos, D.A, eds., Tectonic Evolution of South America: Río de Janeiro, 31st International Geological Congress, p. 453–480.

Allmendinger, R.W., Strecker, M., Eremchuk, J.E., and Francis, P., 1989, Neotectonic deformation of the southern Puna Plateau, northwestern Argentina: Journal of South American Earth Sciences, v. 2, p. 111–130, doi: 10.1016/0895-9811(89)90040-0.

Allmendinger, R.W., Jordan, T.E., Kay, S.M., and Isacks, B.L., 1997, The evolution of the Altiplano-Puna Plateau of the central Andes: Annual Review of Earth and Planetary Sciences, v. 25, p. 139–174, doi: 10.1146/annurev.earth.25.1.139.

Aspden, J.A., and Litherland, M., 1992, The geology and Mesozoic collisional history of the Cordillera Real, Ecuador: Tectonophysics, v. 205, p. 187–204, doi: 10.1016/0040-1951(92)90426-7.

Aspden, J.A., and McCourt, W.J., 1986, Mesozoic oceanic terrane in the Central Andes of Colombia: Geology, v. 14, p. 415–418, doi: 10.1130/0091-7613(1986)14<415:MOTITC>2.0.CO;2.

Aspden, J.A., McCourt, W.J., and Brook, M., 1987, Geometrical control of subduction-related magmatism: The Mesozoic and Cenozoic plutonic history of western Colombia: Journal of the Geological Society of London, v. 144, p. 893–905, doi: 10.1144/gsjgs.144.6.0893.

Aspden, J.A., Fortey, N., Litherland, M., Viteri, F., and Harrison, S.M., 1992a, Regional S-type granites in the Ecuadorian Andes: Possible remnants of the breakup of western Gondwana: Journal of South American Earth Sciences, v. 6, p. 123–132, doi: 10.1016/0895-9811(92)90002-G.

Aspden, J.A., Harrison, S.H., and Rundle, C.C., 1992b, New geochronological control for the tectono-magmatic evolution of the metamorphic basement: Cordillera Real and El Oro Province of Ecuador: Journal of South American Earth Sciences, v. 6, p. 77–96.

Astini, R.A., 2003, The Ordovician proto-Andean basins, *in* Benedetto, J.L., ed., Ordovician Fossils of Argentina: Córdoba, Secretaría de Ciencia y Tecnología, Universidad Nacional de Córdoba, p. 1–74.

Astini, R.A., Benedetto, J.L., and Vaccari, N.E., 1995, The early Paleozoic evolution of the Argentina Precordillera as a Laurentian rifted, drifted, and collided terrane: A geodynamic model: Geological Society of America Bulletin, v. 107, p. 253–273, doi: 10.1130/0016-7606(1995)107<0253:TEPEOT>2.3.CO;2.

Astini, R.A., Ramos, V.A., Benedetto, J.L., Vaccari, N.E., and Cañas, F.L., 1996, La Precordillera: Un terreno exótico a Gondwana, *in* 13th Congreso Geológico Argentino y 3rd Congreso Exploración de Hidrocarburos: Actas, v. 5, p. 293–324.

Atherton, M.P., and Webb, S., 1989, Volcanic facies, structure and geochemistry of the marginal basin rocks of central Perú: Journal of South American Earth Sciences, v. 2, p. 241–261, doi: 10.1016/0895-9811(89)90032-1.

Auboin, J., and Borrello, V.A., 1966, Chaînes alpines et chaines Andines: Regard sur la géologie de la cordillère des Andes au parallele de l'Argentine moyenne: Bulletin de la Société Géologique de France, Série 7, v. 8, p. 1050–1070.

Auboin, J.A., Borrello, A.V., Cecione, G., Charrier, R., Chotin, P., Frutos, J., Thiele, R., and Vicente, J.C., 1973, Esquisse paleogeographique et structurale des Andes meridionales: Revue de Géographie Physique et de Géologie Dynamique, v. 15, p. 11–71.

Audemard, F.E., and Audemard M., F.A., 2002, Structure of the Mérida Andes, Venezuela: Relation with the South America–Caribbean geodynamic interaction: Tectonophysics, v. 345, p. 299–327, doi: 10.1016/S0040-1951(01)00218-9.

Audemard M., F.A., Singer P.A., and Soulas, J.-P., 2006, Fallas y campo de esfuerzos cuaternarios de Venezuela: Revista de la Asociación Geológica Argentina, v. 61, p. 480–491.

Baby, P., Moretti, I., Guillier, B., Limachi, R., Méndez, E., Oller, J., and Specht, M., 1995, Petroleum system of the northern and central Bolivian Subandean zone, *in* Tankard, A.J., Suárez Soruco, R., and Welsink, H.J., eds., Petroleum Basins of South America: American Association of Petroleum Geologists Memoir 62, p. 445–458.

Baeza, L., 1984, Petrography and Tectonics of the Plutonic and Metamorphic Complexes of Limón Verde and Mejillones Península, Northern Chile [Ph.D. thesis]: Tübingen, University of Tübingen, 205 p.

Baeza, L., and Pichowiak, S., 1988, Ancient crystalline basement provinces in the North Chilean Central Andes—Relicts of continental crust development since the mid-Proterozoic: The southern Central Andes: Lecture Notes in Earth Sciences, v. 17, p. 3–24.

Bahlburg, H., 1990, The Ordovician basin in the Puna of NW Argentina and N Chile: Geodynamic evolution from back-arc to foreland basin: Geotektonische Forschungen, v. 75, p. 1–107.

Bahlburg, H., and Hervé, F., 1997, Geodynamic evolution and tectonostratigraphic terranes of northwestern Argentina and northern Chile: Geological Society of America Bulletin, v. 109, p. 869–884, doi: 10.1130/0016-7606(1997)109<0869:GEATTO>2.3.CO;2.

Baldo, E., Casquet, C., Pankhurst, R.J., Galindo, C., Rapela, C.W., Fanning, C.M., Dahlquist, J., and Murra, J., 2006, Neoproterozoic A-type magmatism in the Western Sierras Pampeanas (Argentina): Evidence for Rodinia break-up along a proto-Iapetus rift?: Terra Nova, v. 18, p. 388–394, doi: 10.1111/j.1365-3121.2006.00703.x.

Barazangi, M., and Isacks, B., 1976, Spatial distribution of earthquakes and subduction of the Nazca plate beneath South America: Geology, v. 4, p. 686–692, doi: 10.1130/0091-7613(1976)4<686:SDOEAS>2.0.CO;2.

Barazangi, M., and Isacks, B., 1979, Subduction of the Nazca plate beneath Perú; evidence from spatial distribution of earthquakes: Geophysical Journal of the Royal Astronomical Society, v. 5, p. 537–555.

Barker, P.F., 2001, Scotia Sea regional tectonic evolution: Implications for mantle flow and palaeocirculation: Earth-Science Reviews, v. 55, p. 1–39, doi: 10.1016/S0012-8252(01)00055-1.

Barrero, D., 1979, Geology of the central Western Cordillera, west of Buga and Roldanillo, Colombia: Ingeominas: Publicación Geológica Especial, v. 4, p. 1–75.

Basei, M.A.S., Brito Neves, B.B., Varela, R., Teixeira, W., Siga, O., Jr., Sato, A.M., and Cingolani, C., 1999, Isotopic dating on the crystalline basement rocks of the Bariloche region, Río Negro, Argentina, *in* 2nd South American Symposium on Isotope Geology, Segemar: Anales, v. 34, p. 15–18.

Beck, S.L., and Zandt, G., 2002, The nature of orogenic crust in the central Andes: Journal of Geophysical Research, v. 107, doi: 10.1029/2000JB000124.

Benedetto, J.L., 2003, The Ordovician brachiopod fauna of Argentina: Chronology and biostratigraphic succession, *in* Albanesi, G.L., Beresi, M.S., and Peralta, S.H., eds., Ordovician from the Andes: INSUGEO (Instituto Superior de Geología de la Universidad Nacional de Tucumán), Serie Correlación Geológica, v. 17, p. 87–106.

Blisniuk, P.N., Stern, L.A., Chamberlain, C.T., Zeitler, T.K., Ramos, V.A., Sobel, E.R., Haschke, M., Strecker, M.R., and Warkus, F., 2006, Links between mountain uplift, climates, and surface processes in the southern Patagonian Andes, *in* Oncken, O., et al., eds., The Andes—Active Subduction Orogeny: Berlin, Springer, Frontiers in Earth Sciences Series 1, p. 429–440.

Bordonaro, O., 1992, El Cámbrico de Sudamérica, *in* Gutiérrez Marco, J.C., Saavedra, J., and Rábano, I., eds., Paleozoico Inferior de Ibero-América: Mérida, Universidad de Extremadura, p. 69–84.

Bosch, D., Gabriele, P., Lapierre, H., Malfere, J.L., and Jaillard, E., 2002, Geodynamic significance of the Raspas metamorphic complex (SW Ecuador): Geochemical and isotopic constraints: Tectonophysics, v. 345, p. 83–102, doi: 10.1016/S0040-1951(01)00207-4.

Bourgois, J., Toussaint, J.F., Gonzalez, H., Azema, J., Calle, B., Desmet, A., Murcia, L.A., Acevedo, A.P., Parra, E., and Tournon, J., 1987, Geological history of the Cretaceous ophiolitic complexes of northwestern South America (Colombian Andes): Tectonophysics, v. 143, p. 307–327, doi: 10.1016/0040-1951(87)90215-0.

Burke, K., 1988, Tectonic evolution of the Caribbean: Annual Review of Earth and Planetary Sciences, v. 16, p. 201–230, doi: 10.1146/annurev.ea.16.050188.001221.

Cáceres, C., Cediel, F., and Etayo, F., 2003, Mapas de Distribución de Facies Sedimentarias y Armazón Tectónico de Colombia a Través del Proterozoico y del Fanerozoico: Bogotá, Ingeominas, 40 p.

Caffe, P.J., Trumbull, R.B., Coira, B.L., and Romer, R., 2002, Petrogenesis of early Neogene magmatism in the northern Puna; implications for magma genesis and crustal processes in the central Andean Plateau: Journal of Petrology, v. 43, p. 907–942, doi: 10.1093/petrology/43.5.907.

Cahill, T., and Isacks, B.L., 1992, Seismicity and the shape of the subducted Nazca plate: Journal of Geophysical Research, v. 97, p. 17,503–17,529, doi: 10.1029/92JB00493.

Cardona, A., Cordani, U.G., and Nutman, A.P., 2005a, U-Pb SHRIMP zircon, ^{40}Ar/^{39}Ar geochronology and Nd isotopes from granitoid rocks of the Illescas Massif, Peru: A southern extension of a fragmented late Paleozoic orogen?: Proceedings of the 6th South American Symposium on Isotope Geology: Punta del Este, Uruguay, Facultad de Ciencias, p. 78.

Cardona, A., Cordani, U.G., Ruiz, J., Valencia, V., Nutman, A.P., and Sánchez, A.W., 2005b, U/Pb detrital zircon geochronology and Nd isotopes from Paleozoic metasedimentary rocks of the Marañon Complex: Insights on the proto-Andean tectonic evolution of the eastern Peruvian Andes: Proceedings of the 5th South American Symposium on Isotope Geology: Punta del Este, Uruguay, Facultad de Ciencias, p. 208–11.

Cardona, A., Cordani, U.G., and Sánchez, A.W., 2007, Metamorphic, geochronological and geochemical constraints from the pre-Permian basement of the eastern Peruvian Andes (10°S): A Paleozoic extensional-accretionary orogen?: *in* 20th Colloquium Latin American Earth Sciences: Kiel, Germany, Deutsche Forschungsgemeinschaft, p. 29–30.

Cardona, A., Cordani, U.G., and Nutman, A.P., 2008, U.Pb SHRIMP circón, ^{40}Ar/^{39}Ar geochronology and Nd isotopes from granitoid rocks of the Illescas Massif, Peru: A southern extension of a fragmented Late Paleozoic orogen?: *in* 6th South American Symposium on Isotope Geology, Proceedings: Bariloche, Argentina, Abstracts, p. 78.

Carlotto, V., 2002, Evolution andine et raccourcissement au niveau de Cusco (13–16°S) Pérou: Marseille, Institut de Recherche pour le Développement, Geologie Alpine, Mémoire H.S. 39, p. 1–203.

Case, J.E., Holcombe, T.L., and Martin, R.G., 1984, Map of the geologic provinces in the Caribbean region, *in* Bonini, W.E., Hardgraves, R.B., and Shagam, R., eds., The Caribbean–South America Plate Boundary and Regional Tectonics: Geological Society of America Memoir 162, p. 1–30.

Casquet, C., Pankhurst, R.J., Rapela, C.W., Galindo, C., Fanning, C.M., Chiaradia, M., Baldo, E., González-Casado, J.M., and Dahlquist, J.A., 2008, The Mesoproterozoic Maz terrane in the Western Sierras Pampeanas, Argentina, equivalent to the Arequipa–Antofalla block of southern Peru? Implications for West Gondwana margin evolution: Gondwana Research, v. 13, p. 163–175, doi: 10.1016/j.gr.2007.04.005.

Cawood, P.A., 2005, Terra Australis orogen: Rodinia breakup and development of the Pacific and Iapetus margins of Gondwana during the Neoproterozoic and Paleozoic: Earth-Science Reviews, v. 69, p. 249–279, doi: 10.1016/j.earscirev.2004.09.001.

Cediel, F., Shaw, R.P., and Cáceres, C., 2003, Tectonic assembly of the northern Andean block, *in* Bartolini, C., Buffler, R.T., and Blickwede, J., eds., The Circum-Gulf of Mexico and Caribbean: Hydrocarbon Habitats, Basin Formation, and Plate Tectonics: American Association of Petroleum Geologists Memoir 79, p. 815–848.

Charrier, R., 1973, Interruptions of spreading and the compressive tectonic phases of Meridional Andes: Earth and Planetary Science Letters, v. 20, p. 242–249, doi: 10.1016/0012-821X(73)90164-7.

Chernicoff, C.J., Zappettini, E.O., Santos, J.O.S., Beyer, E., and McNaughton, N.J., 2008, Foreland basin deposits associated with Cuyania accretion in La Pampa Province, Argentina: Gondwana Research, v. 13, p. 189–203, doi: 10.1016/j.gr.2007.04.006.

Chew, D.M., Schaltegger, U., Košler, J., Whitehouse, M.J., Gutjahr, M., Spikings, R.A., and Miškovíc, A., 2007, U-Pb geochronologic evidence for the evolution of the Gondwanan margin of the north-central Andes: Geological Society of America Bulletin, v. 119, p. 697–711, doi: 10.1130/B26080.1.

Coira, B.L., Davidson, J.D., Mpodozis, C., and Ramos, V.A., 1982, Tectonic and magmatic evolution of the Andes of northern Argentina and Chile: Earth-Science Reviews, v. 18, p. 303–332, doi: 10.1016/0012-8252(82)90042-3.

Coira, B., Mahlburg Kay, S., and Viramonte, J., 1993, Upper Cenozoic magmatic evolution of the Argentine Puna—A model for changing subduction geometry: International Geology Review, v. 35, p. 677–720.

Coira, B.L., Mahlburg Kay, S., Pérez, B., Woll, B., Hanning, M., and Flores, P., 1999, Magmatic sources and tectonic setting of Gondwana margin Ordovician magmas, northern Puna of Argentina and Chile, *in* Ramos, V.A., and Keppie, D., eds., Laurentia Gondwana Connections before Pangea: Geological Society of America Special Paper 336, p. 145–170.

Cordani, U.G., Cardonna, A., Jímenez, D.M., Liu, D., and Nutran, A.P., 2005, Geochronology of Proterozoic basement inliers from the Colombian Andes: Tectonic history of remnants from a fragmented Grenville belt, *in* Vaughan, A.P.M., Leat, P.T., and Pankhurst, R.J., eds., Terrane Processes at the Margins of Gondwana: Geological Society of London Special Publication 246, p. 329–46.

Cosma, L., Lapierre, H., Jaillard, E., Laubacher, G., Bosch, D., Desmet, A., Mamberti, M., and Gabriele, P., 1998, Petrographie and geochimie de la Cordillère Occidentale du Nord de l'Equateur (0°30′): Implications tectoniques: Bulletin de la Société Géologique de France, v. 169, p. 739–751.

Costa, C.H., Audemard M., F.A., Bezerra, F.H.R., Lavenu, A., Machette, M.N., and París, G., 2006a, Una perspectiva sobre las principales deformaciones cuaternarias de América del Sur: Revista de la Asociación Geológica Argentina, v. 61, p. 461–479.

Costa, C.H., Smalley, R., Jr., Schwartz, D.P., Stenner, H.D., Ellis, M., Ahumada, E.A., and Velasco, M.S., 2006b, Paleoseismic observations of an onshore transform boundary: The Magallanes-Fagnano fault, Tierra del Fuego Argentina: Revista de la Asociación Geológica Argentina, v. 61, p. 647–657.

Cunningham, D.W., Dalziel, I.A.W., Tung-Yi, L., and Lawver, L.A., 1995, Southernmost South America–Antarctic Peninsula relative plate motions since 84 Ma: Implications for the tectonic evolution of the Scotia Arc region: Journal of Geophysical Research, v. 100, no. B5, p. 8257–8266, doi: 10.1029/95JB00033.

Dalmayrac, B., Laubacher, G., Marocco, R., Martinez, C., and Tomasi, P., 1980, La Chaine Hercynienne d'Amerique du Sud. Structure et evolution d'une orogène intracratonique: Geologische Rundschau, v. 69, p. 1–21, doi: 10.1007/BF01869020.

Daly, M., 1989, Correlations between Nazca/Farallon plate kinematics and forearc evolution in Ecuador: Tectonics, v. 8, p. 769–790, doi: 10.1029/TC008i004p00769.

Dalziel, I.W.D., 1991, Pacific margins of Laurentia and East-Antarctica–Australia as a conjugate rift pair: Evidence and implications for an Eocambrian supercontinent: Geology, v. 19, p. 598–601, doi: 10.1130/0091-7613(1991)019<0598:PMOLAE>2.3.CO;2.

Dalziel, I.W.D., 1994, Precambrian Scotland as a Laurentia-Gondwana link: The origin and significance of cratonic promontories: Geology, v. 22, p. 589–592, doi: 10.1130/0091-7613(1994)022<0589:PSAALG>2.3.CO;2.

Dalziel, I.W.D., 1997, Neoproterozoic-Paleozoic geography and tectonics: Review, hypothesis, environmental speculation: Geological Society of America Bulletin, v. 109, p. 16–42, doi: 10.1130/0016-7606(1997)109<0016:ONPGAT>2.3.CO;2.

Dalziel, I.W., and Forsythe, R.D., 1985, Andean evolution and the terrane concept: Tectonostratigraphic terranes of the circum-Pacific region: Earth Sciences Series, v. 1, p. 565–581.

Dalziel, I.W.D., de Wit, M.F., and Palmer, K.F., 1974, Fossil marginal basin in the Southern Andes: Nature, v. 250, p. 291–294, doi: 10.1038/250291a0.

Damm, K.W., Pichowiak, S., and Todt, W., 1986, Geochimie, Petrologie und Geochronologie der Plutonite und des Metamorphen Grundgebirges in Nordchile: Berliner Geowissenschaftliche Abhandlungen, ser. A, v. 66, p. 73–146.

Damm, K.W., Pichowiak, S., Harmon, R.S., Todt, W., Kelley, S., Omarini, R., and Niemeyer, H., 1990, Pre-Mesozoic evolution of the central Andes; the basement revisited, *in* Kay, S.M., and Rapela, C.W., eds., Plutonism from Antarctica to Alaska: Geological Society of America Special Paper 241, p. 101–126.

DeCelles, P.G., and Horton, B.K., 2003, Early to Middle Tertiary foreland basin development and the history of Andean crustal shortening in Bolivia: Geological Society of America Bulletin, v. 115, p. 58–77, doi: 10.1130/0016-7606(2003)115<0058:ETMTFB>2.0.CO;2.

Dengo, G., 1983, Mid-America: Tectonic setting for the Pacific margin from southern Mexico to northern Colombia: Guatemala, Guatemala, Centro de Estudios Geológicos de América Central, p. 1–56.

de Silva, S.L., 1989, Altiplano-Puna volcanic complex of the central Andes: Geology, v. 17, p. 1102–1106, doi: 10.1130/0091-7613(1989)017<1102:APVCOT>2.3.CO;2.

Dewey, J.F., and Bird, J., 1970, Mountain belts and the new global tectonics: Journal of Geophysical Research, v. 75, p. 2625–2647, doi: 10.1029/JB075i014p02625.

Dewey, J.F., and Pindell, J.L., 1986, Reply to Amos Salvador: Tectonics, v. 5, p. 703–705, doi: 10.1029/TC005i004p00703.

Díaz-Martínez, E., Sempere, T., Isaacson, P.E., and Grader, G.W., 2000, Paleozoic of Western Gondwana active margin (Bolivian Andes): *in* 31st International Geological Congress, Pre-Congress Field Trip 27: Rio de Janeiro, 31st International Geological Congress, 31 p.

Dickinson, W.F., and Snyder, W.S., 1979, Geometry of subducted slabs related to San Andres transform: The Journal of Geology, v. 88, p. 619–638.

Dorbath, C., Granet, M., Poupinet, G., and Martínez, C., 1993, A teleseismic study of the Altiplano and the Eastern Cordillera and northern Bolivia: New constraints on a lithospheric model: Journal of Geophysical Research, v. 98, p. 9825–9844, doi: 10.1029/92JB02406.

Dorbath, L., Dorbath, C., Jimenez, E., and Rivera, L., 1991, Seismicity and tectonic deformation in the Eastern Cordillera and the sub-Andean zone of central Perú: Journal of South American Earth Sciences, v. 4, p. 13–24, doi: 10.1016/0895-9811(91)90015-D.

Duque-Caro, H., 1990, The Choco block in the NW corner of South America; structural, tectonostratigraphic and paleogeographic implications: Journal of South American Earth Sciences, v. 3, p. 71–84, doi: 10.1016/0895-9811(90)90019-W.

Durand, F.R., 1993, Las icnofacies del basamento metasedimentario en el Noroeste argentino: Significado cronológico y aspectos paleogeográficos, *in* 12th Congreso Geológico Argentino y 2nd Congreso de Exploración de Hidrocarburos: Actas, v. 2, p. 260–267.

Eagles, G., Livermore, R., and Morris, P., 2006, Small basins in the Scotia Sea: The Eocene Drake Passage gateway: Earth and Planetary Science Letters, v. 242, p. 343–353, doi: 10.1016/j.epsl.2005.11.060.

Escayola, M.P., Pimentel, M., and Armstrong, R., 2007, Neoproterozoic back-arc basin: Sensitive high-resolution ion microprobe U-Pb and Sm-Nd isotopic evidence from the Eastern Pampean Ranges, Argentina: Geology, v. 35, p. 495–498, doi: 10.1130/G23549A.1.

Etayo-Serna, F., and Barrero, D., 1983, Mapa de Terrenos Geológicos de Colombia: Ingeominas Publicación Especial 14-I, 235 p.

Feininger, T., 1978, Geologic Map of Western El Oro Province: Quito, Politécnica Nacional, scale 1:50,000.

Feininger, T., 1982, Glaucophane schist in the Andes at Jambaló, Colombia: Canadian Mineralogist, v. 20, p. 41–48.

Feininger, T., 1986, Allochthonous terranes in the Andes of Ecuador and northwestern Perú: Geological Survey of Canada Contribution, v. 22686, p. 266–278.

Feininger, T., 1987, Allochthonous terranes in the Andes of Ecuador and northwestern Peru: Canadian Journal of Earth Sciences, v. 24, p. 266–278.

Figari, E.G., Courtade, S.F., and Constantini, L.A., 1996, Stratigraphy and tectonics of Cañadón Asfalto basin, lows of Gastre and Gan Gan, north of Chubut province, Argentina, *in* Riccardi, A.C., ed., Advances in Jurassic Research: GeoResearch Forum, v. 1–2, p. 359–368.

Finney, S.C., 2007, The parautochthonous Gondwanan origin of Cuyania (greater Precordillera) terrane of Argentina: A reevaluation of evidence used to support an allochthonous Laurentian origin: Geologica Acta, v. 5, p. 127–158.

Folguera, A., Ramos, V., and Melnick, D., 2003, Recurrencia en el desarrollo de cuencas de intraarco. Colapso de estructuras orogénicas. Cordillera Neuquina (37°30′): Revista de la Asociación Geológica Argentina, v. 58, p. 3–19.

Folguera, A., Introcaso, A., Giménez, M., Ruiz, F., Martínez, P., Tunstall, C., García Morabito, E., and Ramos, V.A., 2007, Crustal attenuation in the southern Andean retroarc determined from gravimetric studies (38º–39º30′S): The Lonco-Luán asthenospheric anomaly: Tectonophysics, v. 439, p. 129–147, doi: 10.1016/j.tecto.2007.04.001.

Folguera, A., Bottesi, G., Zapata, T., and Ramos, V.A., 2008, Crustal collapse in the Andean back-arc since 2 Ma: Tromen volcanic plateau, Southern Central Andes (36º40 –37º30 S): Tectonophysics, Special Issue on Andean Geodynamics, v. 459, nos. 1–4, p. 140–160, doi: 10.1016/j.tecto.2007.12.013.

Forsythe, R.D., Davidson, I., Mpodozis, C., and Jesinskey, C., 1993, Paleozoic relative motion of the Arequipa block and Gondwana; paleomagnetic evidence from Sierra de Almeida of northern Chile: Tectonics, v. 12, p. 219–236, doi: 10.1029/92TC00619.

Franz, G., Lucassen, F., Kramer, W., Trumbull, R.B., Romer, R.L., Wilke, H.-G., Viramonte, J.G., Becchio, R., and Siebel, W., 2006, Crustal evolution at the central Andean continental margin: A geochemical record of crustal growth, recycling and destruction, *in* Oncken, O., et al., eds., The Andes—Active Subduction Orogeny: Berlin, Springer-Verlag, p. 45–64.

Frutos, J., 1981, Andean tectonics as a consequence of sea floor spreading: Tectonophysics, v. 72, p. 21–32, doi: 10.1016/0040-1951(81)90082-2.

Fuck, R.A., Brito Neve, B.B., and Schobbenhaus, C., 2008, Rodinia descendants in South America: Precambrian Research, v. 160, p. 108–126, doi: 10.1016/j.precamres.2007.04.018.

Gana, P., Wall, R., and Yáñez, G., 1994, Evolución geotectónica de la Cordillera de la Costa de Chile central (33–34°S): Control geológico y geofísico, *in* 7th Congreso Geológico Chileno: Actas, v. 1, p. 38–42.

Gansser, A., 1973, Facts and theories on the Andes: Journal of the Geological Society of London, v. 129, p. 93–131, doi: 10.1144/gsjgs.129.2.0093.

Garzione, C.N., Molnar, P., Libarkin, J.C., and MacFadden, B.J., 2006, Rapid late Miocene rise of the Bolivian Altiplano: Evidence for removal of mantle lithosphere: Earth and Planetary Science Letters, v. 241, p. 543–556, doi: 10.1016/j.epsl.2005.11.026.

Ghidella, M.E., Yañez, G., and LaBrecque, H.L., 2002, Revised tectonic implications for the magnetic anomalies of the Western Weddell Sea: Tectonophysics, v. 347, p. 65–86, doi: 10.1016/S0040-1951(01)00238-4.

Ghidella, M.E., Lawver, L.A., Marenssi, S., and Gahagan, L.M., 2007, Plate kinematic models for Antarctica during Gondwana break-up: A review: Revista de la Asociación Geológica Argentina, v. 62, p. 636–646.

Ghiglione, M.C., and Cristallini, E.O., 2007, Have the southernmost Andes been curved since Late Cretaceous time? An analog test for the Patagonian orocline: Geology, v. 35, p. 13–16, doi: 10.1130/G22770A.1.

Ghiglione, M.C., and Ramos, V.A., 2005, Progression of deformation in the southernmost Andes: Tectonophysics, v. 405, p. 25–46, doi: 10.1016/j.tecto.2005.05.004.

Ghiglione, M.C., Yagupsky, D., Ghidella, M., and Ramos, V.A., 2008, Continental stretching preceding the opening of the Drake Passage: Evidence from Tierra del Fuego: Geology, v. 36, p. 643–646, doi: 10.1130/G24857A.1.

Gómez, E., Jordan, T.E., Allmendinger, R.W., and Cardozo, N., 2005, Development of the Colombian foreland-basin system as a consequence of diachronous exhumation of the northern Andes: Geological Society of America Bulletin, v. 117, p. 1272–1292, doi: 10.1130/B25456.1.

González Bonorino, F., and Aguirre, L., 1970, Metamorphic facies series of the crystalline basement of Chile: Geologische Rundschau, v. 59, p. 979–994, doi: 10.1007/BF02042280.

Götze, H.J., and Krause, S., 2002, The Central Andean Gravity High, a relic of an old subduction complex?: Journal of South American Earth Sciences, v. 14, p. 799–811, doi: 10.1016/S0895-9811(01)00077-3.

Götze, H.J., Lahmeyer, B., Schmidt, S., and Strunk, S., 1994, The lithospheric structure of the central Andes (20°–25°S) as inferred from quantitative interpretation of regional gravity, *in* Reutter, K.J., Scheuber, E., and Wigger, P.J., eds., Tectonic of Southern Central Andes: Structure and Evolution of an Active Continental Margin: Berlin, Springer-Verlag, p. 23–48.

Groeber, P., 1951, La Alta Cordillera entre las latitudes 34° y 29°30′: Instituto Investigaciones de las Ciencias Naturales: Museo Argentino de Ciencias Naturales Bernardino Rivadavia: Revista, v. 1, no. 5, p. 1–352.

Gubbels, T.L., Isacks, B.L., and Farrar, E., 1993, High level surfaces, plateau uplift and foreland development, central Bolivian Andes: Geology, v. 21, p. 695–698, doi: 10.1130/0091-7613(1993)021<0695:HLSPUA>2.3.CO;2.

Gutscher, M.A., Spakman, W., Bijwaard, H., and Engdahl, E.R., 2000, Geodynamics of flat subduction: Seismicity and tomographic constraints from the Andean margin: Tectonics, v. 19, p. 814–833, doi: 10.1029/1999TC001152.

Harrington, H.J., and Kay, M., 1951, Cambrian and Ordovician faunas of eastern Colombia: Journal of Paleontology, v. 25, p. 655–668.

Haschke, M.R., Scheuber, E., Gunther, A., and Reutter, K.-J., 2002, Evolutionary cycles during the Andean orogeny: Repeated slab breakoff and flat subduction?: Terra Nova, v. 14, p. 49–55, doi: 10.1046/j.1365-3121.2002.00387.x.

Hervé, F., Munizaga, F., Marinovic, N., Kawashita, K., Brook, M., and Snelling, N., 1985, Geocronología Rb-Sr y K-Ar del basamento cristalino de Sierra Limón Verde, Antofagasta, Chile, *in* 4th Congreso Geológico Chileno: Actas, v. 3, p. 435–453.

Hervé, F., Demant, A., Ramos, V.A., Pankhurst, R.J., and Suárez, M., 2000, The Southern Andes, *in* Cordani, U.J., Milani, E.J., Thomaz Filho, A., and Campos, D.A., eds., Tectonic Evolution of South America: Río de Janeiro, 31st International Geological Congress, p. 605–634.

Hervé, F., Fanning, C.M., and Pankhurst, R.J., 2003, Detrital zircon age patterns and provenance of the metamorphic complexes of southern Chile: Journal of South American Earth Sciences, v. 16, p. 107–123, doi: 10.1016/S0895-9811(03)00022-1.

Hervé, F., Fanning, C.M., Mpodozis, C., and Pankhurst, R.J., 2008, Aspects of the Phanerozoic evolution of southern Patagonia as suggested by detrital zircon age patterns, *in* 6th South American Symposium on Isotope Geology Abstracts: Bariloche, Argentina, Instituto Nacional de Geología Isotópica, p. 3.

Hoffman, P.F., 1991, Did the breakout of Laurentia turn Gondwanaland inside-out?: Science, v. 252, p. 1409–1412, doi: 10.1126/science.252.5011.1409.

Hughes, R.A., and Pilatasig, L.F., 2002, Cretaceous and Tertiary block accretion in the Cordillera Occidental of the Andes of Ecuador: Tectonophysics, v. 345, p. 29–48, doi: 10.1016/S0040-1951(01)00205-0.

Irving, E., 1971, La evolución estructural de los Andes más septentrionales de Colombia: Ingeominas: Boletín Geológico, v. 19, p. 1–89.

Isacks, B., 1988, Uplift of the Central Andean plateau and bending of the Bolivian orocline: Journal of Geophysical Research, v. 93, p. 3211–3231, doi: 10.1029/JB093iB04p03211.

Jácome, M.I., Kusznir, N., Audemard, F., and Flint, S., 2003, Formation of the Maturín foreland basin, eastern Venezuela: Thrust sheet loading or subduction dynamic topography: Tectonics, v. 22, p. 1-1 to 1-17, doi: 10.1029/2002TC001381.

Jaillard, E., Hérail, G., Monfret, T., Díaz-Martínez, E., Baby, P., Lavenu, A., and Dumon, J.F., 2000, Tectonic evolution of the Andes of Ecuador, Peru, Bolivia, and northernmost Chile, *in* Cordani, U., Milani, E.J., Thomas Filho, A., and Campos, D.A., eds., Tectonic Evolution of South America: Rio de Janeiro, Brazil, 31st International Geological Congress, p. 481–559.

James, D.E., and Sacks, S., 1999, Cenozoic formation of the Central Andes: A geophysical perspective, *in* Skinner, B., ed., Geology and Mineral Deposits of Central Andes: Society of Economic Geology Special Publication 7, p. 1–25.

Jiménez, N., and López-Velásquez, S., 2008, Magmatism in the Huarina belt, Bolivia, and its geotectonic implications: Tectonophysics, v. 459, nos. 1–4, p. 85–106, doi: 10.1016/j.tecto.2007.10.012.

Jordan, T., 1995, Retroarc foreland and related basins, *in* Spera, C., and Ingersoll, R.V., eds., Tectonics of Sedimentary Basins: Cambridge, Massachusetts, Blackwell Scientific, p. 331–362.

Jordan, T., and Allmendinger, R., 1986, The Sierras Pampeanas of Argentina: A modern analogue of Laramide deformation: American Journal of Science, v. 286, p. 737–764.

Jordan, T., Isacks, B., Allmendinger, R., Brewer, J., Ando, C., and Ramos, V.A., 1983a, Andean tectonics related to geometry of subducted plates: Geological Society of America Bulletin, v. 94, p. 341–361, doi: 10.1130/0016-7606(1983)94<341:ATRTGO>2.0.CO;2.

Jordan, T., Isacks, B., Ramos, V.A., and Allmendinger, R.W., 1983b, Mountain building model: The Central Andes: Episodes, v. 1983, no. 3, p. 20–26.

Jordan, T.E., Flemings, P.B., and Beer, J.A., 1988, Dating thrust-fault activity by use of foreland-basin strata, *in* Kleinspehn, K.L., and Paola, C., eds., New Perspectives in Basin Analysis: New York, Springer-Verlag, p. 307–330.

Jordan, T.E., Allmendinger, R.W., Damanti, J.F., and Drake, R.E., 1993, Chronology of motion in a complete thrust belt: The Precordillera, 30–31°S, Andes Mountains: Journal of Geology, v. 101, p. 137–158.

Jordan, T.E., Blanco, N., Dávila, F.M., and Tomlinson, A., 2006, Seismostratigrafía de la cuenca Calama (22°–23°LS), Chile, *in* 11th Congreso Geológico Chileno, Antofagasta: Actas, v. 2, p. 53–56.

Jordan, T.E., Mpodozis, C., Muñoz, N., Blanco, N., Pananont, P., and Gardeweg, M., 2007, Cenozoic subsurface stratigraphy and structure of the Salar de Atacama Basin, northern Chile: Journal of South American Earth Sciences, v. 23, p. 122–146, doi: 10.1016/j.jsames.2006.09.024.

Kay, S.M., and Abbruzzi, J.M., 1996, Magmatic evidence for Neogene lithospheric evolution of the Central Andean flat-slab between 30°S and 32°S: Tectonophysics, v. 259, p. 15–28, doi: 10.1016/0040-1951(96)00032-7.

Kay, S.M., and Coira, B.L., 2009, this volume, Shallowing and steepening subduction zones, continental lithospheric loss, magmatism, and crustal flow under the Central Andean Altiplano-Puna Plateau, *in* Kay, S.M., Ramos, V.A., and Dickinson, W.R., eds., Backbone of the Americas: Shallow Subduction, Plateau Uplift, and Ridge and Terrane Collision: Geological Society of America Memoir 204, doi: 10.1130/2009.1204(11).

Kay, S.M., and Gordillo, C.E., 1994, Pocho volcanic rocks and the melting of depleted continental lithosphere above a shallowly dipping subduction zone in the Central Andes: Contributions to Mineralogy and Petrology, v. 117, p. 25–44, doi: 10.1007/BF00307727.

Kay, R.W., and Mahlburg Kay, S., 1993, Delamination and delamination magmatism: Tectonophysics, v. 219, p. 177–189, doi: 10.1016/0040-1951(93)90295-U.

Kay, S.M., and Mpodozis, C., 2002, Magmatism as a probe to the Neogene shallowing of the Nazca plate beneath the modern Chilean flat-slab: Journal of South American Earth Sciences, v. 15, p. 39–57, doi: 10.1016/S0895-9811(02)00005-6.

Kay, S.M., Maksaev, V., Moscoso, R., Mpodozis, C., and Nasi, C., 1987, Probing the evolving Andean lithosphere: Mid-late Tertiary magmatism in Chile (29°–30°30′S) over the modern zone of subhorizontal subduction: Journal of Geophysical Research, v. 92, no. B7, p. 6173–6189, doi: 10.1029/JB092iB07p06173.

Kay, S.M., Maksaev, V., Moscoso, R., Mpodozis, C., Nasi, C., and Gordillo, C.E., 1988, Tertiary Andean magmatism in Chile and Argentina between 28°S and 33°S: Correlation of magmatic chemistry with a changing Benioff zone: Journal of South American Earth Sciences, v. 1, p. 21–38, doi: 10.1016/0895-9811(88)90013-2.

Kay, S.M., Ramos, V.A., Mpodozis, C., and Sruoga, P., 1989, Late Paleozoic to Jurassic silicic magmatism at the Gondwanaland margin: Analogy to the Middle Proterozoic in North America?: Geology, v. 17, p. 324–328, doi: 10.1130/0091-7613(1989)017<0324:LPTJSM>2.3.CO;2.

Kay, S.M., Mpodozis, C., Ramos, V.A., and Munizaga, F., 1991, Magma source variations for mid to late Tertiary volcanic rocks erupted over a shallowing subduction zone and through a thickening crust in the Main Andean Cordillera (28–33°S), *in* Harmon, R.S., and Rapela, C., eds., Andean Magmatism and Its Tectonic Setting: Geological Society of America Special Paper 265, p. 113–137.

Kay, S.M., Orrell, S., and Abbruzzi, J.M., 1996, Zircon and whole rock Nd-Pb isotopic evidence for a Grenville age and a Laurentian origin for the Precordillera terrane in Argentina: The Journal of Geology, v. 104, p. 637–648.

Kay, S.M., Mpodozis, C., and Coira, B., 1999, Neogene magmatism, tectonism, and mineral deposits of the Central Andes (22°S to 33°S), *in* Skinner, B., ed., Geology and Mineral Deposits of Central Andes: Society of Economic Geology Special Publication 7, p. 27–59.

Kay, S.M., Godoy, E., and Kurtz, A., 2005, Episodic arc migration, crustal thickening, subduction erosion, and magmatism in the south-central Andes: Geological Society of America Bulletin, v. 117, p. 67–88, doi: 10.1130/B25431.1.

Kay, S.M., Burns, W.M., Copeland, P., and Mancilla, O., 2006, Upper Cretaceous to Holocene magmatism and evidence for transient Miocene shallowing of the Andean subduction zone under the northern Neuquén Basin, *in* Kay, S.M., and Ramos, V.A., eds., Evolution of an Andean Margin: A Tectonic and Magmatic View from the Andes to the Neuquén Basin (35°–39°S lat.): Geological Society of America Special Paper 407, p. 19–60.

Kennett, J.P., 1977, Cenozoic evolution of Antarctic glaciation, the circum-Antarctic ocean, and their impact on global paleoceanography: Journal of Geophysical Research, v. 82, p. 3843–3860, doi: 10.1029/JC082i027p03843.

Keppie, J.D., and Dostal, J., 2007, Rift-related basalts in the 1.2–1.3 Ga granulites of the northern Oaxacan Complex, southern Mexico: Evidence for a rifted arc on the northwestern margin of Amazonia: Proceedings of the Geologists' Association, v. 118, p. 63–74.

Keppie, J.D., and Ortega-Gutiérrez, F., 1995, Provenance of Mexican terranes: Isotopic constraints: International Geology Review, v. 37, p. 813–824.

Keppie, J.D., and Ramos, V.A., 1999, Odyssey of terranes in the Iapetus and Rheic Oceans during the Paleozoic, *in* Ramos, V.A., and Keppie, D., eds., Laurentia Gondwana Connections before Pangea: Geological Society of America Special Paper 336, p. 267–276.

Kerr, A.C., and Tarney, J., 2005, Tectonic evolution of the Caribbean and northwestern South America: The case for accretion of two Late Cretaceous oceanic plateaus: Geology, v. 33, p. 269–272, doi: 10.1130/G21109.1.

Kerr, A.C., Tarney, J., Nivia, A., Marriner, G.F., and Saunders, A.D., 1998, The internal structure of oceanic plateaus: Inferences from obducted Cretaceous terranes in western Colombia and the Caribbean: Tectonophysics, v. 292, p. 173–188, doi: 10.1016/S0040-1951(98)00067-5.

Kerr, A.C., Aspden, J.A., Tarney, J., and Pilatasig, L.F., 2002, The nature and provenance of accreted oceanic terranes in western Ecuador: Geochemical and tectonic constraints: Journal of the Geological Society of London, v. 159, p. 577–594, doi: 10.1144/0016-764901-151.

Kerr, A.C., White, R.V., Thompson, P.M.E., Tarney, J., and Saunders, A.D., 2003, No oceanic plateau—No Caribbean plate? The seminal role of an oceanic plateau in Caribbean plate evolution, *in* Bartolini, C., Buffler, R.T., and Blickwede, J.F., eds., The Gulf of Mexico and Caribbean Region: Hydrocarbon Habitats, Basin Formation and Plate Tectonics: American Association of Petroleum Geologists Memoir 79, p. 126–168.

Klepeis, K.A., 1994a, The Magallanes and Deseado fault zones: Major domains of the South American–Scotia transform plate boundary in southernmost

South America, Tierra del Fuego: Journal of Geophysical Research, v. 99, p. 22,001–22,014, doi: 10.1029/94JB01749.

Klepeis, K.A., 1994b, Relationship between uplift of the metamorphic core of the southernmost Andes and shortening in the Magallanes foreland fold and thrust belt, Tierra del Fuego, Chile: Tectonics, v. 13, p. 882–904, doi: 10.1029/94TC00628.

Kontak, D.J., 1985, The Magmatic and Metallogenic Evolution of a Craton-Orogen Interface: The Cordillera de Carabaya, Central Andes, SE Peru [Ph.D. thesis]: Kingston, Queen's University, 713 p.

Kontak, D.J., Clark, A.H., Farrar, E., Archibald, D.A., and Baadsgaard, H., 1990, Late Paleozoic–early Mesozoic magmatism in the Cordillera de Carabaya, Puno, southeastern Peru: Geochronology and petrochemistry: Journal of South American Earth Sciences, v. 3, p. 213–230, doi: 10.1016/0895-9811(90)90004-K.

Lapierre, H., Dupuis, V., Bosch, D., Polvé, M., Maury, R.C., Hernández, J., Monié, P., Yéghicheyan, D., Jaillard, E., Tardy, M., Mercier de Lépinay, B., Mamberti, M., Desnet, A., Keller, F., and Sénebier, F., 2000, Multiple plume events in the genesis of the peri-Caribbean Cretaceous Oceanic Plateau Province: Journal of Geophysical Research, v. 105, p. 8403–8421, doi: 10.1029/1998JB900091.

Lawver, L.A., and Gahagan, L.M., 2003, Evolution of Cenozoic seaways in the circum-Antarctic region: Palaeogeography, Palaeoclimatology, Palaeoecology, v. 198, p. 11–37, doi: 10.1016/S0031-0182(03)00392-4.

Leal, P.R., Hartmann, L.A., Santos, O., Miró, R., and Ramos, V.A., 2004, Volcanismo postorogénico en el extremo norte de las Sierras Pampeanas Orientales: Nuevos datos geocronológicos y sus implicancias tectónicas: Revista de la Asociación Geológica Argentina, v. 58, p. 593–607.

Lebras, N., Mégard, F., Dupuy, C., and Dostal, J., 1987, Geochemistry and tectonic setting of pre-collision Cretaceous and Paleogene volcanic rocks of Ecuador: Geological Society of America Bulletin, v. 99, p. 569–578, doi: 10.1130/0016-7606(1987)99<569:GATSOP>2.0.CO;2.

Levi, B., and Aguirre, L., 1981, Ensialic spreading—Subsidence in the Mesozoic and Paleogene Andes of Central Chile: Journal of the Geological Society of London, v. 138, p. 75–81, doi: 10.1144/gsjgs.138.1.0075.

Levi, B., Nystrom, J.O., Thiele, R., and Aberg, G., 1985, Geochemical variations in Mesozoic-Tertiary volcanic rocks from the Andes in central Chile and implications for the tectonic evolution: Comunicaciones, v. 35, p. 125–128.

Litherland, M., Klinck, B.A., O'Connor, E.A., and Pitfield, P.E.J., 1985, Andean-trending mobile belts in the Brazilian Shield: Nature, v. 314, p. 345–348, doi: 10.1038/314345a0.

Litherland, M., Annells, R.N., Darbyshire, D.P.F., Fletcher, C.J.N., Hawkins, M.P., Klinck, B.A., Mitchell, W.I., Connor, E.A., Pitfield, P.E.J., Power, G., and Webb, B.C., 1989, The Proterozoic of eastern Bolivia and its relationship to the Andean mobile belt: Precambrian Research, v. 43, p. 157–174, doi: 10.1016/0301-9268(89)90054-5.

Litherland, M., Aspden, J.A., and Jemielita, R.A., 1994, The metamorphic belts of Ecuador: British Geological Survey Overseas Memoir 11, p. 1–146.

Livermore, R.A., Nankivell, A.P., Eagles, G., and Morris, P., 2005, Paleogene opening of Drake Passage: Earth and Planetary Science Letters, v. 236, p. 459–470, doi: 10.1016/j.epsl.2005.03.027.

Llambías, E., and Sato, A.M., 1995, El batolito de Colangüil: Transición entre orogénesis y anorogénesis: Revista de la Asociación Geológica Argentina, v. 50, p. 111–131.

Llambías, E.J., Varela, R., Basei, M., and Sato, A.M., 2002, Deformación dúctil y metamorfismo neopaleozoico en Yaminué y su relación con la fase orogénica San Rafael, *in* 15th Congreso Geológico Argentino: Actas, v. 3, p. 123–128.

Lodolo, E., Menichetti, M., Bartole, R., Ben-Avraham, Z., Tassone, A., and Lippai, H., 2003, Magallanes-Fagnano continental transform fault (Tierra del Fuego, southernmost South America): Tectonics, v. 22, p. 1076, doi: 10.1029/2003TC001500.

Lodolo, E., Donda, F., and Tassone, A., 2006, Western Scotia Sea margins: Improved constraints on the opening of the Drake Passage: Journal of Geophysical Research, v. 111, p. B06101, doi: 10.1029/2006JB004361.

Loewy, S.L., Connelly, J.N., Dalziel, I.W.D., and Gower, C.F., 2003, Eastern Laurentia in Rodinia: Constraints from whole-rock Pb and U/Pb geochronology: Tectonophysics, v. 375, p. 169–197, doi: 10.1016/S0040-1951(03)00338-X.

Loewy, S.L., Connelly, J.N., and Dalziel, I.W.D., 2004, An orphaned basement block: The Arequipa-Antofalla basement of the central Andean margin of South America: Geological Society of America Bulletin, v. 116, p. 171–187, doi: 10.1130/B25226.1.

López Ramos, E., and Barrero, D., 2003, Upper Crust Models of Colombia: Bogotá, Ingeominas, 64 p.

Lucassen, F., Franz, G., and Laber, A., 1999, Permian high pressure rocks—The basement of the Sierra de Limón Verde in northern Chile: Journal of South American Earth Sciences, v. 12, no. 2, p. 183–199, doi: 10.1016/S0895-9811(99)00013-9.

Luzieux, L.D.A., Heller, F., Spikings, R., Vallejo, C.F., and Winkler, W., 2006, Origin and Cretaceous tectonic history of the coastal Ecuadorian forearc between 1°N and 3°S: Paleomagnetic, radiometric and fossil evidence: Earth and Planetary Science Letters, v. 249, p. 400–414, doi: 10.1016/j.epsl.2006.07.008.

Mamaní, M., Tassara, A., and Wörner, G., 2007, Crustal domains in the Central Andes and their control on orogenic structures: 20th Colloquium Latin American Earth Sciences: Kiel, Germany, Deutsche Forschungsgemeinschaft, p. 27–28.

Marocco, R., Lavenu, A., and Baudino, R., 1995, Intermontane late Paleogene–Neogene basins of the Andes of Ecuador and Perú: Sedimentologic and tectonic characteristics, *in* Tankard, A.J., Suarez, S.R., and Welsink, H.J., eds., Petroleum Basins of South America: American Association of Petroleum Geologists Memoir 62, p. 597–613.

Martens, U.C., and Dunlap, W.J., 2003, Características del metamorfismo Cretácico del terreno Tahamí como se infiere a partir de edades Ar-Ar obtenidas de las anfibolitas de Medellín, Cordillera Central de Colombia, *in* 9th Congreso Colombiano de Geología: Actas: Medellín, Colombia, Fundación para la Promoción de la Investigación y la Tecnología, p. 47–48.

Martina, F., Astini, R.A., Becker, T.P., and Thomas, W.A., 2005, Granitos grenvilianos milonitizadas en la faja de deformación de Jagüé, noroeste de La Rioja, *in* 16th Congreso Geológico Argentino: Actas, v. 4: La Plata, Argentina, p. 591–594.

Martínez, A., 2005. El Volcanismo del Choiyoi en la Región del Manzano Histórico, Tunuyán, Mendoza [Ph.D. thesis]: Buenos Aires, Universidad de Buenos Aires, 274 p.

Martin-Gombojav, N., and Winkler, W., 2008, Recycling of Proterozoic crust in the Andean Amazon foreland of Ecuador: Implications for orogenic development of the Northern Andes: Terra Nova, v. 20, p. 22–31.

Mathalone, J.M.P., and Montoya, M., 1995, Petroleum geology of the Subandean basins of Peru, *in* Tankard, A.J., Suarez, S.R., and Welsink, H.J., eds., Petroleum Basins of South America: American Association of Petroleum Geologists Memoir 62, p. 423–444.

Matteini, M., Mazzuoli, R., Omarini, R., Cas, R., and Maas, R., 2002, The geochemical variations of the upper Cenozoic volcanism along the Calama–Olacapato–El Toro transversal fault system in central Andes (24°S): Petrogenetic and geodynamic implications: Tectonophysics, v. 345, p. 211–227, doi: 10.1016/S0040-1951(01)00214-1.

McCourt, W.J., Aspden, J.A., and Brook, M., 1984, New geological and geochronological data for the Colombian Andes: Continental growth by multiple accretion: Journal of the Geological Society of London, v. 141, p. 831–845, doi: 10.1144/gsjgs.141.5.0831.

McQuarrie, N., 2002, The kinematic history of the central Andean fold-thrust belt, Bolivia: Implications for building a high plateau: Geological Society of America Bulletin, v. 114, p. 950–963, doi: 10.1130/0016-7606(2002)114<0950:TKHOTC>2.0.CO;2.

Mégard, F., 1987, Cordilleran Andes and Marginal Andes: A review of Andean geology north of the Arica elbow (18°S), *in* Monger, J.M.H., and Francheteau, J., eds., Circum-Pacific Orogenic Belts and Evolution of the Pacific Ocean Basin: International Lithosphere Program Contribution, Geodynamic Series 18, p. 71–96.

Mégard, F., 1989, The evolution of the Pacific Ocean margin in South America north of Arica Elbow (18°S), *in* Ben-Avraham, Z., ed., The Evolution of the Pacific Margins: New York, Oxford University Press, p. 208–231.

Michaud, F., Witt, C., and Royer, J.-Y., 2009, Influence of the subduction of the Carnegie volcanic ridge on Ecuadorian geology: Reality and fiction, *in* Kay, S.M., Ramos, V.A., and Dickinson, W.R., eds., Backbone of the Americas: Shallow Subduction, Plateau Uplift and Ridge and Terrane Collision: Geological Society of America Memoir 204, doi: 10.1130/2009.1204(10).

Montgomery, D.R., Balco, G., and Willett, S.D., 2001, Climate, tectonics, and the morphology of the Andes: Geology, v. 29, p. 579–582, doi: 10.1130/0091-7613(2001)029<0579:CTATMO>2.0.CO;2.

Moores, E.M., 1991, Southwest U.S.–East Antarctic (SWEAT) connection: A hypothesis: Geology, v. 19, p. 425–428, doi: 10.1130/0091-7613(1991)019<0425:SUSEAS>2.3.CO;2.

Moreno-Sánchez, M., 2004, Devonian Plants from Colombia: Geologic Framework and Paleogeographic Implications [Ph.D. thesis]: Liège,

Université de Liège, Faculté des Sciences, Département de Géologie, 206 p.

Mosquera, A., and Ramos, V.A., 2006, Intraplate deformation in the Neuquén Basin, *in* Kay, S.M., and Ramos, V.A., eds., Evolution of an Andean Margin: A Tectonic and Magmatic View from the Andes to the Neuquén Basin (35°–39°S lat): Geological Society of America Special Paper 407, p. 97–124.

Mourier, T., Laj, C., Mégard, F., Roperch, P., Mitouard, P., and Farfan, M., 1988, An accreted continental terrane in northwestern Peru: Earth and Planetary Science Letters, v. 88, p. 182–192, doi: 10.1016/0012-821X(88)90056-8.

Moya, M.C., Malanca, S., Hongn, F.D., and Bahlburg, H., 1993, El Tremadoc temprano en la Puna Occidental Argentina, *in* 12th Congreso Geológico Argentino y 2nd Congreso Exploración de Hidrocarburos: Actas, v. 2: Mendoza, Argentina, p. 20–30.

Mpodozis, C., and Cornejo, P., 1997, El rift Triásico-Sinemuriano de Sierra Exploradora, Cordillera de Domeyko (25–26°S): Asociación de facies y reconstrucción tectónica, *in* 8th Congreso Geológico Chileno: Actas, v. 1: Antofagasta, Chile, p. 550–553.

Mpodozis, C., and Kay, S.M., 1992, Late Paleozoic to Triassic evolution of the Gondwana margin: Evidence from Chilean Frontal Cordilleran Batholiths (28°–31°S): Geological Society of America Bulletin, v. 104, p. 999–1014, doi: 10.1130/0016-7606(1992)104<0999:LPTTEO>2.3.CO;2.

Mpodozis, C., and Ramos, V.A., 1989, The Andes of Chile and Argentina, *in* Ericksen, G.E., Cañas Pinochet, M.T., and Reinemud, J.A., eds., Geology of the Andes and Its Relation to Hydrocarbon and Mineral Resources: Circumpacific Council for Energy and Mineral Resources, Earth Sciences Series 11, p. 59–90.

Mpodozis, C., Arriagada, C., Basso, M., Roperch, P., Cobbold, P., and Reich, M., 2005, Late Mesozoic to Paleogene stratigraphy of the Salar de Atacama Basin, Antofagasta, northern Chile: Implications for the tectonic evolution of the Central Andes: Tectonophysics, v. 399, p. 125–154, doi: 10.1016/j.tecto.2004.12.019.

Murphy, J.B., Gutiérrez-Alonso, G., Nance, R.D., Fernández-Suárez, J., Keppie, J.D., Quesada, C., Strachan, R.A., and Dostal, J., 2006, Origin of the Rheic Ocean: Rifting along a Neoproterozoic suture?: Geology, v. 34, p. 325–328, doi: 10.1130/G22068.1.

Niemeyer, R.H., 1989, El complejo ígneo-sedimentario del Cordón de La Lila, Región de Antofagasta: Estratigrafía y significado tectónico: Revista Geológica de Chile, v. 16, p. 163–182.

Nivia, A., Marriner, G.F., Kerr, A.C., and Tarney, J., 2006, The Quebradagrande Complex: A Lower Cretaceous ensialic marginal basin in the Central Cordillera of the Colombian Andes: Journal of South American Earth Sciences, v. 21, p. 423–436, doi: 10.1016/j.jsames.2006.07.002.

Olivero, E.B., 1998, Mesozoic-Paleogene geology of the marginal-Austral basin of Tierra del Fuego, *in* Mesozoic-Paleogene geology of the marginal-Austral basin of Tierra del Fuego: Third Annual Conference, IGCP Project 381, South Atlantic Mesozoic Correlations, Field Trip Guide: Ushuaia, Argentina, Consejo Nacional de Investigaciones Científicas y Técnicas, p. 1–42.

Omarini, R.H., Sureda, R.J., Götze, H.J., Seilacher, A., and Plüger, F., 1999, Puncovicana fold belt in northwestern Argentina: Testimony of Late Proterozoic Rodinia fragmentation and pre-Gondwana collisional episodes: International Journal of Earth Sciences, v. 88, p. 76–97, doi: 10.1007/s005310050247.

Oncken, O., Chong, G., Franz, G., Giese, P., Götze, H.-J., Ramos, V.A., Strecker, M.R., and Wigger, P., eds., 2006, The Andes—Active Subduction Orogeny: Berlin, Springer, Frontiers in Earth Sciences Series 1, 570 p.

Ordóñez-Carmona, O., Restrepo, J.J., Álvarez, A., and Pimentel, M.M., 2006, Geochronological and isotopical review of pre-Devonian crustal basement of the Colombian Andes: Journal of South American Earth Sciences, v. 21, p. 372–382, doi: 10.1016/j.jsames.2006.07.005.

Palma, M.A., Parica, P., and Ramos, V.A., 1986, El Granito Archibarca: Su edad y significado tectónico: Revista de la Asociación Geológica Argentina, v. 41, p. 414–418.

Pankhurst, R.J., and Rapela, C.W., eds., 1998, The Proto-Andean Margin of Gondwana: Geological Society of London Special Publication 142, 382 p.

Pankhurst, R.J., Rapela, C., Saavedra, J., Baldo, E., Dahlquist, J., Pascua, I., and Fanning, C.M., 1998, The Famatinian magmatic arc in the Central Sierras Pampeanas: An early to mid-Ordovician continental arc on the Gondwana margin, *in* Pankhurst, R.J., and Rapela, C.W., eds., The Proto-Andean Margin of Gondwana: Geological Society of London Special Publication 142, p. 343–367.

Pankhurst, R.J., Rapela, C., and Loske, W.P., 2003, Chronological study of pre-Permian basement rocks of southern Patagonia: Journal of South American Earth Sciences, v. 16, p. 27–44, doi: 10.1016/S0895-9811(03)00017-8.

Pankhurst, R.J., Rapela, C.W., Fanning, C.M., and Márquez, M., 2006, Gondwanide continental collision and the origin of Patagonia: Earth-Science Reviews, v. 76, p. 235–257, doi: 10.1016/j.earscirev.2006.02.001.

Pardo Casas, F., and Molnar, P., 1987, Relative motion of the Nazca (Farallon) and South American plates since Late Cretaceous time: Tectonics, v. 6, p. 233–248, doi: 10.1029/TC006i003p00233.

Parnaud, F., Gou, Y., Pascual, J.C., Truskowski, I., Gallango, O., and Passalacqua, H., 1995, Petroleum geology of the central part of the eastern Venezuela basin, *in* Tankard, A.J., Suarez, S.R., and Welsink, H.J., eds., Petroleum Basins of South America: American Association of Petroleum Geologists Memoir 62, p. 741–756.

Pennington, W.D., 1981, Subduction of the Eastern Panama Basin and seismotectonics of northwest South America: Journal of Geophysical Research, v. 86, p. 10,753–10,770, doi: 10.1029/JB086iB11p10753.

Pereira, E., Ortiz, F., and Prichard, H., 2006, Contribution to the knowledge of amphibolites and dunitas of Medellín (Complejo Ofiolítico of Aburrá): Dyna: Revista Facultad Nacional de Minas, v. 73, p. 17–30.

Pérez, O.J., and GPS Group of Universidad Simón Bolívar, 2006, Geodesia GPS 1994–2006 en el sur del Caribe y Venezuela: Caracterizando y cuantificando las deformaciones corticales de la región: 13th Congreso Venezolano de Geofísica, Digital Archives: Caracas, Venezuela, 10 p.

Petford, N., Atherton, M.P., and Halliday, A.N., 1996, Rapid magma production rate, underplating and remelting in the Andes: Isotopic evidence from northern-central Peru (9°–11°S): Journal of South American Earth Sciences, v. 9, p. 69–78, doi: 10.1016/0895-9811(96)00028-4.

Pindell, J.L., and Tabbutt, K.D., 1995, Mesozoic-Cenozoic Andean paleogeography and regional controls on hydrocarbon systems, *in* Tankard, A.J., Suarez, S.R., and Welsink, H.J., eds., Petroleum Basins of South America: American Association of Petroleum Geologists Memoir 62, p. 101–128.

Pindell, J.L., Higgs, R., and Dewey, J.F., 1998, Cenozoic palinspastic reconstruction, paleogeographic evolution and hydrocarbon setting of the northern margin of South America, *in* Pindell, J.L., and Drake, C.L., eds., Paleogeographic Evolution and Non-Glacial Eustasy, Northern South America: SEPM (Society of Economic Paleontologists and Mineralogists) Special Publication 58, p. 45–86.

Pindell, J.L., Kennan, L., Stanek, K.P., Maresch, W.V., and Draper, G., 2006, Foundations of Gulf of Mexico and Caribbean evolution: Eight controversies resolved: Geologica Acta, v. 4, p. 303–341.

Pratt, T.T., Duque, P., and Ponce, M., 2005, An autochthonous geological model for the eastern Andes of Ecuador: Tectonophysics, v. 399, p. 251–278, doi: 10.1016/j.tecto.2004.12.025.

Quenardelle, S., and Ramos, V.A., 1999, The Ordovician western Sierras Pampeanas magmatic belt: Record of Precordillera accretion in Argentina, *in* Ramos, V.A., and Keppie, D., eds., Laurentia Gondwana Connections before Pangea: Geological Society of America Special Paper 336, p. 63–86.

Quinteros, J., Jacovkis, P., and Ramos, V.A., 2006, Evolution of the upper crustal deformation in subduction zones: Journal of Applied Mechanics, v. 73, p. 984–994, doi: 10.1115/1.2204962.

Ramos, V.A., 1983, Evolución tectónica y metalogénesis de la Cordillera Patagónica, *in* 2nd Congreso Nacional Geología Económica: Actas, v. 1: San Juan, Argentina, p. 108–124.

Ramos, V.A., 1984, Patagonia: ¿Un continente paleozoico a la deriva?, *in* 9th Congreso Geológico Argentino: Actas, v. 2: Bariloche, Argentina, p. 311–325.

Ramos, V.A., 1986, El diastrofismo oclóyico: Un ejemplo de tectónica de colisión durante el Eopaleozoico en el noroeste Argentino: Revista Instituto Ciencias Geológicas, v. 6, p. 13–28.

Ramos, V.A., 1988a, Tectonics of the Late Proterozoic–early Paleozoic: A collisional history of southern South America: Episodes, v. 11, no. 3, p. 168–174.

Ramos, V.A., 1988b, The tectonics of the Central Andes: 30° to 33°S latitude, *in* Clark, S., and Burchfiel, D., eds., Processes in Continental Lithospheric Deformation: Geological Society of America Special Paper 218, p. 31–54.

Ramos, V.A., 1999, Plate tectonic setting of the Andean Cordillera: Episodes, v. 22, no. 3, p. 183–190.

Ramos, V.A., 2000, Evolución tectónica de la Argentina, *in* Caminos, R., ed., Geología Argentina: Instituto de Geología y Recursos Minerales, Anales, v. 29, p. 715–784.

Ramos, V.A., 2002, Evolución tectónica, *in* Haller, M.J., ed., Geología y Recursos Naturales de Santa Cruz: Calafate, Argentina, 15th Congreso Geológico Argentino, Relatorio, p. 365–387.

Ramos, V.A., 2004, Cuyania, an exotic block to Gondwana: Review of a historical success and the present problems: Gondwana Research, v. 7, p. 1009–1026, doi: 10.1016/S1342-937X(05)71081-9.

Ramos, V.A., 2005, Ridge collision and topography: Foreland deformation in the Patagonian Andes: Tectonophysics, v. 399, p. 73–86, doi: 10.1016/j.tecto.2004.12.016.

Ramos, V.A., 2008a, The basement of the Central Andes: The Arequipa and related terranes: Annual Review of Earth and Planetary Sciences, v. 36, p. 289–324, doi: 10.1146/annurev.earth.36.031207.124304.

Ramos, V.A., 2008b, Patagonia: A Paleozoic continent adrift?: Journal of South American Earth Sciences, v. 26, no. 3, p. 235–251.

Ramos, V.A., and Alemán, A., 2000, Tectonic evolution of the Andes, *in* Cordani, E.U.J., Milani, E.J., Thomaz Filho, A., and Campos, D.A., eds., Tectonic Evolution of South America: Río de Janeiro, 31st International Geological Congress, p. 635–685.

Ramos, V.A., and Basei, M.A., 1997, Gondwanan, Perigondwanan, and exotic terranes of southern South America, *in* South American Symposium on Isotope Geology, Extended Abstracts: Campos do Jordao, Brazil, Universidade de São Paulo, p. 250–252.

Ramos, V.A., and Coira, B.L., 2008, Evolución tectónica preandina de la provincia de Jujuy y áreas aledañas, *in* Coira, B.L., and Zappettini, E.O., eds., Geología y Recursos Naturales de la Provincia de Jujuy: 17th Congreso Geológico Argentino: Jujuy, Relatorio, Congreso, p. 401–417.

Ramos, V.A., and Folguera, A., 2005, Tectonic evolution of the Andes of Neuquén: Constraints derived from the magmatic arc and foreland deformation, *in* Veiga, G.D., Spalletti, L.A., Howell, J.A., and Swarz, E., eds., The Neuquén Basin: A Case Study in Sequence Stratigraphy and Basin Dynamics: Geological Society of London Special Publication 252, p. 15–35.

Ramos, V.A., and Folguera, A., 2009, Andean flat slab subduction through time, *in* Murphy, B., ed., Ancient Orogens and Modern Analogues: Geological Society of London Special Publication (in press).

Ramos, V.A., and Ghiglione, M.C., 2008, Tectonic Evolution of the Patagonian Andes, *in* Rabassa, J., ed., Late Cenozoic of Patagonia and Tierra del Fuego: Amsterdam, the Netherlands, Elsevier B.V., Developments in Quaternary Sciences 11, p. 57–71.

Ramos, V.A., and Kay, S.M., 1991, Triassic rifting and associated basalts in the Cuyo Basin, central Argentina, *in* Harmon, R.S., and Rapela, C.W., eds., Andean Magmatism and Its Tectonic Setting: Geological Society of America Special Paper 265, p. 79–91.

Ramos, V.A., and Kay, S.M., 2006, Overview of the tectonic evolution of the southern Central Andes of Mendoza and Neuquén (35°–39°S latitude), *in* Kay, S.M., and Ramos, V.A., eds., Evolution of an Andean Margin: A Tectonic and Magmatic View from the Andes to the Neuquén Basin (35°–39°S lat): Geological Society of America Special Paper 407, p. 1–18.

Ramos, V.A., and Keppie, D., eds., 1999, Laurentia Gondwana Connections before Pangea: Geological Society of America Special Paper 336, 276 p.

Ramos, V.A., and Vujovich, G.I., 1993, The Pampia craton within western Gondwanaland, *in* Ortega Gutiérrez, F., ed., First Circum-Pacific and Circum-Atlantic Terrane Conference (Guanajuato) Proceedings: México, México, Universidad Nacional Autónoma de México, Instituto de Geología, p. 113–116.

Ramos, V.A., Jordan, T.E., Allmendinger, R.W., Mpodozis, C., Kay, S., Cortés, J.M., and Palma, M.A., 1986, Paleozoic terranes of the Central Argentine-Chilean Andes: Tectonics, v. 5, p. 855–880.

Ramos, V.A., Vujovich, G., Kay, S.M., and McDonough, M.R., 1993, La orogénesis de Grenville en las Sierras Pampeanas Occidentales: La Sierra de Pie de Palo y su integración al supercontinente proterozoico, *in* 12th Congreso Geológico Argentino y 2nd Congreso de Exploración de Hidrocarburos: Actas, v. 3: Mendoza, Argentina, p. 343–357.

Ramos, V.A., Escayola, M., Mutti, D., and Vujovich, G.I., 2000, Proterozoic–early Paleozoic ophiolites in the Andean basement of southern South America, *in* Dilek, Y., Moores, E.M., Elthon, D., and Nicolas, A., eds., Ophiolites and Oceanic Crust: New Insights from Field Studies and Ocean Drilling Program: Geological Society of America Special Paper 349, p. 331–349.

Ramos, V.A., Cristallini, E., and Pérez, D.J., 2002, The Pampean flat-slab of the Central Andes: Journal of South American Earth Sciences, v. 15, p. 59–78, doi: 10.1016/S0895-9811(02)00006-8.

Rapalini, A.E., López de Luchi, M., Croce, F., Lince Klinger, F., Tomezzoli, R., and Gímenez, M., 2008, Estudio geofísico del complejo plutónico Navarrte: Implicancias para la evolución tectónica de Patagonia en el Paleozoico Tardío: 5th Simposio Argentino del Paleozoico Superior, Resúmenes: Buenos Aires, Argentina, p. 34.

Rapela, C.W., Pankhurst, R.J., Casquet, C., Baldo, E., Saavedra, J., Galindo, C., and Fanning, C.M., 1998, The Pampean orogeny of the southern proto-Andes: Cambrian continental collision in the Sierras de Córdoba, *in* Pankhurst, R.J., and Rapela, C.W., eds., The Proto-Andean Margin of Gondwana: Geological Society of London Special Publication 142, p. 181–217.

Rapela, C.W., Pankhurst, R.J., Casquet, C., Fanning, C.M., Baldo, E.G., González-Casado, J.M., Galindo, C., and Dahlquist, J., 2007, The Río de la Plata craton and the assembly of SW Gondwana: Earth-Science Reviews, v. 83, p. 49–82, doi: 10.1016/j.earscirev.2007.03.004.

Restrepo, J.J., 2003, Edad de generación y emplazamiento de ofiolitas en la Cordillera Central: Un replanteamiento, *in* 9th Congreso Colombiano de Geología: Actas: Medellín, Colombia, Fundación para la Promoción de la Investigación y la Tecnología, p. 48–49.

Restrepo, J.J., and Toussaint, J.-F., 1974, Obducción Cretácea en el occidente Colombiano: Anales de la Facultad de Minas, v. 58, p. 73–105.

Restrepo, J.J., and Toussaint, J.-F., 1982, Metamorfismos superpuestos en la Cordillera Central de Colombia, *in* 5th Congreso Latinoamericano de Geología: Actas, v. 3: Buenos Aires, Argentina, Servicio Geológico Nacional, p. 505–512.

Restrepo, J.J., and Toussaint, J.-F., 1988, Terranes and continental accretion in the Colombian Andes: Episodes, v. 11, p. 189–193.

Restrepo, J.J., Dunlap, J., Martens, U., Ordoñez-Carmona, O., and Correa, A.M., 2008, Ar-Ar ages of the amphibolites from the Central Cordillera of Colombia and their implications for tectonostratigraphic terrane evolution in the northwestern Andes: 6th South American Symposium on Isotope Geology, Abstracts: Bariloche, Argentina, Instituto Nacional de Geología Isotópica, p. 92.

Restrepo-Pace, P.A., 1993, Petrotectonic characterization of the Central Andean terrane: A Paleozoic (?) polydeformed back-arc turbidite complex in the Central Andes of Colombia, *in* Ortega Gutiérrez, F., ed., First Circum-Pacific and Circum-Atlantic Terrane Conference Proceedings: México, México, Universidad Nacional Autónoma de México, p. 123.

Restrepo-Pace, P.A., Ruiz, J., Gehrels, G., and Cosca, M., 1997, Geochronology and Nd isotopic data of Grenville-age rocks in the Colombian Andes: New constraints for Late Proterozoic–early Paleozoic paleocontinental reconstructions of the Americas: Earth and Planetary Science Letters, v. 150, p. 427–441, doi: 10.1016/S0012-821X(97)00091-5.

Reynaud, C., Jaillard, E., Lapiere, H., Mamberti, M., and Mascle, G., 1999, Oceanic plateau and island arcs of southwestern Ecuador: Their place in the geodynamic evolution of northwestern South America: Tectonophysics, v. 307, p. 235–254, doi: 10.1016/S0040-1951(99)00099-2.

Riller, U., Petrinovic, I.A., Ramelow, J., Strecker, M., and Oncken, O., 2001, Late Cenozoic tectonism, collapse caldera and plateau formation in the central Andes: Earth and Planetary Science Letters, v. 58, p. 1–13.

Roeder, D.H., 1988, Andean-age structure of Eastern Cordillera (Province of La Paz, Bolivia): Tectonics, v. 7, p. 23–40, doi: 10.1029/TC007i001p00023.

Roeder, D.H., and Chamberlain, R.L., 1995, Structural geology of Sub-Andean fold and thrust belt in northwestern Bolivia, *in* Tankard, A., Suárez, S.R., and Welsink, H., eds., Petroleum Basins of South America: American Association of Petroleum Geologists Memoir 62, p. 459–479.

Russo, R.M., and Silver, P.G., 1996, Cordillera formation, mantle dynamics, and the Wilson cycle: Geology, v. 24, no. 6, p. 511–514, doi: 10.1130/0091-7613(1996)024<0511:CFMDAT>2.3.CO;2.

Sadowski, G.R., and Bettencourt, J.S., 1996, Mesoproterozoic tectonic correlations between eastern Laurentia and the western border of the Amazon craton: Precambrian Research, v. 76, p. 213–227, doi: 10.1016/0301-9268(95)00026-7.

Salfity, J.A., and Marquillas, R.A., 1994, Tectonic and sedimentary evolution of the Cretaceous-Eocene Salta Group Basin, Argentina, *in* Salfity, J., ed., Cretaceous Tectonics in the Andes: Vieweg, Braunschweig, Earth Evolution Sciences, p. 266–315.

Sallarés, V., Charvis, P., Flueh, E.R., and Bialas, J., 2003, Seismic structure of Cocos and Malpelo volcanic ridges and implications for hot spot-ridge interaction: Journal of Geophysical Research, v. 108, no. B12, p. 2564, doi: 10.1029/2003JB002431.

Sánchez, J., Palacios, O., Feininger, T., Carlotto, V., and Quispesivana, L., 2006, Puesta en evidencia de granitoides Triásicos en los Amotapes-Tahuín: Deflexión de Huancabamba: 13th Congreso Peruano de Geología, Resúmenes Extendidos: Lima, Perú, Sociedad Geológica del Perú, p. 312–315

Sandeman, H.A., Clark, A.H., and Farrar, E., 1995, An integrated tectono-magmatic model for the evolution of the Southern Peruvian Andes (13°–20°S) since 55 Ma: International Geology Review, v. 37, p. 1039–1073.

Sarmiento-Rojas, L.F., Van Wess, J.D., and Cloetingh, S., 2006, Mesozoic transtensional basin history of the Eastern Cordillera, Colombian Andes: Inferences from tectonic models: Journal of South American Earth Sciences, v. 21, p. 383–411, doi: 10.1016/j.jsames.2006.07.003.

Schurr, B., Rietbrock, A., Asch, G., Kind, R., and Oncken, O., 2006, Evidence for lithospheric detachment in the central Andes from local earthquake tomography: Tectonophysics, v. 415, p. 203–223, doi: 10.1016/j.tecto.2005.12.007.

Segerstrom, K., and Turner, J.C.M., 1972, A conspicuous flexure in regional structural trend in the Puna of northwestern Argentina: U.S. Geological Survey Professional Paper 800, p. 205–209.

Sempere, T., 1995, Phanerozoic evolution of Bolivia and adjacent regions. *in* Tankard, A., Suárez, S.R., and Welsink, H., eds., Petroleum Basins of South America: American Association of Petroleum Geologists Memoir 62, p. 207–230.

Sempere, T., Carlier, G., Soler, P., Fornary, M., Carlotto, V., Jacay, J., Arispe, O., Nereudeau, D., Cárdenas, J., Rosas, S., and Jiménez, N., 2002, Late Permian–Middle Jurassic lithospheric thinning in Peru and Bolivia, and its bearing on Andean-age tectonics: Tectonophysics, v. 345, p. 153–181, doi: 10.1016/S0040-1951(01)00211-6.

Smalley, R., Jr., Pujol, J., Regnier, M., Ming Chiu, J., Chatelain, J.L., Isacks, B., Araujo, M., and Puebla, N., 1993, Basement seismicity beneath the Andean Precordillera thin-skinned thrust belt and implications for crustal and lithospheric behaviour: Tectonics, v. 12, p. 63–76, doi: 10.1029/92TC01108.

Smalley, R., Jr., Dalziel, I.W.D., Bevis, M.G., Kendrick, E., Stamps, D.S., King, E.C., Taylor, F.W., Lauría, E., Zakrajsek, A., and Parra, H., 2007, Scotia arc kinematics from GPS geodesy: Geophysical Research Letters, v. 34, p. L21308, doi: 10.1029/2007GL031699.

Söllner, F., Miller, H., and Hervé, F., 2000, An early Cambrian granodiorite age from the pre-Andean basement of Tierra del Fuego (Chile): The missing link between South America and Antarctica?: Journal of South American Earth Sciences, v. 13, p. 163–177, doi: 10.1016/S0895-9811(00)00020-1.

Somoza, R., 1996, Geocinemática de América del Sur durante el Cretácico: Su relación con la evolución del margen pacífico y la apertura del Atlántico Sur, *in* 13th Congreso Geológico Argentino: Actas, v. 2: Buenos Aires, Argentina, p. 401–402.

Somoza, R., 1998, Updated Nazca (Farallon)–South America relative motions during the last 40 My: Implications for mountain building in the Central Andean region: Journal of South American Earth Sciences, v. 11, p. 211–215, doi: 10.1016/S0895-9811(98)00012-1.

Somoza, R., and Zaffarana, C.B., 2008, Mid-Cretaceous polar standstill of South America, motion of the Atlantic hotspots and the birth of the Andean Cordillera: Earth and Planetary Science Letters, v. 271, p. 267–277, doi: 10.1016/j.epsl.2008.04.004.

Stern, C.R., 1991, Role of subduction erosion in the generation of Andean magmas: Geology, v. 19, p. 78–81, doi: 10.1130/0091-7613(1991)019<0078:ROSEIT>2.3.CO;2.

Stern, C.R., and Mpodozis, C., 1991, Geologic evidence for subduction erosion along the west coast of Central and Northern Chile, *in* 6th Congreso Geológico Chileno: Actas: Santiago, Chile, p. 205–207.

Strecker, M.R., Alonso, R.N., Bookhagen, B., Carrapa, B., Hilley, G.E., Sobel, E.R., and Trauth, M.H., 2007, Tectonics and climate of the southern Central Andes: Annual Review of Earth and Planetary Sciences, v. 35, p. 747–787, doi: 10.1146/annurev.earth.35.031306.140158.

Suárez Soruco, R., 2000, Compendio de geología de Bolivia: Revista Técnica de Yacimientos Petrolíferos Fiscales Bolivianos, v. 18, p. 1–166.

Taboada, A., Rivera, L.A., Fuenzalida, A., Cisternas, A., Philip, H., Bijwaard, H., Olaya, J., and Rivera, C., 2000, Geodynamics of the northern Andes: Subductions and intracontinental deformation (Colombia): Tectonics, v. 19, p. 787–813, doi: 10.1029/2000TC900004.

Teixeira, W., Tassinari, C.C.G., Cordani, U.G., and Kawashita, K., 1989, A review of the geochronology of the Amazonian craton: Tectonic implications: Precambrian Research, v. 42, p. 213–227, doi: 10.1016/0301-9268(89)90012-0.

Thomas, W.A., and Astini, R.A., 2003, Ordovician accretion of the Argentine Precordillera terrane to Gondwana: A review: Journal of South American Earth Sciences, v. 16, p. 67–79, doi: 10.1016/S0895-9811(03)00019-1.

Thornburg, T., and Kulm, L.D., 1981, Sedimentary basins of the Peru continental margin: Structure, stratigraphy, and Cenozoic tectonics from 6°S to 16°S latitude, *in* Kulm, D.M., Dymond, J., Dasch, E.J., and Husson, D.M., eds., Nazca Plate: Crustal Formation and Andean Convergence: Geological Society of America Memoir 154, p. 393–422.

Tosdal, R.M., 1996, The Amazon-Laurentian connection as viewed from the Middle Proterozoic rocks in the Central Andes, western Bolivia and northern Chile: Tectonics, v. 15, p. 827–842, doi: 10.1029/95TC03248.

Toussaint, J.-F., and Restrepo, J.J., 1994, The Colombian Andes during Cretaceous times, *in* Salfity, J., ed., Cretaceous Tectonics in the Andes: Vieweg, Braunschweig, Earth Evolution Sciences, p. 61–100.

Trenkamp, R., Kellogg, J.N., Freymueller, J.T., and Mora, H.P., 2002, Wide plate margin deformation, southern Central America and northwestern South America, CASA GPS observations: Journal of South American Earth Sciences, v. 15, p. 157–171, doi: 10.1016/S0895-9811(02)00018-4.

Trümpy, E., 1943, Pre-Cretaceous of Colombia: Geological Society of America Bulletin, v. 54, p. 1281–1304.

Uliana, M.A., and Biddle, K.T., 1988, Mesozoic-Cenozoic paleogeographic and geodynamic evolution of southern South America: Revista Brasileira de Geociencias, v. 18, p. 172–190.

Vallejo, C., 2007, Evolution of the Western Cordillera in the Andes of Ecuador (Late Cretaceous–Paleogene) [Ph.D. thesis]: Zürich, Swiss Federal Institute of Technology Zürich, 215 p.

Vallejo, C., Spikings, R.A., Winkler, W., Luzieux, L., Chew, D., and Page, L., 2006, The early interaction between the Caribbean Plateau and the NW South American plate: Terra Nova, v. 18, p. 264–269.

Vallejo, C., Winkler, W., Spikings, R.A., Luzieux, L., Heller, F., and Bussy, F., 2009, this volume, Mode and timing of terrane accretion in the forearc of the Andes in Ecuador, *in* Kay, S.M., Ramos, V.A., and Dickinson, W.R., eds., Backbone of the Americas: Shallow Subduction, Plateau Uplift, and Ridge and Terrane Collision: Geological Society of America Memoir 204, doi: 10.1130/2009.1204(09).

Varela, R., Basei, M., Sato, A., Siga, O., Jr., Cingolani, C., and Sato, K., 1998, Edades isotópicas Rb/Sr y U/Pb en rocas de Mina Gonzalito y Arroyo Salado. Macizo Norpatagónico Atlántico, Río Negro, Argentina, *in* 10th Congreso Latinoamericano de Geología y 6th Congreso Nacional de Geología Económica: Actas, v. 1: Buenos Aires, Argentina, p. 71–76.

Vicente, J.C., 1970, Reflexiones sobre la porción meridional del sistema peripacífico oriental, *in* Conferencia sobre Problemas de la Tierra Sólida: Buenos Aires, Proyecto Internacional del Manto Superior, v. 37-I, p. 162–188.

Vinasco, C., 2004, Evolucao Crustal e História Tectònica Dos Granitoides Permo-Triásicos Dos Andes Do Norte [Ph.D. thesis]: São Paulo, Universidade de São Paulo, 121 p.

Vinasco, C.J., Cordani, U.G., González, H., Weber, M., and Pelaez, C., 2006, Geochronological, isotopic, and geochemical data from Permo-Triassic granitic gneisses and granitoids of the Colombian Central Andes: Journal of South American Earth Sciences, v. 21, p. 355–371, doi: 10.1016/j.jsames.2006.07.007.

Viramonte, J.M., Becchio, R.A., Viramonte, J.G., Pimentel, M.M., and Martino, R.D., 2007, Ordovician igneous and metamorphic units in southeastern Puna: New U-Pb and Sm-Nd data and implications for the evolution of northwestern Argentina: Journal of South American Earth Sciences, v. 24, p. 167–183, doi: 10.1016/j.jsames.2007.05.005.

Vujovich, G.I., Van Staal, C.R., and Davis, W., 2004, Age constraints on the tectonic evolution and provenance of the Pie de Palo Complex, Cuyania composite terrane, and the Famatinian orogeny in the Sierra de Pie de Palo, San Juan, Argentina: Gondwana Research, v. 7, p. 1041–1056, doi: 10.1016/S1342-937X(05)71083-2.

Wasteneys, H.A., Clark, A.H., Ferrar, E., and Langridge, R.J., 1995, Grenvillian granulite facies metamorphism in the Arequipa Massif, Peru: A Laurentia Gondwana link: Earth and Planetary Science Letters, v. 132, p. 63–73, doi: 10.1016/0012-821X(95)00055-H.

Wörner, G., Hammerschmidt, K., Henjes-Kunt, F., Lezaun, J., and Wilke, H., 2000a, Geochronology (^{40}Ar/^{39}Ar, K-Ar, and He-exposure ages) of Cenozoic magmatic rocks from northern Chile (18 22°S): Implications for magmatism and tectonic evolution of the Central Andes: Revista Geológica de Chile, v. 27, p. 205–240.

Wörner, G., Lezaun, J., Beck, A., Heber, V., Lucassen, F., Zinngrebe, E., Rössling, R., and Wilcke, H.G., 2000b, Geochronology, petrology, and geochemistry of basement rocks from Belen (N. Chile) and C. Uyarani (W. Bolivian Altiplano): Implication for the evolution of the basement: Journal of South American Earth Sciences, v. 13, p. 717–737, doi: 10.1016/S0895-9811(00)00056-0.

Zandt, G., Leidig, M., Chmielowski, J., Baumont, D., and Yuan, X., 2003, Seismic detection and characterization of the Altiplano-Puna magma body, Central Andes: Pure and Applied Geophysics, v. 160, p. 789–807, doi: 10.1007/PL00012557.

Manuscript Accepted by the Society 5 December 2008

Printed in the USA

The Geological Society of America
Memoir 204
2009

Neogene collision and deformation of convergent margins along the backbone of the Americas

R. von Huene
U.S. Geological Survey, 345 Middlefield Road, Menlo Park, California 94025, USA, and *Department of Geology, University of California–Davis, Davis, California 95616, USA*

C.R. Ranero
Institució Catalana de Recerca i Estudis Avançats (ICREA) at Instituto de Ciencias del Mar-CMIMA, Consejo Superior de Investigaciones Científicas (CSIC), Pg Maritím de la Barceloneta 37-49, 08003, Barcelona, Spain

ABSTRACT

Along Pacific convergent margins of the Americas, high-standing relief on the subducting oceanic plate "collides" with continental slopes and subducts. Features common to many collisions are uplift of the continental margin, accelerated seafloor erosion, accelerated basal subduction erosion, a flat slab, and a lack of active volcanism. Each collision along America's margins has exceptions to a single explanation. Subduction of an ~600 km segment of the Yakutat terrane is associated with >5000-m-high coastal mountains. The terrane may currently be adding its unsubducted mass to the continent by a seaward jump of the deformation front and could be a model for docking of terranes in the past. Cocos Ridge subduction is associated with >3000-m-high mountains, but its shallow subduction zone is not followed by a flat slab. The entry point of the Nazca and Juan Fernandez Ridges into the subduction zone has migrated southward along the South American margin and the adjacent coast without unusually high mountains. The Nazca Ridge and Juan Fernandez Ridges are not actively spreading but the Chile Rise collision is a triple junction. These collisions form barriers to trench sediment transport and separate accreting from eroding segments of the frontal prism. They also occur at the separation of a flat slab from a steeply dipping one. At a smaller scale, the subduction of seamounts and lesser ridges causes temporary surface uplift as long as they remain attached to the subducting plate. Off Costa Rica, these features remain attached beneath the continental shelf. They illustrate, at a small scale, the processes of collision.

INTRODUCTION

When ridges and oceanic features with thicker than normal crust subduct along continental margins, the resulting tectonic process has been termed "collision." If the collision involves large tracts of thick crust that are only partially underthrust, the unsubducted portion will add to the continent. The collision of oceanic ridges or moderate-size plateaus commonly results in subduction of the entire body, and the associated subduction erosion leaves a gap in the geologic record. Collisions occur locally adjacent to particularly high coastal mountains, and a relation between collision and mountain building is commonly inferred but incompletely understood. We illustrate and summarize the general features observed from six current collisions along Pacific margins

von Huene, R., and Ranero, C.R., 2009, Neogene collision and deformation of convergent margins along the backbone of the Americas, *in* Kay, S.M., Ramos, V.A., and Dickinson, W.R., eds., Backbone of the Americas: Shallow Subduction, Plateau Uplift, and Ridge and Terrane Collision: Geological Society of America Memoir 204, p. 67–83, doi: 10.1130/2009.1204(03). For permission to copy, contact editing@geosociety.org.

of the Americas (Fig. 1), including the Yakutat terrane and Cocos Ridge, which subduct along North America, and collisions along South America where the Carnegie, Nazca, and Juan Fernandez Ridges and the South Chile Rise subduct. At Nazca Ridge, the marine data are comparatively less extensive, not withstanding the clear uplift of coastal areas.

Subduction of the Yakutat terrane is truly collisional at a continental scale because it adds significantly to the continent, it forms the coastal part of the central Gulf of Alaska, and it is associated with coastal mountains locally more than 5000 m high. Where the Cocos Ridge has collided for 4–5 m.y., the coastal mountains of Central America reach heights of more than 3000 m. Collisions of oceanic ridges that migrate along a continental margin are commonly associated with local coastal uplift and accelerated plate interface erosion, but after these features migrate away from an area, no significant mass addition to the front of the margin is evident; instead, there is a mass deficit (Hampel et al., 2004). At a smaller scale, colliding ridges and seamounts erode the continental slope and the underside of the upper plate, but obvious surface effects soon become obscured.

The collision zones of subducting ridges locally divide an accreting convergent margin segment from an eroding one, e.g., the Juan Fernandez Ridge. The terms erosion and accretion are applied to convergent margins in a variety of ways. Conceptually, accretion adds lower-plate material to the upper plate, and when a net volume is lost from the upper to the lower plate, it is erosional. The period considered commonly includes many seismic cycles in a long-term tectonic environment. However, interplate material transfer can only seldom be quantified along the whole subduction zone, especially at seismogenic depths. So, it is commonly assumed that material transfer observed at the front of the margin indicates its tectonism, which is inferred from landward-dipping reflections. However, dipping reflections in the frontal prism can occur not only in those composed of trench sediment, but also in nonaccretionary frontal prisms composed of repositioned slope sediment. As an example, the erosional Costa Rican margin has a frontal prism structured with landward-dipping reflections like an accretionary prism, but drilling has recovered only repositioned slope sediment (Kimura et al., 1997). This margin's middle slope clearly subsided from basal subduction erosion, as shown by the paleontology and lithology of drill cores (Vannucchi et al., 2003, 2004). Structure of the frontal prism alone is an unreliable indicator of material transfer unless the accretion of trench sediment slices to the frontal prism is imaged. Nonaccretionary frontal prisms, associated with collision, occur off Central and South America, as described later herein.

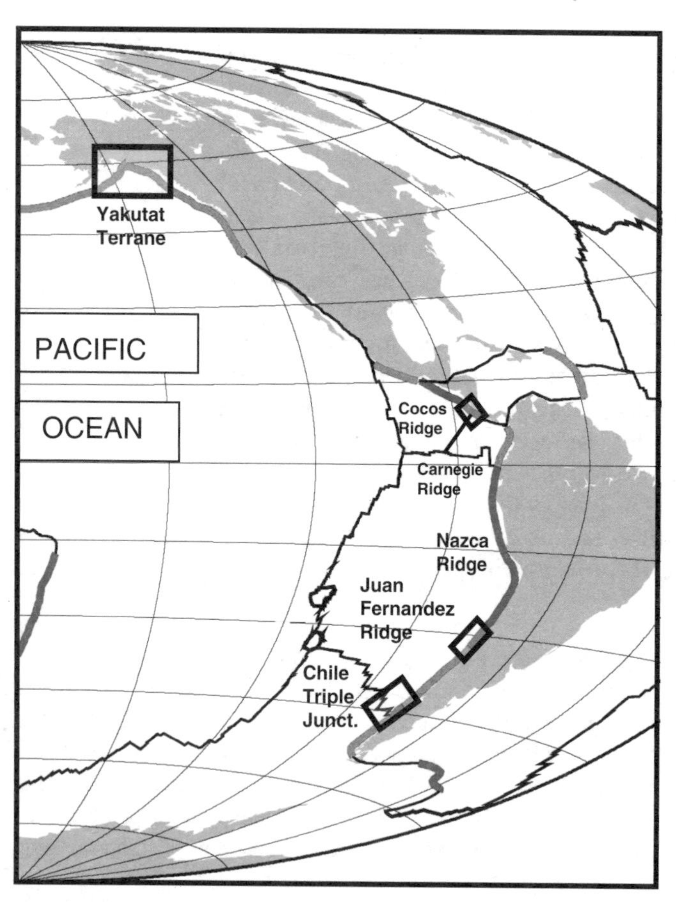

Figure 1. Location of collisions along the American continents. Maps of subsequent figures are indicated by rectangles.

As used in this review, the convergent margin wedge is resolved into a frontal, middle, and inner segment following Kimura et al. (2007), Scholl and von Huene (2007), and von Huene et al. (2009). Both accreting and eroding margin frontal prisms can contain landward-dipping reflections. The rapidly deforming frontal prism is separated by a splay fault or transitional contact from a moderately stable middle prism, which in turn grades into an even more stable inner prism with coherent margin framework structure. This tripartite scheme is obscured in older seismic images, but it characterizes most seismic images acquired with large modern systems and depth-processed data, especially when integrated with multibeam bathymetry. The frontal prism is tectonically most active, whereas the middle prism's permanent deformation is much less. When it consists of margin framework rock, the dominant tectonic process is erosion. In some erosional seismic images, a continuation of framework rock structure extends to the frontal prism, and local drill sampling confirms this (i.e., Vannucchi et al., 2003, 2004). When >1-km-thick sediment fills the trench axis, it is usually associated with frontal prism accretion; when sediment is <1 km thick, the margin is commonly erosional (Clift and Vannucchi, 2004). Sediment abundance is a main controlling factor in accreting or eroding tectonic systems along the convergent margins of the Americas. However, collision overwhelms most other processes, and its effects on sedimentation dominate tectonics for hundreds of kilometers on either side of a ridge axis (Scholl and von Huene, 2008). So, our use of "accrete" and "erode" is based mainly on middle prism character, the margin's multibeam morphology, and trench sediment abundance over a million years or more.

Here, we review and summarize mostly studies over the past two decades as interpreted in a current understanding of convergent margin processes. The descriptions of each collision zone are focused summaries of previous work, in particular, the compilation of data along the Chilean margin by Ranero et al. (2006) and the extensive work offshore Central America by von Huene et al. (2000) and Ranero et al. (2006). A summary of the Yakutat terrane collision includes several recent publications as of this writing, as cited later.

COLLISION OF THE YAKUTAT TERRANE

The Yakutat terrane forms the central Gulf of Alaska coastal mountains and lowlands. It is the most recent addition of terranes to the continent (Figs. 1 and 2) and a focus of studies since the great Alaska earthquake of 1964. Earlier work is summarized in the *Decade of North American Geology* (DNAG) volume edited by Plafker and Berg (1994).

Observations

Underlying the coastal mountainous of the central Gulf of Alaska, the Yakutat terrane is a composite oceanic and continental block of 15- to 20-km-thick crust (Fuis et al., 2008) that is moving northward along the Fairweather fault with the Pacific plate (Fig. 2). This thick crust entered the paleo–Aleutian Trench and continues to subduct beneath North America. Its thick sedimentary cover forms a foreland fold-and-thrust belt that has received 5-km-thick synorogenic sediment for ~5.5 m.y. along Alaska's high coastal mountains (Lagoe et al., 1993; Rea and Snoeckx, 1995; Berger et al., 2008; Wallace, 2009). The Alaska convergent margin north of the Yakutat terrane has a history of subducting vast tracts of oceanic plates beneath North America during Cenozoic time (cf. Howell et al., 1987; Plafker et al., 1994). Central Alaska is composed of terrane segments that migrated from southern latitudes (i.e., Jones, Silberling; Hillhouse, 1977). The Yakutat terrane, as the currently colliding terrane, is perhaps a model of past terrane collision. The subducted leading edge is at least 550 km north of the trailing edge (Transition fault; Fig. 2), where seismic images show an anomalous crustal slab extending to the Alaska Range (Fuis et al., 2008; Eberhart-Phillips et al., 2006; Ferris et al., 2003).

The subducted area of the Yakutat terrane is known from seismic refraction (Brocher et al., 1994; Fuis et al., 2008) and from analyses of earthquake records (Ferris et al., 2003; Eberhart-Phillips et al., 2006). From compilations in reports of these studies, the terrane has an area of ~1600–2015 km^2 or roughly twice the area of the Iceland large igneous province. The subducted segment has a flat-slab–configured segment associated with an absence of volcanism between the Aleutian chain of volcanoes and the Wrangell volcanic field. Of the exposed Yakutat terrane, the western two-thirds consist of a large igneous province faulted against a continental terrane of metamorphosed rock in the eastern one-third. The terrane, an anomalous low-velocity

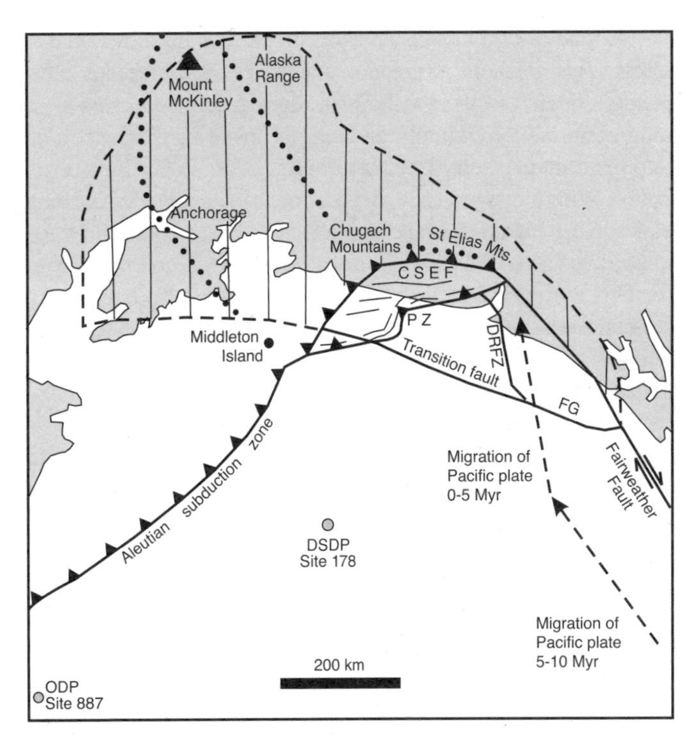

Figure 2. Map of the central Gulf of Alaska simplified to show main structure and the exposed and subducted (vertical lined area) sections of the Yakutat terrane. Dashed lines show subducted area estimated by Eberhart-Phillips et al. (2006), and dotted lines show that estimated by Fuis et al. (2008). Deformation of Yakutat terrane between the Panplona zone (PZ) and the Chugach–St. Elias fault (CSEF) has accelerated during the past 3.5 m.y. Current deformation at Yushin Ridge and at the foot of Fairweather Ground (FG) may show beginning of plate reorganization. Shifting the terrane to the 10 Ma point positions its leading edge near the current coast. Core records that show the increased sedimentation from uplift of coastal mountains are Deep Sea Drilling Project (DSDP) Site 178 and Ocean Drilling Program (ODP) Site 887. DRFZ—Dangerous River fault zone.

unit from 8 to 12 km thick (Brocher et al., 1994), was recently interpreted as a 15–20-km-thick doubled layer (from other refraction data; Fuis et al., 2008). At its northern subducted end, Ferris et al. (2003) estimated an 11- to 22-km-thick low-velocity zone that subducted to 150 km depth. Such structure is similar in velocity and thickness to the larger large igneous provinces and appears to be similar in structure to an off-axis ocean plateau (Coffin et al., 2006). Eberhart-Phillips et al. (2006) and Ferris et al. (2003) indicated subduction of the low-velocity body roughly 600 km inland from the continental slope. The outline of the subducted Yakutat terrane as interpreted by Eberhart-Phillips et al. (2006) is based on inversion of selected earthquake data (Fig. 1), and that of Fuis et al. (2008) is the smallest of other area estimates (Wallace, 2009). The subducted extent of the Yakutat terrane roughly matches the mountains of coastal Alaska. Its crustal thickness and the ~10-km-thick sediment deposited on it indicate that its southern part was once a shallow-water area (Plafker, 1987; Wallace, 2009).

A time of initial collision at ca. 10 Ma can be shown in different ways. First, an assumption that the Yakutat terrane is completely coupled to the Pacific plate during collision provides a minimum value by simply dividing the plate convergence rate into the terrane extent (Eberhart-Phillips et al., 2006; Fuis et al., 2008). With a convergence of ~6 mm/yr, the subducted leading edge of the Yakutat terrane crossed the current position of the margin at ca. 10 Ma. That estimate does not account for contraction during collision. Wallace (2009) carefully balanced a structure section in the central St. Elias Mountains that shows only ~36 km of shortening in 3 m.y., or ~1/7 of the plate convergence. He argued that the surface contractile structure is insufficient to accommodate plate convergence and can be explained by duplexing at depth. Berger et al. (2008) estimated erosion of an exhumed mass of 193 km^2 across the coastal mountains, which, if mass balanced, would account for one-third of the ~6 mm/yr plate convergence, assuming a Yakutat terrane coupled to the Pacific plate during collision. Frontal erosion of the upper plate during subduction was not added to the estimates. Therefore, the simple geometric time estimate might be 1 m.y. short.

The estimated time of collision is also shown by analyses of cores recovered during scientific drilling at two sites (Deep Sea Drilling Project [DSDP] Site 173 and Ocean Drilling Program [ODP] Site 887) in the Gulf of Alaska (Fig. 1). Uplift of coastal mountains resulting from collision that produced tidewater glaciers is indicated by erratic clasts in sediment of the gulf. The source area is constrained to only the collision zone by the Aleutian Trench, which traps sediment from the west. The deep-sea channels south of the Yakutat terrane lead away from the drill sites to the east. Erratic clasts are recorded in cores from both drill sites at ca. 10 Ma (Rea and Snoeckx, 1995). The resolution at Site 887 allowed separation of glacial and interglacial sedimentation. Increased numbers of dropstones and sand released from melting icebergs appear in cored material dated between 5 Ma and 3 Ma, followed by a dramatic increase (Rea and Snoeckx, 1995). In Site 887 cores, the interglacial periods have sufficient resolution for the accelerated nonglacial sedimentation from increased mountain relief to be separated from the glacially accelerated sedimentation. Increased sedimentation is seen in all other cores from scientific drill sites, although ages were not as sharply constrained by biostratigraphy prior to 1971 (DSDP Sites 173, 178, and 183). The sharp increase in sedimentation is consistent with the estimates of earliest age for the Yakataga Formation, which is interpreted to mark the beginning of mountain uplift at 5.9 Ma (Lagoe et al., 1993). Rising coastal relief and tidewater glaciers are explained by collision perhaps beginning ~10 m.y. ago, which at 6 Ma resulted in high mountains and accelerated mountain building at 3 Ma.

Ideas Regarding Tectonism

Motion on the Transition fault has been argued for 20 yr, involving whether the Yakutat terrane was coupled to the Pacific plate during the past 10 m.y. or is decoupled by partial plate motion along the terrane's trailing edge (Plafker 1987; Bruns, 1985, 1984, 1983; Pavlis et al., 2004; Gulick et al., 2007). Arguments for little plate motion on the Transition fault during the past 10 m.y. are as follows. A consistent record of lateral displacement on the trailing edge contact is not evident in seismic dip lines along the margin (Bruns, 1985). In multibeam bathymetry, only two small areas of recent deformation at the base of the slope are evident (Fig. 2) (Gulick et al., 2007). Locally, thick undeformed sediment covers the trace of the Transition fault structure (Jaeger et al., 1998). At its eastern junction with the Fairweather fault, no evidence of interaction between two lateral fault systems has been resolved in either seismic or bathymetric data (Gulick et al., 2007). In the juncture area, the through-going Fairweather transform fault consists of two or three linear faults for 100 km northwest (Bruns and Carlson, 1987). Similarly, at its western subducted end, a projection of the Transition fault shows no corresponding transcurrent faults at the seafloor or in seismic images and no current seismicity from a tear fault in earthquake compilations (Ratchkovski and Hansen, 2002; Eberhart-Phillips et al., 2006). Lateral displacement is not evident in the Yakutat terrane foreland fold-and-thrust belt (Pavlis et al., 2004). Pavlis et al. (2004) suggested that the undeformed sediment across the fault may be too young to show deformation, but they admit that this requires uppermost limits on rates of sedimentation, rates many times greater than those in the nearby Alaska/Aleutian Trench. We consider the evidence favoring coupling of the Yakutat terrane with the Pacific plate for the past 10 m.y. to be most reasonable, thus resulting in a rate of subduction essentially equivalent to the plate convergence rate.

Seismic and sidescan images (Bruns, 1997), multibeam bathymetry (Gulick et al., 2007), and historic seismicity along the Transition fault show current local tectonism (Fig. 2). To explain this activity, we propose that an ever-increasing resistance to terrane subduction, as more of the Yakutat terrane subducts, strains the Transition fault and that, eventually, the interface fault could jump seaward from the Chugach–St. Elias area to the Transition fault. Fisher et al. (2005) have shown a seismic image indicating faulting within the subducted Yakutat terrane near the epicentral area of the great 1964 Alaska earthquake, and Fuis et al. (2008) interpreted fragmentation and underplating of the subducted Yakutat terrane in the same area. So, strain is great enough to break the lower plate, and duplexing has also been proposed (Wallace, 2009).

We propose that the Yakutat terrane migrated north coupled to the Pacific plate at least during the past 10 m.y. and began to subduct beneath Alaska. The increasing resistance to subduction as greater areas of more buoyant crust were underthrust over 10 m.y. caused deformation of the subducting plate down-dip of the Panplona zone (Fig. 2). Resistance to subduction causes interplate faulting to propagate toward segments of the Transition fault as part of a current plate reorganization, as proposed by Gulick et al. (2007). Subduction of the Yakutat terrane has generated a flat slab, a gap in the volcanic arc, and the high coastal mountains surrounding the central Gulf of Alaska. If the plate interface jumps seaward, then in the future, a terrane boundary that dips landward (Berger et al., 2008) or seaward (i.e., Fuis et al., 2008; Wallace, 2009) could be exhumed, and when erosion reaches greater depth, a series of imbricate structures containing metamorphosed subducted material that is currently beneath the foreland fold-and-thrust belt could become part of the exposed geology. Such a geology is similar to the Cretaceous rocks exposed on the Kodiak group of islands (Sample and Moore, 1987) and is known elsewhere along margins of North America.

SUBDUCTION OF COCOS RIDGE BENEATH COSTA RICA

Cocos Ridge extends from the Galapagos hotspot to Osa Peninsula (Figs. 1 and 3). We summarize our discussion in a compilation of Central American studies (Ranero et al., 2007) and then summarize the extensively studied vertical motion on land associated with this collision (cf. Gardner et al., 1992).

Observations

At the base of the Costa Rican margin, the 20-km- to 25-km-thick Cocos Ridge crust was produced ~14 m.y. ago by intrusion of magma and extrusion of volcanic materials from the Galapagos hotspot (Werner et al., 1999). Hotspot magmas were underplated and vented through an ocean crust that formed at the Cocos-Nazca spreading center (Barckhausen et al., 2001). Near the base of the margin, the ridge crest is a graben trending parallel to the ridge with up to 1-km-high normal fault scarps (at bottom of Fig. 4). The crestal graben indicates extension normal to the ridge axis, and it can be traced 200 km seaward toward the Galapagos Islands. Sediment in the graben is ~1 km thick adjacent to the slope and is little deformed (Hinz et al., 1996). On its northwest flank, the seafloor has small ridges, seamounts, and the Quepos Plateau (Fig. 4). Samples from seamounts and the plateau that were isotopically dated are 15–13 m.y. old (Werner et al., 1999) and were emplaced on 21–18 m.y. old lithosphere (Barckhausen et al., 2001). Heat flow recorded with multiple probe measurements shows an elevated temperature on the ridge compared with the host lithosphere (Grevemeyer et al., 2004). The age and geochemistry of the seamounts are essentially the same as the rock forming the Cocos Ridge crest, indicating coeval Galapagos hotspot emplacement. Laterally adjacent to the ridge, the normal 7–8-km-thick ocean crust (Ye et al., 1996; Walther, 2003) bends into the trench more steeply than thickened crust of the ridge crest. This explains the extensional tectonics forming the graben parallel to the ridge axis. Where Cocos Ridge subducts beneath the margin, the base of slope is 1 to 2 km deep, whereas 200 km northwest, it reaches 5 km depth. Ridge collision has introduced a broad swell ~2 to 3 km high into the subduction zone. However, the area of significantly thickened crust represents only part of the ocean crust affected by hotspot magma, and a sill of Galapagos composition was found at ODP Site 1039 some 250 km northwest of the crest (Kimura et al., 1997). Despite the lack of clear topographic expression, the hotspot magma forming the ridge has affected a much broader area

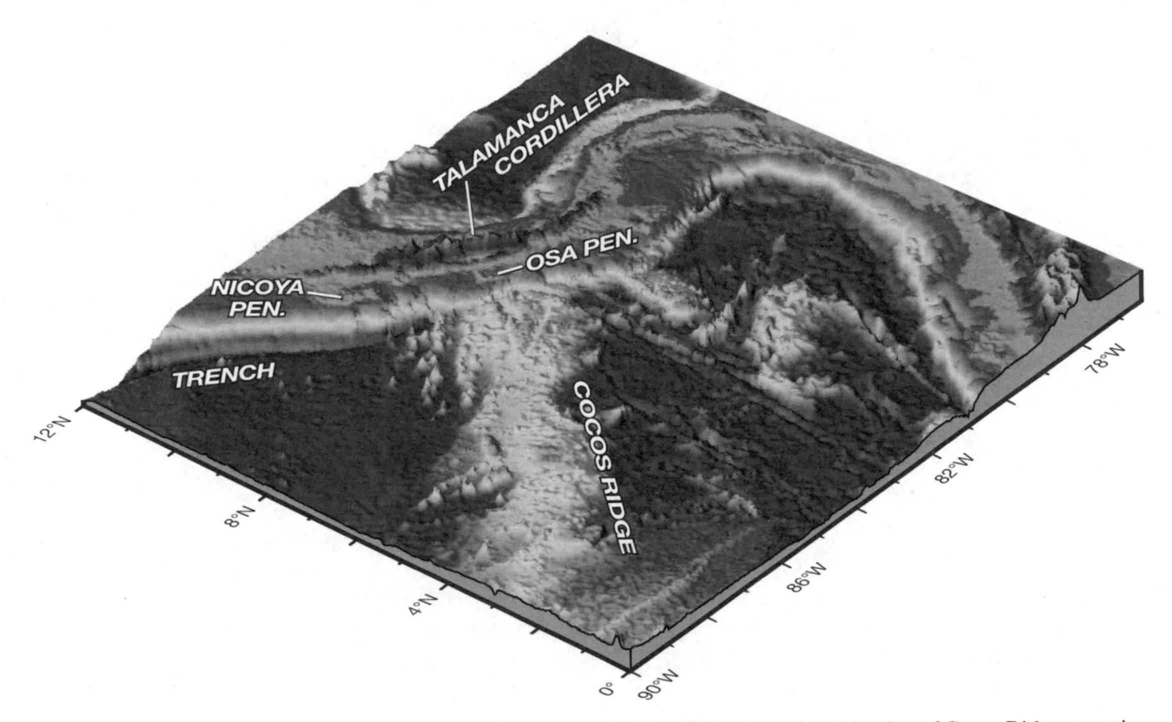

Figure 3. Regional topography and bathymetry off Costa Rica looking ENE showing subduction of Cocos Ridge opposite Osa Peninsula. Cocos Ridge and its associated seamounts are in the foreground. The high mountains opposite the colliding ridge (Talamanca Cordillera) are landward of the collision area.

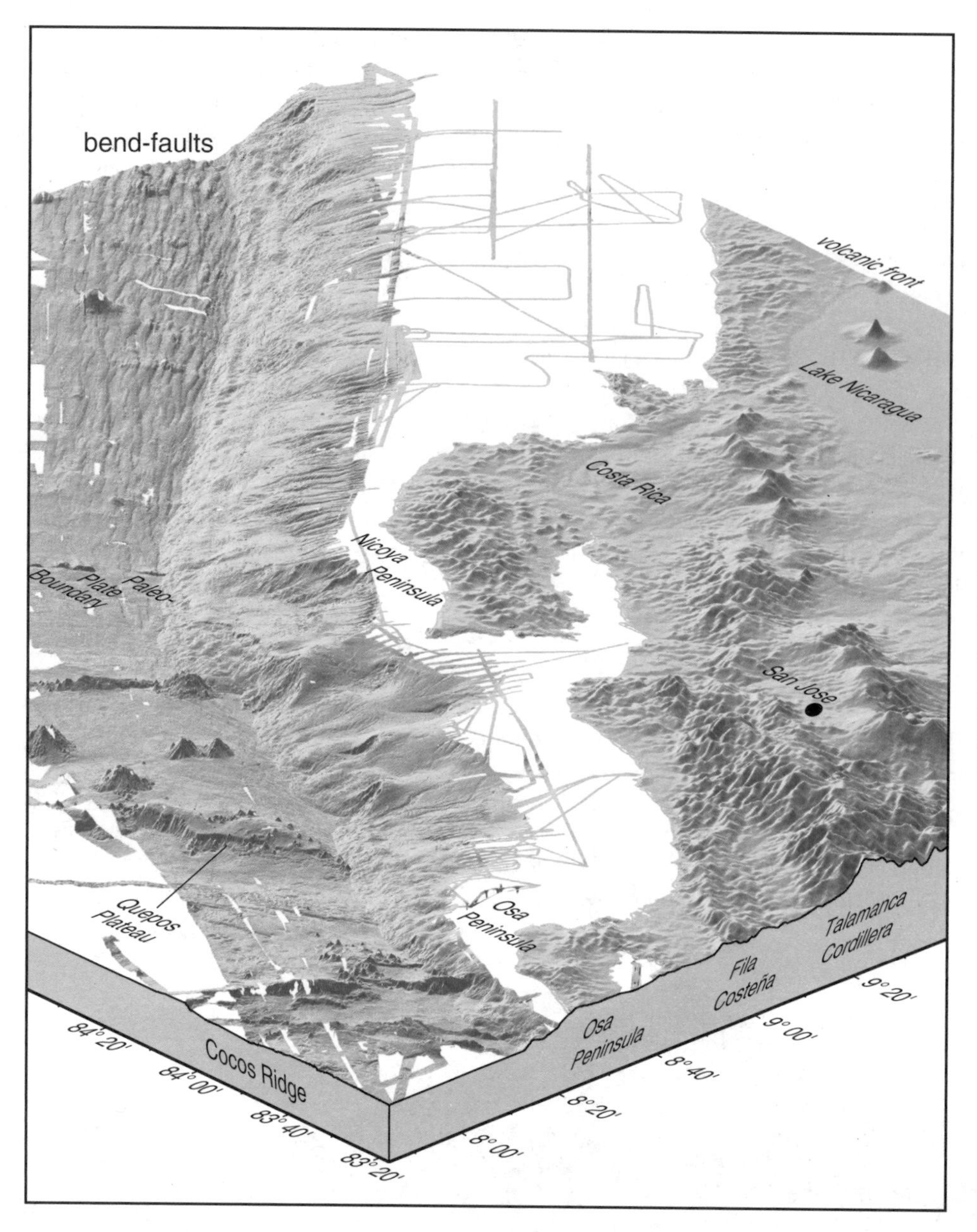

Figure 4. Perspective of multibeam bathymetry and topography looking NW. This shows the retreat of the continental slope where abundant relief associated with the Cocos Ridge subducts as compared to the NW area of horst and graben, where little relief is subducted. The Cocos Ridge crest is marked by the sediment-filled graben paralleling the ridge axis. Since convergence is essentially parallel to the ridge, it has not migrated significantly in the past 3–5 m.y.

of the adjacent seafloor than is apparent from morphology and geophysical data. Cocos Ridge in its present form is estimated to have arrived at the southern Costa Rica margin ~3–5 m.y. ago based on paleontological and geochemical studies of unroofing and uplift (Gräfe et al., 2002; Collins et al., 1995). Because it trends at only a small angle to the convergence vector, the area of collision has been essentially stationary for roughly 3–5 m.y.

Seamounts on the ocean plate associated with the ridge are nominally 20 km across and 2 to 2.5 km high. Seismic images of seamounts subducted to 10 km depth are not resolved well enough to give a precise height, but the seamount's relief appears to be roughly equivalent to that of the exposed ones. Earthquakes and their tightly clustered aftershocks have been recorded above subducted seamounts and on a projection of the subducting Quepos Plateau, indicating that ocean relief remains attached to the subducting Cocos Ridge at least to the shore (DeShon et al., 2003). An extension of the plateau remains attached in the seismogenic zone, as do seamounts such as the one that was the asperity for the 1992 Cobano earthquake (Bilek et al., 2003). Ocean floor relief constructed on crust with only thin pelagic sediment as off Costa Rica could be more firmly attached to the lower plate than relief built on thick ocean floor sediment. Most seamounts sampled returned largely pyroclastic materials, and subduction of an edifice constructed of such weak materials indicates a weak interplate fault (von Huene et al., 2000).

On land, studies of the Quaternary geology along the coast document uplift above subducted seamounts, and a very well-documented example is the uplift at the southern end of the

Nicoya Peninsula (Gardner et al., 2001). Uplift rates of up to 8 mm/yr near the inferred top of the seamount have been reported. Tilted wave-cut erosional terraces are produced even where the upper plate is 25 km thick; however, the tilt is attenuated because of the upper-plate thickness above the subducted relief.

Above the subducted Cocos Ridge, there is the broad uplifted arch of the Osa Peninsula and its adjacent seafloor (von Huene et al., 2000). Uplift is also well defined by uplifted wave-cut terraces on land (Fisher et al., 2004; Gardner et al., 1992; Sak et al., 2004; Sitchler et al., 2007, 2009). A topographic lineament across the peninsula on a projection of the crestal graben reveals that large features of the ridge crest, despite attenuation, are perhaps still visible in the topography of the peninsula (Fig. 4, bottom). Extensive identification and dating of raised wave-cut platforms on Osa Peninsula show uplift rates of 8 mm/yr, as well as a period of subsidence (Sak et al., 2004). Uplift and erosion expose the Osa mélange (cf. Vannucchi et al., 2006), which was buried at depths where temperatures are 110 °C. Rapid uplift began ca. 3 Ma and is greatest where the crest of Cocos Ridge subducts. About two-thirds of the 3- to 5-km-thick uplifted sections have been eroded, exhuming the mélange. Rates of uplift are greatest on a projection of the subducting crest of the ridge.

Landward of the peninsula and opposite the full width of the hotspot-affected ocean crust, the Fila Costeña fold-and-thrust belt displays active contraction along as many as five thrust slices. The Fila Costeña extends parallel to the regional trend and rises against the Talamanca Cordillera (Figs. 3 and 4). The Talamanca mountains reach 3700 m elevation (Gardner et al., 1992) and expose midcrustal granodiorite plutons and abundant Miocene intrusive rocks (de Boer et al., 1995). The maximum elevation of the Talamanca, the maximum deformation of the Fila Costeña, and the highest crest of Osa Peninsula are aligned with the crest of Cocos Ridge. Arc volcanism is now almost completely inactive. Large earthquakes of Mw 7–7.5 have nucleated here, but in the past 20 yr, interseismic earthquakes are fewer than in adjacent segments (Protti et al., 1995). As would be expected, subduction of thickened crust is associated with more rapid uplift than above the thinner crust under the lower seafloor relief on the northwest flank.

An ~230-km-long wide-angle seismic transect across southern Costa Rica just north of Osa Peninsula extends from the Pacific basin to the Caribbean slope (Stavenhagen et al., 1998). Margin wedge seismic velocities increase landward in the common fashion. A moderate velocity reversal occurs within the wedge above the plate interface, probably from a subduction channel containing fluid-rich material (Ranero et al., 2008). The plate interface can only be followed to a depth of nearly 40 km beneath the Talamanca Cordillera. The Moho could not be identified unambiguously. The refraction section (Stavenhagen et al., 1998) and earthquakes (Protti et al., 1995) show no evidence of a flat slab as occurs along other margins where ridges subduct, and a volcanic front is absent. A receiver function study (Dzierma et al., 2007) is consistent with the refraction interpretation. However, beyond the 40-km-deep plate interface where the junction between the Fila Costeña and Talamanca Cordillera occurs, plate interface dip increases from 16° to 30° and is imaged to 170 km depth. Under these conditions, the usual explanation for extinguished volcanism from a flat slab is problematic.

Ideas Regarding Tectonism

The pronounced landward shift in the continental slope along which Cocos Ridge subducts indicates accelerated retreat of the central Costa Rica margin opposite the subducting ridge flank and its associated seamount-studded seafloor. If so, the margin has retreated ~25 km more than the Central America margin to the northwest in ~3–5 m.y. (Fig. 4). Coeval with accelerated erosion, uplift of Osa Peninsula above sea level has occurred. Before that uplift, the peninsula probably had a cover represented by the section now exposed on the peninsula. The cover of younger sediment and other materials has been eroded, exposing Miocene to Eocene rocks (Vannucchi et al., 2006); however, the depth to which these rocks were once buried is still debated (Buchs and Baumgartner, 2003). The Cocos Ridge crest subduction appears to be accommodated by block tilting without obvious compressional faulting and folding of the margin slope, except for a small fold at the deformation front where graben sediment subducts. The collision area of the margin opposite Cocos Ridge is disrupted less than that where seamounts subduct, indicating a low-friction environment (Fig. 4). The thick low-velocity layer, thought to be extrusive volcanic material, may have higher than normal porosity. One explanation for low friction is an abundance of fluid.

The lower-plate relief subducting beneath the Costa Rica margin is of various heights and lengths. Smaller features show the beginning stages of collision. Subducting seamounts leave a trail of local uplift and subsidence across the slope and shelf that are quickly obscured by sedimentation. They uplift the land surface even where the upper plate is 15–20 km thick, and attenuation of deformation above a subducting seamount reduces the surface uplift to ~1/10 of the seamount height. Subducted seamounts associated with the Cocos Ridge form asperities for earthquakes up to M 6.5 (Bilek et al., 2003).

Ridges also form asperities for earthquakes like the 1999 M6.9 earthquake north of the Osa Peninsula (DeShon et al., 2003). This earthquake nucleated on the probable subducted extension of Quepos Plateau. The plateau subducts without much disruption of the slope and causes local slides at the shelf edge associated with an indentation indicating loss of material from subsurface tectonism. On the shelf, it raises the seafloor in a very low ridge that is not visible with conventional bathymetry.

The crestal graben morphology of Cocos Ridge subducts easily at least at the scale of multibeam bathymetry (~100 m), and its morphology is reflected in topography across the peninsula where the upper plate is relatively thin. The ridge's thick axial crust bends less than thinner ridge flank crust as they enter the subduction zone, as shown by the shallower dip beneath the crest than the flank (Protti et al., 1995). Farther inland, compressional deformation of the upper plate is greater above the ridge crest than above its flanks. Compressional structure begins at the Fila Costeña, indicating where subduction zone friction becomes strong enough to significantly deform the upper plate. As this collision indicates, when positive ocean relief has subducted in one spot for 3–5 m.y., significant upper-plate deformation and mountain building results, as long as the ridge does not leave the area.

CARNEGIE RIDGE COLLISION

Carnegie Ridge collides with the South American margin off central Ecuador where the margin bows westward (Fig. 1). Carnegie and Cocos Ridges both formed at the Galapagos hotspot, but their rough ridge morphologies differ in detail. Carnegie's rough seafloor crest contains scattered volcanoes not found on the Cocos Ridge crest, and it lacks the axial graben and wide zone of flanking seamounts. Clear embayments into the lower slope were probably formed by subducting seamounts on the ridge. Where the rough seafloor subducts, the adjacent steep and narrow continental slope is correspondingly rough compared with a smoother slope opposite the ridge flanks (Sage et al., 2006; Calahorrano et al., 2008). The adjacent coast is bordered by a chain of low coastal mountains in the area of collision. Where the ridge axial area subducts, the Andes reach heights over 2000 m. Active volcanoes along the Andes are an exception to the inactive volcanism generally associated with collision.

Seismic images with outstanding resolution and multibeam bathymetry have recently been reported (Sage et al., 2006; Calahorrano, 2008). This erosional margin has a frontal prism structured with landward-dipping reflectors indicating compressional deformation. That prism structure contrasts with the structure in the middle prism, where normal faults that cut the entire plate show extension (Sage et al., 2006; Calahorrano et al., 2008). The plate interface produces a strong reflection, which indicates concentrated fluids in the fault zone. Beneath the plate interface, a stratified layer of uniform thickness represents the subduction channel. Where the subducting plate has a rough seafloor, the subduction channel thickness is irregular, and thick pockets of sediment are separated by stretches where little sediment is resolved (Sage et al., 2006). It is proposed that in the ridge axis area, a plate-interface fault surface of variable sediment thickness produces a patchwork of various frictional behaviors. The area of collision is shown to have a relatively large release of strain during earthquakes compared to the areas that flank it (Gutscher et al., 2000). The narrow area of flattened slab shown by Gutscher et al., (2000) contrasts with the absence of one associated with Cocos Ridge collision (Dzierma et al., 2007), despite a common origin at the Galapagos hotspot.

NAZCA RIDGE COLLISION

The association of no arc volcanism, a flat slab, and ridge collision was noticed along the Peruvian margin in the 1970s. Nazca Ridge subducts obliquely beneath Peru, and we briefly summarize marine studies spanning ~30 yr (cf. Kulm et al., 1981; Hussong and Wipperman, 1981). Early studies focused on accretion or nonaccretion of trench sediment, and interpretation of tectonic structure comes principally from early 1970s seismic acquisition, followed in 1986 with drilling on ODP Leg 112 (Suess et al., 1988). Scientific drilling settled the accretion or nonaccretion controversy and provided data to quantify material flux and rates of elevation change (cf. von Huene et al., 1998; Clift et al., 2003). In 2000, a part of the subducting Nazca Ridge and the lower slope was mapped with digital multibeam bathymetry and single channel seismic records that showed ridge and slope sediment thickness (Hampel et al., 2004).

Drilling showed the rapid uplift and subsidence from ridge subduction superimposed on an erosional margin (von Huene et al., 1996). Subsequently, Clift et al. (2003) quantified seafloor and plate-boundary erosion. The subducted segment of Nazca Ridge is inferred from its mirror image in the Tuamotu Ridge, which formed on the opposite side of the spreading East Pacific Rise. The inferred subducted ridge segment constrained the southward migration history of collision (von Huene et al., 1996; Clift et al., 2003). The 2–4-km-high subducting ridge axis has migrated 800 km since middle Miocene time, leaving a continental shelf and slope that are wider north of the subducting ridge than to the south. The collision first uplifted the margin and eroded both the frontal prism and the base of the upper plate, and when the ridge had left an area, the margin subsided. That subsidence has been associated with development of elongate trench-parallel forearc basins. The frontal prism destroyed by collision was rebuilt in 4 m.y., but it grew only to its original width (von Huene et al., 1996).

The Nazca Ridge axis has crust 17 km thick (Hampel et al., 2004). Even in conventional bathymetry, the ridge outline is irregular, and its rough morphology has been resolved in a small area of digital multibeam bathymetry (Hampel et al., 2004). It shows a rough topography with isolated volcanoes, as expected from a hotspot origin, covered by up to 400 m of sediment. As the ridge bends into the trench, it develops trench-parallel normal faults with scarps reaching 200 m (Hampel et al., 2004). The ~2.5-km-high relief of the ridge adjacent to the trench decreases to 1.5 km in the trench axis, and since trench-axis sediment transport is away from the ridge crest, turbidite trench fill is very sparse. The adjacent continental slope steepens during collision, and a 50 km point of land develops over the subducting ridge relative to the precollision slope. Seismic velocities indicate that the coherent continental framework basement extends to within 15 km of the trench. Subsurface structural deformation is not well imaged, but from multibeam bathymetry, ridge subduction appears to be accommodated readily, as at Cocos Ridge. Like the slope, the coast here protrudes 50 km seaward and is steep and rugged. Its uplift is shown by exposed young marine sediment sections that are correlated with those offshore on the basis of a distinctive siliceous and phosphatic lithology (Dunbar et al., 1990). Fossil marine terraces are up to 350 m high (Hsu, 1992), and subaerial uplift extends ~75 km inland. It has even been proposed that uplift extends into the Amazon Basin (Espurt et al.,

2007). However, an absence of coastal uplift to the north where the ridge once passed shows that subsidence follows uplift that is consistent with the 50-km-wide coastal indentation affected by ridge subduction. In one drill core (ODP Site 682), an unconformity over upwelling sediment characteristic of water depths <500 m indicates ~3 km of subsidence after uplift to wave base (Suess et al., 1988).

JUAN FÉRNANDEZ RIDGE SUBDUCTION

The Juan Férnandez Ridge morphology consists of a gentle swell crested by isolated large seamounts. It was formed by a hotspot that is now ~900 km west of the Chile Trench. The chain of aligned seamounts trends ~85°E, and a 250-km-wide gap between seamounts separates the distant younger group of seamounts and the older ones near the trench. There, the O'Higgins seamounts have ages of ca. 8 Ma and sit on ocean crust ca. 39 Ma in age (Fig. 5). The ridge first collided with the Chile margin in the north at ca. 22 Ma, after which it migrated to its current collision point located at roughly 32–33°S (Yañez et al., 2002). Subduction of this ridge has a spatial correlation with a flat slab and a lack of active volcanism. Its current area of subduction contains Valparaiso Basin, which is the only significant deep-water forearc basin along the central Chilean margin (Laursen et al., 2002).

Observations

Several surveys to acquire multibeam bathymetry have resulted in a high-resolution map of morphotectonic and sedimentary structure developed during ridge subduction (Fig. 5). Structure associated with the ridge strikes NE and is accentuated by vertical displacement on long faults where the ocean crust bends into the trench. The faults parallel the broad swell of the ridge and are assumed to have originated from development of the ridge. The ridge structural trend continues beneath the trench axis and then across the entire continental slope to the shelf. Structure of the collision, subduction of Juan Fernández Ridge, and slope readjustment as the ridge moved southward are also imaged seismically (Laursen et al., 2002; Ranero et al., 2006). However, Juan Fernandez Ridge is not associated with an anomalously thick crust, but only with a small local area of thicker than normal crust (Kopp et al., 2004).

At the current area of collision, the Juan Fernández Ridge erodes the frontal prism and uplifts the margin as it subducts beneath the continent (von Huene et al., 1997). The ridge is a topographic barrier to northward sediment transport along the trench axis (Fig. 5). This barrier separates a trench with more than 2.5-km-thick turbidites to the south from one with turbidites <1 km thick the north (cf. Schweller et al., 1981; Ranero et al., 2006). Thick trench sediment is associated with subduction accretion, and thin sediment is associated with subduction erosion.

Seaward of the trench axis, the structural fabric imparted to the oceanic crust during generation at a spreading ridge, and the superimposed hotspot intrusion, are revealed in multibeam bathymetry (Fig. 5). The morphology of the oceanic plate on either side of the ridge and its seamounts has a NW fabric. Near the trench axis, this seafloor-spreading fabric is crossed locally by faults that formed during bending into the trench prior to subduction and that roughly parallel the trench. In contrast, the area affected by hotspot magmatism has few short trench-parallel scarps. The ridge has numerous small volcanoes in addition to the two large O'Higgins seamounts and a NE-elongated swell adjacent to the trench axis. The multibeam bathymetry shows how hotspot volcanism can modify the seafloor-spreading fabric and structurally overprint older crust (Fig. 5).

Bend faulting of the ridge begins ~100 km from the trench axis, and vertical displacement is up to 1 km at the trench axis (Fig. 5). The swath of hotspot-modified lithosphere is ~150 km wide and trends ~55° from the trench axis. On the slope where the ridge axis has subducted, its trend is distinct and appears to heighten after subduction (von Huene et al., 1997). Seamounts, including one subducted seamount that is clearly defined by magnetic anomalies at the lower slope, are aligned with the diagonal ridge trend (Yañez et al., 2002). Microseismicity on the ridge's diagonal trend projects inland from the coast (von Huene et al., 1997; Engdahl et al., 1998).

The ridge has been a barrier to northward axial turbidite transport, which was most abundant during glaciations of the past 1.5–3 m.y. (Oncken et al., 2006). The influence of trench-sediment distribution, shallow subduction angle, and southward ridge migration has dominated tectonism across the continental slope (Fig. 5). Trench sediment abundance south of the ridge promotes accretion, whereas the meager fill north of the ridge promotes subduction erosion.

The subduction of Juan Fernandez Ridge is spatially associated with the flat slab and a lack of active arc volcanism. The central valley of Chile begins south of the subducted ridge beneath the continent. The geology of the flat slab is discussed elsewhere in the volume, and it is clear that collision is important in the structure of the margin.

Ideas Regarding Tectonism

The association of ridge subduction, flat slab subduction, and termination of volcanism is the basis for various hypotheses discussed in this volume by others. Development of the central valley of Chile is also attributed to subduction of the Juan Fernandez Ridge.

As the hotspot-modified Mesozoic ocean crust bends into the trench, the NE-trending faults are reactivated and break into scarps as much as 1 km high. That relief erodes the frontal prism along the lower slope of the continental margin. The uplifted ridge across the margin clearly marks the subducted relief (Fig. 5), and increased lower-plate displacement once it subducts is suggested (von Huene et al. 1997). Bend faulting continues in the subduction zone, and there the fluid flow is impeded by the upper plate. The elevated pore pressure is likely to reduce friction and facilitate lower-plate fault displacement after subduction. The faults

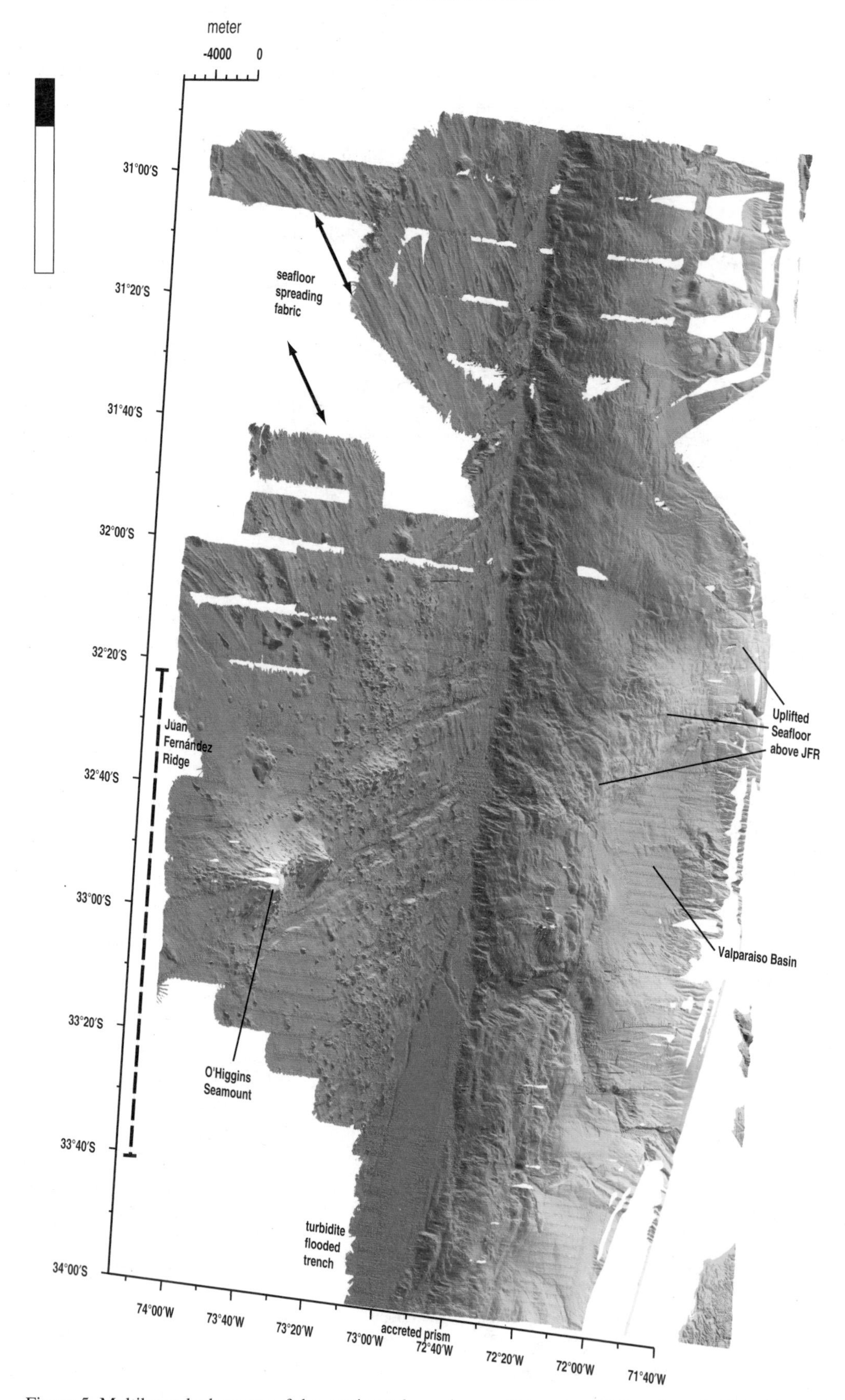

Figure 5. Multibeam bathymetry of the continental margin near Valparaiso, Chile, where the Juan Fernandez Ridge (JFR) collides and subducts. The ridge blocks axial transport of sediment, causing ponding of trench turbidite. Across the slope, the subducting ridge uplifts the overlying seafloor. The zone of collision migrates slowly to the south, where thick sediment is accreted, whereas in the north, the thin trench sediment is associated with subduction erosion.

are also conduits for migration of water into the crust (Kopp et al., 2004).

Disruption of axial sediment transport by the ridge has resulted in a long sediment-starved margin, along with its small sediment supply because of regional aridity since Miocene time. Axial sediment ponding supplies the trench axis south of the ridge with thick turbidite sediment originating from the southern Andes. This division corresponds with accretion in the south and erosion in the north. The erosional margin character of the north is an example of the margin configuration prior to Quaternary accretion south of the Juan Fernandez Ridge barrier. In pre-Quaternary time, the entire Chilean margin is likely to have been erosional.

In the area of collision, interpretation of tectonic history is problematical without temporal control. Ranero et al. (2006) proposed that an accretionary prism was destroyed by Juan Fernandez Ridge collision. After collision and 3–4 m.y. of erosion, the slope morphology and structure reflect the removal of a middle-slope prism (Ranero et al., 2006). A broad diagonal ridge across the margin above the subducted Juan Fernandez Ridge changed structure and morphology on either side as it migrated southward. Since the ridge controlled accretion, the margin has only been accretionary south of the ridge during the past 1.5–3 m.y. Here, middle- and lower-slope morphology consists of margin-parallel ridges that are inferred to consist of trench sediment added to the margin upper plate, forming a 40-km-wide frontal and middle prism. North of the diagonal ridge, the accreted middle- and lower-slope prisms are missing, and the depressed middle slope, its normal faults, and slope failure display a mature erosional margin morphology (Ranero et al., 2006). Alternatively, Laursen et al. (2002) proposed a period of compressional tectonism from subduction of seamounts that reactivated an older landward-dipping sequence of faults to form the outer flank of Valparaiso Basin. Subduction of the ridge also eroded the front of the margin framework. An area along the plate interface eroded and caused subsidence to form the basin depocenter. These scenarios differ between a long or short tectonic history that formed the basin, but a common feature to both scenarios is concentrated deformation over the subducting ridge and subduction erosion.

The modest crustal thickening of the crust beneath Juan Fernandez Ridge is inadequate to provide the significant buoyancy that has been proposed to help in developing the flat slab (Kopp et al., 2004). Wide-angle seismic data indicate an anomalously low upper-mantle velocity that is restricted to the area of reactivated NE-trending fault scarps. The low velocities may indicate mineral alteration by hydration of the upper mantle from water migrating down faults to mantle depths. Up to 20% of the mantle rock may be serpentinized (Kopp et al., 2004), and that would contribute the buoyancy necessary in conversion of steeply dipping subduction zones to a flat-slab configuration.

SOUTH CHILE RISE AND TRIPLE JUNCTION

The South Chile triple junction differs from previous examples in containing an actively spreading ridge that collides with the margin (Cande et al., 1987; Behrmann et al., 1994; Ranero et al., 2006). On either side of the triple junction, plate convergence changes from 20 mm/yr in the south to 90 mm/yr in the north. This difference in convergence is associated with the change in margin tectonism from highly accretionary to erosional (Fig. 6). Farther to the north, the development of accretion from an erosional margin is poorly constrained because prestack depth-migrated seismic data are lacking. Across the subducting triple junction, the change from accretion to erosion occurs without an obvious change in sediment supply from the adjacent glaciated coastal mountains. The trench axis is constricted by the ridge in the area of collision, and sediment flows away from that area both north and south.

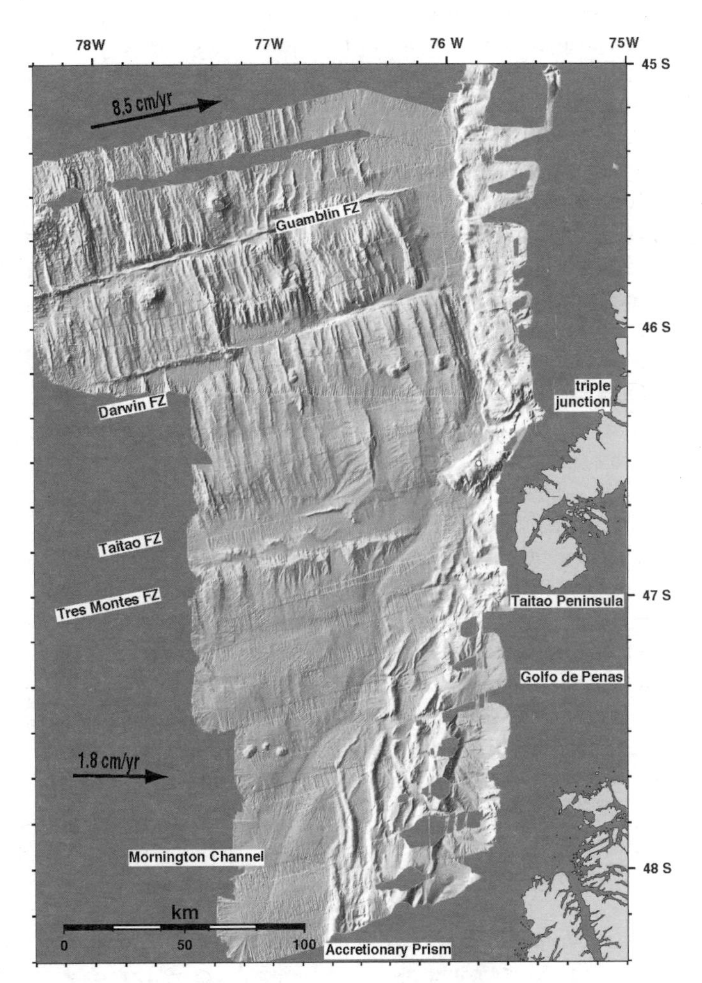

Figure 6. Multibeam bathymetry where the Chile Rise spreading ridge collides and subducts. This map view combines data from cruises of *L'Atalant* and *Sonne* (Bourgois et al., 2000; Ranero et al., 2006). North of the collision area, the seafloor fabric is much less sedimented than to the south because it converges more rapidly and trench sediment is completely subducted. South of the collision, sediment deposition is at the same rate, but the lower plate subducts at one-third the rate. This allows a longer period of sedimentation, and it masks original seafloor fabric morphology. Thicker sediment and slower convergence are associated with accretion, whereas more rapid convergence is associated with erosion. FZ—fracture zone.

A clear picture of the collision tectonics is imaged in the shaded-relief map of multibeam bathymetry (Fig. 6) (to avoid covering morphology in the figure, the triple junction is annotated on the right side). North of the triple junction, ocean crustal fabric is sharp, and transform faults are unmistakably clear. The deformation front north of the constricted collision area is bounded landward by a continuous narrow ridge, which we interpret as the frontal prism, and it fades away as the trench shallows and sediment becomes sparse over the subducting ridge. South of the junction, the ocean crustal fabric is muted by sediment, and the trench is dominated by a turbidity current channel. Low ridges develop in the trench-axis sediment pond, showing the growth of fault-cored folds in the beginning stages of developing an accretionary frontal prism as the triple junction migrates north.

The differing morphology and the different tectonics it indicates on either side of the subducting ridge are instructive (Ranero et al., 2006; Bourgois et al., 1996, 2000). The fully developed accreted terrain (Fig. 6) is 30–40 km wide. Accreted sediment backs against a slope sediment-covered basement. An environment of slow subduction is conducive to accretion because the slow convergence allows more time for deposition of trench sediment, resulting in a thick sediment input to the subduction zone (Behrmann et al., 1994). North of the triple junction, the margin framework basement reaches within 12 km of the deformation front. Convergence is rapid, which is conducive to subduction erosion, and the residence time for sediment in the trench is relatively short. The spreading ridge fills the trench axis for roughly 50 km because the spreading axis and trench axis trends differ by only 10°.

The addition of a spreading center makes the triple junction area a unique collision zone. Contrasting rates of convergence north and south of the junction are the predominant cause of the contrasting changes in the character of the margin. The slow convergence in the south results in a sediment-flooded margin that accretes, whereas to the north, the rapid convergence efficiently subducts a similar sediment volume supplied to the base of the slope. Where the Chile Rise subducts, the margin is severely disrupted, and on shore, young plutons are exposed. However, the exposed young plutons indicate very rapid surface and subsurface erosion (Bourgois et al., 1996).

DISCUSSION

The term collision calls to mind an image of ships smashing into each other and inflicting severe damage. However, in all examples along the Americas, the dominant processes are subduction of ocean floor relief, local uplift of the upper plate, and subduction erosion. In the previous examples, 2–4-km-high relief on a subducting plate is initially accommodated by "aseismic" underthrusting. Underthrusting is facilitated by a high-porosity, low-velocity plate interface zone that loses net fluid volume, thereby progressively increasing interplate friction to a level where it becomes seismogenic (Ranero et al., 2008). As the Costa Rican seamounts show, on entering the subduction zone, they erode an embayment into the thin frontal prism apex and then tunnel underneath the upper plate. Although much material is initially pushed aside and eroded, the principal process is subduction, but since collision is frequently used for this process, we retain it here.

When thousands of square kilometers are subducted, as with the Yakutat terrane, interplate shear stress may build until it exceeds the strength of the lower-plate subduction channel. Lower-plate faulting and duplexing that underplates and uplifts coastal mountains add to upper-plate loading. If loading produces greater interplate friction, the interplate fault may propagate beneath the lower-plate subduction channel to the outboard edge of the anomalously thick crustal feature. Such propagation of a plate boundary has been proposed for the Yakutat terrane (Gulick et al., 2007). In this manner, a large part of the lower plate can transfer to the outboard edge of the upper plate and become its new continental margin. Similar geologic histories could have occurred during past terrane docking around the Gulf of Alaska.

Examples of collision presented here have both common and unique features. Commonly, arc volcanism ends opposite collision zones, except opposite the collision of Carnegie Ridge. Most colliding ridges are associated with a flat slab, except the Cocos Ridge, despite subduction beneath a single area for ~3 to 5 m.y. Perhaps the volume of thick crust is too small to have the positive buoyancy required to form a flat slab; further work to explain Cocos Ridge subduction is in progress (cf. Dzierma et al., 2007). Gutscher et al. (2000) concluded that overthickened oceanic crust, when subducted, is sufficiently buoyant to cause a flat slab to develop. However, that is not possible at Juan Fernandez Ridge, and, instead, it may be mantle hydration and serpentinization that provide buoyancy (Kopp et al., 2004). Another unique collision zone exception is the association of forearc intrusion at the South Chile triple junction.

A discovery from multibeam bathymetric imaging is that hotspot intrusion affects a much wider zone than can be resolved with conventional bathymetry. The Juan Fernandez Ridge in conventional bathymetry is about one-fourth the width of the surface features associated with the ridge in multibeam images (Ranero et al., 2006). Similarly, magma with Cocos Ridge geochemistry has migrated 125 km beyond its region of thick crust (Kimura et al., 1997). Despite the absence of extra thick crust beneath the northwest flank of Cocos Ridge, the recently uplifted mountains on land extend to the same width as hotspot-intruded crust of normal thickness (Fig. 3). Elevated terraces along the coast indicate how far coastal uplift extends beyond the ridge's topographic crest (Gardner et al., 2001). No abrupt shift in depth of the seismogenic zone has been identified except where the subducted ridge flanks end (Protti et al., 1995). At the crest, the zone of interplate microseismicity and heat flow becomes shallower and the up-dip limit of plate-boundary seismogenesis is ~20 km seaward of its position to the northwest, indicating increased interplate friction compared to adjacent areas (Ranero et al., 2006). It seems rational for subducting thick crust of seafloor ridges to be associated with increased interplate friction similar to the local microseismicity associated with subducted seamounts (Bilek et al., 2003). The zone affected by collision is probably much wider than the width of a hotspot ridge in conventional bathymetry.

The wide continental slope of Peru is implied to result from subsidence, where, during its migration, the Nazca Ridge subducted and eroded an 800-km-long segment of the margin (von Huene et al., 1996). Slope retreat is an indicator of the volume of material lost to subduction erosion during a collision and whether the point of collision was stationary or migrated. In Costa Rica, the many subducting seamounts are associated with slope retreat, which is implied to show highly accelerated subduction erosion (Fig. 4) (Ranero et al., 2007).

Although both Juan Fernandez and the South Chile Ridge divide accreting and eroding frontal prisms, the coastal mountains show no differences in character corresponding to differing subduction zones. Apparently, accretion or erosion make less difference to coastal mountain building than crustal thickness of the colliding feature that has divided the trench into sediment-flooded and sediment-starved segments. It seems that where subduction of a colliding feature remains in one place for a longer period, the enlarged mountains are more likely compared to areas above the Cocos Ridge and the Yakutat terrane. The duration of tectonic conditions in one place seems to correlate best with mountain building.

At most collision zones, a dominance of compressional tectonic deformation is generally assumed. However, at the Cocos and Carnegie Ridge collisions, middle-slope extension is revealed in detailed multibeam bathymetry and depth-migrated seismic images (Sage et al., 2006; Ranero et al., 2007; Calahorrano et al., 2008). Frontal prism structure of landward-dipping reflections indicates compressional deformation at the beginning of subduction. The prism materials deform, whereas below the décollement, the lower-plate subduction channel strata remain coherent, except where continued bend faulting occurs (cf. Sage et al., 2006; Calahorrano et al., 2008). The thickened frontal prism mass is presumed to increase pore fluid pressure along the décollement and reduce friction enough for an entire soft trench sediment section to underthrust an intensely faulted and folded prism. In the frontal prism region, the presumed relatively weak upper- and lower-plate materials indicate a weak décollement (von Huene et al., 2009).

Upslope in the middle prism, normal faults cut the entire upper plate (Sage et al., 2006; Ranero et al., 2007). The well-imaged normal faults displacing the slope apron and upper plate indicate that during the history of subduction at collision zones, there can be periods of extension deformation across the middle slope (Wang and Hu, 2006). The initial introduction of anomalous ocean basin features into subduction zones is into the low-friction environment indicated by upper-plate extension and weak microseismicity. At the Costa Rican margin, the surface expression of compressional deformation first begins at the thrust-faulted Fila Costeña above the seismogenic zone.

Putting aside the complexities of differences between collision zones, we summarize observations with a generic model (Fig. 7). In cross section, the colliding feature can be a ridge or a feature of much greater area. As the front of a ridge enters the seaward slope of a trench where the oceanic plate bends downward, its thick crust resists bending, and it forms a shallow segment of the trench axis. During initial collision, the lower-plate dip begins to shallow beneath the frontal prism, and, as the slope erodes, the prism taper increases because the prism seafloor becomes steeper. Initial ridge contact with the lower slope uplifts and erodes the frontal prism, much like the embayment observed during seamount subduction. As collision progresses, erosion of the frontal prism at the foot of the slope continues until the ridge crest can tunnel into the margin. Frontal erosion extends laterally along the foot of the slope as the ridge flanks subduct, and the erosional clastic debris and slope sediment transported to the trench flows away from the elevated subducting ridge crest by axial sediment transport. The elevated ridge becomes a barrier that impedes axial sediment transport, and a wedge of thick trench sediment forms on the upstream side. During ridge

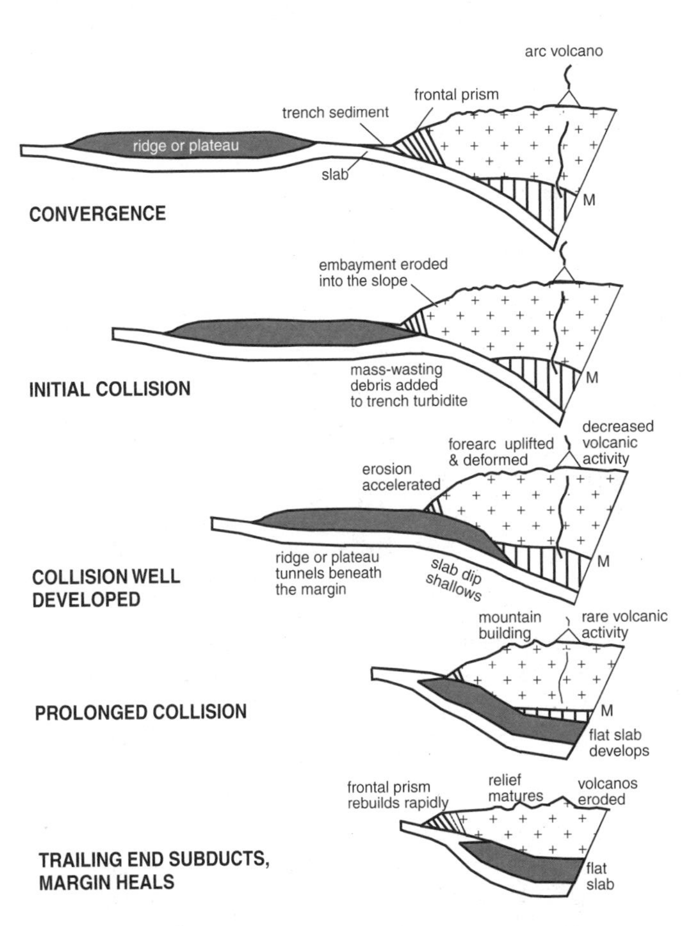

Figure 7. A simplified diagram of major tectonic processes during collision. Accretionary margins become erosional as the colliding body begins to subduct. After 1 or 2 m.y. of collision, uplifted subaerial ground erodes, sedimentation increases, and volcanism declines as a flat slab develops. Once the colliding body is subducted, the trench-axis high disappears, but erosion of high ground continues, and sediment accretes if it is sufficiently thick. A forearc can return to its previous configuration in 2–4 m.y., but temporal geology can leave bits and pieces of evidence. M—Moho.

subduction, margin uplift can locally raise the seafloor above sea level to produce short-lived islands at the shelf edge. With prolonged ridge subduction in one area, the coastal uplift may form a peninsula. If the ridge migrates along the margin, the slope left unsupported by the migrating ridge crest subsides. This is one explanation for the uplift and subsidence of the Peruvian middle and upper slope, which was at neritic depths long enough to have experienced surf zone erosion and receive upwelling sediment as the Nazca Ridge migrated southward beneath it (von Huene et al., 1996). Continued ridge subduction in one area and increasing resistance to subduction may produce a zone of contractile deformation along the plate interface to form mountains landward of the shoreline. With prolonged collision in one area, mountain building continues, but if the collision area migrates rapidly along the margin, a mountain range associated with the collision may not be apparent. The debris from eroding mountain ranges enters the trench, where it can travel down the trench axis or along deep-sea channels for thousands of kilometers to increase the abundance of subducted sediment well away from the collision. Collision of ocean plateaus with a large area intensifies the processes observed in smaller scale at subducting ridges and seamounts. Once the ridge has moved on, the margin returns to its previous configuration, assuming plate-tectonic conditions have remained the same as prior to collision.

CONCLUSIONS

Current collision along the Americas involves subduction of ocean floor relief that uplifts the seafloor and land surface above the colliding feature. The process also involves accelerated subduction erosion and seafloor erosion from steepening of the continental slope. Relief on a subducting plate is initially accommodated in the subduction zone by "aseismic" underthrusting. Among the processes facilitating underthrusting, high fluid pressure along the plate interface reduces friction and allows the zone's aseismic character. A net volume of the fluid in the aseismic zone is lost as it travels deeper down the subduction zone, thereby progressively increasing interplate friction to a level where seismogenic behavior occurs (Ranero et al., 2008). As thickened crustal masses continue to subduct, the increase in friction associated with subduction of thick crust will increase the resistance to subduction. That resistance is inferred to increase beyond the strength of the terrane, i.e., Yakutat, and cause thrusting along the plate interface.

Development of coastal mountains from collision appears to require subduction of large ocean floor relief in one place for 2–3 m.y. In central Alaska, a 600-km-long subducted segment of the Yakutat terrane is associated with coastal mountains >5000 m high. At Cocos Ridge in central America, ~300 km of the ridge subducted in one area, and coeval uplift inland produced >3000-m-high mountains. These are the only two margins along the Americas where collision is clearly associated with exceptionally high coastal mountains. Where ridge subduction migrates rapidly along a margin, no significant permanent change in mountain morphology is evident opposite the collision. Along Peru, short-term uplift during collision was followed by subsidence, including the formation of basins in the middle and upper slopes, and extensional deformation (Clift et al., 2003). Seamount subduction beneath Costa Rica produces a modest local coastal uplift and minor permanent deformation (Gardner et al., 2001; Fisher et al., 2004).

The increased resistance to subduction caused by the 1600–2000 km^2 of subducted Yakutat terrane crust may be causing a seaward jump of the deformation front. Such a jump would attach a segment of the subducting feature to the continental slope, and it may be a model for past terrane docking in Alaska. Deformation of the subducting plate has been indicated by reflection and refraction geophysics (Fuis et al., 2008). Current speculation regarding the mechanics of coastal mountain building invokes either an underplating by duplexing or tectonic thickening of the subducted slab (Wallace, 2009). As resistance to subduction of anomalously thick crust increases, the unsubducted crust becomes a foreland fold-and-thrust belt, and the subducted part may be thrust on itself (Fuis et al., 2008). Once erosion of the mountains contributes large volumes of clastic materials to the forearc, trench axial transport may supply a distant erosional margin with sufficient sediment to become accretionary, despite being far away from any significant sediment source area (Scholl and von Huene, 2007).

Although subducting ridges appear to involve elevated interplate friction, clear images of normal faults through an entire middle prism present the paradox of collision and yet low interplate friction. The increased buoyancy of ridges, whether from high heat flow or less dense materials (Kopp et al., 2004), is commonly associated spatially with the development of a flat slab and cessation of arc volcanism. It is also proposed to cause increased seismicity and explain the greater energy released where some ridges collide with South America (Gutscher et al., 2000). Constraints will come as earthquake records span full earthquake cycles.

ACKNOWLEDGMENTS

We are especially grateful for in-depth reviews by Peter Clift, David Scholl, and Donna Eberhart-Phillips, which greatly improved the content and readability of our manuscripts. We have benefited greatly from their insights, even when we did not agree with them.

REFERENCES CITED

Barckhausen, U., Ranero, C.R., von Huene, R., Cande, S.C., and Roeser, H.A., 2001, Revised tectonic boundaries in the Cocos plate off Costa Rica: Implications for the segmentation of the convergent margin and for plate tectonic models: Journal of Geophysical Research, v. 106, p. 19,207–19,220, doi: 10.1029/2001JB000238.

Behrmann, J.H., Lewis, S.D., and Cande, S.C., 1994, Tectonics and geology of spreading ridge subduction at the Chile triple junction—A synthesis of results from Leg 141 of the Ocean Drilling Program: Geologische Rundschau, v. 83, no. 4, p. 832–852, doi: 10.1007/BF00251080.

Berger, A.L., Spotilla, J.A., Chapman, J.B., Pavlis, T.L., Enkelmann, E., Ruppert, N.A., and Buscher, J.T., 2008, Architecture, kinematics, and exhumation of a convergent orogenic wedge: A thermochronological investigation of tectonic-climatic interactions within the central St. Elias orogen, Alaska: Earth and Planetary Science Letters, v. 270, p. 13–24, doi: 10.1016/j.epsl.2008.02.034.

Bilek, S.L., Schwartz, S.Y., and DeShon, H.R., 2003, Control of seafloor roughness on earthquake rupture behavior: Geology, v. 31, p. 455–458, doi: 10.1130/0091-7613(2003)031<0455:COSROE>2.0.CO;2.

Bourgois, J.H., Martin, Lagabrielle, Y., Le Moigne, J., and Frutos, J., 1996, Subduction erosion related to spreading-ridge subduction: Taitao peninsula (Chile margin triple junction area): Geology, v. 24, p. 723–726.

Bourgois, J., Guivel, C., Lagabrielle, Y., Calmus, T., Boulegue, J., and Daux, V., 2000, Glacial-interglacial trench supply variation, spreading-ridge subduction, and feedback controls on the Andean margin development at the Chile triple junction area (45–48°S): Journal of Geophysical Research, v. 105, p. 8355–8386, doi: 10.1029/1999JB900400.

Brocher, T.M., Fuis, G.S., Fisher, M.A., Plafker, G., Moses, M.J., Taber, J.J., and Christensen, N.I., 1994, Mapping the megathrust beneath the northern Gulf of Alaska using wide-angle seismic data: Journal of Geophysical Research, v. 99, p. 11,663–11,685, doi: 10.1029/94JB00111.

Bruns, T.R., 1983, Model for the origin of the Yakutat block, an accreted terrane in the northern Gulf of Alaska: Geology, v. 11, p. 718–721, doi: 10.1130/0091-7613(1983)11<718:MFTOOT>2.0.CO;2.

Bruns, T.R., 1984, Reply concerning "Model for the origin of the Yakutat block, an accreting terrane in the northern Gulf of Alaska": Geology, v. 12, no. 9, p. 563–564, doi: 10.1130/0091-7613(1984)12<563b:CAROMF>2.0.CO;2.

Bruns, T.R., 1985, Tectonics of the Yakutat Block, an Allochthonous Terrane in the Northern Gulf of Alaska: U.S. Geological Survey Open-File Report 85-13, 112 p.

Bruns, T.R., 1997, *in* Groome, M.G., Gutmacher, C.E, and Stevenson, A.J, eds., Atlas of GLORIA Sidescan-Sonar Imagery of the Exclusive Economic Zone of the United States: EEZ-View Open-File Report 97-540, CD.

Bruns, T.R., and Carlson, P.R., 1987, Geology and petroleum potential of the southeast Alaska continental margin, *in* Scholl, D.W., et al., eds., Geology and Resource Potential of the Continental Margin of Western North America and Adjacent Ocean Basins: Earth Science Series, Volume 6: Houston, Texas, Circum-Pacific Council for Energy and Mineral Resources, p. 269–282.

Buchs, D.M., and Baumgartner, P.O., 2003, The Osa-Cano accretionary complex (southern Costa Rica): Sedimentary processes in the Middle American Trench recorded in an emerged Eocene-Miocene accretionary prism, *in* 11th Swiss Meeting of Sedimentologists, Fribourg, Programme and Abstracts: International Association of Sedimentologists and Swiss Geological Society, p. 35.

Calahorrano B.A., Sallarès, V., Collot, J.-Y., Sage, F., and Ranero, C.R., 2008, Nonlinear variations of the physical properties along the southern Ecuador subduction channel: Results from depth-migrated seismic data, Earth and Planetary Science Letters 267, p. 453–467, doi: 10.1016/j.epsl.2007.11.061.

Cande, S.C., Leslie, R.B., Parra, J.C., and Hobart, M., 1987, Interaction between the Chile Ridge and the Chile Trench: Geophysical and geothermal evidence: Journal of Geophysical Research, v. 92, p. 495–520, doi: 10.1029/JB092iB01p00495.

Clift, P.D., and Vannucchi, P., 2004, Controls on tectonic accretion versus erosion in subduction zones: Implications for the origin and recycling of the continental crust: Reviews of Geophysics, v. 42, p. RG2001, doi: 10.1029/2003RG000127.

Clift, P.D., Pecher, I., Kukowski, N., and Hampel, A., 2003, Tectonic erosion of the Peruvian forearc, Lima Basin, by subduction and Nazca Ridge collision: Tectonics, v. 22, p. 1023, doi: 10.1029/2002TC001386.

Coffin, M.F., Duncan, R.A., Coffin, M.F., Duncan, R.A., Eldholm, O., Fitton, J.G., Frey, F.A., Larsen, H.C., Mahoney, J.J., Saunders, A.D., Schlich, R., and Wallace, P.J., 2006, Large igneous provinces and scientific ocean drilling: Oceanography, Special Issue, v. 9, no. 4, p. 150–160.

Collins, L.S., Coates, A.G., Jackson, J.B.C., and Obando, J.A., 1995, Timing and rates of emergence of the Limon and Bocas del Toro Basins: Caribbean effects of Cocos Ridge subduction?, *in* Mann, P., ed., Geologic and Tectonic Development of the Caribbean Plate Boundary in Southern Central America: Geological Society of America Special Paper 295, p. 362–289.

de Boer, J.Z., Drummond, M.S., Bordelon, M.J., Defant, M.J., Bellon, H., and Maury, R.C., 1995, Cenozoic magmatic phases of the Costa Rican island arc (Cordillera de Talamanca), *in* Mann, P., ed., Geologic and Tectonic Development of the Caribbean Plate Boundary in Southern Central America: Geological Society of America Special Paper 295, p. 35–55.

DeShon, H.R., Schwartz, S.Y., Bilek, S.L., Dorman, L.M., Gonzalez, V., Protti, J.M., Flueh, E.R., and Dixon, T.H., 2003, Seismogenic zone structure of the southern Middle America Trench, Costa Rica: Journal of Geophysical Research, v. 108, no. B10, p. 2491, doi: 10.1029/2002JB002294.

Dunbar, R.B., Marty, R.C., and Baker, P.A., 1990, Cenozoic marine sedimentation in the Sechura and Pisco Basins, Peru: Palaeogeography, Palaeoclimatology, Palaeoecology, v. 77, p. 235–261, doi: 10.1016/0031-0182(90)90179-B.

Dzierma, Y., Thorwart, M., Rabbel, W., Flueh, E., Mora, M., and Alvarado, G., 2007, Imaging Cocos Ridge subduction in southern Costa Rica with receiver functions: Eos (Transactions, American Geophysical Union), v. 52, Fall Meeting supplement, abstract T41C 0703.

Eberhart-Phillips, D., Christensen, D.H., Brocher, T.M., Hansen, R., Ruppert, N.A., Haeussler, P., and Abers, G.A., 2006, Imaging the transition from Aleutian subduction to Yakutat collision in central Alaska, with local earthquakes: Journal of Geophysical Research, v. 111, p. B11303, doi: 10.1029/2005JB004240.

Engdahl, E.R., van der Hilst, R.D., and Buland, R., 1998, Global teleseismic earthquake relocation with improved travel times and procedures for depth relocation: Bulletin of the Seismological Society of America, v. 88, p. 722–743.

Espurt, N., Baby, P., Brusset, S., Roddaz, M., Hermoza, W., Regard, V., Antoine, P.O., Salas-Gismondi, R., and Bolanos, R., 2007, How does the Nazca Ridge subduction influence the modern Amazonian foreland basin?: Geology, v. 35, no. 6, p. 515–518, doi: 10.1130/G23237A.1.

Ferris, A., Abers, G.A., Christensen, D.H., and Veenstra, E., 2003, High resolution image of the subducted Pacific (?) plate beneath central Alaska, 50–150 km depth: Earth and Planetary Science Letters, v. 214, p. 575–588, doi: 10.1016/S0012-821X(03)00403.

Fisher, D.M., Gardner, T.W., Sak, P., Sanchez, J.D., Murphy, K., and Vannucchi, P., 2004, Active thrusting in the inner forearc of an erosive convergent margin, Pacific coast, Costa Rica: Tectonics, v. 23, doi: 10.1029/2002TC001464, 13 p.

Fisher, M.A., Ruppert, N.A., Eberhart-Phillips, D.M., Brocher, T.M., Wells, R.E., Blakely, R.J., and Sliter, R.W., 2005, Tectonics and seismicity of the Alaskan continental margin near the collision zone of the Yakutat terrane and the epicenter of the 1964 Great Alaska Earthquake: Geological Society of America Abstracts with Programs, v. 37, no. 7, p. 79.

Fuis, G.S., Moore, T.E., Plafker, G., Brocher, T.M., Fisher, M.A., Mooney, W.D., Nokleberg, W.J., Page, R.A., Beaudoin, B.C., Christensen, N.I., Levander, A.R., Lutter, W.J., Saltus, R.W., and Ruppert, N.A., 2008, Trans-Alaska Crustal Transect and continental evolution involving subduction underplating and synchronous foreland thrusting: Geology, v. 36, no. 3, p. 267–270, doi: 10.1130/G24257A.1.

Gardner, T.W., Verdonck, D., Pinter, N., Slingerland, R., Furlong, K., Bullard, T.F., and Wells, S.G., 1992, Quaternary uplift astride the aseismic Cocos Ridge, Pacific coast of Costa Rica: Geological Society of America Bulletin, v. 104, p. 219–232, doi: 10.1130/0016-7606(1992)104<0219:QUATAC>2.3.CO;2.

Gardner, T.W., Marshall, J., Merritts, D., Bee, B., Burgette, R., Burton, E., Cooke, J., Kehrwald, N., Protti, M., Fisher, D., and Sak, P., 2001, Holocene forearc deformation in response to seamount subduction, Peninsula de Nicoya, Costa Rica: Geology, v. 29, p. 151–154, doi: 10.1130/0091-7613(2001)029<0151:HFBRIR>2.0.CO;2.

Gräfe, K., Frisch, W., Villa, I.M., and Meschede, M., 2002, Geodynamic evolution of southern Costa Rica related to low-angle subduction of the Cocos Ridge: Constraints from thermochronology: Tectonophysics, v. 348, p. 187–204, doi: 10.1016/S0040-1951(02)00113-0.

Grevemeyer, I., Kopf, A.J., Fekete, N., Kaul, N., Villinger, H.W., Heesemann, M., Wallmann, K., Spieß, V., Gennerich, H.H., Müller, M., and Weinrebe, W., 2004, Fluid flow through active mud dome Mound Culebra offshore Nicoya Peninsula, Costa Rica: Evidence from heat flow surveying: Marine Geology, v. 207, no. 1–4, p. 145–157.

Gulick, S.P., Lowe, L.A., Pavlis, T.L., Gardner, J.V., and Mayer, L.A., 2007, Geophysical insights into Transition fault debate: Propagating strike-slip in response to stalling Yakutat block subduction in the Gulf of Alaska: Geology, v. 35, p. 763–766, doi: 10.1130/G23585A.1.

Gutscher, M.-A., Spakman, W., Bijwaard, H., and Engdahl, E.R., 2000, Geodynamics of flat subduction: Seismicity and tomographic constraints from the Andean margin: Tectonics, v. 19, no. 5, p. 814–833, doi: 10.1029/1999TC001152.

Hampel, A., Kukowski, N., Bialas, J., Huebscher, C., and Heinbockel, R., 2004, Ridge subduction at an erosive margin: The collision zone of the Nazca Ridge in southern Peru: Journal of Geophysical Research–Solid Earth, v. 109, p. B02101, doi: 10.1029/2003JB002593.

Hinz, K., von Huene, R., Ranero, C.R., and the PACOMAR Working Group, 1996, Tectonic structure of the convergent Pacific margin offshore Costa Rica from multichannel seismic reflection data: Tectonics, v. 15, p. 54–66, doi: 10.1029/95TC02355.

Howell, D.G., Moore, G.W., and Wiley, T.J., 1987, Tectonics and basin evolution of western North America—An overview, *in* Scholl, D.W., Grantz, A., and Vedder, J.G., eds., Geology and Resource Potential of the Continental Margin of Western North America and Adjacent Ocean Basins—Beaufort Sea to Baja California: Earth Science Series, Volume 6: Houston, Texas, Circum-Pacific Council for Energy and Mineral Resources, p. 1–17.

Hsu, J.T., 1992, Quaternary uplift of the Peruvian coast related to the subduction of the Nazca Ridge: 13.5 to 15.6 degrees south latitude: Quaternary International, v. 15/16, p. 87–97.

Hussong, D.M., and Wipperman, L.K., 1981,Vertical movement and tectonic erosion of the continental wall of the Peru-Chile Trench near 11 degrees 30°S latitude, *in* Kulm, L.D., et al., eds., Nazca Plate: Crustal Formation and Andean Convergence: Geological Society of America Memoir 154, p. 509–524.

Jaeger, J.M., Nittrouer, C.A., Scott, N.D., and Milliman, J.D., 1998, Sediment accumulation along a glacially impacted mountainous coastline, northeast Gulf of Alaska: Basin Research, v. 10, p. 155–173, doi: 10.1046/j.1365-2117.1998.00059.x.

Jones, D.L., Silberling, N.J., and Hillhouse, J.W., 1977, Wrangellia—A displaced terrane in northwestern North America: Canadian Journal of Earth Sciences, v. 14, p. 2565, doi: 10.1139/e77-222.

Kimura, G., Silver, E.A., Blum, P., et al., 1997, in Proceedings of the Ocean Drilling Program, Initial Reports, Volume 170: College Station, Texas, Ocean Drilling Program, doi: 10.2973/odp.proc.ir.170.1997.

Kimura, G., Kitamura, Y., Hashimoto, Y., Yamaguchi, A., Shibata, T., Ujiie, K., and Okamoto, S., 2007, Transition of accretionary wedge structures around the up-dip limit of the seismogenic subduction zone: Earth and Planetary Science Letters, v. 255, p. 471–484, doi: 10.101j.epsl.2007.01.005.

Kopp, H., Flueh, E.R., Papenberg, C., and Klaeschen, K., 2004, Seismic investigations of the O'Higgins Seamount Group and Juan Fernandez Ridge; aseismic ridge emplacement and lithosphere hydration: Tectonics, v. 23, p. TC2009, doi: 10.1029/2003TC001590.

Kulm, L.D., Prince, R.A., French, W., Johnson, S., and Masias, A., 1981, Crustal structure and tectonics of the central Peru continental margin and trench, *in* Kulm, L.D., et al., eds., Nazca Plate: Crustal Formation and Andean Convergence: Geological Society of America Memoir 154, p. 445–468.

Lagoe, M.B., Eyles, C.H., Eyles, N., and Hale, C., 1993, Timing of late Cenozoic tidewater glaciation in the far north Pacific: Geological Society of America Bulletin, v. 105, p. 1542–1560, doi: 10.1130/0016-7606(1993)105<1542:TOLCTG>2.3.CO;2.

Laursen, J., Scholl, D.W., and von Huene, R., 2002, Neotectonic deformation of the central Chile margin: Deepwater forearc basin formation in response to hot spot ridge and seamount subduction: Tectonics, v. 21, no. 1038(5), doi: 10.1029/2001TC901023.

Oncken, O., Hindle, D., Kley, J., Elger, K., Victor, P., and Schemmann, K., 2006, Deformation of the Central Andean upper plate system—Facts, fiction, and constraints for plateau models; the Andes, *in* Oncken, O., Chong, G., Franz, G., Giese, P., Goetze, H.-J., Ramos, V.A., Strecker, M.R., and Wigger, P., eds., Frontiers in Earth Sciences, Volume 1: Berlin, Springer, p. 3–27.

Pavlis, T.L., Picornell, C., Serpa, L., Bruhn, R.L., and Plafker, G., 2004, Tectonic processes during oblique convergence: Insights from the St. Elias orogen, northern North American Cordillera: Tectonics, v. 23, p. TC3001, doi: 10.1029/2003TC001557.

Plafker, G., 1987, Regional geology and petroleum potential of the northern Gulf of Alaska continental margin, *in* Scholl, D.W., Grantz, A., and Vedder, J.G., eds., Geology and Resource Potential of the Continental Margin of Western North America and Adjacent Ocean Basins—Beaufort Sea to Baja California: Earth Science Series, Volume 6: Houston, Texas, Circum-Pacific Council for Energy and Mineral Resource, p. 229–268.

Plafker, G., and Berg, H.C., 1994, Overview of the geology and tectonic evolution of Alaska, *in* Plafker, G., and Berg, H.G., eds., The Geology of Alaska: Boulder, Colorado, Geological Society of America, Geology of North America, v. G-1, p. 989–1021.

Plafker, G., Moore, J.C., and Winkler, G.R., 1994, Geology of the southern Alaska margin, *in* Plafker, G., and Berg, H.C., eds., The Geology of Alaska: Boulder, Colorado, Geological Society of America, Geology of North America, v. G-1, p. 389–449.

Protti, M., Guendel, F., and McNally, K., 1995, Correlation between the age of the subducting Cocos plate and the geometry of the Wadati-Benioff zone under Nicaragua and Costa Rica, *in* Mann, P., ed., Geologic and Tectonic Development of the Caribbean Plate Boundary: Geological Society of America Special Paper 295, p. 309–340.

Ranero, C.R., von Huene, R., Weinrebe, W., and Reichert, C., 2006, Tectonic processes along the Chile convergent margin, *in* Onken, O., et al., eds., The Andes: Active Subduction Orogeny: Berlin, Springer-Verlag, p. 91–122.

Ranero, C.R., von Huene, R., Weinrebe, W., and Barckhausen, U., 2007, Convergent margin tectonics: A marine perspective, *in* Bundschuh, J., and Alvarado, G.E., eds., Central America: Geology, Resources and Hazards: London, Taylor and Francis, p. 239–265.

Ranero, C.R., Grevemeyer, I., Sahling, H., Barckhausen, U., Hensen, C., Wallmann, K., Weinrebe, W., Vannucchi, P., von Huene, R., and McIntosh, K., 2008, The hydrogeological system of erosional convergent margins and its influence on tectonics and interplate seismogenesis: Geochemistry, Geophysics, Geosystems, v. 9, p. Q03S04, doi: 10.1029/2007GC001679.

Ratchkovski, N.A., and Hansen, R.A., 2002, New evidence for segmentation of the Alaska subduction zone: Seismological Society of America Bulletin, v. 92, p. 1754–1765, doi: 10.1785/0120000269.

Rea, D.K., and Snoeckx, H., 1995, Sediment fluxes in the Gulf of Alaska paleoceanographic record from Site 887 on the Patton-Murray Seamount Platform, *in* Rea, D.K., and Basov, I., co-chiefs, and Janecek, T., ODP scientist, Proceedings of the Ocean Drilling Program. Scientific Results, Volume 145: College Station, Texas, Ocean Drilling Program, p. 247–256.

Sage, F., Collot, J.Y., and Ranero, C.R., 2006, Interplate patchiness and subduction-erosion mechanisms: Evidence from depth-migrated seismic images at the central Ecuador convergent margin: Geology, v. 34, no. 12, p. 997–1000, doi: 10.1130/G22790A.1.

Sak, P., Fisher, D., and Gardner, T., 2004, Effects of subducting seafloor roughness on upper plate vertical tectonism: Osa Peninsula, Costa Rica: Tectonics, v. 23, doi: 10.1029/2002TC001474, 16 p.

Sample, J.C., and Moore, J.C., 1987, Structural style and kinematics of an underplated slate belt, Kodiak Islands, Alaska: Geological Society of America Bulletin, v. 99, p. 7–20, doi: 10.1130/0016-7606(1987)99<7:SSAKOA>2.0.CO;2.

Scholl, D.W., and von Huene, R., 2007, Crustal recycling at modern subduction zones applied to the past—Issues of growth and preservation of continental basement, mantle geochemistry, and supercontinent reconstruction, *in* Hatcher, R.D., Jr., Carlson, M.P., McBride, J.H., and Martínez-Catalán, J.R., eds., 4-D Framework of Continental Crust: Geological Society of America Memoir 200, p. 9–32.

Schweller, W.J., Kulm, L.D., and Prince, R.A., 1981, Tectonics, structure, and sedimentary framework of the Perú-Chile Trench, *in* Kulm, L.D., et al., eds., Nazca Plate: Crustal Formation and Andean Convergence: Geological Society of America Memoir 154, p. 323–349.

Sitchler, J.C., Fisher, D.M., Gardner, T.W., and Protti, M., 2007, Constraints on inner forearc deformation from balanced cross sections, Fila Costeña thrust belt, Costa Rica: Tectonics, v. 26, TC6012, doi: 10.1029/2006TC001949.

Stavenhagen, A.U., Flueh, E.R., Ranero, C., McIntosh, K.D., Shipley, T., Leandro, G., Schulze, A., and Dañobeitia, J.J., 1998, Seismic wide-angle investigations in Costa Rica: A crustal velocity model from the Pacific to the Caribbean coast: Zeitblatt Geologica Palaeontology, Part I, p. 393–408.

Suess, E., von Huene, R., Emeis, K.-C., Bourgois, J., Cruzado Castaneda, J. del C., De Wever, P., Eglinton, G., Garrison, R., Greenberg, M., Paz, E.H., Hill, P.R., Ibaraki, M., Kastner, M., Kemp, A.E.S., Kvenvolden, K.A., Langridge, R., Lindsley-Griffin, N., Marsters, J., Martini, E., McCabe, R., Ocola, L., Resig, J., Sanchez Fernandez, A.W., Schrader, H.-J., Thornburg, T.M., Wefer, G., and Yamano, M., 1988, Proceedings of the Ocean Drilling Program, Initial Reports, Volume 112: College Station, Texas, Ocean Drilling Program.

Vannucchi, P., Ranero, C.R., Galeotti, S., Straub, S.M., Scholl, D.W., and McDougall, R.D., 2003, Fast rates of subduction erosion along the Costa

Rica Pacific margin: Implications for non–steady state rates of crustal recycling at subduction zones: Journal of Geophysical Research, v. 108, doi: 10.1029/2002JB002207.

Vannucchi, P., Galeotti, S., Clift, P.D., Rantero, V.R., and von Huene, R., 2004, Long-term subduction erosion along the Middle America Trench offshore Guatemala: Geology, v. 32, p. 617–620, doi: 10.1130/G20422.1.

Vannucchi, P., Fisher, D.M., Bier, S., and Gardner, T.W., 2006, From seamount accretion to tectonic erosion: Formation of the Osa mélange and the effects of Cocos Ridge subduction in southern Costa Rica: Tectonics, v. 25, TC2004, doi: 1029/2005TC001855, 19 p.

von Huene, R., Pecher, I.A., and Gutscher, M.A., 1996, Development of the accretionary prism along Peru and material flux after subduction of Nazca Ridge: Tectonics, v. 15, p. 19–33, doi: 10.1029/95TC02618.

von Huene, R., Corvalan, J., Flueh, E.R., Hinz, K., Korstgard, J., Ranero, C.R., Weinrebe, W., and C.O.N.D.O.R. Scientists, 1997, Tectonic control of the subducting Juan Fernández Ridge on the Andean margin near Valparaiso, Chile: Tectonics, v. 16, p. 474–488, doi: 10.1029/96TC03703.

von Huene, R., Klaeschen, D., Gutscher, M., and Fruehn, J., 1998, Mass and fluid flux during accretion at the Alaskan margin: Geological Society of America Bulletin, v. 110, p. 468–482, doi: 10.1130/0016-7606(1998)110<0468:MAFFDA>2.3.CO;2.

von Huene, R., Ranero, C.R., Weinrebe, W., and Hinz, K., 2000, Quaternary convergent margin tectonics of Costa Rica, segmentation of the Cocos plate, and Central American volcanism: Tectonics, v. 19, p. 314–334, doi: 10.1029/1999TC001143.

von Huene, R., Ranero, C.R., and Scholl, D.W., 2009, Convergent margin structure in high quality geophysical images and current kinematic and dynamic models, *in* Lallemand, S., and Funiciello, F., eds., Subduction Zone Geodynamics: Berlin, Springer-Verlag, doi: 10.1007/978-3-540-87974-9_8.

Wallace, W.K., 2009, Yakataga fold-and-thrust belt: Structural geometry and tectonic implications of a small continental collision zone, *in* Haeussler, P., Freymueller, J., Wesson, R., and Ekstrom, G., eds., Active Tectonics and Seismic Potential of Alaska: American Geophysical Union Monograph (in press).

Walther, C., 2003, The crustal structure of the Cocos Ridge off Costa Rica: Journal of Geophysical Research, v. 108, doi: 10.1029/2001JB000888.

Wang, K., and Hu, Y., 2006, Accretionary prisms in subduction earthquake cycles: The theory of dynamic Coulomb wedge: Journal of Geophysical Research, v. 111, p. B06410, doi: 10.1029/2005JB004094.

Werner, R., Hoernle, K., van den Bogaard, P., Ranero, C.R., von Huene, R., and Korich, D., 1999, Drowned 14 m.y. old Galapagos archipelago off the coast of Costa Rica: Implications for tectonic and evolutionary models: Geology, v. 27, p. 499–502, doi: 10.1130/0091-7613(1999)027<0499:DMYOGP>2.3.CO;2.

Yañez, G., Cembrano, J., Pardo, M., Ranero, C.R., and Selles, D., 2002, The Challenger–Juan Fernández–Maipo major tectonic transition of the Nazca-Andean subduction system at 33°–34°S: Geodynamic evidences and implications: Journal of South American Earth Sciences, v. 15, p. 23–38, doi: 10.1016/S0895-9811(02)00004-4.

Ye, S., Bialas, J., Flueh, E.R., Stavenhagen, A., von Huene, R., Leandro, G., and Hinz, K., 1996, Crustal structure of the Middle American Trench off Costa Rica from wide-angle seismic data: Tectonics, v. 15, p. 1006–1021, doi: 10.1029/96TC00827.

Manuscript Accepted by the Society 5 December 2008

Printed in the USA

The Geological Society of America
Memoir 204
2009

Relation of flat subduction to magmatism and deformation in the western United States

Eugene Humphreys
Department of Geological Sciences, University of Oregon, Eugene, Oregon 97403, USA

ABSTRACT

Flat subduction of the Farallon plate beneath the western United States during the Laramide orogeny was caused by the combined effects of oceanic plateau subduction and unusually great suction in the mantle wedge, the latter of which was caused by the shallowing slab approaching the North American craton root. Once in contact with basal North America, the slab cooled and hydrated the lithosphere. Upon removal, asthenospheric contact with lithosphere resulted in magma production that was especially intense where the basal lithosphere was fertile (in what now is the Basin and Range Province), and this heating weakened the lithosphere and made it convectively unstable. Small-scale convection has since affected many areas. With slab sinking and unloading, the western United States elevated into a broad plateau, and the weak portion gravitationally collapsed. With development of a transform plate boundary, the western part of the weak zone became entrained with the Pacific plate, and deformation there is dominated by shear.

INTRODUCTION

Tectonic and magmatic activity within the western United States has been unusually vigorous during the last 150 m.y., at times extending into the continent 1000–2000 km from the plate margin (Figs. 1–3). Such non-plate-like behavior, although not common at any time and place, is important through time in forming the geologic and continental structure and the continent itself. The generally held account of western U.S. activity is one of progressively intense subduction coupling at the western plate margin and slab flattening during the Sevier-Laramide orogeny (Livaccari and Perry, 1993), followed by an intense magmatic flare-up over much of the tectonically modified area during post-orogenic collapse (e.g., Burchfiel et al., 1992), the waning stages of which we observe today. The occurrence of Laramide-age "flat slab" subduction of the Farallon slab beneath the western United States is accepted by most (including myself) as being related to the cause of the deeply penetrating tectonics. This is to say, the tectonic and magmatic evolution of the last 150 m.y. in the western United States is attributed to plate-tectonic processes. With Farallon slab flattening and contact with North America and subsequent slab removal in mind, my goal is to account for the general aspects of western U.S. tectonism and magmatism over the last 150 m.y., i.e., during the Sevier-Laramide orogeny and the following episode of postorogenic collapse. My desire is to understand what caused the western United States to behave as it did. In particular, how did plate tectonics create the western United States and what non–plate tectonic processes contributed to the tectonic and magmatic activity? In some regards, interpretations for this activity are generally agreed upon, and in other regards, the relationship between Farallon subduction and the geologic record is quite ambiguous. The following represents my view, most of which is the commonly accepted understanding. My treatment is divided into the time intervals pre-Laramide, Laramide, and post-Laramide, and then, with this history in mind, the current physical state.

Humphreys, E., 2009, Relation of flat subduction to magmatism and deformation in the western United States, *in* Kay, S.M., Ramos, V.A., and Dickinson, W.R., eds., Backbone of the Americas: Shallow Subduction, Plateau Uplift, and Ridge and Terrane Collision: Geological Society of America Memoir 204, p. 85–98, doi: 10.1130/2009.1204(04). For permission to copy, contact editing@geosociety.org.

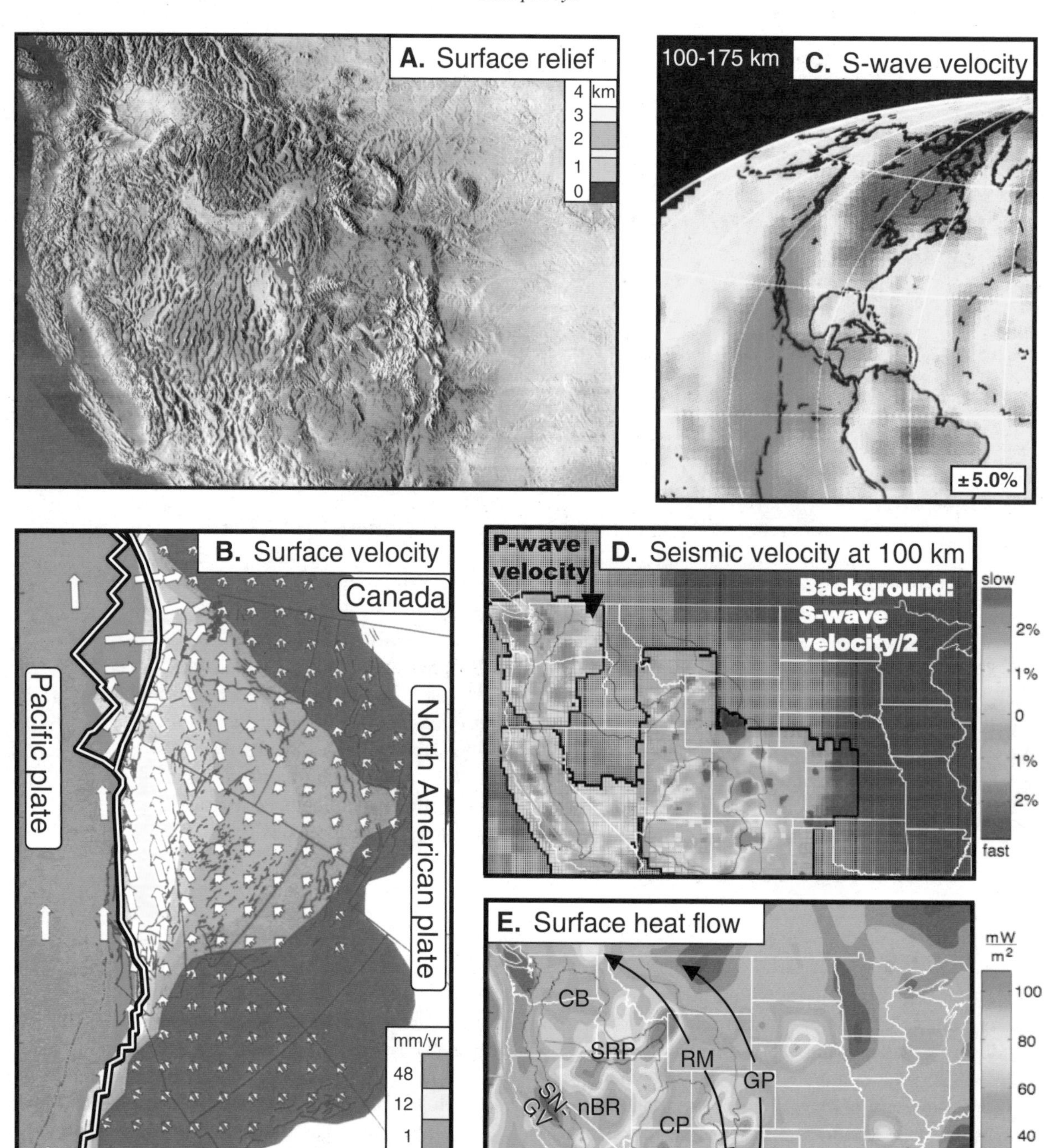

Figure 1. Geophysical observations of the western United States region. (A) Surface relief (Simpson and Anders, 1992). Along the length of the Cordillera, the region of uplift is unusually wide within the western United States. (B) Surface velocity field, modeled using global positioning system (GPS) data. Projection is oblique Mercator about Pacific–North America pole. In Mexico and Canada, accommodation of plate motion is confined nearly to the major plate-bounding faults (white-on-black lines), whereas it is distributed broadly over the western portion of the uplifted western United States. (C) Seismic S-wave velocity perturbations (Grand, 1994) averaged in the 100–175 km depth range (velocity range shown is 5%). The contrast between the fast North America craton and the slow western United States–East Pacific Rise volume is as great as any on Earth. (D) Seismic P-wave and rescaled S-wave velocity perturbations (Humphreys et al., 2003) at 100 km depth. The contrasts seen within the western United States are as great as those seen between the craton and the western United States. High-velocity mantle probably is continental or subducted ocean lithosphere, and low-velocity mantle probably is partially molten. (E) Surface heat flow (Humphreys et al., 2003). Most areas of heat flow greater than 70 mW/m^2 have experienced young magmatism, although water-flow effects, apparent in the Great Plains east of Wyoming and in south-central Nevada, also affect this map (for state locations, see Fig. 6). Abbreviations indicate tectonic provinces discussed in the text: CB—Columbia Basin, CP—Colorado Plateau, GP—Great Plains, nBR—northern Basin and Range, sBR—southern Basin and Range, RM—Rocky Mountains, SN-GV—Sierra Nevada–Great Valley, SRP—Snake River Plain.

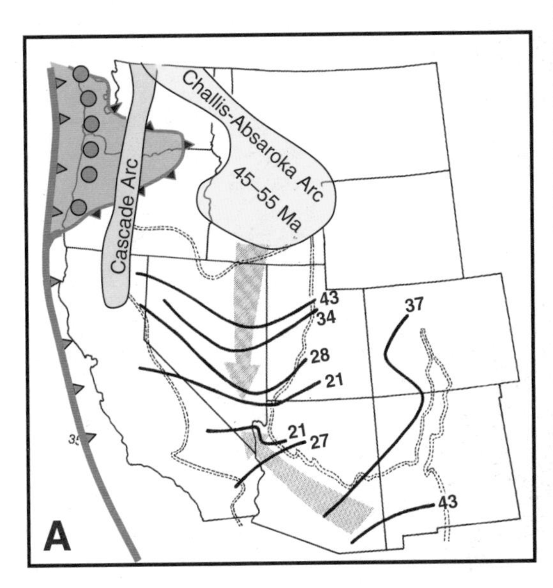

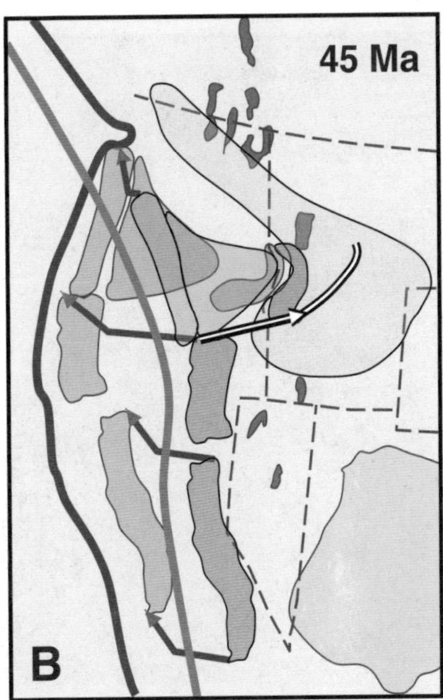

Figure 2. Simplified western United States evolution from 55 to 20 Ma. (A) Major volcanic activity. Green area shows the oceanic Siletzia terrane (with dark green seamounts indicated), which accreted at ca. 48 Ma (Madsen et al., 2006). Following accretion, the subduction zone (blue lines) and arc-related volcanism (yellow areas) jumped west to the Cascadia subduction zone and the Cascade arc, and the ignimbrite flare-up initiated and propagated to the south across the northern Basin and Range and NW across the southern Basin and Range (age of initial magmatism indicated). (B) Map showing post-Laramide deformation of the western United States (modified from Dickinson, 2002). Red areas indicate the extent of Mesozoic accreted and plutonic terranes of the Sierra Nevada, Klamath and Blue Mountains, which are used as indicators of deformation in the continental interior. The green Siletzia terrane plays a similar role as a kinematic indicator (shown as broken at the Cascade arc; offshore portion not indicated). Current positions of these terranes are shown in the background (in dark gray). Yellow shows the Challis-Absaroka arc, and darkest gray shows the locations of the major metamorphic core complexes in the region. White-on-black line represents the southern edge of the slab window created by Siletzia accretion (at ca. 45 Ma, 3 m.y. after accretion).

Much has been written about the geologic history of the western United States, and it cannot be reviewed comprehensively here. Instead, I focus on the lithospheric-scale seismic and geodynamic aspects of western U.S. evolution. Many good reviews have been written about the geologic, tectonic, and magmatic history (e.g., Burchfiel et al., 1992; DeCelles, 2004; Christiansen and Yeats, 1992), to which the reader should refer for more detail.

PRE-LARAMIDE SEVIER

Evidence for shallowing Farallon slab dip during the course of the Sevier orogeny is provided by an eastward migration of arc magmatism into Nevada (e.g., DeCelles, 2004), an inferred increase of subduction zone coupling evidenced by fold-and-thrust contraction across a wide belt extending east from the magmatic front to the Colorado Plateau and western Wyoming (e.g., DeCelles, 2004; Burchfiel et al., 1992), the intense transpressional truncation of the Pacific Northwest continental margin (Giorgis et al., 2005), and the dynamic subsidence of the continental interior creating the Cretaceous Interior Seaway (Mitrovica et al., 1989). The widespread interior subsidence and intensifying tectonism may have resulted from an “avalanche” or rapid subduction through the 660 km discontinuity of the Farallon slab that previously was laid out in the transition zone. Alternatively, a more steady subduction of the Farallon slab laid out along the 660 km discontinuity could have provided similar dynamic subsidence but with a less punctuated subsidence history.

The suggestion that the Farallon slab laid out at the top of the upper mantle is based on the continuation of arc magmatism near the California-Nevada border during Sevier time, (implying the subducting slab dipped into and exposed itself to asthenosphere near eastern California, i.e., it was not in direct contact with North America east of California), and yet the dynamic effects of the slab influenced continental subsidence as far east as the Great Lakes, suggesting that the Farallon slab was near enough to the surface beneath the north-central United States to pull the surface down dynamically (Mitrovica et al., 1989; Gurnis, 1993). A shallow dip and a vertically thin asthenospheric wedge extending across half a continent seem to be dynamically unreasonable. It is more likely that the slab subducted to and was supported by the viscosity increase of the lower mantle and the endothermic

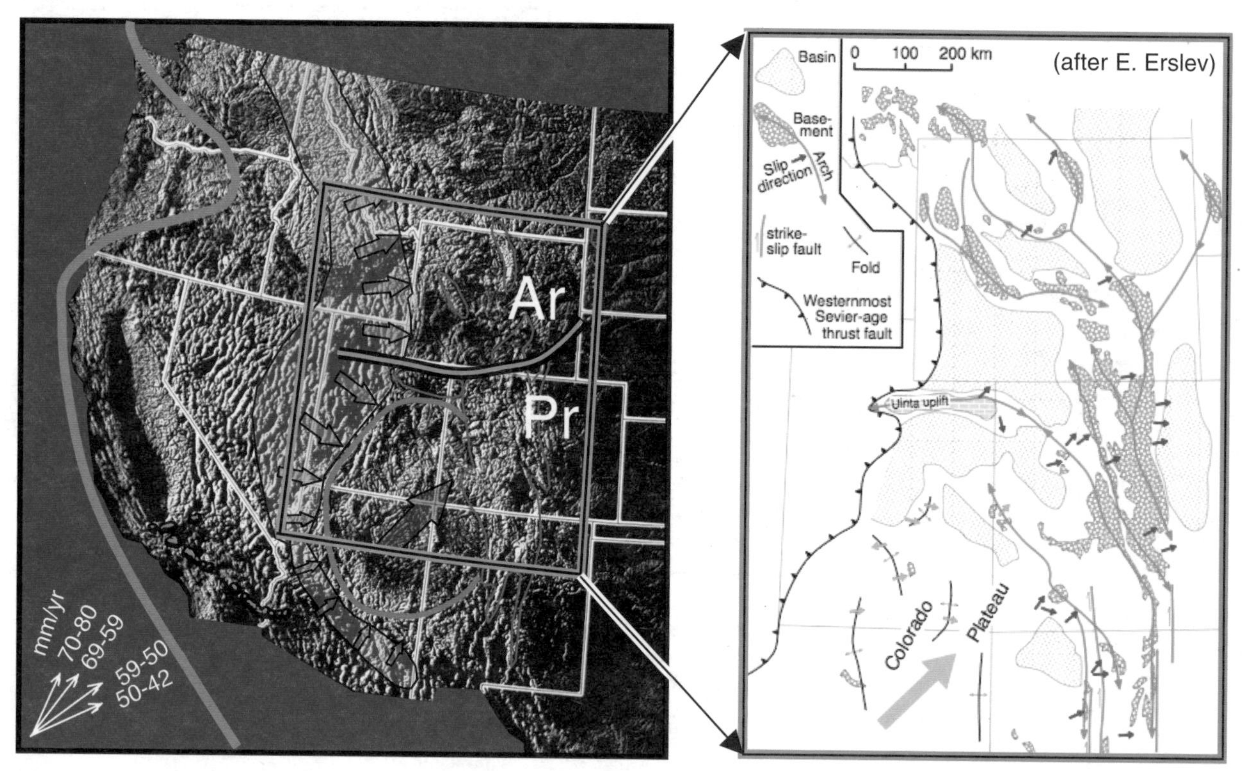

Figure 3. Laramide contraction in relation to other structures, and relation of Laramide uplifts (thin blue outlines) to the Colorado Plateau (lavender outline) and the thickened crust of the Sevier crustal welt (light green). The crustal welt would push the Colorado Plateau toward the east (lavender arrows), whereas subduction is toward the NE (white arrows, with ages in m.y.; from Saleeby, 2003), in the direction that the Colorado Plateau moved during the Laramide orogeny (large blue arrow). Archean (Ar) and Proterozoic (Pr) lithosphere is separated by the blue-green line; note the difference in tectonic character of Laramide uplifts in each region. Paleo–subduction zone is shown with thick blue line, and Pelona-type schist outcrops are shown in red. Enlargement to right shows shortening directions (short arrows) in the southern Rocky Mountains and Colorado Plateau (dark blue, Erslev, 2005; light blue, Bump and Davis, 2003; green, strike-slip faults from Karlstrom and Daniel, 1994; Cather et al., 2006).

phase transition at the 660 km discontinuity (e.g., Billen, 2008), in a manner similar to that imaged beneath SE Asia (Bijwaard et al., 1998). The slab may have then avalanched (e.g., Tackley et al., 1993) or rapidly sunk into the lower mantle during the Sevier orogeny. With avalanching, only minor dynamic subsidence would occur prior to avalanching, followed by a strong suction in the upper mantle beneath the western and central United States, which would pull the continent down (Pysklywec and Mitrovica, 1997) during the Sevier-Laramide orogeny.

The presence of a wide and thick North America craton, by restricting asthenospheric flow into the volume evacuated by subduction entrainment, would act to enhance the magnitude of the suction and its geodynamic effects (Cadek and Fleitout, 2003). Increased suction beneath the western United States would pull the subducted slab near the western plate margin in an eastward direction and pull the craton westward toward the subduction zone, and the west-directed force acting on the craton would increase North America absolute velocity and pull North America over the subduction zone, greatly intensifying compression of the western United States in the process (O'Driscoll et al., 2009).

LARAMIDE

The Laramide phase of the Sevier-Laramide orogeny is limited to the period ca. 75 to ca. 45 Ma, and it is confined to the latitude of the western United States. It is characterized by a low rate of magmatic production and strong tectonic activity reaching far into the continent. The quiescence of arc magmatism presumably resulted from slab flattening against the base of North America (Dumitru et al., 1991; English et al., 2003). These observations suggest that slab contact with the North American interior did not occur prior to the Laramide orogeny, and the extent of Laramide slab flattening involved only a portion of the subducted slab. In particular, during the Laramide orogeny, normal arc magmatism continued in Canada and extended SE from the north Cascades across eastern Washington, most of Idaho, western Montana, and NE Wyoming as the Challis-Absaroka volcanic trend (Christiansen and Yeats, 1992; Fig. 2). I view this portion of the arc to be transitional between the region of normal subduction beneath Canada and flat subduction beneath the Laramide uplifts that extend from SW Montana to westernmost Texas.

Slab Flattening

I suspect that Laramide slab flattening resulted from the combined effects of plateau subduction beneath southern California (Livacarri et al., 1981; Saleeby, 2003), the more regional subduction dynamics associated with the enhanced mantle-wedge suction (O'Driscoll et al., 2009), and the decreasing age of subducted Farallon plate (van Hunen et al., 2002). The region affected by plateau subduction was narrow (Saleeby, 2003) compared to the width of slab that eventually flattened from northern California and Nevada to perhaps central Mexico, as evidenced by an eastward sweep of the magmatic front, quiescence of normal arc magmatism, and subsequent magmatic flare-up over the broad area (i.e., as introduced by Coney and Reynolds, 1977). Ferrari et al. (1999) and Ferrari (2006) argued for a similar slab flattening beneath Mexico. While slab flattening may have occurred beneath most of Mexico, the Laramide style of basement-cored uplifts was limited to the area backed by thick continental lithosphere, which does not include Mexico (Fig. 1C).

Laramide Deformation

With slab flattening during the Laramide orogeny, Colorado Plateau compression against North America (e.g., Hamilton, 1989; Saleeby, 2003) drove NE- to ENE-directed shortening (Fig. 3; Varga, 1993; Erslev, 2005) across a relatively narrow north-trending belt in central New Mexico and Colorado and across a wide area in Wyoming and adjoining states. This direction of shortening was similar to the relative motion of the subducting Farallon slab (Fig. 3), suggesting that the tractions applied by this slab to the base of the Colorado Plateau supplied the most important force driving the Laramide contraction. The alternative—a crustal welt created by earlier Sevier contraction of Great Basin crust—would have pushed the Colorado Plateau in a more easterly direction (Fig. 3).

As illustrated in Figures 4C and 4D, Wyoming and Colorado lithosphere is ~200 km thick, tapering to the SW (~140 km beneath the Four Corners [Smith, 2000] and ~0 km at the Pelona-type schist outcrops of southern California; Figs. 4B and 4E). Tomographic imaging (Fig. 4C; Humphreys et al., 2003) and receiver function imaging (Fig. 4D; Dueker et al., 2001) indicate that the mantle beneath the area of crustal contraction in Colorado and New Mexico is slow to depths of ~200 km (Colorado) and 120 km (New Mexico). This mantle probably is slow because it is partially molten, which suggests that it has been modified relatively recently. Considering that the Colorado Plateau has acted as a strong block (Fig. 4), has low heat flow, and has seismically high-velocity lithosphere (Figs. 1D and 1E), it seems reasonable that Laramide shortening in the Proterozoic lithosphere of Colorado and New Mexico (Fig. 3) occurred directly beneath the zone of crustal shortening, in the area now imaged as seismically slow (Figs. 1D and 4C). In contrast, beneath Wyoming, the mantle appears to be strong everywhere, based on flexural modeling (Lowry and Smith, 1994), low heat flow, and seismic imaging (Figs. 1D and 1E). For these reasons, I assume that the Wyoming lithosphere did not deform greatly during the Laramide orogeny, implying that the upper-crustal shortening distributed broadly over this Archean lithosphere (Fig. 3) was accommodated by a lower-crustal detachment that was rooted somewhere west of Wyoming (as suggested by Erslev, 2005).

POST-LARAMIDE

End of Laramide and Start of Ignimbrite Flare-Up

Laramide termination and initiation of the ignimbrite flare-up that followed were caused by removal of the flat slab and exposure of the thinned and hydrated lithosphere to the infilling asthenosphere (Humphreys, 1995; Humphreys et al., 2003). The Laramide orogeny ended earlier in the northern part of the western United States, and coincided in time with the accretion of a large fragment of oceanic lithosphere (Siletzia and adjoining lithosphere; green area in Fig. 2) to the Pacific Northwest at ca. 48 Ma (Madsen et al., 2006). Accretion of this lithosphere filled the Columbia Embayment and caused subduction to jump west of the accreted lithosphere, initiating the Cascadia subduction zone. Rapidly, the Challis-Absaroka arc ceased volcanic activity, and the Cascade arc became active across Oregon and Washington (Christiansen and Yeats, 1992; Madsen et al., 2006), signaling the establishment of a subduction zone of more typical slab dip by ca. 45 Ma north of California.

The subducting Farallon plate must have torn at the southern margin of the accreted lithosphere (in central Oregon), separating a Cascadia slab of rather normal dip from the Laramide-related flat slab to the south. This opened a northern "window" in the slab. The white-on-black line in Figure 2B shows the southern edge of the window created in 3 m.y. This window opened to the east as the Farallon slab continued subducting. Considering that the Farallon slab at this latitude was associated with the Challis-Absaroka volcanic arc, this window opened within the asthenosphere beneath North America lithosphere. An exception may have been the exposure of forearc to asthenosphere in central-to-northeastern Oregon, which could account for the Clarno volcanic rocks of this age.

Once created, the southern edge of the window propagated from north to south across the Great Basin, progressively exposing the hydrated lithosphere to asthenosphere and causing the ignimbrite flare-up (Fig. 2A). Figure 5 shows my preferred means of slab removal, although a north-to-south rollback of the slab edge (Dickinson, 2002) is also possible. My preference for a buckling style is based on seismic imaging (which shows a high-velocity feature where expected for this style of slab removal; Fig. 6 at 710 km) and my sense for the mechanical difficulty of peeling slab off of the continent (especially when the slab is moving). As subduction of ocean lithosphere continued south of the Siletzia terrane, the northern portion of this slab steepened in dip, as evidenced by the southward propagation of the Cascade arc as far as Lake Tahoe. I am unaware of any post-Laramide

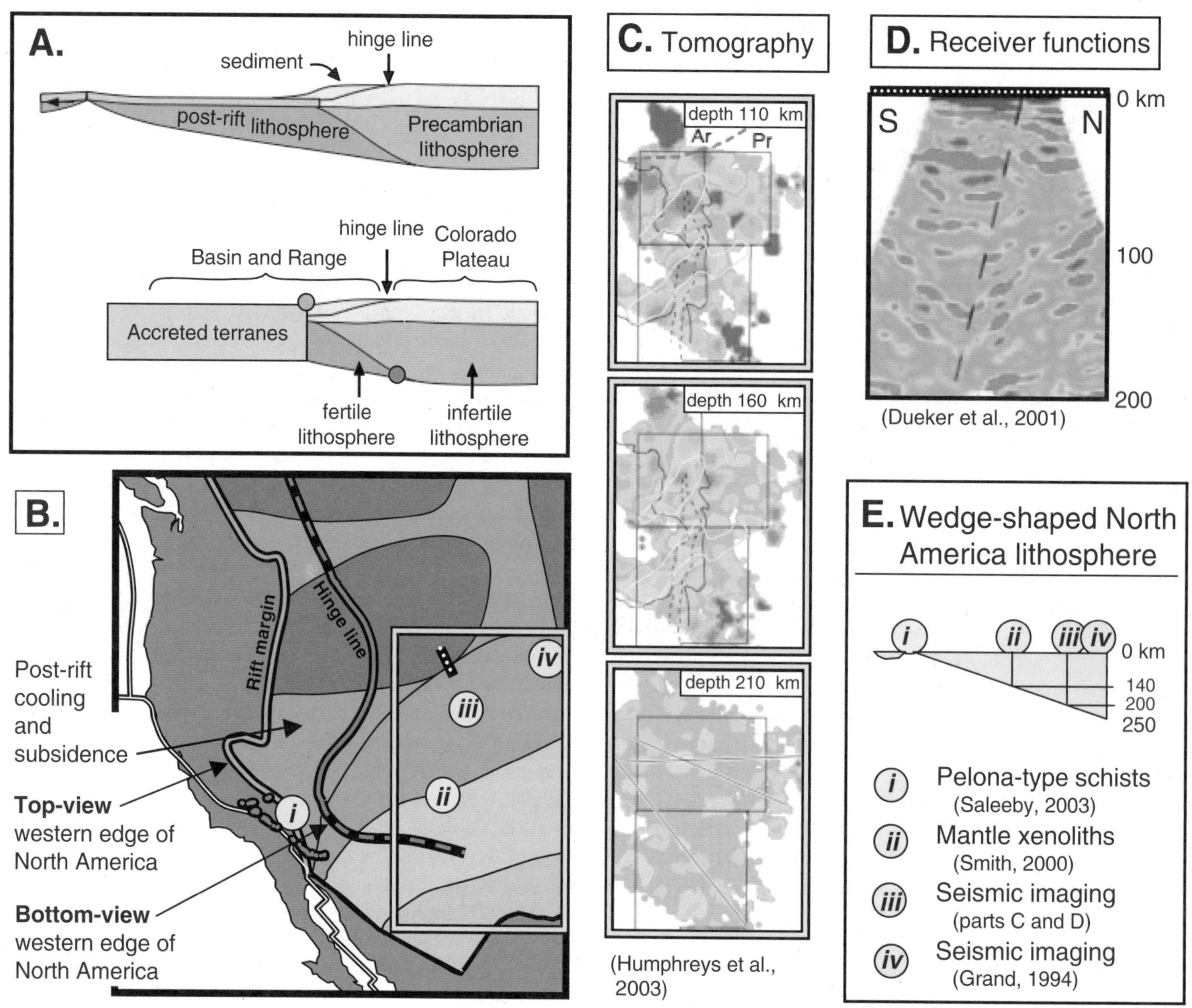

Figure 4. Major inherited features of western North America lithosphere. (A) Creation of fertile lithosphere beneath the Basin and Range Province (from Humphreys et al., 2003). Postrift lithospheric cooling and subsidence west of the hinge line during the lower Paleozoic involved accretion of fertile asthenosphere to the base of the lithosphere (in contrast to the infertile Precambrian lithosphere to the east). This contrast between young (fertile) and old (infertile) North America occurs at the lavender dot; the rift margin usually discussed (blue dot) is farther west. (B) Western U.S. lithospheric elements. The lavender and blue lines correspond to the colored dots in A. Blue areas represent Archean (Ar) lithosphere (dark cratons and light mobile zones), and green areas represent Proterozoic (Pr) lithosphere (accreted arcs that young to the SE). Gray area is Phanerozoic accreted continent, with the current plate-boundary faults shown (white-on-black line). Red areas represent Pelona-type schist that underplated North America during the Laramide orogeny. Short dotted line refers to cross section shown in C, and yellow square outlines area shown in D. Numerals refer to part E. (C) Tomographic image of Rocky Mountain lithosphere (Humphreys et al., 2003). Rectangles show Colorado and New Mexico borders. Yellow lines indicate major areas of Cenozoic volcanic activity, which correlate approximately with the low-velocity (and presumably partly molten) upper-mantle lithosphere. Seismic velocity contrast beneath Colorado extends to ~200 km depth, suggesting that inherited North America lithosphere extends to this depth. The Southern Rocky Mountains are outlined with a solid black line, and the Rio Grande rift is shown with a dashed line. (D) Receiver function image of Rocky Mountain lithosphere (Dueker et al., 2001). Layer structure beneath the region of Proterozoic-Archean suture extends to ~200 km depth. Dashed line shows inferred suture at depth. (E) Representation of wedge-shaped lithosphere of southwestern United States, from southern California to central United States. This is the corridor under which the subducted oceanic plateau is thought to have passed during the Laramide orogeny (Livacarri et al., 1981; Saleeby, 2003). Each numbered location on this figure corresponds with a numbered location in B where a lithospheric thickness estimate exists.

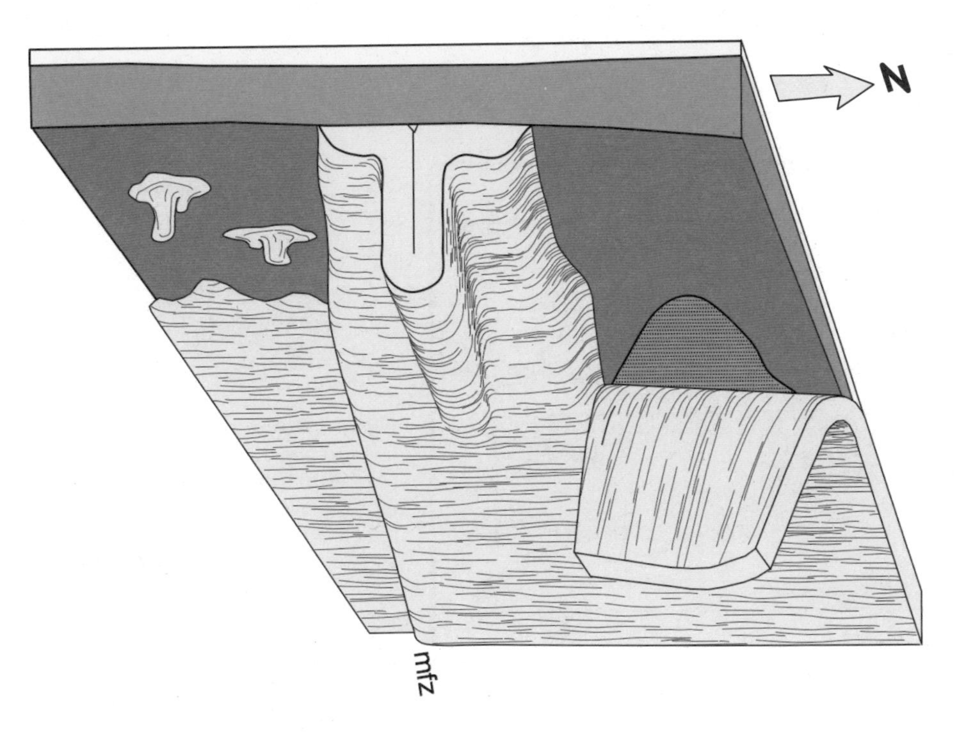

Figure 5. Buckling option for post-Laramide removal of the Farallon slab, representing the period ca. 40 Ma. Accretion of Siletzia (stippled pattern) caused tear in the Farallon slab, which then subducted steeply beneath northern United States. This tear has enabled the flat-slab portion to buckle and its northern edge to propagate to the south, exposing the hydrated lithosphere to asthenosphere. Similarly, the Mendocino fracture zone (mfz) propagates north, leaving hot thin slab beneath Mexico and the southwestern United States; thin slab is shown dripping off of the base of North America. By using the propagation of magmatism during the ignimbrite flare-up (Fig. 2), downwelling occurred along an axis extending WNE from southern Nevada. Plate motion in the 20 m.y. since the ignimbrite flare-up ended leaves this buckled slab where high-velocity mantle is imaged (Fig. 6, 710-km-depth frame; arrow shows 20 m.y. of North America absolute motion).

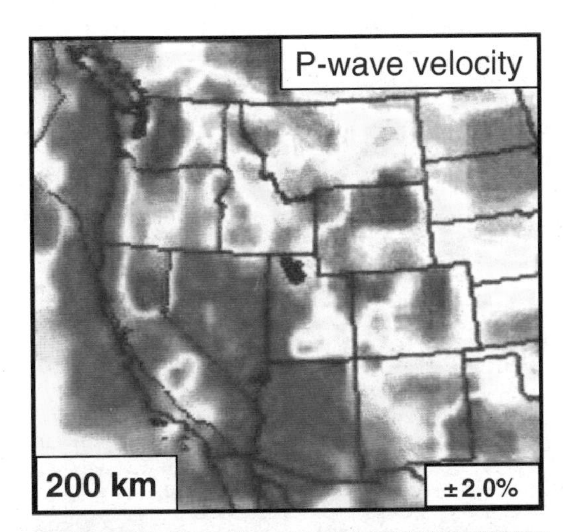

Figure 6. Seismic P-wave velocity perturbations (Bijwaard et al., 1998) at indicated depths. Blue is fast, and range of imaged variations is indicated in each frame. High velocities below 200 km represent slab subducted beneath western North America.

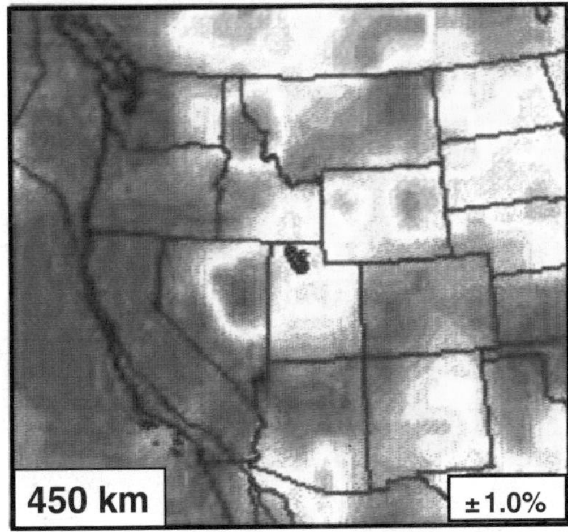

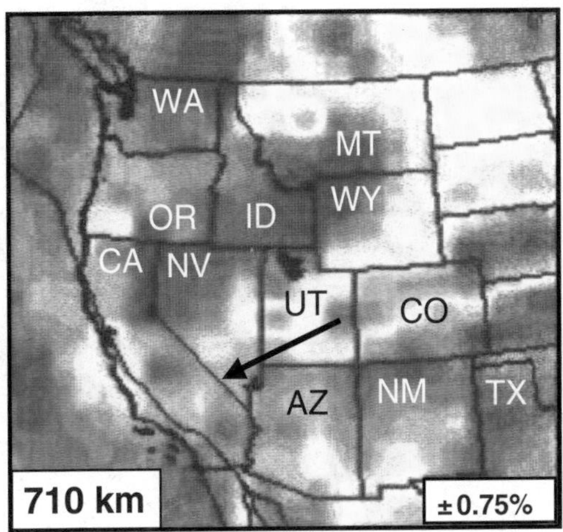

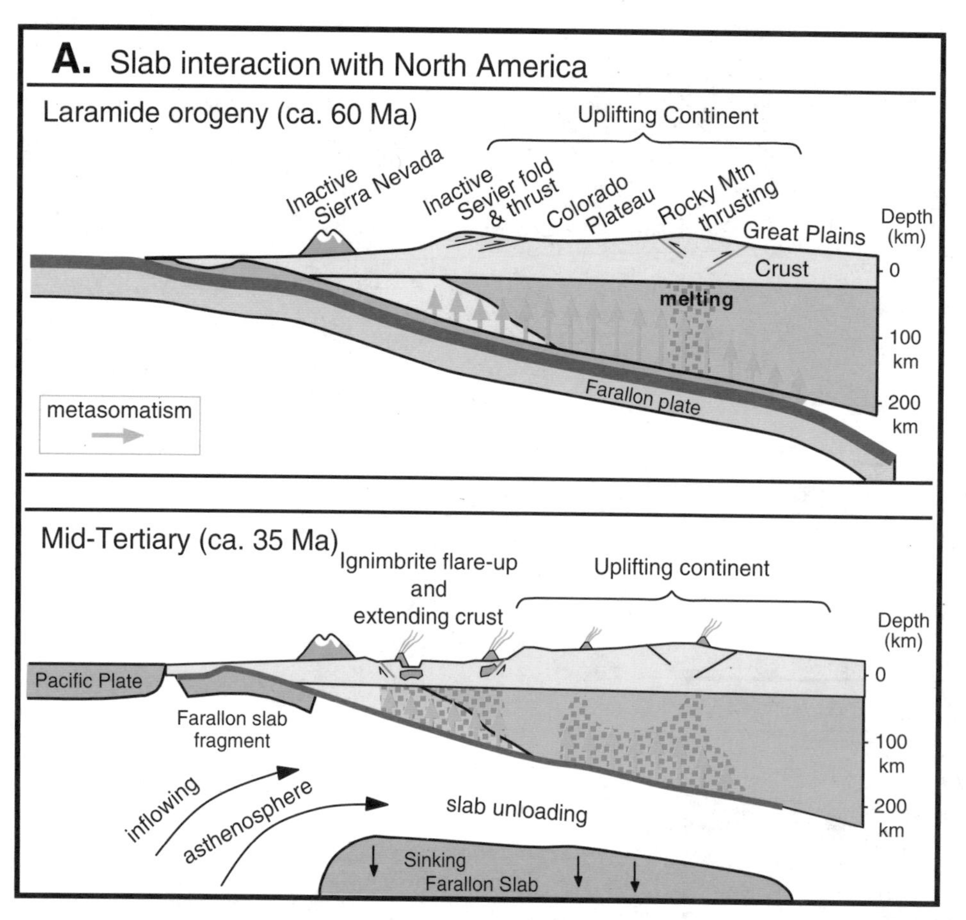

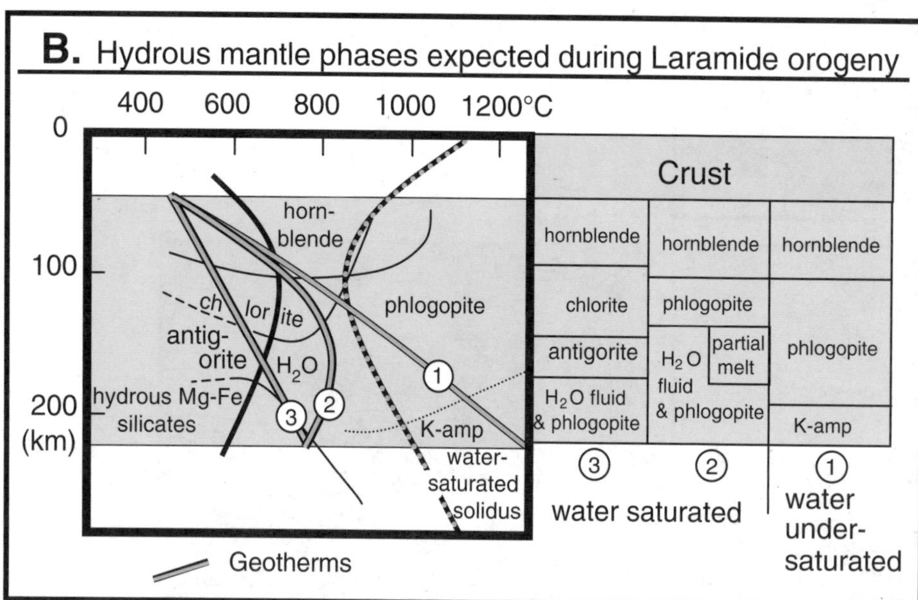

Figure 7. Laramide lithospheric hydration and post-Laramide mantle melting. (A) Relation of mantle processes to surface geologic activity. Top frame represents the Laramide orogeny and shows the western margin of North America with subducted slab in contact with (and hydrating; green arrows) North America. The mantle types are from Figure 4A. The bottom frame represents the ignimbrite flare-up and shows the magmatism of variable intensity (red for more intense; pink for less intense). (B) Hydrous mantle phases under the water-saturated and cool conditions expected during Laramide orogeny (phase diagram after Bromiley and Pawley, 2003; Elkins-Tanton, 2007; Kawamoto and Holloway, 1997). Lithosphere that is 150 km thick can hold 0.1 km thickness of water in olivine. Pink dotted line is the wet solidus. Green lines represent hypothetical geotherms. Prior to and during initial slab contact (geotherm 1), the infusion of water would cause melting of basal lithosphere and some magmatism as the Farallon slab flattened. After cooling caused by slab contact (geotherm 2), free-phase water (at greater depth) and hydrous phases (at shallower depths) would be stable. Geotherm 3 shows temperature profile at conductive equilibrium for basal temperature of 750 °C (for reference). To the right, illustration of lithosphere shows possible hydrous phases for the three geotherms shown.

arc activity in the United States south of the Lake Tahoe region (Henry et al., 2006), suggesting that a normal subduction zone never was reestablished beneath the southwestern United States after the Laramide orogeny. Assuming that the flat slab was beneath the Rocky Mountains, slab removal would have brought asthenosphere in contact with areas that were not involved in the ignimbrite flare-up (i.e., the Rocky Mountains, Colorado Plateau, and perhaps the Great Plains). I attribute the remarkable volume of magma production from the Basin and Range Province (and not the areas interior to this) to this lithosphere being thinner (with greater volumes of melt created at the lower pressures) and more fertile (see Figs. 4A, 4B, and 7A).

Great Basin lithosphere, being thermally weakened and increasing in elevation (as a result of slab removal and heating), gravitationally collapsed by expanding over the subducting Farallon plate (Sonder et al., 1987; Jones et al., 1996). Within the deforming western United States, Siletzia and the Sierra Nevada–Great Valley blocks (the two terranes composed largely of oceanic lithosphere) retained sufficient strength to avoid deformation. As a result, Great Basin extension involved a westward drift of the Sierra Nevada–Great Valley block and a clockwise rotation of the Siletzia terrane, as illustrated in Figure 2B.

Transition to Shear-Dominated Deformation

North America encountered the Pacific plate near the later part of the ignimbrite flare-up, which initiated the transform margin and a second slab window (that discussed by Dickinson and Snyder, 1979). Sierra Nevada–Great Valley motion away from North America became regulated by the rate at which the Pacific plate moved away from North America, and Basin and Range extension slowed to this rate.

A change in Pacific–North America plate motion from NNW to WNW at 7–8 Ma (Atwater and Stock, 1998) caused the Pacific plate to move approximately parallel to the North America plate margin. This further inhibited North America extension and caused North America deformation to reorganize. Southern Basin and Range extension became nearly inactive, and significant Gulf of California opening began, which incorporated extension by stepping into the continent and placing the transform margin (the San Andreas fault) into California (e.g., Holt et al., 2000). The Sierra Nevada–Great Valley block changed its motion from WNW to NNW (Wernicke et al., 1988), nearly parallel to Pacific–North America relative motion (shown by the kink in the arrows in Fig. 2B). This block motion was accommodated by an interior shear zone, the Eastern California–Walker Lane shear zone (Fig. 8), which has accommodated ~ 1 cm/yr of right-lateral shear strain during the last ~7 m.y. (Faulds et al., 2005). The oceanic Siletzia terrane remained too strong to deform significantly, and it accommodated right-lateral shear through continued clockwise block rotation. To accommodate the width of this block, the interior shear zone broadened across NW Nevada and SE Oregon (Fig. 8). This broadening resulted in faults of a releasing orientation, and the shear zone was then integrated with Basin and Range extension that includes the central Nevada seismic zone (Hammond and Thatcher, 2004). North of the Siletzia terrane (green area in Fig. 8), the shear zone narrows as it trends to the north by stepping westward toward the plate margin. The resulting restraining geometry results in north-south contraction in the Yakima fold-and-thrust belt (along the northern margin of the Siletzia terrane, Fig. 8), continuing to the Seattle area (Wells et al., 1998).

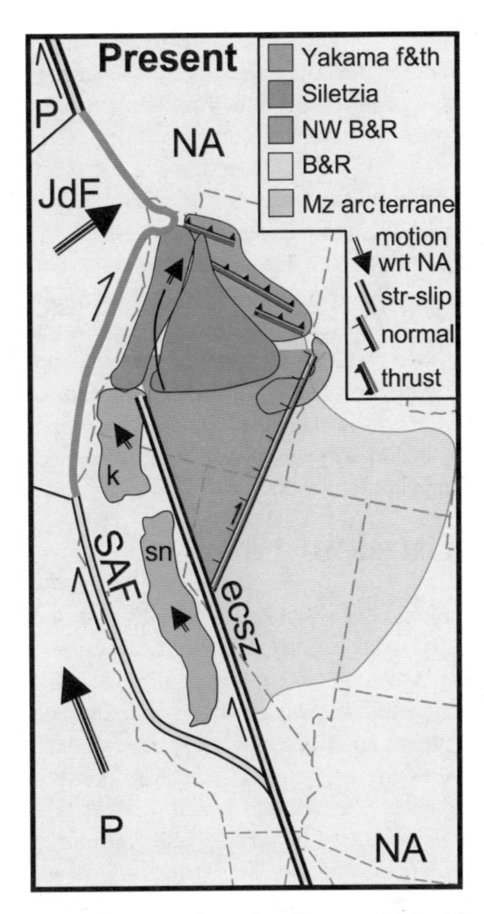

Figure 8. Current tectonic setting of western United States, emphasizing the shear margin. At the southern and northern ends, Pacific–North America (P-NA) relative motion is accommodated on narrow transform systems (yellow-on-black line for strike-slip fault). At the latitude of California, the San Andreas fault (SAF) and Eastern California shear zone (ecsz) accommodate Pacific–North America relative motion, and the Sierra Nevada–Great Valley (sn) and Klamath (k) blocks occupy a transitional setting (motion shown with respect to North America). Northern motion of the Klamath block is accommodated by rotation of the Siletzia block (green), shown as broken and separating at the Cascade graben (Priest, 1990). The shear zone broadens across the width of the rotating Siletzia block, necessitating extension of the NW Basin and Range (B&R) and contraction across the Yakima fold-and-thrust belt (f&th). Siletzia rotation is consistent with the oblique convergence of the Juan de Fuca plate (JdF) south of Canada; no shear deformation occurs where JdF subduction is normal.

Extension in the northern Basin and Range and NNW motion of the transform-entrained Sierra Nevada–Great Valley block each require space, and the Pacific plate does not permit continental growth in a westward direction. Instead, accommodation occurs by south Cascadia rollback. Northern Basin and Range faulting has reoriented so that extension is toward southern Cascadia, and the Sierra Nevada–Great Valley block moves to the NNW over the subduction zone. Overall, shear strain dominates the marginal ~300 km of the continent, dilatational strain dominates areas to the east where deviatoric stress is still controlled by high gravitational potential energy (Flesch et al., 2000; Humphreys and Coblentz, 2007), and contraction is common in the northwest United States; where strength is low, deformation occurs at geologically significant rates (Fig. 1B) (Kreemer and Hammond, 2007).

CURRENT PHYSICAL STATE

The geologic history of the western United States has created a density structure and strength distribution that largely dictate current tectonic activity by contributing driving forces and controlling the distribution of deformation (Flesch et al., 2000; Humphreys and Coblentz, 2007). Knowledge of this evolution provides the context with which to constrain thinking about the physical changes that created the density and strength variations. Through seismic imaging and geodetic and geologic mapping of young and ongoing deformation, we recognize and better understand the connections between what has been inherited and its controls activity; we also find structure in these fields that is surprising and requires some new thinking. In this section, I describe my view of the current western U.S. physical state, how prior events created this, and how the physical state controls current activity.

Uplift and Buoyancy

Most of the lithosphere above which the flat slab is thought to have contacted North America lithosphere is, on average, seismically slow (Figs. 1C and 1D), and the surface has been elevated from near sea level to ~2 km (Fig. 1A). The western portion of the uplifted area has been and continues to be deformed (Fig. 1B) and magmatically modified. These indicate a rather widespread and profound increase in lithospheric buoyancy and temperature and a decrease in strength. The areas that have been uplifted involve a diverse set of distinctive geomorphic and tectonic provinces, including the mildly contracted Colorado Plateau, the more strongly contracted Rocky Mountains, the highly extended Great Basin, and the tilted Great Plains (Fig. 1A). The broad nature of uplift over this diversity of tectonic provinces suggests a broad-scale and deep underlying cause. Potential causes for the increased buoyancy include Laramide and post-Laramide contributions, and because the timing of uplift is poorly understood, the sources of buoyancy are poorly understood.

The amount of Laramide crustal shortening (Erslev, 1993; Hamilton, 1989) accounts for the 0.5–1 km of elevation of the Rocky Mountains above their surrounding area. Most other potential buoyancy contributions are more regional. Pre-Laramide dynamic subsidence owing to the negative buoyancy of the Farallon slab would continue as an isostatic depression during the flat-slab Laramide-age subduction. (In either case, the depression is attributed to the weight of the slab.) As the age of the flat slab became younger, the influence of slab negative buoyancy diminished, and some uplift resulted. Simultaneously, hydration (Humphreys et al., 2003) and mechanical erosion (Spencer, 1996) of basal North America lithosphere by the flat slab would contribute buoyancy, as would post-Laramide heating associated with magmatic invasion of the lithosphere.

Within the uplifted western United States, areas of little magmatism occur where the lithosphere is thick. Lithospheric erosion, therefore, probably is not the primary cause of uplift in these areas. I infer that hydration-related buoyancy was widely important for holding up the western United States region, and that in areas of high-volume magmatism, this buoyancy was replaced by thermal and compositional (basalt depletion) buoyancy. Thus, I view lithospheric hydration under the cool conditions of flat-slab contact as fundamental to post-Laramide western U.S. tectonic and magmatic activity (Humphreys et al., 2003). Most directly, it created buoyancy through the production of low-density minerals. Figure 7B illustrates that under the hydrous and low-temperature conditions created by flat-slab subduction, free-phase water is stable and hence available to migrate upward, where hornblende and chlorite would form above ~100 km depth. With lithospheric cooling, serpentine (antigorite) becomes stable at all depths (assuming sufficient cooling; Fig. 7B). Upon slab removal and heating, the hydrated and fertile lithosphere produced large volumes of magma (especially from the thin lithosphere of the compositionally fertile Basin and Range Province; Figs. 4A and 4B). This mantle is expected to be dehydrated now. The corresponding loss of hydration-related buoyancy probably is compensated by the addition of thermal and depletion buoyancy (Humphreys and Dueker, 1994).

Below the Lithosphere

The subducting Gorda–Juan de Fuca plate retains a gap at the location of the inherited tear in central Oregon (that resulted with Siletzia accretion; Fig. 2). The slab south of this gap now involves only the Gorda plate. This subduction is a continuation of long-lived Farallon subduction (Engebretson et al., 1985; Stock and Molnar, 1988), and the width of the subduction zone has diminished as the Mendocino triple junction has migrated north (Atwater and Stock, 1998). This subducted slab is imaged in the mantle to extend beneath northern California to northern Nevada and beyond (Fig. 6 at 450 km). The north-propagating southern edge of the Gorda slab is imaged where predicted by plate reconstruction models, extending from the Mendocino triple junction to central Nevada. North of central Oregon, the subducting Juan de Fuca slab is imaged to extend beneath Washington and northern Idaho. This implies slab that has been subducted at the Cascadia subduction zone since Siletzia accretion. The gap between the northern and southern slab segments is imaged starting at a depth of ~150 km beneath central Oregon (Rasmussen and Humphreys, 1988; Bostock and Vandecar, 1995) and extending NE across Oregon, central Idaho, and Montana (Fig. 6). The persistence of this tear in the slab is consistent with modeled slab behavior. When slabs subduct, they tend to buckle and narrow in width as they sink (Piromello et al., 2006), especially near their edges (Schellart et al., 2007). This would lead to the two sides of the slab to pull away from each other, widening the gap.

The weight of the sinking Gorda–Juan de Fuca slab drives a toroidal flow of asthenosphere from beneath the slab around its southern edge to above the slab (Piromello et al., 2006; Stegman et al., 2006; Zandt and Humphreys, 2008). The tectonic effects of this flow, if any, are not clear. Flow beneath the Great Basin will apply north-directed tractions that may be important in driving Great Basin lithosphere northward. It may also supply a means of sweeping away the base of the cooling lithosphere, thereby keeping the Great Basin hot, elevated, and magmatically active (in contrast to the southern Basin and Range of Arizona and western Sonora). The WNW-directed flow across northern California and southern Oregon appears to have a more obvious effect on the magmatic propagation of Newberry and Medicine Lake volcanic activity, which is toward the WNW (Draper, 1991).

Seismic tomography of the upper mantle beneath Yellowstone images a plume-like structure dipping ~75°WNW to a depth of ~450 km (Yuan and Dueker 2005; Waite et al., 2006), and receiver function imaging finds a 410 km discontinuity deflected downward (consistent with anomalously high temperatures) at the location of the plume-like structure (Fee and Dueker, 2004). Considering that Yellowstone magma has anomalously high $^3He/^4He$ values (Hearn et al., 1990), which suggest a lower-mantle source, and that excessively hot mantle probably has an origin in a lower thermal boundary layer, it seems reasonable to suggest that Yellowstone plume originated from the lower mantle. However, neither the plume-like structure nor the 660 km discontinuity indicates anomalous structure near or below this depth. It appears that Yellowstone magmatism involves anomalously hot mantle that ascends from the transition zone, but that its connection to the lower mantle (if any) is not a simple plume-like structure. The apparent location of the Yellowstone "plume" at 400–450 km depth is in the gap between the Gorda and Juan de Fuca slabs (Fig. 6), which suggests that the negative buoyancy of the Gorda–Juan de Fuca slab drives a flow of lower mantle through this gap that is anomalously hot and elevated in $^3He/^4He$. This would make Yellowstone magmatism fundamentally an interaction between subducted slab and a hot volume of upper mantle.

Small-Scale Convection Everywhere

Following post-Laramide slab removal, activity in the upper mantle appears to have been dominated by small-scale convection of various forms (Fig. 1D), including systems driven by positively buoyant asthenosphere and others driven by negatively buoyant lithosphere. Upwellings include: Yellowstone and low-velocity mantle left in its track (Yuan and Dueker, 2005); the central Colorado low-velocity zone associated with the large San Juan volcanic field (Dueker et al., 2001); the central Utah Springerville volcanic field beneath the Colorado Plateau (Dueker and Humphreys, 1990) and Colorado Plateau–fringing areas of low velocity; and the Salton Trough (Humphreys et al., 1984). Except for the Salton Trough, these low-velocity volumes appear to be active (i.e., they are ascending under the influence of their positive buoyancy). Downwellings include: the Transverse Ranges "drip" (Bird and Rosenstock, 1984; Humphreys and Hager, 1990; Billen and Houseman, 2004) (lying beneath Laramide-age subduction complex, this probably is a sinking fragment of the abandoned Farallon slab); southern Sierra Nevada delamination, which owes its negative buoyancy as much to its composition as its temperature (Ducea and Saleeby, 1996, 1998); the apparent delamination to the east of the Rio Grande Rift (Gao et al., 2004); Wallowa delamination beneath NE Oregon (Hales et al., 2005), which is a past event associated with the Columbia River flood basalt eruptions; and subduction of the Gorda and Juan de Fuca slabs. All these cases are active, and except for sinking ocean slab, all involve destabilized North America mantle lithosphere; most appear to be related to the initiation of focused strain zones.

Individually these constitute a set of small-scale processes that are only partially understood but are important to local tectonics and magmatic activity. Collectively, they represent a poorly understood "mesoscale" event that clearly is important to the mass and heat flux through the lithosphere, and to the construction and modification of continental lithosphere. A major goal in the next decade is to understand how the active small-scale convective processes relate to one another, to prior and current mantle conditions, and how they act as an integrated whole beneath the western United States to create continental lithosphere.

SUMMARY

In the western United States, many processes related more-or-less directly to subduction have destabilized what was a stable plate, resulting in strength and density reduction, which in turn have activated plate disintegration on scales ranging from small-scale convection to lithosphere-scale extension. Forces responsible for deformation have resulted from the plate-tectonic boundary conditions and from the elevated land and resulting gravitational potential energy of the western United States; over the western half of the uplifted western United States, a regional loss of lithospheric strength has enabled deformation. Magmatism has been both a result of and an agent for modification through its effects on temperature, strength, buoyancy, and composition.

Although most western U.S. tectonic and magmatic activity over the last 150 m.y. has been related to plate-tectonic processes in general, and to subduction in particular, much of this activity is related to processes not typically thought of as plate tectonic in nature. Non–plate tectonic activities include: basal tractions acting on North America (e.g., dynamic downwarping and strong contraction caused by slab sinking); regional increases in lithospheric buoyancy (e.g., resulting from mantle lithosphere hydration and basaltic depletion) and the uplift and gravitational collapse it drives; loss of lithospheric strength (e.g., caused directly by hydration or by hydration-related magmatic heating); small-scale convection and the resulting loss of strength and density; thermal-mechanical effects of the asthenosphere flow around the edge of sinking Gorda–Juan de Fuca slab; and mantle flow driven by processes related to Yellowstone magmatism.

This view has plate tectonics playing a dominant role in shaping the western United States, but it incorporates more mechanical coupling with the interior of Earth and a more important role for vertical tectonics and far-from-boundary effects of plate interaction than most geologists usually consider. Once this region cools, the continent will stabilize (through the increase in strength and reduction in gravitational potential energy) as a largely reconstructed volume of continental lithosphere. Lithosphere that remains strongly hydrated or depleted will retain a component of long-term buoyancy and elevation.

ACKNOWLEDGMENTS

This synthesis is the result of free and enthusiastic conversations and interactions with many researchers and students, for which I am most grateful. I would like to thank Sue Kay for handling this manuscript and for her time and effort (along with Víctor Ramos) in organizing the outstanding "Backbone of the Americas" meeting. Steve Grand and Walt Snyder provided useful and thoughtful reviews. I thank Mark Hemphill-Haley for Figure 1B. The research involved with this synthesis was supported by National Science Foundation (NSF) grants EAR-0642487, EAR-0545404, EAR-509965, EAR-511000, and EAR454489.

REFERENCES CITED

Atwater, T., and Stock, J.M., 1998, Pacific–North America plate tectonics of the Neogene southwestern United States: An update: International Geology Review, v. 40, p. 375–402.

Bijwaard, H., Spakman, W., and Engdahl, E.R., 1998, Closing the gap between regional and global travel time tomography: Journal of Geophysical Research, v. 103, p. 30,055–30,078, doi: 10.1029/98JB02467.

Billen, M.I., 2008, Modeling the dynamics of subducting slabs: Annual Review of Earth and Planetary Sciences, v. 36, p. 325–356, doi: 10.1146/annurev.earth.36.031207.124129.

Billen, M.I., and Houseman, G.A., 2004, Lithospheric instability in obliquely convergent margins: San Gabriel Mountains, southern California: Journal of Geophysical Research, v. 109, doi: 10.1029/2003JB002605.

Bird, P., and Rosenstock, R.W., 1984, Kinematics of present crust and mantle flow in southern California: Geological Society of America Bulletin, v. 95, p. 946–957, doi: 10.1130/0016-7606(1984)95<946:KOPCAM>2.0.CO;2.

Bostock, M.G., and Vandecar, J.C., 1995, Upper mantle structure of the northern Cascadia subduction zone: Canadian Journal of Earth Sciences, v. 32, p. 1–12.

Bromiley, G.D., and Pawley, A.R., 2003, The stability of antigorite in the systems $MgO-SiO_2-H_2O$ (MSH) and $MgO-Al_2O_3-SiO_2-H_2O$ (MASH): The effects of Al^{3+} substitution on high-pressure stability: The American Mineralogist, v. 88, p. 99–108.

Bump, A.P., and Davis, G.H., 2003, Late Cretaceous to early Tertiary Laramide deformation of the northern Colorado Plateau, Utah and Colorado: Journal of Structural Geology, v. 25, p. 421–440, doi: 10.1016/S0191-8141(02)00033-0.

Burchfiel, B.C., Cowan, D.S., and Davis, G.A., 1992, Tectonic overview of the Cordilleran orogen in the western United States, *in* Burchfiel, B.C., Lipman, P.W., and Zoback, M.L., eds., The Cordilleran Orogen: Conterminous U.S.: Boulder, Colorado, Geological Society of America, Geology of North America, v. G-3, p. 407–480.

Cadek, O., and Fleitout, L., 2003, Effect of lateral viscosity variations in the top 300 km on the geoid and dynamic topography: Geophysical Journal International, v. 152, p. 566–580, doi: 10.1046/j.1365-246X.2003.01859.x.

Cather, S.M., Karlstrom, K.E., Timmons, J.M., and Heizler, M.T., 2006, Palinspastic reconstruction of Proterozoic basement-related aeromagnetic features in north-central New Mexico: Implications for Mesoproterozoic to late Cenozoic tectonism: Geosphere, v. 2, no. 6, p. 299–323, doi: 10.1130/GES00045.1.

Christiansen, R.L., and Yeats, R.L., 1992, Post-Laramide geology of the U.S. Cordilleran region, *in* Burchfiel, B.C., Lipman, P.W., and Zoback, M.L., eds., The Cordilleran Orogen: Conterminous U.S.: Boulder, Colorado, Geological Society of America, Geology of North America, v. G-3, p. 261–406.

Coney, P.J., and Reynolds, S.J., 1977, Cordilleran Benioff zones: Nature, v. 270, p. 403–406, doi: 10.1038/270403a0.

DeCelles, P.G., 2004, Late Jurassic to Eocene evolution of the Cordilleran thrust belt and foreland basin system, western U.S.A.: American Journal of Science, v. 304, p. 105–168, doi: 10.2475/ajs.304.2.105.

Dickinson, W.R., 2002, The Basin and Range Province as a composite extensional domain: International Geology Review, v. 44, p. 1–38, doi: 10.2747/0020-6814.44.1.1.

Dickinson, W.R., and Snyder, W.S., 1979, Geometry of subducted slabs related to the San Andreas transform: The Journal of Geology, v. 87, p. 609–627.

Draper, D., 1991, Late Cenozoic bimodal volcanism in the northern Basin and Range: Journal of Volcanology and Geothermal Research, v. 47, p. 299–328, doi: 10.1016/0377-0273(91)90006-L.

Ducea, M.N., and Saleeby, J.B., 1996, Buoyancy sources for a large, unrooted mountain range, the Sierra Nevada, California; evidence from xenolith thermobarometry: Journal of Geophysical Research, v. 101, p. 8229–8244, doi: 10.1029/95JB03452.

Ducea, M.N., and Saleeby, J.B., 1998, A case for delamination of the deep batholithic crust beneath the Sierra Nevada, California: International Geology Review, v. 40, p. 78–93.

Dueker, K., and Humphreys, E., 1990, Upper-mantle velocity structure of the Great Basin: Geophysical Research Letters, v. 17, p. 1327–1330, doi: 10.1029/GL017i009p01327.

Dueker, K.G., Yuan, H., and Zurek, B., 2001, Thick-structured Proterozoic lithosphere of the Rocky Mountains: GSA Today, v. 11, no. 12, p. 4–9, doi: 10.1130/1052-5173(2001)011<0004:TSPLOT>2.0.CO;2.

Dumitru, T.A., Gans, P.B., Foster, D.A., and Miller, E.L., 1991, Refrigeration of the western Cordilleran lithosphere during Laramide shallow-angle subduction: Geology, v. 19, p. 1145–1148, doi: 10.1130/0091-7613(1991)019<1145:ROTWCL>2.3.CO;2.

Elkins-Tanton, L.T., 2007, Continental magmatism, volatile recycling, and a heterogeneous mantle caused by lithospheric gravitational instabilities: Journal of Geophysical Research, v. 107, doi: 10.1029/2005JB004072.

Engebretson, D.C., Cox, A., and Gordon, R.G., 1985, Relative Motions between Oceanic and Continental Plates in the Pacific Basin: Geological Society of America Special Paper 206, 59 p.

English, J.M., Johnston, S.T., and Wang, K., 2003, Thermal modelling of the Laramide orogeny: Testing the flat-slab subduction hypothesis: Earth and Planetary Science Letters, v. 214, p. 619–632, doi: 10.1016/S0012-821X(03)00399-6.

Erslev, E.A., 1993, Thrusts, back-thrusts, and detachment of Laramide foreland arches, *in* Schmidt, C.J., Chase, R., and Erslev, E.A., eds., Laramide Basement Deformation in the Rocky Mountain Foreland of the Western United States: Geological Society of America Special Paper 280, p. 339–358.

Erslev, E.A., 2005, 2D Laramide geometries and kinematics of the Rocky Mountain region, *in* Karlstrom, K.E., and Keller, G.R., eds., The Rocky Mountain Region—An Evolving Lithosphere: Tectonics, Geochemistry, and Geophysics: Washington, D.C., American Geophysical Union, p. 7–20.

Faulds, J.E., Henrey, C.D., and Hinz, N.H., 2005, Kinematics of the northern Walker Lane: An incipient transform fault along the Pacific–North American plate boundary: Geology, v. 33, p. 505–508, doi: 10.1130/G21274.1.

Fee, D., and Dueker, K., 2004, Mantle transition zone topography and structure beneath the Yellowstone hotspot: Geophysical Research Letters, v. 31, doi: 10.1029/2004GL020636.

Ferrari, L., 2006, Laramide and Neogene shallow subduction in Mexico: Constraints and contracts, *in* Backbone of the Americas—Patagonia to Alaska: Geological Society of America Abstracts with Programs, p. 23.

Ferrari, L., Martinez, M.L., Diaz, G.A., and Nunez, G.C., 1999, Space-time patterns of Cenozoic arc volcanism in central Mexico; from the Sierra Madre Occidental to the Mexican volcanic belt: Geology, v. 27, p. 303–306, doi: 10.1130/0091-7613(1999)027<0303:STPOCA>2.3.CO;2.

Flesch, L.M., Holt, W.E., Haines, A.J., and Shen-Tu, B., 2000, Dynamics of the Pacific–North American plate boundary in the western United States: Science, v. 287, p. 834–836, doi: 10.1126/science.287.5454.834.

Gao, W., Grand, S., Baldridge, W.S., Wilson, D., West, M., Ni, J., and Aster, R., 2004, Upper mantle convection beneath the central Rio Grande rift imaged by P and S wave tomography: Journal of Geophysical Research, v. 109, p. B03305, doi: 10.1029/2003JB002743.

Giorgis, S., Tikoff, B., and McClelland, W., 2005, Missing Idaho arc: Transpressional modification of the $^{87}Sr/^{86}Sr$ transition on the western edge of the Idaho batholith: Geology, v. 33, p. 469–472, doi: 10.1130/G20911.1.

Grand, S.P., 1994, Mantle shear structure beneath the Americas and surrounding oceans: Journal of Geophysical Research, v. 99, p. 11,591–11,621, doi: 10.1029/94JB00042.

Gurnis, M., 1993, Phanerozoic marine inundation of continents driven by dynamic topography above subducting slabs: Nature, v. 364, p. 589–593, doi: 10.1038/364589a0.

Hales, T.C., Abt, D., Humphreys, E., and Roering, J., 2005, A lithospheric origin for Columbia River flood basalts and Wallowa Mountains uplift in northeast Oregon: Nature, v. 438, p. 842–845, doi: 10.1038/nature04313.

Hamilton, W.B., 1989, Crustal geologic processes of the United States, *in* Pakiser, L.C., and Mooney, W.D., eds., Geophysical Framework of the Continental United States: Geological Society of America Memoir 172, p. 743–781.

Hammond, W., and Thatcher, W., 2004, Contemporary tectonic deformation of the Basin and Range Province, western United States: 10 years of observation with the global positioning system: Journal of Geophysical Research, v. 109, doi: 10.1029/2003JB002746.

Hearn, E.H., Kennedy, B.M., and Truesdell, A.T., 1990, Coupled variations in helium isotopes and fluid chemistry: Shoshone Geyser Basin, Yellowstone National Park: Geochimica et Cosmochimica Acta, v. 54, p. 3103–3113, doi: 10.1016/0016-7037(90)90126-6.

Henry, C.D., Cousns, B.L., Castor, S.B., Faulds, J.E., Garside, L.J., and Timmermans, A., 2004, The Ancestral Cascades arc, northern California/western Nevada: Spatial and temporal variations in volcanism and geochemistry: Eos (Transactions, American Geophysical Union), v. 85, no. 47, V13B-1478.

Holt, J.W., Holt, E.W., and Stock, J.M., 2000, An age constraint on Gulf of California rifting from the Santa Rosalía basin, Baja California Sur, Mexico: Geological Society of America Bulletin, v. 112, p. 540–549, doi: 10.1130/0016-7606(2000)112<0540:AACOGO>2.3.CO;2.

Humphreys, E., 1995, Post-Laramide removal of the Farallon slab, western United States: Geology, v. 23, p. 987–990, doi: 10.1130/0091-7613(1995)023<0987:PLROTF>2.3.CO;2.

Humphreys, E.D., and Coblentz, D.D., 2007, North America dynamics and western U.S. tectonics: Reviews of Geophysics, v. 45, RG3001, doi: 10.1029/2005RG000181.

Humphreys, E., and Dueker, K., 1994, Physical state of the western U.S. upper mantle: Journal of Geophysical Research, v. 99, p. 9635–9650, doi: 10.1029/93JB02640.

Humphreys, E., and Hager, B., 1990, A kinematic model for the late Cenozoic development of southern California crust and upper mantle: Journal of Geophysical Research, v. 95, p. 19,747–19,762, doi: 10.1029/JB095iB12p19747.

Humphreys, E., Clayton, R., and Hager, B., 1984, A tomographic image of mantle structure beneath southern California: Geophysical Research Letters, v. 11, p. 625–627, doi: 10.1029/GL011i007p00625.

Humphreys, E., Hessler, E., Dueker, K., Erslev, E., Farmer, G.L., and Atwater, T., 2003, How Laramide-age hydration of North America by the Farallon slab controlled subsequent activity in the western U.S.: International Geology Review, v. 45, p. 575–595, doi: 10.2747/0020-6814.45.7.575.

Jones, C.H., Unruh, J.R., and Sonder, L.J., 1996, The role of gravitational potential energy in active deformation in the southwestern United States: Nature, v. 381, p. 37–41, doi: 10.1038/381037a0.

Karlstrom, K.E., and Daniel, C.G., 1994, Restoration of Laramide right-lateral strike slip in northern New Mexico by using Proterozoic piercing points: Tectonic implications from the Proterozoic to the Cenozoic: Geology, v. 22, p. 863–864.

Kawamoto, T., and Holloway, J.R., 1997, Melting temperature and partial melt chemistry of H_2O-saturated mantle peridotite to 11 gigapascals: Science, v. 276, p. 240–243, doi: 10.1126/science.276.5310.240.

Kreemer, C., and Hammond, W.C., 2007, Geodetic constraints on areal changes in the Pacific–North America plate boundary zone: What controls Basin and Range extension?: Geology, v. 35, p. 943–946, doi: 10.1130/G23868A.1.

Livaccari, R.F., and Perry, F.V., 1993, Isotopic evidence for preservation of lithospheric mantle during the Sevier-Laramide orogeny, western U.S.: Geology, v. 21, p. 719–722, doi: 10.1130/0091-7613(1993)021<0719:IEFPOC>2.3.CO;2.

Livaccari, R.F., Burke, K., and Şengor, A.M.C., 1981, Was the Laramide orogeny related to subduction of an oceanic plateau?: Nature, v. 289, p. 276–278, doi: 10.1038/289276a0.

Lowry, A.R., and Smith, R.B., 1994, Flexural rigidity of the Basin and Range–Colorado Plateau–Rocky Mountain transition from coherence analysis of gravity and topography: Journal of Geophysical Research, v. 99, p. 20,123–20,140.

Madsen, J.K., Thorkelson, D.J., Friedman, R.M., and Marshall, D.D., 2006, Cenozoic to Recent configurations in the Pacific Basin: Ridge subduction and slab window magmatism in western North America: Geosphere, v. 2, p. 11–34, doi: 10.1130/GES00020.1.

Mitrovica, J.X., Beaumont, C., and Jarvis, G.T., 1989, Tilting of continental interiors by the dynamical effects of subduction: Tectonics, v. 8, p. 1079–1094, doi: 10.1029/TC008i005p01079.

O'Driscoll, L.J., Humphreys, E.D., and Saucier, F., 2009, Subduction adjacent to deep continental roots: Enhanced negative pressure in the mantle wedge, mountain building and continental motion: Earth and Planetary Science Letters, doi: 10.1016/j.epsl.2009.01.020 (in press).

Piromello, C., Becker, T.W., Funiciello, F., and Faccenna, C., 2006, Three-dimensional instantaneous mantle flow induced by subduction: Geophysical Research Letters, v. 33, L08304, doi: 10.1029/2005GL025390.

Priest, G.R., 1990, Volcanic and tectonic evolution of the Cascade volcanic arc, central Oregon: Journal of Geophysical Research, v. 95, p. 19,583–19,599, doi: 10.1029/JB095iB12p19583.

Pysklywec, R.N., and Mitrovica, J.X., 1997, Mantle avalanches and the dynamic topography of continents: Earth and Planetary Science Letters, v. 148, p. 447–455, doi: 10.1016/S0012-821X(97)00045-9.

Rasmussen, J., and Humphreys, E., 1988, Tomographic image of the Juan de Fuca plate beneath Washington and western Oregon using teleseismic P-wave travel times: Geophysical Research Letters, v. 15, p. 1417–1420, doi: 10.1029/GL015i012p01417.

Saleeby, J., 2003, Segmentation of the Laramide slab—Evidence from the southern Sierra Nevada region: Geological Society of America Bulletin, v. 115, p. 655–668, doi: 10.1130/0016-7606(2003)115<0655:SOTLSF>2.0.CO;2.

Schellart, W.P., Freeman, J., Stegman, D., Moresi, L., and May, D., 2007, Evolution and diversity of subduction zones controlled by slab width: Nature, v. 446, p. 308–311, doi: 10.1038/nature05615.

Simpson, D.W., and Anders, M.H., 1992, Tectonics and topography of the western U.S.—An example of digital map making: GSA Today, v. 2, no. 6, p. 118–121.

Smith, D., 2000, Insights into the evolution of the uppermost continental mantle from xenolith localities on and near the Colorado Plateau and regional comparisons: Journal of Geophysical Research, v. 105, p. 16,769–16,781, doi: 10.1029/2000JB900103.

Sonder, L.J., England, P.C., Wernicke, B.P., and Christiansen, R.L., 1987, Extension in the Basin and Range Province and East Pacific margin: A physical model for Cenozoic extension of western North America: Geological Society of London Special Publication 28, p. 187–201.

Spencer, J.E., 1996, Uplift of the Colorado Plateau due to lithosphere attenuation during Laramide low-angle subduction: Journal of Geophysical Research, v. 101, p. 13,595–13,609, doi: 10.1029/96JB00818.

Stegman, D.R., Freeman, J., Schellart, W.P., Moresi, L., and May, D., 2006, Influence of trench width on subduction hinge retreat rates in 3-D models of slab rollback: Geochemistry, Geophysics, Geosystems, v. 7, p. Q03012, doi: 10.1029/2005GC001056.

Stock, J., and Molnar, P., 1988, Uncertainties and implications of the Late Cretaceous and Tertiary position of North America relative to the Farallon, Kula, and Pacific plates: Tectonics, v. 7, p. 1339–1384, doi: 10.1029/TC007i006p01339.

Tackley, P.J., Stevenson, D.J., Glatzmaier, G.A., and Schubert, G., 1993, Effects of an endothermic phase transition at 670 km depth in a spherical model of convection in the Earth's mantle: Nature, v. 361, p. 699–704, doi: 10.1038/361699a0.

van Hunen, J., van den Berg, A.P., and Vlaar, N.J., 2002, On the role of subducting oceanic plateaus in the development of shallow flat subduction: Tectonophysics, v. 352, p. 317–333.

Varga, R.J., 1993, Rocky Mountain foreland uplifts: Products of a rotating stress field or strain partitioning?: Geology, v. 21, p. 1115–1118, doi: 10.1130/0091-7613(1993)021<1115:RMFUPO>2.3.CO;2.

Waite, G.P., Smith, R.B., and Allen, R.M., 2006, *VP* and *VS* structure of the Yellowstone hot spot from teleseismic tomography: Evidence for an upper mantle plume: Journal of Geophysical Research, v. 111, doi: 10.1029/2005JB003867.

Wells, R.E., Weaver, C.S., and Blakely, R.J., 1998, Fore-arc migration in Cascadia and its neotectonic significance: Geology, v. 26, p. 759–762, doi: 10.1130/0091-7613(1998)026<0759:FAMICA>2.3.CO;2.

Wernicke, B., Axen, G.J., and Snow, J.K., 1988, Basin and Range extensional tectonics at the latitude of Las Vegas, Nevada: Geological Society of America Bulletin, v. 100, p. 1738–1757, doi: 10.1130/0016-7606(1988)100<1738:BARETA>2.3.CO;2.

Yuan, H., and Dueker, K., 2005, Teleseismic P-wave tomogram of the Yellowstone plume: Geophysical Research Letters, v. 32, p. L07304, doi: 10.1029/2004GL022056.

Zandt, G., and Humphreys, E., 2008, Toroidal mantle flow through the western U.S. slab window: Geology, v. 36, p. 295–298, doi: 10.1130/G24611A.1.

Manuscript Accepted by the Society 5 December 2008

The Geological Society of America
Memoir 204
2009

Structural geologic evolution of the Colorado Plateau

George H. Davis
Department of Geosciences, University of Arizona, Tucson, Arizona 85721, USA

Alex P. Bump
BP Exploration and Production Technology, Houston, Texas 77079, USA

ABSTRACT

The Colorado Plateau is composed of Neoproterozoic, Paleozoic, and Mesozoic sedimentary rocks overlying mechanically heterogeneous latest Paleoproterozoic and Mesoproterozoic crystalline basement containing shear zones. The structure of the plateau is dominated by ten major basement-cored uplifts and associated monoclines, which were constructed during the Late Cretaceous through early Tertiary Laramide orogeny. Structural relief on the uplifts ranges up to 2 km. Each uplift is a highly asymmetric, doubly plunging anticline residing in the hanging wall of a (generally) blind crustal shear zone with reverse or reverse/oblique-slip displacement. The master shear zones are rooted in basement, and many, if not most, originated along reactivated, dominantly Neoproterozoic shear zones. These can be observed in several of the uplifts and within basement exposures of central Arizona that project "down-structure" northward beneath the Colorado Plateau. The basement shear zones, which were reactivated by crustal shortening, formed largely as a result of intracratonic rifting, and thus the system of Colorado Plateau uplifts is largely a product of inversion tectonics. The overall deformational style is commonly referred to as "Laramide," and this is how we use this term here.

In order to estimate the dip of basement faults beneath uplifts, we applied trishear modeling to the Circle Cliffs uplift and the San Rafael swell. Detailed and repeated applications of trishear inverse- and forward-modeling for each of these uplifts suggest to us that the uplifts require initiation along a low-angle shear zone (between ~20° and ~40°), an initial shear-zone tip well below the basement-cover contact, a propagation (*p*) to slip (*s*) ratio that is higher for mechanically stiffer rocks and lower for mechanically softer rocks, a planar shear-zone geometry, and a trishear angle of ~100°. These are new results, and they demonstrate that the basement uplifts, arches, and monoclines have cohesive geometries that reflect fault-propagation folding in general and trishear fault-propagation in particular. Expressions of the shear zones in uppermost basement may in some cases be neoformed shear zones that broke loose as "footwall shortcuts" from the deeper reactivated zones.

Structural analysis of outcrop-scale structures permitted determination of horizontal-shortening directions in the Paleozoic and Mesozoic sedimentary cover of the uplifts. These arrange themselves into two groupings of uplifts, one that reveals NE/SW-directed shortening, and a second that reveals NW/SE shortening. Because the

Davis, G.H., and Bump, A.P., 2009, Structural geologic evolution of the Colorado Plateau, *in* Kay, S.M., Ramos, V.A., and Dickinson, W.R., eds., Backbone of the Americas: Shallow Subduction, Plateau Uplift, and Ridge and Terrane Collision: Geological Society of America Memoir 204, p. 99–124, doi: 10.1130/2009.1204(05). For permission to copy, contact editing@geosociety.org.

strain in cover strata is localized to the upward projections of the blind shear zones, and because the measured shortening directions are uniform across a given uplift (independent of variations in the strike of the bounding monocline), it seems clear that the regional stresses ultimately responsible for deformation were transmitted through the basement at a deeper level. Thus, the stresses responsible for deformation of the cover may be interpreted as a reflection of basement strain. The basement strain (expressed as oblique shear displacements *into* cover driven by reactivations of dominantly Neoproterozoic normal-shear zones) was a response to regional tectonic stresses and, ultimately, plate-generated tectonic stresses.

The driving mechanism for the Sevier fold-and-thrust belt was coupled to subduction of the Farallon plate, perhaps enhanced by slab flattening and generation of higher traction along the base of the lithosphere. However, the disparate shortening directions documented here suggest that two tectonic drivers may have operated on the Colorado Plateau: (1) gravitational spreading of the topographically high Sevier thrust belt on the northwest side of the plateau adjacent to an active Charleston-Nebo salient of the Sevier thrust belt, which imparted northwest-southeast shortening; and (2) northeast-southwest shortening driven by the flat slab. The effect of the two drivers tended to "crumple" part of the region in a bidirectional vise, creating added complications to structural patterns. Testing of this idea will require, among other things, very precise age determinations of progressive deformation of the Colorado Plateau within the latest Cretaceous to early Tertiary time window, and sophisticated finite-element modeling to evaluate the nature of the deformation *gradients* that would be induced by the two drivers.

The Raplee anticline, Monument Uplift, Utah. Photograph courtesy of Peter Kresan.

INTRODUCTION

Viewed in the context of the "Backbone of the Americas Conference" and the papers presented within this volume on the Cordillera (from "top" to "bottom"), the Colorado Plateau appears as a small "outcrop." Yet, it is a phenomenal outcrop from the standpoint of scenic beauty and tectono-structural insight, and it is an informative outcrop in disclosing how distinct geologic structures and tectonic provinces reflect the interplay of mechanical stratigraphy, rheology, shear zone deformation, and tectonic loading. Even though the Colorado Plateau was affected by and responded to the major plate-generated movements and stresses, it was never overwhelmed. The fact that this region is marked by superb exposures of rocks and structures, and it resides as an island of structural coherence surrounded by a sea of strain, provides special opportunities in interpreting structure and tectonics.

Variously motivated by oil, uranium, and academic curiosity, distinguished, now classic, geologic work has been carried out on the Colorado Plateau for over a century (Powell, 1873; Gilbert, 1876; Dutton, 1882; Baker, 1935; Strahler, 1948; Eardley, 1949; Gilluly, 1952; Kelley, 1955a, 1955b; Kelley and Clinton, 1960). This work was done in a pre–plate tectonic era, and the results have had a relative lasting impact on interpretation because of the relative hiatus of investigations on the Colorado Plateau when work along or near plate margins held such attraction. The early work, among other things, elucidated the highly variable trends of Laramide folds, faults, and uplifts within the Colorado Plateau, an observation that has proven to be an enduring thorn for scientists trying to come up with a unifying tectonic model, and it is especially frustrating because the geology looks so simple. Models that form the early core of tectonic interpretations variously emphasized compressional shortening (Baker, 1935), differential vertical uplift (Stearns, 1978), compressional shortening with discrete shifts in regional stress directions (Kelley, 1955a), compressional shortening with incremental rotation of regional stress fields (Gries, 1990), progressive rotation of regional strain within a province-broad zone of distributed progressive transpression (Sales, 1968), and one self-described "outrageous" hypothesis (Wise, 1963).

P.B. King's (1969) *Tectonic Map of North America* presents the Colorado Plateau in an artistic and scientifically informative way. The simplified black-and white version here (Fig. 1) captures the lightly deformed cratonic assemblage of the Colorado Plateau, bordered by the more substantial deformation of the Wyoming Province to the north (Brown, 1988, 1993), the Rocky Mountains to the east and east-southeast (Tweto, 1979), the Rio Grande rift on the southeast, and the Sevier fold-and-thrust belt on the west. Basin and Range faulting marks adjacent tectonic provinces to the south and west of the Colorado Plateau, and examples of this faulting include the several major normal faults that define the Western High Plateaus of southwestern Utah (Fig. 1). Basin and Range faulting and volcanism in the Basin and Range tectonic province proper are superimposed on intense older tectonic products of deformation, including superposed rifting, thrusting, magmatism, core complex development, detachment faulting, and Basin and Range faulting.

The apparent cohesiveness of the Colorado Plateau derives importantly from the sharp tectonic boundaries on the west and south. On the west, the boundary is not simply one of abutting "against" the Basin and Range, but dropping off the Cordilleran hinge line from craton sediments on the east to passive-margin basin sediments on the west. To the south, there are the northwest-trending Mogollon Highlands, which are fundamentally controlled by a tectonic fabric that originated at least as far back as the Jurassic. Boundaries to the north and east are more transitional and are related to abrupt increases in structural relief of the Laramide-style basement cored uplifts (Wyoming uplifts, Rocky Mountain uplifts).

The tectonic and structural characteristics of the Colorado Plateau reflect a combination of the initial character of pre–Upper Cretaceous lithotectonic assemblages, including the Precambrian basement; the heterogeneous nature of the basement, including the presence of crustal shear zones; a distinctive combination of loading conditions; and the changed rheologic conditions brought about by plate dynamics in the late Mesozoic and early Tertiary. We review and address these characteristics in this paper, with the goal of explaining, both from structural and tectonic perspectives, the variability in orientation of the uplifts, and the shifts in vergence from some groupings of uplifts to others. These fundamental descriptive facts have posed major difficulties in achieving coherent tectonic deformation plans to explain them. There is irony in this, for this "outcrop" seems tectonically so simple! To achieve our goals, we not only review pertinent literature, but also contribute some new observations and findings that have emerged from our recent work.

BASEMENT-CORED UPLIFTS OF THE COLORADO PLATEAU

The uplifts and folds of the Colorado Plateau were first studied in detail by Kelley (1955a, 1955b), who carried out comprehensive analyses of structures in the Colorado Plateau in relation to minerals (notably uranium), energy, and Laramide-style tectonics. His work not only features the Colorado Plateau system of uplifts and monoclines (Fig. 2), but also folds and fracture patterns (Kelley and Clinton, 1960). Kelly's descriptions of the uplifts were informed importantly through assembling structure-contour maps of the Colorado Plateau tectonic province (Kelley, 1955b). Mapping reveals that the Colorado Plateau, Laramide-style, basement-cored uplifts are not simply rectilinear blocks with sharp monocline margins, but they instead are more nuanced doubly-plunging asymmetric anticlinoria that typically have conspicuous monoclines along their steep flanks (e.g., see detailed descriptions by Davis, 1999) (Figs. 3 and 4). Over the entire span of the Colorado Plateau, shortening accommodated by these uplifts is less than several percent; in fact, Davis (1978) ran a rough calculation showing that shortening achieved by monoclines in the Arizona part of the Colorado Plateau is no more than

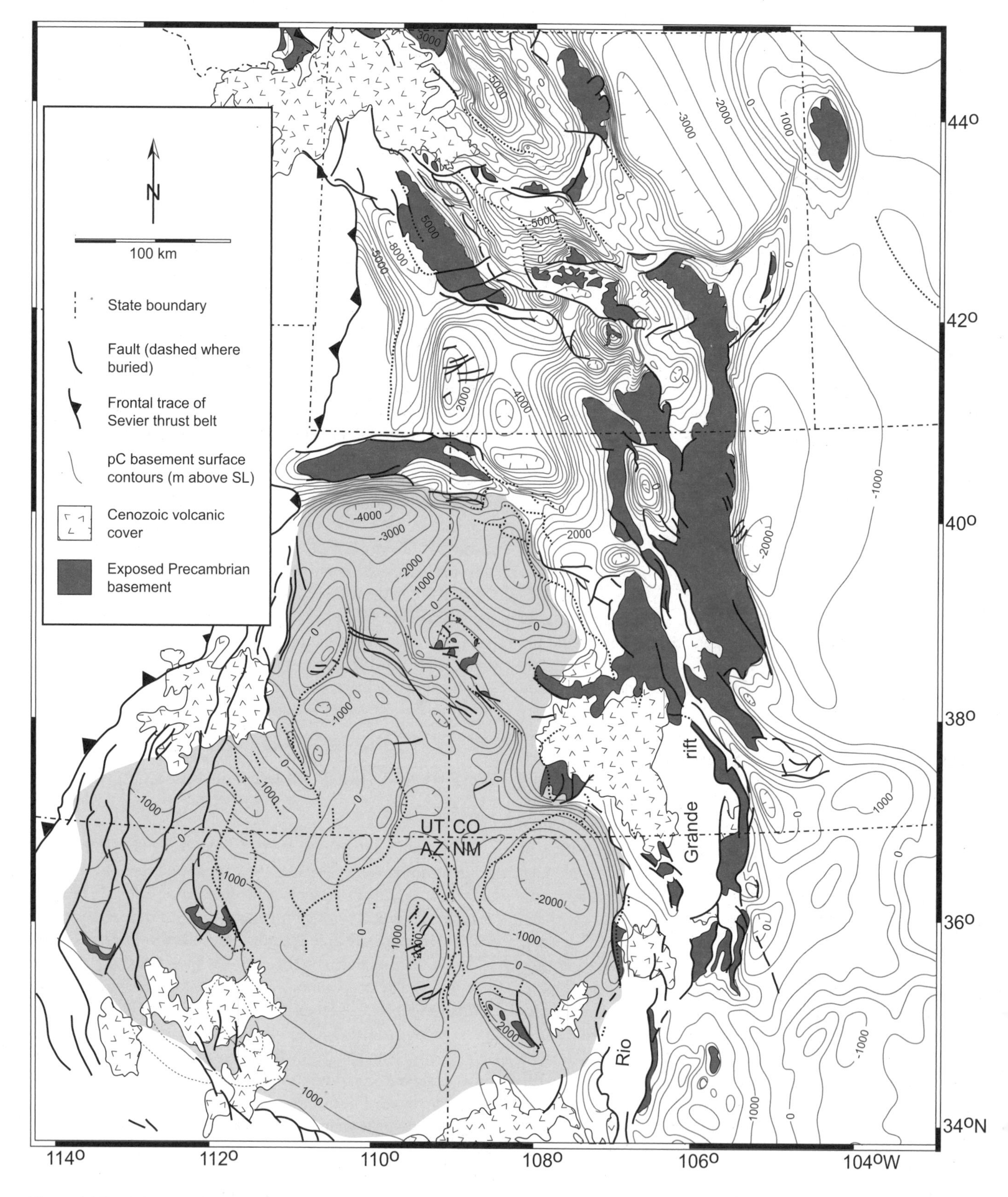

Figure 1. The Colorado Plateau tectonic province (light gray) in relation to the Wyoming province to the north, the Rocky Mountains to the east, the Rio Grande rift system to the southeast, and the front of the Sevier fold-and-thrust belt to the west. Not shown is the Basin and Range province, which borders the Colorado Plateau on the south and west. (Several Basin and Range faults that encroach upon the Colorado Plateau are shown in the lower left, where they demarcate the Western High Plateaus of Utah.) Map is from P.B. King (1969). pC—pre-Cambrian, SL—sea level.

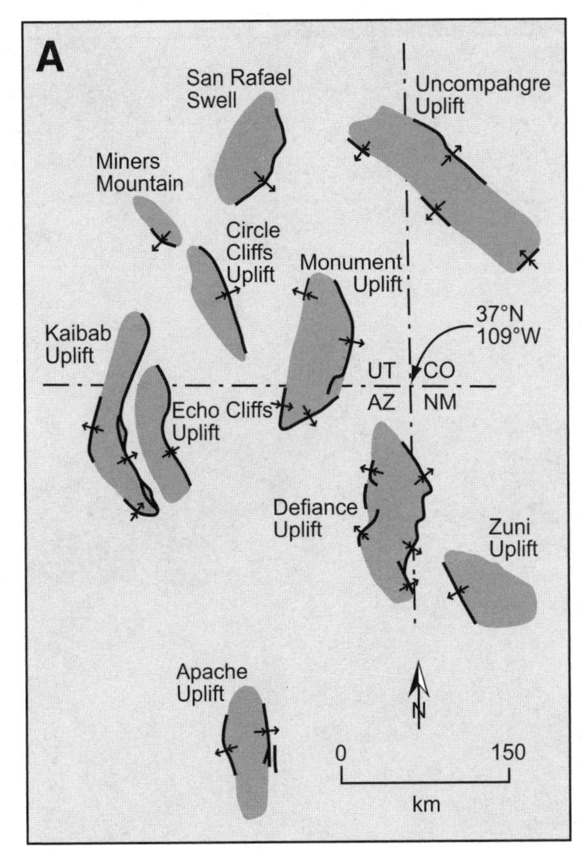

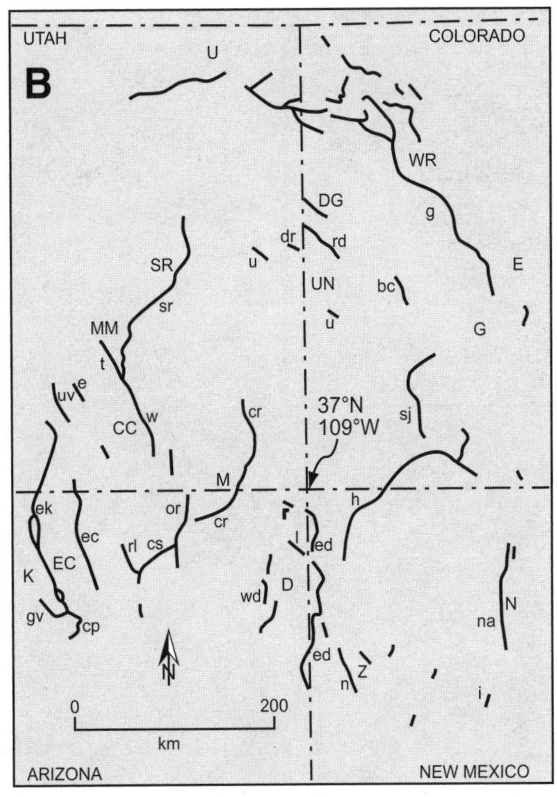

Figure 2. (A) Map of Colorado Plateau uplifts (from Davis et al., 1981, Fig. 42, p. 94; originally adapted from Kelley, 1955b). Reproduced by permission of the Arizona Geological Survey. Kelley did not include "Miners Mountain" as an uplift per se, though he did show the Teasdale faulted monocline on the southwest margin of this uplift. Furthermore, Kelley (1955b) was not aware of the "Apache uplift" (see text). (B) Map of monoclines of the Colorado Plateau (from Davis, 1978, Fig. 5, p. 222; originally adapted from Kelley, 1955b). Monoclines: bc—Book Cliffs; cp—Coconino Point; cr—Comb Ridge; cs—Cow Springs; dr—Davis Ranch; e—Escalante; ec—Echo Cliffs; ed—East Defiance; ek—East Kaibab; g—Grand; gv—Grandview; h—Hogback; i—Ignacio; l—Lukachukai; n—Nutria; na—Nacimiento; or—Organ Rock; r—Rattlesnake; rd—Redlands; rl—Red Lake; sj—San Juan; sr—San Rafael; t—Teasdale; u—Uncompahgre; uv—Upper Valley; w—Waterpocket; wd—West Defiance. Uplifts: CC—Circle Cliffs; D—Defiance; DG—Douglas; E—Elk; EC—Echo Cliffs; G—Gunnison; K—Kaibab; M—Monument; MM—Miners Mountain; N—Nacimiento; SR—San Rafael; U—Uinta; UN—Uncompahgre; WR—White River; Z—Zuni.

1%. The long back-limbs of the uplifts tend to be homoclinal or almost-undetectably curviplanar, with dips ranging from 0.5° to several degrees, whereas the short forelimbs of the uplifts tend to be curviplanar, with maximum dips as low as 10° (or less) and as high as 85° overturned. Individual uplifts tend to trend NNW, but there are N-trending and NNE-trending uplifts as well.

The Colorado Plateau uplifts are generally greater than 100 km in strike length and broader than 30 km. Structural relief associated with the uplifts is as great as 2000 m (e.g., the Circle Cliffs uplift!) (see Figs. 2 and 3). Kelley (1955a) noted that the system of Colorado Plateau uplifts can be subdivided into two systems. In the western part of the Colorado Plateau, the uplifts are asymmetric toward the east, and in the eastern part of the Colorado Plateau, the uplifts are asymmetric toward the west (see Fig. 2A). The western group consists of the San Rafael, Circle Cliffs, Kaibab, Monument, Echo Cliffs, and Defiance uplifts. The eastern group consists of the Uncompahgre, Zuni, and Naciemiento uplifts. The monoclines themselves are more variable in orientation than the uplifts (see Fig. 2B). NNW-trending monoclines are the most abundant, but some are NE- and N- to NNE-trending within the system as a whole.

Based on a variety of observations, the Colorado Plateau uplifts are interpreted as underlain by faulted basement. In the cases of the Kaibab and Uncompahgre uplifts, faulted basement is directly exposed (Lohman, 1963; Cashion, 1973; Huntoon, 1971; Huntoon and Sears, 1975; Stern, 1992; Huntoon, 1993). In the case of the San Rafael Swell, Allmendinger et al. (1987) demonstrated through the Consortium for Continental Reflection Profiling (COCORP) seismic data the presence of a basement high beneath this uplift. Furthermore, Cook et al. (1991) showed that Colorado Plateau uplifts express themselves as gravity highs, which they interpreted as expressions of uplifted basement.

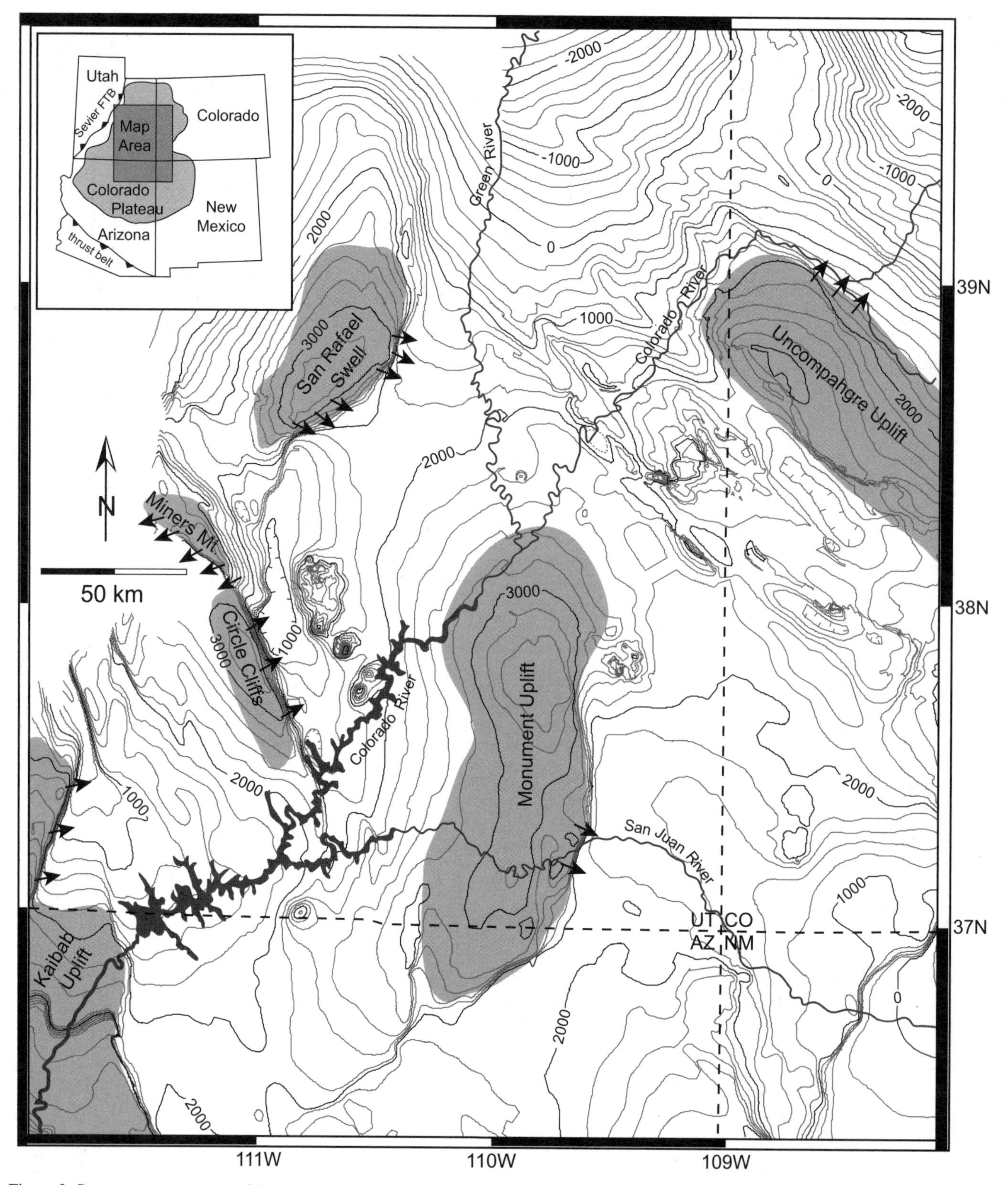

Figure 3. Structure contour map of the northern Colorado Plateau showing the approximate areas of the uplifts (shaded gray) in this study and their local shortening directions (black arrows) (from Bump, 2004, Fig. 1). Contours are drawn at 200 m intervals on the base of the Cretaceous Dakota sandstone. Source for contours, modified after Baker (1935), O'Sullivan (1963), Williams (1964), Williams and Hackman (1971), Haynes et al. (1972), Cashion (1973), Hackman and Wyant (1973), and Haynes and Hackman (1978). Structural elevations are given in meters above sea level. The Colorado River and its tributaries (dark gray) are shown for reference. FTB—fold-and-thrust belt. Reprinted from Bump, A.P., 2004, Three-dimensional Laramide deformation of the Colorado Plateau: Competing stresses from the Sevier thrust belt and the flat Farallon slab: Tectonics, v. 23, doi: 10.1029/2002TC001424, with permission from the American Geophysical Union.

Finally, the Colorado Plateau uplifts are similar in style to uplifts in Wyoming and Colorado, which are known to be basement-cored. Where exposed, these basement faults are ancient features that show multiple episodes of slip, further complicating the kinematic picture (e.g., Huntoon and Sears, 1975; Stone, 1977).

The Colorado Plateau was modified by faulting related to (Miocene to present) regional extension to the west and to the east, but the uplifts shown in Kelly's regional structure map (see Fig. 2A) for the most part escaped this superposed deformation. One exception within the Colorado Plateau proper is the Kaibab uplift, which is cut by the Paunsaugunt and Sevier faults of Basin and Range origin (Davis, 1999). One more Colorado Plateau uplift completely escaped Kelly's detection because of the effects of post-Laramide superposed extension. This uplift, the Apache uplift (see Fig. 2A), lies in the Arizona transition zone between the Colorado Plateau and Basin and Range. Davis et al. (1981) named this uplift, basing its presence on the work of Finnell (1962), Granger and Raup (1969), Peirce et al. (1979), Peirce (1981, personal commun.), and their own structural mapping and analysis along the Salt River Canyon in central Arizona. Much of this uplift lies within the Fort Apache Indian Reservation, which is thus the context for its naming.

Figure 4. North-directed photograph of the East Kaibab monocline (from Davis, 1999, Fig. 35A, p. 39). Gently dipping Navajo Sandstone (Jurassic) caps the top of the monocline, on the far left. More readily eroded Carmel Formation (Jurassic), Dakota Sandstone, and Tropic Shale (Lower Cretaceous) crop out in the foreground. Straight Cliffs Formation (Upper Cretaceous) is on the upper far right. Breadth of view is ~1 km.

SHEAR ZONES IN COLORADO PLATEAU BASEMENT

It has been known for decades that many monoclines in the Grand Canyon "root" into Laramide reverse shear zones that are reactivated Upper Proterozoic normal faults. Noble (1914) recognized this in the Shinumo quadrangle, Maxson (1961) recognized this in the Bright Angel quadrangle, and Huntoon and Sears (1975) recognized it in their analysis of structures in the eastern part of the Grand Canyon. In particular, Huntoon (1974) concluded that the abrupt shifts in trend of individual monoclines in the eastern Grand Canyon are expressions of parts of the original Precambrian fault-trace geometry in the underlying basement. Huntoon (1974) emphasized that the basement faults formed originally as normal faults during the latest Mesoproterozoic and Neoproterozoic, but later, during the Laramide (late Mesozoic through early Tertiary), they were reactivated with a reverse sense of displacement to form monoclines involving Neoproterozoic, Paleozoic, and Mesozoic strata (Fig. 5). In the footwall immediately beneath the Butte fault, there is outcrop-scale evidence of shortening in the form of thrust faults and associated folding (Fig. 6).

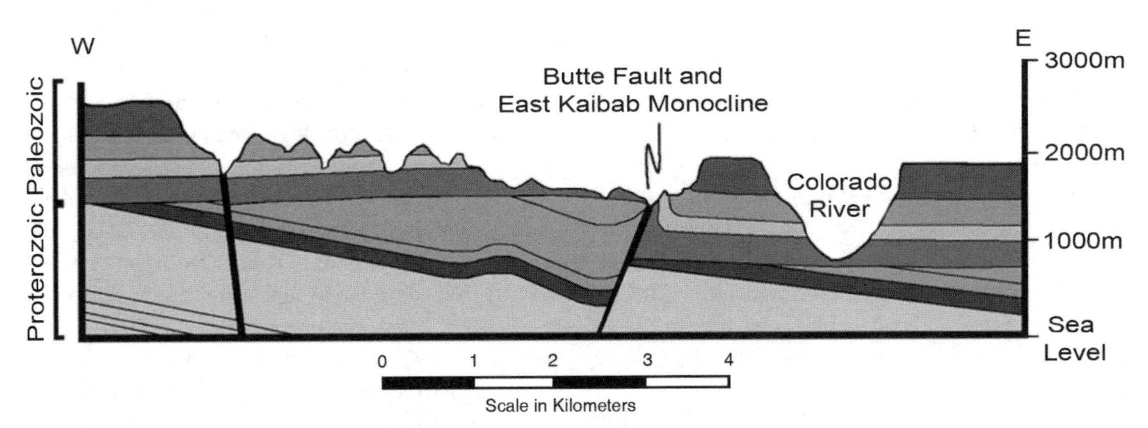

Figure 5. Structure section showing the Butte fault in relation to basement and cover. During the Neoproterozoic, movement on the Butte fault created normal displacement. During the latest Cretaceous and early Tertiary, reactivation of the Butte fault produced reverse displacement, tipping out upward into monoclinal folding. The magnitude of reverse throw was less than Neoproterozoic normal throw, and thus net offset of Mesoproterozoic basement remains of normal displacement. From Tindall (2000a, Fig. A5, p. 54), used with permission of Sarah Tindall.

Figure 6. Photograph of outcrop-scale thrust fault in Neoproterozoic strata (Uncar Formation) in the immediate footwall of the Butte fault. This is an expression of Laramide compressional deformation during reactivation of the Butte fault. Photograph by Sarah Tindall.

Indeed, the structural history of the Butte fault, which is exposed in the Grand Canyon directly below monoclinally folded Paleozoic and Mesozoic strata, permits the salient relationships between monoclines and basement faults to be understood. There are well documented west-side-down offsets along the west-dipping Butte fault in the Grand Canyon Supergroup (Walcott, 1890; Maxson, 1961; Huntoon, 1969, 1993; Huntoon and Sears, 1975; Tindall, 2000b). Reactivation of this fault, beginning presumably in the latest Cretaceous, caused west-side-up reactivation of the Butte fault during the creation of the Kaibab uplift and East Kaibab monocline (Huntoon, 1993; Huntoon and Sears, 1975). As reemphasized by Tindall (2000b, p. 632): "The magnitude of reverse offset must have been smaller than the magnitude of ancient normal offset, because normal separation is still preserved at the Precambrian level."

Davis (1978, p. 215) expanded the fault-specific conclusions reached by Noble (1914), Maxson (1961), Huntoon (1974), and Huntoon and Sears (1975) (Fig. 7) and concluded that the monocline fold pattern *as a whole* in the Colorado Plateau reflects the expression of many elements of a provincewide basement-fault system. Davis further suggested that the shapes of many of the Colorado Plateau uplifts are the muted expressions of underlying basement-block edges (Davis, 1978, p. 225). These conclusions were not entirely new. Case and Joesting (1972), based on analysis of gravity and magnetic gradients evident in Precambrian basement, interpreted a "fundamental" fracture pattern of Precambrian age in Precambrian basement beneath a part of the Colorado Plateau.

We suspect that those who have studied the faults into which the uplifts and monoclines root would agree that major deep-seated "faults" associated with Colorado Plateau uplifts and monoclines are "fault zones." Strictly speaking, these "fault zones" are most accurately described as "brittle, semibrittle, or ductile shear zones," depending upon depth-level of observation, rheology, and superposition of fabrics, reflecting histories of activation and reactivation (Davis and Reynolds, 1996). Thus, throughout the remainder of this paper, we have chosen to replace the term "fault" with "shear zone" in *our* descriptions and interpretations, though in referencing past work of other workers, we will continue to use the term "fault" in the same way used by them in the literature.

NATURE OF BASEMENT SHEAR ZONES ADJACENT TO THE PLATEAU

Given the limited exposures of Precambrian rocks and structures within the Colorado Plateau, observations regarding Precambrian shear zones in the central Arizona "transition zone" between the Colorado Plateau (to the north) and the Basin and Range (to the south) become very important. We here introduce this kind of analysis as a variation on Mackin's (1950) "down-structure" method to examine a geologic map as if it were a cross section and, in this case, to visualize Precambrian shear zones beneath the southern edge of the Colorado Plateau. The transition zone lends itself to this evaluation because this part of Arizona was uplifted during the late Mesozoic and early Tertiary in such a way that Paleozoic and Mesozoic strata were eroded, extensively exposing the Precambrian basement (Kamilli and Richard, 1998). Furthermore, the region was tilted slightly northeastward (by ~1° to 2°). According to Peirce et al. (1979), and as summarized in Faulds (1986, p. 126–127), the tilting was accomplished in two episodes, the first between the Late Jurassic and Late Cretaceous, and the second during the latest Cretaceous and Eocene. Because of this history, a very low-plunging northward viewing of the transition zone in relation to the Colorado Plateau becomes an approximation of a structure-section.

The framework for being able to identify shear-zone boundaries within the Precambrian of southwestern North America was built by Lee Silver and his students and colleagues, based upon extensive geologic and geochronologic surveying (e.g., Cooper and Silver, 1964; Silver, 1965, 1967, 1969, 1978; Anderson et al., 1971; Anderson and Silver, 1976; Anderson and Silver, 2005). This work led to the understanding that the late Paleoproterozoic and early Mesoproterozoic of southwestern North America grew through accretionary crust-forming events, expressed importantly as NE-SW–trending provinces of cogenetic suites of volcanic and plutonic rocks (Anderson and Silver, 2005). Recognition of these distinctive provinces was based upon crystallization ages derived from developing and using the U-Pb system on zircon (e.g., Silver, 1963; Silver and Deutsch, 1961, 1963). Anderson and Silver (2005, p. 12, their Figs. 5A–5C) nicely summarized the progressive delineation of two dominant provinces in southern Arizona and southern New Mexico: the Pinal Province (to the southeast), marked by crystallization ages of ca. 1.7–1.6 Ga, and the Yavapai Province (bordering the Pinal Province on the northwest), marked by crystallization ages of ca. 1.8–1.7 Ga (Conway et al., 1987).

Karlstrom and Humphreys (1998) and Nourse et al. (2005) presented interpretations of Proterozoic crustal provinces complementary to the aforementioned Yavapai-Pinal accretion map, but they used a "Yavapai-Mazatzal" province taxonomy (Fig. 8) and split out relationships in more detail based on the work of Karlstrom and Bowring (1988), Karlstrom and Williams (1998), Wooden and DeWitt (1991), Bender et al. (1993), Karlstrom (1993), Karlstrom and Daniel (1993), Ilg et al. (1996), and Eisele and Isachsen (2001). The conclusion remains: the Paleoproterozoic of central Arizona was assembled by tectonic shortening between ca. 1.8 and ca. 1.6 Ga (Karlstrom, 1993; Conway et al., 1987; Karlstrom and Humphreys, 1998; Anderson and Silver, 2005).

The principal tectonic grain resulting from this tectonic accretion is northeast-oriented (see Fig. 8), as expressed by the trends of the provinces themselves (e.g., the Mojave province, the Mojave-Yavapai transition, the Yavapai province, the Yavapai-Mazatzal transition, and the Mazatzal province), but also by the "thrust-sense shear zones" (Karlstrom and Humphreys, 1998, p. 162) that separate each province from one another. From northwest to southeast, these late Paleoproterozoic–early Mesoproterozoic shear zones are the Gneiss Canyon shear zone, the Crystal shear zone, the Moore Gulch shear zone, and the Slate Creek shear zone (see Fig. 8). As emphasized by Karlstrom and Humphreys (1998), these shear zone boundaries influenced the distribution and character of Laramide magmatism and metallogenesis but were *not* reactivated in any conspicuous way during the formation of Laramide basement-cored uplifts. However, Karlstrom and Humphreys (1998) pointed out that it is possible for northeast grain to have been reactivated from place to place in the form of transfer zones or accommodation zones oriented parallel to the direction of Laramide shortening, i.e., NE/SW. We point out here that this may have occurred in fashioning the NE-trending Cow Springs monocline near Kayenta, Arizona, just southwest of the Monument uplift (see Fig. 2B) (Davis and Kiven, 1975; Davis, 1978, his Figs. 4 and 5).

It was the late Mesoproterozoic and Neoproterozoic shear zones of normal-sense shear displacement that were reactivated preferentially during the formation of the Laramide-style basement-cored uplifts (Karlstrom and Humphreys, 1998). The best examples of such normal-sense displacement shear zones are the Canyon Creek and Butte faults (Karlstrom and Humphreys, 1998) (Fig. 9). Continental-scale rifting created these faults between 1.1 Ga and 700 Ma, which coincided with syntectonic diabase intrusions and rift-basin sediment accumulations (Silver, 1960; Shride, 1967; Silver, 1978; Granger and Raup, 1969; Elston, 1979). Indeed, the Canyon Creek "fault," as a Paleoproterozoic shear zone, controlled the emplacement and distribution of dikes and sills of Neoproterozoic diabase (Finnell, 1962; Shride, 1967; Granger and Raup, 1969), which were dated by Silver (1960, 1978) as 1.1 Ga. This overall timing of rifting makes sense in the context of plate reconstructions for the Neoproterozoic (Burke and Dewey, 1973; Stewart, 1976; Stewart and Suczek, 1977; Dickinson, 1977).

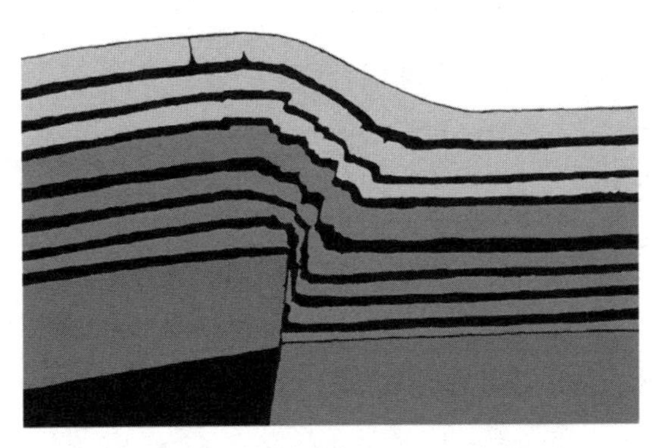

Figure 7. Experimental simulation of the relationship of monoclines to faulted basement (from Davis, 1978, Fig. 1, p. 216). In this case, the strata are alternating layers of powdery kaolin clay and modeling clay, the basement is a wooden board, and the preexisting basement fault is a saw cut. The monocline was produced by compressional end-loading of the board.

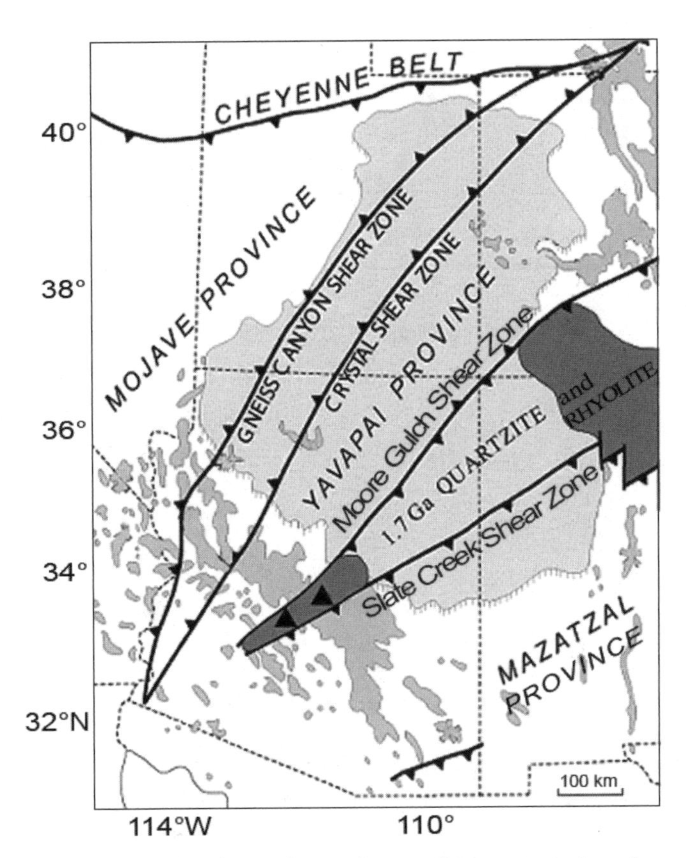

Figure 8. Map showing major northeast-striking tectonic boundaries of latest Paleoproterozoic and earliest Mesoproterozoic age (1.8–1.65 Ga), as projected beneath the Colorado Plateau. Note that the tectonic boundaries are major shear zones. The Moore Gulch shear zone coincides with the north edge of 1.65 Ga deformation. The Slate Creek shear zone coincides with the south edge of pre–1.7 Ga basement. Figure is based on Karlstrom and Humphreys (1998, *Rocky Mountain Geology*, v. 33, no. 2, Fig. 1, p. 163).

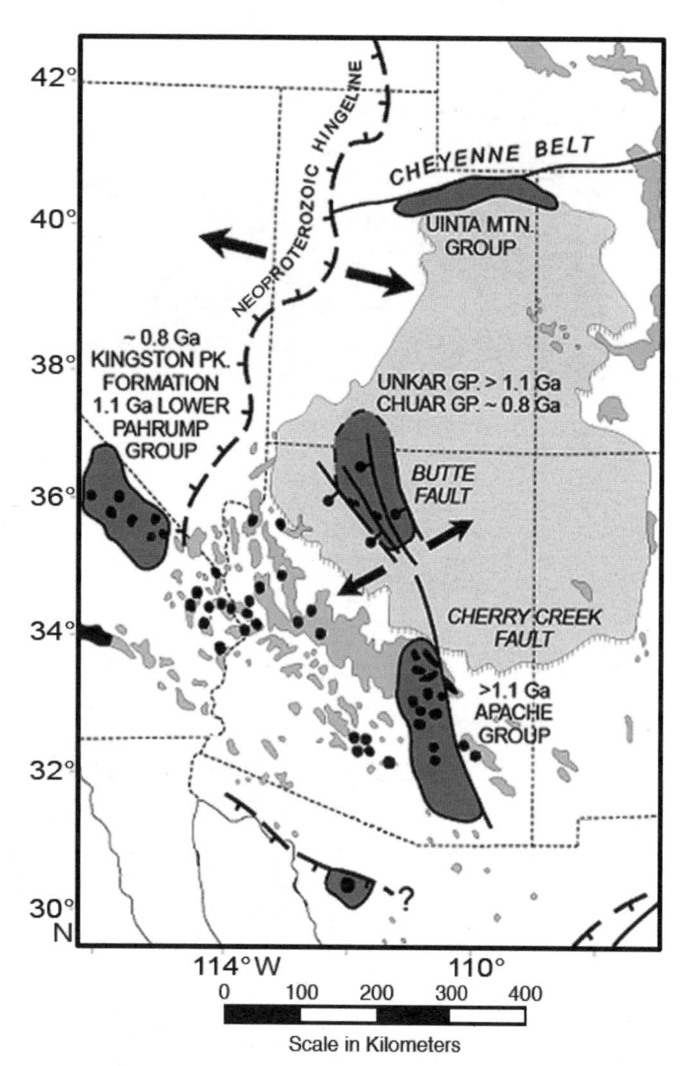

Figure 9. Map showing major NNW-striking shear zones of latest Mesoproterozoic and Neoproterozoic age (1.1–0.6 Ga), as projected beneath the Colorado Plateau. Figure is based on Karlstrom and Humphreys (1998, *Rocky Mountain Geology*, v. 33, no. 2, Fig. 3, p. 168).

The map relationships of the Apache uplift and its bounding shear zones (Canyon Creek shear zone on the east, Cherry Creek shear zone on the west) are quite pertinent to understanding Laramide-style basement-cored uplifts (Fig. 10), and thus we will discuss them here in some detail, drawing together some dispersed literature that has tended to be "off the radar." Finnell (1962) and Granger and Raup (1969) concluded decades ago that the Canyon Creek fault zone coincides with a major Precambrian shear zone. The presence of this shear zone was felt tectonically in Neoproterozoic time, for it exerted control on the emplacement of ca. 1.1 Ga diabase sills and dikes. Finnell (1962) recognized that the Canyon Creek fault was the site of a major east-facing monocline that deformed Neoproterozoic and Phanerozoic strata. Peirce et al. (1979), more specifically, determined that the Canyon Creek (Precambrian) fault accommodated at least two major reactivations in the Phanerozoic. During the latest Cretaceous and early Tertiary, the Canyon Creek fault experienced eastward-directed reverse displacement of at least 1.5 km (Peirce et al., 1979). Subsequently, as the result of extensional faulting in middle to late Tertiary time, this structural relief was countermanded when the Canyon Creek fault accommodated westward-directed normal displacement on the order of 1.5 km. Because the vertical component of the (older) reverse faulting exceeded the offset achieved by the superposed normal faulting, Tertiary rocks to the west of the Canyon Creek fault are structurally higher than equivalents to the east. Thus, the requisite throw related to reverse faulting must have exceeded 1.5 km of structural relief (H.W. Peirce, 1981, personal commun.).

Faulds (1986, p. 231) placed the up-on-the-west reverse throw at 1650–1750 m and calculated a down-to-the-west normal displacement of 750+ m. The Cherry Creek shear zone, which marks the western margin of the Apache uplift (see Fig. 10), also coincides with a Paleoproterozoic shear zone, a ca. 1.1 Ga reactivation history during diabase sill and dike emplacement, a shortening-induced reactivation history of down-to-the-west monocline development, an extension-induced down-to-the-west Oligocene normal faulting, and post–14 Ma normal faulting of Basin and Range tectonic origin (Faulds, 1986). Overall, as emphasized by Davis et al. (1981), the Apache uplift is less obvious because it lacks the stripped structural form so characteristic of typical Colorado Plateau uplifts. However, its breadth (~16 km), length (~100 km), and structural relief (at least ~1650 m) are quite comparable to that of other Colorado Plateau uplifts, such as the Circle Cliffs uplift (Davis et al., 1981, p. 94).

The development of the Apache basement-cored uplift was followed by extensive erosion, as revealed by the nonconformity atop the Apache uplift between late Paleoproterozoic and Mesoproterozoic basement beneath Tertiary sedimentary and volcanic strata above (see Fig. 10). Monoclines associated with the Apache uplift are shown in Figure 11. The orientations and locations of the monoclines associated with the Apache uplift are nicely compatible with the East Kaibab system of monoclines (see Fig. 11), and they convincingly match the Butte–Canyon Creek fault system (Fig. 9) as presented in Karlstrom and Humphreys (1998).

OBLIQUE SLIP ALONG REACTIVATED BASEMENT SHEAR ZONES

As noted earlier, reactivation of the Neoproterozoic Butte shear zone within the Grand Canyon is well documented, and it contributed importantly to the development of the East Kaibab monocline along the eastern margin of the Kaibab uplift. Prior to the work of Davis and Tindall (1996), interpretations developed for the Butte shear zone exclusively reported *vertical* throw component(s). It is important to recognize that some shear-zone reactivations associated with Colorado Plateau uplifts have oblique-shear expressions. A case in point is the NNE stretch

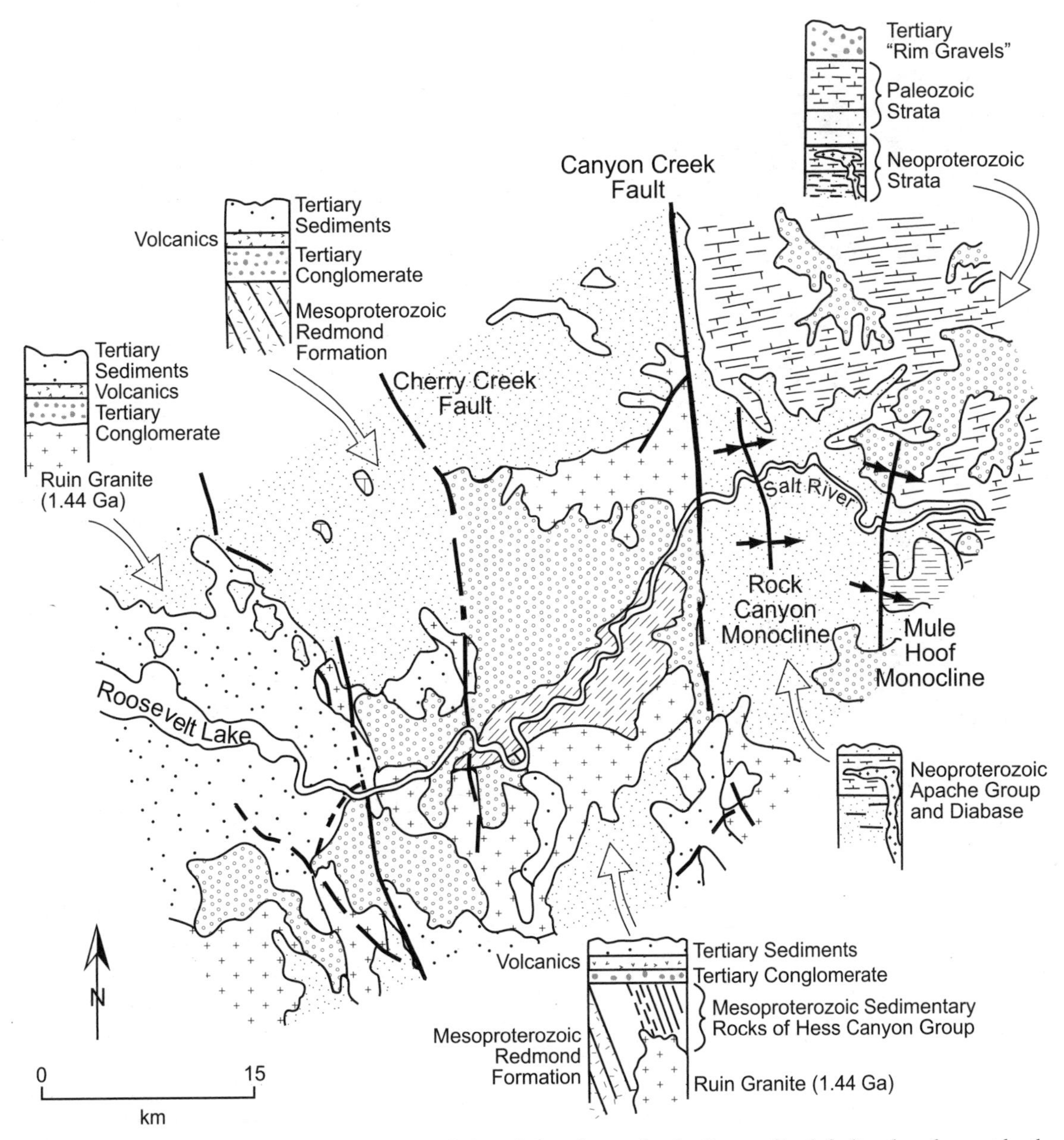

Figure 10. Geologic map showing the Apache uplift, bounded on the east by the Canyon Creek fault and on the west by the Cherry Creek fault. Schematic geological columns underscore that all Paleozoic sedimentary rocks as well as Neoproterozoic sedimentary rocks and associated diabase intrusions are completely missing atop the Apache uplift, where Tertiary sedimentary and volcanic rocks rest nonconformably upon late Paleoproterozoic and Mesoproterozoic basement. Figure is modified from Davis et al. (1981, Fig. 9, p. 58, and Fig. 27, p. 78). Reproduced by permission of the Arizona Geological Survey.

of the East Kaibab monocline in southern Utah (Babenroth and Strahler, 1945; Kelley, 1955a, 1955b; Davis, 1978, 1999).

Mesozoic strata within the East Kaibab monocline (see Fig. 4) display an elegant, penetrative system of NNE-striking and WNW-striking faults first identified by Sargent and Hansen (1982). Based upon detailed structural analysis and geologic mapping, these fault sets are recognized as right-handed and left-handed oblique slip faults, respectively (Davis and Tindall, 1996; Tindall and Davis, 1998; Davis, 1999; Tindall, 2000a, 2000b). Sarah Tindall comprehensively mapped the faulting along the full 60 km stretch of the East Kaibab shear zone within Utah (Fig. 12) (Tindall, 2000a). The faults and fault sets are synthetic and antithetic with respect to a major, overall, reverse right-handed strike-slip shearing along this N20°E-trending segment of the East Kaibab monocline (Davis, 1999; Tindall and Davis, 1999; Tindall, 2000a, 2000b). Slickenlines and grooves along the major shear zone consistently rake 30°S to 40°S, and the shear zone surfaces dip steeply (~75°) westward. As first emphasized by Davis and Tindall (1996), this oblique right-handed strike-slip shearing is the natural consequence of reactivation of a

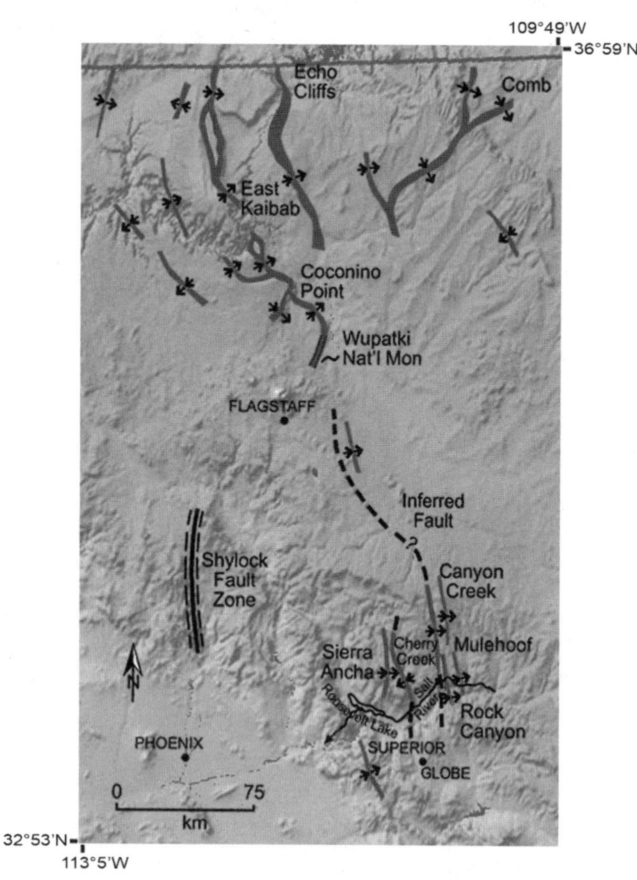

Figure 11. Map of monoclines in the Salt River Canyon region (after Granger and Raup, 1969; Davis et al., 1981) and their relation to monoclines of the Colorado Plateau of northern Arizona (after Davis and Kiven, 1975). Map is from Davis et al. (1981, Fig. 23, p. 74). Reproduced by permission of the Arizona Geological Survey.

NNE-striking basement shear zone in response to NE-SW shortening (Davis, 1999; Tindall and Davis, 1999). A N65°E-trending horizontal-shortening direction acting across a N10°E-striking, steeply W-dipping basement shear zone is ideally suited to reactivate the shear zone in right-handed strike-slip fashion (Fig. 13).

Reverse right-handed oblique slip also characterizes the NNE-trending stretch of the East Kaibab monocline north of Flagstaff, Arizona, within Wupatki National Monument (see Fig. 11). There, strata of the Permian Kaibab and Triassic Moenkopi Formation exhibit abundant low-raking slickenlines on slickenlined fault surfaces. Where individual fault zones tip out, there are relays that conform to right-handed (oblique) strike-slip shearing.

Another good example of oblique-slip shearing associated with monocline folding was recognized along the southwest margin of Miners Mountain in Utah. This Colorado Plateau uplift is situated just north of the Circle Cliffs uplift (see Fig. 3). A doubly plunging NW-trending Colorado Plateau uplift, Miners Mountain is bordered on the southwest by the N65°W-trending Teasdale faulted monocline, the structural relief of which is 150 m (Smith et al., 1963; Billingsley et al., 1987; Davis, 1999). Anderson and Barnhard (1986) documented that this faulted monocline accommodated left-handed strike-slip movement. Bump et al. (1997) later studied this relationship and determined independently that the Teasdale faulted monocline is a reverse, left-handed transpressive structure, with perhaps 1–2 km of left-lateral offset.

COLORADO PLATEAU UPLIFTS AND INVERSION TECTONICS

The evidence is abundant to support the hypothesis that individual Neoproterozoic shear zones were reactivated as reverse faults and/or transpressive reverse/oblique-slip shear zones. These observations have led to an important provincewide perspective, namely, that Colorado Plateau uplifts are an expression of "inversion tectonics," and in particular "intracratonic rift inversion" (Marshak et al., 2000, p. 736). Marshak et al. (2000) made a compelling argument for this, building on some of their previous work (Marshak and Paulsen, 1996; Karlstrom and Humphreys, 1998; Timmons et al., 2001). Marshak et al. (2000) inserted an important and necessary concept into the story of structural evolution of the Colorado Plateau uplifts in particular.

The mechanical basis for understanding reactivation of normal faults and shear zones as reverse/thrust faults is long established (e.g., Donath, 1961), and it is perhaps best summarized in Byerlee (1978), the contribution for which "Byerlee's law" was coined. Byerlee's law describes the conditions necessary to cause slip on a preexisting fault, namely, by assessing the product of the coefficient of sliding friction and normal stress acting on a given preexisting fault, and comparing that product to the sum of the shear strength and cohesive strength of the body of rock, where unfaulted. Certain products of the first faulting can reduce the coefficient of sliding friction, including breccia, gouge, and other fault-rock products; weak hydrothermally altered mineral assemblages; and finer-grained and/or highly foliated rocks (e.g., in shear zones) (Etheridge, 1986). The final necessary condition for reactivation is suitability of orientation of the preexisting fault surface (or zone) within the prevailing stress field (Donath and Cranwell, 1981; Etheridge, 1986). Etheridge (1986) emphasized that reverse (thrust) reactivation of normal faults is the most likely circumstance of reactivation, since normal fault systems and thrust systems may have very similar geometries (see, for example, Cooper and Williams, 1989; Boyer and Elliott, 1982).

Etheridge (1986) applied the conceptual framework of fault reactivation to deformation systems, using primarily examples from southeastern Australia. He emphasized that insightful models for lithospheric stretching (McKenzie, 1978; Le Pichon and Sibuet, 1981; and Dewey, 1988) provide a basis for understanding the formation of primary extensional fault systems in basement and cover, thus setting up conditions for reactivation during subsequent compressional deformation (Etheridge, 1986, p. 185).

Marshak et al. (2000) emphasized that intracratonic rift inversion is expressed in uplift structures in the Rocky Mountains, Colorado Plateau, and Midcontinent regions of North

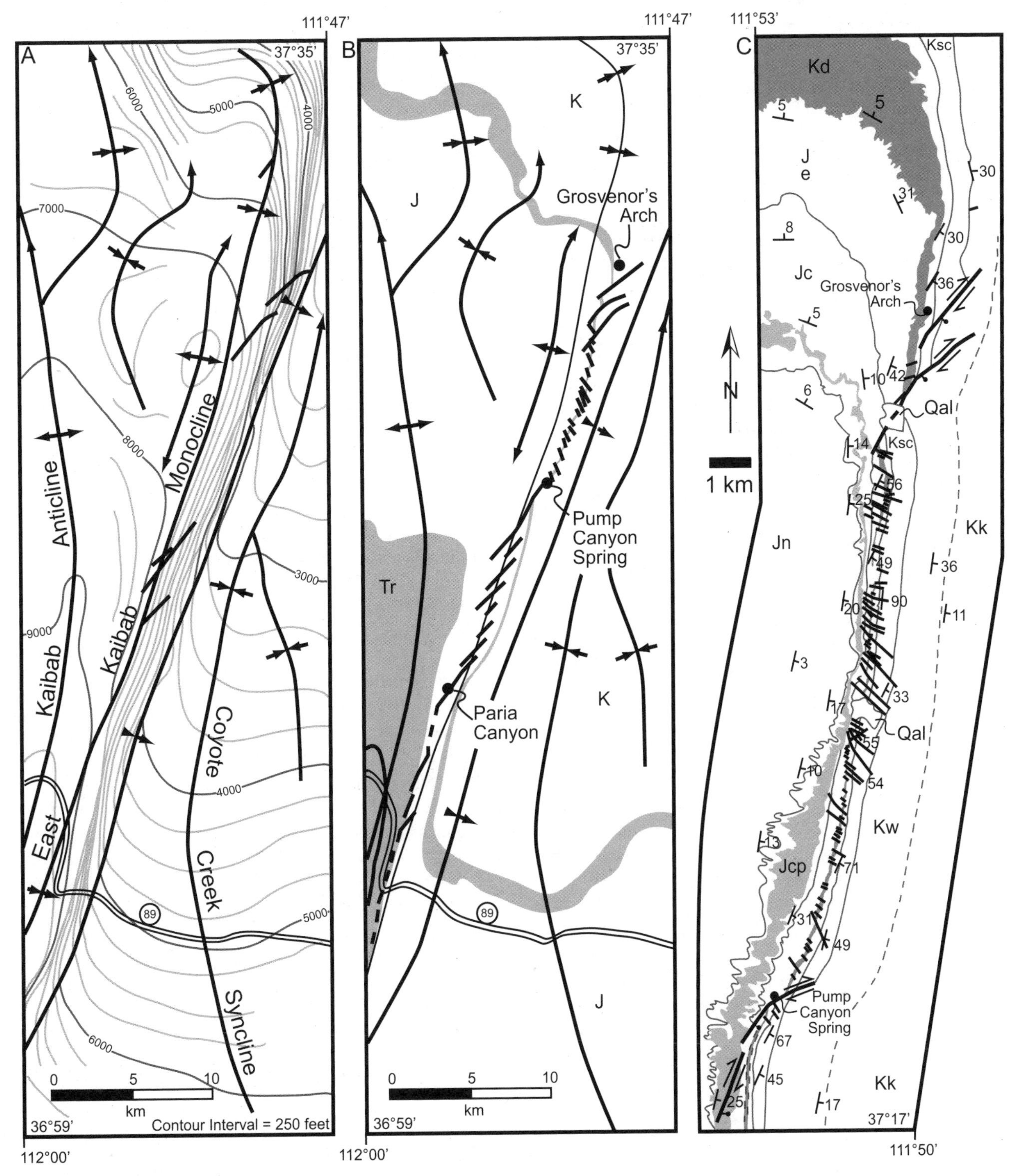

Figure 12. Northern part of East Kaibab monocline. (A) Structural contour map. Structure contours (ft) are drawn on the base of the Cretaceous Dakota Sandstone. From Tindall and Davis (1999, Fig. 2, p. 1305). (B) Simplified geologic map. Faulting and folding in the steep limb move from older stratigraphic units in the south into higher stratigraphic units northward. NE-striking faults are right lateral; NW-striking faults are left lateral. From Tindall and Davis (1999, Fig. 3, p. 1306). (C) Structural details. Short, NW-striking, NE-dipping faults are left lateral. Near Grosvenor's Arch and Pump Canyon Spring, NE-striking faults dip NW and accommodate reverse right-handed offset. Jurassic stratigraphy from oldest to youngest: Navajo Sandstone (Jn), Page Sandstone Member (Jcp), Carmel Formation (Jc), and Entrada Formation (Je). Cretaceous stratigraphy from oldest to youngest: Dakota Formation (Kd), Straight Cliffs Formation (Ksc), Wahweap Formation (Kw), Kaiparowits Formation (Kk), and Qal—Quaternary alluvium. Figures 12A, 12B, and 12C reprinted from Tindall, S., and Davis, G.H., 1999, Monocline development by oblique-slip fault-propagation folding: The East Kaibab monocline, Colorado Plateau, Utah: Journal of Structural Geology, v. 21, no. 10, Figure 6, p. 1308, with permission from Elsevier.

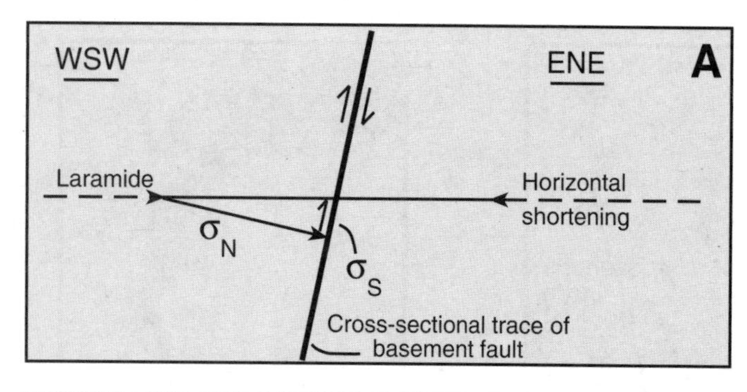

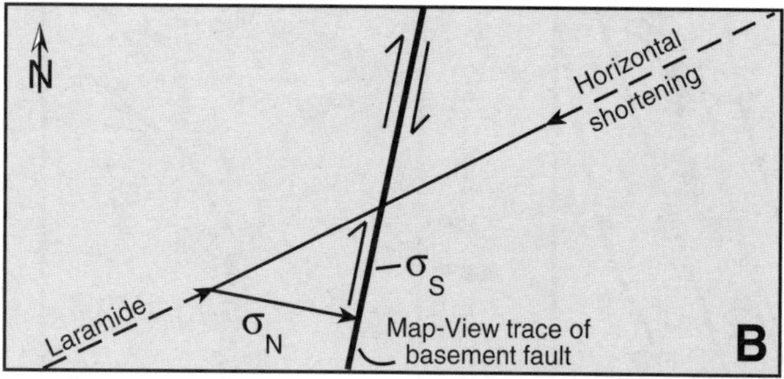

Figure 13. (A) Cross-sectional diagram showing Laramide horizontal compressive stress resolved on a steeply dipping basement fault. σ_S is shear component; σ_N is normal component. (B) Map-view diagram showing resolved shear stress on a steeply dipping basement fault, given N65°E-trending horizontal compressive stress. This configuration favors right-handed transpressive strike slip. Figure is from Davis (1999, Fig. 39, p. 42).

America, and differences between uplifts across these provinces are only a matter of scale (Fig. 14). They reiterated that the Proterozoic normal faults and shear zones, some of which later became reactivated to form the Ancestral Rockies and Laramide-style basement-cored uplifts, were created during rifting of continental crust (Fig. 15). The cratonwide Proterozoic fault orientations, according to the analysis by Marshak et al. (2000), were marked by two dominant sets: N to NE, and W to NW. Marshak et al. (2000) suggested that these rifting-related faults and shear zones evolved during the Mesoproterozoic and Neoproterozoic, specifically in the interval 1.5 to 0.7 Ga (Marshak and Paulsen, 1996; Timmons et al., 2000). Of course, different kinematics could develop on different faults at the same time, depending on the orientation of the faults (Marshak et al., 2003).

Reactivation of Proterozoic normal faults and shear zones into reverse-slip and reverse/oblique-slip structures created monocline fold vergence reflecting "antecedent fault dips" of the basement structures (Marshak et al., 2000, p. 735). Absence of Proterozoic rift strata in the hanging walls of many of the uplift-bounding shear zones was not viewed as a problem by Marshak et al. (2000, p. 738)—they pointed out that regional erosion, now expressed in the Great Unconformity, removed at least 10 km of crust from mid-Proterozoic basement.

Davis et al. (1981, p. 94–95) subscribed to the conclusions reached by Silver (1978) as to why the Neoproterozoic shear zones were prone to reactivation. Silver (1978) argued that intrusion of the 1.41 Ga, regionally extensive, Mesoproterozoic granitic suite was a "cratonization process" that imparted a greater rigidity to the basement. Thus, when basement was subjected to horizontal compressive shortening, neither the relatively incompressible granitic batholiths could shorten, nor could the mechanically softer pendants of metamorphic rocks insulated by the granite. As a result, crustal shortening was accommodated selectively on major, wide-spaced shear-zone discontinuities.

SUBSTANTIALLY DIFFERING VIEWS OF ORIGIN OF THE FOLDS

The structural geologic literature underscores how difficult it has been, over the decades, to work out the deformation mechanism(s) responsible for the formation of the monoclines and the regional anticlines (the uplifts!) with which they are associated. For the longest time, there was no agreement that the fundamental origin was due to crustal shortening, although this conclusion had been reached early and very compellingly by workers such as Baker (1935). End-member compressional deformation mechanisms for folding of regional tectonic significance are "free folding" and "forced folding." Products of free folding have geometric profiles with characteristics that reflect the physical and mechanical properties of the rock layers themselves, especially thickness, stiffness, ductility contrast between layers, and cohesion along layer boundaries (Biot, 1957, 1959; Ramberg, 1959, 1962; Biot et al., 1961; Currie et al., 1962; Johnson, 1977; Davis and Reynolds, 1996, p. 414).

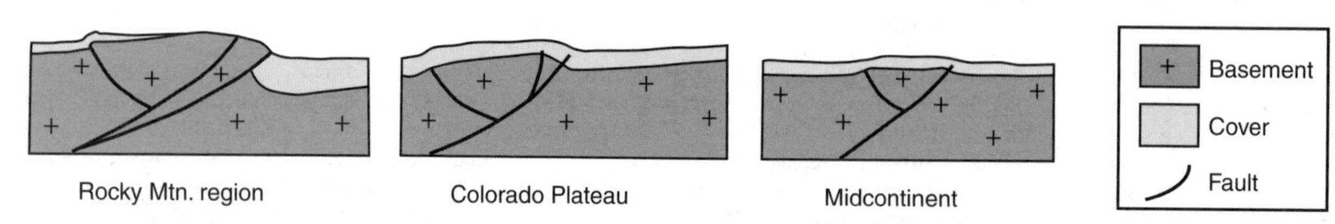

Figure 14. Cross-sectional sketches that emphasize the similarity of style of basement uplifts in the Rocky Mountain region, the Colorado Plateau, and the Midcontinent. Differences are fundamentally a matter of scale. Figure is from Marshak et al. (2000, Fig. 1B, p. 735). Used with permission of the Geological Society of America.

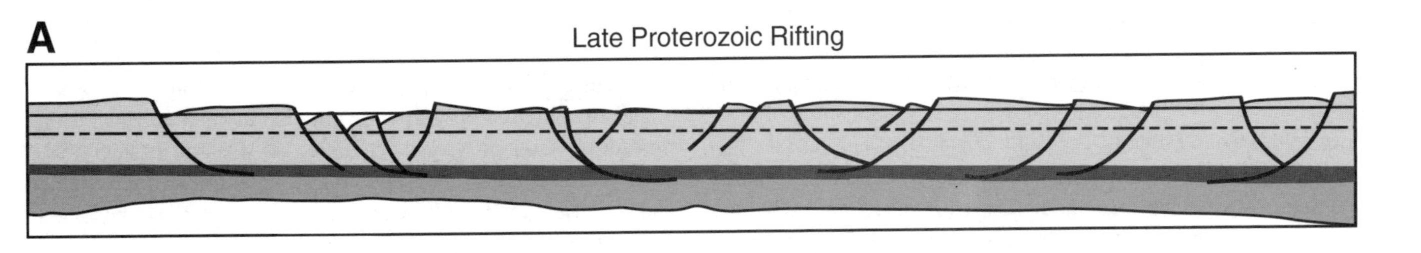

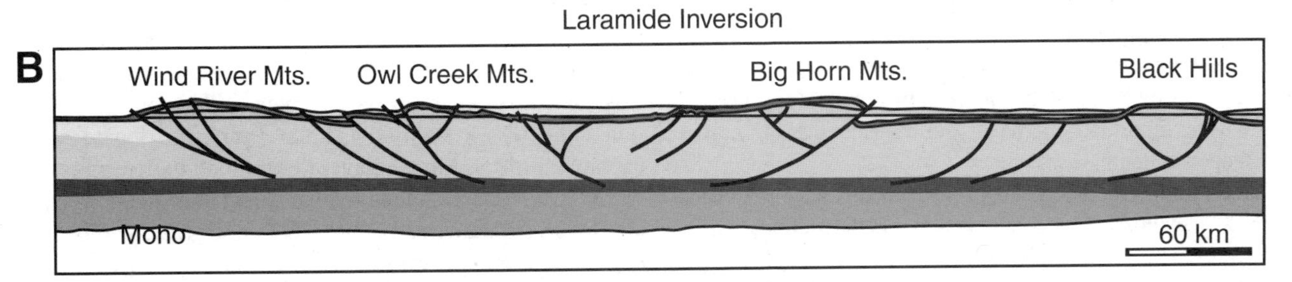

Figure 15. Illustrations showing intracratonic rift inversion. (A) Schematic cross section of Rocky Mountains: GRB—Green River Basin, WRB—Wind River Basin, BHB—Bighorn Basin, PRB—Powder River Basin. (B) Schematic E-W cross section illustrating how the Rocky Mountain section (in A) may have looked prior to Phanerozoic inversion. Figures are from Marshak et al. (2000, Figs. 3A and 3B, p. 737). Published with permission from the Geological Society of America.

Reches and Johnson (1978) emphasized, for example, that buckling or (large-scale) kink folding are both "viable options" for the formation of monoclines on the Colorado Plateau. Based upon both mechanical and experimental modeling, they concluded that the asymmetry—so pronounced in the case of monoclines!—can be produced by layer-parallel shortening and layer-parallel shearing, with or without differential movement along high-angle faults.

Yin (1994) emphasized free folding as well, modeling the upper crust of the entire Colorado Plateau as an elastic thin plate (10 km thick) and having it become deflected into a broad NNW-trending antiform by the combination of horizontal compression and vertical edge loading. The structural relief of the antiform was modeled as 1.5 km, which corresponds to limb-dip inclinations of ~1° or less. The bending of the crust was, according to Yin (1994), accompanied by layer-parallel shear on the antiform's broad, regional, western and eastern limbs, each of which would have been ~500 km wide. Yin proposed that the layer-parallel shear caused the monoclines to form as giant asymmetrical drag folds.

We doubt that sufficient layer-parallel shear could have been generated to achieve the final condition. Ramsay (1967, p. 392–393) has demonstrated that the actual amount of (flexural) slippage along the top of any layer within a flexurally folded sequence can be determined by multiplying the thickness of the folded layer and the dip of the layer in radians (where 1° = 0.175 radian). In Yin's model, the folded basement layer within the "thin elastic plate" is ~7 km thick, and it resides beneath a 3-km-thick cover of Paleozoic, Mesozoic, and Cenozoic strata. If, for Yin's model, *all* of the slip concentrated itself at the basement-cover interface, only 122 m of slip could be generated along the basement-cover interface during bending and formation of the antiform. If, more realistically, this flexural slip were distributed throughout the cover at the boundaries of major, thick, stiff formations, the impact would be negligible. In fact, field observations show that away from the monoclines proper, there is no physical evidence for flexural slip along bedding planes.

Tikoff and Maxson (2001) elevated the scale of application of free folding even higher and proposed that the initiation of the arches and uplifts in the Colorado Plateau, Rocky Mountains, and Wyoming Province reflected buckling at a lithospheric scale. In particular, they envisioned that the Cordilleran foreland was laterally stressed in such a way that the entire lithosphere experienced a buckling instability, and that the "stiff layer," which controlled the dominant wavelength (~190 km), was the lithospheric mantle. The wavelength data reported by Tikoff and Maxson (2001) are marked by a very high standard deviation, perhaps unacceptably high in relation to the conclusions reached. They reported distances between arches (measured along east-west transects) as 140 km, 140 km, 80 km, 60 km, 300 km, and 230 km.

In contrast to free folding, products of forced folding have geometric profiles with characteristics that reflect the form of the faults, and the amount of displacement along the faults with which the folding is associated. Stearns (1971, 1978) was a strong proponent of forced folding in interpreting deformation associated with basement-cored uplifts in the central Rocky Mountains, emphasizing the expression "drape folding" in describing this deformation mechanism. Strata were "draped" over the edges of basement blocks, which had differentially vertically moved with respect to one another by faulting. Descriptions of the structural geometries and rock properties by Stearns and his students and colleagues, through both field and laboratory work, are abundantly detailed (e.g., Jamison, 1979; Jamison and Stearns, 1982; Couples et al., 1994). Stearns (1978) emphasized that the specific fold geometries reflected such factors as ductility contrast between basement and cover work, absolute rheologies, and presence or absence of detachment between basement and cover. Stearns' work sparked great debate, not so much in relation to the fold geometries and rock properties, but in relation to his emphasis that the fault movements associated with basement uplifts were not generated by layer-parallel compression and shortening, but by "vertical tectonics." His emphasis on vertical tectonics was importantly derived from his view that the master faults steepen at depth.

Forced folding geometries and mechanisms have now been examined in excruciating detail, following the delineation of the two dominant classes of forced folding: fault-bend folding (Boyer and Elliott, 1982; Suppe, 1983; Mitra, 1990), and fault-propagation folding (Suppe, 1983, 1985; Mitra and Mount, 1998; Mosar and Suppe, 1992; Poblet and McClay, 1996). "Fault-bend folding" is germane to understanding the Sevier fold-and-thrust belt west of the Colorado Plateau, where, for example, DeCelles and Coogan (2006) reported 220 km of total shortening based on their studies in central Utah. As is evident in the classical literature on structure-tectonics of the Canadian Rockies, the Appalachian Mountains, and the Sevier fold-and-thrust belt (including the Wyoming-Idaho thrust belt), thick miogeoclinal sequences lend themselves to deformation by fault-bend folding. As emphasized in the following section, fault-propagation folding is the mechanism directly applicable to understanding the uplifts of the Colorado Plateau.

TRISHEAR DEVELOPMENT OF COLORADO PLATEAU MONOCLINES

As is apparent, divergent views on the origin of the monoclines and anticlines have commonly been inseparable from diverging views of the orientations of the faults with which these folds are associated. Advances in structural geology in the past two decades have made it crystal clear that, almost always, the form of a major fold reflects the form of the major fault with which it is associated (often "blind" and at depth), and that folds of the type we are addressing are the products of incremental progressive development over the course of thousands of earthquake cycles. This recognition, not understood at the time when the "classical" Colorado Plateau studies were being carried out, creates leverage in determining the orientations and shapes of faults at depth, even in the absence of subsurface data.

It is fault-propagation folding that best applies to formation of Laramide-style uplifts and associated monoclines within the Colorado Plateau. Such folding takes place above the tip of the blind shear zone as it propagates upward and laterally through basement and into cover. Erslev (1991) took issue with the practice of modeling fault-propagation folding as if the fault-propagation fold was a migrating kink band (Suppe and Medwedeff, 1990), establishing its shape early (as a function of fault-ramp angle) and then simply growing in size as displacement along the master fault progressively increased. Instead Erslev (1991, p. 617) emphasized that broad zones of folding in cover tighten and constrict downward toward narrow shear zones in basement, and that the fold geometry throughout affected cover tends to be triangular in cross section.

Erslev (1991) went on to provide the "trishear" kinematic model for fault-propagation folding, and this has proven to be a powerful contribution. "Trishear" (Erslev, 1991, p. 617–618) was coined for the triangular geometry of the zone of deformation within which strain-compatible shear is distributed (see, for example, Fig. 7). Allmendinger (1998) took the precepts of trishear deformation, and the mathematical analysis of "trishear" developed by Hardy and Ford (1997), and created software for both inverse and forward numerical modeling of trishear fault-propagation folding (see also Cristallini and Allmendinger, 2001). The profile geometries that emerge from forward modeling bear a very close relationship to the profiles of Colorado Plateau monoclines. One feature in particular stands out to us, both in the models and in field observations, namely, Erslev's (1991, p. 617) emphasis that the synclinal hinges of monoclines are especially *angular*. In our experience, the synclinal hinges are seldom well exposed in the Colorado Plateau monoclines, but where we have seen them (e.g., the Rock Canyon monocline on the eastern margin of the Apache uplift, and the Nutria monocline along the Zuni uplift), the synclinal hinges are strikingly angular.

We take the opportunity in this paper to insert our own modeling of Colorado Plateau folding using the trishear programs of Allmendinger (1998), using very carefully rendered field profiles of bedding dip and contact locations across uplifts for control (Cardozo, 2005). The trishear modeling itself tracks five variables: initial x and y locations of the fault tip, fault dip (ramp angle), total fault slip, trishear angle, and propagation-to-slip (p/s) ratio (Allmendinger, 1998). A common homogeneous trishear movement plan (a potential sixth variable) was assumed throughout (Erslev, 1991).

An example of our findings is revealed in the results of forward modeling of the Waterpocket monocline, which is the eastern margin of the Circle Cliffs uplift (Fig. 16). Bedding dip data were collected by University of Arizona students Darren Green and Hillary Brown, who mapped a transect across the fold, measured dips using a meter-long digital level (leveling with respect to expansive normal-profile dip exposures), and compiled well logs made available by the Utah Geological Survey. Local stratigraphic thickness data were compiled from well logs as well as from the measured sections of Smith et al. (1963), Billingsley et al. (1987), and Hintze (1988).

We ran nearly 200 forward models of the Waterpocket monocline, experimenting with different trishear angles, ramp angles, and p/s ratios. The best fit for the Waterpocket fold is one

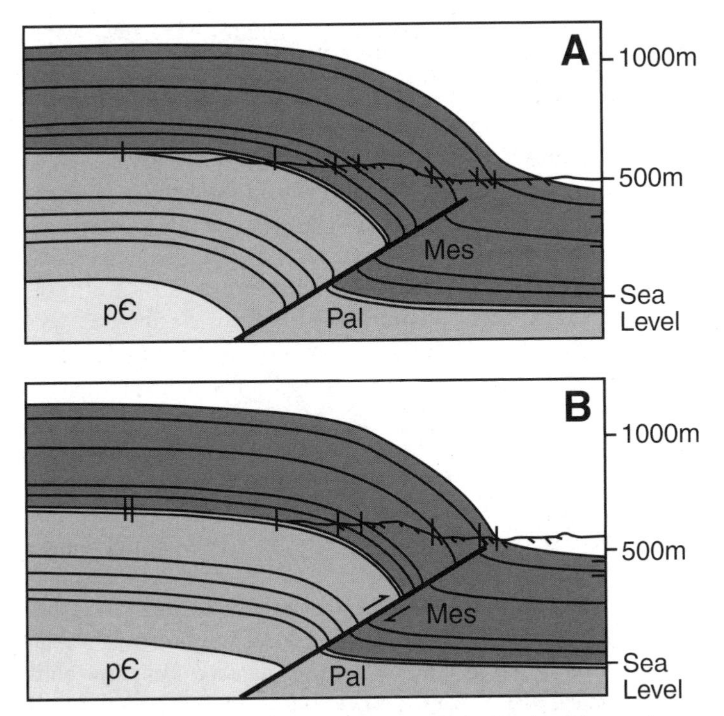

Figure 16. Best-fit inverse and forward models for the Waterpocket monocline (east margin of Circle Cliffs uplift). For the inverse model, note the mismatches in both locations and dips of contacts at topographic surface. Mes—Mesozoic; Pal—Paleocene; pЄ—pre-Cambrian.

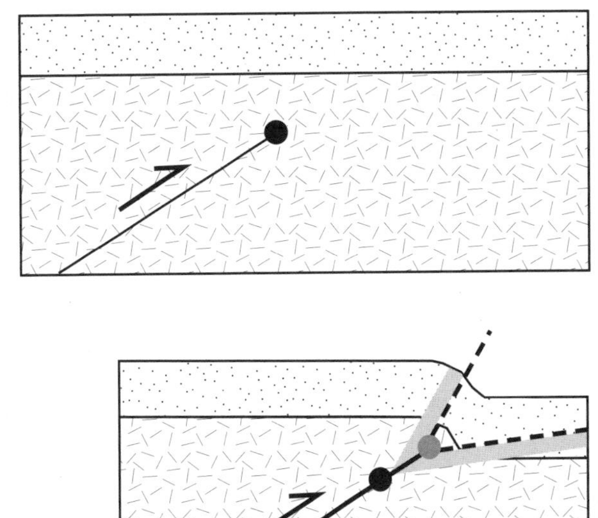

Figure 17. Model of fault tip propagating from within basement into shallower basement, resulting in folding of basement and cover. Figure was modified from García and Davis (2004, Fig. 3b, p. 1260). Published with permission of the American Association of Petroleum Geologists.

with a fault ramp angle of 30°, a trishear angle of 100°, and an initial fault tip 2.3 km below the basement-cover contact. In our best solutions, we used a *p*/*s* ratio of 6.0 within basement, reducing it to 2.1 in cover for the remainder of the total 3.5 km of fault displacement (Fig. 16).

Where basement is found to be folded, the fault tip of the shear zone to be reactivated must have started out below the basement-cover contact (Fig. 17), perhaps originating as a footwall shortcut on an inverted listric shear zone. Insight into the folding of granite, through trishear fault-propagation folding, was presented by García and Davis (2004, p. 1274), based on their analysis of the Sierra de Hualfín basement-cored uplift in the Sierras Pampeanas:

The very propagation of the tip of an advancing basement fault creates, beyond the tip, a physical ground preparation that not only may more readily accommodate further tip advance, but may also accommodate folding of the material, even granite, through which the tip may advance. Thus, in the trishear example, the trishear angle, combined with slip magnitude, influences the amplitude dimensions of the eventual basement-cover folded interface. The damage in the trishear zone, in combination with predeformation anisotropy, renders even granite a macroductility. In short, before imagining folds as fault-propagation folds, it was impossible to imagine a set of mechanisms that would incrementally, first, damage the rock appropriately, and second, impart systematic movements of all constituent parts to create a folded form, even in granite.

REGIONAL STRESS AND STRAIN DIRECTIONS

With few exceptions, the observations and interpretations reported up to this point in this paper have skirted the challenge of inferring shortening directions, and then perhaps stress orientations, on the basis of the structures themselves. It is one thing to call attention to the geometry and deformation plan of the Colorado Plateau basement-cored uplifts, but quite another to invert products of the strain history in order to evaluate shortening directions and causal stresses. Methods have been available to invert fault-slip data to regional stress (Angelier, 1979, 1990, 1994; Suppe, 1985). There are limitations in applying these approaches to the Colorado Plateau uplifts. Although fracturing, largely jointing, is penetrative and pervasive within and across all of the uplifts, mesoscale faulting is essentially absent, except in the deformed zones in close proximity to the monoclines. There are a number of exceptions, e.g., the Chimney Rock area of the San Rafael Swell (Krantz, 1986, 1988), the steep limbs of the Miner's Mountain and Circle Cliffs uplifts, and the West Kaibab fault zone on the margin of the Kaibab uplift (Strahler, 1948), but the mesofaulting record far afield from the deformed zones is too spotty for comprehensive analysis.

Nevertheless, Anderson and Barnhard (1986) applied Angelier's (1979) fault-inversion technique to mesofaults in the sedimentary cover of the Waterpocket and Teasdale monoclines. Although these are not the major uplift-bounding basement faults, they gave a maximum compressive stress direction of N65°E/S65°W.

Kelley and Clinton (1960) studied the regional joint patterns within the entire Colorado Plateau and yet were not able to extract from those data substantive conclusions about the Laramide stress field. Ziony (1966) described joints in the southeastern part of the Monument uplift, near the Four Corners region. Based on the orientations of regional sets of shear joints, he interpreted a Laramide minimum compressive stress direction of ~N20°E/S20°W, approximately parallel to the strike of the Comb Ridge monocline, which bounds the Monument uplift.

Bergerat et al. (1992) reported a detailed study of joints in selected areas of the Colorado Plateau. They identified both shear and tensile joints, from which they interpreted a progressive rotation of maximum compressive stress from 65° to 115°. They cautioned, however, that although jointing could be a useful indicator of tectonic stress directions, it was difficult to use alone.

Swanberg (1999) analyzed joints in the Wingate Sandstone across well-exposed cliff outcrops of the doubly plunging Circle Cliffs uplift. Her work also suggested that it may be productive to invert joint system data to interpret regional compressive stress (Swanberg and Davis, 1999). The joints appear to "emerge" from strain energy stored from the time of folding, for there is a tight relationship between the geometry of the Circle Cliffs anticline and the geometry of the jointing. To date, there have been no such systematic analyses across all of the many uplifts of the Colorado Plateau.

Davis (1999) evaluated the systems of deformation band shear zones in the Colorado Plateau region of southern Utah, with an initial objective (among others) of determining if deformation band shear zones related to the Laramide-style uplifts were evident across the uplifts, and not just within the deformed zones along the boundaries of uplifts. Deformation bands and deformation band shear zones typically are distinguished by millimeter- to centimeter- to meter-scale bands or zones of cataclasis within porous sandstones (of ~20% porosity). They are shear phenomena and fault-like in their structural and kinematic significance. They "work harden" during their development and, unlike fractures and faults, do not become reactivated, thus creating a continuous record of evolving strain, which is a distinct advantage in stress-inversion studies. He found that deformation band shear zones of such origin are fundamentally restricted to the deformed zones, and although they contain an important local record of progressive structural deformation, the regional tectonic signals are much less clear.

Varga (1993) emphasized the challenge in using any mesoscale structural data within the region of Rocky Mountain foreland uplifts to interpret regional stresses within deformed zones, such as along the monoclines and faulted monoclines. He noted, on the one hand, that the regional stress signature may be masked by the local stresses inverted from the local strain conditions. On the other hand, he cautioned that if the deformed zones are weak zones, the far-field regional stresses will be rotated into principal planes parallel and perpendicular to the deformed zone(s). In both instances, local conditions rule, and regional stress signatures remain elusive.

In spite of these collective challenges, Bump (2001) and Bump and Davis (2002) analyzed, for purposes of stress inversion, mesoscale penetrative structures within and near the steep limb of the San Rafael, Miners Mountain, Monument, Uncompahgre, Circle Cliffs, and Kaibab uplifts. A goal was to resolve the challenge of whether penetrative structures observed are products of monoclinal folding and/or manifestations of regional stresses that formed the monoclines in the first place. The approach was to gather structural data at diverse locations along each deformed zone and then determine whether the preferred orientations of the small-scale structures shifted in ways that conformed to comparable shifts in the local strike of the monocline, or whether they remained fixed in preferred orientation even where the trend of the monocline itself shifted significantly.

The conventional mesostructures that proved useful were tectonic stylolites, en echelon vein arrays, deformation bands, and mesoscale slickenlined fault surfaces, including crystal-fiber lineations (Bump and Davis, 2002). We also used a relatively unconventional mesostructure, namely, Eshelby joints (Eidelman and Reches, 1992). These are millimeter- to centimeter-spaced parallel, planar fractures that occur in stiff "inclusions" (stress concentrators such as chert nodules) within soft layers (such as limestone).

For tectonic stylolites, we assumed that the orientation of maximum compressive stress (σ_1) was parallel to the preferred orientation of teeth and cones. Within certain strata in the Monument uplift, the presence of conjugate semibrittle shear zones, composed of nested en echelon crystal-fiber gash veins (calcite) and en echelon tectonic stylolites (Fig. 18), revealed the direction of maximum compressive stress (i.e., perpendicular to the line of intersection of the two conjugate sets, and bisecting the acute angle between the sets). Deformation band shear zones of tectonic origin and slickenlined mesoscale faults commonly occur in conjugate sets as well, with demonstrable expression of sense-of-shear for each set, permitting the direction of maximum compressive stress to be deduced. We assumed that the strike azimuth of Eshelby joints (in the form of penetrative parallel joints in chert nodules within limestone) was the direction of maximum compressive stress.

Inferred directions of maximum compressive stress were found to be consistently oriented within each uplift, irrespective of changes in trends of monoclines associated with the uplift. The directions of maximum compressive stress were thus inferred to show the principal shortening directions (Table 1). Viewed as a system, the data reveal that there are two groupings of the Colorado Plateau uplifts, one that reveals a NE-SW principal shortening direction and a second that reveals a NW-SE principal shortening direction. The first group is composed of the Miners Mountain, Circle Cliffs, Kaibab, and Uncompahgre uplifts. The second group is composed of the San Rafael Swell and the Monument uplift. The conclusions reached regarding the N60°W/S60°E direction of maximum compressional stress (σ_1) in the San Rafael Swell are in good agreement with the work

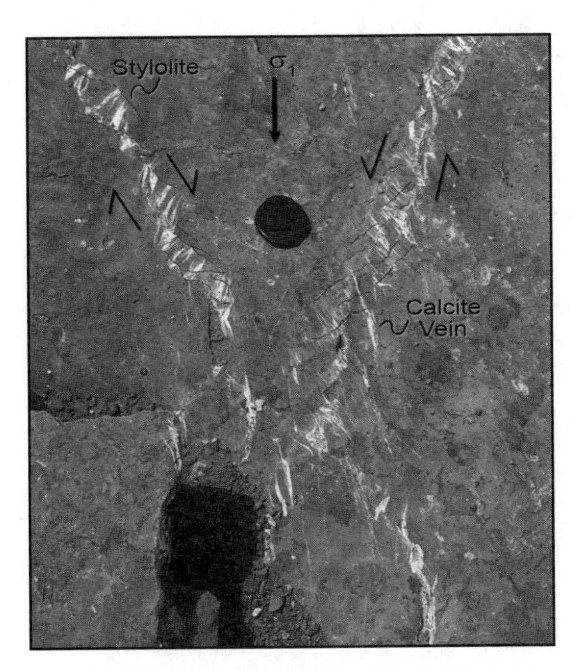

Figure 18. Photograph of semiductile conjugate shear zones composed of nested en echelon crystal-fiber gash veins (calcite) and en echelon tectonic stylolites, in the Honaker Trail Formation, Monument uplift. Note lens cap for scale. Photograph by G.H. Davis.

of Davis (1999) and Christensen and Fischer (2000). The N60°-70°E/S60°-70°W maximum compressive stress (σ_1) determined for the Miners Mountain uplift is similar to observations by Anderson and Barnhard (1986). The N70°W/S70°E determination of maximum compressional stress (σ_1) for the Monument uplift is consistent with conclusions reached by Ziony (1966). The conclusions reached regarding a N50°-55°E/S50°-55°W orientation of maximum compressional stress (σ_1) for the Uncompahgre uplift are in general agreement with the work of Jamison and Stearns (1982). A N55°E/S55W interpretation of maximum compressional stress (σ_1) for the Kaibab uplift is consistent with the findings of Reches (1978), Davis (1999), and Tindall and Davis (1999), and Tindall (2000a, 2000b). Finally, the interpretation of maximum compressional stress (σ_1) of N55°-65°E/S55°-65°W for the Circle Cliffs uplift is consistent with the work of Anderson and Barnhard (1986) and Swanberg (1999).

Bump and Davis (2002) concluded that the compressive stresses identified in their study, and those inferred to have been at work during the formation of the Laramide-style uplifts, were a direct manifestation of basement strain. "[T]he direction of greatest shortening in the basement was parallel to the maximum compressive stress direction in the cover" (Bump and Davis, 2002, p. 436). Stated differently, "[T]he interpreted cover *stress* directions can be viewed as basement *strain* directions" (Bump and Davis, 2002, p. 436).

More broadly, the interpretation of ~65°-directed maximum compressive stress is consistent with the findings of many studies in Wyoming and Colorado (Erslev, 1993; Erslev and Rogers, 1993; Erslev and Wiechelman, 1997). Our interpretation—that some uplifts shortened in an ~110° direction—is not consistent with these studies. Explanations for the origins of two shortening directions for uplifts in the Colorado Plateau, and why only one of these is seen elsewhere, require examination of the timing of the uplifts and consideration of inferred tectonic conditions west of and beneath the Colorado Plateau during the Cretaceous and early Tertiary.

TABLE 1. SUMMARY OF STRESS INVERSIONS

Uplift	Inferred direction of σ_1
San Rafael	N60°W/S60°E
Miners Mountain	N60°-70°E/S60°-70°W
Monument	N70°W/S70°E
Uncompahgre	N45°-55°E/S45°-55°W
Kaibab	N60°-70°E/S60°-70°W

TIMING OF UPLIFTS

Over the years, there have been several attempts to determine the exact timing of the uplifts in the Colorado Plateau. These have involved several approaches, from sedimentologic analysis to apatite fission-track analysis to (U-Th)/He thermochronology. The former have been most successful, but none has demonstrated any clear difference in timing among the different uplifts (Lawton, 1983, 1986; Dumitru et al., 1994; Goldstrand, 1994; Stockli et al., 2002).

Lawton (1983, 1986) examined fluvial deposits of the Upper Cretaceous Mesa Verde Group and Paleocene North Horn Formation in the Uinta Basin on the northern flank of the San Rafael Swell. Starting in the latest Campanian, he documented local thinning of stratigraphy over the northern end of the San Rafael Swell and the diversion of paleodrainages off its flank. By the Maastrichtian, sediments were also ponding against the western flank of the swell. Based on these observations, Lawton interpreted the onset of uplift in the latest Campanian. Uplift was initially rapid, waning later, and it continued until the late Paleocene.

Goldstrand (1994) examined deposits of the Upper Cretaceous Canaan Peak Formation in the Kaiparowits adjacent to the Kaibab and Circle Cliffs uplifts. Like Lawton, he documented stratal thinning onto the uplifts and diversion of the main axial drainages, starting in the latest Campanian or earliest Maastrichtian. The cessation of deformation was recorded by the presence of middle Eocene lacustrine deposits of the Claron Formation, which onlap and overtop Laramide paleotopographic highs (Goldstrand, 1994).

Dumitru et al. (1994) collected apatite-bearing samples spanning ~4 km of stratigraphy, starting with the Precambrian in the bottom of the Grand Canyon, continuing up to the Permian at the rim of the canyon, then shifting laterally to the Circle Cliffs uplift and continuing up through the Cretaceous. Fission-track analysis showed that Permian samples reached a temperature of 90–100 °C and began cooling at 75 Ma. Dumitru et al. (1994)

interpreted this as reflective of erosion triggered by the rise of the Kaibab and Circle Cliffs uplifts. Stockli et al. (2002) hoped to apply the (then) new technique of (U-Th)/He thermochronology to the problem of dating structural deformation of many of the Colorado Plateau uplifts. To that end, they collected and analyzed samples from stratigraphic profiles on the Circle Cliffs, San Rafael, Monument, and Kaibab uplifts. Unfortunately, all of the samples yielded ages from 33 to 11 Ma, consistent with uplift and denudation of the Colorado Plateau, but too young to be referencing the rise of individual Laramide uplifts (Stockli et al., 2002).

Though the constraints on timing are poorly defined, thermochronologic and sedimentologic observations suggest that the uplifts examined all began to rise concurrently in the latest Campanian. Notably, this group includes the uplifts that developed in response to NE-SW shortening and those that developed in response to NW-SE shortening. More broadly, these results are consistent with results of similar studies in Wyoming and Colorado, namely, a rapid onset of deformation in the Campanian followed by quick eastward propagation of shortening, with little or no north-south diachroneity (Bird, 1988; Dickinson et al., 1988; Roberts and Kirschbaum, 1995; Crowley et al., 2002).

ORIGIN AND DEVELOPMENT OF THE SYSTEM OF UPLIFTS

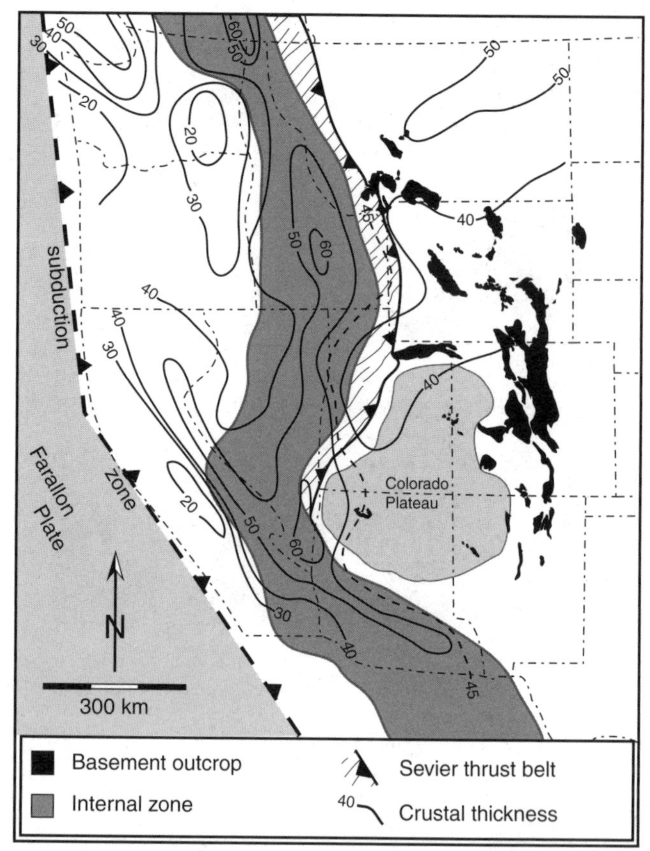

Figure 19. Tectonic map of the western United States in the Late Cretaceous and early Tertiary. Crustal thickness contours and state boundaries are after Coney and Harms (1984) based on their palinspastic restorations. The "internal zone" was defined by Hodges and Walker (1992) on the basis of Cretaceous anatexis and the locations of metamorphic core complexes. Basement outcrops are shown only for the region east of the Sevier belt and serve to highlight the locations of major Laramide uplifts. Location of the subduction zone is approximate. Note the geometry of the Sevier belt and the "internal zone." Stresses generated by them, perpendicular to strike, would be southeast-directed in the Colorado Plateau and east- or northeast-directed elsewhere in the foreland.

From the Late Cretaceous to the early Tertiary, coincident with construction of the uplifts described here, the Colorado Plateau was subjected to compressive stress from two distinctly different sources. To the west and northwest, the thin-skinned Sevier thrust belt, over the course of the preceding 80 m.y., evolved into an enormous topographic edifice (Fig. 19) (DeCelles and Coogan, 2006). Balanced cross sections (Coogan et al., 1995; DeCelles et al., 1995), pressure-time-temperature histories of midcrustal rocks now exposed in the hinterland (Hodges and Walker, 1992; Camilleri and Chamberlain, 1997; Lewis et al., 1999), flexural modeling of the foreland basin (Jordan, 1981; Currie, 1998), kinematic restorations of postorogenic extension (Coney and Harms, 1984), and paleofloral data (Chase et al., 1998) all suggest an average regional elevation of 3–4 km in western Utah. Like all thrust belts, or, for that matter, like all topographic highs, the Sevier thrust belt exerted a compressive stress operating generally perpendicular to the thrust front (Elliott, 1976; Davis et al., 1983; Molnar and Lyon-Caen, 1988; Bada et al., 2001). The Charleston-Nebo salient of the Sevier belt strikes NE-SW (see Fig. 19), and thus the local topography-derived stress was probably directed southeastward into the Colorado Plateau (Bump, 2004).

The second source of compressive stress acting on the Colorado Plateau was related to subduction of the Farallon plate, enhanced by slab flattening and attendant generation of higher traction along the base of the lithosphere. A subduction zone had sat off of the western coast of North America for much of the Phanerozoic, with the slab plunging steeply into the mantle for much or all of that time. Starting in the Late Cretaceous, however, the subducting slab shallowed, sending a pulse of magmatism sweeping far into the foreland (Coney and Reynolds, 1977; Dickinson and Snyder, 1978; Humphreys, 1995; Sterne and Constenius, 1997). Forward and inverse modeling (Bird, 1984, 1998), together with investigations of modern analogs (Jordan and Allmendinger, 1986; Gutscher et al., 2000; Gutscher, 2002), indicates that low-angle subduction exerts a viscous shear on the base of the overriding plate in a direction parallel to the relative plate motion, northeast in the case of western North America and the Farallon plate (Coney, 1978; Engebretson et al., 1985; Cole, 1990; Severinghaus and Atwater, 1990). Additionally, the subducting slab weakened the North American crust and any lithospheric mantle lid, both by hydration and magmatic heating (Humphreys et al., 2003).

In the Late Cretaceous and early Tertiary, the Colorado Plateau was thus caught in a tectonic squeeze, compressed NW-SE by the encroaching thrust belt, and compressed NE-SW by the shallowly subducting Farallon plate and cratonic North America. Paleostress magnitudes are very difficult to assess, but they were probably similar in magnitude. As Molnar and Lyon-Caen (1988) pointed out, topographic highs are crude pressure gauges, in that the weight of the topography must be supported by an equal lateral force at its margins. Assuming a bulk rock density of 2550 kg/m^3, a 3–4-km-high thrust belt would exert a lateral stress of 75–100 MPa on the surrounding region (Bump, 2004). Plate-interaction stress magnitudes are less well known, but estimates range from tens to hundreds of MPa (Govers et al., 1992; Richardson, 1992; Zoback and Healy, 1992; Zoback et al., 1993; Richardson and Coblentz, 1994; Coblentz and Richardson, 1996), about the same order of magnitude as those results for a thrust belt. Farther to the east, the thrust-belt derived stress would have waned due to friction on the basal detachment (be that within the crust or at the base of the lithosphere). Stress from the flat slab, on the other hand, would have increased to the east as the viscous coupling acted over greater distances (Bird, 1988).

In the context of the Cordillera, the Colorado Plateau thus sat in a unique location, subjected to unique stresses. By virtue of the northeastward-subducting Farallon slab below and the southeastward-propagating thrust belt to the west, the plateau was squeezed in two directions simultaneously. Furthermore, by virtue of its position adjacent to the thrust belt, the Colorado Plateau was subjected to the greatest possible stress from the thrust belt and a relatively low stress from the flat slab, such that the two stresses were uniquely similar in magnitude (Bump, 2004). Caught in this constrictional stress field, existing weaknesses in the plateau deformed in a dominantly reverse-slip sense (Bump, 2004). NE-striking faults, such as those bounding the San Rafael and Monument uplifts, slipped to the southeast, while NW-striking faults, such as those bounding the Waterpocket, Miners Mountain, and Uncompahgre uplifts, slipped to the northeast. The Kaibab uplift, which is bounded by a north-northeast–striking fault that slipped obliquely to the northeast (Tindall and Davis, 1999), did not slip to the southeast like those bounding the San Rafael and Monument uplifts because it lies well to the south of the Charleston-Nebo salient, out of the line of southeast-directed thrust-belt stress (Bump, 2004).

CONCLUSIONS

Our present understanding of the Colorado Plateau, its structure, and its tectonic evolution is built on the shoulders of giants and more than 100 yr of geologic field work, analysis, and reanalysis. Beginning with the very first explorations by Powell (Powell, 1873), Gilbert (Gilbert, 1876), and Dutton (Dutton, 1882), structural knowledge of the Colorado Plateau has progressed from basic (and elegant) description of the monoclines and their underlying shear zones, to confusion and debate over the tectonic significance of the range in orientations (Baker, 1935; Kelley, 1955b; Wise, 1963; Sales, 1968; Stearns, 1978; Gries, 1990; Yin, 1994), and to yet another attempt at synthesis, which undoubtedly will be debated and lead to other discussions and understandings. We infer that monoclines bounding the ten major uplifts of the Colorado Plateau are rooted downward into ancient basement shear zones with a long history of reactivation. These shear zones have their origins in different tectonic events and consequently span a broad range of orientations. Like most other workers, we have focused on reactivation of basement shear zones that were originally created in the Proterozoic, but Marshak et al. (2003) rightly pointed out that faults created during Ancestral Rocky Mountain deformation may have been reactivated as well. The most likely areas for this are in the eastern part of the Colorado Plateau (e.g., Paradox basin) and perhaps in the Four Corners region (e.g., Defiance uplift). During the formation of the Laramide-style uplifts, some shear zones were reactivated in pure reverse slip; others were reactivated obliquely. The San Rafael and Monument uplifts shortened toward the southeast. The Circle Cliffs, Kaibab, and Uncompahgre uplifts shortened toward the northeast. These two simultaneous, nearly orthogonal shortening directions were set up by a constrictional strain field that was the product of the northeastward-subducting flat slab and the southeastward-advancing Sevier thrust belt. This constrictional stress field was unique to the Colorado Plateau. Elsewhere in the Rocky Mountains (e.g., in Wyoming), the directions of thrust-belt propagation and plate convergence were more nearly parallel.

An improved understanding of the origin of Laramide-style basement-cored uplifts in the Colorado Plateau will simultaneously improve understanding of the plate-tectonic evolution of continental interiors, including active tectonic phenomena far removed from plate margins. In terms of testing the synthesis presented here, there is an urgent need for a nuanced, sophisticated, very precise understanding of the timing and movement histories of individual uplifts, and for finite-element modeling that would elucidate likely gradients of deformation across the plateau.

It is natural to think of the Colorado Plateau region as strong, but we believe it is more illuminating to think of its basement as riddled with weak shear zones and tectonic displacements that made it quite unnecessary for the shear-zone–bounded blocks themselves, each of which occupies thousands of square kilometers, to exercise layer-parallel strain. It was just a matter of time before the Colorado Plateau thus deformed because the "tip zone" of the Sevier fold-and-thrust belt had been marching steadily eastward toward the Cordilleran hinge line, ultimately advancing eastward beyond the last remaining passive-margin basin sediments. The encounter with "new" crustal structure capped by basement-supported craton sediments triggered a "new" deformation mechanism, one that exploited crustal heterogeneities, including preexisting shear-zone weaknesses, in order to achieve requisite shortening.

ACKNOWLEDGMENTS

We thank the conveners of the "Backbone of the Americas" symposium, Suzanne Mahlburg Kay (Cornell University) and Víctor A. Ramos (Universidad de Buenos Aires), for inviting us to submit this paper to the symposium volume. The observations and ideas presented are an outgrowth of multiple research projects over time. Research support came from The University of Arizona through a number of sources, notably the Department of Geosciences, the Laboratory of Geotectonics (with a special grant from the SOHIO Petroleum Company); the GeoStructure Partnership (sponsors of which include Mobil, Conoco, British Petroleum, British Petroleum/Amoco Exploration, ExxonMobil Upstream Research Company, GeoMap, Inc., and Midland Valley Exploration); and the Office of Arid Lands Studies (through which George Davis received a National Aeronautics and Space Administration grant directed toward analysis of monoclines in the Arizona sector of the Colorado Plateau).

Multiple grants were awarded to George Davis by the Tectonics Program of the Earth Sciences Division of the National Science Foundation (NSF). These include NSF grant EAR-9406208 (use of deformation bands as a guide to regional tectonic stress patterns within the southwestern Colorado Plateau); grant EAR-9706028 (structural analysis of Laramide strike-slip faulting along the eastern margin of the Kaibab uplift, Colorado Plateau); grant INT-9908622 (basement-cored systematics, a structural study of Sierra de Haulfin and adjacent uplifts, Sierra Pampeanas, NW Argentina, with Pilar García as co–principal investigator [PI]); grant INT-9907204 (the making of a high plateau, a field-based determination of tectonic strain and displacement across the Altiplano of Bolivia and implications for modeling Andean plateau uplift, with Nadine McQuarrie as co-PI); and grant EAR-0001222 (inverse and forward modeling of Utah system of Colorado Plateau basement–cored uplifts, with Alex Bump as co-PI).

Furthermore, grants were also awarded to George Davis by the American Chemical Society's Petroleum Research Fund (ACS-PRF). These include ACS-PRF support for "Thin-skinned deformation in the Bryce-Zion region of southern Utah," and "Characterization of deformation-band controlled connectivity and compartmentalization in sandstone."

George Davis thanks former University of Arizona graduate students who participated with him in structural geological research on the Colorado Plateau, and these include Steve Ahlgren, Greg Benson, Pilar García, Lucy Harding, Bob Krantz, Steve Lingrey, Chuck Kiven, Eric Lundin, Lisa McCallmot, Steve Naruk, Scott Showalter, Karen Swanberg, Sarah Tindall, and Alex Bump. Undergraduate students who contributed to advancement of structural geologic research in the program include Greg Belinski, Shari Christofferson, Erin Collin, David Boardman, Scott Grasse, Jessica Graybeal, Angela Smith, and Danielle Vanderhorst. I also call attention to colleagues Nick Nicelsen, Olivier Merle, Fred Cropp, and Gayle Pollock, with whom I benefited in participating together in Colorado Plateau research.

We have benefited over the years from close interaction with special colleagues in the Department of Geosciences, in particular, the late Peter Coney, Bill Dickinson, and Pete DeCelles.

Certain mapping projects in the past were advanced through the excellent support of Western Mapping (formerly GeoMap, Inc.), headed by Jim Holmlund. Collaborations with industry were invaluable, and I particularly acknowledge Chuck Kluth in this regard. Susie Gillatt prepared final figures and illustrations excellently.

Finally, we thank John Geissman and Stephen Marshak for reviewing this manuscript, for providing helpful editorial suggestions, and for providing valuable appraisals of strengths and limitations of our conclusions and our arguments. Our paper is better as a result, but we take full responsibility for the shortcomings in this final product, and we frankly remain humbled by how such "simple" geology can be so difficult to "wire."

REFERENCES CITED

Allmendinger, R.W., 1998, Inverse and forward numerical modeling of trishear fault-propagation folds: Tectonics, v. 17, p. 640–656, doi: 10.1029/98TC01907.

Allmendinger, R., Hauge, T., Hauser, E., Potter, C., Klemperer, S., Nelson, K., Kneupfer, P., and Oliver, J., 1987, Overview of the COCORP 40°N transect, western United States: The fabric of an orogenic belt: Geological Society of America Bulletin, v. 98, p. 308–319, doi: 10.1130/0016-7606(1987)98<308:OOTCNT>2.0.CO;2.

Anderson, C.A., and Silver, L.T., 1976, Yavapai Series—A greenstone belt: Arizona Geological Society Digest, v. 10, p. 13–26.

Anderson, C.A., Blacet, P.M., Silver, L.T., and Stern, T.W., 1971, Revision of Precambrian stratigraphy in the Prescott-Jerome area, Yavapai County, Arizona: U.S. Geological Survey Bulletin 1323-C, p. C1–C16.

Anderson, R.E., and Barnhard, T.P., 1986, Genetic relationship between faults and folds and determination of Laramide and neotectonic paleostress, western Colorado Plateau transition zone: Tectonics, v. 5, no. 2, p. 335–357, doi: 10.1029/TC005i002p00335.

Anderson, T.H., and Silver, L.T., 2005, The Mojave-Sonora megashear—Field and analytical studies leading to the conception and evolution of the hypothesis, *in* Anderson, T.H., Nourse, J.A., McKee, J.W., and Steiner, M.B., eds., The Mojave-Sonora Megashear Hypothesis: Geological Society of America Special Paper 393, p. 1–50.

Angelier, J., 1979, Determination of the mean principal directions of stresses for a given fault population: Tectonophysics, v. 56, p. T17–T26, doi: 10.1016/0040-1951(79)90081-7.

Angelier, J., 1990, Inversion of field data in fault tectonics to obtain the regional stress. III. A new rapid direct inversion method by analytical means: International Journal of Geophysics, v. 103, p. 363–376, doi: 10.1111/j.1365-246X.1990.tb01777.x.

Angelier, J., 1994, Fault slip analysis and palaeostress reconstruction, *in* Hancock, P.L., ed., Continental Deformation: Oxford, Pergamon Press, p. 53–100.

Babenroth, D.L., and Strahler, A.N., 1945, Geomorphology of the East Kaibab monocline, Arizona and Utah: Geological Society of America Bulletin, v. 56, p. 107–150, doi: 10.1130/0016-7606(1945)56[107:GASOTE]2.0.CO;2.

Bada, G., Horvath, F., Cloetingh, S., Coblentz, D., and Toth, T., 2001, Role of topography-induced gravitational stresses in basin inversion: The case study of the Pannonian Basin: Tectonics, v. 20, no. 3, p. 343–363, doi: 10.1029/2001TC900001.

Baker, A.A., 1935, Geologic structure of southeastern Utah: American Association of Petroleum Geologists Bulletin, v. 19, p. 1472–1507.

Bender, E.E., Morrison, J., Anderson, J.L., and Wooden, J.L., 1993, Early Proterozoic ties between two suspect terranes and the Mojave crustal block of the southwestern U.S.: The Journal of Geology, v. 101, p. 715–728.

Bergerat, F., Bouroz-Weil, C., and Angelier, J., 1992, Paleostresses inferred from macrofractures, Colorado Plateau, western USA: Tectonophysics, v. 206, p. 219–243, doi: 10.1016/0040-1951(92)90378-J.

Billingsley, G.H., Huntoon, P.W., and Breed, W.J., 1987, Geologic Map of Capitol Reef National Park and Vicinity, Emery, Garfield, Millard, and Wayne Counties, Utah: Utah Geological and Mineral Survey, colored geologic map, scale 1:62,500, 5 sheets.

Biot, M.A., 1957, Folding instability of a layered viscoelastic medium under compression: Royal Society of London Proceedings, ser. A, v. 242, p. 211–228.

Biot, M.A., 1959, On the instability of folding deformation of a layered visco-elastic medium under compression: Journal of Applied Mechanics, v. 26, p. 393–400.

Biot, M.A., Ode, H., and Roever, W.L., 1961, Experimental verification of the folding of stratified viscoelastic media: Geological Society of America Bulletin, v. 72, p. 1621–1630, doi: 10.1130/0016-7606(1961)72[1621:EVOTTO]2.0.CO;2.

Bird, P., 1984, Laramide crustal thickening event in the Rocky Mountain foreland and Great Plains: Tectonics, v. 3, p. 741–758, doi: 10.1029/TC003i007p00741.

Bird, P., 1988, Formation of the Rocky Mountains, western United States: A continuum computer model: Science, v. 239, p. 1501–1507, doi: 10.1126/science.239.4847.1501.

Bird, P., 1998, Kinematic history of the Laramide orogeny in latitudes 35°–49°N, western United States: Tectonics, v. 17, p. 780–801, doi: 10.1029/98TC02698.

Boyer, S.E., and Elliott, D., 1982, Thrust systems: American Association of Petroleum Geologists Bulletin, v. 66, p. 1196–1230.

Brown, W.G., 1988, Deformation style of Laramide uplifts in the Wyoming foreland, *in* Schmidt, C.J., and Perry, W.L.J., eds., Interaction of the Rocky Mountain Foreland and the Cordilleran Thrust Belt: Geological Society of America Memoir 171, p. 1–25.

Brown, W.G., 1993, Structural style of Laramide basement-cored uplifts and associated folds, *in* Snoke, A.W., Steidtmann, J.R., and Roberts, S.M., eds., Geology of Wyoming: Geological Survey of Wyoming Memoir 5, p. 312–371.

Bump, A.P., 2001, Kinematics, Dynamics, and Mechanics of Laramide Deformation, Colorado Plateau, Utah and Colorado [Ph.D. dissertation]: Tucson, University of Arizona, 149 p.

Bump, A. P., 2004, Three-dimensional Laramide deformation of the Colorado Plateau: Competing stresses from the Sevier thrust belt and the flat Farallon slab: Tectonics, v. 23, doi: 10.1029/2002TC001424.

Bump, A., and Davis, G.H., 2002, Forward and inverse trishear modeling of the San Rafael and Waterpocket monoclines, Colorado Plateau, Utah: Geological Society of America Abstracts with Programs, v. 34, no. 6, p. 40.

Bump, A.P., and Davis, G.H., 2003, Late Cretaceous to early Tertiary Laramide deformation of the northern Colorado Plateau, Utah and Colorado: Journal of Structural Geology, v. 25, p. 421–440, doi: 10.1016/S0191-8141(02)00033-0.

Bump, A., Ahlgren, S.G., and Davis, G.H., 1997, A tale of two uplifts: Waterpocket fold, Capitol Reef National Park: Eos (Transactions, American Geophysical Union), v. 78, no. 46 (Fall Meeting), p. F701.

Bump, A., Davis, G.H., Bilinski, G.E., Swiders, A.E., and Kidder, S., 2000, Progressive development of brittle and semi-brittle structures on the Monument uplift, Utah, and their implications for regional strain: Geological Society of America Abstracts with Programs, v. 32, no. 5, p. A-4.

Burke, K., and Dewey, J.G., 1973, Plume generated triple junctions: Key indicators in applying plate tectonics to old rocks: The Journal of Geology, v. 81, p. 406–433.

Byerlee, J.D., 1978, Friction of rocks: Pure and Applied Geophysics, v. 116, p. 615–626, doi: 10.1007/BF00876528.

Camilleri, P.A., and Chamberlain, K.R., 1997, Mesozoic tectonics and metamorphism in the Pequop Mountains and Woods Hills region, northwest Nevada: Implications for the architecture and evolution of the Sevier orogen: Geological Society of America Bulletin, v. 109, no. 1, p. 74–94, doi: 10.1130/0016-7606(1997)109<0074:MTAMIT>2.3.CO;2.

Cardozo, N., 2005, Trishear modeling of fold bedding data along a topographic profile: Journal of Structural Geology, v. 27, p. 495–502, doi: 10.1016/j.jsg.2004.10.004.

Case, J.E., and Joesting, H.E., 1972, Regional Geophysical Investigations in the Central Colorado Plateau: U.S. Geological Survey Professional Paper 736, 31 p.

Cashion, W.B., 1973, Geologic and Structure Map of the Grand Junction Quadrangle, Colorado and Utah: U.S. Geological Survey Miscellaneous Investigations Series I-0736, scale 1:250,000.

Chase, C.G., Gregory, K.M., Parrish, J.T., and DeCelles, P.G., 1998, Topographic history of the western Cordillera of North America and the etiology of climate, *in* Crowley, T.J., and Burke, K., eds., Tectonic Boundary Conditions for Climate Reconstructions: Oxford Monographs on Geology and Geophysics 39, p. 73–99.

Christensen, R.D., and Fischer, M.P., 2000, Using mesoscopic structural analysis to interpret the deformation history of the San Rafael Swell, east-central Utah: Geological Society of America Annual Meeting Abstracts with Programs, v. 32, no. 7, p. 233.

Coblentz, D.D., and Richardson, R.M., 1996, Analysis of the South American intraplate stress field: Journal of Geophysical Research, v. 101, p. 8643–8657, doi: 10.1029/96JB00090.

Cole, G.L., 1990, Models of Plate Kinematics along the Western Margin of the Americas: Cretaceous to Present [Ph.D. dissertation]: Tucson, University of Arizona, 460 p.

Coney, P.J., 1978, Mesozoic-Cenozoic Cordilleran plate tectonics, *in* Smith, R.B., and Eaton, G.P., eds., Cenozoic Tectonics and Regional Geophysics of the Western Cordillera: Geological Society of America Memoir 152, p. 33–50.

Coney, P.J., and Harms, T.A., 1984, Cordilleran metamorphic core complexes: Cenozoic extensional relics of Mesozoic compression: Geology, v. 12, p. 550–554, doi: 10.1130/0091-7613(1984)12<550:CMCCCE>2.0.CO;2.

Coney, P.J., and Reynolds, S.J., 1977, Cordilleran Benioff zones: Nature, v. 270, p. 403–406, doi: 10.1038/270403a0.

Conway, C.M., Karlstrom, K.E., Silver, L.T., and Wrucke, C.T., 1987, Tectonic and magmatic contrasts across a two-province Proterozoic boundary in central Arizona, *in* Davis, G.H., and VandenDolder, E.M., eds., Geologic Diversity of Arizona and Its Margins: Excursions to Choice Areas (Geological Society of America Field Trip Guidebook 1987 Annual Meeting, Phoenix): Arizona Bureau of Geology and Mineral Technology Special Paper 5, p. 158–175.

Coogan, J.C., DeCelles, P.G., Mitra, G., and Sussman, A.J., 1995, New regional balanced cross-section across the Sevier Desert region and central Utah thrust belt: Geological Society of America Abstracts with Programs, v. 27, no. 4, p. 7.

Cook, K.L., Mabey, D.R., and Bankey, V., 1991, The new gravity map of Utah—An overview of its usefulness for geologic studies: Utah Geologic and Mineral Survey Notes, v. 24, p. 12–21.

Cooper, J.R., and Silver, L.T., 1964, Geology and Ore Deposits of the Dragoon Quadrangle, Cochise County, Arizona: U.S. Geological Survey Professional Paper 416, 196 p.

Cooper, M.A., and Williams, G.D., eds., 1989, Inversion Tectonics: Geological Society of London Special Publication 44, 375 p.

Couples, G.D., Stearns, D.W., and Handin, J.W., 1994, Kinematics of experimental forced folds and their relevance to cross-section balancing: Tectonophysics, v. 233, p. 193–213, doi: 10.1016/0040-1951(94)90241-0.

Cristallini, E., and Allmendinger, R., 2001, Pseudo 3-D modeling of trishear fault-propagation folding: Journal of Structural Geology, v. 23, p. 1883–1900, doi: 10.1016/S0191-8141(01)00034-7.

Crowley, P.D., Reiners, P.W., Reuter, J.M., and Kaye, G.D., 2002, Laramide exhumation of the Bighorn Mountains, Wyoming: An apatite (U-Th)/He thermochronologic study: Geology, v. 30, p. 27–30, doi: 10.1130/0091-7613(2002)030<0027:LEOTBM>2.0.CO;2.

Currie, B.S., 1998, Jurassic-Cretaceous Evolution of the Central Cordilleran Foreland-Basin System [Ph.D. dissertation]: Tucson, University of Arizona, 239 p.

Currie, J.B., Patnode, A.W., and Trump, R.P., 1962, Development of folds in sedimentary strata: Geological Society of America Bulletin, v. 73, p. 655–674, doi: 10.1130/0016-7606(1962)73[655:DOFISS]2.0.CO;2.

Davis, D.M., Suppe, J., and Dahlen, F.A., 1983, Mechanics of fold-and-thrust belts and accretionary wedges: Journal of Geophysical Research, v. 88, no. B2, p. 1153–1172, doi: 10.1029/JB088iB02p01153.

Davis, G.H., 1978, The monocline fold pattern of the Colorado Plateau, *in* Matthews, V., ed., Laramide Folding Associated with Basement Block Faulting in the Western United States: Geological Society of America Memoir 151, p. 215–233.

Davis, G.H., 1999, Structural Geology of the Colorado Plateau Region of Southern Utah, with Special Emphasis on Deformation Band Shear Zones: Geological Society of America Special Paper 342, 157 p.

Davis, G.H., and Kiven, C.W., 1975, Structure Map of Folds in Phanerozoic Rocks, Colorado Plateau Tectonic Province of Arizona: Phoenix, Arizona, Arizona Oil and Gas Conservation Commission, scale 1:500,000.

Davis, G.H., and Reynolds, S.J., 1996, Structural Geology of Rocks and Regions: New York, John Wiley and Sons, 776 p.

Davis, G.H., and Tindall, S.E., 1996, Discovery of major right-handed Laramide strike-slip faulting along the eastern margin of the Kaibab Uplift, Colorado Plateau, Utah: Eos (Transactions, American Geophysical Union), v. 77, 1996 Fall Meeting, p. F641–F642.

Davis, G.H., Showalter, S.R., Benson, G.S., McCalmont, L.C., and Cropp, F.W., 1981, Guide to the geology of the Salt River Canyon, Arizona: Arizona Geological Society Digest, v. 13, p. 48–97.

DeCelles, P.G., and Coogan, J.C., 2006, Regional structure and kinematic history of the Sevier fold-thrust belt, central Utah: Geological Society of America, v. 118, p. 841–864, doi: 10.1130/B25759.1.

Dewey, J.F., 1988, Lithospheric stress, deformation, and tectonic cycles; the disruption of Pangaea and the closure of Tethys, *in* Audley-Charles, M.G., and Hallam, A., eds., Gondwana and Tethys: Geological Society of London Special Publication 37, p. 23–40.

Dickinson, W.R., 1977, Paleozoic plate tectonics and the evolution of the Cordilleran continental margin, *in* Stewart, J.H., and Stevens, C.H., eds., Paleozoic Paleogeography of the Western United States: Society of Economic Paleontologists and Mineralogists, Pacific Section Symposium I, Los Angeles, p. 137–155.

Dickinson, W.R., and Snyder, W.S., 1978, Plate tectonics of the Laramide orogeny, *in* Matthews, V., III, ed., Laramide Folding Associated with Basement Block Faulting in the Western United States: Geological Society of America Memoir 11, p. 355–366.

Dickinson, W.R., Klute, M.A., Hayes, M.J., Janecke, S.U., Lundin, E.R., McKittrick, M.A., and Olivares, M.D., 1988, Paleogeographic and paleotectonic setting of Laramide sedimentary basins in the central Rocky Mountain region: Geological Society of America Bulletin, v. 100, p. 1023–1039, doi: 10.1130/0016-7606(1988)100<1023:PAPSOL>2.3.CO;2.

Donath, F.A., 1961, Experimental study of shear failure in anisotropic rocks: Geological Society of America Bulletin, v. 72, p. 985–989, doi: 10.1130/0016-7606(1961)72[985:ESOSFI]2.0.CO;2.

Donath, F.A., and Cranwell, R.M., 1981, Probabilistic treatment of faulting in geologic media, *in* Carter, N.L., Friedman, M., Logan, J.M., and Stearns, D.W., eds., Mechanical Behavior of Crustal Rocks: The Handin Volume: Geophysical Monograph, v. 24, p. 231–241.

Dumitru, T., Duddy, I., and Green, P., 1994, Mesozoic-Cenozoic burial, uplift, and erosion of the west-central Colorado Plateau: Geology, v. 22, p. 499–502, doi: 10.1130/0091-7613(1994)022<0499:MCBUAE>2.3.CO;2.

Dutton, C.E., 1882, The physical geology of the Grand Canyon district: U.S. Geological Survey Second Annual Report: Washington, D.C., p. 49–166.

Eardley, A.J., 1949, Structural evolution of Utah, *in* Hansen, G.H., and Bell, M.M., eds., The Oil and Gas Possibilities of Utah: Salt Lake, Utah Geological and Mineralogical Survey, p. 10–23.

Eidelman, A., and Reches, Z., 1992, Fractured pebbles—A new stress indicator: Geology, v. 20, p. 307–310, doi: 10.1130/0091-7613(1992)020<0307:FPANSI>2.3.CO;2.

Eisele, J., and Isachsen, C.E., 2001, Crustal growth in southern Arizona: U-Pb geochronological and Sm-Nd isotopic evidence for addition of the Paleoproterozoic Cochise block to the Mazatzal Province: American Journal of Science, v. 301, p. 773–797, doi: 10.2475/ajs.301.9.773.

Elliott, D., 1976, The energy balance and deformation mechanism of thrust sheets: Philosophical Transactions of the Royal Society of London, v. 283, p. 289–312, doi: 10.1098/rsta.1976.0086.

Elston, D.P., 1979, Late Precambrian Sixtymile Formation and Orogeny at the Top of the Grand Canyon Supergroup, Northern Arizona: U.S. Geological Survey Professional Paper 1092, 20 p.

Engebretson, D.C., Cox, A., and Gordon, R.G., 1985, Relative Motions between Oceanic and Continental Plates in the Pacific Basin: Geological Society of America Special Paper 206, 59 p.

Erslev, E.A., 1991, Trishear fault propagation folding: Geology, v. 19, p. 617–620, doi: 10.1130/0091-7613(1991)019<0617:TFPF>2.3.CO;2.

Erslev, E., 1993, Thrusts, back-thrusts, and detachments of Rocky Mountain foreland arches, *in* Schmidt, C.J., Chase, R.B., and Erslev, E.A., eds., Laramide Basement Deformation in the Rocky Mountain Foreland of the Western United States: Geological Society of America Special Paper 280, p. 339–358.

Erslev, E.A., and Rogers, J.L., 1993, Basement-cover geometry of Laramide fault-propagation folds, *in* Schmidt, C.J., Chase, R.B., and Erslev, E.A., eds., Laramide Basement Deformation in the Rocky Mountain Foreland of the Western United States: Geological Society of America Special Paper 280, p. 125–146.

Erslev, E.A., and Wiechelman, D., 1997, Fault and fold orientations in the central Rocky Mountains of Colorado and Utah, *in* Fractured Reservoirs: Characterization and Modeling Guidebook–1997: Denver, Rocky Mountain Association of Geologists, p. 131–136.

Etheridge, M.A., 1986, On the reactivation of extensional fault systems: Royal Society of London Philosophical Transactions, ser. A, v. 317, p. 179–194.

Faulds, J.E., 1986, Tertiary Geologic History of the Salt River Canyon Region, Gila County, Arizona [M.S. thesis]: Tucson, University of Arizona, 319 p.

Finnell, T.L., 1962, Recurrent movement on the Canyon Creek Fault, Navajo County, Arizona: U.S. Geological Survey Professional Paper 450-D, art. 143, p. 80–82.

García, P.E., and Davis, G.H., 2004, Evidence and mechanism for folding of granite, in Sierra de Haulfín basement-cored uplift, northwest Argentina: American Association of Petroleum Geologists Bulletin, v. 88, no. 9, p. 1255–1276.

Gilbert, G.K., 1876, The Colorado Plateau province as a field for geological study: American Journal of Science, v. 12, p. 16–24, 85–103.

Gilluly, J., 1952, Connection between orogeny and epeirogeny as deduced from the history of the Colorado Plateau and the Great Basin: Geological Society of America Bulletin, v. 63, p. 1329–1330.

Goldstrand, P.M., 1994, Development of Upper Cretaceous to Eocene strata of southwestern Utah: Geological Society of America Bulletin, v. 106, p. 145–154, doi: 10.1130/0016-7606(1994)106<0145:TDOUCT>2.3.CO;2.

Govers, R., Wortel, M.J.R., Cloetingh, S.A.P.I., and Stein, C.A., 1992, Stress magnitude estimates from earthquakes in oceanic plate interiors: Journal of Geophysical Research, v. 97, no. B8, p. 11,749–11,759, doi: 10.1029/91JB01797.

Granger, H.C., and Raup, R.B., 1969, Geology of Uranium Deposits in the Dripping Spring Quartzite, Gila County, Arizona: U.S. Geological Survey Professional Paper 595, 108 p.

Gregory, H.E., and Moore, R.C., 1931, The Kaiparowits Region—A Geographic and Geological Reconnaissance of Parts of Utah and Arizona: U.S. Geological Survey Professional Paper 164, 161 p.

Gries, R.R., 1990, Rocky Mountain foreland structures: Changes in compression direction through time, *in* Letouzey, J., ed., Petroleum and Tectonics in Mobile Belts: Paris, Editions Technip, p. 129–148.

Gutscher, M.-A., 2002, Andean subduction styles and their effect on thermal structure and interplate coupling: Journal of South American Earth Sciences, v. 15, p. 3–10, doi: 10.1016/S0895-9811(02)00002-0.

Gutscher, M.-A., Spakman, W., Bijwaard, H., and Engdahl, E.R., 2000, Geodynamics of flat subduction: Seismicity and tomographic constraints from the Andean margin: Tectonics, v. 19, no. 5, p. 814–833, doi: 10.1029/1999TC001152.

Hackman, R.J., and Wyant, D.G., 1973, Geology, Structure, and Uranium Deposits of the Escalante Quadrangle, Utah and Arizona: Washington, D.C., U.S. Geological Survey, 2 sheets.

Hardy, S., and Ford, M., 1997, Numerical modeling of trishear fault propagation folding: Tectonics, v. 16, p. 841–854, doi: 10.1029/97TC01171.

Haynes, D.D., Vogel, J.D., and Wyant, D.G., 1972, Geology, Structure, and Uranium Deposits of the Cortez Quadrangle, Utah and Colorado: Washington, D.C., U.S. Geological Survey Miscellaneous Investigations Series I-629, scale 1:250,000, 2 sheets.

Hintze, L.F., 1980, Geologic Map of Utah: Salt Lake City, Utah Geological and Mineral Survey.

Hintze, L.F., 1988, Geologic History of Utah: Brigham Young University Geology Studies Special Publication 7, 202 p.

Hodges, K.V., and Walker, J.D., 1992, Extension in the Cretaceous Sevier orogen, North American Cordillera: Geological Society of America Bulletin, v. 104, p. 560–569, doi: 10.1130/0016-7606(1992)104<0560:EITCSO>2.3.CO;2.

Humphreys, E.D., 1995, Post-Laramide removal of the Farallon slab, western United States: Geology, v. 23, p. 987–990, doi: 10.1130/0091-7613(1995)023<0987:PLROTF>2.3.CO;2.

Humphreys, E., Hessler, E., Dueker, K., Farmer, G., Erslev, E., and Atwater, T., 2003, How Laramide-age hydration of North American lithosphere by the Farallon slab controlled subsequent activity in the Western United States: International Geology Review, v. 45, no. 7, p. 575–595, doi: 10.2747/0020-6814.45.7.575.

Huntoon, P.W., 1969, Recurrent movements and contrary bending along the West Kaibab fault zone: Plateau, v. 42, p. 66–74.

Huntoon, P.W., 1971, The deep structure of the monoclines in eastern Grand Canyon, Arizona: Plateau, v. 43, p. 148–158.

Huntoon, P.W., 1974, Synopsis of Laramide and post-Laramide structural geology of the eastern Grand Canyon, Arizona, *in* Eastwood, R.L., et al., eds., Geology of Northern Arizona: Part I. Regional Studies: Flagstaff, Northern Arizona University, p. 317–335.

Huntoon, P.W., 1993, Influence of inherited Precambrian basement structure on the localization and form of Laramide monoclines, Grand Canyon, Arizona, *in* Schmidt, C.J., Chase, R.B., and Erslev, E.A., eds., Laramide Basement Deformation in the Rocky Mountain Foreland of the Western United States: Geological Society of America Special Paper 280, p. 243–256.

Huntoon, P.W., and Sears, J.W., 1975, Bright Angel and Eminence faults, eastern Grand Canyon, Arizona: Geological Society of America Bulletin, v. 86, p. 465–472, doi: 10.1130/0016-7606(1975)86<465:BAAEFE>2.0.CO;2.

Ilg, B.R., Karlstrom, K.E., Hawkins, D.P., and Williams, M.L., 1996, Tectonic evolution of Paleoproterozoic rocks in the Grand Canyon: Insights into middle-crustal processes: Geological Society of America Bulletin, v. 108, p. 1149–1166, doi: 10.1130/0016-7606(1996)108<1149:TEOPRI>2.3.CO;2.

Jamison, W.R., 1979, Laramide deformation of the Wingate Sandstone, Colorado National Monument; a study of cataclastic flow [PhD thesis]: Texas A&M University, College Station, Texas, 181 p.

Jamison, W.R., and Stearns, D.W., 1982, Tectonic deformation of the Wingate Sandstone, Colorado National Monument: American Association of Petroleum Geologists Bulletin, v. 66, p. 2584–2608.

Johnson, A.M., 1977, Styles of Folding: Mechanics and Mechanisms of Folding of Natural Elastic Materials: Amsterdam, Elsevier Scientific Publishing Company, 406 p.

Jordan, T.E., 1981, Thrust loads and foreland basin evolution, Cretaceous, western United States: American Association of Petroleum Geologists Bulletin, v. 65, no. 12, p. 2506–2520.

Jordan, T.E., and Allmendinger, R.W., 1986, The Sierras Pampeanas of Argentina: A modern analogue of Rocky Mountain foreland deformation: American Journal of Science, v. 286, p. 737–764.

Kamilli, R.J., and Richard, S.M., eds., 1998, Geologic Highway Map of Arizona: Tucson, Arizona, Arizona Geological Society and Arizona Geological Survey, scale 1:1,000,000.

Karlstrom, K.E., 1993, Proterozoic orogenic history of Arizona, *in* Van Schmus, W.R., Bickford, M.E., and 23 others, Transcontinental Proterozoic provinces, *in* Reed, J.C., Jr., Bickford, M.E., Houston, R.S., Link, P.K., Rankin, D.W., Sims, P.K., and Van Schmus, W.R., eds., Precambrian: Conterminous U.S.: Boulder, Colorado, Geological Society of America, Geology of North America, v. C-2, p. 188–211.

Karlstrom, K.E., and Bowring, S.A., 1988, Early Proterozoic assembly of tectonostratigraphic terranes in southwestern North America: The Journal of Geology, v. 96, p. 561–576.

Karlstrom, K.E., and Daniel, C.G., 1993, Restoration of Laramide right-lateral strike slip in northern New Mexico by using Proterozoic piercing points: Tectonic implications from the Proterozoic to the Cenozoic: Geology, v. 21, p. 1139–1142, doi: 10.1130/0091-7613(1993)021<1139:ROLRLS>2.3.CO;2.

Karlstrom, K.E., and Humphreys, E.D., 1998, Persistent influence of Proterozoic accretionary boundaries in the tectonic evolution of southwestern North America: Interaction of cratonic grain and mantle modification events: Rocky Mountain Geology, v. 33, no. 2, p. 161–179.

Karlstrom, K.E., and Williams, M.L., 1998, Heterogeneity of the middle crust: Implications for strength of continental lithosphere: Geology, v. 26, p. 815–818, doi: 10.1130/0091-7613(1998)026<0815:HOTMCI>2.3.CO;2.

Kelley, V.C., 1955a, Monoclines of the Colorado Plateau: Geological Society of America Bulletin, v. 66, p. 789–804, doi: 10.1130/0016-7606(1955)66[789:MOTCP]2.0.CO;2.

Kelley, V.C., ed., 1955b, Regional Tectonics of the Colorado Plateau and Relationship to the Origin and Distribution of Uranium: New Mexico University Publications in Geology 5, 120 p.

Kelley, V.C., and Clinton, N.J., eds., 1960, Fracture Systems and Tectonic Elements of the Colorado Plateau: New Mexico University Publications in Geology 6, 104 p.

King, P.B., 1969, Tectonic Map of North America: Washington, D.C., American Geological Institute, scale 1:5,000,000, 2 sheets.

Krantz, R.W., 1986, The Odd-Axis Model: Orthorhombic Fault Patterns and Three-Dimensional Strain Fields [Ph.D. dissertation]: Tucson, University of Arizona, 97 p.

Krantz, R.W., 1988, Multiple fault sets and three-dimensional strain: Theory and application: Journal of Structural Geology, v. 10, p. 225–237, doi: 10.1016/0191-8141(88)90056-9.

Lawton, T.F., 1983, Late Cretaceous fluvial systems and the age of foreland uplifts in central Utah, *in* Lowell, J.D., and Gries, R.R., eds., Rocky Mountain Foreland Basins and Uplifts: Denver, Rocky Mountain Association of Geologists, p. 181–199.

Lawton, T.F., 1986, Fluvial systems of the Upper Cretaceous Mesa Verde Group and Paleocene North Horn Formation, central Utah; a record of transition from thin-skinned to thick-skinned deformation in the foreland regions, *in* Peterson, J., ed., Paleotectonics and Sedimentation in the Rocky Mountain Region, United States: American Association of Petroleum Geologists Memoir 41, p. 423–442.

Le Pichon, X., and Sibuet, J., 1981, Passive margins: A model of formation: Journal of Geophysical Research, v. 86, no. B5, p. 3708–3720.

Lewis, C.J., Wernicke, B.P., Selverstone, J., and Bartley, J.M., 1999, Deep burial of the footwall of the northern Snake Range décollement: Geological Society of America Bulletin, v. 111, p. 39–51, doi: 10.1130/0016-7606(1999)111<0039:DBOTFO>2.3.CO;2.

Lohman, S.W., 1963, Geologic map of the Grand Junction area, Colorado: U.S. Geological Survey Miscellaneous Geologic Investigations Map I-0404, N38°30′00″–N39°20′00″ and W109°05′00″–W107°20′00″.

Mackin, J.H., 1950, The down-structure method of viewing geologic maps: The Journal of Geology, v. 58, p. 55–72.

Marshak, S., and Paulsen, T., 1996, Midcontinent U.S. fault and fold zones: A legacy of Proterozoic intracratonic extensional tectonism?: Geology, v. 24, p. 151–154, doi: 10.1130/0091-7613(1996)024<0151:MUSFAF>2.3.CO;2.

Marshak, S., Karlstrom, K., and Timmons, J.M., 2000, Inversion of Proterozoic extensional faults: An explanation for the pattern of Laramide and Ancestral Rockies intracratonic deformation: Geology, v. 28, p. 735–738, doi: 10.1130/0091-7613(2000)28<735:IOPEFA>2.0.CO;2.

Marshak, S., Nelson, W.J., and McBride, J., 2003, Strike-slip faulting in the continental interior of North America, *in* Storty, F., Holdsworth, R.E., and Salvine, F., eds., Intraplate Strike-Slip Deformation Belts: Geological Society of London Special Publication 210, p. 171–196.

Maxson, J.H., 1961, Geologic Map of the Bright Angel Quadrangle, Grand Canyon National Park, Arizona: Grand Canyon, Arizona, Grand Canyon Natural History Association, scale 1:48,000.

McKenzie, D.P., 1978, Some remarks on the development of sedimentary basins: Earth and Planetary Science Letters, v. 40, p. 25–32.

Mitra, S., 1990, Geometry and kinematic evolution of inversion structures: American Association of Petroleum Geologists Bulletin, v. 77, p. 1159–1191.

Mitra, S., and Mount, V.S., 1998, Foreland basement involved structures: American Association of Petroleum Geologists Bulletin, v. 82, p. 70–109.

Molnar, P., and Lyon-Caen, H., 1988, Some simple physical aspects of the support, structure, and evolution of mountain belts, *in* Clark, S.P.J., Burchfiel, B.C., and Suppe, J.C., eds., Processes in Continental Lithospheric Deformation: Geological Society of America Special Paper 218, p. 179–207.

Mosar, J., and Suppe, J., 1992, Role of shear in fault-propagation folds, *in* McClay, K., ed., Thrust Tectonics: London, Chapman & Hall, p. 123–132.

Noble, L.F., 1914, The Shinumo Quadrangle, Grand Canyon District, Arizona: U.S. Geological Survey Bulletin 549, 100 p.

Nourse, J.A., Premo, W.R., Iriondo, A., and Stahl, E., 2005, Contrasting Proterozoic basement complexes near the truncated margin of Laurentia, northwestern Sonora–Arizona international border region, *in* Anderson, T.H., Nourse, J.A., McKee, J.W., and Steiner, M.B., eds., The Mojave-Sonora Megashear Hypothesis: Geological Society of America Special Paper 393, p. 123–182.

O'Sullivan, R.B., 1963, Geology, Structure, and Uranium Deposits of the Shiprock Quadrangle, New Mexico and Arizona: Washington, D.C., U.S. Geological Survey Miscellaneous Investigations Series I-345, scale 1:250,000, 2 sheets.

Peirce, H.W., Damon, P.E., and Shafiqullah, M., 1979, An Oligocene (?) Colorado Plateau edge in Arizona: Tectonophysics, v. 61, p. 1–24, doi: 10.1016/0040-1951(79)90289-0.

Poblet, J., and McClay, K., 1996, Geometry and kinematics of single-layer detachment folds: American Association of Petroleum Geologists Bulletin, v. 80, p. 1085–1109.

Powell, J.W., 1873, Some remarks on the geological structure of a district of country lying to the north of the Grand Canyon of the Colorado: American Journal of Science, v. 5, p. 456–465.

Ramberg, H., 1959, Evolution of ptygmatic folding: Norsk Geologisk Tidsskrift, v. 39, p. 99–151.

Ramberg, H., 1962, Contact strain and folding instability of a multilayered body under compression: Geologische Rundschau, v. 51, p. 405–439, doi: 10.1007/BF01820010.

Ramsay, J.G., 1967, Folding and Fracturing of Rocks: New York, McGraw-Hill Book Company, 560 p.

Reches, Z., 1978, Development of monoclines: Part 1. Structure of the Palisades Creek branch of the East Kaibab monocline, Grand Canyon, Arizona, *in* Mathews, V., III, ed., Laramide Folding Associated with Basement Block Faulting in the Western United States: Geological Society of America Memoir 151, p. 235–272.

Reches, Z., and Johnson, A.M., 1978, Development of monoclines. Part II. Theoretical analysis of monoclines, *in* Matthews, V., III, ed., Laramide Folding Associated with Basement Block Faulting in the Western United States: Geological Society of America Memoir 151, p. 273–311.

Richardson, R.M., 1992, Ridge forces, absolute plate motions, and the intraplate stress field: Journal of Geophysical Research, v. 97, no. B8, p. 11,739–11,748, doi: 10.1029/91JB00475.

Richardson, R.M., and Coblentz, D.D., 1994, Stress modeling in the Andes: Constraints on the South American intraplate stress magnitudes: Journal of Geophysical Research, v. 99, no. B11, p. 22,015–22,025, doi: 10.1029/94JB01751.

Roberts, N.L.R., and Kirschbaum, M.A., 1995, Paleogeography of the Late Cretaceous of the Western Interior of the Middle of North America—Coal Distribution and Sediment Accumulation: U.S. Geological Survey Professional Paper 1561, 115 p.

Sales, J.K., 1968, Cordilleran foreland deformation: American Association of Petroleum Geologists Bulletin, v. 52, p. 2000–2015.

Sargent, K.A., and Hansen, D.E., 1982, Bedrock Geologic Map of the Kaiparowits Coal-Basin Area, Utah: U.S. Geological Survey Miscellaneous Investigations Map I-1033-I, scale 1:125,000.

Severinghaus, J., and Atwater, T., 1990, Cenozoic geometry and thermal state of the subducting slabs beneath western North America, *in* Wernicke, B.P., ed., Basin and Range Extensional Tectonics near the Latitude of Las Vegas, Nevada: Geological Society of America Memoir 176, p. 1–22.

Shride, A.F., 1967, Younger Precambrian Geology in Southern Arizona: U.S. Geological Survey Professional Paper 566, 89 p.

Silver, L.T., 1960, Age determinations on Precambrian diabase differentiates in Sierra Ancha, Gila County, Arizona: Geological Society of America Bulletin, v. 71, p. 1973–1974.

Silver, L.T., 1963, The use of cogenetic uranium-lead isotope systems in zircons in geochronology: Radioactive dating: Athens, International Atomic Energy Agency, p. 279–285.

Silver, L.T., 1965, Mazatzal orogeny and tectonic episodicity, *in* Abstracts for 1964: Geological Society of America Special Paper 82, p. 185–186.

Silver, L.T., 1967, Apparent age relations in the older Precambrian stratigraphy of Arizona [abs.], *in* Burwash, R.A., and Morton, R.D., eds., Geochronology of Precambrian Stratified Rocks: Edmonton, University of Alberta, p. 87.

Silver, L.T., 1969, Precambrian batholiths of Arizona, *in* Abstracts for 1968: Geological Society of America Special Paper 121, p. 558–559.

Silver, L.T., 1978, Precambrian formations and Precambrian history in Cochise County, southeastern Arizona, *in* Callender, J.F., Wilt, J.C., and Clemons, R.E., eds., Land of Cochise-Southeastern Arizona: New Mexico Geological Society Guidebook for 29th Field Conference: Socorro, New Mexico, p. 157–163.

Silver, L.T., and Deutsch, S., 1961, Uranium-lead method on zircons: New York Academy of Science Annual Reports, v. 91, p. 279–283, doi: 10.1111/j.1749-6632.1961.tb35460.x.

Silver, L.T., and Deutsch, S., 1963, Uranium-lead isotope variations in zircons: A case study: The Journal of Geology, v. 71, p. 721–728.

Smith, J.F., Lyman, C., Hinrichs, E.N., and Luedke, R.G., 1963, Geology of the Capitol Reef Area, Wayne and Garfield Counties, Utah: U.S. Geological Survey Professional Paper 363, 102 p.

Stearns, D.W., 1971, Mechanisms of drape folding in the Wyoming province: Wyoming Geological Association, 23rd Annual Field Conference Guidebook: Casper, Wyoming, p. 125–143.

Stearns, D.W., 1978, Faulting and forced folding in the Rocky Mountain foreland, *in* Matthews, V., III, ed., Laramide Folding Associated with Basement Block Faulting in the Western United States: Geological Society of America Memoir 151, p. 1–37.

Stern, S.M., 1992, Geometry of Basement Faults Underlying the Northern Extent of the East Kaibab Monocline, Utah [M.S. thesis]: Chapel Hill, University of North Carolina at Chapel Hill, 33 p.

Sterne, E.J., and Constenius, K.N., 1997, Space-time relationships between magmatism and tectonism in the western United States between 120 Ma and 10 Ma: A regional context for the Front Range of Colorado, *in* Bolyard, D.W., and Sonnenberg, S.A., eds., Geologic History of the Colorado Front Range: Denver, Rocky Mountain Association of Geologists, 1997 RMS-AAPG Field Trip 7, p. 85–100.

Stewart, J.H., 1976, Late Pre-Cambrian evolution of North American—Plate tectonic implications: Geology, v. 4, no. 1, p. 11–15, doi: 10.1130/0091-7613(1976)4<11:LPEONA>2.0.CO;2.

Stewart, J.H., and Suczek, C.A., 1977, Cambrian and latest Precambrian paleogeography and tectonics in the western United States, *in* Stewart, J.H., Stevens, C.H., and Fritsche, A.E., eds., Paleozoic Paleogeography of the Western United States: Society of Economic Paleontologists and Mineralogists, Pacific Section Symposium: Los Angeles, California, p. 1–17.

Stockli, D.F., Bump, A.P., Davis, G.H., and Farley, K.A., 2002, Preliminary (U-Th)/He thermochronological constraints on the timing and magnitude of early Tertiary deformation of the Colorado Plateau: Geological Society of America Abstracts with Programs, v. 34, no. 6, p. 321.

Stone, D.S., 1977, Tectonic history of the Uncompahgre uplift, *in* Veal, H.K., ed., Exploration Frontiers of the Central and Southern Rockies: Rocky Mountain Association of Geologists, 1977 Symposium: Denver, Colorado, p. 23–30.

Strahler, A.N., 1948, Geomorphology and structure of the West Kaibab fault zone and Kaibab plateau, Arizona: Geological Society of America Bulletin, v. 59, p. 513–540, doi: 10.1130/0016-7606(1948)59[513:GASOTW]2.0.CO;2.

Suppe, J., 1983, Geometry and kinematics of fault-bend folding: American Journal of Science, v. 283, p. 684–721.

Suppe, J., 1985, Principles of Structural Geology: Englewood Cliffs, New Jersey, Prentice-Hall, 537 p.

Suppe, J., and Medwedeff, D.A., 1990, Geometry and kinematics of fault-propagation folding: Eclogae Geologicae Helvetiae, v. 83, p. 409–454.

Swanberg, K., 1999, Analysis of Spacing and Orientation of Fractures in Wingate Sandstone, Circle Cliffs, Colorado Plateau [M.S. thesis]: Tucson, University of Arizona, 49 p.

Swanberg, K.A., and Davis, G.H., 1999, Analysis of spacing and orientation of jointing in Wingate Sandstone, Circle Cliffs, Colorado Plateau: Geological Society of America Abstracts with Programs, v. 31, no. 4, p. A58.

Tikoff, B., and Maxson, J., 2001, Lithospheric buckling of the Laramide foreland during Late Cretaceous to Paleogene, western United States: Rocky Mountain Geology, v. 36, p. 13–35, doi: 10.2113/gsrocky.36.1.13.

Timmons, J.M., Karlstrom, K.E., Dehler, C.M., Geissman, J.W., and Heizler, M.T., 2001, Proterozoic multistage (ca. 1.1 and ca. 0.8 Ga) extension in the Grand Canyon Supergroup and establishment of northwest and north-south tectonic grains in the southwestern United States: Geological Society of America Bulletin, v. 113, p. 163–181, doi: 10.1130/0016-7606(2001)113<0163:PMCAGE>2.0.CO;2.

Tindall, S., 2000a, Development of Oblique-Slip Basement-Cored Uplifts: Insights from the Kaibab Uplift and from Physical Models [Ph.D. dissertation]: Tucson, University of Arizona, 261 p.

Tindall, S., 2000b, The Cockscomb segment of the East Kaibab monocline: Taking the structural plunge, *in* Sprinkle, D.A., Chidsey, T.C., Jr., and Anderson, P.B., eds., Geology of Utah's Parks and Monuments: Utah Geological Association Publication 28, p. 629–643.

Tindall, S.E., and Davis, G.H., 1998, Partitioning of faulting within a reverse, right-lateral shear zone along the East Kaibab monocline, Colorado Plateau, Utah: Geological Society of America Abstracts with Programs, v. 30, no. 6, p. 38.

Tindall, S.E., and Davis, G.H., 1999, Monocline development by oblique-slip fault-propagation folding: The East Kaibab monocline, Colorado Plateau Utah: Journal of Structural Geology, v. 21, p. 1303–1320, doi: 10.1016/S0191-8141(99)00089-9.

Tweto, O., 1979, Geologic Map of Colorado: U.S. Geological Survey: Reston, Virginia, scale 1:500,000.

Varga, R.J., 1993, Rocky Mountain foreland uplifts: Products of a rotating stress field or strain partitioning?: Geology, v. 21, p. 1115–1118, doi: 10.1130/0091-7613(1993)021<1115:RMFUPO>2.3.CO;2.

Walcott, C.D., 1890, Study of line displacement in the Grand Canyon of the Colorado in northern Arizona: Geological Society of America Bulletin, v. 1, p. 49–64.

Williams, P.L., 1964, Geology, Structure, and Uranium Deposits of the Moab Quadrangle, Colorado and Utah: Washington, D.C., U.S. Geological Survey Miscellaneous Investigations Series I-360, scale 1:250,000, 2 sheets.

Williams, P.L., and Hackman, R.J., 1971, Geology, Structure, and Uranium Deposits of the Salina Quadrangle, Utah: Washington, D.C., U.S. Geological Survey Miscellaneous Investigations Series I-591, scale 1:250,000, 2 sheets.

Wise, D., 1963, An outrageous hypothesis for the tectonic pattern of the North American Cordillera: Geological Society of America Bulletin, v. 74, p. 357–362.

Wooden, J.L., and DeWitt, E., 1991, Pb isotopic evidence for the boundary between the Early Proterozoic Mojave and Central Arizona crustal provinces in western Arizona, *in* Karlstrom, K.E., ed., Proterozoic Geology and Ore Deposits of Arizona: Arizona Geological Society Digest, v. 19, p. 27–50.

Yin, A., 1994, Mechanics of monoclinal systems in the Colorado Plateau during the Laramide orogeny: Journal of Geophysical Research, v. 99, p. 22,043–22,058, doi: 10.1029/94JB01408.

Ziony, J.I., 1966, Analysis of Systematic Jointing in Part of the Monument Upwarp, Southeastern Utah [Ph.D. thesis]: Los Angeles, University of California, 112 p.

Zoback, M.D., and Healy, J.H., 1992, In situ stress measurements of 3.5 km depth in the Cajon Pass scientific research borehole: Implications for the mechanics of crustal faulting: Journal of Geophysical Research, v. 97, p. 5039–5057.

Zoback, M.D., Apel, R., Baumgaertner, J., Brudy, M., Emmermann, R., Engeser, B., Fuchs, K., Kessels, W., Rischmueller, H., Rummel, F., and Vernik, L., 1993, Upper-crustal strength inferred from stress measurements to 6 km depth in the KTB borehole: Nature, v. 365, p. 633–635, doi: 10.1038/365633a0.

Manuscript Accepted by the Society 5 December 2008

The Geological Society of America
Memoir 204
2009

Three-dimensional kinematics of Laramide, basement-involved Rocky Mountain deformation, USA: Insights from minor faults and GIS-enhanced structure maps

Eric A. Erslev*
Nicole V. Koenig†
Department of Geosciences, Colorado State University, Fort Collins, Colorado 80523, USA

ABSTRACT

Contractional, basement-involved foreland deformation in the Rocky Mountains of the conterminous United States occurred during the latest Cretaceous to Paleogene Laramide orogeny. Current kinematic hypotheses for the Laramide orogeny include single-stage NE- to E-directed shortening, sequential multidirectional shortening, and transpressive deformation partitioned between NW-striking thrust and N-striking strike-slip faults. In part due to this kinematic uncertainty, the links between Laramide deformation and plate-margin processes are unresolved, and proposed driving forces range from external stresses paralleling plate convergence to internal stresses due to gravitational collapse of the Cordilleran thrust belt.

To determine the tectonic controls on Laramide deformation, kinematic data from minor faults (*n* = 21,129) were combined with a geographic information system (GIS) database quantifying Rocky Mountain structural trends. Minor fault data were collected from a variety of pre-Laramide units to calculate average Laramide slip (N67E-01) and maximum compressive stress (N67E-02) directions for the Rocky Mountains. These largely unimodal, subhorizontal slip and compression directions vary slightly in space; more E-W directions occur in the southern and eastern Rockies, and more NE-SW directions are found near the Colorado Plateau.

This kinematic framework was extended to the entire orogen using map data for faults, folds, arches, and Precambrian fabrics from Wyoming, Colorado, northern New Mexico, southeastern Utah, and northeastern Arizona. Vector mean calculations and length-weighted rose diagrams show that Precambrian fabrics are at a high angle to most larger Phanerozoic structures but are commonly reactivated by smaller structures. Ancestral Rocky Mountain structures are subparallel to Laramide structures, suggesting similar tectonic mechanisms.

Laramide faults, defined by their involvement of Mesozoic and Paleogene strata but not Neogene strata, are complex, and preexisting weaknesses and minor strain

*Current address: Department of Geology and Geophysics, University of Wyoming, Laramie, Wyoming 82071, USA; eerslev@uwyo.edu.
†Current address: 1587 S. Xenon Ct., Lakewood, Colorado 80228, USA.

Erslev, E.A., and Koenig, N.V., 2009, Three-dimensional kinematics of Laramide, basement-involved Rocky Mountain deformation, USA: Insights from minor faults and GIS-enhanced structure maps, *in* Kay, S.M., Ramos, V.A., and Dickinson, W.R., eds., Backbone of the Americas: Shallow Subduction, Plateau Uplift, and Ridge and Terrane Collision: Geological Society of America Memoir 204, p. 125–150, doi: 10.1130/2009.1204(06). For permission to copy, contact editing@geosociety.org.

components commonly predominate. In contrast, Laramide fold (avg. N24W) and arch (avg. N23W) axis trends are oriented perpendicular to minor fault slip and compression directions due to their generation by thrust-related folding.

Laramide deformation shows the primary external influence of ENE-directed shortening paralleling published convergence vectors between the North American and Farallon plates. Slightly radial shortening directions, from more NE-directed to the north to more E-directed to the south, suggest focused contraction originating near current-day southern California. A slight clockwise rotation of shortening directions going from west to east is consistent with proposed changes in Farallon–North American plate trajectories as the orogen expanded eastward. Additional complexities caused by localized preexisting weaknesses and impingement by the adjoining Cordilleran thrust belt have provided structural diversity within the Laramide province. Obliquities between convergence directions and the northern and southeastern boundaries of the Laramide province have resulted in transpressive arrays of en echelon folds and arches, not major through-going strike-slip faults.

INTRODUCTION

The kinematics and controls on basement-involved, compressive foreland orogens remain critical unknowns in the geosciences. The Rocky Mountain foreland of the conterminous United States provides two excellent examples of basement-involved foreland orogens: the Pennsylvanian–early Permian Ancestral Rocky Mountain orogen and the latest Cretaceous through Paleogene Laramide orogen (Fig. 1). This paper tests kinematic hypotheses for these orogens and the subsequent Rio Grande extension using minor fault data (e.g., Erslev et al., 2004) and a geographic information system (GIS) database of Rocky Mountain map-scale structures (Bolay-Koenig and Erslev, 2003). The primary focus of the paper is the Laramide orogeny, which is the most extensive and best documented Rocky Mountain orogeny.

There is a growing consensus that the Laramide orogeny was dominated by horizontal shortening and compression, although the relative importance of strike-slip and thrust faulting remains contested (e.g., Fankhauser and Erslev, 2004; Erslev, 2004; Cather, 1999; Cather et al., 2006; Wawrzyniec et al., 2007). Hypotheses explaining the diversity of Laramide basement-involved contractional trends can be grouped into two categories, those involving external controls and those involving internal controls. Proposed external controls on Laramide structural variations include variable plate convergence directions (Gries, 1983; Engebretson, 1985; Bird, 1998), encroachment and/or gravitational collapse of flanking orogenic belts (Hamilton, 1988; Livaccari, 1991), and the inherent geometric complexity of thick-skinned foreland deformation (Paylor and Yin, 1993; Erslev, 1993). Proposed internal controls on Laramide structural variations generally center on the reactivation of preexisting weaknesses (Chamberlin, 1945; Osterwald, 1961; Hodgson, 1968; Blackstone, 1980; Maughan and Perry, 1986), including those caused by reactivation of old fault systems (Marshak et al., 2000; Timmons et al., 2001) and lithologic contacts (Tonnson, 1986; Sims et al., 2001). Examples of clear internal (e.g., reactivation of faults bounding the Proterozoic Uinta Supergroup; Stone, 1986) and external controls (e.g., parallelism between relative plate convergence vectors and Laramide shortening directions; Engebretson, 1985; Saleeby, 2003) make it clear that both external and internal controls exist. The real question is their relative importance over the entire Rocky Mountains and analogous orogens.

This paper addresses the kinematics of the Phanerozoic Rocky Mountain orogenies by attempting to quantify the structural heterogeneity of the Rocky Mountains. The greater variability of structural trends in basement-involved orogens relative to more classical thin-skinned thrust belts requires accurate, unbiased methods. Kinematic patterns revealed by minor fault data are then generalized over most of the Rocky Mountains using a GIS database integrating fault, fold, and Precambrian fabric patterns from U.S. Geological Survey (USGS) and petroleum industry maps. The combined results are used to evaluate the relative contributions of internal and external controls on basement-involved Laramide structures.

PROPOSED EXTERNAL AND INTERNAL CONTROLS ON ROCKY MOUNTAIN DEFORMATION

Many external and internal controls on Rocky Mountain deformation have been proposed in the literature (see Table 1 for a subset of published hypotheses). In many cases, particularly those invoking local controls by preexisting weaknesses, published hypotheses are valid explanations of local features, but whether the proposed process provides an essential regional control remains uncertain.

This paper will not dwell on the past Laramide controversy between advocates of vertical tectonics and horizontal contraction (Stearns, 1971; Lowell, 1983) because modern geological, geophysical, and kinematic data clearly show the dominance of horizontal slip and compression (e.g., Stone, 2005; Erslev, 2005; Erslev and Larson, 2006). One common element of current hypotheses for Laramide deformation (Table 1) is their prediction of NW- to N-trending fold axes and thrust fault strikes. Some hypotheses for external controls, such as the proposed causes of

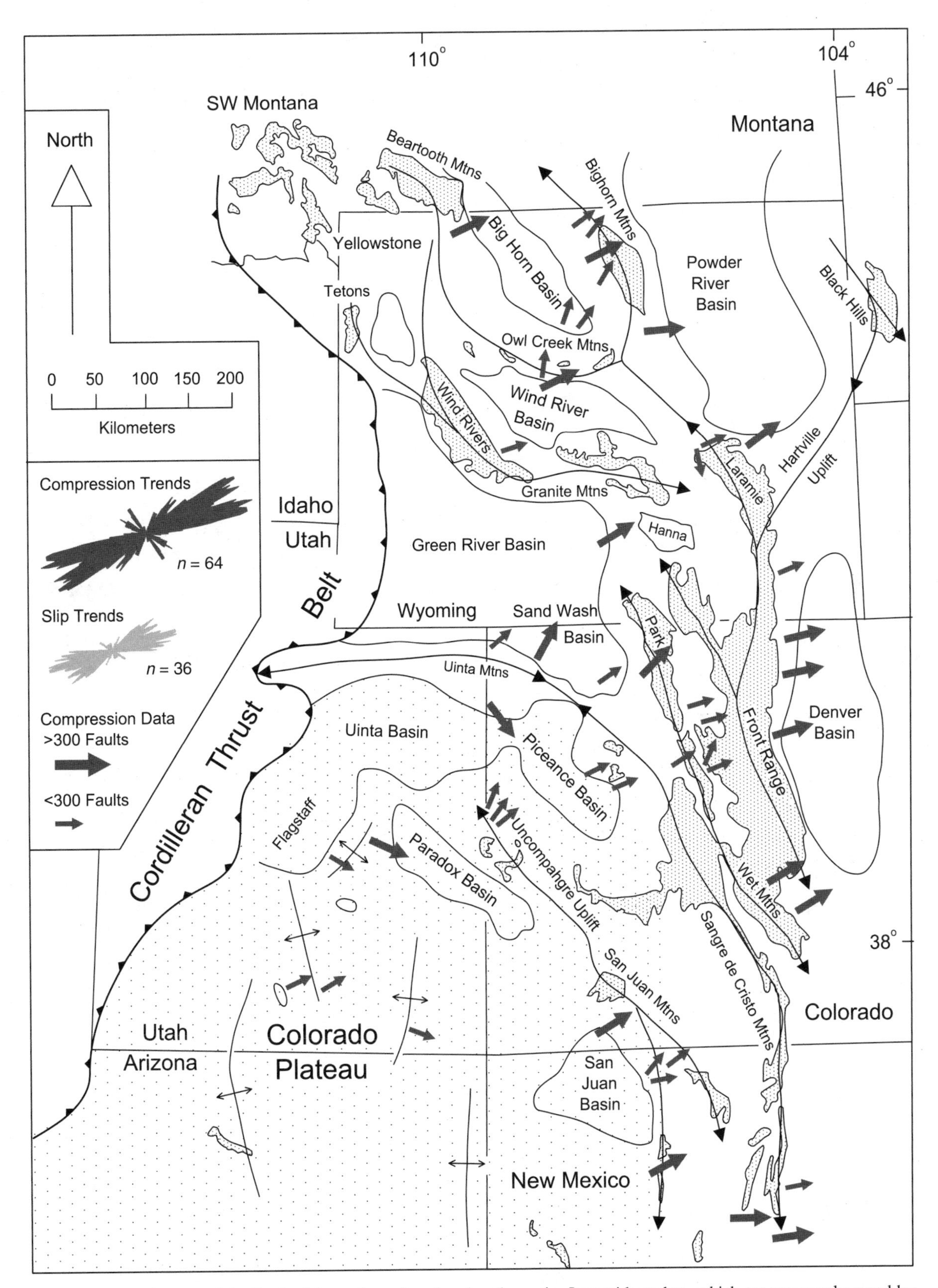

Figure 1. Tectonic map of the Rocky Mountain region showing the major Laramide arches, which are commonly cored by Precambrian crystalline basement exposures (fine stipple), as well as the adjoining Colorado Plateau (coarse stipple) and Cordilleran thrust belt. Average compression directions from minor faults are shown as arrows; smoothed (10°) rose diagrams show all compression and slip directions from Table 2.

TABLE 1. PROPOSED CONTROLS ON PHANEROZOIC ROCKY MOUNTAIN OROGENIES

Ancestral Rocky Mountain orogeny		
External controls	Trend of predicted structures	Recent author(s)
NNW-directed escape tectonics from continental collision to SE	NW and N-S transpressional faults and folds	Kluth and Coney (1981); Kluth (1986, 1998)
ENE to NE shortening due to Andean collision to SW	NNW to NW thrust faults and folds	Ye et al. (1996)
Internal controls	Trend of predicted structures	Recent author(s)
Reactivation of WNW-trending Wichita megashear	WNW sinistral strike-slip faulting and N-S–trending thrust faulting and folding	Budnik (1986)
Reactivation of orthogonal, NW- and NE-trending fracture systems	NW and NE strike-slip faults and folds	Stevenson and Baars (1986)
N-S–trending zones of lithospheric weakness	N-S transpressional dextral faults and NW folds	Yang and Dorobek (1995)
Inversion of Proterozoic N- to NE- and W- to NW-striking extensional fault systems	Bimodal N and NW faults and folds	Marshak et al. (2000); Timmons et al. (2001)
Laramide orogeny		
External controls	Trend of predicted structures	Recent author(s)
ENE-directed subcrustal shear from low-angle subduction during oblique Andean collision	NNW thrust faults and folds	Dickinson and Snyder (1978); Bird (1984, 1988)
ENE-directed crustal shortening during oblique Andean collision	NNW thrust faults and folds	Sales (1968); Stone (1969); Lowell (1983); Oldow et al. (1990); Blackstone (1990); Erslev (1993); Egan and Urquart (1993)
NE-directed crustal shortening during oblique Andean collision	NW thrust faults and folds	Engebretson et al. (1985); Brown (1988); Yin and Ingersol (1997)
Punctuated changes in shortening directions from E-W to NE to N-S	Range of N-S to E-W thrust faults and folds	Gries (1983, 1990); Bergh and Snoke (1992)
Punctuated change in shortening directions from ENE to NE	WNW and NW thrust faults and folds	Chapin and Cather (1981); Erslev (2001)
Punctuated change in shortening directions from ENE to E-W	NNW and N-S thrust faults and folds	Bird (1998)
Punctuated changes in shortening directions from ENE to NW to NE	NNW, NE, and NW thrust faults and folds	Erslev (1997); Ruf (2000)
Punctuated changes in shortening directions from N65E to N95E to N115E	NNW, N-S, and NNE thrust faults and folds	Bergerat et al. (1992)
Rotational convergence of the Colorado Plateau	Thrust faults and folds paralleling Colorado Plateau margin	Hamilton (1988)
Northward translation of the Colorado Plateau	N-S transpressional folds and faults east of plateau, NW thrust faults and folds north of plateau	Chapin and Cather (1981); Chapin (1983); Cather (1999)
NE translation of the Colorado Plateau	NW thrust faults and folds	Woodward (1976)
Gravitational collapse of Cordilleran thrust belt	Thrust faults and folds paralleling thrust belt margin	Laubach et al. (1992); Livaccari (1991)
Internal controls	Trend of predicted structures	Recent author(s)
Slip on optimally oriented Precambrian contacts	Faults parallel Precambrian contacts	Tonnson (1986)
Inversion of E-striking Uinta normal fault system	E-W faults and folds	Stone (1986)
Transpressive partitioning on preexisting N-S faults	N-S strike-slip faults, thrust faults, and folds	Karlstrom and Daniel (1993); Cather (1999)
Pervasive N30-35W– and N65-70W–trending weaknesses in basement	Bimodal N30-35W and N65-70W faults and folds	Blackstone (1990)
Inversion of Proterozoic N- to NE- and W- to NW-striking extensional fault systems	Bimodal N and NW faults and folds	Marshak et al. (2000); Timmons et al. (2001)
Detachment directions guided by corrugations on brittle-ductile transition	Arches occur where Precambrian lithologic changes are oblique to slip	Erslev et al. (2001)
Rio Grande orogeny		
External controls	Trend of predicted structures	Recent author(s)
Extension parallel to successive ENE and E-W to ESE Basin and Range extension	WNW cut by N-S to NNE normal faults	Zoback et al. (1981); Aldrich et al. (1986); Keller and Baldridge (1999)
Clockwise rotation of Colorado Plateau	Normal fault strikes form great circles and strike-slip faults form small circles about pole of rotation	Chapin and Cather (1994)
Internal controls	Trend of predicted structures	Author(s)
Inversion of Laramide thrust faults	Rio Grande normal faults parallel Laramide faults	Kellogg (1999)
Reactivation of N-S strike-slip faults	N-S normal faults	Chapin and Seager (1975); Karlstrom and Humphreys (1998); Erslev (2001); Timmons et al. (2001)

Laramide shortening (e.g., detachment of the crust, pure shear flattening, or subcrustal shear), cannot be simply differentiated on the basis of map patterns because they predict similar near-surface deformation. Deeper information, the fragmentary beginnings and implications of which were summarized in Erslev (2005), is now being collected in EarthScope's USArray project.

Still, many hypotheses make predictions that can be tested by near-surface minor faults and structural trends. For instance, Laramide hypotheses indicating one stage of ENE-WSW shortening differ from those suggesting punctuated episodes of shortening with different orientations (see Table 1 for references). More uniform fold-axis and fault-slip orientations would be expected if unidirectional ENE-WSW shortening had occurred. In contrast, multimodal fault-slip and fold trends of regional extent would be predicted if punctuated episodes of differently oriented shortening had occurred. In addition, some hypotheses can be tested by the geographical location of associated structures. For instance, Laramide hypotheses invoking gravitational collapse of the Cordilleran thrust belt or rotational convergence of the Colorado Plateau predict that folds and thrust faults should parallel the margins of these regions and have slip directions radial to their margins. Also, widespread control by orthogonal or conjugate preexisting weaknesses should result in consistent bimodal distributions of fold and fault orientations, whereas other hypotheses (Tonnson, 1986) suggest systematic relationships between Phanerozoic structures and Precambrian lithologic contacts.

MINOR FAULT METHODS

Minor faults provide an important tool for determining shortening and paleostress directions. Moreover, minor faults give the optimal orientation(s) for constructing and restoring cross sections. Deflections of regional shortening directions can indicate vertical-axis rotations in areas of localized strike-slip deformation (Erslev and Larson, 2006; Tetreault et al., 2008).

Slickensided minor faults are common in sandstone units throughout the Rocky Mountains, particularly in units adjacent to Laramide faults and folds. In siliceous sandstones, minor faults typically form 1–10-mm-thick zones of striated, quartz-rich cataclasite that lack clay gouge and crop out either as erosionally resistant shear bands or as discrete outcrop surfaces. In carbonate-rich rocks, calcite growth fibers are commonly seen on fault planes. Both types of slickensides typically occur as crosscutting thrust and/or strike-slip conjugate sets. Most slickensided faults in the Rockies indicate layer-parallel shortening consistent with Laramide horizontal shortening, where the acute bisector of conjugate thrust and strike-slip faults is contained within the bedding plane (Erslev et al., 2004). Where Laramide thrust and strike-slip faults occur together, they usually share a common acute bisector, suggesting one direction of maximum shortening and compression.

Slip and orientation data from minor faults were collected from data stations in 49 discrete regions throughout the Rockies (see references in Table 2 and average orientations plotted in Fig. 1). Slip sense was determined in the field using the criteria of Petit (1987). Strain calculations were limited to calculating average slickenline orientations because the total slip on individual minor fault planes, so critical to the determination of absolute elongations, is rarely apparent.

Stress orientations were mostly calculated using the ideal Compton (1966) σ_1 method, where an ideal compression direction is calculated for each fault using an α angle (half the acute bisector of conjugate faults) of 20°, which most closely fits the conjugate faulting patterns of rock units within the region (e.g., Erslev, 2001; Erslev et al., 2004; Neely, 2006; Erslev and Larson, 2006). In most cases, faults paralleling bedding and preexisting planar weaknesses were not sampled. Angelier (1990) calculations of least squares reduced stress tensors from the literature and from earlier Colorado State University (CSU) studies are also reported. Comparisons of stress inversion methods used by Compton (1966) and Angelier (1990) show that least squares averaging in the Angelier method can exaggerate the influence of distal data points, reducing that method's precision (Erslev et al., 2004). Average slickenline and ideal σ_1 orientations for each CSU data set were calculated using eigenvector analysis, and eigenvalues were used to quantify data clustering (Table 2). In most cases, minor fault patterns were analyzed in their current orientations as well as with bedding restored to horizontal.

MINOR FAULT RESULTS

Due to the biasing effect of large numbers of measurements near Colorado State University, fault data (Table 2) are reported by regions. In areas of detailed CSU analyses, both average and preferred σ_1 orientations are tabulated (see the references in Table 2 for detailed rationales). For instance, in the northeastern Front Range, average slip and ideal Compton (1966) σ_1 orientations north of Fort Collins are more E-W than those south of Fort Collins. Because the forelimbs of NE-trending folds north of Fort Collins show clockwise rotations of faults and paleomagnetic poles (Holdaway, 1998; Tetreault et al., 2008), a distributed zone of right-lateral shear has been hypothesized to be responsible for this deflection (Erslev and Larson, 2006).

In another example from the southern plunge of the Bighorn Arch in central Wyoming, minor faults associated with adjacent anticline-syncline pairs with different fold-axis trends show different slip directions (Fisher, 2003). Faults associated with a N17W-trending anticline-syncline pair indicate thrust shortening normal to fold axes, whereas in an adjacent N35W-trending anticline-syncline pair, shortening directions differ with location. Within the syncline, shortening was uniformly perpendicular to the fold axis. Within the anticline, however, high-angle faults segmenting the anticlinal axis indicate a substantial strike-slip component parallel to the fold axis. This sort of fold-slip partitioning was demonstrated by surface deformation in the Chi-Chi earthquake of Taiwan (Angelier, 2003), and it suggests that fault measurements need to be taken over a wider area than just the synclinal region if regional shortening directions are to be determined.

TABLE 2. SUMMARY OF LARAMIDE MINOR FAULT INFORMATION FOR THE ROCKY MOUNTAINS

Locations	No. of faults	Avg. slickenline trend-plunge, E_1*		Avg. ideal s_1 trend-plunge, E_1*		Interp. s_1 axis	Primary source
Colorado							
Front Range						075–00[§]	Erslev and Larson (2006)
Lyons to N. Fort Collins	2233	080–13	.7232	079–11	.8397		Holdaway (1998)
Fort Collins to Wyoming	2973	093–07	.7027	093–06	.8028	077–00	Erslev and Larson (2006)
Golden to Boulder	784	087–66	.5863	075–25	.5583		Selvig (1994)
Canon City Embayment	642	076–06	.6697	075–04	.7393	060–00	Jurista (1996)
Webster Park area	543	231–06	.6416	233–01	.7635	060–00	Fryer (1996)
Western Front Range	612	250–06	.6485	253–02	.7189		Erslev et al. (2004)
Dillon	208	237–03	.7261	234–05	.7473		Erslev et al. (2004)
Stage 1	130			249–06	.8745		Erslev et al. (2004)
Stage 2	78			025–00	.8491		Erslev et al. (2004)
Hot Sulphur Springs	217	252–12	.5990	254–03	.6838		Erslev et al. (2004)
Ute Pass	46	254–10	.6215	256–02	.7543		Erslev et al. (2004)
Laramie River Valley	49					077–03[†]	E. Erslev (2008)
Western Gore Range							
NE set	128	049–02	.9160	058–02	.8606		Copfer (2005)
NW set (post-Laramide?)	28	322–09	.8445	323–14	.8415		Copfer (2005)
Northern San Juan Basin	842						
NE set		237–03	.8155	239–06	.9076		Ruf (2000)
NW set (post-Laramide?)		123–04	.7982	120–08	.8500		Ruf (2000)
Durango area							
NE set	72					215–06[†]	Erslev (1997)
NW set (post-Laramide?)	39					123–00[†]	
Craig	48	236–01	.8138	056–04	.9649		C. Gillett (2008, personal commun.)*
Steamboat Springs	369	085–05	.6061			239–05[§]	Ehrlich (1999)
NE Set						046–00	Ehrlich (1999)
NW Set (post-Laramide?)						146–00	Ehrlich (1999)
Uinta Mountains							
Northern flank	72	043–09	.8087			050–04[†]	Gregson and Erslev (1997)
Southern flank	1193	228–65	.3877(162–06[§])	344–21	.4363	322–04[§]	Gregson and Erslev (1997)
NE corner	722	017–37	.6667	029–34	.7414		J. Detring (2008, personal commun.)
Colorado Nat. Monument	63	030–23	.7258	038–09	.7852		E. Erslev (2008)
Colorado Nat. Monument	143					050–00 to 055–00	Bump and Davis (2003)
Fruita	41	022–06	.6542	024–19	.6454		E. Erslev (2008)
Rifle	47	252–28	.7134	245–24	.7561		E. Erslev (2008)
New Castle	22	260–29	.8286	248–34	.8673		E. Erslev (2008)
Utah							
San Rafael Swell	341					294–00[§]	Fischer and Christensen (2004)
San Rafael Swell	151 deformation bands and slickenlines					120–00	

(*Continued*)

TABLE 2. SUMMARY OF LARAMIDE MINOR FAULT INFORMATION FOR THE ROCKY MOUNTAINS (*Continued*)

Locations	No. of faults	Avg. slickenline trend-plunge, E_1*		Avg. ideal s_1 trend-plunge, E_1*		Interp. s_1 axis	Primary source
Utah (*Continued*)							
Monument Uplift	47					110–00	Bump and Davis (2003)
N, Capital Reef Nat. Park	218 deformation bands					060–00 to 070–00	Bump and Davis (2003)
Circle Cliffs						060–00	Bump and Davis (2003)
New Mexico							
Eastern San Juan Basin	415	242–05	.7165	63–00	.8005		Erslev (2001)
NE rim (San Juan Basin)	144	244–03	.7573	62–00	.8120		Erslev (2001)
Stage 1						260–02	Erslev (2001)
Stage 2						041–01	Erslev (2001)
Heron Dam	106	069–02	.7705	054–02	.8209		Erslev (2001)
Santa Fe to Las Vegas	533	084–02	.7293	083–00	.8289		Erslev (2001)
Glorieta to Canoncito	1081	268–01	.7172	269–01	.8301		Fankhauser (2006)
Las Vegas to Moria	234	119–30	.7449	109–29	.7718	080–00	Magnani et al. (2005)
Truth or Consequences	533	085–06	.5515	074–09	.6135		E. Erslev (2008)
Wyoming							
East Laramie Range	40					068–24†	Varga (1993)
Douglas	606	055–25	.7534	68–12	.8043	068–12	D. Cooley (2008, personal commun.)
Rawlins	786	076–09	.6364			059–02§	Ehrlich (1999)
Bighorn Mtns., Lovell	93	234–55	.8750	234–46	.8788		Hager (2001)
Bighorn Mtns., Five Springs	27					040–13†	Varga (1993)
Bighorn Mtns., Shell Creek	24					212–02†	Varga (1993)
Owl Creek Mtns.	32					188–08	Varga (1993)
Casper Mountain	28					168–06	Varga (1993)
West of Kaycee	686	071–14	.4994	076–14	.5918	085–00	Fisher (2003)
Cody	1581					065–00	Neely (2006)
Lander	17	075–14	.8150	073–10	.9961		
Thermopolis							
Red Canyon	20					038–00	E. Erslev (2008)
Hot Springs Park	20					020–00	E. Erslev (2008)
Greybull	450	241–04	.7950	234–04	.9294	065–00	S. Smaltz (2008, personal commun.)
Casper Mountain	62	064–00				067–03†	Molzer and Erslev (1995)
Owl Creek Mountains	718	227–03				245–02†	Molzer and Erslev (1995)
Summary of above data (not including probable post-Laramide data)							
Avg. lineation + stress data	(99 values)			064–02	.8130		
Subsets w/ full eigendata	30	067–01	.8244	067–02	.8140		
Subsets w/ lineation data	35	066–01	.8368				
Subsets w/ stress data	64			063–02	.8005		

*E_1 is the eigenvalue of the first eigenvector for the data set. The sum of all three eigenvalues equals 1.0; clustering of the line and axis orientations occurs toward the first eigenvector orientation, increasing as E_1 approaches 1.0.

†Angelier (1990) reduced stress tensor inversion.

§Average of sites.

At Rocky Mountain localities that omit possible post-Laramide fault sets and contain full eigendata (*n* = 30, Table 2), fault slip and ideal Compton (1966) σ_1 orientations are nearly identical, having average orientations of N67E-01 and N67E-02, respectively. Likewise, the average slip direction from all sources (*n* = 35; N66E-01) is closely aligned with the average σ_1 orientations from all sources (*n* = 64; N63E-02). The overall average of slip directions and compressive stress directions, N64E-02, contains data from both the Rockies and the Colorado Plateau, and it is probably a good average slip and compression direction for the entire region.

In general, slip and compression orientations appear to be more E-W to the south and east, and more NE-SW to the north and west (Fig. 1). The absence of consistent obliquity between slip directions and σ_1 orientations suggests a lack of regional strike-slip. Few localities indicate multidirectional Laramide compression (see Table 2). Our previous interpretations of a distinct phase of NW-SE compression (Ruf and Erslev, 2005; Erslev et al., 2004) due to gravitational collapse of the Cordilleran thrust belt (Livaccari, 1991; Bump, 2004) have been contradicted by recent work in NW Colorado. In this area, C. Gillett (2008, personal commun.) has shown that the NW-SE strike-slip faults that dominate these data sets are more consistent with post-Laramide extensional systems cutting Miocene rocks. Thus, minor fault measurements are consistent with one phase of ENE-WSW Laramide deformation, where there is minor fanning of orientations to more easterly orientations from west to east, and from north to south.

GIS METHODS

We extended the kinematic framework established with minor fault data by using more extensive and uniformly distributed map-scale structures. Phanerozoic fault and fold data were combined with Precambrian contact, shear zone, and foliation data from Wyoming, Colorado, New Mexico, and parts of eastern Utah and Arizona in a GIS database (Bolay-Koenig, 2001; Bolay-Koenig and Erslev, 2003). The boundaries of the area, somewhat arbitrary due to time constraints and limited data sources, extended south from the Montana-Wyoming border to central New Mexico and east from eastern Utah and Arizona to the eastern edges of Colorado and New Mexico (Fig. 2), including most of the Colorado Plateau. Thin-skinned Sevier thrust belt structures were excluded in our attempt to focus on basement-involved deformation.

Sources for fault and fold orientations include 19 USGS 1° × 2° geologic quadrangle maps (listed in references separately) covering all but north-central New Mexico and southern Wyoming and 22 basin structure contour maps on Early Cretaceous and late Paleozoic formations provided by the Rocky Mountain Map Company. Areas not covered by these data sets were covered using the Wyoming basement contour map (Blackstone, 1993), larger-scale USGS geologic maps (Colton, 1978; Houston and Karlstrom, 1992), Kelley's (1955) tectonic map of the Colorado Plateau, and large-scale maps over the Sierra Nacimiento (Woodward, 1987), southern Sangre de Cristo (Miller et al., 1963), and Sandia (Kelley and Northrop, 1975) mountains of northern New Mexico. The pattern of anastomosing Laramide arch trends (Erslev, 1993) was defined by examining regional structure contour maps. Most exposed Precambrian contacts and shear zone traces were acquired from USGS 1° × 2° geologic quadrangle maps. Subsurface Precambrian contacts were acquired from basement maps of Sims et al. (2001), who extrapolated surface contacts to the subsurface using core and aeromagnetic data. Colorado Precambrian foliation data were collected from an unpublished foliation map compiled by Ogden Tweto and loaned to us by John C. Reed Jr.

Acetate overlays were generated and digitized for each source map, and data were subdivided by structure type and the age of involved strata. Fault and fold data were subdivided into four age-based subsets (PRE, ARM, LAR, and RIO), which were determined by the age of the youngest strata involved in the structure. PRE structures only involve Precambrian rocks, ARM structures involve pre-Permian rocks, LAR structures involve Permian to Paleogene rocks, and RIO structures involve Neogene rocks.

These age-based subsets assured that no older structures were attributed to younger orogens. It should be noted, however, that younger structures can and were attributed to older orogens. For instance, the PRE fault subset contains Phanerozoic faults that happen to be limited to Precambrian exposures. In addition, the certainty of a structure's age varies with the type of structure. The LAR arch data set is more reliably Laramide in age than the LAR fault data set, which probably contains numerous post-Laramide faults. PRE contact (crystalline rock contacts), shear zone, and foliation data are more certain to be Precambrian because they formed or were modified by ductile processes that, for currently exposed rocks in the Rocky Mountains, were largely restricted to the Precambrian.

The raw data (Table 3) were digitized from the acetate overlays into ARCINFO™ and transferred into ARCVIEW™ software. In order to quantify and compare the relationships between structural fabrics on geologic maps, a subroutine (script) broke each arc into individual line segments. This was necessary because if an arc's orientation is solely calculated from its end points, there would be no difference between straight and highly curved arcs.

Individual structure data sets were divided into geographic subsets by outlining an area for analysis using ARCVIEW™'s selection capabilities or by using another script to create an evenly spaced array of subdomains. For each subdomain, subsets can be created to include all line segments having center points within a user-specified distance from central grid points.

Once a subset was selected, vector mean and dispersion values were calculated. Dispersion is a measure of clustering; a dispersion of 0.0 indicates only one orientation within the data, and a dispersion of 1.0 indicates no single preferred orientation. The vector mean was calculated for the subset's line segments using the 2θ vector mean method for lines (Krumbein, 1939; Davis,

Figure 2. Composite of all map data for the Rocky Mountains and adjacent areas used in this paper.

TABLE 3. SUMMARY OF MAP DATA DIGITIZED INTO ARCINFO™

Data set	Subset*	Number of arcs	Number of line segments	Total length of segments (km)
PRE contacts		2033	12,906	20,433
PRE shear zones		202	603	3127
PRE foliations		250	622	2319
Major LAR arches		27	110	3514
Folds	RIO	364	1209	2124
	LAR	2096	6691	23,894
	ARM	144	397	1599
	PRE	89	290	412
Faults	RIO	5576	15,903	18,112
	LAR	21,049	48,107	39,742
	ARM	2166	6082	7659
	PRE	1737	4936	7839
TOTALS:		35,733	97,856	130,674

*Four age-based subsets were determined by the age of the youngest strata involved in the structure. PRE—Precambrian rocks; ARM—pre-Permian rocks; LAR—Permian to Paleogene rocks; RIO—Neogene rocks.

1986), which essentially (1) doubles each line angle, (2) does a standard vector mean calculation of the doubled angles (here weighted for line length), and (3) divides the final value by two to determine the mean line orientation.

Length-weighted rose diagrams of line segment orientations were created to provide a more direct visualization of line distributions. The total line segment length for every 1° increment was calculated and smoothed using a window of 10°, where the extent of an individual petal from the center is proportional to the sum of the lengths of lines within 5° of that petal's orientations. Rose diagrams are scaled to efficiently fill available space and are not scaled to the total amount of data.

SUMMARY OF ROCKY MOUNTAIN STRUCTURAL TRENDS

Rose diagrams of line segments were created for all (1) PRE contacts, shear zones, and foliations, (2) major LAR arches, and (3) faults and folds for the RIO, LAR, ARM, and PRE data sets (Fig. 3). In addition, structure maps for the RIO, LAR, ARM, and PRE categories were combined with rose diagrams (Figs. 4–8; see Table 4) for three latitudinal subregions encompassing (1) Wyoming, (2) Colorado and southeastern Utah (referred to as Colorado-Utah), and (3) New Mexico and eastern Arizona (referred to as New Mexico–Arizona). Vector means, dispersions, and total lengths in kilometers for each data set are shown in Table 4.

In this section, we discuss each data set to test hypotheses of Rio Grande and Ancestral Rocky Mountain deformation. Hypotheses for Laramide deformation are tested in the next, more detailed section.

RIO Data Sets

RIO faults and folds (Figs. 3H and 4; Table 4) can be unambiguously resolved because all of these structures involve post-Eocene, and thus post-Laramide, rocks. The average strike of RIO faults is N06W, and the trimodal distribution is made up of NNW-SSE, N-S, and less abundant NE-SW orientations. The more sparse RIO fold data display similar trends, averaging N13W (Fig. 3H), and an additional WNW-ESE petal due to folds in north-central Colorado. The parallelism between these data sets indicates that the folds are fault-related, perhaps due to reverse drag folding adjacent to curved normal faults or extensional fault-propagation folds above blind normal faults.

RIO faults in Wyoming consist of a random-looking pattern around the Yellowstone magmatic system, adjacent N-S striking faults (e.g., the Teton fault east of the Teton Range), and Laramide arch-parallel faults, particularly in more E-W arches like the Granite Mountains. RIO faults in Colorado-Utah and New Mexico–Arizona are better clustered and slightly bimodal about NNW-SSE and N-S directions; average strike orientations are N15W and N14E. RIO folds have average orientations of N29W and N04W, but the amount of data for these rose diagrams (Table 4) is very limited.

Faults and folds within the RIO data sets were formed during mid-Tertiary and Rio Grande deformation. Folds in north-central Colorado that are oblique to adjacent faults may represent a mid-Tertiary pulse of strike-slip deformation (Wawrzyniec et al., 2002, 2007; Erslev, 2001). The overall and state-to-state bimodality of fault rose diagrams supports dual-stage extension histories, as proposed by Zoback et al. (1981), Aldrich et al. (1986), and Keller and Baldridge (1999). Alternatively, fault bimodality could be due to reactivation of preexisting N-S (Chapin and Seager, 1975; Erslev, 2001) and NNW-SSE Laramide faults (Kellogg, 1999). These hypotheses could be tested by seeing if fault extension directions are bidirectional, as predicted by dual-stage extension hypotheses and supported by preliminary fault analyses (Finnan and Erslev, 2001), or more unidirectional, which would be predicted by reactivation hypotheses.

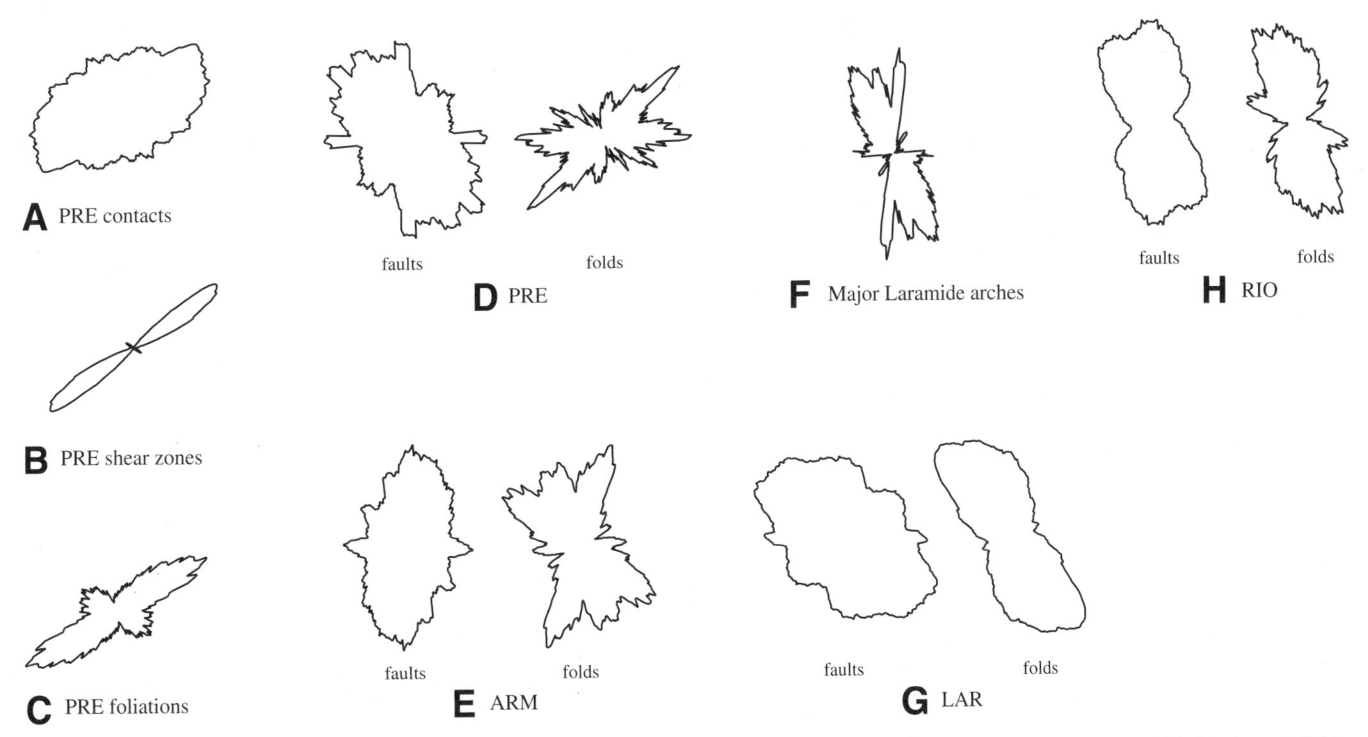

Figure 3. Rose diagrams for all (A) PRE contacts, (B) PRE shear zones, (C) PRE foliations, (D) PRE faults and folds, (E) ARM faults and folds, (F) Major LAR arches, (G) LAR faults and folds, and (H) RIO faults and folds.

LAR Data Sets

LAR arch, fault, and fold data sets represent structures involving Mesozoic but not post-Eocene rocks (Figs. 3F, 3G, and 5; Table 4). Inclusion of RIO structures in LAR data sets is probably not a major problem because LAR structures are much more abundant and many RIO normal faults are easily differentiated from LAR contractional faults.

LAR arches show the first-order deformation of the Laramide foreland; they have a trimodal distribution of trends between NNW-SSE to just east of N-S and an overall vector mean of N23W (Fig. 3F). The Colorado-Utah and New Mexico–Arizona data sets (Fig. 5) are unimodal, having average trends of N29W and N03W, respectively. The Wyoming arch data set has an average trend of N35W, and it is trimodal, characterized by a dominant NNW-SSE mode and secondary WNW-ESE (Owl Creek and Granite Mountains) and NE-SW (Hartville uplift) modes. The excellent correlation between these arches and the major thrust faults that bound them (Blackstone, 1993) suggests that arches are a good proxy for the major Laramide thrust and transpressive strike-slip faults.

LAR folds have an average trend of N25W and a dominant and relatively uniform NNW-SSE mode in all rose diagrams. Dispersion values for the fold data sets (Table 4) are lower than dispersion values for fault data sets, indicating a more limited range of orientations. Fold rose diagrams for Wyoming and Colorado-Utah are similar, with an average trend of N29W. The New Mexico–Arizona subregion has a parallel primary mode of NNW-SSE fold orientations and secondary NNE-SSW and ENE-WSW modes, giving a more northerly average trend (N07W).

LAR faults have an average orientation of N59W and are quite variable in orientation. The average orientation of faults changes radically from one region to another: WNW-ESE orientations in Wyoming (N80W average), NW-SE orientations in the Colorado-Utah region (N53W average), and more N-S orientations in the New Mexico–Arizona region (N14E average). These complex data will be addressed later by examining more localized subsets.

ARM Data Set

Faults and folds identified in the ARM data sets (Figs. 3E and 6; Table 4) are those that cut Paleozoic units but do not cut post-Pennsylvanian units. Unfortunately, many Laramide and post-Laramide structures in areas of Paleozoic bedrock are mislabeled as ARM in our database, and thus these data sets may have limited validity as evidence for Ancestral Rocky Mountain processes.

ARM folds from Colorado-Utah and New Mexico–Arizona differ only slightly from Laramide fold patterns. ARM folds in New Mexico–Arizona are strongly N-S in orientation, paralleling major Laramide arches and the N-S mode of LAR folds. In Colorado-Utah, NW-SE and NNW-SSE modes roughly parallel the region's LAR folds. An additional NNE-SSW mode in ARM folds is lacking in the LAR folds data set, but they are from the Laramide San Rafael Swell and, thus, are Laramide in age. The

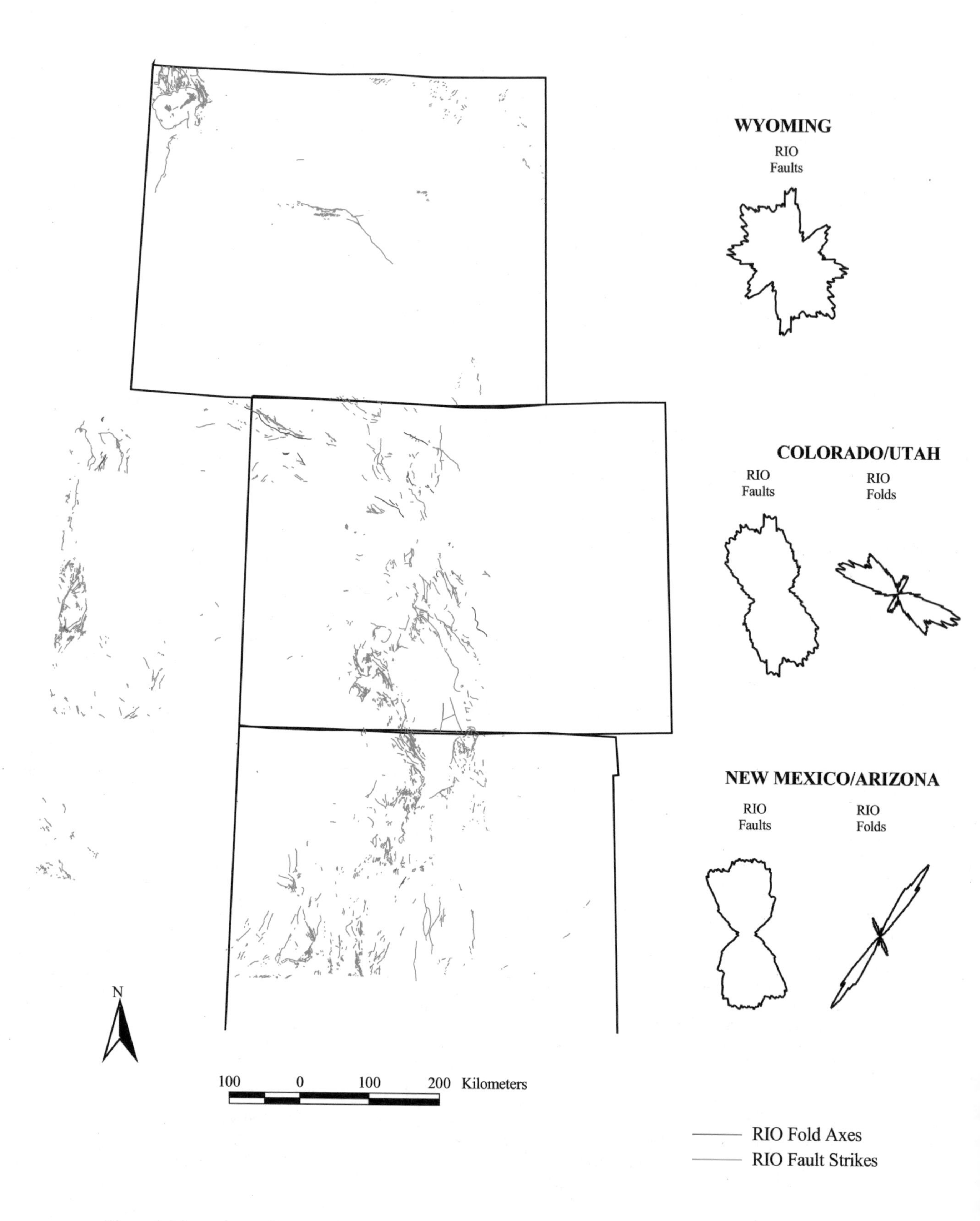

Figure 4. Map and rose diagrams of RIO faults and folds in Wyoming, Colorado–Utah, and New Mexico–Arizona.

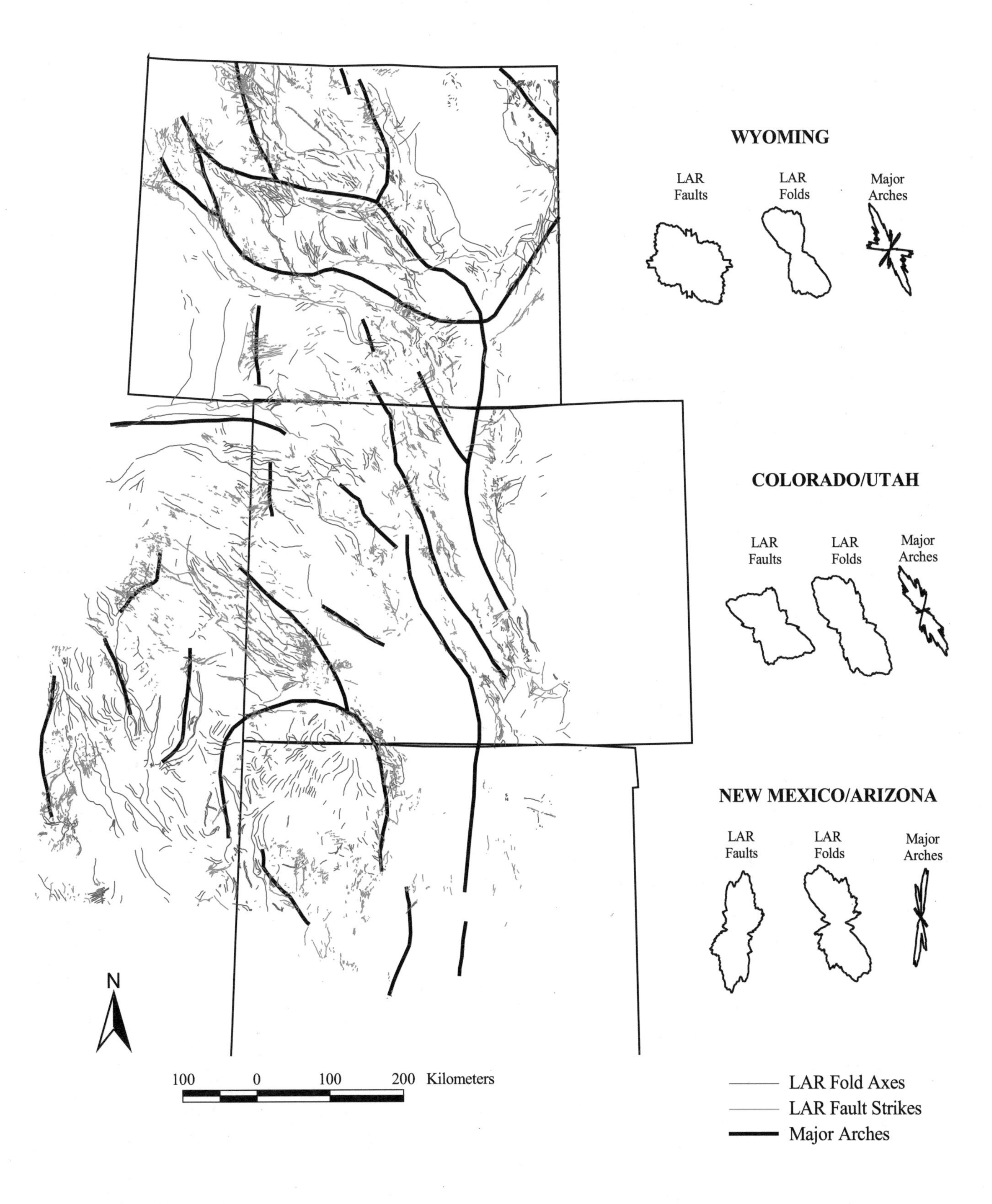

Figure 5. Map and rose diagrams of LAR arches, faults, and folds in Wyoming, Colorado-Utah, and New Mexico–Arizona.

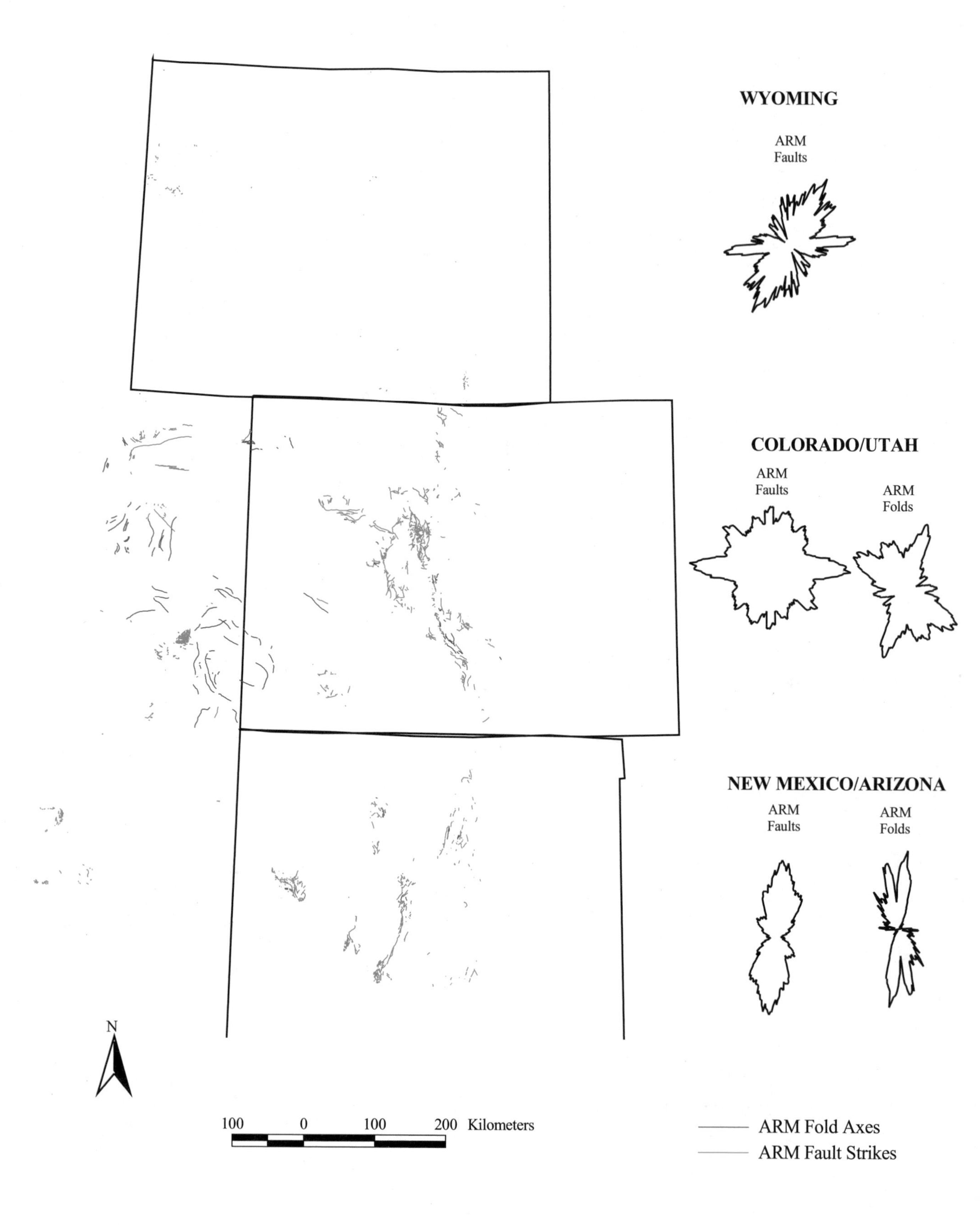

Figure 6. Map and rose diagrams of ARM faults and folds in Wyoming, Colorado-Utah, and New Mexico–Arizona.

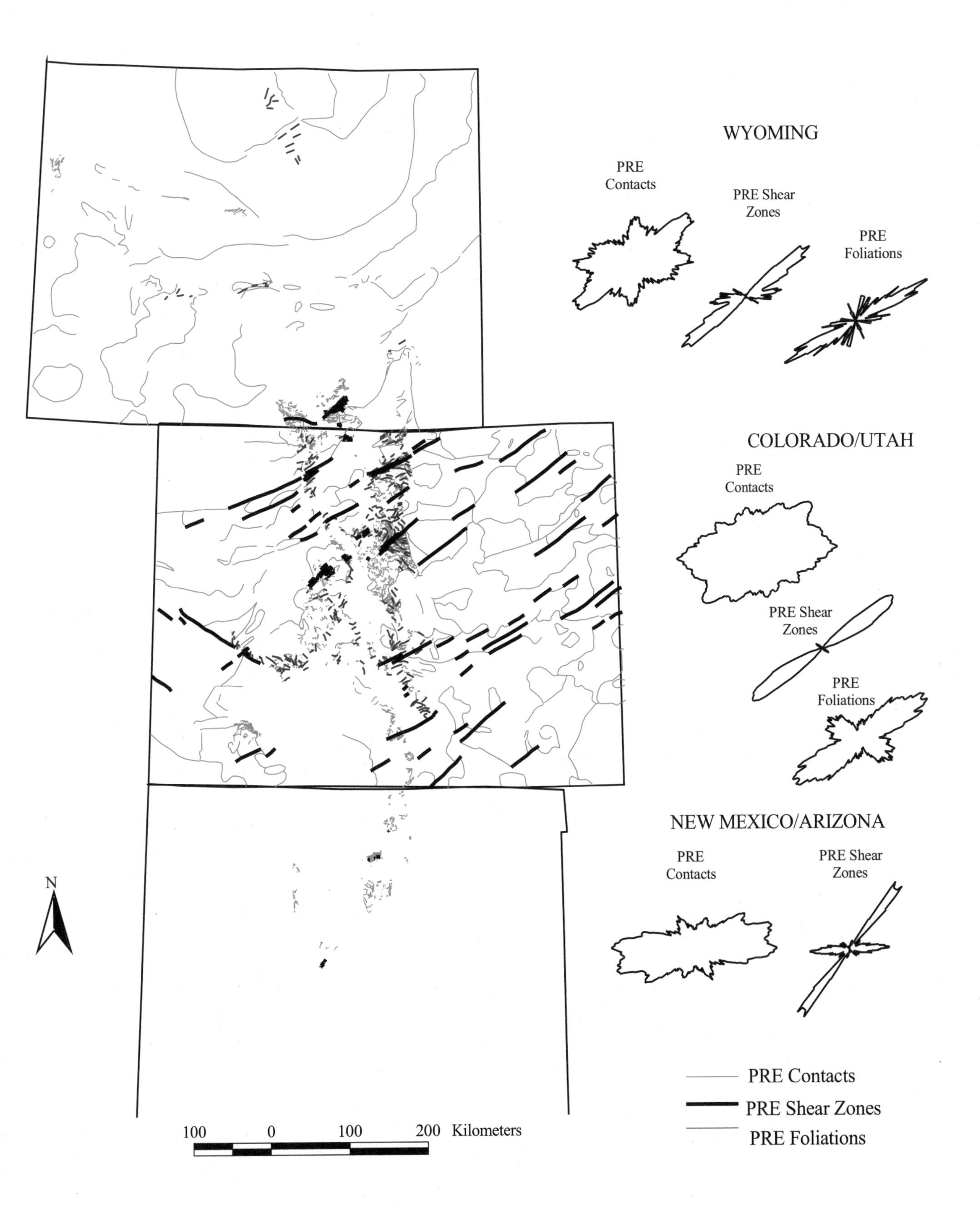

Figure 7. Map and rose diagrams of PRE contacts, shear zones, and foliations in Wyoming, Colorado-Utah, and New Mexico–Arizona.

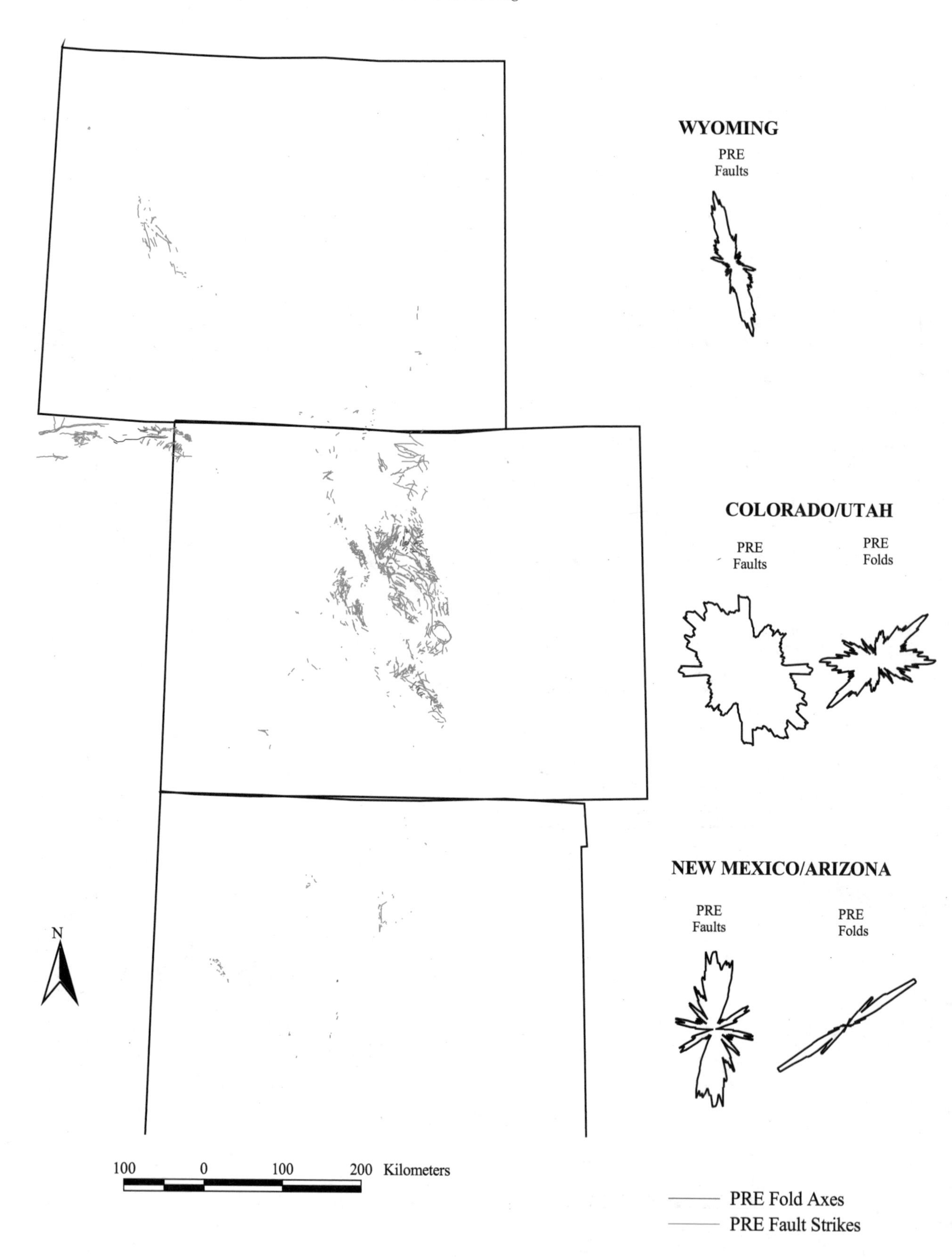

Figure 8. Map and rose diagrams of PRE faults and folds in Wyoming, Colorado-Utah, and New Mexico–Arizona.

TABLE 4. VECTOR MEAN, DISPERSION, AND TOTAL LENGTH FOR EACH DATA SET

	Total			Wyoming			Colorado-Utah			New Mexico–Arizona		
Data set	Vector mean (°)	Disp.	Length (km)	Vector mean (°)	Disp.	Length (km)	Vector mean (°)	Disp.	Length (km)	Vector mean (°)	Disp.	Length (km)
PRE shear zones	63.03	0.32	3127	59.41	0.49	182	58.82	0.20	2918	86.19	0.40	27
PRE foliations	66.67	0.73	2319	57.64	0.39	229	67.60	0.75	2089	NA	NA	0
PRE contacts	71.18	0.88	20,433	65.05	0.83	5894	69.80	0.92	13,565	77.43	0.72	1001
Major LAR arches	156.67	0.53	5660	144.68	0.53	2204	150.96	0.33	1964	177.07	0.21	865
RIO folds	137.25	0.69	429	NA	NA	0	132.15	0.57	411	22.20	0.33	180
RIO faults	173.13	0.73	18,638	154.65	0.87	2679	164.83	0.73	7388	0.37	0.62	8606
LAR folds	155.92	0.69	23,894	151.18	0.61	8700	151.01	0.71	9635	175.85	0.69	5432
LAR faults	133.49	0.88	40,426	124.51	0.90	16,072	127.44	0.72	15,586	13.59	0.73	8099
ARM folds	162.49	0.77	1599	NA	NA	0	159.91	0.80	1471	170.53	0.46	128
ARM faults	11.07	0.86	7659	40.93	0.78	127	50.42	0.98	4920	6.79	0.58	2610
PRE folds	59.83	0.71	412	NA	NA	0	60.16	0.75	404	58.47	0.12	8
PRE faults	148.30	0.86	7839	157.44	0.65	394	141.09	0.88	7198	173.54	0.63	238

Note: Age-based subsets were determined by the age of the youngest strata involved in the structure: Disp—dispersion; PRE—Precambrian rocks; ARM—pre-Permian rocks; LAR—Permian to Paleogene rocks; RIO—Neogene rocks.

nearly orthogonal modes of fold axes in Colorado-Utah are suggestive of orthogonal fracture control, but the folds occur in separate regions, not as a crisscrossing grid of folds, as envisioned by Stevenson and Baars (1986).

ARM faults in Wyoming are minimal in number and length (127 km) due to the lesser Ancestral Rocky Mountain deformation. In Colorado-Utah, uniformly distributed fault strikes are difficult to interpret; the most prominent mode trends in a nearly E-W direction, a direction that is not well represented in LAR faults from this region. In New Mexico–Arizona, RIO, LAR, and ARM faults are subparallel, suggesting that some LAR and RIO faults were incorrectly identified as ARM faults and/or ARM faults were reactivated by Laramide and Rio Grande faulting.

In conclusion, ARM fold and fault trends are largely indistinguishable from Laramide trends. If this is not just due to misclassification of Laramide structures, it suggests that stress orientations responsible for Ancestral Rocky Mountain deformation were analogous to those during the Laramide orogeny, supporting NE-SW ARM shortening models (e.g., Ye et al., 1996; Poole et al., 2005). Inversion of Precambrian E-W normal faults, locally demonstrated by Stone (1986) and hypothesized for the region by Marshak et al. (2000), could account for the E-W mode of fault strikes. The inherent uncertainties of this analysis, however, indicate that more detailed analyses are needed to determine the structural patterns of the Ancestral Rocky Mountain orogeny.

PRE Data Sets

The northeast-trending PRE contact, shear zone, and foliation data sets (Figs. 3A, 3B, 3C, and 7; Table 4) represent ductile deformation that is nearly certain to be Precambrian in age. The map-scale basement folds in the PRE fold data set (Figs. 3D and 8) are also probably Precambrian because near-surface basement deformation during Phanerozoic orogenies was dominated by faulting and gentle arching. These PRE data sets have vector mean orientations within 11° of each other (Table 4: N60E for folds, N63E for contacts, N71E for shear zones, and N67E for foliations). Within these data sets, diffuse NW-SE modes are seen in PRE contacts, shear zones, and foliations. Regionally, contacts show a slight change from NE-SW orientations in Wyoming to nearly E-W orientations in New Mexico–Arizona. Shear zones, which disrupt contacts at numerous localities, have nearly uniform strikes in Wyoming and Colorado-Utah. The similarity of the PRE fault data set to RIO, ARM, and LAR fault data sets suggests that PRE faults largely represent post-Precambrian faulting.

INTERNAL AND EXTERNAL CONTROLS ON LARAMIDE DEFORMATION

If average fault, arch, and fold orientations are perpendicular to shortening and compression directions, then LAR data sets provide powerful tests of Laramide hypotheses (Table 1). The vector mean average orientations of these data sets are discordant, however; LAR faults predict N31E-S31W shortening and compression, LAR arches predict N67E-S67W shortening and compression, and LAR folds predict N66E-S66W shortening and compression.

LAR faults (Figs. 3F, 5, and 9A) show diffuse patterns that probably reflect the inclusion of differently oriented thrust, dextral, sinistral, and even normal faults (due to localized outer-arc extension during Laramide folding) in the LAR fault data set. A histogram of LAR fault strikes (Fig. 9A) has a broad peak at N45W and a vector mean orientation of N59W, suggesting N45E and N31E shortening directions, oblique to minor fault-slip and compression directions.

In contrast, the more clustered LAR fold-axis trends (Fig. 9B) have a skewed distribution with a tighter peak at N40W and a N24W vector mean orientation. These LAR folds and arches are thrust-related folds, either fault-propagation, detachment, or fault-bend folds, and thus they represent a thrust fault subset of Laramide structures. As such, they are excellent indicators

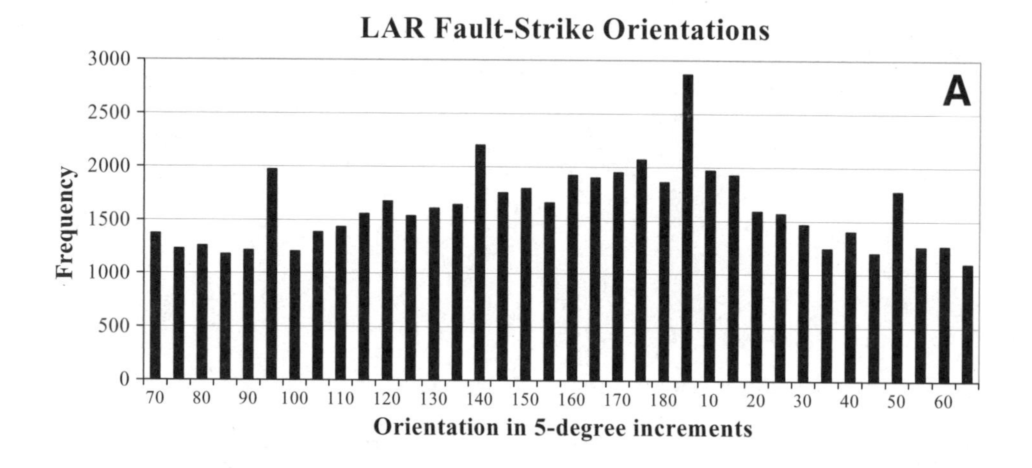

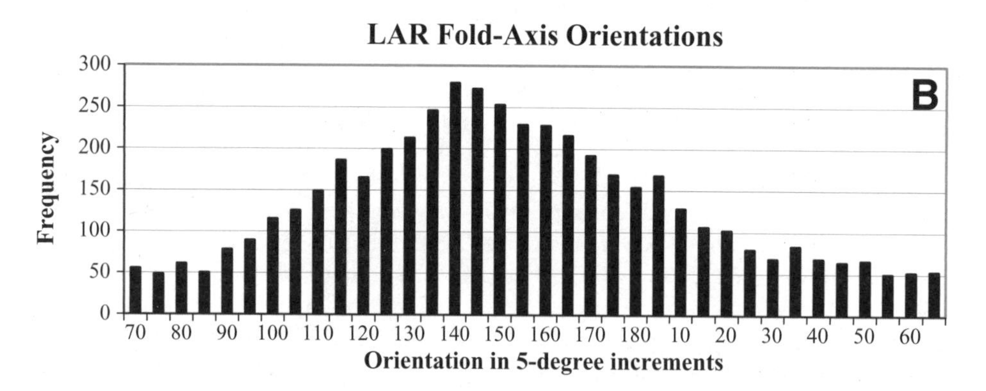

Figure 9. Histograms for all digitized LAR faults and folds from 0° to 180° in 5° increments.

of Laramide horizontal shortening and compression directions. Laramide thrusts are underrepresented by surface fault data because thrust faults commonly go blind by fully transferring their slip into folding and cryptic layer-parallel slip. For example, exposed faults in the NE Bighorn Basin are mostly small, E- to NE-striking, high-angle faults, not the large, NNW-striking blind thrusts that dominate fault slip at the basement level (Stanton and Erslev, 2004). In addition, LAR arch and fold data sets include fewer post-Laramide structures because post-Laramide extension was dominated by faulting, not folding. Thus, we interpret the overall Laramide GIS and minor fault data sets as indicating ENE-WSW (N66-67E–S66-67W) shortening and compression during the Laramide orogeny.

Hypotheses invoking multidirectional deformation, either due to multiple stages of differently oriented compression (Chapin and Cather, 1981; Gries, 1983, 1990) or due to pervasive multidirectional structural weaknesses (Marshak et al., 2000; Timmons et al., 2001), predict that Laramide structures should be systematically multimodal. This is not supported by regional minor fault or LAR fold data, which show dominantly unimodal distributions (Figs. 1, 3G, and 5). The LAR arch rose diagram (Fig. 3F) is multimodal, but non-NNW-SSE arches are geographically restricted—N-S arches largely come from New Mexico–Arizona, and E-W arches largely come from Wyoming, where they follow major Precambrian zones of weakness. However, rose diagrams of LAR faults (Figs. 3G and 5) can be interpreted as showing NNW-SSE and N-S modes, consistent with the previous hypotheses.

To further test the possibility of multimodal structural trends of regional extent, LAR fault and fold rose diagrams were created on a 75 km grid with a 75 km radius of analysis (Figs. 10 and 11). These rose diagrams show that N-S petals come mostly from eastern subsets, NNW-SSE petals come mostly from western subsets, and relatively few local rose diagrams are truly bimodal. Thus, indications of multidirectional Laramide deformation appear to result from the combination of local unimodal domains, not regionally pervasive bimodal distributions.

To aid in visualizing regional patterns, fault (Fig. 10) and fold (Fig. 11) orientations were subdivided into six distinct domains based on LAR fold (primarily) and fault orientations: (1) Wyoming, (2) Green River Basin, (3) Uinta Mountain, (4) Colorado, (5) Colorado Plateau, and (6) New Mexico (Fig. 12; Table 5). The Green River Basin, Uinta Mountain, and Colorado Plateau domain boundaries occur where changes in fold rose diagrams correspond with major structural boundaries. The Colorado domain, which contains the northeastern corner of the Colorado Plateau, captures the systematic change in fold orientations across Colorado in a single domain. The Wyoming domain was differentiated from the Colorado domain due to changes in fault patterns at the Colorado-Wyoming border. The multimodality of

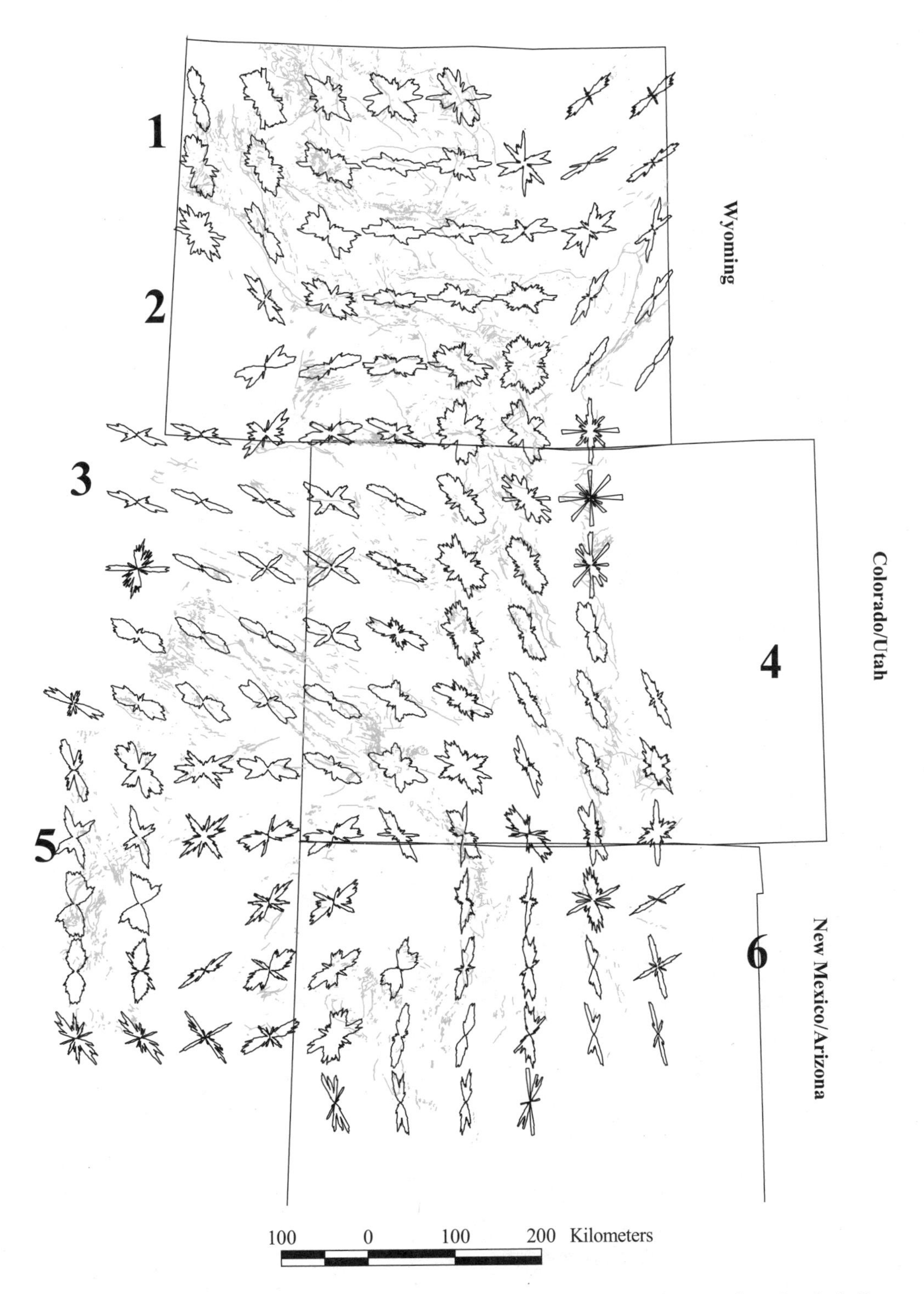

Figure 10. Rose diagrams of LAR faults using a 75 km analysis grid and 75 km for the radius of analysis. Rose diagrams are separated into the (1) Wyoming, (2) Green River, (3) Uinta, (4) Colorado, (5) Colorado Plateau, and (6) New Mexico domains.

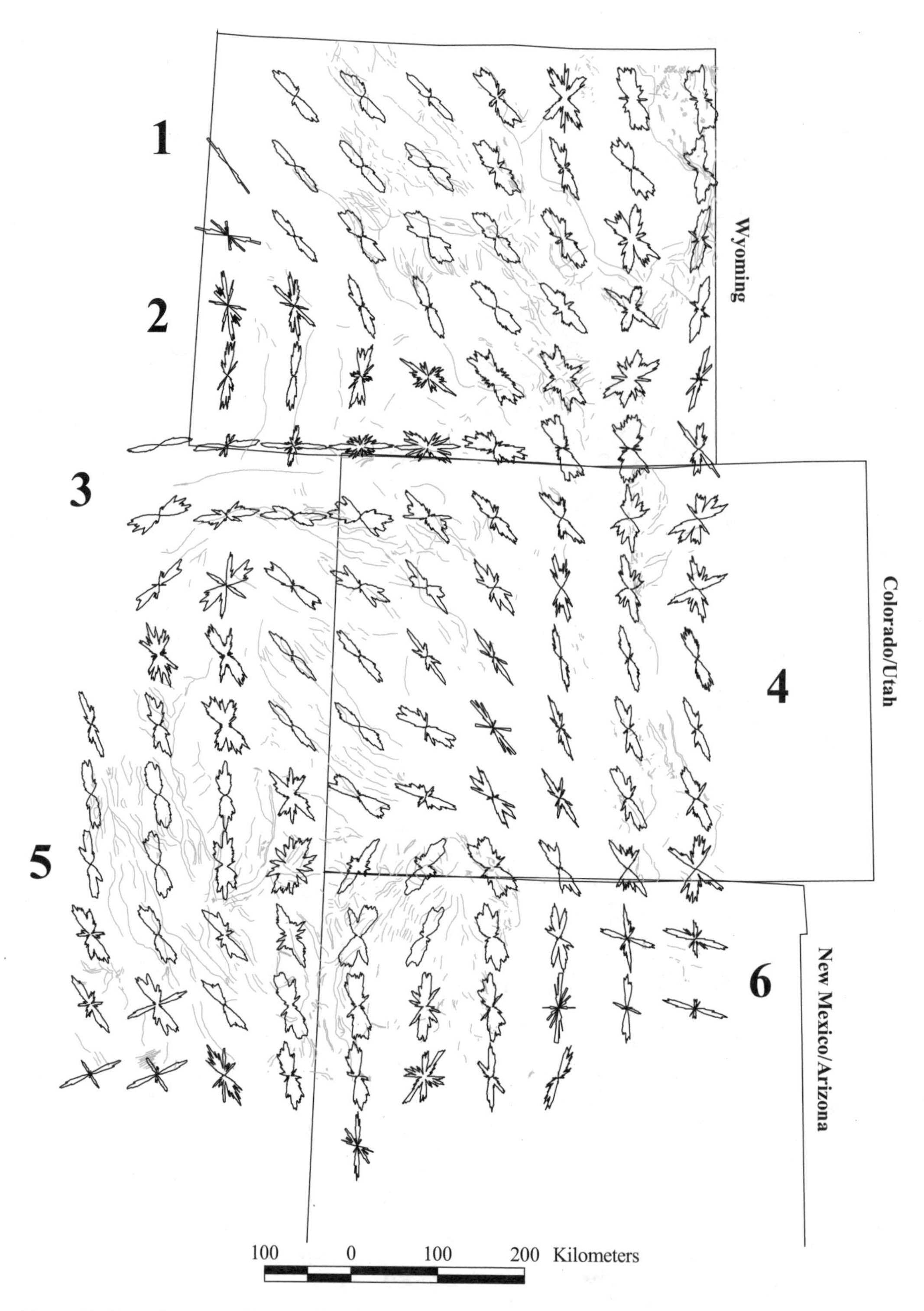

Figure 11. Rose diagrams of LAR folds using a 75 km analysis grid and 75 km for the radius of analysis. Rose diagrams are separated into the (1) Wyoming, (2) Green River, (3) Uinta, (4) Colorado, (5) Colorado Plateau, and (6) New Mexico domains.

Figure 12. LAR fault and fold rose diagrams for the (A) Green River, (B) Wyoming, (C) Uinta, (D) Colorado, (E) Colorado Plateau, and (F) New Mexico domains.

TABLE 5. VECTOR MEAN, DISPERSION (DISP.), AND TOTAL LENGTH FOR THE WYOMING, GREEN RIVER, UINTA, COLORADO PLATEAU, COLORADO, AND NEW MEXICO DOMAINS

Domain	Data set	Vector mean (°)	Disp.	Length (km)
Wyoming	Fold	151.18	0.60	8.028
	Fault	108.57	0.94	12.774
Greater Green River Basin	Fold	172.68	0.35	373
	Fault	76.37	0.60	1183
Uinta Mountain	Fold	101.63	0.62	2114
	Fault	110.68	0.68	2477
Colorado Plateau	Fold	172.76	0.32	9413
	Fault	170.86	0.93	11,661
Colorado	Fold	148.84	0.57	3745
	Fault	134.02	0.73	10,019
New Mexico	Fold	176.19	0.82	292
	Fault	10.86	0.69	540

eastern New Mexico structures and the SE edge of the Colorado Plateau define the New Mexico domain.

Differences in structural orientations between individual domains (Figs. 10, 11, and 12) require multiple explanations.

(1) The parallelism of anomalously E-W faults and folds with the Proterozoic Uinta fault system in the Uinta domain supports local control by basement faults (Stone, 1986).

(2) The N-S structural grain evident in the southern part of the Colorado Plateau and New Mexico domains supports control by local basement weaknesses (Timmons et al., 2001).

(3) Faults in the Wyoming domain parallel exposed and extrapolated (from aeromagnetic patterns: Sims et al., 2001) Precambrian contacts, consistent with Tonnson's (1986) hypothesis that faults reactivated Precambrian contacts.

(4) Faults and folds paralleling the Sevier thrust belt are limited to the western margin of the Laramide province, suggesting that gravitational collapse of the thrust belt (Livaccari, 1991; Laubach et al., 1992) was only important immediately adjacent to the thrust belt.

(5) The discordance of many structures in the Wyoming and Colorado domains with Colorado Plateau margins shows that the plateau was not the primary control on Laramide structural trends.

Perhaps the most striking change in Laramide structural fabric is the systematic change in fold and arch trends from north to south. The NW-SE trends in Wyoming and southern Montana change to NNW-SSE trends in Colorado and N-S trends in New Mexico, defining a concentric arc that is concave to the west. Similarly, in Wyoming and Colorado, folds start at NW-SE orientations to the west and progressively become more N-S toward the east, where they approach the orientation of the eastern boundary of the Laramide province.

These regional shifts in structural orientations eastward across the Laramide orogen have some interesting potential implications. If preexisting systems of Precambrian weaknesses controlled Laramide deformation, structural orientations would not be expected to be sensitive to position within the orogen. In addition, while there are obvious examples of control by preexisting weaknesses (e.g., E-W Uinta arch), there are other obvious pre-Laramide weaknesses, like the Ancestral Rocky Mountain thrust bounding the Paradox Basin, that were minimally reactivated during the Laramide (C. Kluth, 2007, personal commun.). These observations, along with the highly discordant relationships between Precambrian and Phanerozoic structures, indicate that preexisting weaknesses were not the primary control on Laramide structural orientations. We agree with Rodgers' (1987) summary of basement-involved foreland orogens, which concluded that "many details of the pattern of the ranges are controlled by older structures, but the overall pattern was not."

CONCLUSIONS AND GENERAL IMPLICATIONS TO FORELAND DEFORMATION

We combined Laramide minor faults with a GIS database of structural trends to determine the tectonic controls on Rocky Mountain basement-involved foreland structures. GIS analyses show that post-Eocene (RIO) structures are dominated by bimodal fault orientations consistent with dual-stage NE-SW and E-W extension during Browns Park and Rio Grande extension. Ancestral Rocky Mountain (ARM) structures generally parallel Laramide structures, supporting a collisional push from the southwest for the Ancestral Rocky Mountain orogeny (e.g., Ye et al., 1996; Poole et al., 2005).

Unlike the analogous Sierras Pampeanas of southern Argentina, where active Laramide-style structures parallel Paleozoic structures, Rocky Mountain Precambrian (PRE) ductile folds, contacts, shear zones, and foliations dominantly trend NE-SW, at high angles to most Phanerozoic structures. Thus, our data suggest that reactivation of pre-orogenic ductile fabrics has played only localized, nonessential roles in determining structural trends within basement-involved foreland orogens.

Preexisting major faults may or may not have been reactivated, perhaps depending on their position relative to basement-involved arches. Local control by preexisting faults occurs throughout the Rockies (e.g., Brown, 1993), particularly in the E-W Uinta arch. Hypotheses suggesting that Laramide faults parallel a widespread grid of Precambrian faults (Marshak et al., 2000; Timmons et al., 2001) are not supported by our data: the fault modes commonly occur independently and systematically change their orientation within the orogen. Thus, reactivation of regionally extensive fault sets was not the major control on Laramide deformation.

Impingement and/or collapse of the Colorado Plateau and the Cordilleran thrust belt also were not primary controls on Laramide structural trends. For the Colorado Plateau, fault and fold orientations sweep systematically across the eastern plateau boundary in western Colorado as if the boundary had minimal mechanical significance. For the thrust belt, Laramide deformation only parallels thin-skinned thrust belt deformation immediately adjacent to its boundary.

For syn-Laramide minor faults, average slip (N67E-01) and compression (N67E-02) directions indicate unimodal, ENE-WSW shortening and compression, in excellent agreement with shortening directions predicted by Laramide arches (N67E) and folds (N66E), assuming shortening perpendicular to fold axes. The lack of consistent multidirectional fold trends appears to negate hypotheses invoking major, punctuated rotations of compression directions due to changing plate boundaries (Gries, 1983; Bird, 1998, 2002). This result is consistent with syn-Laramide igneous activity near the U.S.-Canada border that suggests only one oceanic plate, the Farallon plate, underlay the Rocky Mountains during the Laramide orogeny (Haeussler et al., 2003; Madsen et al., 2006).

The eastward change from NW-SE to N-S fold trends in Wyoming and Colorado appears to be proportional to distance from the eastern limit of the Laramide orogen. This could be construed as evidence for the partitioning of NE-SW Laramide shortening into N-S dextral strike-slip and E-W thrust motion at the eastern boundary of Rocky Mountain deformation (Karlstrom and Daniel, 1993; Cather, 1999; Cather et al., 2006). However, continuous Precambrian ductile structures in southern Wyoming (Fankhauser, 2006; Sims et al., 2001) show no evidence of through-going, N-S dextral strike-slip faults. The lack of large strike-slip systems in the analogous Sierras Pampeanas (Allmendinger et al., 1990; Ramos et al., 2002) also indicates analogous horizontal contraction without orogen-scale strain partitioning. In addition, the nearly identical average Laramide slip and compression directions suggest no large-scale, regional strike-slip shear in the Rockies. The obliquity of en echelon Laramide arches in New Mexico (N-S trends) and northern Wyoming (more E-W trends) to province boundaries appears to result from transpression required by three-dimensional strain compatibility, where the southeastern orogen margin needs a component of dextral motion and the northern orogen margin needs a mirroring component of sinistral motion.

The close spatial correlations between the present geometry of the low-angle segment of the Andean Benioff zone, the Juan Fernandez Ridge on the Nazca plate, and basement-involved foreland arches of the Sierras Pampeanas show a clear causal relationship (Jordan et al., 1983; Ramos et al., 2002; Alvarado et al., 2005). These observations support the hypothesis of Saleeby (2003), which states that Laramide deformation of the cratonic interior and the disruption of the southern Sierra Nevada batholith were both due to the subduction of an oceanic plateau on the Farallon plate. Certainly, the arcuate Laramide front, with its concentric fold and radial slip directions, is consistent with a primary impact occurring in the vicinity of modern-day southern California. Gradual west-to-east changes in Laramide fold-axis orientations from NW-SE to N-S may reflect rotation of Farallon–North American convergence directions. In Wyoming, where deformation progressed from southwest to northeast (Brown, 1988; Perry et al., 1992), this predicts a clockwise rotation of shortening and compression directions consistent with rotations of plate convergence vectors hypothesized by Saleeby (2003).

The actual connection between subduction processes and basement-involved deformation remains uncertain. End loading of the North American plate by continental collision during the Pennsylvanian (Ancestral Rocky Mountain orogeny) and by plateau subduction during the Cretaceous-Paleogene (Laramide orogeny) could explain both orogens. The unusually thick crust of the Rocky Mountain region (Keller et al., 2005) may have allowed the remarkable extent of these orogens by providing hotter, and thus weaker, lower crust. For the Laramide orogeny, a lithospheric thrust wedge defined by the subducting plate below and a crustal detachment above is consistent with current models of Andean deformation (Jordan and Allmendinger, 1986; Ramos et al., 2002, 2004). We anticipate that the deep seismic data from EarthScope investigations will test current Laramide models invoking thrust imbrication of the entire lithosphere (McQueen and Beaumont, 1989), lithospheric buckling (Tikoff and Maxson, 2001), lithospheric hydration (Humphreys et al., 2003), and crustal buckling and detachment (Fletcher, 1984; Erslev, 2005).

In conclusion, the consistency of Laramide minor fault, arch, and fold trends supports hypotheses invoking one stage of ENE-WSW shortening and compression due to subduction-driven Laramide contraction. Local complications due to reactivation of preexisting weaknesses and external impingement of the Cordilleran thrust belt were superimposed on this primary ENE-WSW contractional pattern. In general, our view of Rocky Mountain structural trends is analogous to looking through a thoroughly fractured window. Local complexities inherited from a long history of prior events have blurred and distorted Laramide patterns, requiring careful integration to reveal their essential systematics. The minor fault and GIS methods presented here can help integrate complex deformation patterns into coherent, information-rich images.

ACKNOWLEDGMENTS

This research was made possible through generous contributions of geologic and geographic information systems (GIS) expertise by John C. Reed Jr., Paul Sims, Carol Finn, Victoria Rystrom, Melinda Laituri, Dave Theobald, Tim Wawrzyniec, John Geissman, Karl Kellogg, David Lageson, Steve Cather, Karl Karlstrom, and many others. Colorado State University (CSU) fault data were mostly collected by CSU students Bjorn Selvig, Phillip Molzer, David Hager, Joe Gregson, Stephanie O'Meara, Branislav Jurista, Tim Ehrlich, Jason Ruf, Mandy Fisher, Steve Holdaway, Melissa Copfer, Seth Fankhauser, Thomas Neely, Cyrus Gillett, John Detring, and Scott Larson. Rocky Mountain Map Company, a division of Barlow and Haun, donated structure contour maps of the Rocky Mountain basins. Funding came from the Petroleum Research Fund of the American Chemical Society, the Continental Dynamics Program of the National Science Foundation, Colorado Geological Survey, and Edward Warner of Expedition Oil Company. We gratefully acknowledge a review by David Lageson of this paper.

REFERENCES CITED

Aldrich, M.J., Jr., Chapin, C.E., and Laughlin, A.W., 1986, Stress history and tectonic development of the Rio Grande rift, New Mexico: Journal of Geophysical Research, v. 91, p. 6199–6211, doi: 10.1029/JB091iB06p06199.

Allmendinger, R.W., Figueroa, D., Snyder, D., Beer, J., Mpodozis, C., and Isacks, B.L., 1990, Foreland shortening and crustal balancing in the Andes at 30°S latitude: Tectonics, v. 9, p. 789–809, doi: 10.1029/TC009i004p00789.

Alvarado, P., Beck, S., Zandt, G., Araujo, M., and Triep, E., 2005, Crustal deformation in the south-central Andes backarc terranes as viewed from regional broad-band seismic waveform modeling: Geophysical Journal International, v. 163, p. 580–598, doi: 10.1111/j.1365-246X.2005.02759.x.

Angelier, J., 1990, Inversion of field data in fault tectonics to obtain the regional stress: Part 3. A new rapid direct inversion method by analytical means: Geophysical Journal International, v. 103, no. 2, p. 363–379, doi: 10.1111/j.1365-246X.1990.tb01777.x.

Angelier, J., 2003, Three-dimensional deformation along the rupture trace of the September 21st, 1999, Taiwan earthquake; a case study in the Kuangfu School: Journal of Structural Geology, v. 25, p. 351–370, doi: 10.1016/S0191-8141(02)00039-1.

Bergerat, F., Bouroz-Weil, C., and Angelier, J., 1992, Paleostresses inferred from macrofractures, Colorado Plateau, Western U.S.A.: Tectonophysics, v. 206, p. 219–243, doi: 10.1016/0040-1951(92)90378-J.

Bergh, S.G., and Snoke, A.W., 1992, Polyphase Laramide deformation in the Shirley Mountains, south central Wyoming foreland: The Mountain Geologist, v. 29, p. 85–100.

Bird, P., 1984, Laramide crustal thickening event in the Rocky Mountain foreland and Great Plains: Tectonics, v. 3, p. 741–758, doi: 10.1029/TC003i007p00741.

Bird, P., 1988, Formation of the Rocky Mountains, western United States: A continuum computer model: Science, v. 239, p. 1501–1507, doi: 10.1126/science.239.4847.1501.

Bird, P., 1998, Kinematic history of the Laramide orogeny in latitudes 35°–49°, western United States: Tectonics, v. 17, p. 780–801, doi: 10.1029/98TC02698.

Bird, P., 2002, Stress direction history of the western United States and Mexico since 85 Ma: Tectonics, v. 21, doi: 10.1029/2001gc000252.

Blackstone, D.L., 1980, Foreland deformation: Compression as a cause: Contributions to Geology (Copenhagen), v. 18, p. 83–100.

Blackstone, D.L., 1990, Rocky Mountain foreland structure exemplified by the Owl Creek Mountains, Bridger Range and Casper arch, central Wyoming: Wyoming Geological Association Guidebook, v. 41, p. 151–166.

Blackstone, D.L., 1993, Precambrian basement map of Wyoming: Outcrop and structural configuration, *in* Mayer, L., ed., Extensional Tectonics of the Southwestern United States: A Perspective on Processes and Kinematics: Geological Society of America Special Paper 208, p. 335–337.

Bolay-Koenig, N.V., 2001, Phanerozoic Deformation in the Southern Rocky Mountains, North America: Insights from a Kinematic Study of Minor Faults in North-Central New Mexico and Regional GIS Analyses [M.S. thesis]: Fort Collins, Colorado State University, 159 p.

Bolay-Koenig, N.V., and Erslev, E.A., 2003, Internal and External Controls on Phanerozoic Rocky Mountain Structures, U.S.A.: Insights from GIS-enhanced tectonic maps, *in* Raynolds, R.G., and Flores, R.M., eds., Cenozoic Systems of the Rocky Mountain Region: Denver, Colorado, Rocky Mountain section, SEPM (Society for Sedimentary Geology, Denver Colorado), p. 33–63.

Brown, W.G., 1988, Deformational style of Laramide uplift in the Wyoming foreland, *in* Schmidt, C.J., and Perry, W.J., Jr., eds., Interaction of the Rocky Mountain Foreland and the Cordilleran Thrust Belt: Geological Society of America Memoir 171, p. 1–25.

Brown, W.G., 1993, Structural style of Laramide basement-cored uplifts and associated folds, *in* Snoke, A.W., Steidmann, J.R., and Roberts, S.M., eds., Geology of Wyoming: Casper, Wyoming Geological Survey of Wyoming Memoir 5, v. 1, p. 312–371.

Budnik, R.T., 1986, Left-lateral intraplate deformation along the Ancestral Rocky Mountains: Implications for late Paleozoic plate motions: Tectonophysics, v. 132, p. 195–214, doi: 10.1016/0040-1951(86)90032-6.

Bump, A.P., 2004, Three-dimensional Laramide deformation of the Colorado Plateau; competing stresses from the Sevier thrust belt and the flat Farallon slab: Tectonics, v. 23, doi: 10.1029/2001TC001329.

Bump, A.P., and Davis, G.H., 2003, Late Cretaceous–early Tertiary Laramide deformation of the northern Colorado Plateau, Utah and Colorado: Journal of Structural Geology, v. 25, p. 421–440, doi: 10.1016/S0191-8141(02)00033-0.

Cather, S.M., 1999, Implications of Jurassic, Cretaceous, and Proterozoic piercing lines for Laramide oblique-slip faulting in New Mexico and rotation of the Colorado Plateau: Geological Society of America Bulletin, v. 111, p. 849–868, doi: 10.1130/0016-7606(1999)111<0849:IOJCAP>2.3.CO;2.

Cather, S.M., Karlstrom, K.E., Timmons, J.M., and Heizler, M.T., 2006, Palinspastic reconstruction of Proterozoic basement-related aeromagnetic features in north-central New Mexico: Implications for Mesoproterozoic to late Cenozoic tectonism: Geosphere, v. 2, no. 6, p. 299–323, doi: 10.1130/GES00045.1.

Chamberlin, R.T., 1945, Basement control in Rocky Mountain deformation: American Journal of Science, v. 243-A, p. 98–116.

Chapin, C.E., 1983, An overview of Laramide wrench faulting in the southern Rocky Mountains with emphasis on petroleum exploration, *in* Lowell, J.D., ed., Rocky Mountain Foreland Basins and Uplifts: Denver, Colorado, Rocky Mountain Association of Geologists, p. 169–179.

Chapin, C.E., and Cather, S.M., 1981, Eocene tectonics and sedimentation in the Colorado Plateau–Rocky Mountain area, *in* Dickinson, W.R., and Payne, M.D., eds., Rocky Mountain Foreland Basins and Uplifts: Denver, Colorado, Rocky Mountain Association of Geologists, p. 173–198.

Chapin, C.E., and Cather, S.M., 1994, Tectonic setting of the axial basins of the northern and central Rio Grande rift, *in* Keller, G.R., and Cather, S.M., eds., Basins of the Rio Grande Rift: Structure, Stratigraphy and Tectonic Setting: Geological Society of America Special Paper 291, p. 5–25.

Chapin, C.E., and Seager, W.R., 1975, Evolution of the Rio Grande rift in the Socorro and Las Cruces area: New Mexico Geological Society Guidebook, v. 26, p. 297–321.

Colton, R.B., 1978, Geologic Map of the Boulder–Fort Collins–Greeley Area, Colorado: U.S. Geological Survey Miscellaneous Investigations Series I-855-G, scale 1:100,000.

Compton, R.R., 1966, Analyses of Pliocene-Pleistocene deformation and stresses in northern Santa Lucia Range, California: Geological Society of America Bulletin, v. 77, p. 1361–1380, doi: 10.1130/0016-7606(1966)77[1361:AOPDAS]2.0.CO;2.

Copfer, M.N.N., 2005, Kinematic Analysis of the Gore Fault and Surrounding Region, Eagle County, Colorado [M.S. thesis]: Fort Collins, Colorado State University, 85 p.

Davis, J.C., 1986, Statistics and Data Analysis in Geology: New York, John Wiley & Sons, 646 p.

Dickinson, W.R., and Snyder, W.S., 1978, Plate tectonics of the Laramide orogeny, *in* Matthews, V., III, ed., Laramide Folding Associated with Basement Block Faulting in the Western United States: Geological Society of America Memoir 151, p. 355–366.

Egan, S.S., and Urquart, J.M., 1993, Numerical modeling of lithosphere shortening; application to the Laramide orogenic province, western U.S.A.: Tectonophysics, v. 221, p. 385–411, doi: 10.1016/0040-1951(93)90170-O.

Ehrlich, T.K., 1999, Fault Analysis and Regional Balancing of Cenozoic Deformation in Northwest Colorado and South-Central Wyoming [M.S. thesis]: Fort Collins, Colorado State University, 116 p.

Engebretson, D.C., 1985, Conjectures on the birth, life, death, and posthumous effects of the Kula plate: Eos (Transactions, American Geophysical Union), v. 66, p. 863.

Engebretson, D.C., Cox, A., and Thompson, G.A., 1985, Relative motions between oceanic and continental plates in the Pacific Basin: Geological Society of America Special Paper 206, 59 p.

Erslev, E.A., 1993, Thrusts, back-thrusts and detachment of Rocky Mountain foreland arches, *in* Schmidt, C.J., Chase, R., and Erslev, E.A., eds., Laramide Basement Deformation in the Rocky Mountain Foreland of the Western United States: Geological Society of America Special Paper 280, p. 339–358.

Erslev, E.A., 1997, Multi-directional Laramide compression in the Durango area—Why?, *in* Close, J.C., and Casey, T.A., eds., Natural Fracture Systems in the Southern Rockies: Durango, Colorado, Four Corners Geological Society, p. 1–5.

Erslev, E.A., 2001, Multistage, multidirectional Tertiary shortening and compression in north-central New Mexico: Geological Society of America Bulletin, v. 113, p. 63–74, doi: 10.1130/0016-7606(2001)113<0063:MMTSAC>2.0.CO;2.

Erslev, E.A., 2005, 2D Laramide geometries and kinematics of the Rocky Mountains, western U.S.A., *in* Karlstrom, K.E., and Keller, G.R., eds., The Rocky Mountain Region—An Evolving Lithosphere: Tectonics, Geochemistry, and Geophysics: American Geophysical Union Geophysical Monograph 154, p. 7–20.

Erslev, E.A., and Larson, S.M., 2006, Testing Laramide hypotheses for the Colorado Front Range arch using minor faults, *in* Raynolds, R., and Sterne, E., eds., Special Issue on the Colorado Front Range: Mountain Geologist, v. 43, p. 45–64.

Erslev, E.A., Bolay, N.V., Ehrlich, T.K., and Karlstrom, K.E., 2001, Mechanisms for Precambrian basement control of Laramide and Ancestral Rocky Mountain structures: Geological Society of America Abstracts with Programs, v. 33, no. 5, p. A51.

Erslev, E.A., Holdaway, S.M., O'Meara, S.A., Jurista, B., and Selvig, B., 2004, Laramide minor faulting in the Colorado Front Range: New Mexico Bureau of Geology and Mineral Resources Bulletin, v. 160, p. 181–203.

Fankhauser, S.D., 2006, The Picuris-Pecos Fault System, Southern Sangre de Cristo Mountains, New Mexico: Evidence for Major Precambrian Slip Followed by Multiphase Reactivation [M.S. thesis]: Fort Collins, Colorado State University, 178 p.

Fankhauser, S.D., and Erslev, E., 2004, Unconformable and cross-cutting relationships indicate major Precambrian faulting on the Picuris-Pecos fault system, southern Sangre De Cristo Mountains, New Mexico, *in* Lucas, S.G., Zeigler, K.E., Lueth, V.W., and Owen, D.E., eds., Geology of the Chama Basin: Socorro, New Mexico Geological Society, 55th Field Conference Guidebook, p. 121–133.

Finnan, S., and Erslev, E., 2001, Post-Laramide deformation in central Colorado: Geological Society of America Abstracts with Programs, v. 33, no. 5, p. A7.

Fischer, M.P., and Christensen, R.D., 2004, Insights into the growth of basement uplifts deduced from a study of fracture systems in the San Rafael monocline, east-central Utah: Tectonics, v. 23, doi: 10.1029/2002TC001470.

Fisher, A.B., 2003, Geometry and kinematics of Laramide basement-involved anticlines: Southeastern Bighorn Mountains, Wyoming [M.S. thesis]: Fort Collins, Colorado State University, 208 p.

Fletcher, R.C., 1984, Instability of lithosphere undergoing shortening: A model for Laramide foreland structures: Geological Society of America Abstracts with Programs, v. 16, no. 2, p. 83.

Fryer, S.L., 1996, Laramide Faulting Associated with the Ilse Fault System, Northern Wet Mountains, Colorado [M.S. thesis]: Fort Collins, Colorado State University, 120 p.

Gregson, J., and Erslev, E.A., 1997, Heterogeneous deformation in the Uinta Mountains, Colorado and Utah, *in* Hoak, T.E., Klawitter, A.L., and Bloomquist, P.K., eds., Fractured Reservoirs: Characterization and Modeling: Denver, Colorado, Rocky Mountain Association of Geologists, 1997 Guidebook, p. 137–154.

Gries, R., 1983, North-south compression of Rocky Mountain foreland structures, *in* Lowell, J.D., ed., Rocky Mountain Foreland Basins and Uplifts: Denver, Colorado, Rocky Mountain Association of Geologists, p. 9–32.

Hackman, R.J., and Wyant, D.G., 1973, Geology, Structure, and Uranium Deposits of the Escalante Quadrangle, Utah and Arizona: U.S. Geological Survey Miscellaneous Investigations Series I-744, scale 1:250,000.

Haeussler, P.J., Bradley, D.C., Wells, R.E., and Miller, M.L., 2003, Life and death of the Resurrection plate: Evidence for its existence and subduction in the northeastern Pacific in Paleocene-Eocene time: Geological Society of America Bulletin, v. 115, p. 867–880, doi: 10.1130/0016-7606(2003)115<0867:LADOTR>2.0.CO;2.

Hager, D.H., Jr., 2001, Geometry and Kinematics of Fault-Propagation Folds in the Northwest Flank of the Bighorn Mountains, Wyoming [M.S. thesis]: Fort Collins, Colorado State University, 126 p.

Hamilton, W., 1988, Laramide crustal shortening, *in* Schmidt, C.J., and Perry, W.J., Jr., eds., Interaction of the Rocky Mountain Foreland and the Cordilleran Thrust Belt: Geological Society of America Memoir 171, p. 27–39.

Hodgson, R.A., 1968, Genetic and geometric relations between structures in basement and overlying sedimentary rocks, with examples from Colorado Plateau and Wyoming: American Association of Petroleum Geologists Bulletin, v. 49, p. 935–949.

Holdaway, S.M., 1998, Laramide Deformation of the Northeastern Front Range, Colorado: Evidence for Deep Crustal Wedging during Horizontal Compression [M.S. thesis]: Fort Collins, Colorado State University, 146 p.

Houston, R.S., and Karlstrom, K.E., 1992, Geologic Map of Precambrian Metasedimentary Rocks of the Medicine Bow Mountains, Albany and Carbon Counties, Wyoming: U.S. Geological Survey Miscellaneous Investigations Series I-2280, scale 1:50,000.

Humphreys, G., Hessler, E., Dueker, K., Farmer, L., Erslev, E., and Atwater, T., 2003, How Laramide-aged hydration of the North American lithosphere by the Farallon slab controls subsequent activity in the western U.S.: International Geology Review, v. 45, p. 575–595, doi: 10.2747/0020-6814.45.7.575.

Johnson, R.B., 1969, Geologic Map of the Trinidad Quadrangle, South-Central Colorado: U.S. Geological Survey Miscellaneous Investigations Series I-558, scale 1:250,000.

Jordan, T.E., and Allmendinger, R.W., 1986, The Sierras Pampeanas of Argentina—A modern analog of Rocky-Mountain foreland deformation: American Journal of Science, v. 286, p. 737–764.

Jordan, T.E., Isacks, B.L., Allmendinger, R.W., Brewer, J.A., Ramos, V.A., and Ando, C.J., 1983, Andean tectonics related to the geometry of the subducted Nazca plate: Geological Society of America Bulletin, v. 94, p. 341–361, doi: 10.1130/0016-7606(1983)94<341:ATRTGO>2.0.CO;2.

Jurista, B.K., 1996, East-Northeast Laramide Compression and Shortening of the Canon City Embayment, South-Central Colorado [M.S. thesis]: Fort Collins, Colorado State University, 147 p.

Karlstrom, K.E., and Daniel, C.G., 1993, Restoration of Laramide right-lateral strike slip in northern New Mexico by using Proterozoic piercing points: Tectonic implications from the Proterozoic to the Cenozoic: Geology, v. 21, p. 1139–1142, doi: 10.1130/0091-7613(1993)021<1139:ROLRLS>2.3.CO;2.

Karlstrom, K.E., and Humphreys, E.D., 1998, Persistent influence of Proterozoic accretionary boundaries in the tectonic evolution of southwestern North America; interaction of cratonic grain and mantle modification events: Rocky Mountain Geology, v. 33, p. 161–179.

Keller, G.R., and Baldridge, W.S., 1999, The Rio Grande rift: A geological and geophysical overview: Rocky Mountain Geology, v. 34, p. 121–130, doi: 10.2113/34.1.121.

Keller, G.R., Karlstrom, K.E., Williams, M.L., Miller, K.C., Andronicos, C.L., Levander, A., Snelson, C.M., and Prodehl, C., 2005, The dynamic nature of the continental crust-mantle boundary: Crustal evolution in the southern Rocky Mountain region as an example, *in* Karlstrom, K.E., and Keller, G.R., eds., Lithospheric Structure and Evolution of the Rocky Mountain Region: American Geophysical Union Monograph 154, p. 403–420.

Kelley, V.C., 1955, Tectonic Map of the Colorado Plateau Showing Uranium Deposits: University of New Mexico Publications in Geology 5, scale 1:50,000.

Kelley, V.C., and Northrop, S.A., 1975, Geology of Sandia Mountains and Vicinity, New Mexico: New Mexico Bureau of Mines and Mineral Resources Memoir 29, 136 p.

Kellogg, K.S., 1999, Neogene basins of the northern Rio Grande rift—Partitioning and asymmetry inherited from Laramide and older uplifts: Tectonophysics, v. 305, p. 141–152, doi: 10.1016/S0040-1951(99)00013-X.

Kluth, C.F., 1986, Plate tectonics of the Ancestral Rocky Mountains, *in* Peterson, J.A., ed., Paleotectonics and Sedimentation: American Association of Petroleum Geologists Memoir 41, p. 353–370.

Kluth, C.F., 1998, Late Paleozoic deformation of interior North America: The greater Ancestral Rocky Mountain: American Association of Petroleum Geologists Bulletin, v. 82, p. 2272–2276.

Kluth, C.F., and Coney, P.J., 1981, Plate tectonics of the Ancestral Rocky Mountains: Geology, v. 9, p. 10–15, doi: 10.1130/0091-7613(1981)9<10:PTOTAR>2.0.CO;2.

Krumbein, W.C., 1939, Graphic presentation and statistical analysis of sedimentary data: Sedimentary Petrology, v. 6, p. 558–591.

Laubach, S.E., Tyler, R., Ambrose, W.A., Tremain, C.M., and Grout, M.A., 1992, Preliminary map of fracture patterns in coal in the western United States: Casper, Wyoming Geological Association Guidebook, p. 253–267.

Livaccari, R.F., 1991, Role of crustal thickening and extensional collapse in the tectonic evolution of the Sevier-Laramide orogeny, western United States: Geology, v. 19, p. 1104–1107, doi: 10.1130/0091-7613(1991)019<1104:ROCTAE>2.3.CO;2.

Lowell, J.D., 1983, Foreland detachment deformation: American Association of Petroleum Geologists Bulletin, v. 67, p. 1349.

Madsen, J.K., Thorkelson, D.J., Friedman, R.M., and Marshall, D.D., 2006, Cenozoic to Recent plate configurations in the Pacific Basin: Ridge subduction and slab window magmatism in western North America: Geosphere, v. 2, p. 11–34, doi: 10.1130/GES00020.1.

Magnani, M.B., Levander, A., Erslev, E.A., Bolay-Koenig, N., and Karlstrom, K., 2005, Listric thrust faulting in the Laramide front of north-central New Mexico guided by Precambrian basement anisotropies, *in* Karlstrom, K.E., and Keller, G.R., eds., The Rocky Mountain Region—An Evolving Lithosphere: Tectonics, Geochemistry, and Geophysics: American Geophysical Union Geophysical Monograph 154, p. 239–252.

Marshak, S., Karlstrom, K.E., and Timmons, J.M., 2000, Inversion of Proterozoic extensional faults: An explanation for the pattern of Laramide and Ancestral Rockies intracratonic deformation, United States: Geology, v. 28, p. 735–738, doi: 10.1130/0091-7613(2000)28<735:IOPEFA>2.0.CO;2.

Maughan, E.K., and Perry, W.J., Jr., 1986, Lineaments and their tectonic implications in the Rocky Mountains and adjacent plains region, *in* Peterson, J.A., ed., Paleotectonics and Sedimentation: American Association of Petroleum Geologists Memoir 41, p. 41–53.

McQueen, H.W.S., and Beaumont, C., 1989, Mechanical models of tilted block basins, *in* Price, R.A., ed., Origin and Evolution of Sedimentary Basins and Their Energy and Mineral Resources: American Geophysical Union Geophysical Monograph, v. 48, p. 65–71.

Miller, J.P., Montgomery, A., and Sutherland, P.K., 1963, Geology of Part of the Southern Sangre de Cristo Mountains, New Mexico: New Mexico Bureau of Mines and Mineral Resources Memoir 11, 106 p.

Molzer, P.C., and Erslev, E.A., 1995, Oblique convergence during northeast-southwest Laramide compression along the east-west Owl Creek and Casper Mountain arches, central Wyoming: American Association of Petroleum Geologists Bulletin, v. 79, p. 1377–1394.

Neely, T.G., 2006, Three-Dimensional Strain at Foreland Arch Transitions: Structural Modeling of the Southern Beartooth Arch Transition Zone, Northwest Wyoming [M.S. thesis]: Fort Collins, Colorado State University, 107 p.

Oldow, J.S., Bally, A.W., and Lallemant, H.G., 1990, Transpression, orogenic float, and lithospheric balance: Geology, v. 18, p. 991–994, doi: 10.1130/0091-7613(1990)018<0991:TOFALB>2.3.CO;2.

Osterwald, F.W., 1961, Critical review of some tectonic problems in Cordilleran foreland: American Association of Petroleum Geologists Bulletin, v. 45, p. 219–237.

Paylor, E.D., II, and Yin, A., 1993, Left-slip evolution of the North Owl Creek fault system, Wyoming, during Laramide shortening, *in* Schmidt, C.J., Chase, R., and Erslev, E.A., eds., Laramide Basement Deformation in the Rocky Mountain Foreland of the Western United States: Geological Society of America Special Paper 280, p. 229–242.

Perry, W.J., Jr., Nicols, D.J., Dyman, T.S., and Haley, C.J., 1992, Sequential Laramide deformation in Montana and Wyoming, *in* Thorman, C.H., ed., Application of Structural Geology to Mineral and Energy Resources of the Central and Western United States: U.S. Geological Survey Bulletin 2012, p. C1–C14.

Petit, J.P., 1987, Criteria for the sense of movement on fault surfaces in brittle rocks: Journal of Structural Geology, v. 9, p. 597–608, doi: 10.1016/0191-8141(87)90145-3.

Pierce, W.G., 1997, Geologic Map of the Cody 1° × 2° Quadrangle, Northwestern Wyoming: U.S. Geological Survey Miscellaneous Investigations Series I-2500, scale 1:250,000.

Poole, F.G., Perry, W.J., Jr., Madrid, R.J., and Amaya-Martinez, R., 2005, Tectonic synthesis of the Ouachita-Marathon-Sonoran orogenic margin of southern Laurentia: Stratigraphic and structural implications for timing of deformation events and plate-tectonic model, *in* Anderson, T.H., Nourse, J.A., McKee, J.W., and Steiner, M.B., eds., The Mojave-Sonora Megashear Hypothesis: Development, Assessment and Alternatives: Geological Society of America Special Paper 393, p. 543–596, doi: 10.1130/2005.2393(21).

Ramos, V.A., Cristallini, E.O., and Perez, D.J., 2002, The Pampean flat-slab of the Andes: Journal of South American Earth Sciences, v. 15, p. 59–78, doi: 10.1016/S0895-9811(02)00006-8.

Ramos, V.A., Zapata, T., Cristallini, E., and Introcasa, A., 2004, The Andean thrust system—Latitudinal variations in structural styles and orogenic shortening, *in* McClay, K.R., ed., Thrust Tectonics and Hydrocarbon Systems: American Association of Petroleum Geologists Memoir 82, p. 30–50.

Rodgers, J., 1987, Chains of basement uplifts within cratons marginal to orogenic belts: American Journal of Science, v. 287, p. 661–692.

Ruf, J.C., 2000, Origin of Laramide to Holocene Fractures in the Northern San Juan Basin, Colorado and New Mexico [M.S. thesis]: Fort Collins, Colorado State University, 167 p.

Ruf, J.C., and Erslev, E.A., 2005, Origin of Late Mesozoic to Holocene fractures in the northern San Juan Basin, Colorado and New Mexico: Rocky Mountain Geology, v. 40, p. 91–114.

Saleeby, J., 2003, Segmentation of the Laramide slab-evidence from the southern Sierra Nevada region: Geological Society of America Bulletin, v. 115, p. 655–668, doi: 10.1130/0016-7606(2003)115<0655:SOTLSF>2.0.CO;2.

Sales, J.K., 1968, Crustal mechanisms of Cordilleran foreland deformation: A regional and scale-model approach: American Association of Petroleum Geologists Bulletin, v. 52, p. 2016–2044.

Scott, G.R., Taylor, R.B., Epis, R.C., and Wobus, R.A., 1978, Geologic Map of the Pueblo 1° × 2° Quadrangle, South-Central Colorado: U.S. Geological Survey Miscellaneous Investigations Series I-1022, scale 1:250,000.

Selvig, B.W., 1994, Kinematics and Structural Models of Faulting Adjacent to the Rocky Flats Plant, Central Colorado [M.S. thesis]: Fort Collins, Colorado State University, 133 p.

Sims, P.K., Finn, C.A., and Rystrom, V.L., 2001, Preliminary Precambrian basement map of Wyoming showing geologic-geophysical domains, Wyoming: U.S. Geological Survey Open-File Report 01-199, http://pubs.usgs.gov/of/2001/ofr-01-0199/ (last accessed 8 April 2009).

Stanton, H.I., and Erslev, E.A., 2004, Sheep Mountain: Backlimb tightening and sequential deformation in the Bighorn Basin, Wyoming, *in* 53rd Field Conference Guidebook: Casper, Wyoming Geological Association, p. 75–87.

Stearns, D.W., 1971, Mechanisms of drape folding in the Wyoming Province, *in* Renfro, A.R., ed., Symposium on Wyoming Tectonics and their Economic Significance: 23rd Annual Field Conference Guidebook: Casper, Wyoming Geological Association, p. 125–143.

Steven, T.A., Lipman, P.W., Hail, W.J., Jr., Barker, F., and Luedke, R.G., 1974, Geologic Map of the Durango Quadrangle, Southwestern Colorado: U.S. Geological Survey Miscellaneous Investigations Series I-764, scale 1:250,000.

Stevenson, G.M., and Baars, D.L., 1986, The Paradox: A pull-apart basin of Pennsylvanian age, *in* Peterson, J.A., ed., Paleotectonics and Sedimentation: American Association of Petroleum Geologists Memoir 41, p. 513–539.

Stone, D.S., 1969, Wrench faulting and Rocky Mountain tectonics: The Mountain Geologist, v. 6, p. 67–79.

Stone, D.S., 1986, Geometry and kinematics of thrust-fold structures in central Rocky Mountain foreland: American Association of Petroleum Geologists Bulletin, v. 70, p. 1057.

Stone, D.S., 2005, On illogical interpretation of geological structures in the Rocky Mountain foreland province: The Mountain Geologist, v. 42, p. 159–186.

Tetreault, J., Jones, C.H., Erslev, E., Hudson, M., and Larson, S., 2008, Paleomagnetic and structural evidence for oblique slip fold, Grayback monocline, Colorado: Geological Society of America Bulletin, v. 120, p. 877–892, doi: 10.1130/B26178.1.

Tikoff, B., and Maxson, J., 2001, Lithospheric buckling of the Laramide foreland during Late Cretaceous and Paleogene, western United States: Rocky Mountain Geology, v. 36, p. 13–35, doi: 10.2113/gsrocky.36.1.13.

Timmons, J.M., Karlstrom, K.E., Dehler, C.M., Geissman, J.W., and Heizler, M.T., 2001, Proterozoic multistage (ca. 1.1 and 0.8 Ga) extension recorded in the Grand Canyon Supergroup and establishment of northwest- and north-trending tectonic grains in the southwestern United States: Geological Society of America Bulletin, v. 113, p. 163–181, doi: 10.1130/0016-7606(2001)113<0163:PMCAGE>2.0.CO;2.

Tonnson, J.J., 1986, Influence of tectonic terranes adjacent to the Precambrian Wyoming Province on Phanerozoic stratigraphy in the Rocky Mountain region, *in* Peterson, J.A., ed., Paleotectonics and Sedimentation: American Association of Petroleum Geologists Memoir 41, p. 21–40.

Varga, R.J., 1993, Rocky Mountain foreland uplifts: Products of a rotating stress field or strain partitioning?: Geology, v. 21, p. 1115–1118, doi: 10.1130/0091-7613(1993)021<1115:RMFUPO>2.3.CO;2.

Wawrzyniec, T.F., Geissman, J.W., Melker, M.D., and Hubbard, M., 2002, Dextral shear along the eastern margin of the Colorado Plateau: A kinematic link between Laramide contraction and Rio Grande rifting (ca. 75–13 Ma): The Journal of Geology, v. 110, p. 305–324, doi: 10.1086/339534.

Wawrzyniec, T.F., Geissman, J.W., Ault, A.K., Erslev, E.A., and Fankhauser, S.D., 2007, Paleomagnetic dating of fault slip in the southern Rocky Mountains, USA, and its importance to an integrated Laramide foreland strain field: Geosphere, v. 3, p. 16–25, doi: 10.1130/GES00066.1.

Williams, P.L., 1964, Geology, Structure, and Uranium Deposits of the Moab Quadrangle, Colorado and Utah: U.S. Geological Survey Miscellaneous Investigations Series I-360, scale 1:250,000.

Woodward, L.A., 1976, Laramide deformation of Rocky Mountain foreland: Geometry and mechanics, *in* Woodward, L.A., and Northrop, S.A., eds., Tectonics and Mineral Resources of Southwestern North America: New Mexico Geological Society Special Publication 6, p. 11–17.

Woodward, L.A., 1987, Geology and Mineral Resources of Sierra Nacimiento and Vicinity, New Mexico: New Mexico Bureau of Mines and Mineral Resources Memoir 42, 84 p.

Yang, K., and Dorobek, S.L., 1995, The Permian Basin of West Texas and New Mexico; flexural modeling and evidence for lithospheric heterogeneity across the Marathon foreland, *in* Dorobek, S.L., and Ross, G.M., eds., Stratigraphic Evolution of Foreland Basins: Society for Sedimentary Geology Special Publication 52, p. 37–50.

Ye, H., Royden, L., Burchfiel, C., and Schuepbach, M., 1996, Late Paleozoic deformation of interior North America: The greater Ancestral Rocky Mountains: American Association of Petroleum Geologists Bulletin, v. 80, p. 1397–1432.

Yin, A., and Ingersol, R.V., 1997, A model for evolution of Laramide axial basins in the southern Rocky Mountains, U.S.A.: International Geology Review, v. 39, p. 1113–1123.

Zoback, M.L., Anderson, R.E., and Thompson, G.A., 1981, Cainozoic evolution of the state of stress and style of tectonism of the Basin and Range Province of the western United States: Philosophical Transactions of the Royal Society of London, ser. B, Biological Sciences, v. 300, p. 407–434, doi: 10.1098/rsta.1981.0073.

Manuscript Accepted by the Society 5 December 2008

Printed in the USA

The Geological Society of America
Memoir 204
2009

Cretaceous–Eocene magmatism and Laramide deformation in southwestern Mexico: No role for terrane accretion

Michelangelo Martini
Luca Ferrari*
Centro de Geociencias, Universidad Nacional Autónoma de México, Campus Juriquilla, 76230 Querétaro, Mexico

Margarita López-Martínez
Laboratorio de Geocronología, División de Ciencias de la Tierra, CICESE (Centro de Investigación Cientifica y Educación Superior de Ensenada), Ensenada, Baja California, Mexico

Mariano Cerca-Martínez
Centro de Geociencias, Universidad Nacional Autónoma de México, Campus Juriquilla, 76230 Querétaro, Mexico

Víctor A. Valencia
Department of Geosciences, University of Arizona, Tucson, Arizona 85721, USA

Lina Serrano-Durán
Centro de Geociencias, Universidad Nacional Autónoma de México, Campus Juriquilla, 76230 Querétaro, Mexico

ABSTRACT

In southwestern Mexico, Late Cretaceous to Early Tertiary deformation has been generally associated with the Laramide orogeny of the Cordillera. Several alternative models consider the deformation to result from the accretion of the Guerrero terrane, formed by the Zihuatanejo, Arcelia, and Teloloapan intraoceanic island arcs, to the continental margin of the North American plate. Here, we present a detailed geologic and structural study and new $^{40}Ar/^{39}Ar$ and U-Pb ages for a broad region in the central-eastern part of the Guerrero terrane that allow the accretion model to be tested. In the Huetamo–Ciudad Altamirano part of the region, an almost complete Cretaceous-Paleogene succession records the transition from an early Cretaceous shallow-marine environment to continental conditions that began in Santonian times, followed by the development of a major continental Eocene magmatic arc. Folding of the marine and transitional successions signifies a shortening episode between the late Cenomanian and the Santonian, and a subsequent, out-of-sequence, coaxial refolding event in Maastrichtian-Paleocene time amplified the previous structures. A major left-lateral shear zone postdates the contractional deformation, and it passively controlled the geographic distribution of Eocene silicic volcanism. Minor transcurrent faulting followed.

*luca@geociencias.unam.mx

Martini, M., Ferrari, L., López-Martínez, M., Cerca-Martínez, M., Valencia, V.A., and Serrano-Durán, L., 2009, Cretaceous–Eocene magmatism and Laramide deformation in southwestern Mexico: No role for terrane accretion, *in* Kay, S.M., Ramos, V.A., and Dickinson, W.R., eds., Backbone of the Americas: Shallow Subduction, Plateau Uplift, and Ridge and Terrane Collision: Geological Society of America Memoir 204, p. 151–182, doi: 10.1130/2009.1204(07). For permission to copy, contact editing@geosociety.org.

Our results indicate that the Huetamo–Ciudad Altamirano region, which has been considered part of the Zihuatanejo subterrane, was in proximity to a continent during most of the Mesozoic. We found continental recycled material at various stratigraphic levels of the Huetamo Cretaceous succession and Grenvillian inherited ages in zircons from the ca. 120 Ma Placeres del Oro pluton. More importantly, detrital zircon ages from the pre-Cretaceous basement of the Huetamo succession (Tzitzio metaflysch) and the pre–Early Jurassic basement of the Arcelia subterrane (Tejupilco suite) yield very similar Late Permian and Ordovician age peaks. These ages are typical of the Acatlán complex, onto which the Guerrero terrane has been proposed to have been accreted in the Late Cretaceous. Similarly, Paleozoic and Precambrian ages are reported in detrital zircons from the volcano-sedimentary successions of the Zihuatanejo, Arcelia, and Teloloapan subterranes. Models considering this part of the Guerrero terrane as having formed by intraoceanic island arcs separated by one or more subduction zones cannot explain the ubiquitous presence of older continental material in the Mesozoic succession. We favor a model in which most of the Guerrero terrane consisted of autochthonous or parautochthonous units deposited on the thinned continental margin of the North American plate and where the Mesozoic magmatic and sedimentary record is explained in the framework of an enduring west-facing migrating arc and related extensional backarc and forearc basins.

The results presented here exclude the accretion of allochthonous terranes as the cause for Laramide deformation and require an alternative driving force to explain the generation of the Late Cretaceous–early Tertiary shortening and shearing on the southern margin of the North American plate.

INTRODUCTION

Geologists have long wondered about the mechanisms responsible for orogenesis. In the case of the North America Cordillera, ridge collision and flat-slab subduction have been recognized to be among the main processes producing orogenic deformation. Both processes are driven by the arrival of buoyant lithospheric sections of an oceanic plate in the subduction zone. Aseismic ridges, island arcs, oceanic plateaus, and clusters of seamounts can be too buoyant to be subducted and can eventually be accreted to the upper plate as allochthonous terranes. This process results in the growth of the continental crust and complex deformation of the margin. Terrane accretion has shaped part of the Cordillera, which is now formed by the juxtaposition of different blocks that have independent tectono-magmatic histories prior to the accretion (Coney et al., 1980). A model entrenched in the geologic literature of southern Mexico considers the Laramide orogeny to be the result of the accretion of Mesozoic intraoceanic arcs to the Pacific margin of the North American continental plate (Campa and Coney, 1983; Lapierre et al., 1992; Tardy et al., 1994; Talavera-Mendoza and Guerrero-Suástegui, 2000; Keppie et al., 2004; Talavera-Mendoza et al., 2007). However, the link between accretion of allochthonous terranes and the Laramide orogeny has not been discussed in detail, probably because the timing and spatial extent of this deformational event has been poorly defined in southern Mexico. However, stratigraphic, structural, and geochemical data allow scenarios that do not require the accretion of allochthonous terranes separated by a subduction zone from mainland Mexico (Cabral-Cano et al., 2000a, 2000b; Elías-Herrera et al., 2000; Busby, 2004; Busby et al., 2005; Centeno-García et al., 2008). In this paper, we address these questions with new stratigraphic and structural data, supported by new $^{40}Ar/^{39}Ar$ and U-Pb geochronologic ages, for a key area in the central part of the Guerrero terrane in southwestern Mexico. Based on our results and a review of previous work, we discuss the applicability of the terrane accretion model in explaining Late Cretaceous and early Tertiary deformation in southwestern Mexico.

GEOLOGIC SETTING

The western and southern parts of Mexico have been considered to consist of allochthonous terranes accreted to the continental margin of North America in the Mesozoic and earliest Cenozoic (see Dickinson and Lawton, 2001; Keppie, 2004, for a review). The Guerrero represents the largest terrane of Mexico, and it extends from the states of Sinaloa and Zacatecas, north of the Trans-Mexican volcanic belt, to Zihuatanejo and Taxco in southern Mexico (Fig. 1) (Centeno-García et al., 2008). It is formed by Mesozoic volcano-sedimentary successions, which have been traditionally interpreted as intraoceanic arc(s) sequences accreted to nuclear Mexico during the latest Mesozoic or the early Tertiary (Campa and Coney, 1983; Lapierre et al., 1992; Tardy et al., 1994; Talavera-Mendoza and Guerrero-Suástegui, 2000). The Guerrero is considered to be a composite terrane, made up of distinct subterranes, the name, number, and geographic extents of which have varied in the literature. Following Talavera-Mendoza

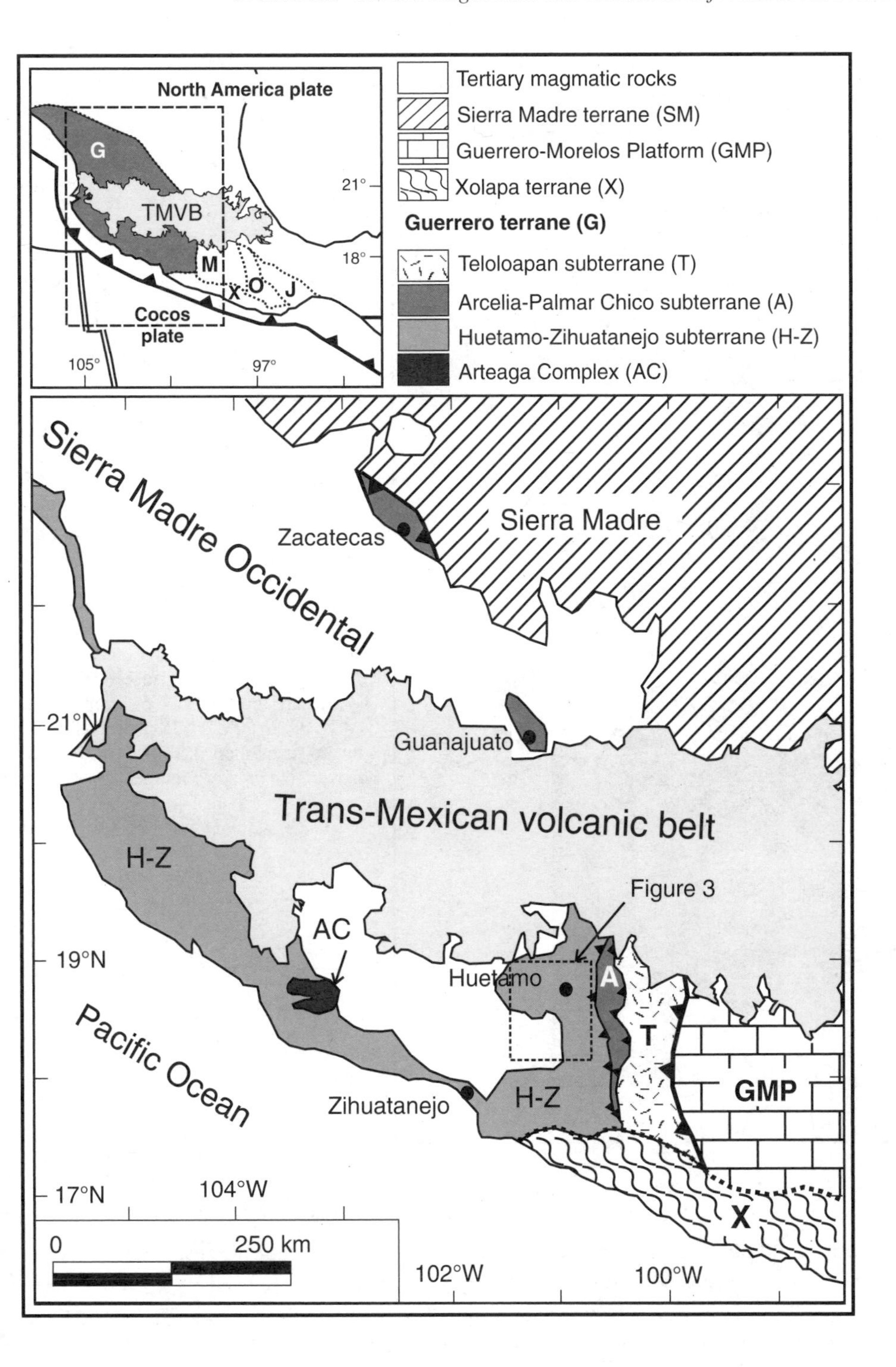

Figure 1. Schematic tectonic map of the central and southern part of the Guerrero terrane and its subterranes as defined in Talavera-Mendoza and Guerrero-Suástegui (2000). Inset shows the present plate boundaries, the terrane subdivisions of southern Mexico, and other tectonic elements described in the text. G—Guerrero terrane; M—Mixteco terrane; O—Oaxaca terrane; J—Juarez terrane; X—Xolapa terrane; TMVB—Trans-Mexican volcanic belt.

and Guerrero-Suástegui (2000) and Talavera-Mendoza et al. (2007), the Guerrero consists of three subterranes (Figs. 1 and 2) from east to west:

(1) The Teloloapan subterrane is characterized by an island-arc volcanic and sedimentary rock association, the petrological and geochemical features of which have been described in Talavera-Mendoza et al. (1995). The age of the succession is Early Cretaceous based on paleontologic (radiolarian) data (Talavera-Mendoza et al., 1995) and volcanic rocks, which yield U/Pb zircon ages in the range ca. 130–146 Ma (Campa-Uranga and Iriondo, 2003; Mortensen et al., 2008).

(2) The Arcelia–Palmar Chico subterrane includes pillow lavas, pillow breccias, and hyaloclastites capped by a marine sedimentary cover, which is considered to represent either a primitive intraoceanic island arc with a backarc ocean basin (Ortíz-Hernández et al., 1991; Talavera-Mendoza and Guerrero-Suástegui,

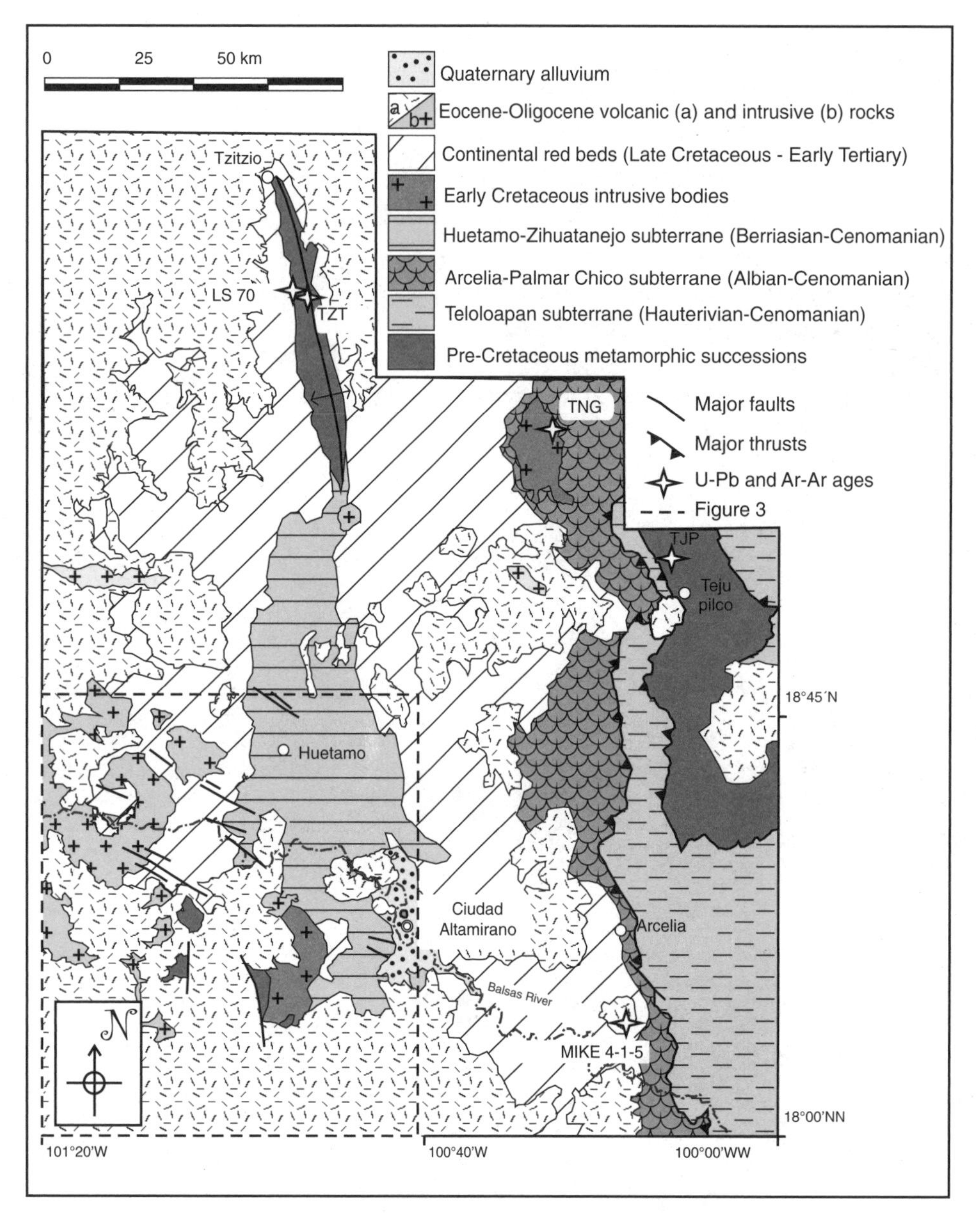

Figure 2. Geologic map of the Huetamo, Arcelia–Palmar Chico, and Teloloapan subterranes of the Guerrero superterrane (based on our geologic mapping and reinterpretation of Servicio Geológico Mexicano [2002] 1:250,000 scale geologic maps). Metaflysch exposed in the core of the Tzitzio anticline is the basement of the Huetamo subterrane, whereas the metaflysch of the Tejupilco area represents the basement of the Arcelia–Palmar Chico subterrane. Also shown are the locations of samples dated by U-Pb (TZT, TJP, and TNG) and by Ar-Ar method (LS-70, MIKE 4-1-5) outside the area of Figure 3.

2000; Elías-Herrera et al., 2000) or a backarc basin only (Elías-Herrera and Ortega-Gutiérrez, 1998). The age of the sequence is Albian-Cenomanian based on $^{40}Ar/^{39}Ar$ ages of the lavas and paleontological determinations (Dávila-Alcocer and Guerrero-Suástegui, 1990; Elías-Herrera et al., 2000).

(3) The Huetamo-Zihuatanejo subterrane is interpreted as a complex assemblage made up of a Cretaceous island-arc sequence (Zihuatanejo sequence) and a backarc basin sequence (Huetamo sequence) (Talavera-Mendoza and Guerrero-Suástegui, 2000; Talavera-Mendoza et al., 2007; Centeno-García et al., 2008).

In most cases, the proposed subterrane boundaries are buried beneath Tertiary volcanic and sedimentary deposits. Where exposed, they appear as low-angle thrust faults affecting the Mesozoic volcano-sedimentary successions. A Triassic accretionary complex (Arteaga and Las Ollas Complexes) has been considered to be the basement of the Huetamo-Zihuatanejo subterrane (Centeno-García et al., 1993, 2008), whereas the Late Triassic–Early Jurassic Tejupilco metamorphic suite is considered to form the basement of the Arcelia–Palmar Chico subterrane (Elías-Herrera et al., 2000). The region studied here is located in the Huetamo-Zihuatanejo subterrane (Figs. 1 and 2), and it was chosen because of its large and complete stratigraphic record, which spans most of the Cretaceous and Paleogene (Figs. 2 and 3). We dated key rock units exposed to the north and to the east that have important implications for the geology of the studied area. These data allow us to constrain the age of the deformation and to discuss the hypothesis of terrane accretion as the cause of the Laramide orogeny. Our methodology is described in the Appendix.

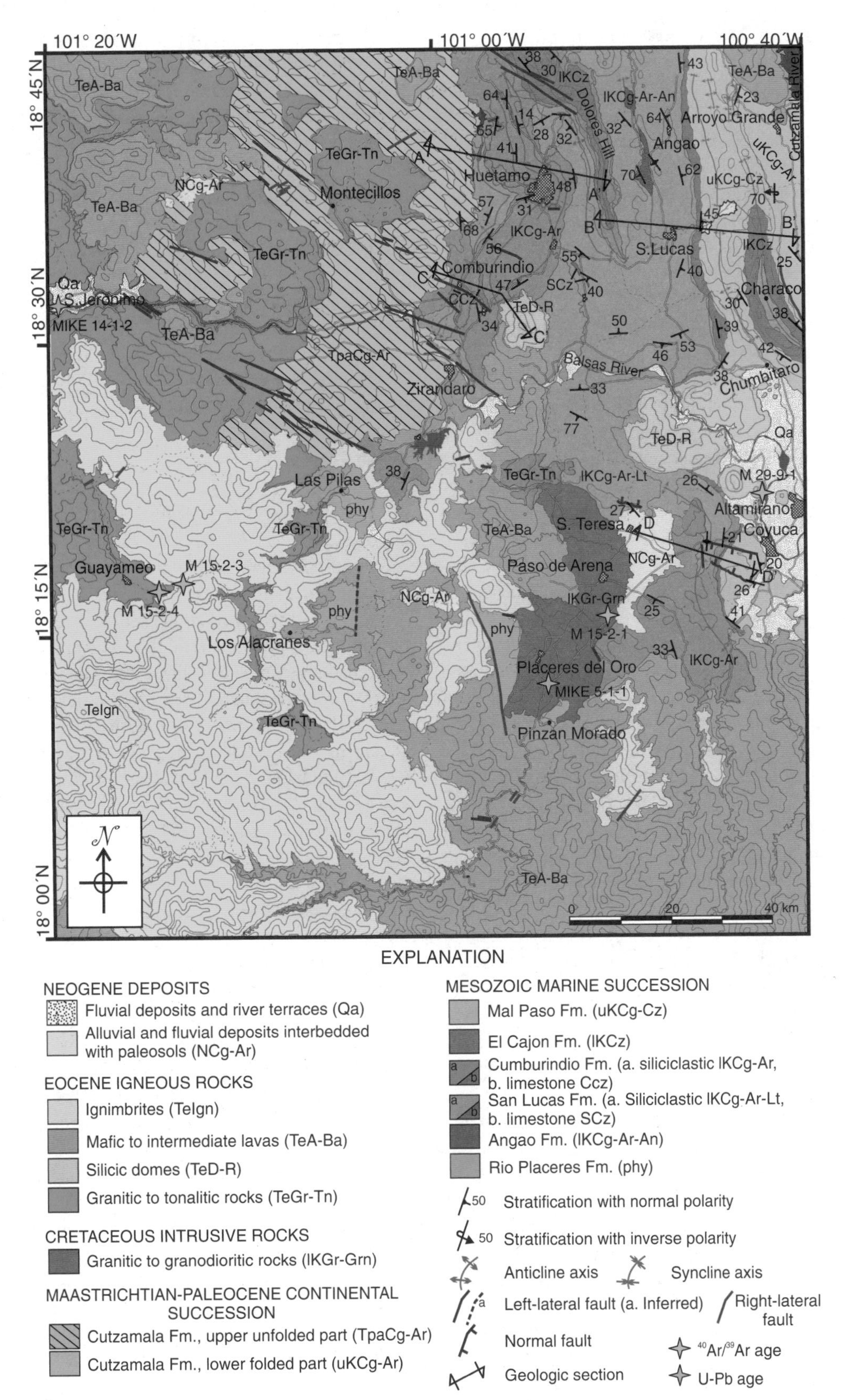

Figure 3. Detailed geologic map of the Huetamo–Ciudad Altamirano region showing the Mesozoic to Tertiary units and the main structures described in the text. The location of dated samples is also indicated.

STRATIGRAPHIC FRAMEWORK

Stratigraphic and geologic data for the Huetamo–Ciudad Altamirano area are scarce in the literature (Pantoja-Alór, 1959; Campa-Uranga and Ramírez, 1979; Guerrero-Suástegui, 1997; Altamira-Areyán, 2002; Pantoja-Alór and Gómez-Caballero, 2003), and the nomenclature and age of the main stratigraphic units are controversial. Most previous studies have focused on relatively small areas, defining stratigraphy at a very local scale and proposing local names for the different formations recognized. In some cases, the same name was used to refer to different formations and vice-versa. Although the stratigraphic contacts and the geometric relationships between the main stratigraphic units were correctly established in most cases, the lack of stratigraphic correlations between successions in different places has led to an incomplete understanding of the Mesozoic-Cenozoic depositional evolution of the Huetamo Basin. Comprehensive geologic mapping of the region and modern isotopic age determinations were also lacking.

In our study, we recognized a general shallowing-upward succession deposited at the western side of a sedimentary basin, characterized by an active depositional history during the Cretaceous and the early Tertiary. Based on the presence of volcanic or volcaniclastic material at several levels, this Cretaceous–early Tertiary succession was likely deposited in a backarc basin. The lateral variation of the sedimentary facies shows that the marginal basin deposits are localized in the western part of the study area, whereas pelagic rocks are distributed at the eastern side, indicating that the Cretaceous depocenter was located toward the east. Next, we present a description of the geologic units that make up the stratigraphic succession of the Huetamo–Ciudad Altamirano region. Although we used most of the stratigraphic units already recognized in the literature, we discuss and revise their use according to our field observations. We present three stratigraphic columns and a section that cuts across the basin margin from the outer (western) to the inner (eastern) part, and it illustrates the lateral variation of the sedimentary facies and their spatial migration according to the relative fluctuation of the sea level (Fig. 4).

Río Placeres Formation

This formation was introduced by Pantoja-Alór (1990) to designate a sequence of metasedimentary rocks that crop out discontinuously between the small villages of Pinzan Morado, Los Alacranes, and Las Pilas in the central part of the region (Fig. 3). It consists of gray to violet slates and phyllites, alternating with quartz-rich metasandstones and few decimetric-scale levels of recrystallized black limestones. Sandstones and phyllites have developed a foliated texture, shown by elongated quartz ribbons and the iso-orientation of white micas and tremolite-actinolite amphiboles along surfaces parallel to the bedding planes. Quartz clasts are characterized by undulate extinction, subgrain domains, and pervasive grain boundary migration, suggesting significant plastic deformation of the crystals' lattice and recrystallization. Foliation developed as a continuous cleavage in the finest lithologic terms and is space disjunctive in the coarser parts, varying from anastomosing to moderately rough using the Twiss and Moore (1992) classification. The original stratigraphic position of this formation is unknown, because a 2000-m-thick sequence of volcanic rocks covers any relation with the Cretaceous succession. Pantoja-Alór and Gómez-Caballero (2003) suggested a pre–Late Jurassic age for this formation based on a tentative stratigraphic correlation with the Varales lithofacies of the Arteaga Complex, which is exposed ~150 km to the west (Fig. 1), and which has a reported Triassic age (Campa-Uranga et al., 1982). In their review of the Guerrero terrane, Centeno-García et al. (2008) grouped the Río Placeres Formation and the low-grade metamorphic flysch in the core of the Tzitzio anticline with the Arteaga Complex. Ages for detrital zircons from the Tzitzio metaflysch reported in Talavera-Mendoza et al. (2007) and herein support this correlation.

Angao Formation

Pantoja-Alór (1959) originally introduced this formation to describe a sequence of marine clastic rocks made up of sandstone, shale, and polymictic varicolored conglomerate, cropping out in the core of the San Lucas anticline (Fig. 3). Subsequently, Campa-Uranga (1977) reported some basaltic lava with pillowed structure interbedded in the clastic sequence of the Angao Formation. Guerrero-Suástegui (1997) presented a detailed lithostratigraphic description of the Angao Formation, which is described as being composed of conglomerate, sandstone, and shale, with lesser amount of volcanic rocks. The clastic fraction is dominated by volcanic fragments (basalt, andesite, and minor tuff and dacite), with lesser amounts of clasts, limestone, and metamorphic clasts, including gneiss, schist, and metapelite (Guerrero-Suástegui, 1997). Two metric levels of pillow lavas, interbedded with the siliciclastic sediments, have been reported in the middle part of the sequence (Guerrero-Suástegui, 1997). Pillows are 60–80 cm in diameter and are composed of aphyric basalt with tholeiitic affinity (Talavera-Mendoza, 1993). The lithologic assemblage of the Angao Formation was interpreted by Guerrero-Suástegui (1997) to represent proximal to distal apron deposits, in agreement with the fossil content (Pantoja-Alór, 1959; Guerrero-Suástegui, 1997). Since the base of the Angao Formation is not exposed in the region, the total thickness is unknown. Pantoja-Alór (1959) reported a Late Jurassic age for the Angao Formation, based on the occurrence of some species of ammonites and molluscs characteristic of the Kimmeridgian-Tithonian. Later, Guerrero-Suástegui (1997) recognized that this Jurassic fauna was reworked and proposed an Early Cretaceous age based on late Valanginian rudists in the upper part of the sequence. The Angao Formation grades transitionally upward into the siliciclastic pelagic deposits of the lower member of the San Lucas Formation.

San Lucas Formation

The San Lucas Formation was originally defined by Pantoja-Alór (1959) as a marine turbiditic succession composed of

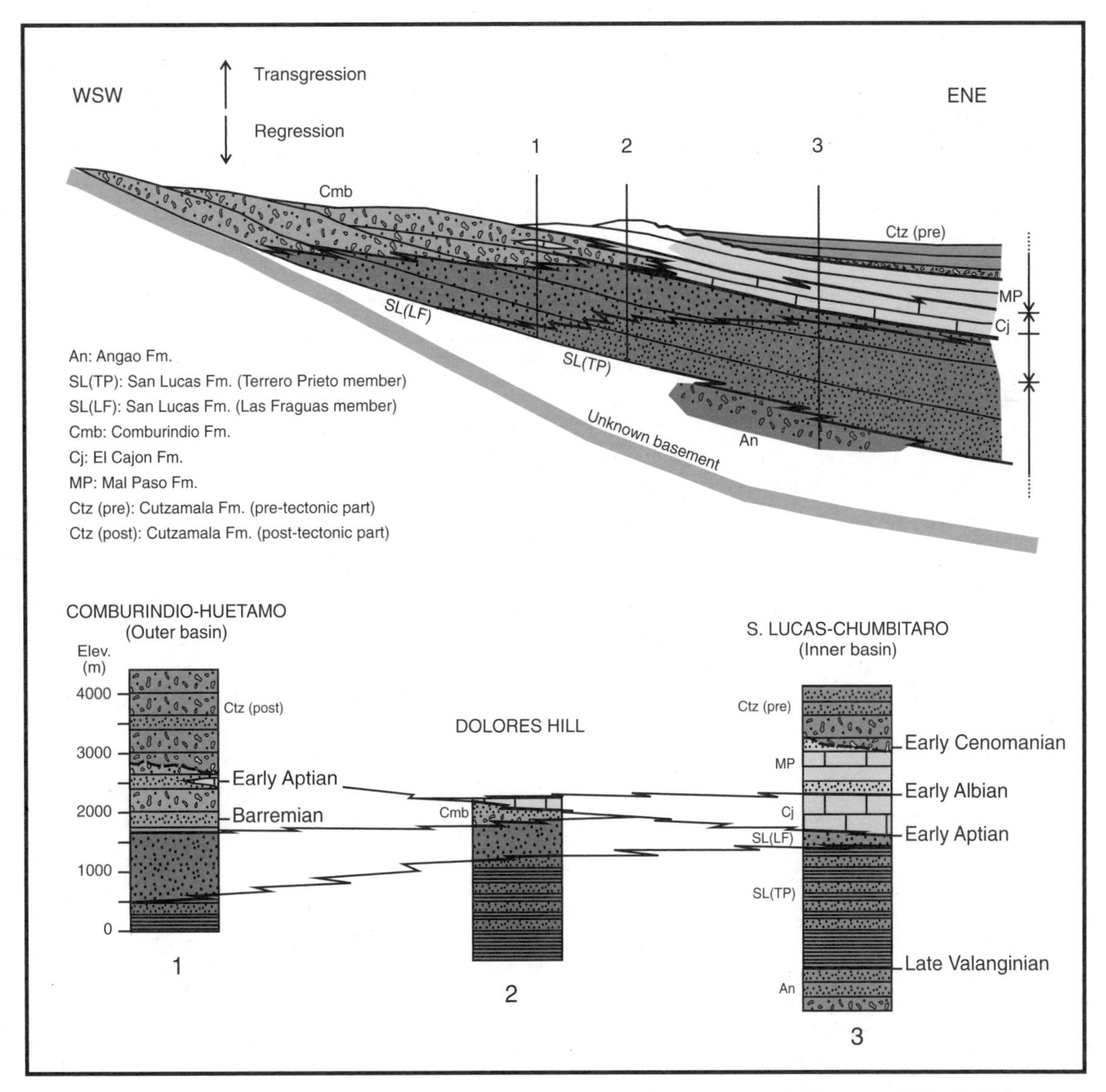

Figure 4. Stratigraphic columns for the study area and reconstruction of the Huetamo Basin across a WNW-ESE section.

mudstone, shale, sandstone, conglomerate, and some interbedded biostromic limestone banks that crop out between the town of San Lucas and the Turitzio Hill, east of Zirandaro village and west of Huetamo (Fig. 3). Based on its sedimentary characteristics, this formation was subdivided by Pantoja-Alór (1990) and Pantoja-Alór and Gómez-Caballero (2003) into the lower Terrero Prieto member and the upper Las Fraguas member. The lower member consists of a rhythmic sequence of mudstone, shale, and sandstone, in strata variable from 2 to 30 cm. The clastic fraction is composed principally of plagioclase monocrystals, with lesser amounts of quartz and volcanic lithics. The lithology and faunal association of the Terrero Prieto member suggest a pelagic deposit in an open-marine environment (Pantoja-Alór, 1990; Pantoja-Alór and Gómez-Caballero, 2003; Omaña-Pulido et al., 2005). The upper member is composed of thick feldspathic sandstone, and lesser amounts of brown to green mudstone and conglomerate, which conformably cover the lower member. Petrographic analysis shows that the dominant clastic component consists of plagioclase and minor quartz monocrystals. Scarce lithic fragments of andesitic and metamorphic rocks also occur. Metamorphic fragments consist of polycrystalline quartz that shows clear evidences of plastic deformation and recrystallization (e.g., grain boundary migration, undulate extinction, subgrain domains), which could be from quartzite or quartz-schist. The lithologic association and the faunal assemblage of the Las Fraguas member suggest deposition in a distal deltaic fan environment (Pantoja-Alór, 1990; Pantoja-Alór and Gómez-Caballero, 2003).

The sandstones of the Las Fraguas member transition upward into the coastal deposits of the Comburindio Formation in the western part of the mapped region. In the eastern part, the early Aptian limestone of the El Cajón Formation directly overlies the Las Fraguas member in a transitional contact (Fig. 4). Paleontological age determinations are reported from different parts of the San Lucas Formation at many sites (Pantoja-Alór, 1959; Campa-Uranga, 1977; Gómez-Luna et al., 1993; Guerrero-Suástegui, 1997; Omaña-Pulido et al., 2005). Based on these data, Pantoja-Alór and Gómez-Caballero (2003) proposed a Berriasian-Barremian age for this formation, whereas Guerrero-Suástegui (1997) proposed a late Valanginian–late Aptian age. Recent U-Pb radiometric ages of detrital zircons from the Las Fraguas member in the eastern part span from ca. 542 to ca. 118 Ma, with a peak of ca. 126 Ma (sample Gro-09; Talavera-Mendoza et al., 2007). Taking into account the stratigraphic position, the vertical and lateral variations in sedimentary facies, and the available paleontologic and radiometric data, the top and basal contacts of the San Lucas Formation seem to be diachronic surfaces, as is typical in aggradational and retrogradational basin systems where sedimentation is influenced by sea-level variations. In the eastern part of the region, the top of the San Lucas Formation was deposited during the early Aptian, as indicated by the transitional contact with the overlying limestones of the El Cajon Formation. In the western part, the distal deltaic sedimentation of the Las Fraguas member is replaced by the nearshore and shoreface deposits of the Comburindio Formation, which has a Barremian age (Alencaster and Pantoja-Alór, 1998; Pantoja-Alór and Gómez-Caballero, 2003). This sequence clearly suggests a progressive rejuvenation of the upper contact of the San Lucas Formation toward the depocenter of the basin controlled by the lateral shift of the coastline toward the inner part of the basin during a prolonged period of marine regression (Fig. 4). The basal contact of the San Lucas Formation crops out in a restricted area, north of the village of San Lucas, where a late Valanginian age has been established on paleontological grounds (Guerrero-Suástegui, 1997). As a late Valanginian transgressive event is required to explain the progressive deepening of the sedimentary environment, it is reasonable to expect a gradual rejuvenation of the lower contact of the San Lucas Formation toward the outer part of the basin, making the age of the San Lucas Formation variable throughout the region. In the inner part of the basin, the sedimentation age is well constrained to the late Valanginian–early Aptian interval; in the outer part, it varies from the Barremian to, probably, the Hauterivian (Fig. 4).

Comburindio Formation

This name has been controversially used in previous works to define different stratigraphic units in the Huetamo–Ciudad Altamirano region. Salazar (1973) introduced the term Comburindio Limestone to define some biostromic calcareous rocks of Aptian age exposed in the surroundings of the town of Comburindio, southwest of Huetamo (Fig. 3). Subsequently, Campa-Uranga and Ramirez (1979) documented limestone intercalated with the clastic succession and used the name Comburindio Formation to designate biostromic bodies and to differentiate them from the Albian-Cenomanian limestone of the Morelos Formation. In their revision of the stratigraphy of the Huetamo region, Pantoja-Alór and Gómez-Caballero (2003) redefined the Comburindio Formation as a thick succession of deltaic clastic sediments with some interbedded biostromic limestone banks and faunal assemblages of early Aptian age. We follow the definition of Pantoja-Alór and Gómez-Caballero (2003) because it best agrees with our field observations. The Comburindio Formation is composed of a thick sequence of quartz-feldspathic sandstone, volcaniclastic and calcareous conglomerate, and lesser amounts of shale. The clastic fraction is mostly represented by plagioclase monocrystals and lithic fragments, mainly derived from andesitic lavas. Lesser amounts of quartz crystals and calcareous and metamorphic lithic clasts are also observed. The latter are polycrystalline quartz fragments characterized by a well-developed foliated texture and evidence of recrystallization, suggesting a quartzite or quartz-schist source rock. Three biostromic limestone banks interbedded with the upper part of the siliciclastic sequence crop out near Comburindio. They bear abundant faunas, including rudists, corals, ammonites and nerineids, and form outstanding morphologic rims that make them easy to distinguish from the surroundings clastic deposits. Some fossil wood has been found in the siliciclastic deposits (Pantoja-Alór and Gómez-Caballero, 2003). One of the most characteristic features of the Comburindio Formation is the dominant reddish color of the clastic part of the sequence, suggesting prolonged subaerial exposure. The faunal assemblage of the Comburindio Formation indicates a Barremian–early Aptian age (Alencaster and Pantoja-Alór, 1998; Pantoja-Alór and Gómez-Caballero, 2003).

El Cajon Formation

Originally introduced by Pantoja-Alór (1990) to designate a thick sequence of platform and reefal limestone previously known as the Morelos Formation (Pantoja-Alór, 1959), the El Cajon Formation crops out in a N-S–oriented belt in the eastern part of the map area (Fig. 3). East of the village of San Lucas, the El Cajon Formation conformably overlies the San Lucas Formation and is overlain by the Mal Paso Formation in a conformable contact. Along Dolores Hill, in the northeastern part of the map area, its base conformably covers the coastal deposits of the Comburindio Formation, and its upper contact has been removed by erosion. The El Cajon Formation consists of sandy limestone, coquina of orbitolinids, bioclastic limestone with corals, rudists, and gastropods, and massive limestone with shale interlayered at the top of the succession. The carbonates contain a rudist fauna similar to that of the biostromic banks of the Comburindio Formation and caprinids of early Aptian age (Skelton and Pantoja-Alór, 1999). Omaña-Pulido and Pantoja-Alór (1998) reported benthic foraminifera of Aptian age in calcareous beds east of the town of San Lucas. The early Albian fauna reported for the lowest beds of the overlying Mal Paso Formation (Buitrón-Sánchez and Pantoja-Alór, 2002) confirms that the El Cajon Formation is entirely within the Aptian time interval in the eastern portion of the map area. No precise paleontological data are available for the El Cajon Formation at Dolores Hill. The lower part is considered to represent a heteropic lateral variation of the upper beds of the Comburindio Formation (Pantoja-Alór and Gómez-Caballero, 2003), as strongly supported by paleontological data and by the increase of the calcareous fraction in the coastal deposits during the early Aptian.

Mal Paso Formation

The Mal Paso Formation was introduced by Pantoja-Alór (1959) to describe a succession of marine clastic rocks and carbonates that conformably overlie the platform and reef limestones of the El Cajon Formation in the eastern part of the mapped area. Based on the lithologic assemblage and sedimentary features, Buitrón-Sánchez and Pantoja-Alór (1998) subdivided the Mal Paso Formation into a lower deltaic clastic member and an upper reef and lagoonal member. The lower member consists of a sequence of medium- to thick-bedded quartzo-feldspathic and lithic sandstone and massive polymictic conglomerate, mostly composed of cobbles and boulders of limestone. Toward the top, the sequence grades into medium-bedded red claystone, siltstone, and sandstone and an interbedded biostromic bank with abundant *Toucasia* (Pantoja-Alór, 1959). Buitrón-Sánchez and Pantoja-Alór (1998) reported scarce igneous and metamorphic clasts in the conglomerate levels without specifying their lithologic type. Fossil wood and logs have been reported at various stratigraphic levels (Pantoja-Alór and Gómez-Caballero, 2003). The reef and lagoonal member consists of calcareous quartzo-feldspathic, medium-bedded gray sandstone and some intercalations of siltstone, claystone, and limestone. Toward the top, the sandstones gradually change to marl and thin-bedded argillaceous limestone, with abundant corals, gastropods, echinoids, ammonites, bivalves, and biostromic rudists. Several fossil faunal assemblages have been reported, constraining the age of the Mal Paso Formation to the early Albian–early Cenomanian (Pantoja-Alór, 1959; García-Barrera and Pantoja-Alór, 1991; Buitrón-Sánchez and Pantoja-Alór, 1998; Pantoja-Alór and Skelton, 2000; Filkorn, 2002).

Cutzamala Formation

The Cutzamala Formation was introduced by Campa-Uranga and Ramirez (1979) to describe a thick succession of continental red beds widely exposed along the Cutzamala River basin, east of the village of San Lucas (Fig. 3). It consists of a thick succession of reddish continental deposits that developed in fluvial and floodplain environments (Altamira-Areyán, 2002). An important shortening episode during deposition makes it possible to subdivide the Cutzamala Formation into a lower folded part and an upper unfolded part. The stratigraphy of the Cutzamala Formation is intrinsically complex because deposition took place in a continental fluvial plain environment where the drainage system was affected by lateral migration, and ongoing deformation significantly affected the distribution of the sedimentation centers. Detailed chronostratigraphic and sedimentologic studies are preliminary and restricted to several areas north and east of the studied region (Altamira-Areyán, 2002).

We use the deformational fabrics to establish the relative stratigraphy of the Cutzamala Formation, which is defined to consist of a thick succession of coarse- to medium-grained continental deposits, made up of reddish to gray conglomerate, sandstone, and lesser amounts of shale, with some intercalation of andesitic levels increasing toward the top of the sequence. Volcanic clasts, which are principally andesitic in composition, with lesser amounts of dacite and tuff, dominate throughout the entire succession. Shale, sandstone, limestone, and phyllite clasts derived from the underlying stratigraphic units occur in the lower part (Altamira-Areyán, 2002), whereas clasts in the undeformed upper part are characteristically from the lower part. The upper part is cut by extensive intrusive bodies that produce high-temperature–low-pressure (HT-LP) metamorphic aureoles. Recoil effects in the country rocks frequently produced quartzite characterized by grain boundary area reduction and static recrystallization of quartz.

The lower part of the Cutzamala Formation crops out in the easternmost portion of the map region, along the western margin of the Cutzamala River basin, whereas the upper part disconformably covers the folded Cretaceous succession in the northwestern portion, indicating that a significant migration of the depocenter occurred after the deformation event. In the eastern portion of the region, deposits of the Cutzamala Formation overlie the calcareous sediments of the Mal Paso Formation. In our study area, the basal stratigraphic contact of the continental succession is covered or has been replaced by subvertical left-lateral faults that put it in contact with the Mal Paso Formation. However, further east, Altamira-Areyán (2002) reported that the basal contact of the Cutzamala Formation is an angular unconformity. A thick conglomerate level with limestone boulders and cobbles derived from the Mal Paso Formation at the base of the continental succession indicates that the carbonate deposits were exposed and partially eroded before continental sedimentation began.

The age of the Cutzamala Formation is not well constrained. No ages have been reported for the continental deposits cropping out in the Huetamo–Ciudad Altamirano area. At Los Bonetes canyon, ~30 km north of the map area, Benammi et al. (2005) found well-preserved dinosaur bones at three levels in the folded part of the sequence. They assigned the fossils to the Hadrosauride family, giving a Late Cretaceous age to the lower part. In the same area, Altamira-Areyán (2002) found pollen species of the *Momipites microcoriphaeous* and *Momipites temipolus* groups, the first occurrence of which is Maastrichtian in age (Nicholls, 1973). Mariscal-Ramos et al. (2005) reported a radiometric age of 84 Ma for a lava flow interbedded within the basal beds near the dinosaur site, indicating a possible maximum age of late Santonian. We obtained a $^{40}Ar/^{39}Ar$ radiometric age of ca. 74 Ma for an andesitic clast (Fig. 5) from a gently folded conglomerate of the lower part of the Cutzamala Formation, 50 km north of the study

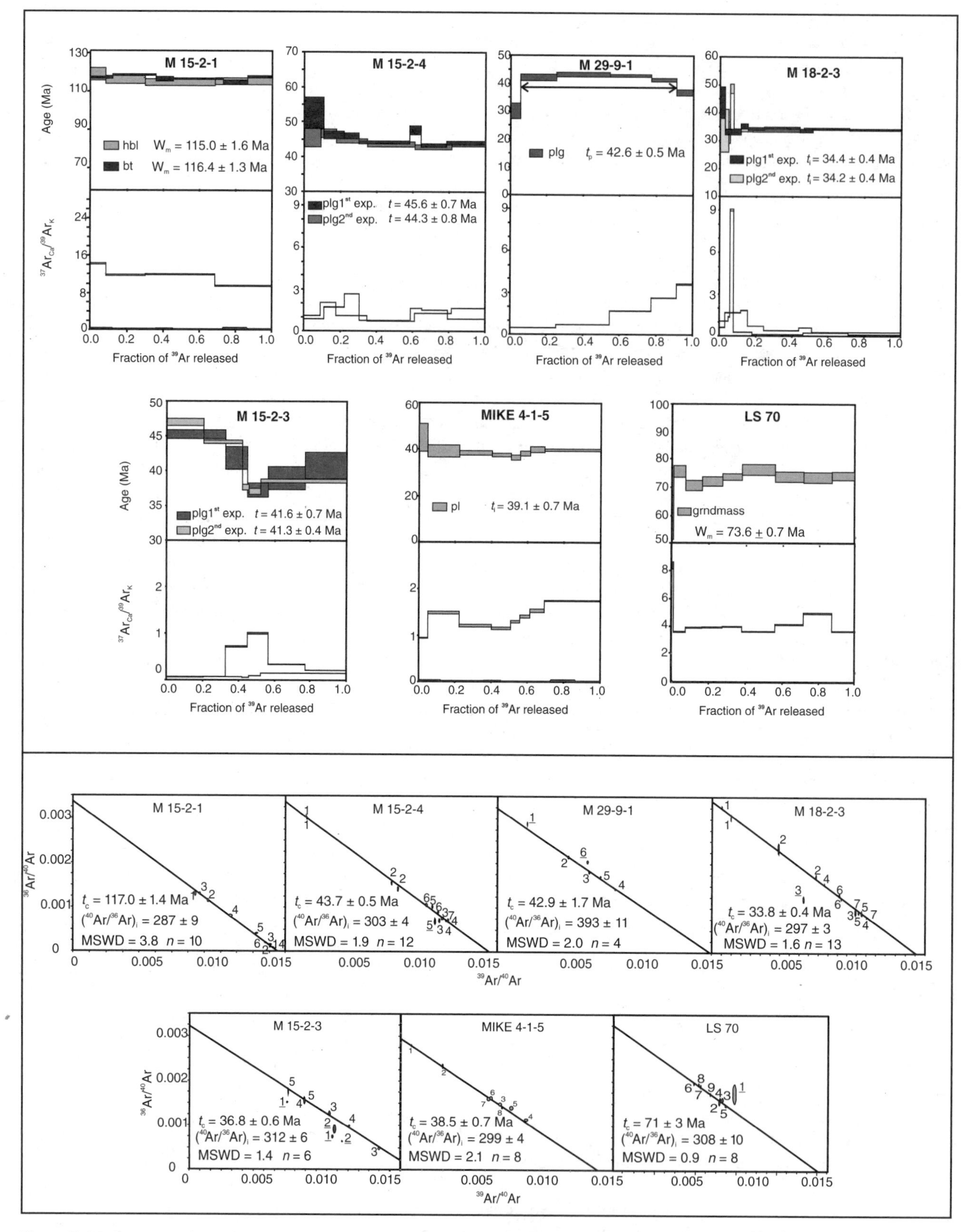

Figure 5. Age spectra and correlation diagrams for samples dated by the $^{40}Ar/^{39}Ar$ method at the CICESE (Centro de Investigación Cientifica y Educación Superior de Ensenada) Geochronologic Laboratory. Sample locations are shown in Figure 2 and 3; coordinates are in Table 1. MSWD—mean square of weighted deviations.

area (Fig. 2), indicating a late Campanian age for the volcanic source (see details in following section). No data are available for the upper part of the Cutzamala Formation, but considering that the youngest age for the folded part of the continental succession is Maastrichtian and that extensive early Eocene intrusive bodies cut these strata, we suggest a Paleocene and/or earliest Eocene age for the upper undeformed part of the Cutzamala Formation.

Magmatic Rocks

The previously described sedimentary successions are intruded by the Placeres del Oro, San Jeronimo–Guayameo, and Montecillos plutons, and they are overlain by Tertiary lavas and ignimbrites. New $^{40}Ar/^{39}Ar$ and U-Pb isotopic ages are presented here for plutonic and volcanic samples that have imprecise previously reported ages or that contrast with geologic observations and for magmatic units that have important stratigraphic and structural relationships. Samples locations are shown in Figures 2 and 3, and listed in Tables 1 and 2.

Plutonic Rocks

The Placeres del Oro pluton crops out in the central portion of the map area, between the villages of Placeres del Oro and Santa Teresa (Fig. 3). It cuts the lower member of the San Lucas Formation, and it is unconformably covered by Eocene lavas and ignimbrites and undeformed postmagmatic continental deposits. The pluton consists of phaneritic, holocrystalline, medium- to coarse-grained granodiorites to diorites, which are mainly composed of plagioclase + quartz + K-feldspar + biotite + hornblende + oxides. Several amphibolite xenoliths, 2–11 cm in size, occur in the granodiorite at Puente del Oro, a few kilometers south of Placeres del Oro (Fig. 3). The pluton appears to be only weakly deformed in thin section, and undulatory extinction in quartz, mechanical twinning in plagioclase, and microkinks in biotite were observed. Reported K/Ar and Rb/Sr ages range from 79 to 115 Ma (Fries, 1962; Salas, 1972; PEMEX, 1987, *in* Lemos-Bustos and Fu-Orozco, 2002; Larsen et al., 1958).

A new $^{40}Ar/^{39}Ar$ age for the Placeres del Oro pluton comes from four one-step and six one-step laser fusions, respectively, on hornblende and biotite multigrain concentrates from sample M15-2-1. Individual ages are statistically undistinguishable, and the weighted mean age from each mineral is shown in Figure 5. To facilitate the comparison of the results of the one-step experiment, pseudo-age spectra are plotted. We consider the $^{40}Ar/^{39}Ar$ age of ca. 115–116 Ma to be the best estimate for the crystallization age of the Placeres del Oro pluton. For comparison, we also obtained U/Pb zircon isotopic ages on sample MIKE 5-1-1, which was collected 15 km SW of sample M15-2-1. Forty-three single grains were analyzed using a laser spot diameter of 35 µm. Thirty-nine yielded concordant ages of 120.2 ± 2.1 Ma, which are interpreted to represent the crystallization age of the zircons. Along with the $^{40}Ar/^{39}Ar$ ages, this age indicates a relatively slow cooling rate for the pluton. Concordant ages of 1109.7 ± 93.2 and 1071.0 ± 28.1 Ma from the core of one zircon provide evidence for an inherited older component assimilated during the Early Cretaceous magmatic event (Fig. 6). Five grains yielding discordant ages that range from 266.1 ± 7.3 Ma to 160.3 ± 12.0 Ma define a well-constrained discordia with intercepts of 120 ± 2 Ma and 1075 ± 58 Ma, which are in good agreement with crystallization and inheritance ages.

The essentially undeformed San Jeronimo–Guayameo pluton, which crops out in the western portion of the map area (Fig. 3), intrudes the unfolded continental red beds of the Cutzamala Formation in the northwestern part of the map area. It is extensively covered by a 1000-m-thick sequence of ignimbrites toward the south. The pluton consists of phaneritic, holocrystalline, coarse- to fine-grained units ranging in composition from diorite to granite. The most common assemblage is plagioclase + quartz + K-feldspar + biotite + hornblende + clinopyroxene + oxides. Pantoja-Alór (1983) reported K/Ar ages of 36 Ma and 47 Ma for the northern and southern part of the San Jeronimo–Guayameo pluton, respectively.

Results from a $^{40}Ar/^{39}Ar$ age determination for sample M 15-2-4 from the southern part of the San Jeronimo–Guayameo pluton, near the village of Guayameo (Fig. 3), are shown in Figure 5. Two step-heating experiments were performed using aliquots of plagioclase concentrates. Six and seven fractions were collected for the first and the second experiment, respectively, and analysis yielded weak reproducible "U-shaped" spectrum

TABLE 1. SUMMARY OF THE $^{40}Ar/^{39}Ar$ AGE RESULTS

Sample	Long. (°W)	Lat. (°N)	Rock type	Material dated	t_i (Ma)	t_p (Ma)	t_c (Ma)
M 15-2-1	305976	2021553	Granodiorite	hbl	115.0 ± 1.6†		
				bt	116.4 ± 1.3†		117.0 ± 1.4 (hbl + bt)
M 29-9-1	320058	2032163	Dacite	plg		42.6 ± 0.5	42.9 ± 1.7
M 15-2-4	265106	2023872	Granodiorite	plg			43.7 ± 0.5
M 15-2-3	262022	2024356	Ignimbrite	plg			36.8 ± 0.6
M 18-2-3	202862	210397	Granodiorite	plg	34.4 ± 0.4 (1st exp.)		33.8 ± 0.4
					34.2 ± 0.4 (2nd exp.)		
MIKE 4-1-5	361081	2014273	Microgranite	plg	39.1 ± 0.7		38.5 ± 0.7
LS 70	302591	2146976	Andesite (clast)	groundmass	73.9 ± 1.1†		71 ± 3

Note: Errors are reported at 1σ level; hbl—hornblende; bt—biotite; plg—plagioclase; t_i—integrated age; t_p—plateau age; t_c—isochron age.
†Weighted mean of one-step fusion experiments.

TABLE 2. U-Pb GEOCHRONOLOGICAL DATA OF INTRUSIVE ROCKS FROM THE HUETAMO-CIUDAD ALTAMIRANO REGION, SOUTHERN MEXICO

			Isotope ratios									Apparent ages (Ma)			
U (ppm)	$^{206}Pb/^{204}Pb$	U/Th	$^{207}Pb^*/^{235}U^*$	± (%)	$^{206}Pb^*/^{238}U$	± (%)	error corr.	$^{206}Pb^*/^{238}U^*$	± (Ma)	$^{207}Pb^*/^{235}U$	± (Ma)	$^{206}Pb^*/^{207}Pb^*$	± (Ma)	Best age (Ma)	± (Ma)
TJP—Tejupilco Quartz-Schist (UTM: 0377346-2094908, h. 1714 [m])															
184	5000	0.8	0.3043	7.2	0.0416	1.8	0.25	263.0	4.7	269.8	17.1	328.9	159.1	263.0	4.7
681	38,080	2.7	2.1860	1.4	0.1963	1.0	0.71	1155.4	10.6	1176.6	9.9	1215.7	19.7	1215.7	19.7
285	14,740	1.1	0.5940	3.5	0.0754	2.6	0.74	468.4	11.7	473.4	13.1	498.1	51.1	468.4	11.7
212	7730	1.1	0.3062	5.3	0.0414	2.2	0.41	261.5	5.6	271.3	12.5	356.6	108.0	261.5	5.6
337	54,205	2.1	4.6414	1.7	0.3182	1.4	0.81	1781.1	21.3	1756.8	14.2	1727.9	18.4	1727.9	18.4
299	9795	2.1	0.3181	3.3	0.0441	1.4	0.41	278.1	3.7	280.5	8.0	300.6	68.2	278.1	3.7
200	12,965	1.2	0.6263	3.2	0.0787	1.9	0.60	488.3	9.0	493.8	12.6	519.0	56.8	488.3	9.0
221	34,850	2.1	1.7335	2.5	0.1725	2.1	0.83	1025.8	19.7	1021.1	16.0	1010.9	27.8	1010.9	27.8
383	35,315	1.5	2.4142	2.4	0.2145	1.3	0.56	1252.7	15.1	1246.8	17.0	1236.7	38.5	1236.7	38.5
244	52,045	4.3	2.2865	3.2	0.2039	2.8	0.86	1196.5	30.6	1208.1	22.9	1229.0	32.2	1229.0	32.2
192	60,815	2.5	3.2760	3.7	0.2534	3.0	0.81	1456.1	39.2	1475.4	28.7	1503.2	40.5	1503.2	40.5
208	16,950	1.2	0.6034	5.3	0.0766	4.0	0.76	475.9	18.4	479.4	20.2	496.1	75.4	475.9	18.4
187	34,755	2.3	2.3249	2.1	0.2050	1.8	0.87	1202.0	20.1	1219.9	14.9	1251.8	20.0	1251.8	20.0
173	8340	1.6	0.2914	10.3	0.0409	2.2	0.21	258.4	5.5	259.6	23.7	271.3	232.5	258.4	5.5
228	31,555	1.3	1.6884	2.0	0.1687	1.4	0.69	1004.8	12.9	1004.2	12.9	1002.8	29.9	1002.8	29.9
381	24,760	1.6	0.7194	2.5	0.0885	2.2	0.90	546.7	11.7	550.3	10.5	565.1	22.9	546.7	11.7
174	21,585	1.3	1.8224	2.0	0.1766	1.2	0.58	1048.1	11.4	1053.6	13.3	1064.8	33.2	1064.8	33.2
141	4980	1.9	0.3065	6.7	0.0411	1.0	0.15	259.8	2.5	271.4	16.1	372.9	150.4	259.8	2.5
216	7990	2.2	0.2848	10.5	0.0399	2.5	0.23	252.5	6.1	254.5	23.6	273.2	234.3	252.5	6.1
250	9870	2.0	0.3003	6.1	0.0410	1.4	0.22	259.3	3.5	266.6	14.4	331.9	135.8	259.3	3.5
182	7025	2.3	0.3050	7.0	0.0427	1.7	0.24	269.5	4.4	270.3	16.5	276.7	154.7	269.5	4.4
414	43,035	3.6	1.4990	2.1	0.1523	1.7	0.81	913.7	14.8	930.0	13.0	968.8	25.3	968.8	25.3
92	3715	3.3	0.3382	12.0	0.0450	3.3	0.27	283.5	9.1	295.8	30.7	394.0	258.6	283.5	9.1
152	5270	1.5	0.3130	9.2	0.0409	2.3	0.25	258.7	5.9	276.5	22.3	430.0	199.4	258.7	5.9
240	10,180	1.7	0.3105	3.4	0.0396	2.2	0.63	250.5	5.3	274.6	8.3	485.5	59.0	250.5	5.3
364	26,940	3.3	0.5977	2.5	0.0765	1.3	0.52	475.0	5.9	475.8	9.4	479.8	46.7	475.0	5.9
212	9165	1.4	0.3047	6.1	0.0422	2.0	0.33	266.5	5.3	270.1	14.5	301.2	131.7	266.5	5.3
104	11,815	2.2	0.6845	7.7	0.0860	2.8	0.37	531.9	14.3	529.5	31.7	519.1	157.2	531.9	14.3
56	1720	5.4	0.2967	18.5	0.0472	1.4	0.08	297.2	4.2	263.8	42.9	–23.0	449.0	297.2	4.2
1003	59,100	17.0	0.5963	1.9	0.0743	1.6	0.85	462.1	7.3	474.9	7.3	537.1	21.9	462.1	7.3
121	9880	1.6	0.3050	5.8	0.0411	2.0	0.34	259.8	5.0	270.3	13.7	361.8	122.4	259.8	5.0
342	23,850	5.1	0.5961	3.8	0.0737	3.5	0.93	458.3	15.5	474.7	14.3	554.8	29.6	458.3	15.5
1351	208,485	19.0	1.5965	3.2	0.1626	3.1	0.95	971.2	27.9	968.8	20.3	963.5	20.4	963.5	20.4
79	28,855	1.0	2.9275	2.5	0.2403	1.5	0.58	1388.3	18.4	1389.1	19.0	1390.2	39.2	1390.2	39.2
176	5290	1.7	0.3713	7.6	0.0448	1.2	0.16	282.7	3.3	320.6	20.9	606.1	162.5	282.7	3.3
132	30,595	2.2	2.2978	3.7	0.2082	3.3	0.88	1219.4	36.8	1211.6	26.5	1197.8	34.5	1197.8	34.5
396	42,285	2.4	0.7195	4.9	0.0875	3.8	0.79	540.9	19.9	550.4	20.7	589.9	65.1	540.9	19.9
876	21,755	4.5	0.3206	2.2	0.0437	1.3	0.60	275.7	3.6	282.3	5.5	337.7	40.4	275.7	3.6
135	7700	1.6	0.6107	4.9	0.0758	2.9	0.60	470.8	13.3	484.0	18.8	547.2	85.6	470.8	13.3
1139	32,050	2.0	0.2922	2.5	0.0407	1.8	0.71	257.4	4.5	260.3	5.7	286.5	40.5	257.4	4.5
301	12,230	3.0	0.3414	4.9	0.0467	3.6	0.73	294.3	10.3	298.2	12.7	329.2	76.1	294.3	10.3
96	4105	1.9	0.2959	16.5	0.0407	2.5	0.15	257.0	6.2	263.2	38.3	319.1	373.2	257.0	6.2
432	62,250	4.6	1.6822	2.6	0.1681	2.3	0.87	1001.7	21.1	1001.8	16.7	1002.2	26.4	1002.2	26.4
823	110,185	4.2	1.7031	3.9	0.1710	3.8	0.97	1017.7	35.4	1009.7	24.9	992.2	20.5	992.2	20.5
196	34,530	1.8	2.1748	2.3	0.1988	1.9	0.80	1168.8	20.0	1173.0	16.2	1180.8	27.3	1180.8	27.3
329	59,800	1.8	1.7956	6.4	0.1734	6.3	0.99	1030.7	59.9	1043.9	41.6	1071.5	20.1	1071.5	20.1
92	11,905	4.0	1.8096	2.8	0.1787	1.4	0.51	1060.1	14.0	1048.9	18.5	1025.8	49.3	1025.8	49.3
255	63,740	2.8	3.5469	2.3	0.2703	2.1	0.90	1542.5	28.3	1537.8	18.1	1531.3	18.8	1531.3	18.8
371	61,000	3.0	1.7877	2.4	0.1782	2.2	0.91	1057.2	21.3	1041.0	15.6	1007.1	20.3	1007.1	20.3
116	13,570	1.3	1.1007	2.8	0.1230	1.0	0.35	748.0	7.1	753.7	15.1	770.6	56.0	748.0	7.1
171	11,120	1.8	0.6470	3.8	0.0815	2.5	0.67	505.0	12.3	506.6	15.1	514.0	62.0	505.0	12.3
173	13,025	1.2	0.6766	3.5	0.0816	1.7	0.49	505.5	8.3	524.7	14.2	609.0	65.1	505.5	8.3
152	5850	1.3	0.2918	10.8	0.0414	1.6	0.15	261.3	4.2	259.9	24.8	247.3	246.7	261.3	4.2
183	9365	1.0	0.6746	4.9	0.0830	1.0	0.20	514.1	4.9	523.5	20.1	564.8	104.7	514.1	4.9

(*Continued*)

TABLE 2. U-Pb GEOCHRONOLOGICAL DATA OF INTRUSIVE ROCKS FROM THE HUETAMO-CIUDAD ALTAMIRANO REGION, SOUTHERN MEXICO (*Continued*)

			Isotope ratios									Apparent ages (Ma)			
U (ppm)	$^{206}Pb/^{204}Pb$	U/Th	$^{207}Pb^*/^{235}U^*$	± (%)	$^{206}Pb^*/^{238}U$	± (%)	error corr.	$^{206}Pb^*/^{238}U^*$	± (Ma)	$^{207}Pb^*/^{235}U$	± (Ma)	$^{206}Pb^*/^{207}Pb^*$	± (Ma)	Best age (Ma)	± (Ma)
TJP—Tejupilco Quartz-Schist (UTM: 0377346-2094908, h. 1714 [m]) (*Continued*)															
398	55,040	2.5	5.4037	2.2	0.3480	1.9	0.85	1925.1	30.8	1885.4	18.7	1842.0	21.0	1842.0	21.0
412	71,845	3.6	2.3374	3.7	0.2109	3.6	0.96	1233.6	40.5	1223.7	26.6	1206.3	19.7	1206.3	19.7
85	4060	1.3	0.6959	7.5	0.0797	4.3	0.57	494.3	20.5	536.3	31.3	719.1	130.7	494.3	20.5
599	57,445	4.2	0.8263	1.7	0.1002	1.2	0.72	615.8	7.1	611.5	7.7	596.0	25.1	615.8	7.1
341	26,425	1.3	0.7669	2.8	0.0960	2.4	0.84	590.8	13.5	578.0	12.5	528.0	33.5	590.8	13.5
133	13,810	1.2	0.6320	4.3	0.0760	2.2	0.52	472.0	10.1	497.3	16.8	615.5	79.3	472.0	10.1
303	21,715	2.5	0.6082	3.0	0.0757	2.2	0.75	470.2	10.1	482.4	11.4	540.7	42.7	470.2	10.1
129	6880	2.7	0.3308	11.5	0.0431	1.5	0.13	272.3	4.1	290.1	29.0	436.3	254.2	272.3	4.1
209	16,055	1.7	0.5951	5.7	0.0757	2.9	0.51	470.3	13.2	474.1	21.6	492.4	108.1	470.3	13.2
132	9295	1.4	1.4665	3.3	0.1473	2.8	0.86	885.8	23.1	916.7	19.7	991.8	34.3	991.8	34.3
97	26,030	1.6	1.9189	3.5	0.1816	1.9	0.56	1075.6	19.0	1087.7	23.1	1111.9	57.4	1111.9	57.4
169	33,340	2.1	1.9540	4.7	0.1872	2.7	0.59	1106.2	27.8	1099.8	31.3	1087.2	75.8	1087.2	75.8
172	17,150	1.3	0.6015	5.5	0.0747	1.8	0.32	464.5	7.9	478.2	20.9	544.2	113.5	464.5	7.9
849	41,920	2.4	0.2904	3.4	0.0414	2.3	0.68	261.2	5.9	258.9	7.7	237.5	57.0	261.2	5.9
131	4935	2.8	0.4184	6.4	0.0533	1.0	0.16	334.8	3.3	354.9	19.1	488.6	139.3	334.8	3.3
74	23,990	2.7	3.2270	2.5	0.2578	1.9	0.77	1478.5	25.4	1463.7	19.3	1442.2	30.1	1442.2	30.1
183	21,030	3.2	1.5857	2.4	0.1607	1.0	0.42	960.5	8.9	964.6	15.0	974.0	44.7	974.0	44.7
129	29,715	2.6	2.3240	2.7	0.2079	1.8	0.68	1217.7	20.4	1219.7	19.2	1223.2	39.1	1223.2	39.1
101	14,235	2.5	1.8287	2.9	0.1792	1.5	0.53	1062.6	15.0	1055.8	18.9	1041.9	49.1	1041.9	49.1
363	20,645	2.5	0.3253	4.7	0.0436	3.6	0.76	275.0	9.7	285.9	11.8	376.3	69.1	275.0	9.7
568	98,140	3.8	0.8394	2.6	0.1015	2.5	0.93	622.9	14.5	618.8	12.3	603.9	21.6	622.9	14.5
41	2575	2.2	0.3207	18.4	0.0404	3.5	0.19	255.5	8.6	282.4	45.3	511.7	399.1	255.5	8.6
94	10,525	1.0	0.7955	7.1	0.0922	5.2	0.73	568.6	28.1	594.3	32.1	693.4	104.6	568.6	28.1
182	8410	2.1	0.3179	8.6	0.0413	4.9	0.56	261.1	12.4	280.3	21.1	443.6	158.4	261.1	12.4
968	52,545	1.3	0.7338	1.7	0.0917	1.4	0.81	565.5	7.5	558.8	7.4	531.2	21.9	565.5	7.5
118	17,085	2.7	1.8682	3.9	0.1818	3.5	0.89	1076.9	34.2	1069.9	25.6	1055.6	35.5	1055.6	35.5
414	13,980	1.1	1.9182	1.8	0.1863	1.2	0.67	1101.5	12.5	1087.4	12.3	1059.3	27.6	1059.3	27.6
69	6525	2.2	1.7444	4.7	0.1752	1.1	0.22	1040.6	10.2	1025.1	30.5	992.2	93.8	992.2	93.8
145	5090	1.2	0.3400	8.3	0.0440	1.3	0.16	277.4	3.6	297.1	21.4	455.4	182.5	277.4	3.6
220	15,865	1.6	0.5691	2.8	0.0760	1.4	0.49	472.4	6.2	457.4	10.2	382.8	54.1	472.4	6.2
97	5090	1.0	0.7656	7.5	0.0873	1.0	0.13	539.3	5.2	577.2	33.2	729.4	158.6	539.3	5.2
223	41,125	2.2	3.2119	1.5	0.2573	1.1	0.72	1476.0	13.9	1460.0	11.2	1436.8	19.1	1436.8	19.1
219	55,135	4.8	2.4356	9.0	0.2116	8.8	0.98	1237.5	99.1	1253.2	64.8	1280.2	35.5	1280.2	35.5
222	9015	1.6	0.3222	6.6	0.0435	1.0	0.15	274.4	2.7	283.6	16.3	360.0	146.9	274.4	2.7
298	14,275	1.1	0.2934	5.0	0.0408	1.3	0.27	258.0	3.4	261.2	11.5	290.4	110.2	258.0	3.4
308	11,815	0.8	0.3070	5.5	0.0420	1.0	0.18	265.2	2.6	271.9	13.1	329.3	122.4	265.2	2.6
65	9265	2.2	2.0231	4.5	0.1904	2.3	0.51	1123.6	23.4	1123.3	30.4	1122.7	77.0	1122.7	77.0
229	7170	1.3	0.3495	9.4	0.0427	2.1	0.22	269.5	5.5	304.4	24.8	581.3	199.8	269.5	5.5
TZT—Tzitzio metasandstone (UTM: 0303231-2144860, h. 818 [m])															
249	36,520	8.0	0.6219	4.9	0.0767	2.4	0.48	476.7	11.0	491.0	19.2	558.5	94.0	476.7	11.0
66	7600	1.4	0.6214	11.0	0.0751	1.0	0.09	466.6	4.5	490.7	42.9	605.1	238.0	466.6	4.5
152	10,015	4.6	0.3575	7.0	0.0454	2.8	0.41	286.4	8.0	310.3	18.7	494.5	141.1	286.4	8.0
177	33,610	5.7	1.6142	3.4	0.1538	2.9	0.86	922.2	25.2	975.7	21.4	1098.2	34.8	1098.2	34.8
288	138,200	3.1	8.5465	2.0	0.3946	1.8	0.87	2144.1	31.9	2290.9	18.3	2424.5	17.0	2424.5	17.0
416	16,935	1.0	0.2973	2.5	0.0406	1.1	0.46	256.5	2.8	264.3	5.7	334.1	49.5	256.5	2.8
217	35,930	5.7	1.8982	1.8	0.1826	1.1	0.58	1081.1	10.6	1080.5	12.2	1079.2	30.0	1079.2	30.0
122	4800	1.5	0.3267	14.2	0.0431	1.0	0.07	271.9	2.7	287.0	35.5	412.3	318.1	271.9	2.7
143	5745	1.5	0.3018	16.5	0.0428	1.4	0.08	270.3	3.6	267.8	38.7	245.8	379.8	270.3	3.6
149	40,720	3.3	1.5919	2.7	0.1603	1.1	0.42	958.3	10.2	967.0	17.1	986.9	50.7	986.9	50.7
399	65,330	3.3	2.2171	1.8	0.2023	1.5	0.83	1187.7	15.8	1186.4	12.4	1184.1	19.8	1184.1	19.8
389	13,345	1.0	0.3008	3.3	0.0429	1.2	0.36	271.0	3.1	267.0	7.6	232.7	70.0	271.0	3.1
196	12,775	3.0	0.6289	3.8	0.0789	1.0	0.27	489.3	4.8	495.4	14.9	523.9	80.3	489.3	4.8
108	3915	1.4	0.3064	16.5	0.0423	1.0	0.06	267.0	2.7	271.4	39.2	309.0	376.4	267.0	2.7

(*Continued*)

TABLE 2. U-Pb GEOCHRONOLOGICAL DATA OF INTRUSIVE ROCKS FROM THE HUETAMO-CIUDAD ALTAMIRANO REGION, SOUTHERN MEXICO (*Continued*)

			Isotope ratios									Apparent ages (Ma)			
U (ppm)	$^{206}Pb/^{204}Pb$	U/Th	$^{207}Pb^{*}/^{235}U^{*}$	± (%)	$^{206}Pb^{*}/^{238}U$	± (%)	error corr.	$^{206}Pb^{*}/^{238}U^{*}$	± (Ma)	$^{207}Pb^{*}/^{235}U$	± (Ma)	$^{206}Pb^{*}/^{207}Pb^{*}$	± (Ma)	Best age (Ma)	± (Ma)
TZT—Tzitzio metasandstone (UTM: 0303231-2144860, h. 818 [m]) (*Continued*)															
725	101,245	3.0	1.6215	1.4	0.1631	1.0	0.71	974.1	9.0	978.6	8.9	988.6	20.3	988.6	20.3
269	67,230	2.6	2.9321	1.8	0.2433	1.5	0.83	1403.8	18.7	1390.2	13.5	1369.4	19.2	1369.4	19.2
61	9380	2.3	1.8928	4.5	0.1838	2.8	0.61	1087.8	27.8	1078.6	30.1	1059.9	72.1	1059.9	72.1
360	64,295	3.1	2.1716	1.5	0.2010	1.2	0.76	1180.8	12.5	1172.0	10.6	1155.8	19.8	1155.8	19.8
508	23,190	2.1	0.3209	6.0	0.0440	1.2	0.20	277.8	3.2	282.6	14.8	322.0	133.2	277.8	3.2
554	111,885	2.9	2.6138	1.9	0.2226	1.6	0.85	1295.6	19.2	1304.5	14.1	1319.2	19.4	1319.2	19.4
74	11,535	1.7	1.9841	3.3	0.1905	2.0	0.62	1124.0	20.9	1110.1	22.3	1083.1	52.2	1083.1	52.2
187	46,275	6.3	2.2334	3.4	0.2018	3.1	0.91	1185.2	33.3	1191.6	23.8	1203.2	27.8	1203.2	27.8
310	14,340	3.3	0.3278	5.3	0.0440	2.2	0.41	277.5	5.9	287.9	13.2	373.2	107.7	277.5	5.9
224	44,985	2.4	2.1980	4.7	0.2033	4.4	0.94	1193.0	48.0	1180.4	32.8	1157.3	32.1	1157.3	32.1
202	39,830	1.7	1.6103	2.5	0.1639	2.3	0.92	978.2	21.1	974.2	15.9	965.4	20.4	965.4	20.4
168	13,910	2.7	2.0703	6.4	0.1740	2.1	0.33	1033.8	20.4	1139.0	44.1	1345.3	117.2	1345.3	117.2
80	6350	1.8	0.3106	17.2	0.0426	4.8	0.28	269.0	12.6	274.6	41.5	322.9	378.5	269.0	12.6
92	24,480	2.9	1.8175	3.2	0.1739	2.2	0.67	1033.5	20.6	1051.8	21.1	1089.9	47.9	1089.9	47.9
132	9325	3.5	0.2959	9.8	0.0415	2.3	0.24	262.4	5.9	263.2	22.7	270.1	218.2	262.4	5.9
532	27,595	1.9	0.3003	3.2	0.0411	1.7	0.54	259.8	4.4	266.6	7.4	326.8	60.4	259.8	4.4
497	57,785	13.2	0.7336	3.0	0.0894	2.4	0.80	551.7	12.6	558.6	12.8	586.9	38.6	551.7	12.6
115	33,040	4.0	1.6542	2.0	0.1672	1.1	0.53	996.8	9.9	991.1	12.7	978.5	34.5	978.5	34.5
571	175,305	2.7	3.3491	1.8	0.2616	1.5	0.82	1497.9	19.4	1492.6	13.8	1485.0	18.9	1485.0	18.9
102	19,895	3.3	1.6716	3.6	0.1645	3.0	0.83	981.7	27.5	997.8	23.2	1033.3	41.3	1033.3	41.3
548	124,110	4.7	2.0884	3.2	0.1901	3.1	0.95	1121.8	31.6	1145.0	22.2	1189.2	19.7	1189.2	19.7
48	12,655	2.1	1.8503	4.3	0.1766	1.5	0.35	1048.6	14.8	1063.5	28.4	1094.2	80.8	1094.2	80.8
250	67,720	4.2	2.1696	2.4	0.1999	2.2	0.91	1174.5	23.5	1171.4	16.8	1165.5	20.2	1165.5	20.2
250	18,175	1.8	0.6111	4.0	0.0783	2.8	0.70	486.1	13.2	484.2	15.4	475.3	63.3	486.1	13.2
607	38,315	15.8	1.6902	1.8	0.1667	1.4	0.81	994.2	13.3	1004.9	11.3	1028.2	21.0	1028.2	21.0
289	47,485	4.6	1.8942	1.6	0.1817	1.0	0.61	1076.1	9.9	1079.0	10.9	1085.0	26.1	1085.0	26.1
131	31,555	2.8	1.7683	2.8	0.1741	2.3	0.81	1034.6	21.9	1033.9	18.4	1032.3	33.6	1032.3	33.6
264	22,505	1.8	0.6211	3.7	0.0790	1.5	0.39	490.4	6.9	490.5	14.6	491.0	76.0	490.4	6.9
923	23,075	1.7	0.3157	9.1	0.0411	1.0	0.11	259.5	2.5	278.6	22.1	442.6	200.8	259.5	2.5
145	10,555	5.5	0.3781	7.8	0.0473	1.3	0.17	297.9	3.8	325.6	21.6	528.9	167.7	297.9	3.8
168	10,125	1.6	0.3248	8.3	0.0430	2.3	0.28	271.5	6.2	285.6	20.7	402.0	179.3	271.5	6.2
535	29,920	2.0	0.2826	3.5	0.0395	2.0	0.55	249.8	4.8	252.7	7.9	280.3	67.1	249.8	4.8
389	23,975	2.0	0.5831	3.3	0.0730	2.7	0.81	454.3	11.8	466.4	12.4	526.4	42.3	454.3	11.8
250	26,330	1.9	0.6134	2.1	0.0773	1.1	0.54	480.1	5.2	485.7	7.9	512.1	37.9	480.1	5.2
199	10,305	2.0	0.2939	8.7	0.0405	5.6	0.65	256.1	14.1	261.6	20.0	311.0	150.6	256.1	14.1
183	12,040	1.9	0.3021	3.4	0.0415	2.3	0.67	262.3	5.9	268.0	8.0	317.9	57.1	262.3	5.9
495	25,220	2.3	0.3370	2.9	0.0456	2.3	0.78	287.4	6.3	294.9	7.3	355.0	40.2	287.4	6.3
237	25,505	9.1	0.5991	4.2	0.0755	3.4	0.81	469.0	15.4	476.7	15.9	513.7	53.7	469.0	15.4
85	22,335	2.6	1.7770	3.9	0.1704	2.7	0.70	1014.1	25.5	1037.1	25.1	1085.9	55.0	1085.9	55.0
108	42,020	1.0	5.4976	1.8	0.3472	1.5	0.83	1921.2	24.6	1900.2	15.3	1877.3	18.0	1877.3	18.0
103	28,570	3.3	1.8782	3.1	0.1786	1.3	0.43	1059.3	12.9	1073.4	20.5	1102.2	56.0	1102.2	56.0
193	11,520	2.2	0.3056	6.0	0.0396	2.5	0.41	250.1	6.0	270.8	14.2	453.4	121.1	250.1	6.0
281	14,840	2.2	0.6198	5.4	0.0733	2.0	0.37	455.9	8.8	489.7	21.1	651.5	108.6	455.9	8.8
99	27,640	2.4	1.6722	3.9	0.1647	1.8	0.47	982.7	16.7	998.0	24.7	1031.9	69.4	1031.9	69.4
503	41,765	2.2	0.3708	3.1	0.0494	2.1	0.68	310.8	6.4	320.3	8.5	389.8	51.2	310.8	6.4
189	9540	1.6	0.2994	7.5	0.0403	1.8	0.24	254.9	4.5	265.9	17.5	364.6	164.1	254.9	4.5
252	48,390	3.5	1.7408	1.5	0.1733	1.0	0.68	1030.2	9.6	1023.8	9.5	1010.0	21.9	1010.0	21.9
146	49,080	1.7	4.2397	2.2	0.2993	1.7	0.76	1687.9	24.8	1681.7	18.1	1674.0	26.4	1674.0	26.4
300	49,965	3.1	1.8537	2.8	0.1804	2.1	0.73	1069.4	20.2	1064.7	18.5	1055.2	38.5	1055.2	38.5
242	30,725	3.6	0.8728	2.4	0.1041	1.0	0.42	638.5	6.1	637.1	11.3	632.0	46.8	638.5	6.1
268	41,365	1.7	1.0374	2.8	0.1184	2.2	0.79	721.5	15.1	722.6	14.4	726.2	36.1	721.5	15.1
232	15,535	2.0	0.5794	4.3	0.0704	2.4	0.55	438.4	10.0	464.1	16.1	593.2	78.3	438.4	10.0
203	17,750	1.0	0.6265	3.9	0.0776	1.2	0.30	481.6	5.4	493.9	15.1	551.4	80.2	481.6	5.4
234	21,170	1.4	0.6082	2.7	0.0770	1.0	0.37	478.4	4.6	482.4	10.4	501.7	55.5	478.4	4.6

(*Continued*)

TABLE 2. U-Pb GEOCHRONOLOGICAL DATA OF INTRUSIVE ROCKS FROM THE HUETAMO-CIUDAD ALTAMIRANO REGION, SOUTHERN MEXICO (*Continued*)

			Isotope ratios									Apparent ages (Ma)			
U (ppm)	$^{206}Pb/^{204}Pb$	U/Th	$^{207}Pb^{*}/^{235}U^{*}$	± (%)	$^{206}Pb^{*}/^{238}U$	± (%)	error corr.	$^{206}Pb^{*}/^{238}U^{*}$	± (Ma)	$^{207}Pb^{*}/^{235}U$	± (Ma)	$^{206}Pb^{*}/^{207}Pb^{*}$	± (Ma)	Best age (Ma)	± (Ma)
TZT—Tzitzio metasandstone (UTM: 0303231-2144860, h. 818 [m]) (*Continued*)															
274	15,205	2.7	0.5589	19.9	0.0669	4.1	0.20	417.6	16.5	450.8	72.5	624.0	423.6	417.6	16.5
122	7080	4.5	0.3341	11.7	0.0466	1.0	0.09	293.7	2.9	292.7	29.8	284.9	267.6	293.7	2.9
1096	9715	11.5	0.3238	4.7	0.0400	2.9	0.63	253.0	7.3	284.8	11.6	555.1	79.0	253.0	7.3
429	154,070	3.9	6.2269	4.7	0.3668	4.6	0.98	2014.4	78.9	2008.2	40.9	2001.9	17.8	2001.9	17.8
241	12,330	2.0	0.2949	7.1	0.0410	3.5	0.50	259.1	9.0	262.4	16.3	292.0	139.9	259.1	9.0
379	14,900	2.1	0.3147	2.9	0.0428	1.5	0.52	270.4	4.0	277.8	7.1	340.0	56.3	270.4	4.0
386	86,320	3.3	2.9797	1.6	0.2459	1.2	0.71	1417.1	14.6	1402.4	12.3	1380.2	21.9	1380.2	21.9
237	14,255	2.9	0.3406	5.5	0.0465	2.1	0.38	293.2	6.1	297.6	14.3	332.1	116.2	293.2	6.1
562	50,625	0.5	0.6070	3.2	0.0777	3.0	0.95	482.1	13.9	481.6	12.1	479.4	22.3	482.1	13.9
123	13,410	2.2	0.6591	4.9	0.0787	2.4	0.48	488.5	11.1	514.0	19.7	629.1	92.1	488.5	11.1
259	11,090	1.3	0.3186	5.2	0.0430	1.0	0.19	271.3	2.7	280.8	12.8	360.9	115.4	271.3	2.7
257	61,175	14.9	2.7632	3.0	0.2363	1.1	0.38	1367.5	13.9	1345.7	22.2	1311.1	53.4	1311.1	53.4
434	45,825	1.9	0.8446	1.7	0.1021	1.0	0.60	626.5	6.0	621.7	7.8	604.1	29.0	626.5	6.0
115	21,945	1.0	1.7468	3.6	0.1740	2.8	0.76	1034.3	26.3	1026.0	23.5	1008.2	48.1	1008.2	48.1
163	7945	2.0	0.2828	7.8	0.0407	2.5	0.32	257.1	6.2	252.8	17.6	213.9	172.6	257.1	6.2
382	114,555	12.8	3.6176	1.6	0.2773	1.2	0.78	1578.0	17.4	1553.4	12.7	1520.2	18.9	1520.2	18.9
297	70,360	1.8	2.6467	1.9	0.2286	1.6	0.85	1327.3	19.2	1313.7	13.9	1291.7	19.5	1291.7	19.5
496	21,665	2.7	0.3219	2.5	0.0447	1.2	0.48	281.9	3.4	283.3	6.3	295.5	50.7	281.9	3.4
110	25,100	4.2	2.3836	1.8	0.2141	1.0	0.58	1250.9	11.7	1237.7	12.8	1214.8	28.8	1214.8	28.8
259	11,845	1.2	0.2957	5.3	0.0406	3.4	0.64	256.6	8.6	263.0	12.4	321.2	92.9	256.6	8.6
173	49,300	1.0	4.6051	1.8	0.3148	1.2	0.67	1764.2	18.7	1750.2	15.0	1733.5	24.4	1733.5	24.4
174	16,050	3.1	0.8317	5.1	0.0953	2.3	0.44	586.8	12.7	614.6	23.4	718.2	96.7	586.8	12.7
201	71,430	7.9	1.9158	2.9	0.1804	2.0	0.69	1069.2	19.9	1086.6	19.4	1121.7	41.9	1121.7	41.9
216	16,755	4.3	0.3356	5.3	0.0448	2.0	0.37	282.4	5.4	293.9	13.6	385.6	111.0	282.4	5.4
158	100,710	3.3	6.7762	2.0	0.3797	1.7	0.86	2075.1	30.0	2082.6	17.4	2090.1	17.6	2090.1	17.6
MIKE 5-1-1—Placeres del Oro Granodiorite (UTM: 0300135-2014990, h. 394 [m])															
42	453	2.8	0.17563	16.4	0.01758	7.1	0.43	112.4	7.9	164.3	24.9	998.7	302.1	112.4	7.9
55	740	1.6	0.19943	17.4	0.01876	6.8	0.39	119.8	8.1	184.6	29.4	1123.4	321.3	119.8	8.1
75	579	1.8	0.18194	9.3	0.01849	3.7	0.40	118.1	4.4	169.7	14.5	968.2	173.9	118.1	4.4
142	1713	2.1	0.15119	5.2	0.01883	2.3	0.44	120.2	2.7	143.0	6.9	538.8	101.7	120.2	2.7
67	1943	1.2	0.36579	5.6	0.04214	2.8	0.50	266.1	7.3	316.5	15.2	706.5	102.9	266.1	7.3
36	334	2.4	0.24562	13.1	0.01782	8.5	0.65	113.9	9.6	223.0	26.3	1623.5	187.3	113.9	9.6
109	9681	1.1	2.08238	5.0	0.19727	1.7	0.35	1160.6	18.2	1143.0	34.1	1109.7	93.2	1109.7	93.2
38	459	2.8	0.22681	18.9	0.01870	8.7	0.46	119.4	10.3	207.6	35.4	1381.9	324.0	119.4	10.3
83	1123	2.2	0.17601	15.2	0.01912	3.2	0.21	122.1	3.8	164.6	23.1	830.4	311.7	122.1	3.8
49	425	2.1	0.19439	18.2	0.01845	4.0	0.22	117.8	4.6	180.4	30.0	1106.1	357.2	117.8	4.6
36	375	2.5	0.20144	19.2	0.01940	9.0	0.47	123.9	11.1	186.3	32.7	1076.8	342.8	123.9	11.1
47	594	2.2	0.18264	13.1	0.01835	8.9	0.68	117.2	10.3	170.3	20.5	990.7	195.6	117.2	10.3
54	362	2.2	0.13583	14.9	0.01819	8.0	0.54	116.2	9.2	129.3	18.1	377.1	283.5	116.2	9.2
143	1700	2.0	0.27469	9.6	0.03447	9.1	0.94	218.5	19.5	246.4	21.0	521.9	69.4	218.5	19.5
86	807	2.2	0.16333	9.0	0.01836	3.5	0.39	117.3	4.0	153.6	12.8	758.6	175.4	117.3	4.0
64	1054	2.0	0.16701	10.1	0.01870	4.0	0.40	119.4	4.8	156.8	14.6	767.0	194.8	119.4	4.8
133	25,449	3.6	1.67624	7.6	0.16189	7.5	0.98	967.3	67.3	999.6	48.5	1071.0	28.1	1071.0	28.1
81	1361	2.6	0.22969	8.2	0.01934	4.3	0.52	123.5	5.2	209.9	15.6	1340.7	136.4	123.5	5.2
40	490	2.3	0.25879	31.1	0.02016	8.4	0.27	128.7	10.7	233.7	65.1	1489.7	580.9	128.7	10.7
63	599	2.4	0.12059	8.8	0.01898	4.8	0.55	121.2	5.8	115.6	9.6	1.9	177.7	121.2	5.8
67	639	2.5	0.14633	10.9	0.01878	4.8	0.44	120.0	5.8	138.7	14.1	472.3	216.2	120.0	5.8
142	1287	2.0	0.12111	9.1	0.01884	1.9	0.21	120.3	2.2	116.1	10.0	30.4	213.4	120.3	2.2
69	728	2.7	0.17110	13.6	0.01936	5.7	0.42	123.6	7.0	160.4	20.1	745.3	260.9	123.6	7.0
68	678	1.8	0.15048	11.5	0.01870	3.9	0.34	119.4	4.7	142.3	15.3	543.3	237.8	119.4	4.7
48	443	2.9	0.19892	20.1	0.01741	4.7	0.23	111.3	5.2	184.2	33.8	1265.8	384.8	111.3	5.2
158	2099	1.0	0.15270	10.7	0.01774	2.8	0.26	113.3	3.1	144.3	14.4	689.2	220.5	113.3	3.1
60	524	2.0	0.16530	20.2	0.01765	8.5	0.42	112.8	9.5	155.3	29.1	866.1	383.5	112.8	9.5

(*Continued*)

TABLE 2. U-Pb GEOCHRONOLOGICAL DATA OF INTRUSIVE ROCKS FROM THE HUETAMO-CIUDAD ALTAMIRANO REGION, SOUTHERN MEXICO (*Continued*)

			Isotope ratios										Apparent ages (Ma)			
U (ppm)	$^{206}Pb/^{204}Pb$	U/Th	$^{207}Pb^*/^{235}U^*$	± (%)	$^{206}Pb^*/^{238}U$	± (%)	error corr.	$^{206}Pb^*/^{238}U^*$	± (Ma)	$^{207}Pb^*/^{235}U$	± (Ma)	$^{206}Pb^*/^{207}Pb^*$	± (Ma)	Best age (Ma)	± (Ma)
MIKE 5-1-1—Placeres del Oro Granodiorite (UTM: 0300135-2014990, h. 394 [m]) (*Continued*)															
78	732	1.9	0.13504	13.1	0.01826	4.2	0.32	116.7	4.8	128.6	15.8	355.3	281.9	116.7	4.8
59	544	2.1	0.14745	8.8	0.01845	2.8	0.32	117.9	3.3	139.7	11.4	528.3	182.2	117.9	3.3
32	386	2.9	0.12540	22.0	0.02016	5.5	0.25	128.7	7.0	120.0	24.9	–49.8	522.4	128.7	7.0
48	1299	2.1	0.22564	14.2	0.01813	6.5	0.46	115.8	7.5	206.6	26.6	1431.5	241.6	115.8	7.5
74	937	2.3	0.16827	9.4	0.01861	4.9	0.52	118.9	5.7	157.9	13.8	792.5	169.1	118.9	5.7
68	1288	3.1	0.14896	9.5	0.01841	5.2	0.55	117.6	6.1	141.0	12.5	555.0	174.1	117.6	6.1
90	1882	2.3	0.23493	4.9	0.02918	2.4	0.50	185.4	4.5	214.3	9.5	544.1	92.7	185.4	4.5
62	847	2.4	0.18405	12.9	0.01791	5.4	0.42	114.5	6.1	171.5	20.4	1055.5	237.3	114.5	6.1
298	3400	1.5	0.19214	7.9	0.02518	7.6	0.96	160.3	12.0	178.5	12.9	425.8	49.9	160.3	12.0
62	1113	2.4	0.16448	11.3	0.01858	5.1	0.45	118.7	5.9	154.6	16.2	748.6	214.3	118.7	5.9
69	967	2.7	0.15554	15.9	0.01885	8.7	0.55	120.4	10.4	146.8	21.7	597.9	288.3	120.4	10.4
64	757	2.1	0.14824	17.6	0.01894	9.7	0.55	120.9	11.6	140.4	23.1	482.4	326.1	120.9	11.6
71	720	1.7	0.16413	20.6	0.01819	6.5	0.31	116.2	7.5	154.3	29.5	788.6	414.2	116.2	7.5
40	698	3.0	0.23663	16.4	0.01930	3.6	0.22	123.3	4.4	215.7	31.8	1402.0	308.0	123.3	4.4
58	659	1.6	0.15992	12.8	0.01950	2.3	0.18	124.5	2.9	150.6	17.9	584.2	274.0	124.5	2.9
37	488	2.6	0.17773	9.4	0.01864	6.4	0.68	119.1	7.5	166.1	14.5	903.1	143.4	119.1	7.5
31	508	2.1	0.21786	19.6	0.01863	8.8	0.45	119.0	10.4	200.1	35.6	1311.3	342.2	119.0	10.4
37	594	2.2	0.21273	16.0	0.01851	6.6	0.41	118.2	7.7	195.8	28.5	1277.6	286.0	118.2	7.7
70	858	2.1	0.18084	7.2	0.01923	2.4	0.34	122.8	2.9	168.8	11.2	874.8	141.2	122.8	2.9
138	2119	1.5	0.15066	9.5	0.01832	3.3	0.34	117.1	3.8	142.5	12.7	590.3	194.2	117.1	3.8
106	1075	2.8	0.16671	11.1	0.01900	3.6	0.33	121.3	4.4	156.6	16.2	729.9	223.6	121.3	4.4
MIKE 14-1-2—San Jeronimo Granodiorite (UTM: 0254399-2051570, h. 198 [m])															
218	547	1.4	0.04932	35.2	0.00616	4.1	0.12	39.6	1.6	48.9	16.8	531.6	788.2	39.6	1.6
581	1995	1.1	0.04461	5.8	0.00609	3.5	0.60	39.1	1.3	44.3	2.5	336.0	105.3	39.1	1.3
647	1316	1.2	0.04311	8.3	0.00623	3.2	0.39	40.0	1.3	42.9	3.5	205.3	176.7	40.0	1.3
1990	2951	0.8	0.04047	4.7	0.00609	2.7	0.57	39.1	1.0	40.3	1.9	109.6	91.3	39.1	1.0
432	1292	1.0	0.04308	9.1	0.00603	2.5	0.27	38.7	1.0	42.8	3.8	278.2	200.6	38.7	1.0
153	385	1.5	0.06235	14.8	0.00616	4.4	0.30	39.6	1.7	61.4	8.8	1025.4	286.7	39.6	1.7
1026	1586	1.0	0.03940	5.7	0.00609	1.7	0.31	39.1	0.7	39.2	2.2	46.8	128.7	39.1	0.7
480	913	1.4	0.05034	6.4	0.00634	2.6	0.41	40.8	1.1	49.9	3.1	512.9	127.4	40.8	1.1
495	1366	1.0	0.04345	8.2	0.00610	2.5	0.31	39.2	1.0	43.2	3.4	271.3	178.0	39.2	1.0
633	4218	1.8	0.05160	4.5	0.00634	2.7	0.60	40.7	1.1	51.1	2.2	568.3	78.7	40.7	1.1
1268	4541	1.1	0.03971	6.8	0.00629	2.8	0.41	40.5	1.1	39.5	2.6	–15.1	150.7	40.5	1.1
498	446	1.0	0.04548	26.6	0.00631	3.0	0.11	40.5	1.2	45.2	11.7	298.4	612.0	40.5	1.2
1741	4640	0.8	0.04138	3.3	0.00613	1.8	0.56	39.4	0.7	41.2	1.3	144.9	64.6	39.4	0.7
450	1431	1.2	0.04814	7.2	0.00616	3.4	0.47	39.6	1.3	47.7	3.4	478.3	140.4	39.6	1.3
226	1213	1.1	0.06193	10.6	0.00612	2.8	0.26	39.4	1.1	61.0	6.3	1023.6	206.6	39.4	1.1
562	363	1.1	0.04045	28.5	0.00607	4.6	0.16	39.0	1.8	40.3	11.2	113.6	675.0	39.0	1.8
304	1399	1.6	0.05765	9.4	0.00633	4.1	0.44	40.7	1.7	56.9	5.2	808.3	176.8	40.7	1.7
442	1266	1.1	0.05176	11.3	0.00621	2.9	0.25	39.9	1.1	51.2	5.6	617.9	236.7	39.9	1.1
456	285	1.5	0.06423	37.4	0.00734	8.1	0.22	47.2	3.8	63.2	22.9	723.2	801.4	47.2	3.8
498	329	1.2	0.05483	9.6	0.00624	3.2	0.33	40.1	1.3	54.2	5.1	732.7	191.8	40.1	1.3
475	665	1.9	0.05486	5.4	0.00715	2.4	0.45	45.9	1.1	54.2	2.8	437.7	107.2	45.9	1.1
210	612	1.2	0.06879	12.3	0.00684	3.0	0.24	43.9	1.3	67.6	8.0	1012.4	242.1	43.9	1.3
534	1256	1.5	0.05122	6.3	0.00632	2.1	0.33	40.6	0.8	50.7	3.1	559.8	129.0	40.6	0.8
571	359	1.2	0.05112	14.5	0.00629	2.7	0.19	40.4	1.1	50.6	7.2	564.0	312.5	40.4	1.1
422	1807	1.1	0.04847	7.9	0.00589	5.7	0.72	37.9	2.1	48.1	3.7	592.2	117.6	37.9	2.1
1772	7466	0.9	0.04186	3.2	0.00621	2.1	0.66	39.9	0.8	41.6	1.3	142.4	56.6	39.9	0.8
1793	14,064	1.5	0.04362	5.3	0.00614	2.4	0.46	39.4	1.0	43.4	2.2	265.2	108.0	39.4	1.0
406	2636	1.4	0.05492	5.0	0.00614	2.5	0.49	39.5	1.0	54.3	2.7	769.7	92.3	39.5	1.0
452	1479	0.8	0.04976	7.0	0.00592	2.4	0.34	38.0	0.9	49.3	3.4	639.5	142.6	38.0	0.9
275	1037	1.6	0.07156	10.2	0.00645	3.8	0.38	41.4	1.6	70.2	6.9	1209.8	186.5	41.4	1.6

(*Continued*)

TABLE 2. U-Pb GEOCHRONOLOGICAL DATA OF INTRUSIVE ROCKS FROM THE HUETAMO-CIUDAD ALTAMIRANO REGION, SOUTHERN MEXICO (*Continued*)

			Isotope ratios								Apparent ages (Ma)			
U (ppm)	$^{206}Pb/^{204}Pb$	U/Th	$^{207}Pb^*/^{235}U^*$	± (%)	$^{206}Pb^*/^{238}U$	± (%)	error corr.	$^{206}Pb^*/^{238}U^*$	± (Ma)	$^{207}Pb^*/^{235}U$	± (Ma)	$^{206}Pb^*/^{207}Pb^*$	± (Ma)	Best age (Ma)	± (Ma)
TNG—Tingambato Granodiorite (UTM: 0350049-2123559, h. 1000 [m]) (*Continued*)															
311	6385	1.4	0.1435	9.0	0.0206	4.8	0.53	131.7	6.2	136.1	11.5	213.8	177.1	131.7	6.2
246	3895	1.1	0.1322	10.6	0.0198	1.9	0.17	126.5	2.3	126.1	12.6	118.7	247.6	126.5	2.3
335	6515	0.7	0.1412	7.1	0.0200	4.0	0.57	127.5	5.0	134.1	8.9	252.3	134.1	127.5	5.0
484	9975	0.7	0.1235	6.6	0.0193	2.3	0.34	123.3	2.8	118.2	7.4	16.8	149.0	123.3	2.8
140	3205	1.8	0.1674	18.4	0.0212	5.5	0.30	135.0	7.3	157.2	26.8	506.0	389.3	135.0	7.3
83	2145	1.5	0.1846	23.7	0.0202	6.6	0.28	129.2	8.5	172.0	37.5	811.2	481.4	129.2	8.5
211	5605	1.1	0.1494	10.3	0.0204	4.6	0.45	130.1	5.9	141.4	13.6	336.0	208.9	130.1	5.9
378	8355	3.7	0.1453	8.9	0.0199	6.9	0.78	126.8	8.6	137.8	11.4	331.2	126.8	126.8	8.6
228	4485	1.8	0.1432	8.9	0.0197	3.4	0.38	126.0	4.3	135.9	11.4	311.7	188.2	126.0	4.3
299	5965	1.4	0.1390	8.5	0.0195	3.5	0.40	124.3	4.2	132.2	10.6	277.0	179.2	124.3	4.2
164	3365	1.4	0.1552	15.5	0.0212	2.1	0.13	135.2	2.7	146.5	21.2	334.2	351.2	135.2	2.7
132	3165	1.3	0.1383	18.7	0.0202	1.7	0.09	129.0	2.2	131.5	23.1	177.6	437.3	129.0	2.2
171	3695	1.2	0.1184	14.2	0.0208	1.6	0.11	132.7	2.0	113.6	15.3	–269.9	360.1	132.7	2.0
375	8655	1.0	0.1427	6.7	0.0194	2.0	0.31	123.9	2.5	135.4	8.5	342.7	144.1	123.9	2.5
83	1385	1.5	0.1670	26.3	0.0202	3.3	0.13	128.9	4.2	156.8	38.2	602.9	573.7	128.9	4.2
186	4255	1.3	0.1239	15.2	0.0204	2.5	0.16	129.9	3.2	118.6	17.1	–103.0	371.6	129.9	3.2
212	4545	1.5	0.1505	11.2	0.0212	2.1	0.19	135.1	2.8	142.4	14.9	265.0	253.1	135.1	2.8
145	3100	1.3	0.1476	15.1	0.0207	4.1	0.27	132.4	5.4	139.8	19.7	267.2	334.1	132.4	5.4
545	11,270	1.1	0.1390	3.2	0.0205	1.0	0.32	131.1	1.3	132.1	3.9	150.1	70.2	131.1	1.3
253	5400	1.5	0.1431	11.9	0.0199	4.1	0.34	126.8	5.1	135.8	15.1	296.1	255.5	126.8	5.1
265	5665	1.0	0.1396	12.8	0.0196	1.8	0.14	124.8	2.2	132.7	15.9	276.2	290.3	124.8	2.2
220	5510	1.1	0.1511	10.3	0.0207	3.9	0.37	132.0	5.0	142.9	13.8	326.6	217.9	132.0	5.0
102	2360	1.4	0.1459	27.2	0.0195	2.0	0.07	124.2	2.4	138.3	35.1	386.7	618.8	124.2	2.4
466	8775	0.9	0.1483	5.9	0.0198	1.3	0.21	126.5	1.6	140.4	7.7	381.2	129.9	126.5	1.6
97	2970	2.1	0.1593	29.7	0.0206	1.0	0.03	131.2	1.3	150.1	41.5	461.4	671.2	131.2	1.3
351	8155	1.1	0.1321	7.3	0.0196	2.4	0.33	125.4	3.0	126.0	8.7	136.4	162.6	125.4	3.0
155	3835	1.1	0.1395	19.9	0.0207	1.4	0.07	132.1	1.8	132.6	24.7	143.1	469.7	132.1	1.8
208	4325	1.2	0.1270	11.1	0.0201	1.7	0.15	128.4	2.1	121.4	12.6	–14.0	264.8	128.4	2.1
140	2525	1.6	0.1472	22.7	0.0204	2.3	0.10	130.0	3.0	139.4	29.6	302.4	521.2	130.0	3.0
257	5095	1.2	0.1476	10.3	0.0201	3.1	0.30	128.3	3.9	139.8	13.5	339.8	223.3	128.3	3.9
250	5435	1.0	0.1489	6.1	0.0213	1.3	0.21	135.6	1.7	141.0	8.0	233.2	136.9	135.6	1.7
428	7400	0.8	0.1277	7.9	0.0196	1.2	0.15	125.0	1.4	122.0	9.1	65.1	185.7	125.0	1.4
220	4965	1.0	0.1415	8.6	0.0208	2.0	0.23	132.7	2.6	134.4	10.8	164.2	195.8	132.7	2.6
350	5685	0.9	0.1569	5.0	0.0207	1.0	0.20	132.3	1.3	148.0	6.8	408.5	108.8	132.3	1.3
50	1370	1.3	0.1674	34.9	0.0207	1.0	0.03	131.8	1.3	157.1	50.8	558.8	781.7	131.8	1.3
93	2160	2.1	0.1502	22.1	0.0198	2.4	0.11	126.7	3.0	142.1	29.3	408.2	496.1	126.7	3.0
685	5830	0.9	0.1515	4.9	0.0193	2.5	0.52	123.1	3.1	143.2	6.5	491.7	91.3	123.1	3.1
61	1450	1.3	0.1510	26.1	0.0205	1.5	0.06	130.9	2.0	142.8	34.7	346.5	597.3	130.9	2.0
96	2470	1.1	0.1429	19.2	0.0203	2.3	0.12	129.6	3.0	135.6	24.4	241.5	443.6	129.6	3.0

Note: Analyses were by laser-ablation multicollector inductively coupled plasma (ICP) mass spectrometry at Laser Chron Laboratory, Department of Geoscience, University of Arizona. All uncertainties are reported at the 1σ level and include only measurement errors. Systematic errors would increase age uncertainties by 1%–2%. U concentration and U/Th were calibrated relative to NIST SRM 610 and are accurate to ~20%. Common Pb correction is from ^{204}Pb, with composition interpreted from Stacey and Kramers (1975). Uncertainties are 1.0 for $^{206}Pb/^{204}Pb$, 0.3 for $^{207}Pb/^{204}Pb$, and 2.0 for $^{208}Pb/^{204}Pb$. U/Pb and $^{206}Pb/^{207}Pb$ fractionation was calibrated relative to fragments of a large Sri Lanka zircon of 564 ± 4 Ma (2σ). U decay constants and composition are as follows: $^{238}U = 9.8485 \times 10^{-10}$, $^{235}U = 1.55125 \times 10^{-10}$, $^{238}U/^{235}U = 137.88$.

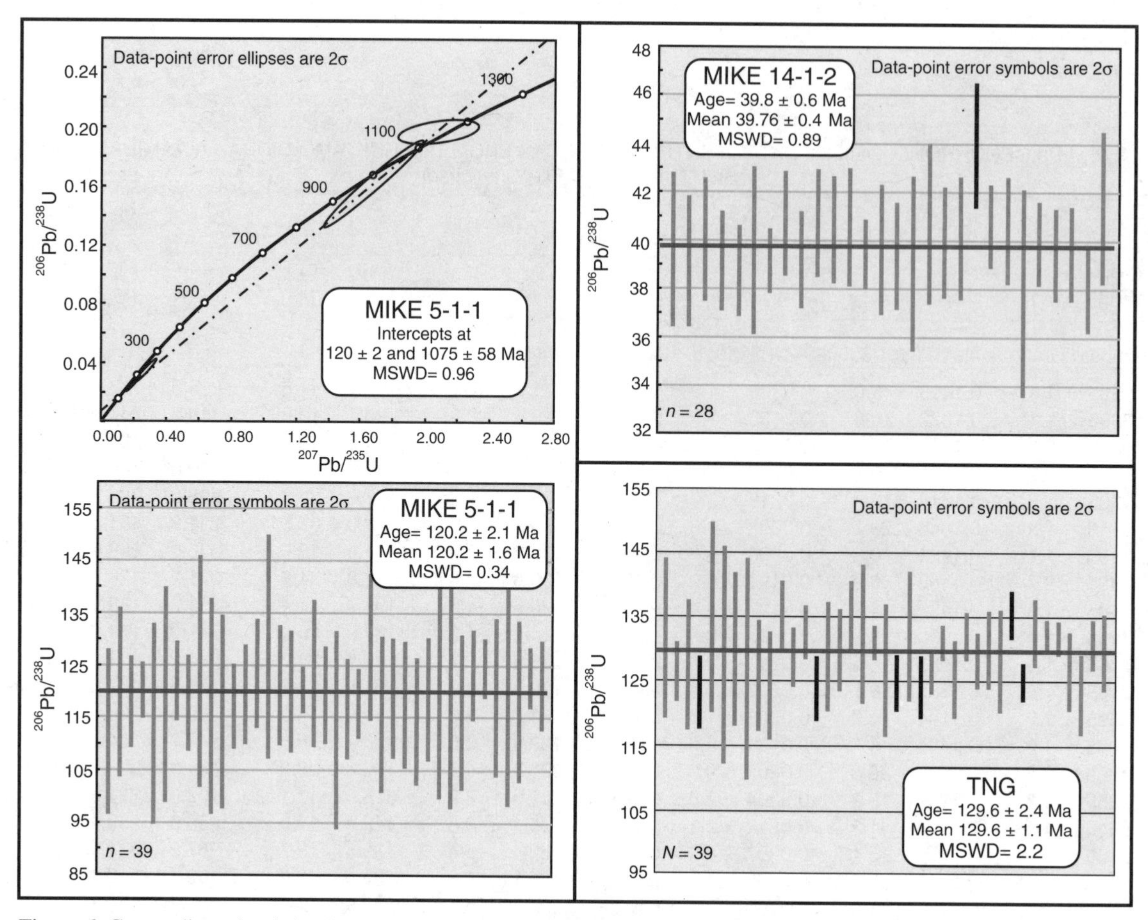

Figure 6. Concordia and weighted mean diagrams for sample dated with the U-Pb method. Crystallization ages are determined from the weighted mean of the $^{206}Pb/^{238}U$ ages of the concordant and overlapping analyses. Analyses that are statistically excluded from the main cluster age are shown in dark gray. Errors, which include contribution for the standard calibration, age of the calibration standard, composition of common Pb, and U decay constant are reported to the 2σ level. Analytical data and sample coordinates are in Table 2. Sample locations are shown in Figures 2 and 3. MSWD—mean square of weighted deviations.

(Fig. 5), the integrated ages of which are considered geologically meaningless. In both experiments, the first fractions yielded ages of ca. 50 Ma, the base of the saddle yielded an age of 43.3 ± 0.5 Ma, and the last fractions when the mineral concentrate was brought to fusion showed a slightly older age. The fifth fraction of the first experiment displayed anomalous behavior that was not reproduced in the second experiment, and it has been ignored in the isochron age calculation. "U-shaped" spectra are typically interpreted as resulting from excess argon (Lanphere and Dalrymple, 1976), when excess argon fractions are released at low and high temperatures, and normal fractions are released at intermediate temperatures at the base (Kaneoka, 1974; Harrison and McDougall, 1981). The resulting ages of 43.3 ± 0.5 Ma and 42.8 ± 0.7 Ma from the base of the saddles for sample M15-2-4 are statistically undistinguishable and in very good agreement with the isochron age of 43.7 ± 0.5 Ma calculated from the combined fractions of the two experiments. An $(^{40}Ar/^{36}Ar)_i$ value of 303 ± 4 supports the interpretation that excess argon was incorporated into the sample. We consider the best estimate for the crystallization age to be ca. 43 Ma. We also analyzed zircons from sample MIKE 14-1-2 from the northern portion of the San Jeronimo–Guayameo pluton, near the village of San Jeronimo (Fig. 3). Twenty-eight single grains were analyzed with a laser spot diameter of 35 μm. All grains yielded concordant analyses that define a mean age of 39.8 ± 0.6 Ma, which is interpreted as the crystallization age for this sample (Fig. 6).

The undeformed Montecillos pluton, which crops out in the north-central part of the map area in the surroundings of the village of Montecillos (Fig. 3), cuts the upper part of the Cutzamala Formation. This pluton is composed of phaneritic, holocrystalline, medium- to fine-grained granite, granodiorite, and quartz-monzonite with plagioclase + quartz + K-feldspar + uralitized clinopyroxene + biotite + oxides. A reported Rb/Sr age of 115 Ma (Garduño-Monroy and Corona-Chavez, 1992) is inconsistent with the pluton intruding the unfolded part of the Cutzamala Formation. We consider this pluton to be Tertiary in age.

The granitic to tonalitic La Huacana pluton, near the village of La Huacana, outside of the map area in Figure 3 intrudes the unfolded continental red beds of the upper part of the Cutzamala Formation. To the southwest, this pluton is largely covered by an extensive Tertiary ignimbrite sequence. A three point Rb/Sr isochron age of 42 Ma has been reported for the eastern portion of the pluton (Schaaf et al., 1995). Here, we present a $^{40}Ar/^{39}Ar$ age from sample M 18-2-3 from the easternmost portion of the pluton (see coordinates in Table 1). Two step-heating experiments were performed using plagioclase concentrates. Reproducible results with statistically undistinguishable age spectra and integrated ages of 34.4 ± 0.4 and 34.2 ± 0.4 Ma (Fig. 5) and a weighted mean age of 34.4 ± 0.4 Ma (SumS/[n – 1] = 0.6) were produced from the last six fractions of the remarkably flat age spectrum of the first run. The second run displayed a roughly flat pattern, except for the third step, which coincided with an abrupt increase in the $^{37}ArCa/^{39}ArK$ ratio. The age of this fraction is not geologically meaningful, suggesting that the plagioclase concentrate was not homogeneous. If we combine all of the fractions, except the outlier in the third step of the second experiment, we obtain an isochron age of 33.8 ± 0.4 Ma, which we interpret to be the best estimate for the cooling age for the Huacana pluton.

The Tingambato batholith (Fig. 2) is a large body within the Arcelia–Palmar Chico subterrane northeast of the study area. Elías-Herrera et al. (2000) reported a K/Ar age of 107 ± 5 Ma for this batholith, which is inconsistent with the ca. 103 and ca. 93 Ma $^{40}Ar/^{39}Ar$ ages of the Arcelia–Palmar Chico succession that they report is intruded by the batholith. To resolve this inconsistency, zircons were analyzed from granitic sample TNG from the northern part of the batholith. The sample contains quartz + alkaline feldspar + plagioclase + biotite + hornblende + oxides. Thirty-nine single grains were analyzed with a laser spot size of 35 µm, obtaining concordant ages defining a mean age of 129.6 ± 1.1 Ma (Fig. 6), which we consider to represent the crystallization of the sample. The age is consistent with the ca. 131–133 Ma U-Pb sensitive high-resolution ion microprobe (SHRIMP) ages obtained by Garza-González et al. (2004) in the northernmost part of the body and indicates that the rocks intruded by the batholith are older than the Arcelia–Palmar Chico succession.

Tertiary Lavas and Ignimbrites

A thick volcanic succession is exposed in the southeastern and northwestern parts of the map area (Fig. 3). This sequence is composed of ~1500 m of mafic to silicic massive lava flows and autoclastic breccias that discordantly cover the Mesozoic folded succession and the Placeres del Oro pluton. In the northwestern part of the area, the volcanic succession is in contact with the San Jeronimo–Guayameo pluton, but the stratigraphic relationship is not clear. The lavas are dominantly mafic to intermediate in composition and have porphyritic to megaporphyritic textures, or are holocrystalline to hypohyaline. Phenocrysts are mainly plagioclase and hornblende in a matrix of plagioclase + oxides ± hornblende ± clinopyroxene ± volcanic glass. Silicic domes with a porphyritic texture characterized by phenocrysts of plagioclase and clinopyroxene in a fine- to very fine–grained matrix of plagioclase + oxides ± clinopyroxene ± volcanic glass are exposed west of Ciudad Altamirano (Fig. 3). Frank et al. (1992) reported K/Ar ages between 43 Ma and 46 Ma for porphyritic andesites along the main road connecting Ciudad Altamirano to Zihuatanejo but did not give the exact locations of the dated samples.

Our new $^{40}Ar/^{39}Ar$ data also indicate an Eocene age for these volcanic rocks. The first sample (M 29-9-1) dated is from a dacitic dome, located 3 km NNW of Ciudad Altamirano (Fig. 3). Six fractions were collected in a step-heating experiment on a plagioclase concentrate. A plateau age of 42.6 ± 0.5 Ma was defined by the four intermediate fractions, which represent 85.7% of the total ^{39}Ar released (Fig. 5). The plateau age is in very good agreement with the isochron age of 42.9 ± 1.7 Ma calculated from the plateau fractions. We take ca. 43 Ma as our best estimate for the crystallization and cooling age of these rocks.

The second dated sample (MIKE 4-1-5) comes from Cerro Alacrán, a silicic cryptodome located ~50 km SE of Ciudad Altamirano (Fig. 2). This subvolcanic body, which is emplaced within the Cutzamala Formation, has the morphology of a dome as the result of accumulation of magma above the neutral buoyancy level. It shows a pseudoporphyritic holocrystalline texture, and it is composed of phenocrysts of plagioclase in a microcrystalline matrix of quartz + plagioclase + oxides. Eight fractions collected from a plagioclase concentrate show a roughly flat age spectrum that yields an integrated age of 39.1 ± 0.7 Ma. This age is statistically undistinguishable from the isochron age of 38.5 ± 0.7 Ma (Fig. 5), and we consider ca. 38.5 Ma to be the best estimation for the crystallization age of this dome.

A 1000-m-thick ignimbrite succession crops out in the southwestern part of the study area (Fig. 3), where it covers Eocene lavas and plutonic rocks and rests in angular unconformity over the Mesozoic folded succession. The ignimbrites are moderately to well consolidated and vary from lithic to crystalline. Crystals, which are frequently broken, are mainly plagioclase + quartz + hornblende ± biotite floating in a pink to white matrix of volcanic ash. Lithics are mostly angular to subrounded fragments of porphyritic lavas and fine- to medium-grained plutonic rocks. Pumice is frequent and, in some cases, flattened and elongated parallel to the stratification, developing an eutaxitic texture.

We determined a $^{40}Ar/^{39}Ar$ age for sample M 15-2-3 from the base of the ignimbrite sequence (Fig. 3). A plagioclase concentrate was separated and step-heated in duplicate experiments in which five fractions were collected. The age spectra are reproducible and display a well-defined saddle shape indicating excess Ar (Fig. 5). The initial age of ca. 47 Ma decreases to ca. 37 Ma, before rising again, indicating a crystallization age near 37 Ma. The intermediate- and high-temperature fractions of the two experiments display a well-correlated array on the $^{36}Ar/^{40}Ar$ versus $^{39}Ar/^{40}Ar$ diagram, yielding an isochron age of 36.8 ± 0.6 Ma, consistent with a best estimate of ca. 37 Ma for the age of the base of the ignimbrite succession in this area.

Postmagmatic Deposits

The continental sedimentation that followed the last Eocene magmatic episode was restricted to small local basins that formed following the deformation and magmatic activity. Five small basins have been recognized in the studied area, the largest of which developed in the surroundings of the village of Santa Teresa (Fig. 3). The postmagmatic sedimentation is represented by moderately consolidated red conglomerates and sandstones derived from older sedimentary and magmatic units that disconformably overlie the Mesozoic-Cenozoic succession in the Huetamo–Ciudad Altamirano region.

ADDITIONAL GEOCHRONOLOGIC CONSTRAINTS

In this section, we present $^{40}Ar/^{39}Ar$ ages and U/Pb ages for sedimentary units that are relevant to the Cretaceous-Tertiary stratigraphic and structural evolution of the Guerrero terrane.

The first age is for an andesitic clast (sample LS 70) from a gently tilted conglomeratic bed in the lower part of the Cutzamala Formation, which forms the western limb of the Tzitzio anticline 50 km north of the study area, (Fig. 2). Nine one-step fusion experiments were conducted on a plagioclase-rich matrix concentrate. The results plot in Figure 5 as a pseudo-age spectra, showing ages in agreement within 1σ for the majority of the analyses. A weighted mean age of 73.6 ± 0.7 Ma was calculated from these analyses. The data cluster on the $^{36}Ar/^{40}Ar$ versus $^{39}Ar/^{40}Ar$ correlation diagram, and the line calculated with eight out of nine points yields an isochron age of 71 ± 3 Ma, which is statistically indistinguishable from the weighted mean age. We consider ca. 74 Ma to be the best estimate for the crystallization of the andesitic clast.

Sample TZT is a muscovite-rich, metamorphic, fine-grained sandstone from a metaturbidite exposed in the core of the Tzitzio anticline west of El Devanadero (Fig. 2). These rocks have been considered to be the basement of the Huetamo subterrane and have been correlated with the Triassic Arteaga complex (Talavera-Mendoza et al., 2007; Centeno-García et al., 2008). Abundant elongated subrounded to elliptical rounded zircons indicate extensive transport. We dated 100 of these zircons, 93 of which produced an acceptable age. The great majority (95%) have U/Th ratios less than 6, as expected for a magmatic origin (Rubatto, 2002). The main population peaks are at 257, 467, and 1173 Ma, with minor Precambrian, Proterozoic, and Archean peaks (Fig. 7). These results support the Triassic age for the Tzitzio metamorphic rocks suggested by Centeno-García et al. (2008). Talavera-Mendoza et al. (2007) apparently dated the same rock unit (their sample GRO-12), although the coordinates of their sample location plot some tens of kilometers to the west of the Tzitzio anticline in a region covered by Tertiary volcanics. The main age peaks for zircons in their sample are comparable to ours, although they also report a peak at ca. 202 Ma.

Sample TJP is a quartz- and biotite-rich metamorphic sandstone from the Tejupilco suite collected ~10 km northwest of Tejupilco. Elías-Herrera et al. (2000) assigned a Late Triassic–Early Jurassic age to this suite and considered it to constitute the basement of the Arcelia–Palmar Chico subterrane. Talavera-Mendoza and Guerrero-Suástegui (2000) considered this suite to be part of the Early Cretaceous Teloloapan subterrane. The dated sample yielded abundant zircons with a morphology similar to those in the TZT sample. Ninety-two of 100 zircons produced acceptable ages. The overwhelming majority of the dated grains were of magmatic origin (U/Th ratios <6 in 95% of the cases). The main age population peaks at 259, 274, 471, 1000, and 1212 Ma are very similar to those of the Tzitzio sample. Talavera-Mendoza et al. (2007) also reported detrital zircons ages from the Arcelia and Teloloapan subterrane that systematically yielded Early Cretaceous age peaks. Thus, our results support the assertion that

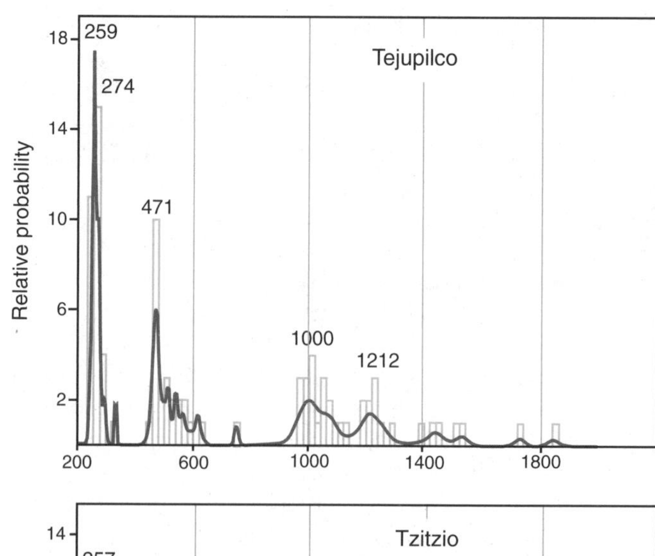

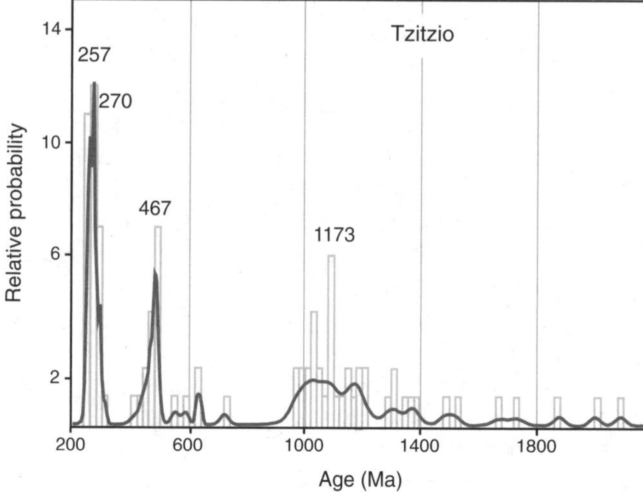

Figure 7. Age probability plots from detrital zircons in the Tejupilco (TJP) and Tzitzio (TZT) metamorphic succession. The plots were constructed using the $^{206}Pb/^{238}U$ age for young (younger than 1.0 Ga) zircons and the $^{206}Pb/^{207}Pb$ age for older (older than 1.0 Ga) zircons. In old grains, ages with >20% discordance or >10% reverse discordance are considered unreliable and were not used. Location of the samples and the complete list of ages are in Table 2.

the Tejupilco metasedimentary succession is the pre-Cretaceous basement of these subterranes, as suggested by Elías-Herrera et al. (2000), but constrain the maximum age as Late Permian.

STRUCTURAL ANALYSIS

Although multiple phases of deformation have been suggested for the study region, complete and well-documented descriptions of the structures have not been presented. For example, large folds in the calcareous beds of the El Cajon and Mal Paso Formations east of the San Lucas village have been associated with the Laramide orogeny (Pantoja-Alór, 1959; Campa-Uranga, 1978; Campa-Uranga and Ramirez, 1979), but no details on their age, kinematics, and rate of shortening have been provided. Several lateral and normal faults have been mapped by Altamira-Areyán (2002) and the Mexican Geologic Survey (Montiel Escobar et al., 2000) in the Huetamo–Ciudad Altamirano area, but the kinematics and rheology of the deformation are not addressed.

Next, we present detailed geometrical descriptions and statistical analysis of the principal structures of the region, which are depicted in the map in Figure 8. The map shows a complex structural pattern characterized by large variations in the trend of the regional structures from ~N-S in the northern part to ~E-W in the central portion. Based on kinematic compatibility, the type of deformation, and available geochronologic data, we identify five main deformational events spanning from late Cenomanian to early Tertiary times.

Deformation 1: Shortening Event (Late Cenomanian–Santonian)

A first shortening event that produced the uplift of the Mesozoic marine succession of the Huetamo region is inferred from an angular unconformity between the Mal Paso and the Cutzamala Formations reported by Altamira-Areyán (2002), and from the lower part of the Cutzamala Formation being directly in contact with marine Cretaceous rocks in the Huetamo area and with Triassic metaturbidites in the Tzitzio region (Fig. 2). Uplift of the Mesozoic marine succession is indicated by the progressive transition from a marine to a continental environment during the early and middle Upper Cretaceous, and by the numerous clasts of marine rocks in the lower part of the Cutzamala Formation. The geometry and style of the deformation producing this uplift are difficult to reconstruct because of the following episodes of folding and faulting. The age of deformation is constrained between the early Cenomanian age of the upper part of the Mal Paso Formation and the 84 Ma age reported for the base of the Cutzamala Formation.

Deformation 2: E-W Shortening Event (Maastrichtian–Paleocene)

The most remarkable structural features of the region are the large folds involving the Cretaceous sedimentary succession of the Huetamo Basin. The four structural profiles in Figure 9 are constructed perpendicular to the principal folds to illustrate the geometry of the deformation. The folds have centimeter to kilometer scales, inducing the formation of a very low penetrative axial plane cleavage. The folds are typically symmetrical upright, and they have vertical to subvertical axial planes and horizontal to subhorizontal axes. The interlimb angles vary from 52° to 118°, and hinges vary from types 2C to 2F and from 1C to 1D in the visual harmonic classification of Hudlestone (1973). Scarce, asymmetrical, and moderately recumbent folds, showing a constant vergence toward the east, are also observed. The recumbent folds have interlimb angles from 45° to 70° and hinge geometries of types 2E and 2D in the classification of Hudlestone (1973). In some cases, overturned folds show local inversion of the polarity of the stratigraphic succession, as along the middle portion of the eastern flank of the Characo anticline (Fig. 8; geologic section B–B′ of Fig. 9), where Cutzamala Formation continental sandstones are overlain in a steep conformable contact by the calcareous sediments of the upper member of the Mal Paso Formation.

The trend of the main structures varies strongly from north to south, allowing two principal structural domains, roughly separated by the Balsas river to be defined:

(1) Domain 1 in the north is characterized by a rather homogeneous structural pattern with a dominant N-S trend. This region is characterized by 10 main kilometric-scale folds, the axes of which strike from N-S to NNW-SSE. A counterclockwise rotation of the N-S structures in the western part of the domain is the result of subsequent transcurrent deformation. Poles to bedding are very similar in all of the main structures and cluster symmetrically along the equator of a stereonet, reflecting the dominant east and west dips of the flanks of the folds. This suggests a homogeneous stress field associated with east-west shortening of the Cretaceous rocks in the Huetamo Basin. Axes of the principal fold structures on the stereographic projection vary from N171 to N186, which are in very good agreement with fold axes measured in the field (Fig. 9).

(2) Domain 2, which extends from the region of the Balsas River south to the edge of the Eocene volcanic cover, displays a heterogeneous structural pattern and significant lateral variations in strike of the structures. The region is characterized by three kilometer-scale folds. The structures display a sinuous trend that defines a sigmoidal geometry in the southern portion of the domain. In the south, the fold axes and the trend of the bedding planes show an average strike of N152, which gradually deflects northward to a mean direction of N015 and reaches a northwest-trending direction south of the Balsas River (N134 on average). An important change in the structural trend occurs near the Balsas River, where near east-west–striking structures abruptly turn north and connect with north-south–striking structures in domain 1. We interpret this lateral variation in strike to result from a strong mechanical contrast between the Early Cretaceous Placeres del Oro pluton and the sedimentary cover. East-west shortening caused the sedimentary succession to behave

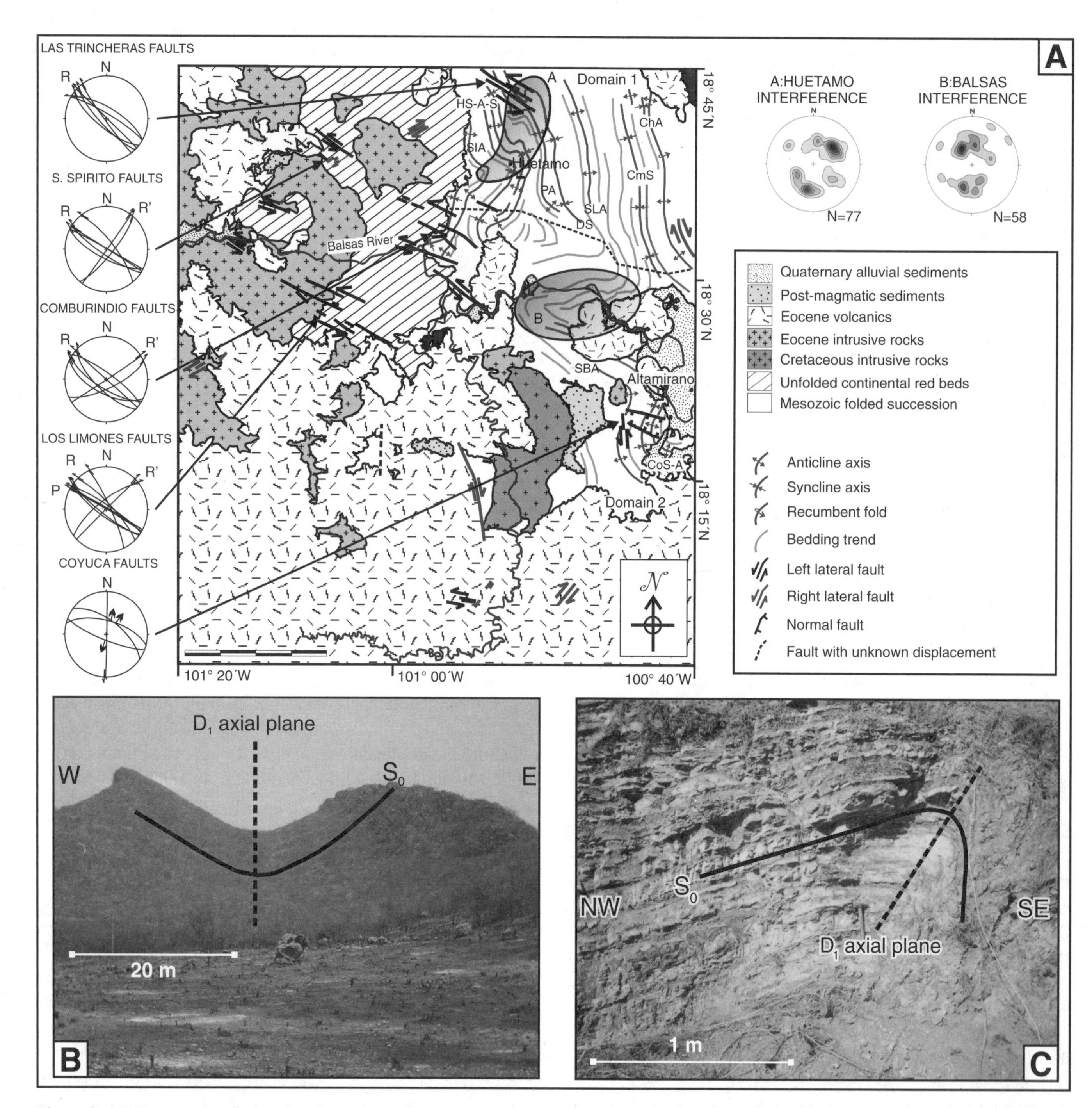

Figure 8. (A) Structural map showing the major structures of the Huetamo–Ciudad Altamirano area. ChA—Characo anticline; CmS—Chumbitaro syncline; SLA—San Lucas anticline; DS—Dolores syncline; PA—La Parota anticline; HS-A-S—Huetamo anticline-syncline; SIA—San Ignacio anticline; SBA—Santa Barbara anticline; CoS-A—Coyuca syncline-anticline. (B) D1 vertical fold in the Mal Paso Formation, southwest of the village of Arroyo Grande (Fig. 2). (C) D_1 moderately recumbent fold in the San Lucas Formation, along the main road between San Lucas and Ciudad Altamirano.

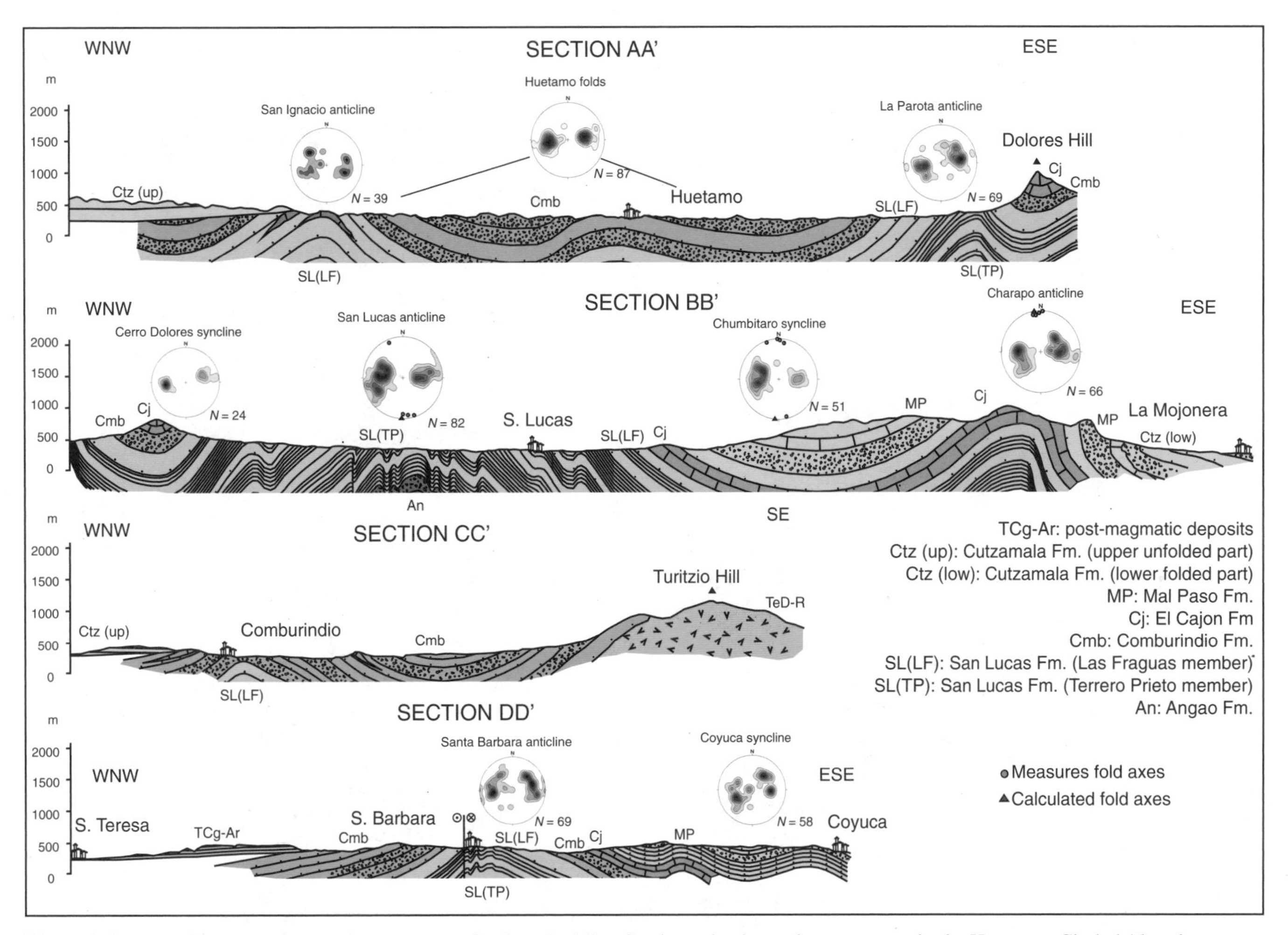

Figure 9. Structural cross sections and stereogram of pole to bedding for the main shortening structures in the Huetamo–Ciudad Altamirano area (D1 deformation phase). Traces of sections are in Figure 3. See text for description and interpretation.

as an incompetent matrix, whereas the intrusive body acted as a rigid bock controlling the regional flow fabric. This interpretation is consistent with a lack of penetrative deformation in the Placeres del Oro pluton, and it implies some decoupling between the intrusive bodies and the sedimentary succession.

The age of shortening can only be roughly constrained. Based on the fossil fauna reported by Benammi et al. (2005) in the folded part of the Cutzamala Formation, we assume a maximum Maastrichtian age for this shortening. A minimum age comes from the 43 Ma ^{40}Ar/^{39}Ar age of the dacitic domes, northwest of Ciudad Altamirano, which cut the folded Cretaceous succession of the Huetamo Basin. This age agrees with the ca. 43 Ma age of the undeformed felsic intrusion at San Jeronimo–Guayameo, which cuts the unfolded part of the Cutzamala Formation. Taking the undeformed part of the Cutzamala Formation as the oldest unit not involved in the D_2 shortening, and using the Paleogene age discussed previously for the base, the D_2 event is bracketed between the Maastrichtian and Paleocene.

Deformation 3: Left-Lateral Ductile Shear (Late Paleocene–Early Eocene)

A second deformation event affected the sedimentary succession of the Huetamo–Ciudad Altamirano region. We recognized a major lineament with a NW-SE orientation running from San José Creek to Comburindio (Fig. 8) that is defined by a 20–50-m-wide, highly deformed belt of localized ductile deformation. In the northwestern portion of the study area, the Montecillos pluton (Fig. 8) cuts and obliterates this lineament. The deformed rocks in the lineament show a well-defined penetrative mylonitic foliation. The foliation along the lineament shows a smooth to moderately rough disjunctive spacing in the siliciclastic rocks to stylolites in the calcareous rocks. A stereographic projection of the poles to the foliation planes highlights the vertical to subvertical attitude and the N130 to N159 strike of the foliation planes (Fig. 10). In thin section, the mylonitic rocks show a shape-preferred crystal orientation that is mostly produced by mechanical rotation and crystal-plastic deforma-

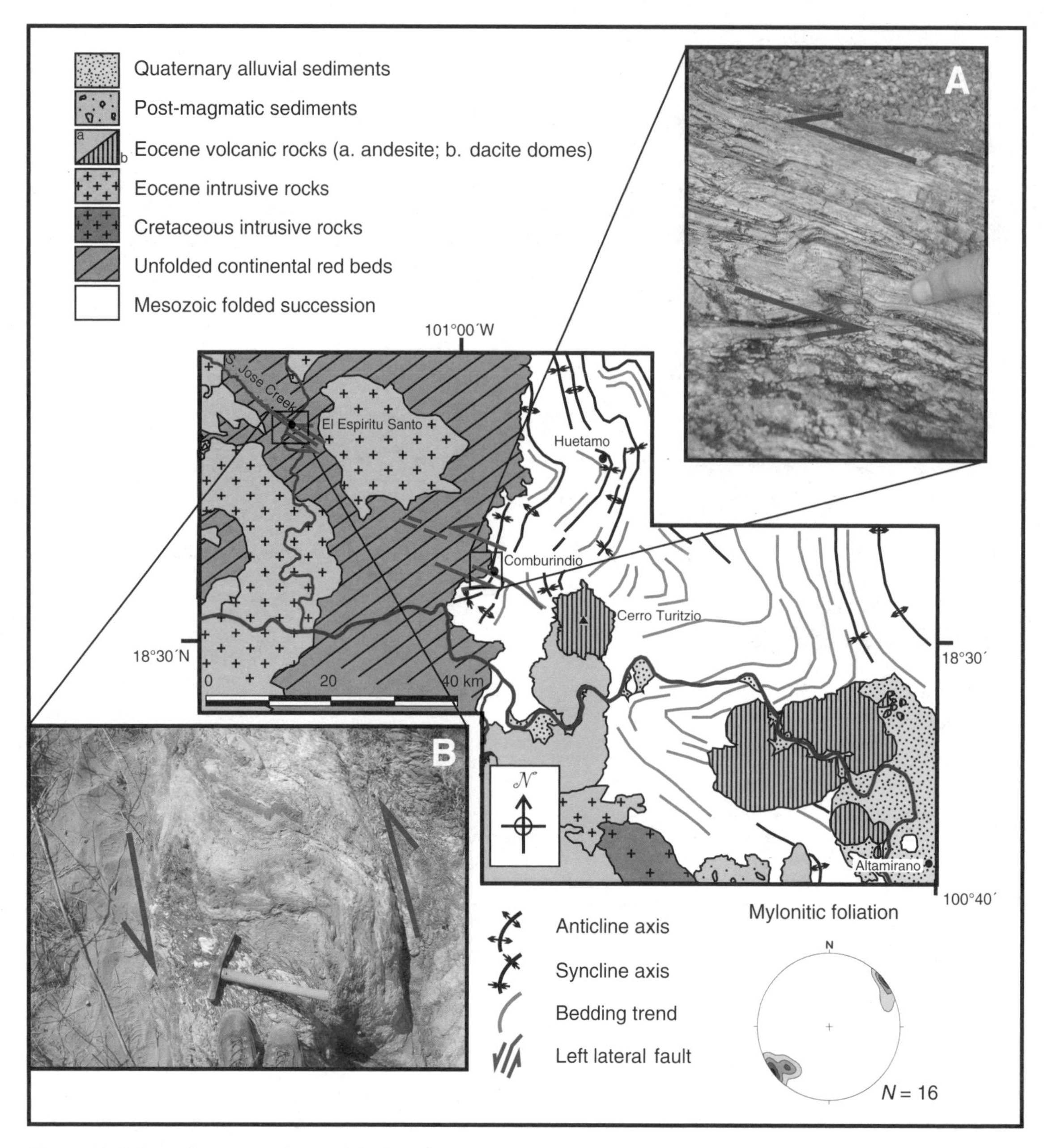

Figure 10. Schematic structural map of the D_2 left-lateral deformation along the Balsas River (northern part of the region covered by Fig. 3). Photographs A and B show horizontal view of two outcrops with ductile indicators of left-lateral displacement along the La Huacana–Villa Hidalgo shear zone. Poles to mylonitic foliation indicate a NW-SE trend of the structure consistent with lineaments observed from aerial and space imagery. See text for discussion.

tion. Grain boundary migration indicates recrystallization during deformation. Stratigraphic offsets and deflections, the trend of the mylonitic foliation, and asymmetric folds and fish structures from kilometer to microscopic scale along the shear zone clearly indicate a left-lateral sense of shear. A dissected limestone bank immediately west of Comburindio suggests a displacement of near 500 m along the shear zone (Fig. 3).

Evidence for this shear zone has not been observed in the rocks east of Cerro Turitzio (Fig. 10), but its southeastward prolongation is strongly suggested by the distribution of Tertiary magmatic rocks and the course of the Balsas River. From the La Huacana intrusion to the ESE, 14 igneous bodies line up along the projected direction of the shear zone, which is called the La Huacana–Villa Hidalgo shear zone (Fig. 11). The alignment of the igneous centers is consistent with a major crustal weakness that facilitated magma ascent. Volcanic rocks covering the magmatic rocks show no evidence of penetrative left-lateral deformation. The drainage system also appears to be controlled by the La

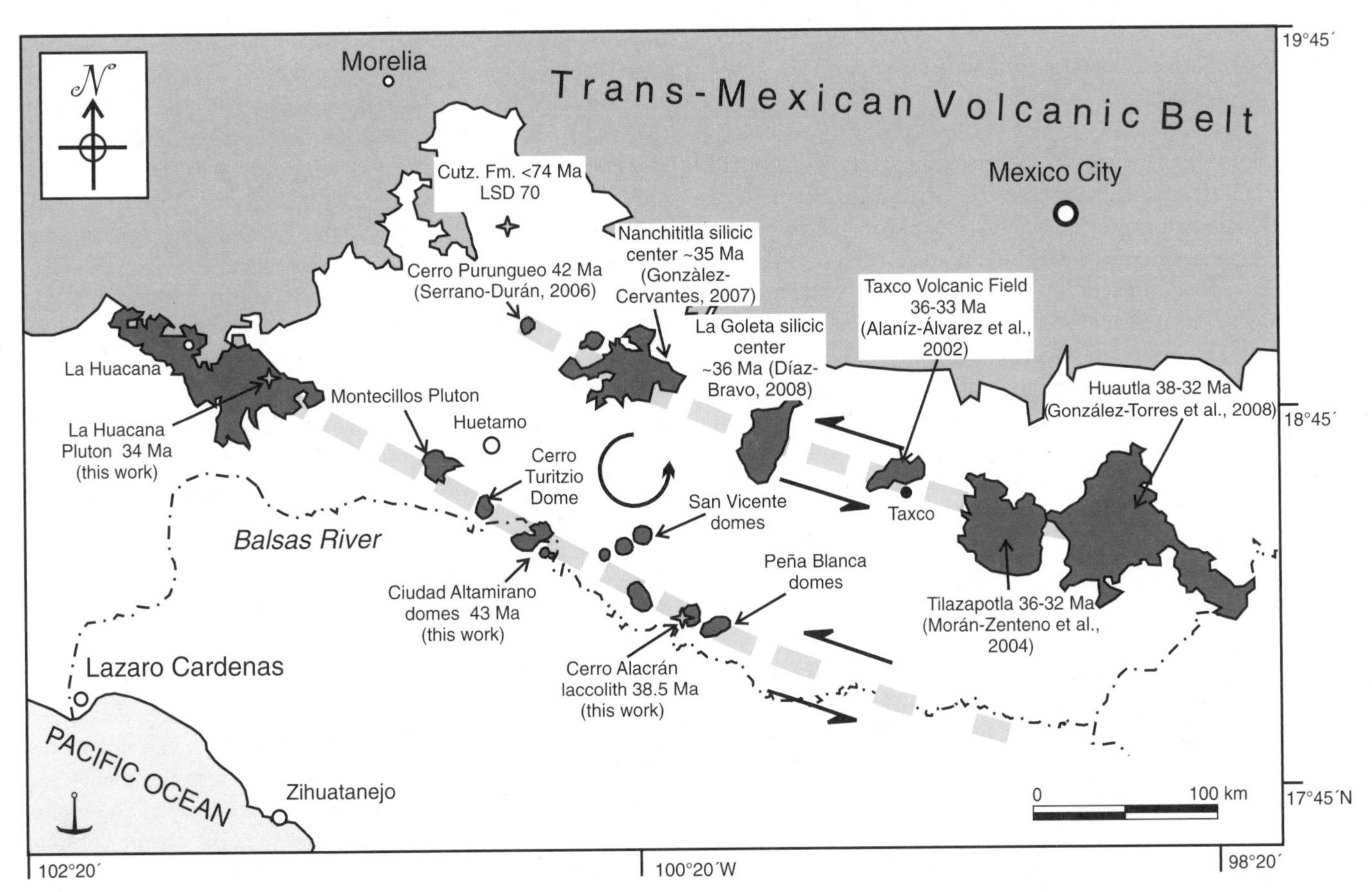

Figure 11. Schematic map of the central part of the Sierra Madre del Sur, showing the two major left-lateral shear zones of La Huacana–Villa Hidalgo (south) and Cerro Purungueo–Huautla (north) and alignment of Tertiary igneous centers (dark gray). Cutz. Fm.—Cutzamala Formation.

Huacana–Villa Hidalgo shear zone. The Balsas River, which is the largest river of southwestern Mexico, runs for some 100 km along the trace of the lineament, before being diverted westward near Cerro Turitzio (Figs. 10 and 11). In the northwestern part of the region, San José creek also trends NW-SE, producing a major erosional valley along the trace of the shear zone (Fig. 10). A kinematically similar lineament has been reported ~70 km northeast of the La Huacana–Villa Hidalgo shear zone, running for ~200 km parallel to the coast from Cerro Purungueo to the Huautla volcanic field (Morán-Zenteno et al., 2004; Ferrari et al., 2004) (Fig. 11).

Left-lateral displacements along these major shear zones can explain the counterclockwise rotation of preexisting structures, development of kilometer-scale sigmoidal geometries near Huetamo, and local bending related to drag along the trace of the shear zone. A subvertical N-S to NNW-SSE right-lateral protomylonitic shear zone along the eastern flank of the Characo anticline has been interpreted as a major structure associated with the transpressive deformation that produced the principal folding of the region (Altamira-Areyán, 2002). However, on the basis of the analyses here, this structure is more likely related to partial decoupling along the interface between the calcareous deposits of the Mal Paso Formation and the continental red beds of the Cutzamala Formation by counterclockwise rotation related to left-lateral shearing. Our ca. 43 Ma $^{40}Ar/^{39}Ar$ age for the oldest undeformed dacitic domes along the shear zone and the Paleocene age inferred for the continental red beds of the Cutzamala Formation, (i.e., youngest stratigraphic unit affected by penetrative mylonitic deformation) constrain this deformation to the late Paleocene or the early Eocene.

Deformation 4: Post–Early Eocene Deformation

Brittle faults affect all of the stratigraphic units in the region. On the basis of kinematics inferred from symmetrical structures on fault surfaces, the faults are subdivided into two groups. Due to the scarcity of post–early Eocene units in the region, the only constraint on the age of these faults is that both groups cut the 37 Ma ignimbrites in the southern part of the study region (Fig. 3).

The first group consists of left-lateral and right-lateral brittle faults that define typical Riedel patterns, consistent with a general left-lateral shear zone parallel to the present coast (Fig. 8). A second set of N-S– to NNW-SSE–trending right-lateral faults in the southern portion of the study area cuts Cretaceous sedimentary rocks of the Huetamo Basin and early Eocene volcanic deposits. Structures in this group are kinematically compatible with transcurrent left-lateral deformation associated with a principal NW elongation.

The second group is represented by NW-SE normal faults and right-lateral reactivations of preexisting left-lateral faults. Normal faults between the villages of Coyuca and Santa Teresa in the southeastern part of the region (Fig. 8) form small grabens and semigrabens oriented parallel to the coast. Right-lateral faults along preexistent left-lateral structures clearly show more than one generation of displacement-related striae. On many fault surfaces, right-lateral kinematic indicators are superimposed on left-lateral ones. Structures in this group are kinematically compatible with right-lateral transtension associated with principal NE elongation.

DISCUSSION

The causes of the Laramide orogeny in southern Mexico are still poorly understood. A main question emerging from the stratigraphic, geochronologic, and structural data presented here is the relation between the Laramide orogeny and the accretion of allochthonous terranes to the continental margin of the North American plate, a relation that has been taken for granted by many authors. Next, we review the timing of Laramide deformation and discuss the applicability of the accretion model to explain the deformation.

Age of the Laramide Orogeny in Southern Mexico

South of the Trans-Mexican volcanic belt, the Laramide deformation involves an ~550-km-wide belt of folds and thrusts that extend from the Pacific coast to the Sierra de Zongolica (Fig. 12). The trends of the structures vary from NW to NE, reflecting a crustal heterogeneity on a regional scale. The chronology and spatial distribution of the deformation show a progressive eastward migration of shortening with time (De Cserna et al., 1980; Nieto-Samaniego et al., 2006; Cerca et al., 2007). Laramide deformation occurred in the middle Eocene in the Zongolica and western Veracruz Basin, before the Bartonian in the Acatlán-Oaxacan block, during the Maastrichtian-Paleocene in the eastern part of the Guerrero-Morelos platform, and in the Santonian-Campanian in the west (Cerca et al., 2007, and references therein). The age of deformation in the Arcelia–Palmar Chico region is poorly defined, but must postdate the ca. 93 Ma $^{40}Ar/^{39}Ar$ age of the Palmar Chico lavas (Elías-Herrera et al., 2000).

The main shortening structures in the Huetamo area are the Maastrichtian-Paleocene folds of the D_2 event. However, earlier late Cenomanian–Santonian D_1 regional deformation and uplift are supported by indirect evidence, as discussed already. Pre-D_2 folds cannot be clearly recognized in the marine Cretaceous succession, suggesting that D_1 and D_2 folds were coaxial. As such, the shortening structures of the study area can be interpreted as the sum of two folding events characterized by parallel axes and axial planes (type 0 of Ramsay, 1967), in which D_2 folding only amplified D_1 structures. The late Cenomanian–Santonian age inferred for the D_1 phase is compatible with eastward migration of the contractile Laramide front. Between the Guerrero-Morelos platform and the Teloloapan subterrane, the beginning of Laramide shortening and uplift has been constrained by the deposition of the syntectonic marine turbidites of the Mexcala Formation (Hernández-Romano et al., 1997; Perrilliat et al., 2000). We

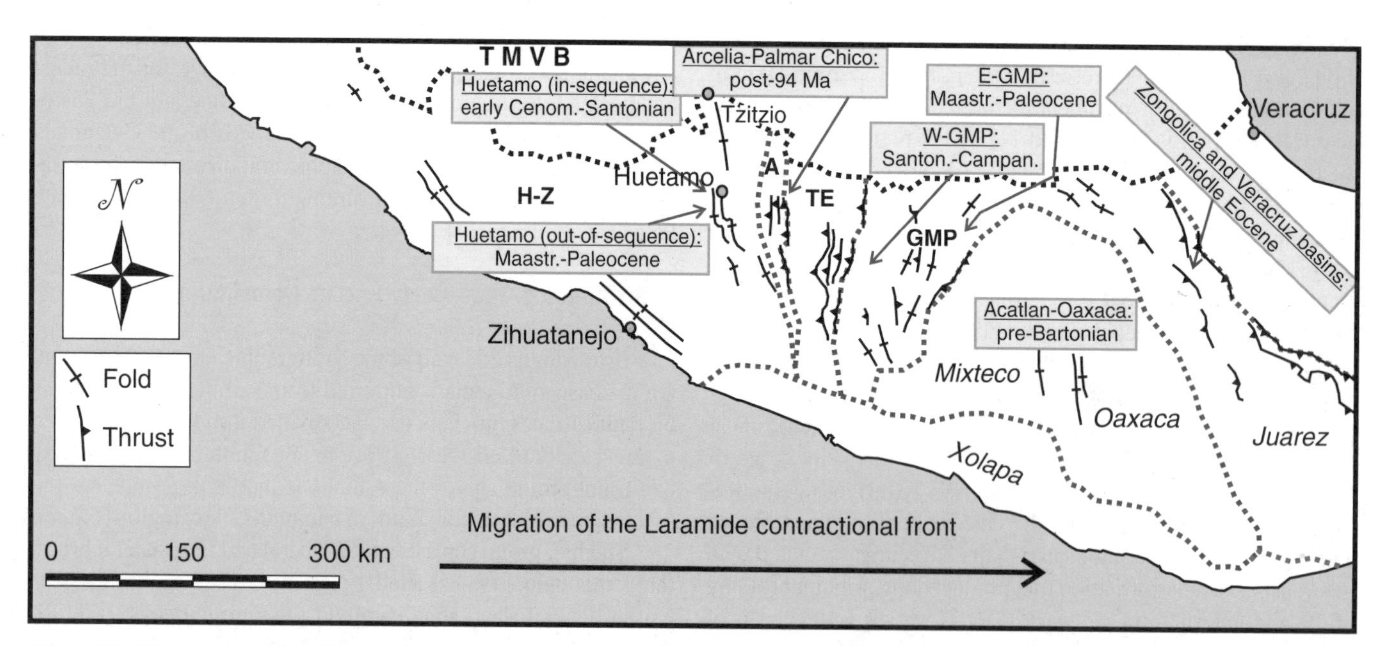

Figure 12. Map summarizing the age of deformation and the main structures of the Laramide orogeny in southern Mexico. Names in italics refer to the terranes in Figure 1. TMVB—Trans-Mexican volcanic belt; H-Z—Huetamo-Zihuatanejo subterrane; A—Arcelia subterrane; TE—Teloloapan subterrane; GMP—Guerrero-Morelos platform.

propose that the continental deposits of the Cutzamala Formation may represent the westward equivalent of the Mexcala Formation, and may date the uplift related to Laramide shortening in the study area. The D_2 folding phase (Maastrichtian–early Paleocene) occurred when the Laramide deformation front was in the eastern part of the Guerrero-Morelos platform, i.e., ~200 km east of Huetamo. In this framework, the D_2 contractile structures are out-of-sequence folds that developed in the hinterland during the eastward migration of the deformation front. In accord with this, Salinas-Prieto et al. (2000) and Cabral-Cano et al. (2000b) observed two, almost coaxial phases of deformation with opposite vergence in the region east of our study area. The deposition of the upper unfolded part of the Cutzamala Formation marks the end of the contractile regime and the beginning of the early Tertiary shearing deformation in the Huetamo area.

Nature of the Guerrero Terrane in Southwestern Mexico

Many previous workers have considered the Mesozoic volcano-sedimentary succession of southwestern Mexico in terms of allochthonous terranes, represented by one or more oceanic island arcs and related sedimentary basins developed concurrently with a more or less prolonged period of subduction toward the west (Campa and Coney, 1983; Lapierre et al., 1992; Tardy et al., 1994; Talavera-Mendoza and Guerrero-Suástegui, 2000; Keppie, 2004). Some models propose two or more subduction zones with opposite vergence during mid-Jurassic to Early Cretaceous times (Dickinson and Lawton, 2001; Talavera-Mendoza et al., 2007). All of these models require the complete consumption of one or more oceanic plate(s) and a change in subduction polarity to arrive at the present plate configuration. The idea of accretion of allochthonous terranes to nuclear Mexico was introduced to explain the apparent lateral incompatibilities of the volcano-sedimentary successions of southwestern Mexico, and it requires first-order diffuse or discrete crustal deformation belts (sutures) at the inferred edges of the accreted terrane. However, evidence for such sutures has not been reported. On the contrary, where exposed, the inferred boundaries of the composite Guerrero terrane are low-angle thrust faults without ophiolite associations. This is the case, for example, in the Teloloapan thrust system, which was originally considered the eastern boundary of the Guerrero terrane (Campa and Coney, 1983). Published geologic cross sections depict these structures as a typical thin-skinned deformation with a subhorizontal geometry just a few kilometers from the surface (Montiel Escobar et al., 1998; Salinas-Prieto et al., 2000; Cabral-Cano et al., 2000a, 200b).

In the model of Talavera-Mendoza and Guerrero-Suástegui (2000), Centeno-García et al. (2003, 2008), and Talavera-Mendoza et al. (2007), the study area would be entirely within the Huetamo-Zihuatanejo subterrane, which would be bounded by the Arcelia–Palmar Chico subterrane a few kilometers east of Ciudad Altamirano (Fig. 2). According to these authors, the subterrane boundary would represent a suture between two different island arcs separated by a subduction zone. This boundary would not be observable, since it would be covered by the post-tectonic Cutzamala Formation and Eocene volcanic rocks (Figs. 1 and 2).

Indirect evidence for a boundary between the Huetamo and Arcelia subterrane has come from geochemical data. Based on major- and trace-element analyses, the magmatic suite of the Arcelia subterrane has been interpreted as a heterogeneous complex composed of an immature intraoceanic island arc and a related backarc basin (Talavera-Mendoza and Guerrero-Suástegui, 2000). The huge granitoid Tingambato batholith that intrudes the westernmost part of the Arcelia–Palmar Chico subterrane (Fig. 2) is clearly the result of arc magmatism. However, the ca. 129 Ma U-Pb age reported here and the ca. 131–133 Ma SHRIMP age of Garza-González et al. (2004) demonstrate that this batholith is part of the Teloloapan arc. Taking these data into account, we envision the Arcelia–Palmar Chico as a backarc basin that opened in the western part of the Teloloapan arc during a westward migration of arc volcanism (see Fig. 13).

If an intraoceanic arc existed in the Arcelia–Palmar Chico subterrane, a subduction zone must have existed to the west in Albian times (e.g., as depicted in Fig. 7 *in* Talavera-Mendoza et al., 2007), which would later become a subterrane boundary. The existence of such a subterrane boundary between Huetamo and Arcelia is difficult to defend in light of our results that show a good match between the ages of detrital zircon populations from the basements of the Huetamo and Arcelia–Palmar Chico subterranes (TZT and TJP samples, respectively, Fig. 7). The main peaks are in the Late Permian, Ordovician, and Grenville. The Ordovician peak is of particular importance as it has been found in the Acatlán complex of the Mixteco terrane (Talavera-Mendoza et al., 2005), which is the continental block immediately to the east of the Guerrero terrane (Fig. 1). These peaks in detrital zircon ages are also similar to those reported by Centeno-García et al. (2005) for the Triassic siliciclastic successions of Arteaga and Zacatecas in the Guerrero terrane, as well as those of nuclear Mexico to the east (Sierra Madre terrane).

There is also abundant evidence of a continental signature in the Late Jurassic to early Tertiary sedimentary and volcanic rocks of the Guerrero terrane. The Cretaceous siliciclastic succession of the Huetamo Basin documents a dominantly volcanic source, likely derived from the arc magmatic rocks that are widely exposed in southwestern Mexico. An older continental source is documented by metamorphic quartz-rich clasts from pervasively deformed and recrystallized rocks in various levels of the stratigraphic succession. Also, the U-Pb ages obtained for some inherited zircons from the Placeres del Oro pluton indicate the assimilation of Grenvillian zircons directly from older continental basement or from recycled material in the Mesozoic sedimentary succession. Similar indications are found in other regions of the Guerrero terrane. Elías-Herrera et al. (1996) and Elías-Herrera and Ortega-Gutierrez (1997) reported peraluminous granulitic and gneiss xenoliths in the Eocene subvolcanic body of Pepechuga, north of Tejupilco, in the western part of the Teloloapan subterrane. Centeno-García et al. (2003) reported clasts derived from polydeformed metamorphosed quartz-rich sandstones,

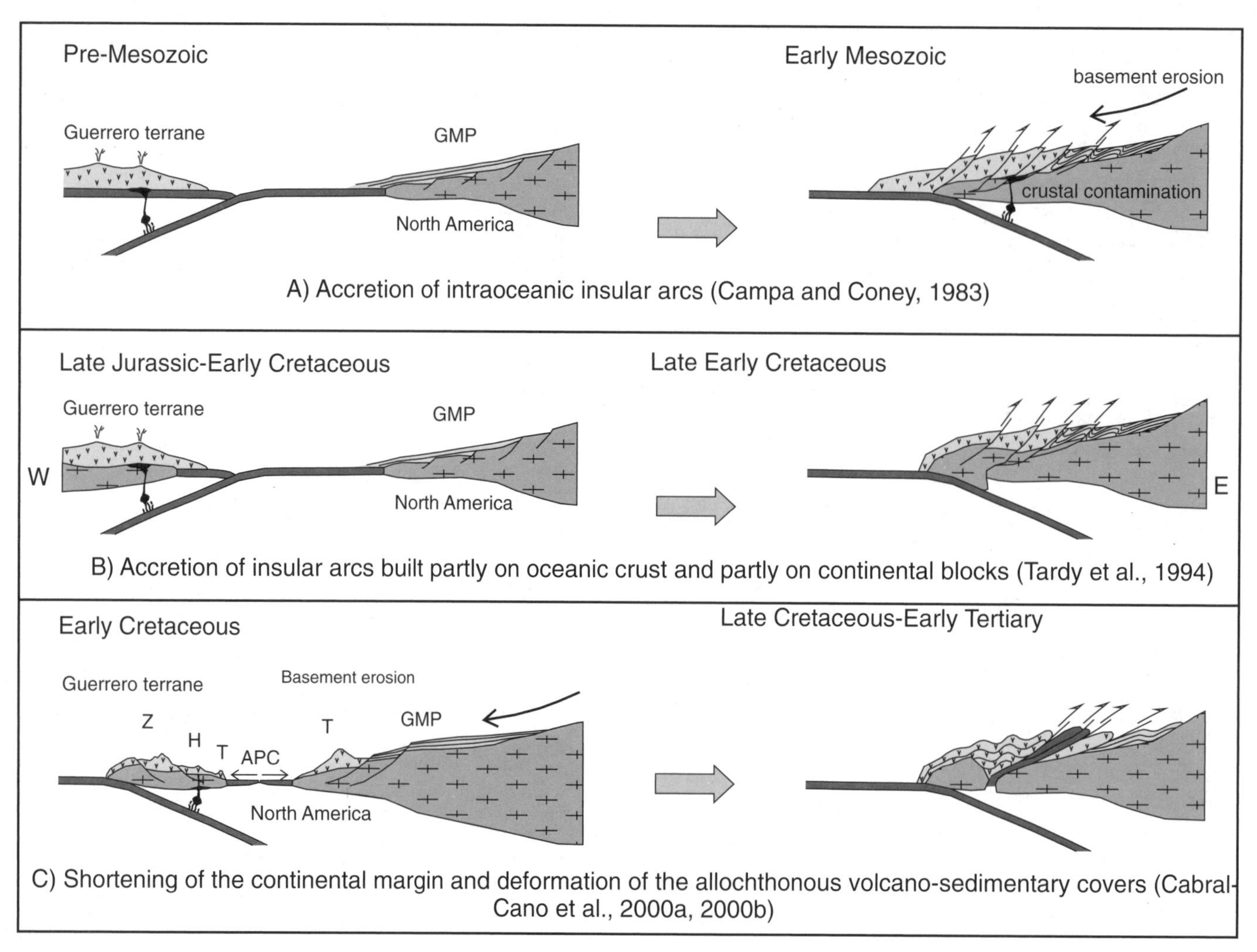

Figure 13. Models proposed in the literature for the evolution of the southern continental margin of the North America plate. Model C is the most consistent with the data presented in this paper. See text for discussion. Zihuatanejo arc; H—Huetamo basin margin; APC—Arcelia–Palmar Chico backarc basin; T—Teloloapan arc; GMP—Guerrero-Morelos platform.

gneiss, and two mica–granitoid rocks in the Posquelite conglomerate along the Pacific coast in the eastern Zihuatanejo subterrane. Talavera-Mendoza et al. (2007) documented Paleozoic and Grenville ages in detrital zircons in samples interbedded with volcanic rocks of the Arcelia, Teloloapan, and Zihuatanejo subterranes (their samples Gro-01, -04, -06, -08 and, -11). In the Zihuatanejo area (sample Gro-11), the age spectrum shows abundant Paleozoic and Precambrian peaks that are only partly similar to the Tzitzio and Tejupilco samples. However, Talavera-Mendoza et al. (2007) recognized the Arteaga complex and its Jurassic intrusive bodies as the source for these zircons, tying the Zihuatanejo area to the same continental source, as documented by Centeno-García et al. (2008). Finally, a Proterozoic and Precambrian inheritance has been reported in magmatic zircons from volcanic rocks from the Teloloapan and Guanajuato subterranes (Mortensen et al., 2008).

The recognition of a pre-Mesozoic continental component in the magmatic and sedimentary successions of southwestern Mexico has important implications for the Mesozoic-Cenozoic reconstruction of the southern margin of the North American plate. The ubiquitous presence of metamorphic clasts and xenoliths, as well as of Paleozoic and Precambrian xenocrystic zircons, often with similar age peaks, in the Huetamo-Zihuatanejo, Arcelia, and Teloloapan subterranes is hard to explain by models that interpret them as the remnants of intraoceanic island arcs separated by one or more trench systems (Campa and Coney, 1983; Lapierre et al., 1992; Tardy et al., 1994; Talavera-Mendoza and Guerrero-Suástegui, 2000; Dickinson and Lawton, 2001; Keppie, 2004; Talavera-Mendoza et al., 2007). In our view, the whole region between the Guerrero-Morelos platform and the Huetamo-Zihuatanejo region was located on or in proximity to the North America continental margin at least since the early Mesozoic. Our data confirm the recent proposal by Centeno-García et al. (2008), who stated that the pre-Jurassic basement of the Guerrero terrane was a thick submarine siliciclastic turbidite succession that accumulated on the western paleocontinental shelf/slope region of the North America plate (Potosi submarine fan).

Terrane Accretion and Laramide Deformation in Southern Mexico

Three possible scenarios may produce a continental signature in the sedimentary and magmatic rocks of southwestern Mexico. In the first scenario, one or more island arcs, which were built on oceanic crust during a prolonged period of west-verging subduction, would be accreted to the continental margin of southwestern Mexico (Fig. 13A). In this case, the continental metamorphic lithics and Precambrian to Paleozoic age zircons would come from tectonic basement acquired by accretion and overthrusting. The erosion of this basement would produce the old continental detritus in the Mesozoic siliciclastic succession, and detrital zircons in the Mesozoic cover could provide the Grenville inherited component in the Placeres del Oro pluton. In this model, the accretion and juxtaposition of the ocean-island arc sequences to the continental margin must be pre-Mesozoic in age, since Paleozoic detrital zircons occur in the Triassic metaflysch of Tejupilco and Tzitzio. This model implies that the Mesozoic volcano-sedimentary successions of southwestern Mexico developed on a thinned North American continental margin or at least within the reach of its turbidites. Accretion in pre-Mesozoic times would preclude any direct relationship between accretion and Laramide deformation.

A variant of the accretion model has been proposed by Tardy et al. (1994), who suggested accretion of an island arc built on both oceanic and continental crust during Laramide time (Fig. 13B). A continental basement beneath the western part of the Mesozoic arc-succession can explain the continental influence in the Mesozoic magmatic and sedimentary units of southwestern Mexico. In this model, the continental basement of the Mesozoic arc would be a proper microcontinent of Grenvillian age, situated in an open-ocean environment, and its accretion to the continental margin of southern Mexico would result in collision between a continental block and a continental craton. This, in turn, calls for the development of a continental suture, characterized by intense deformation, high-grade metamorphism, plutonism, and ophiolites, to separate continental blocks with different precollision histories. Evidence for such a suture has not been documented in southwestern Mexico, and the similarity in detrital zircon ages between the pre-Cretaceous basement of the Huetamo-Zihuatanejo and Arcelia subterranes would appear to rule out such a structure, at least east of Huetamo.

We favor a third scenario that does not require the accretion of allochthonous terranes, at least in the eastern half of the Guerrero terrane. In this view, the Mesozoic volcano-sedimentary rocks of southwestern Mexico are autochthonous or parautochthonous successions that developed on the thinned or rifted continental margin of the North American plate concurrently with the progressive westward migration of a continental magmatic arc, which in turn resulted from the east-verging subduction of the Farallon plate beneath the continental margin (Fig. 13C). The arc rocks could have been partially contaminated at lower and/or middle crustal levels with the continental basement, or at shallower levels with detrital zircons of sedimentary successions derived from erosion of the same pre-Mesozoic continental basement. In this model, there is no need to invoke major crustal boundaries between allochthonous terranes or subterranes with different sedimentary, magmatic, and tectonic histories. In our view, the data indicate that the differences in the Mesozoic successions of southwestern Mexico are best explained by lateral variations of facies as suggested by Cabral-Cano et al. (2000a, 2000b) for the region east of this study. We think that the Late Jurassic to early Tertiary magmatic and sedimentary record of the Guerrero terrane can be explained in the framework of a migrating arc and related extensional backarc and forearc basins, as recently proposed by Mortensen et al. (2008) and Centeno-García et al. (2008) in papers that revise their earlier models (Talavera Mendoza et al., 2007). These arcs and basins were subsequently juxtaposed during the Laramide shortening episode, which, in turn was unrelated to accretion or collision of intraoceanic island arcs from the west. Another driving force is required to explain the generation of the Late Cretaceous–early Tertiary shortening structures of the southern margin of the North American plate.

CONCLUSIONS

Stratigraphic, structural, and geochronologic analyses of the Huetamo–Ciudad Altamirano area show a continental affinity in the Cretaceous magmatic and sedimentary record as well as in the basement. Precambrian and Paleozoic continental signatures in the intrusive and sedimentary rocks are difficult to explain by considering the Cretaceous succession of southwestern Mexico as intraoceanic arcs accreted to the continental margin of Mexico during Laramide times. We cannot exclude the possibility of terrane accretion in southern Mexico, but we show that accretion could only have occurred prior to the Mesozoic. Our results support a model in which the Mesozoic volcano-sedimentary successions of at least part of the Guerrero terrane were deposited directly on a thinned continental margin of the southern part of the North American plate and that the Late Jurassic to early Tertiary magmatic/sedimentary record can be explained by a migrating arc and related extensional backarc and forearc basins. Our results imply that most of the Guerrero terrane is not allochthonous and that the Laramide orogeny is not the result of the accretion of intraoceanic arcs to the North America plate.

ACKNOWLEDGMENTS

This work has been funded by grant CONACyT (Consejo Nacional de Ciencias y Tecnología) SEP 2003-C02–42642. L.F. Michelangelo Martini acknowledges a Ph.D. grant from DGEP-UNAM (Dirección General de Estudio de Posgrado–Universidad Nacional Autónoma de México). M.A. García, S. Rosas, G. Rendón, and V. Pérez assisted with mass spectrometry and sample preparation for ^{40}Ar-^{39}Ar dating. Juan Tomás Vasquez prepared thin sections. We thank Joaquin Ruiz for access to the Arizona LaserChron Center, which is partially supported by National Science Foundation (NSF) grant EAR-0443387. Reviews by Elena Centeno and Dante Morán improved the clarity of the manuscript. Elena Centeno is also thanked for discussions about the Guerrero terrane and for a paper in advance of publication. We thank Víctor Ramos and Suzanne Kay for encouraging us to submit this contribution.

APPENDIX. METHODOLOGY

Geological mapping of six 1:50,000 scale topographic maps (sheets E14-4 A83, A84, A73, A74, A63, and A64 published by INEGI (Instituto Nacional de Estadistica Geografía e Informatica, Mexican Mapping Agency) was carried out during several months of field work between 2005 and 2007. Enhanced Landsat Thematic Mapper images and 1:75,000 scale aerial photos were studied before and after each field season. The petrography of dozens of samples for each geologic unit was studied in thin section. Observations of the contacts and the geometric relationship between the main stratigraphic units, as well as collection of structural data at the mesoscopic scale (e.g., attitude of the stratification and foliation surfaces, fold axes and axial planes, shear surfaces, and related kinematics indicators), were conducted along all existing roads and many riverbeds. Structural data were processed statistically using conventional methods (Ramsay and Huber, 1987; Twiss and Moore, 1992) and were used to produce detailed structural map and geological sections.

Samples for petrographic and microstructural studies were collected at several locations. Minerals were separated from nine volcanic and intrusive rocks samples for dating to produce a good chronological control on the stratigraphy and to better constrain the age of the main deformation events. Approximately 5–10 kg of material were collected for each of the seven samples analyzed by the $^{40}Ar/^{39}Ar$ method. Each sample was crushed and sieved to –25 +45, –45 +60, –60 +80 and –80 +120 mesh size fractions. Mineral separations were performed with a Frantz magnetic separator and handpicking. The mineral concentrates were irradiated in the uranium-enriched research reactor of McMaster University in Hamilton, Ontario, Canada. The argon isotopic analysis was performed at the geochronologic laboratory of Centro de Investigación Cientifica y Educación Superior de Ensenada (CICESE), using a VG5400 mass spectrometer, specifically drowned for the noble gas analysis. The argon extraction line used a Coherent Innova 70C argon-ion laser to heat the samples. Details of the experiments are reported in Figure 5 and Table 1.

Zircons separated from two intrusive and two metasedimentary rock samples were used for the U-Pb analysis. Approximately 20 kg of rock were collected for the intrusive samples and ~6 kg for the metasandstone samples. Mineral separation was carried out at the mineral facility of Centro de Geociencias, Universidad Nacional Autónoma de México, using standard technical procedures, including crushing, sieving to –32 +60, –60 +80, and –80 +120 mesh size fractions, magnetic separation, heavy liquids, and handpicking. Individual zircon U-Pb ages were obtained by laser ablation–multicollector–inductively coupled plasma–mass spectrometry (LA-MC- ICP-MS) at the Laser Chron Center of the University of Arizona at Tucson. Ablation of zircons was performed with a New Wave/Lambda Physik DUV193 Excimer laser, operating at a wavelength of 193 nm (Gehrels et al., 2008). Common Pb correction was accomplished by using the measured ^{204}Pb and assuming an initial Pb composition from Stacey and Kramers (1975). Our measurements of ^{204}Pb were unaffected by the presence of ^{204}Hg because backgrounds were measured on peaks, and because very little Hg was present in the argon gas. Errors are reported to the 2σ level. Details of the experiments are presented in Figures 6 and 7 and in Tables 1 and 2. In the case of metasedimentary rocks, at least 100 zircons were chosen randomly between all the zircons mounted.

REFERENCES CITED

Alaníz-Alvárez, S.A., Nieto-Samaniego, A.F., Morán-Zenteno, D.J., and Alba-Aldave, L., 2002, Rhyolitic volcanism in extension zone associated with strike-slip tectonics in the Taxco region, southern Mexico: Journal of Volcanology and Geothermal Research, v. 118, p. 1–14, doi: 10.1016/S0377-0273(02)00247-0.

Alencaster, G., and Pantoja-Alór, J., 1998, Two new Lower Cretaceous rudists (Bivalvia-Hippuritacea) from the Huetamo region; southwestern Mexico: Geobios, v. 31, p. 15–28, doi: 10.1016/S0016-6995(98)80061-2.

Altamira-Areyán, A., 2002, Las Litofacies y sus Implicaciones de la Cuenca Sedimentaria Cutzamala-Tiquicheo, Estado de Guerrero y Michoacán, México [M.Sc. thesis]: México D.F., Universidad Nacional Autónoma de México, Instituto de Geología, 79 p.

Benammi, M., Centeno-García, E., Martínez-Hernández, E., Morales-Gámez, M., Tolson, G., and Urrutia-Fucugauchi, J., 2005, Presencia de dinosaurios en la Barranca Los Bonetes en el sur de México (Región de Tiquicheo, Estado de Michoacán) y sus implicaciones cronoestratigráficas: Revista Mexicana de Ciencias Geológicas, v. 22, no. 3, p. 429–435.

Buitrón-Sánchez, B.E., and Pantoja-Alór, J., 1998, Albian gastropods of the rudist-bearing Mal Paso Formation, Chumbítaro region, Guerrero, Mexico: Revista Mexicana de Ciencias Geológicas, v. 15, no. 1, p. 14–20.

Cabral-Cano, E., Draper, G., Lang, H.R., and Harrison, C.G.A., 2000a, Constraining the late Mesozoic and early Tertiary tectonic evolution of southern Mexico: Structure and deformation history of the Tierra Caliente region: The Journal of Geology, v. 108, p. 427–446, doi: 10.1086/314414.

Cabral-Cano, E., Lang, H.R., and Harrison, C.G.A., 2000b, Stratigraphic assessment of the Arcelia-Teloloapan area, southern Mexico: Implication for southern Mexico's post-Neocomian tectonic evolution: Journal of South American Earth Sciences, v. 13, p. 443–457, doi: 10.1016/S0895-9811(00)00035-3.

Campa, M.F., 1978, La evolución tectónica de Tierra Caliente, Guerrero: Boletín de la Sociedad Geológica Mexicana, v. 39, no. 2, p. 52–54.

Campa, M.F., and Coney, P.J., 1983, Tectono-stratigraphic terranes and mineral resource distribution in Mexico: Canadian Journal of Earth Sciences, v. 20, p. 1040–1051.

Campa, M.F., and Iriondo, A., 2004, Significado de dataciones Cretácicas de los arcos volcánicos de Taxco, Taxco Viejo y Chapolapa, en la evolución de la plataforma Guerrero-Morelos: Unión Geofísica Mexicana, Reunión Nacional de Ciencias de la Tierra: GEOS, v. 24, n. 2, p. 173.

Campa-Uranga, M.F., 1977, Estudio Tectónico: Prospecto Altamirano-Huetamo: Petróleos Mexicano, Informe Geológico 146 (IGPR-146), 94 p.

Campa-Uranga, M.F., and Ramírez, J., 1979, La Evolución Geológica y la Metalogénesis del Noroccidente de Guerrero: Universidad Autónoma de Guerrero, Serie Técnico-Científica 1, 101 p.

Campa-Uranga, M.F., Ramírez, J., and Bloome, C., 1982, La secuencia volcánico-sedimentaria metamorfizada del Triásico (Ladiniano-Cárnico) de la región de Tumbiscatio, Michoacán, *in* Sociedad Geológica Mexicana, Convención Geológica Nacional Abstracts, v. 6a, p. 48.

Centeno-García, E., Ruiz, J., Coney, P.J., Patchett, P.J., and Ortega-Gutiérrez, F., 1993, Guerrero terrane of Mexico: Its role in the Southern Cordillera from new geochemical data: Geology, v. 21, p. 419–422, doi: 10.1130/0091-7613(1993)021<0419:GTOMIR>2.3.CO;2.

Centeno-García, E., Corona-Chavez, P., Talavera-Mendoza, and O., Iriondo, A., 2003, Geologic and tectonic evolution of the western Guerrero terrane—A transect from Puerto Vallarta to Zihuatanejo, Mexico, *in* Alcayde, M., and Gómez-Caballero, A., eds., Geologic Transects across Cordilleran Mexico, Guidebook for the Field Trips of the 99th Geological Society of America Cordilleran Section Annual Meeting, Puerto Vallarta, Jalisco, Mexico, 4–7 April 2003: Universidad Nacional Autónoma de México, Instituto de Geología, Special Publication 1, p. 201–228.

Centeno-García, E., Gehrels, G., Diaz-Salgado, C., and Talavera-Mendoza, O., 2005, Zircon provenance of Triassic (Paleozoic?) turbidites from central and western Mexico: Implications for the early evolution of the Guerrero Arc: Geological Society of America Abstracts with Programs, v. 37, no. 4, p. 12.

Centeno-García, E., Guerrero-Suástegui, M., and Talvera-Mendonza, O., 2008, The Guerrero composite terrane of western Mexico: Collision and subsequent rifting in a suprasubduction zone, *in* Draut, A.E., Clift, P.D., and Scholl, D.W., eds., Formation and Application of the Sedimentary Record in Arc Collision Zones: Geological Society of America Special Paper 436, p. 279–308, doi: 10.1130/2008.2436(13).

Cerca, M., Ferrari, L., López-Martínez, M., Martiny, B., and Iriondo, A., 2007, Late Cretaceous shortening and early Tertiary shearing in the central Sierra Madre del Sur, southern Mexico: Insights into the evolution of the Caribbean–North America plate interaction: Tectonics, v. 26, doi: 10.1029/2006TC001981.

Coney, P.J., Jones, D.L., and Monger, J.W.H., 1980, Cordilleran suspect terranes: Nature, v. 288, p. 329–333, doi: 10.1038/288329a0.

Dávila-Alcocer, V.M., and Guerrero-Suástegui, M., 1990, Una edad basada en radiolarios para la secuencia volcánicasedimentaria de Arcelia, Estado de Guerrero, *in* Sociedad Geológica Mexicana, X Convención Geológica Nacional (México D.F.): Puebla, México, Libro de Resúmenes, p. 83.

De Cserna, Z., Ortega-Gutiérrez, F., and Palacios-Nieto, M., 1980, Reconocimiento geologico de la parte central de la cuenca del alto Río Balsas, estados de Guerrero y Puebla, *in* Sociedad Geológica Mexicana, Field-Trip Guidebook of Geologic Excursions to the Central Part of the Balsas River Basin: Fifth National Geologic Congress, p. 2–33.

Díaz-Bravo, B.A., 2008, Estratigrafía, petrología y estilo eruptivo del centro volcánico silícico de La Goleta-Sultepec, estados de México y Guerrero [master's thesis]: Universidad Nacional Autónoma de México, México, D.F., 101 p.

Dickinson, W.R., and Lawton, T.F., 2001, Carboniferous to Cretaceous assembly and fragmentation of Mexico: Geological Society of America Bulletin, v. 113, p. 1142–1160, doi: 10.1130/0016-7606(2001)113<1142:CTCAAF>2.0.CO;2.

Elías-Herrera, M., and Ortega-Gutiérrez, F., 1997, Petrology of high-grade metapelitic xenoliths in an Oligocene rhyodacite plug—Precambrian crust beneath the southern Guerrero terrane, Mexico?: Revista Mexicana de Ciencias Geológicas, v. 14, no. 1, p. 101–109.

Elías-Herrera, M., and Ortega-Gutiérrez, F., 1998, The Early Cretaceous Arperos oceanic basin (western Mexico): Geochemical evidence for an aseismic ridge formed near a spreading center: Comment: Tectonophysics, v. 292, p. 321–326.

Elías-Herrera, M., Ortega-Gutiérrez, F., and Cameron, K., 1996, Pre-Mesozoic continental crust beneath the southern Guerrero terrane: Xenolith evidence: GEOS Unión Geofísica Mexicana Abstracts with Programs, v. 16, p. 234.

Elías-Herrera, M., Sánchez-Zavala, J.L., and Macias-Romo, C., 2000, Geologic and geochronologic data from the Guerrero terrane in the Tejupilco area, southern Mexico: New constraints on its tectonic interpretation: Journal of South American Earth Science, v. 13, p. 355–375, doi: 10.1016/S0895-9811(00)00029-8.

Ferrari, L., Cerca-Martinez, M., López-Martínez, M., Serrano-Duran, L., and González-Cervantes, N., 2004, Age of formation of the Tzitzio antiform and structural control of volcanism in eastern Michoacán and western Guerrero, *in* IV Reunion Nacional de Ciencias de la Tierra Abstracts: GEOS Boletín de la Union Geofísica Mexicana, Puerto Vallarta, México, v. 24, p. 165.

Filkorn, H.F., 2002, A new species of Mexicaprina (Caprinidae, Coalcomaninae) and review of the age and paleobiogeography of the genus: Journal of Paleontology, v. 76, p. 672–691, doi: 10.1666/0022-3360(2002)076<0672:ANSOMC>2.0.CO;2.

Frank, M.M., Kratzeisen, M.J., Negendank, J.F.W., and Boehnel, H., 1992, Geología y tectónica en el terreno Guerrero (México-sur), *in* III Congreso Geológico de España and VIII Congreso Latinoamericano de Geología, Salamanca, España: Actas tomo 4, p. 290–293.

García-Barrera, P., and Pantoja-Alór, J., 1991, Equinoides del Albiano tardío de la Formación Mal Paso de la región de Chumbítaro, estados de Guerrero y Michoacán: Revista de la Sociedad Mexicana de Paleontología, v. 4, p. 23–41.

Garduño-Monroy, V.H., and Corona-Chavez, P., 1992, Mapa geológico del estado de Michoacán, Universidad Michoacana de San Nicolas de Hidalgo, Instituto de Investigaciones Metalurgicas, 1 map with illustrative notes.

Garza-González, A., González-Partida, E., Tritlla, J., Levresse, G., Arriaga-García, G., Rosique-Naranjo, F., Medina-Avila, J.J., and Iriondo, A., 2004, Evolución magmática en el pórfido de cobre de Tiámaro, Michoacán: Evidencias del potencial Cu-Au en la sur de México: GEOS, Boletín Unión Geofisica Mexicana, Abstracts of the IV Reunión Nacional de Ciencias de la Tierra, Puerto Vallarta, Mexico, v. 24, n. 2, p. 288.

Gehrels, G.E., Valencia, V.A., and Ruiz, J., 2008, Enhanced precision, accuracy, efficiency, and spatial resolution of U-Pb ages by laser ablation–multi-collector–inductively coupled plasma–mass spectrometry: Geochemistry, Geophysics, Geosystems, v. 9, p. Q03017, doi: 10.1029/2007GC001805.

Gómez-Luna, M.E., Contreras-Montero, B., Guerrero-Suástegui, M., and Ramírez-Espinoza, J., 1993, Ammonitas del Valanginiano superior-Barremiano de la Formación San Lucas en el área de Huetamo, Michoacán: Revista de la Sociedad Mexicana de Paleontología, v. 6, no. 1, p. 57–65.

González-Cervantes, N., 2007, Evolución del centro silícico de la Sierra de Nanchititla, Estado de México y Michoacán [master's thesis]: Universidad Nacional Autónoma de México, Centro de Geociencias, Campus Juriquilla, Qro., 128 p.

González-Torres, E., Morán-Zenteno, D., Chapela-Lara, M., Solé-Viñas, J., Valencia, V., and Pompa-Mera, V., 2008. Geocronología del campo volcánico de Huautla, Morelos y del sector norte-central de la Sierra Madre del Sur y sus implicaciones en el conocimiento de la evolución magmática del Cenozoico: GEOS, Boletin Unión Geofísica Mexicana, v. 27, p. 145.

Guerrero-Suástegui, M., 1997, Depositional History and Sedimentary Petrology of the Huetamo Sequence, Southwestern Mexico [M.Sc. thesis]: El Paso, University of Texas–El Paso, 95 p.

Harrison, T.M., and McDougall, I., 1981, Excess ^{40}Ar in metamorphic rocks from Broken Hill, New South Wales: Implications for ^{40}Ar/^{39}Ar age spectra and thermal history of the region: Earth and Planetary Science Letters, v. 55, p. 123–149, doi: 10.1016/0012-821X(81)90092-3.

Hernández-Romano, U., Aguilera-Franco, N., Martínez-Medrano, M., and Barceló-Duarte, J., 1997, Guerrero-Morelos Platform drowning at the Cenomanian-Turonian boundary, Huitziltepec area, Guerrero State, southern Mexico: Cretaceous Research, v. 18, p. 661–686.

Hudlestone, P.J., 1973, An analysis of single layer folds developed experimentally in viscous media: Tectonophysics, v. 16, p. 17–98.

Kaneoka, I., 1974, Investigation of excess argon in ultramafic rocks from the Kola peninsula by the ^{40}Ar/^{39}Ar method: Earth and Planetary Science Letters, v. 22, p. 145–156, doi: 10.1016/0012-821X(74)90075-2.

Keppie, D.J., 2004, Terranes of Mexico revisited: A 1.3 billion year odyssey: International Geology Review, v. 46, p. 765–794, doi: 10.2747/0020-6814.46.9.765.

Lanphere, M.A., and Dalrymple, G.B., 1976, Identification of excess ^{40}Ar by the ^{40}Ar/^{39}Ar age spectrum technique: Earth and Planetary Science Letters, v. 32, p. 141–148, doi: 10.1016/0012-821X(76)90052-2.

Lapierre, H., Ortiz, L.E., Abouchami, W., Monod, O., Coulon, C., and Zimmermann, J.L., 1992, A crustal section of an intra-oceanic island arc: The Late Jurassic–Early Cretaceous Guanajuato magmatic sequence, central Mexico: Earth and Planetary Science Letters, v. 108, p. 61–77, doi: 10.1016/0012-821X(92)90060-9.

Larsen, E.S., Gottfried, D., Jaffe, H.H., and Waring, C.L., 1958, Lead-alpha ages of the Mesozoic batholiths of Western North America: U.S. Geological Survey Bulletin, n. 1070b, p. 46–47.

Lemos-Bustos, O., and Fu-Orozco, V.M., 2002, Carta Geológico-Minera, Placeres del Oro E14–A84 Guerrero: Servicio Geologico Mexicano, México DF, scale 1:50,000.

Mariscal-Ramos, C., Talavera-Mendoza, O., Centeno-García, E., Morales-Gámez, M., and Benammi, M., 2005, Preliminary magnetostratigraphic study of the Upper Cretaceous dinosaur site from La Barranca Los Bonites, Tiquicheo (Michoacán State, Southern Mexico): Reunión Anual de la Unión Geofísica Mexicana, v. 25, no. 1, p. 57–58.

Montiel-Escobar, J.E., Segura de la Teja, M.A., Estrada-Rodarte, G., Cruz-López, D.E., and Rosales-Franco, E., 2000, Carta geológico-minera Ciudad Altamirano E14-4, Guerrero, Michoacán y Edo de México, Servicio Geológico Mexicano, scale: 1:250.000.

Morán-Zenteno, D.J., Alba-Aldave, L.A., Solé, J., and Iriondo, A., 2004, A major resurgent caldera in southern Mexico: The source of the late Eocene Tilzapotla ignimbrite: Journal of Volcanology and Geothermal Research, v. 136, p. 97–119, doi: 10.1016/j.jvolgeores.2004.04.002.

Mortensen, J.K., Hall, B.V., Bissig, T., Friedman, R.M., Danielson, T., Oliver, J., Rhys, D.A., Ross, K.V., and Gabites, J.E., 2008, Age and paleotectonic setting of volcanogenic massive sulphide deposits in the Guerrero terrane of Central Mexico: Constraints from U-Pb age and Pb isotope studies: Economic Geology and the Bulletin of the Society of Economic Geologists, v. 103, p. 117–140.

Nicholls, D.J., 1973, North American and European species of Momipites ("Engelhardtia") and related genera: Geoscience and Man, v. 7, p. 103–117.

Nieto-Samaniego, A.F., Alaniz-Álvarez, S.A., Silva-Romo, G., Eguiza-Castro, M.H., and Mendoza-Rosales, C.C., 2006, Latest Cretaceous to Miocene deformation events in the eastern Sierra Madre del Sur, Mexico, inferred from the geometry and age of major structures: Geological Society of America Bulletin, v. 118, no. 1/2, p. 238–252, doi: 10.1130/B25730.1.

Omaña-Pulido, L., and Pantoja-Alór, J., 1998, Early Aptian benthic foraminifera from the El Cajón Formation, Huetamo, Michoacán, SW Mexico: Revista Mexicana de Ciencias Geológicas, v. 15, no. 1, p. 64–72.

Omaña-Pulido, L., González-Arreola, C., and Ramírez-Garza, B.M., 2005, Barremian planktonic foraminiferal events correlated with the ammonite zones from the San Lucas Formation, Michoacán (SW Mexico): Revista Mexicana de Ciencias Geológicas, v. 22, no. 1, p. 88–96.

Ortíz-Hernández, H.E., Yta, M., Talavera, O., Lapierre, H., Monod, O., and Tardy, M., 1991, Origine intra-pacifique des formations pluto-volcaniques d'arc du Jurassique supérieur-Crétacé inférieur du Mexique centro-méridional: Comptes Rendus de l'Académie des Sciences (Paris), v. 312, p. 399–406.

Pantoja-Alór, J., 1959, Estudio geológico de reconocimiento de la región de Huetamo, Estado de Michoacán, Consejo de Recursos Naturales no Renovables: Boletin (Instituto de Estudios de Poblacion y Desarrollo [Dominican Republic]), v. 50, p. 3–33.

Pantoja-Alór, J., 1983, Geocronometría del magmatismo Cretácico-Terciario de la Sierra Madre del Sur: Boletín de la Sociedad Geológica Mexicana, v. 47, p. 1–46.

Pantoja-Alór, J., 1990, Redefinición de las unidades estratigráficas de la secuencia mesozoica de la región Huetamo-Cd. Altamirano, estados de Michoacán y Guerrero, *in* Convención Geológica Nacional: Sociedad Geológica Mexicana Memoir 10: Puebla, México, p. 121–123.

Pantoja-Alór, J., and Gómez-Caballero, J.A., 2003, Geologic features and biostratigraphy of the Cretaceous of southwestern México (Guerrero Terrane), *in* Alcayde, M., and Gómez-Caballero, A., eds., Geologic Transects across Cordilleran Mexico, Guidebook for the Field Trips of the 99th Geological Society of America Cordilleran Section Annual Meeting, Puerto Vallarta, Jalisco, Mexico, 4–7 April 2003: Universidad Nacional Autónoma de México, Instituto de Geología, Special Publication 1, p. 229–260.

Pantoja-Alór, J., and Skelton, P.W., 2000, Tepeyacia corrugata Palmer: Rudist of the Polyconitidae family, *in* VII Congreso Nacional de Paleontología y I Simposio Geológico en el Noreste de México Actas: Linares, Nuevo León, México, p. 58–59.

Perrilliat, M.C., Vega, F., and Corona, R., 2000, Early Maastrichtian mollusca from the Mexcala Formation of the State of Guerrero, southern Mexico: Journal of Paleontology, v. 74, no. 1, p. 7–24.

Ramsay, J.G., 1967, Folding and Fracturing of Rocks: New York, McGraw-Hill, 568 p.

Ramsay, J.G., and Huber, M.I., 1987, The Technique of Modern Structural Geology, vol. 1: Strain Analysis: Oxford, Harcourt Brace Jovanovich Publishers, Alden Press, 307 p.

Rubatto, D., 2002, Zircon trace element geochemistry: Partitioning with garnet and the link between U-Pb ages and metamorphism: Chemical Geology, v. 184, p. 123–138, doi: 10.1016/S0009-2541(01)00355-2.

Salazar, M.S., 1973, Prospecto Altamirano-Área Huetamo: Petroleos Mexicanos R-114 Open-File Report, 16 p.

Salinas-Prieto, J.C., Monod, O., and Faure, M., 2000, Ductile deformations of opposite vergence in the eastern part of the Guerrero terrane (SW Mexico): Journal of South American Earth Sciences, v. 13, p. 389–402, doi: 10.1016/S0895-9811(00)00031-6.

Schaaf, P., Morán-Zenteno, D., Hernández-Bernal, M., Solís-Pichardo, G., Tolson, G., and Koehler, H., 1995, Paleogene continental margin truncation in southwestern Mexico: Geochronological evidence: Tectonics, v. 14, no. 6, p. 1339–1350, doi: 10.1029/95TC01928.

Serrano-Durán, L., 2006, Estudio de los Enjambres de Diques y del Fallamiento Terciario en la Región de Tuzantla-Tiquicheo-Nanchititla, Estados de Michoacán, México y Guerrero [Bachelor's thesis]: Juriquilla, Universidad Nacional Autónoma de México, Centro de Geociencias, Campus Juriquilla, Qro., 120 p.

Skelton, P.W., and Pantoja-Alór, J., 1999, Discovery of Coalcomana (Caprinidae) in the Lower Aptian El Cajón Formation in the San Lucas area, Michoacán, SW Mexico, *in* Fifth International Congress on Rudists: Erlanger Geologische Abhandlungen, Erlanger, Sonderband 3, p. 66.

Stacey, J.S., and Kramers, J.D., 1975, Approximation of terrestrial lead isotope evolution by a two-stage model: Earth and Planetary Science Letters, v. 26, p. 207, doi: 10.1016/0012-821X(75)90088-6.

Talavera-Mendoza, O., 1993, Les Formations Orogéniques Mésozoiques du Guerrero (Mexique meridional). Contribution à la Conaissance de l'évolution géogynamique des Cordillères Mexicaines [Ph.D. thesis]: Grenoble, France, Université Joseph Fourier, 462 p.

Talavera-Mendoza, O., and Guerrero-Suástegui, M., 2000, Geochemistry and isotopic composition of the Guerrero terrane (western Mexico): Implication for the tectonomagmatic evolution of southwestern North America during the late Mesozoic: Journal of South American Earth Sciences, v. 13, p. 297–324, doi: 10.1016/S0895-9811(00)00026-2.

Talavera-Mendoza, O., Ramírez-Espinoza, J., and Guerrero-Suástegui, M., 1995, Petrology and geochemistry of the Teloloapan subterrane: A Lower Cretaceous evolved intra-oceanic island arc: Geofísica Internacional, v. 34, p. 3–22.

Talavera-Mendoza, O., Ruiz, J., Gehrels, G.E., Meza-Figueroa, D.M., Vega-Granillo, R., and Campa-Uranga, M.F., 2005, U-Pb geochronology of the Acatlán complex and implication for the Paleozoic paleogeography and tectonic evolution of southern Mexico: Earth and Planetary Science Letters, v. 235, p. 682–699, doi: 10.1016/j.epsl.2005.04.013.

Talavera-Mendoza, O., Ruiz, J., Gehrels, G.E., Valencia, V.A., and Centeno-García, E., 2007, Detrital zircon U/Pb geochronology of southern Guerrero and western Mixteca arc successions (southern Mexico): New insights for the tectonic evolution of the southwestern North America during the late Mesozoic: Geological Society of America Bulletin, v. 119, p. 1052–1065, doi: 10.1130/B26016.1.

Tardy, M., Lapierre, H., Freydier, C., Coulon, C., Gill, J.B., Mercier De Lepinay, B., Beck, C., Martínez, J., Talavera-Mendoza, O., Ortiz, E., Stein, G., Bourdier, J.L., and Yta, M., 1994, The Guerrero suspect terrane (western Mexico) and coeval arc terranes (the Greater Antilles and the Western Cordillera of Colombia): A late Mesozoic intra-oceanic arc accreted to cratonal America during the Cretaceous: Tectonophysics, v. 230, p. 49–73, doi: 10.1016/0040-1951(94)90146-5.

Twiss, R.S., and Moore, E.M., 1992, Structural Geology: New York: W.H. Freeman and Company, 532 p.

Manuscript Accepted by the Society 5 December 2008

The Geological Society of America
Memoir 204
2009

Geochemical evolution of igneous rocks and changing magma sources during the formation and closure of the Central American land bridge of Panama

Gerhard Wörner
Abteilung Geochemie, Geowissenschaftliches Zentrum Göttingen, Universität Göttingen, Goldschmidtstrasse 1, 37077 Göttingen, Germany

Russell S. Harmon
Army Research Office, U.S. Army Research Laboratory, P.O. Box 12211, Research Triangle Park, North Carolina 27709, USA

Wencke Wegner
Abteilung Geochemie, Geowissenschaftliches Zentrum Göttingen, Universität Göttingen, Goldschmidtstrasse 1, 37077 Göttingen, Germany

ABSTRACT

The geological development of Panama's isthmus resulted from intermittent magmatism and oceanic plate interactions over approximately the past 100 m.y. Geochemical data from ~300 volcanic and intrusive rocks sampled along the Cordillera de Panama document this evolution and are used to place it in a tectonic framework. Three distinct trace-element signatures are recognized in the oldest basement rocks: (1) oceanic basement of the Caribbean large igneous province (CLIP basement) displays flat trace-element patterns, (2) CLIP terranes show enriched ocean-island basalt (OIB) signatures, and (3) CLIP rocks exhibit arc signatures. The Chagres igneous complex represents the oldest evidence of arc magmatism in Panama. These rocks are tholeiitic, and they have enriched but highly variable fluid-mobile element (Cs, Ba, Rb, K, Sr) abundances. Ratios of these large ion lithophile elements LILEs) to immobile trace elements (e.g., Nb, Ta, middle and heavy rare earth elements) have a typical, but variably depleted, arc-type character that was produced by subduction below the CLIP oceanic plateau. These early arc rocks likely comprise much of the upper crust of the Cordillera de Panama and indicate that by 66 Ma, the mantle wedge beneath Panama was chemically distinct (i.e., more depleted) and highly variable in composition compared to the Galapagos mantle material, from which earlier CLIP magmas were derived.

Younger Miocene andesites were erupted across the Cordillera de Panama from 20 to 5 Ma, and these display relatively uniform trace-element patterns. High field strength elements (HFSEs) increase from tholeiitic to medium-K arc compositions. The change in mantle sources from CLIP basement to arc magmas indicates that

Wörner, G., Harmon, R.S., and Wegner, W., 2009, Geochemical evolution of igneous rocks and changing magma sources during the formation and closure of the Central American land bridge of Panama, *in* Kay, S.M., Ramos, V.A., and Dickinson, W.R., eds., Backbone of the Americas: Shallow Subduction, Plateau Uplift, and Ridge and Terrane Collision: Geological Society of America Memoir 204, p. 183–196, doi: 10.1130/2009.1204(08). For permission to copy, contact editing@geosociety.org.

enriched sub-CLIP (i.e., plume) mantle material was no longer present in the mantle wedge by the time that subduction magmatism commenced in the area. Instead, a large spectrum of mantle compositions was present at the onset of arc magmatism, onto which the arc fluid signature was imprinted. Arc maturation led to a more homogeneous mantle wedge, which became progressively less depleted due to mixing or entrainment of less-depleted backarc mantle through time.

Normal arc magmatism in the Cordillera de Panama terminated around 5 Ma due to the collision of a series of aseismic ridges with the developing and emergent Panama landmass. Younger heavy rare earth element–depleted magmas (younger than 2 Ma), which still carry a strong arc geochemical signature, were probably produced by ocean-ridge melting after their collision.

INTRODUCTION AND GEOLOGICAL SETTING

The western part of the Central American land bridge between the Santa Elena transform fault in northern Costa Rica, which is the boundary with the continental Chortis terrane of northern Central America, and the Atrato fault zone in northern Colombia, which is a westward extension of the South Caribbean fault, is characterized by a series of mafic complexes (Fig. 1) that form the foundation of the Central American land bridge. These mafic terranes represent the major constructional elements of the Central American crust of Costa Rica and Panama. Similar terranes are known in Colombia and northern Ecuador (Goossens et al., 1977; Reynaud et al., 1999). Goossens et al. (1977) first noted their age range, from the Cretaceous to the Eocene, and described their distribution from northern Costa Rica to the northern Colombian Andes. Based on similar lithologies and common tholeiitic characteristics, Goossens et al. (1977) also proposed their correlation and common origin as accreted oceanic terranes. More recently, Hauff et al. (2000) and Hoernle et al. (2002, 2004) suggested the term Caribbean large igneous province (CLIP) for this region and linked these CLIP rocks to the Galapagos plume and the thickened oceanic crust of the Caribbean plate. They interpreted these mafic terranes as a series of aseismic ridges and oceanic islands from the Galapagos plume, equivalents of which can be observed as aseismic ridges on the Nazca plate (Fig. 1). These authors also confirmed the Cretaceous to Tertiary age for these terranes first proposed by Goossens et al. (1977) based on more precise Ar-Ar dating (139 and 50 Ma; see recent compilation by Hoernle and Hauff, 2007). Still, the origin of these terranes and the Caribbean plate remains a matter of debate and involves a complex and controversial tectonic history (Pindell et al., 2006, and reference therein) for the Caribbean plate, with which they are associated. Hypothetical origins that have been proposed include:

(1) A change from westward-dipping subduction of the Farallon plate to an eastward-facing Caribbean arc system in Aptian time (Pindell et al., 2005);

(2) Formation of the Caribbean large igneous province (CLIP), either subsequent to the development of the Galapagos plume as a consequence of subduction of the proto-Caribbean spreading center and formation of a slab window (Pindell et al., 2006), or by successive accretion of oceanic ridges over some 70 m.y. (Hoernle et al., 2004);

(3) Northwestward movement of the Caribbean plate between North and South America and southwest subduction of the proto-Caribbean oceanic crust and associated magmatism in the Greater Antillean arc;

(4) Consumption of the proto-Caribbean oceanic crust by trench-trench collision with South America during Late Cretaceous–Paleocene time (i.e., 60–70 Ma; Pindell et al., 2006); and

(5) Establishment of a more recent northeastward-verging subduction zone, terrane accretion, and arc magmatism that developed at the western margin of the modern Caribbean plate (Mann and Kolarsky, 1995; de Boer et al., 1988; Pindell and Barret, 1990).

Subaerial volcanic rocks in Costa Rica and Panama were produced by a younger island-arc system that evolved from the Oligocene to the present from a tholeiitic to a more mature arc calc-alkaline character (Alvarado et al., 1992; de Boer et al., 1995; Abratis and Wörner, 2001). Presently active arc volcanism in Costa Rica, westernmost Panama, and northern Colombia is associated with the northwestward (northeastward) subduction of the Cocos and Nazca plates, respectively (Fig. 1).

Two volcanic gaps presently exist in the region, the first where the Cocos Ridge collided with the arc in southern Costa Rica, and the second in central and eastern Panama, where plate convergence has been accommodated since mid-Miocene time by movement along the N-S–trending Panama fracture zone and in the Panama deformed belt of the Caribbean plate (Fig. 1). As a consequence, the convergent plate boundary in central and eastern Panama has locked, arc volcanism has ceased, and related northward displacement of Panama has been accommodated by strike-slip motion, oroclinal bending, and resulting E-W extension (Fig. 1; Escalante, 1990). The result of this complex tectonic situation is a zone of low topography in central Panama where the final closure of the land bridge occurred some 3–2 m.y. ago (Collins et al., 1996; Haug and Tiedemann, 1998; Coates et al., 2000).

The Panama Canal was built in this region, between mountainous landscapes generated by a compressional regime to the west and transpressional deformation in eastern Panama and northern Colombia.

Based on data from the literature, combined with our new analyses, this paper documents changing magma compositions

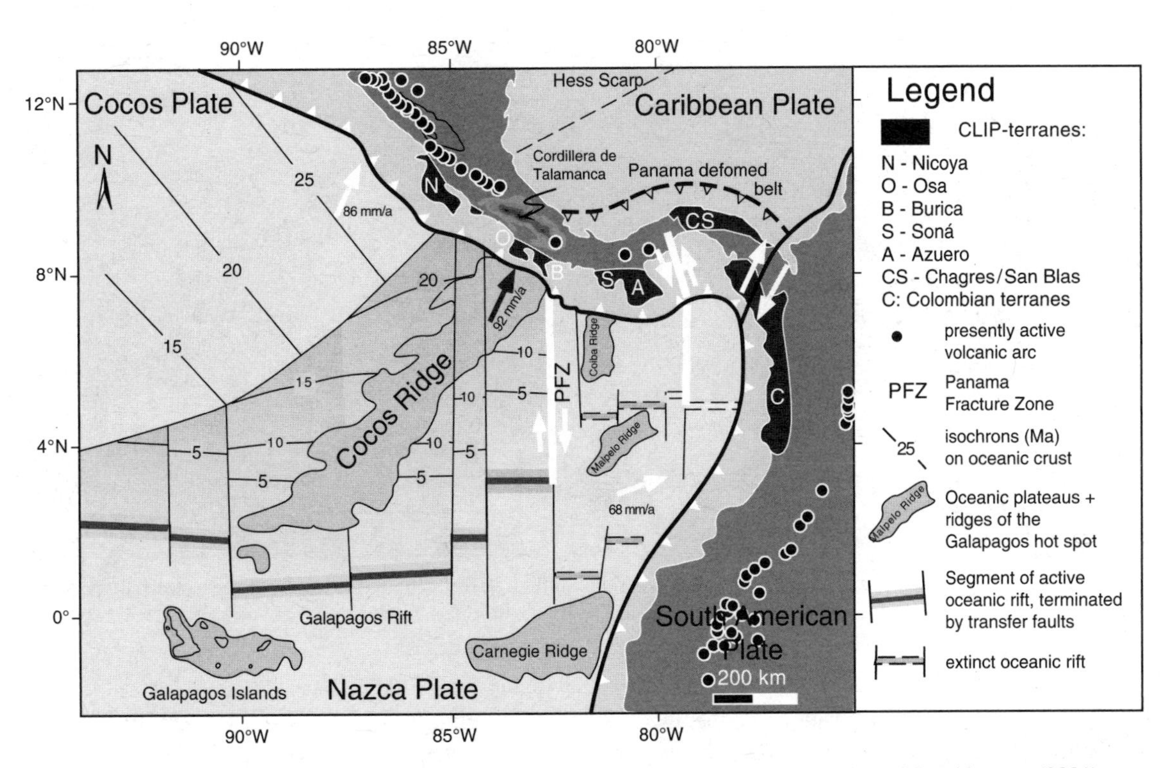

Figure 1. Plate-tectonic setting of the Central American land bridge following Meschede and Barckhausen (2001); accreted mafic complexes are from Goossens et al. (1977). CLIP—Caribbean large igneous province.

over the past 66 m.y. in western-central Panama and traces their changing mantle sources through this time interval. These changes reflect the crustal growth and arc maturation that, together with the juvenile magmatic addition to the upper plate, led to the final closure of the land bridge and fostered its consolidation,

ANALYTICAL METHODS

Some 300 samples of Late Cretaceous– to Quaternary-age igneous rocks were collected at 207 localities along the Cordillera de Panama (Fig. 2). Representative samples of between 1 and 3 kg of the most unaltered rock present at each location were selected for X-ray fluorescence (XRF) spectroscopy and inductively couple plasma–mass spectrometry (ICP-MS) determination of whole-rock chemical composition. For some of the more altered samples, secondary veins were removed by handpicking after initial crushing. About 100 g of this sample material was then ground to powder (<65 μm) in an agate mill.

Major- and minor-element (Si, Ti, Al, Fe, Mn, Mg, Ca, Na, K, P) abundances were measured by XRF on fused glass discs. Measurements were undertaken on a Philips PW 1480 XRF spectrometer. Based on multiple measurements of internal and international reference material, the analytical error for both major and minor elements was ±0.4%–1.8%. Trace-element (Nb, Ta, Be, Cs, Cu, Hf, Li, Y, Pb, Rb, Tl, Th, U, and rare earth element [REE]) abundances were measured by ICP-MS. Samples were digested in pressurized Teflon vessels by dissolving a 100 mg sample in a solution of 3 ml HF and 2 ml HNO_3. Samples were then repeatedly dried down and redissolved in a mixture of 1 ml $HClO_4$ and 1 ml HF. In a final step, an internal standard containing 20 ppb Rh, Re, and In was mixed into a mixture of 750 μl $HClO_4$, 500 μl HF, and 2 ml HNO_3, and the final solution was diluted with deionized water to a volume of 100 ml. A blank solution and standards were prepared for each batch of 18 samples. The spiked solution was analyzed using a FISONS VG PQ STE instrument. Standard JA2 was analyzed every 11 samples. The $\pm 2\sigma$ error of the method is estimated to be <20% for Nb and Ta, <10% for Be, Cs, Cu, Hf, Li, Y, Pb, Rb, Tl, Th, and U, and ~5% for the rare earth elements (REEs).

RESULTS

Geochemical analyses of representative samples are provided in Table 1, where all data have been normalized to volatile-free compositions, and total Fe is reported as $Fe^{2+}O$. A complete data set of major- and trace-element analyses is available in the GSA Data Repository, or, alternatively, it can be obtained in spread-sheet form from the first author upon request.[1]

[1]GSA Data Repository item 2009133, major- and trace-element data for 30 selected and representative samples for different groups of volcanic rocks from Central and Western Panama: CLIP oceanic basement, CLIP-OIB, CLIP-Arc, Early Arc, Miocene Arc, and Adakites, is available at http://www.geosociety.org/pubs/ft2009.htm or by request to editing@geosociety.org.

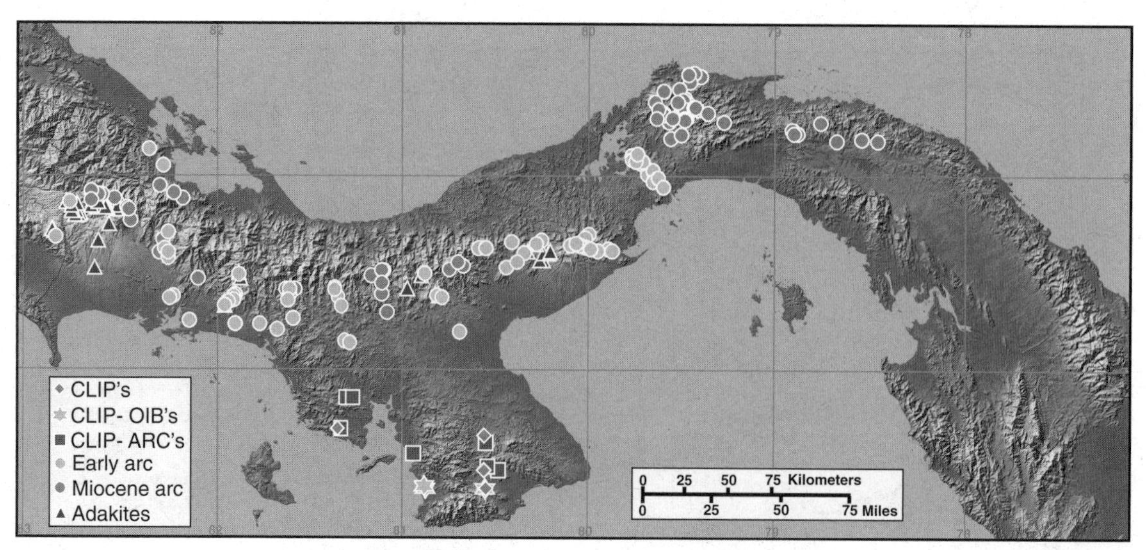

Figure 2. Map of Panama showing the sample locations of this study. The different symbols refer to the tectonic association of each sample. CLIP—Caribbean large igneous province; OIB—ocean-island basalt; ARC—arc rocks of variable age.

Variations in major-element composition for the igneous rocks of the Cordillera de Panama are illustrated in the total alkali–silica diagram (Fig. 3). With few exceptions, all rocks are subalkaline, and the suite of samples spans the entire compositional range from basalt and gabbro to rhyolite and tonalite/granite. Higher $Na_2O + K_2O$ values (i.e., mildly alkaline rocks) could contain an introduced effect resulting from secondary alteration. These samples have been omitted.

The remaining samples were subdivided into four groups based on their age and tectonic setting.

(1) Compositional data from CLIP oceanic basement rocks were compiled from a range of localities throughout Central America (Nicoya, Azuero, Soná Peninsulas and other minor occurrences; Hoernle et al., 2002, 2004; Hauff et al., 2000, 1997; Sinton et al., 1997, 1998) and combined with our data from Soná, Azuero, and Nicoya. CLIP rocks have been dated by Hoernle et al. (2002, 2004). These rocks were further divided on the basis of their trace-element signatures (as discussed later) into: (a) oceanic basement rocks of the Caribbean plate from 139 to 55 Ma (AN8: 139 ± 1 Ma from Hoernle et al. [2004]; OS4: 55 ± 2 Ma from Hoernle et al. [2002]), (b) rocks from accreted ridges and ocean-island complexes as young as 21 Ma (G22: 21 ± 1; Hoernle et al., 2002), and (c) rare samples from the CLIP terranes from Soná and Azuero that have a clear subduction-zone signature of unknown age (this paper; Buchs et al., 2007).

(2) "Early" arc rocks are defined as those between 66 and 42 Ma, which are dominated by samples from the Chagres igneous complex. These mostly represent deeply exposed sections of submarine volcanic and volcaniclastic rocks and intrusions that range in composition from gabbro to tonalite.

(3) Miocene arc rocks are from central and western Panama, and they erupted between 36 and 5 Ma.

(4) Pliocene-Holocene rocks exhibit steep heavy REE (HREE) patterns, which occur in the "magmatic gap" of southeastern Costa Rica and western Panama (previously referred to as "adakites" by Defant et al. [1991a, 1991b]; Abratis and Wörner, 2001). The following sections will concentrate on the trace-element signatures for rocks in these groups in order to identify the change of magma sources through time.

Age, Composition, and Mantle Source of CLIP Terranes

The oldest known rocks in the central portion of the Central American land bridge are in terranes of oceanic basalt from the Nicoya Peninsula in Costa Rica; Fig. 1), which yield ages between 139 and 111 Ma (Hoernle et al., 2004). Younger CLIP ages down to 55 Ma have also been found (Hauff et al., 2000; Hoernle et al., 2004, 2002, and references therein). With respect to the initiation of subduction-zone magmatism (66 Ma, see following), it is interesting to note that the accretion of seamounts from the Galapagos hotspot track (ocean-island basalts) commenced at 66 Ma (data compilation in Hoernle et al., 2004). The oldest CLIP rocks in Panama come from the Azuero Peninsula (Fig. 1), where most ages are between 50 and 66 Ma (Hoernle et al., 2002). One sample from Soná gave 71 ± 2 Ma, which is the oldest dated rock in Panama (Hoernle et al., 2002; see also compilation by Hoernle and Hauff, 2007). Hoernle and other workers (Hoernle et al., 2002; Hauff et al., 2000, 1997; Sinton et al., 1997, 1998) documented the geochemical and isotopic signatures of these rocks. As originally noted by Goossens et al. (1977), the larger basaltic terranes are tholeiitic in character (Hoernle et al., 2004). Younger (20–66 Ma) and smaller occurrences represent intraplate ocean-island basaltic (OIB) rocks (i.e., seamounts) from the hotspot track that have accreted onto the evolving active margin since 66 Ma.

TABLE 1. MAJOR- AND TRACE-ELEMENT COMPOSITION OF REPRESENTATIVE SAMPLES

	Group 1: CLIP oceanic basement					Group 2: CLIP-OIB				
Sample:	PAN-06-197	PAN-06-198	PAN-05-029	PAN-05-018	PAN-05-009	PAN-05-019	PAN-05-028	PAN-05-020	PAN-05-011	PAN-05-026
Region:	Sona	Sona	Azuero	Azuero	Azuero	Azuero	Azuero	Azuero	Azuero	Azuero
EASTING:	459472	460897	503618	538893	546883	510795	509629	510795	539597	510795
NORTHING:	853859	853719	842217	831487	831171	824126	824335	824126	820923	824126
Lithology:	basalt	basalt	basalt	basalt	basalt	basalt	basalt	basalt	basalt	basalt
SiO_2	47.9	48.2	47.9	48.7	48.4	39.9	42.0	45.4	48.3	46.1
TiO_2	1.02	1.08	2.02	1.30	1.99	0.76	1.27	3.46	2.80	3.05
Al_2O_3	14.37	14.17	14.40	14.10	14.90	4.20	6.30	13.40	13.40	13.80
Fe_2O_3	10.44	11.31	13.18	12.73	13.36	12.66	13.85	13.13	11.46	11.76
MnO	0.18	0.18	0.25	0.18	0.22	0.17	0.17	0.17	0.15	0.16
MgO	8.34	8.32	6.83	7.71	4.70	30.91	26.30	5.74	6.80	5.53
CaO	13.24	11.07	10.64	12.47	8.19	5.09	5.08	8.79	9.38	7.86
Na_2O	1.93	3.05	3.22	1.89	4.68	0.01	0.74	3.54	3.82	3.04
K_2O	0.07	0.11	0.17	0.07	0.17	0.02	0.16	1.24	0.64	0.17
P_2O_5	0.077	0.088	0.194	0.101	0.215	0.079	0.134	0.383	0.273	0.383
Total	97.57	97.59	98.80	99.25	96.82	93.80	96.01	95.25	97.03	91.84
Sc	64	57	53	51	43	15	19	30	35	35
V	333	327	357	317	383	96	145	299	287	304
Co	42	45	45	49	41	109	105	43	40	44
Ni	97	103	73	90	28	1463	1155	57	98	83
Zn	70	76	132	100	124	88	117	134	116	120
Rb	2.23	2.4	1.9	0.7	2.5	2.6	5.7	27.7	7.8	3.4
Sr	141	111	198	113	265	21	134	429	427	346
Y	21	24	51	26	42	8	13	36	29	38
Zr	57	70	82	49	86	34	58	178	132	147
Nb	2.88	3.29	3.47	4.04	3.92	4.39	8.41	35.30	25.36	29.09
Ta	0.13	0.16	0.19	0.27	0.25	0.15	0.34	1.94	1.63	1.69
Cd	0.10	0.10	0.13	0.15	0.22	0.05	0.07	0.13	0.11	0.17
Cs	0.03	0.05	0.08	0.005	0.11	0.04	0.17	0.13	0.004	0.05
Ba	48	17	62	21	96	10	30	255	212	78
Hf	1.43	1.70	3.35	2.06	3.52	1.26	2.14	6.53	5.11	5.70
Pb	0.34	0.43	1.64	0.52	2.72	0.79	0.69	1.28	1.15	1.33
Th	0.19	0.21	0.26	0.26	0.49	0.52	0.76	2.43	1.70	1.97
U	0.08	0.08	0.10	0.09	0.16	0.29	0.22	0.68	0.51	0.55
La	2.46	3.08	5.13	3.31	6.31	4.90	7.95	24.08	18.42	20.24
Ce	6.96	8.60	16.36	9.42	17.55	12.34	19.05	55.18	43.43	47.41
Pr	1.16	1.41	2.97	1.57	3.07	1.60	2.65	7.73	6.14	6.67
Nd	6.29	7.39	16.74	8.50	16.63	7.44	12.30	35.05	27.93	30.98
Sm	2.21	2.49	5.54	2.76	5.20	1.80	3.00	8.19	6.54	7.25
Eu	0.84	0.94	1.89	1.04	1.90	0.60	0.98	2.55	2.18	2.38
Gd	2.93	3.35	6.97	3.51	6.34	1.80	2.92	7.98	6.58	7.25
Tb	0.54	0.61	1.28	0.64	1.10	0.27	0.45	1.19	0.98	1.08
Dy	3.52	3.98	8.87	4.45	7.39	1.61	2.66	7.01	5.84	6.55
Ho	0.75	0.86	1.85	0.92	1.56	0.30	0.48	1.28	1.08	1.24
Er	2.22	2.54	5.52	2.68	4.52	0.77	1.22	3.24	2.74	3.31
Tm	0.30	0.35	0.78	0.38	0.65	0.10	0.16	0.43	0.36	0.45
Yb	2.03	2.33	4.99	2.44	4.12	0.62	1.01	2.49	2.14	2.73
Lu	0.30	0.35	0.75	0.37	0.62	0.09	0.14	0.37	0.31	0.40

(*Continued*)

TABLE 1. MAJOR- AND TRACE-ELEMENT COMPOSITION OF REPRESENTATIVE SAMPLES (*Continued*)

	Group 3: CLIP-arc					Group 4: Early arc				
Sample:	PAN-06-204	PAN-06-203	PAN-06-205	PAN-05-004	PAN-05-008	PAN-03-007	PAN-03-004	PAN-05-031	PAN-06-088	PAN-06-206
Region:	Sona	Sona	Sona	Azuero	Azuero	Chagres	Chagres	Santa Fe Traverse	Penonome Loop	Bajano
EASTING:	464621	464621	468175	540343	546883	681638	675034	496224	545793	787121
NORTHING:	870824	870824	870544	846613	831171	1020020	1027177	921091	953830	1017837
Lithology:	ba. andesite	andesite	dacite	andesite	dacite	gabbro	microdiorite	andesite	andesite	granodiorite
SiO_2	54.9	56.6	64.4	57.0	59.4	48.6	53.8	58.3	57.6	61.6
TiO_2	0.38	0.51	0.53	0.25	0.18	0.42	0.71	1.00	0.78	0.42
Al_2O_3	15.81	16.03	14.65	14.60	13.10	19.10	17.90	16.40	15.69	16.20
Fe_2O_3	9.07	8.36	5.71	8.68	7.49	6.37	7.12	8.10	7.01	6.09
MnO	0.18	0.14	0.12	0.16	0.22	0.13	0.11	0.13	0.15	0.16
MgO	5.86	4.63	2.28	5.16	6.45	7.07	4.53	2.50	2.50	2.93
CaO	9.15	8.47	6.71	7.32	4.10	15.13	11.56	6.85	5.67	5.86
Na_2O	2.24	3.08	2.91	2.91	3.38	1.35	2.58	2.83	2.95	3.36
K_2O	0.29	0.40	0.78	1.95	0.46	0.51	0.12	1.92	3.41	1.09
P_2O_5	0.169	0.153	0.117	0.078	0.081	0.156	0.116	0.308	0.302	0.110
Total	98.02	98.40	98.16	98.11	94.86	98.83	98.55	98.34	96.03	97.82
Sc	40	38	32	40	30	53	29	27	19	18
V	278	312	212	236	174	226	180	177	166	137
Co	31	22	13	26	23	37	19	21	16	15
Ni	35	25	10	13	31	43	25	6	6	10
Zn	82	67	61	77	167	49	42	93	80	67
Rb	5.3	8.2	13.6	23.0	4.4	8.8	2.3	39.9	77.6	18.1
Sr	349	578	270	321	359	456	240	428	511	252
Y	23	21	18	12	9	15	31	31	24	16
Zr	32	67	64	21	33	46	27	91	150	73
Nb	0.54	1.58	1.16	1.25	1.14	2.20	1.49	8.08	9.40	1.44
Ta	0.03	0.09	0.07	0.14	0.09	0.14	0.11	0.49	0.50	0.09
Cd	0.10	0.07	0.06	0.07	0.25	0.08		0.10	0.07	0.05
Cs	0.16	1.08	1.38	0.17	0.07	0.19	0.01	0.66	0.91	0.20
Ba	195	242	351	884	295	298	43	934	859	601
Hf	0.83	1.76	1.68	0.99	1.37	1.39	0.94	3.66	3.99	2.05
Pb	1.81	3.61	1.85	2.80	3.38	1.25	0.39	5.57	5.43	1.83
Th	0.38	1.09	0.49	0.70	0.95	0.71	0.48	3.20	4.02	1.41
U	0.15	0.46	0.20	0.34	0.39	0.24	0.17	1.15	1.59	0.46
La	4.60	5.86	4.75	4.12	7.04	8.53	4.55	18.09	20.47	7.80
Ce	7.58	11.43	10.54	9.20	14.02	18.30	10.29	35.72	39.81	15.15
Pr	1.41	1.93	1.79	1.45	2.15	2.30	1.71	5.13	5.19	2.22
Nd	6.65	8.87	8.65	6.53	9.67	10.10	9.12	23.09	21.03	9.56
Sm	1.79	2.44	2.41	1.55	2.14	2.59	3.16	5.46	4.51	2.23
Eu	0.61	0.71	0.77	0.38	0.61	0.83	1.00	1.43	1.23	0.70
Gd	2.23	2.89	2.80	1.66	1.95	2.91	4.50	5.52	4.69	2.30
Tb	0.37	0.48	0.47	0.26	0.25	0.44	0.76	0.88	0.66	0.39
Dy	2.55	3.10	3.03	1.82	1.42	2.58	5.18	5.66	4.03	2.57
Ho	0.61	0.70	0.66	0.41	0.28	0.52	1.11	1.14	0.83	0.55
Er	1.99	2.19	2.01	1.28	0.77	1.42	3.21	3.27	2.43	1.71
Tm	0.29	0.32	0.29	0.20	0.12	0.20	0.44	0.48	0.34	0.26
Yb	2.04	2.23	1.95	1.39	0.78	1.29	2.71	3.11	2.38	1.83
Lu	0.33	0.35	0.30	0.22	0.13	0.20	0.40	0.48	0.36	0.30

(*Continued*)

TABLE 1. MAJOR- AND TRACE-ELEMENT COMPOSITION OF REPRESENTATIVE SAMPLES (*Continued*)

	Group 5: Miocene arc					Group 6: Adakites				
Sample:	PAN-06-114	PAN-05-056	PAN-06-011	PAN-03-032	PAN-06-219	PAN-05-049	PAN-06-097	PAN-06-166	PAN-06-177	PAN-06-180g
Region:	Road to Bocas	Canal-Tour	Santa Clara Loop	Coclecito Traverse	Campana	La Yeguada	El Valle	Baru	Baru	Baru
EASTING:	358387	644264	311164	561246	614431	515419	594961	340254	326350	330650
NORTHING:	1007756	1000475	976475	956056	954900	932671	953725	973213	955330	973680
Lithology:	ba. andesite	ba. andesite	ba. andesite	ba. andesite	dacite	dacite	dacite	andesite	andesite	pumice
SiO_2	52.2	51.4	52.6	55.2	62.6	69.1	64.2	57.2	57.0	54.9
TiO_2	0.84	1.70	0.88	0.76	0.56	0.28	0.22	0.72	0.73	0.72
Al_2O_3	21.50	16.00	17.18	15.60	16.70	16.10	15.72	16.80	16.50	16.30
Fe_2O_3	5.70	11.18	9.08	8.72	5.07	1.56	2.37	6.54	6.28	7.05
MnO	0.20	0.18	0.35	0.17	0.06	0.04	0.06	0.11	0.10	0.12
MgO	1.91	4.19	3.91	4.78	1.86	0.74	0.84	3.44	3.30	6.39
CaO	6.94	8.53	8.65	8.40	5.72	2.96	4.19	7.48	6.64	8.03
Na_2O	4.22	3.55	2.95	2.31	3.62	4.47	4.64	2.95	2.80	3.31
K_2O	3.31	0.71	1.85	2.31	1.56	2.70	1.40	2.40	3.03	1.36
P_2O_5	0.478	0.458	0.290	0.272	0.149	0.110	0.099	0.242	0.289	0.202
Total	97.30	97.89	97.78	98.52	97.90	98.07	93.73	97.87	96.67	98.37
Sc	6	36	27	29	13	2	3	18	9	17
V	147	308	249	245	128	24	37	197	120	165
Co	12	31	28	33	15	4	5	22	15	23
Ni	7	24	46	33	21	5	4	27	29	73
Zn	97	110	92	81	59	33	37	72	72	74
Rb	28.5	13.5	27.3	51.50	18.65	28.57	12.25	43.30	46.00	33.60
Sr	1334	446	1571	585	300	706	670	1500	1250	1030
Y	22	35	17	23	23	5	4	13	8	10
Zr	163	113	123	124	93	36	62	139	121	100
Nb	14.34	10.71	6.58	7.83	4.71	11.92	1.89	5.70	8.53	6.40
Ta	0.66	0.64	0.33	0.41	0.30	0.76	0.11	0.32	0.48	0.36
Cd	0.04	0.17	1.65	0.18	0.05	0.03	0.02	0.03	0.03	0.06
Cs	2.45	0.14	0.40	0.56	0.17	0.72	0.18	0.62	0.59	0.51
Ba	1354	344	850	1100	546	1336	475	1065	1175	856
Hf	3.68	3.93	3.27	2.92	2.78	1.76	1.78	3.78	3.34	2.68
Pb	14.15	5.50	6.22	4.59	2.23	7.85	4.34	7.58	6.60	4.98
Th	5.55	2.10	5.79	2.64	1.82	9.08	2.12	6.36	6.81	4.05
U	1.35	0.66	1.52	1.07	0.51	1.96	0.82	2.01	2.03	1.30
La	34.09	19.41	32.57	18.20	11.88	26.01	6.35	28.10	31.00	21.40
Ce	60.44	39.55	57.06	36.00	20.44	48.72	13.64	54.70	57.70	40.00
Pr	8.93	5.33	7.73	4.34	3.09	5.00	1.69	6.65	6.51	4.74
Nd	35.67	23.72	30.15	18.20	13.26	17.06	6.66	26.10	24.47	18.53
Sm	6.38	5.69	5.31	4.11	2.85	2.40	1.28	4.56	3.87	3.26
Eu	2.06	1.86	1.54	1.30	0.87	0.63	0.42	1.35	1.13	1.02
Gd	5.27	6.19	4.44	4.51	3.12	1.72	1.13	3.74	2.92	2.66
Tb	0.70	0.98	0.53	0.65	0.49	0.19	0.14	0.45	0.33	0.34
Dy	4.07	6.40	2.97	3.84	3.20	0.98	0.76	2.52	1.73	1.96
Ho	0.78	1.29	0.57	0.80	0.70	0.18	0.15	0.46	0.31	0.37
Er	2.12	3.67	1.54	2.29	2.15	0.41	0.39	1.22	0.77	1.00
Tm	0.31	0.52	0.21	0.33	0.31	0.06	0.06	0.17	0.11	0.14
Yb	2.07	3.43	1.44	2.15	2.03	0.42	0.40	1.10	0.69	0.94
Lu	0.31	0.51	0.22	0.34	0.33	0.07	0.06	0.16	0.11	0.14

Note: Major-element data are reported as wt% oxide. Trace-element concentrations are given as ppm. CLIP—Caribbean large igneous province; OIB—ocean-island basalt; ba. andesite—basaltic andesite.

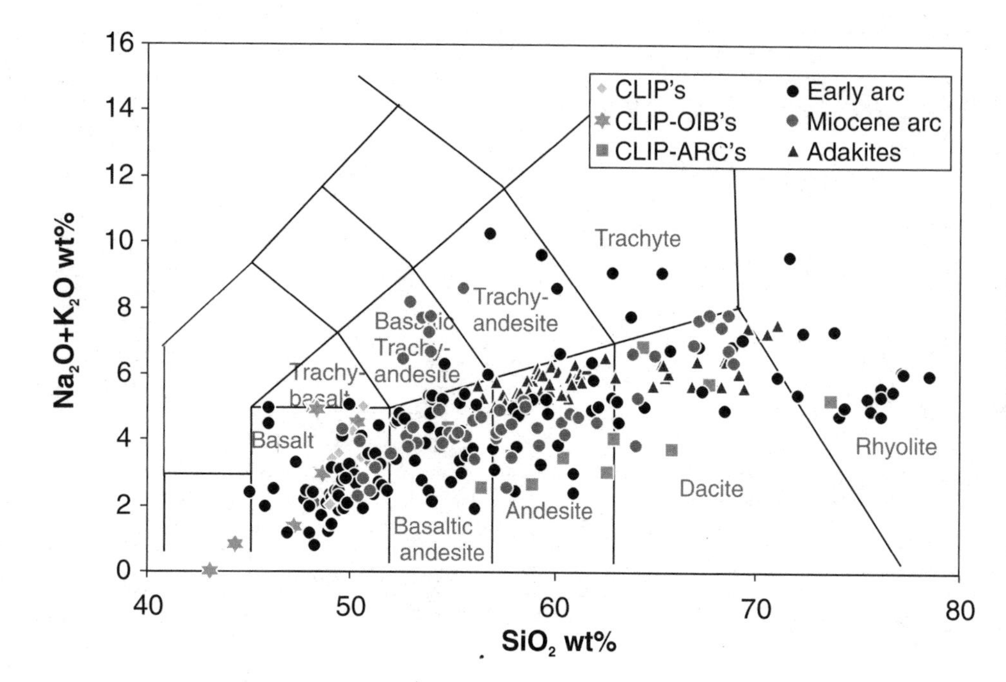

Figure 3. Total alkali–silica diagram for the volcanic and intrusive rocks from central and western Panama analyzed in this study. CLIP—Caribbean large igneous province; OIB—ocean-island basalt; ARC—arc rocks of variable age.

The trace-element patterns of CLIP rocks described by Hauff et al. (2000, 1997) and our new data set from west-central Panama are typically flat or only slightly enriched (Fig. 4A). The scatter in some mobile elements (K, Sr, Ba) is attributed to secondary, low-temperature hydrothermal alteration of these submarine basalts and should not be taken as a subduction-zone signature. Younger (70–20 Ma) accreted terranes are invariably more enriched in incompatible trace elements, akin to an intraplate OIB signature, and these can be related to the Galapagos hotspot track (Hoernle et al., 2002; Fig. 4B). In addition to these CLIP and Galapagos-derived OIB rocks, rocks of undoubted arc signature (12 samples) have been found associated with the Azuero and Soná CLIP terranes (Fig. 4C). Similar arc rocks have also been reported from the Osa Peninsula in SE Costa Rica (Fig. 1; Buchs et al., 2007). These enigmatic rocks may hold significant clues to the earliest volcanic arc development along the southern margin of the Caribbean plate. At this point, however, information about their structural relations with CLIP rocks and ages is lacking due to insufficient outcrop and the absence of datable material.

The "Early" Arc in Central Panama (66–42 Ma)

Our sampling and analytical work on these rocks concentrated on the Chagres igneous complex, located in the mountains east of the Panama Canal (Fig. 1). The deepest and oldest sections of the Central American arc are exposed in this region. As a consequence of an international, multidisciplinary research project on the Rio Chagres watershed (Harmon, 2004), the geology of the region was studied.

The geological basement of the Chagres region consists of hydrothermally altered submarine volcanic rocks represented by highly deformed basalts, basaltic andesites as sheet flows, rare pillow lavas, and volcaniclastic sequences. These are intruded by dike swarms, some of which evolved into intrusive complexes as multiple dike injection inflated the intrusions. Gabbros and diorites with chemically more evolved granodiorites, tonalites, and granites are also common. Younger and relatively undeformed basaltic andesite/andesite dikes and dike complexes crosscut all lithologies. Intercalated volcaniclastic breccias and rare sandstones have also been observed. Rare oxidized scoria in volcanic debris flows provides evidence for subaerial eruptions and their redeposition on the seamount flanks. The Chagres igneous complex thus represents a deeply eroded section of a submarine lava flow–dike complex, associated with overlying and intercalated volcanic breccias and subvolcanic intrusive rocks, dike swarms, and large intrusive complexes of gabbroic composition. Our geological reconstruction envisages a large, mostly submarine seamount and volcanic island complex with large magma chambers represented by the intrusions. The predominance of sheet flows and abundance of dikes and larger intrusive bodies suggest a relatively high eruption rate for these submarine volcanoes. Ultramafic rocks, which would provide evidence for an ultramafic oceanic basement (i.e., mid-ocean-ridge basalt [MORB] crust or oceanic mantle) in the Chagres region, are absent. Most rocks, in particular the volcanics, volcaniclastic sediments, and granites, show evidence of brittle deformation along shear zones, sometimes grading into mylonites. Chemical alteration is ubiquitous in these submarine rocks, but mineralization by sulfides is primarily observed near the fault zones.

On a water-free normalized basis, major-element concentrations vary from (rare) ferro-gabbros with 45% SiO_2 to granites with 78.5% SiO_2. Potassium contents are generally below 1%, often much lower. This attests both to the initially low

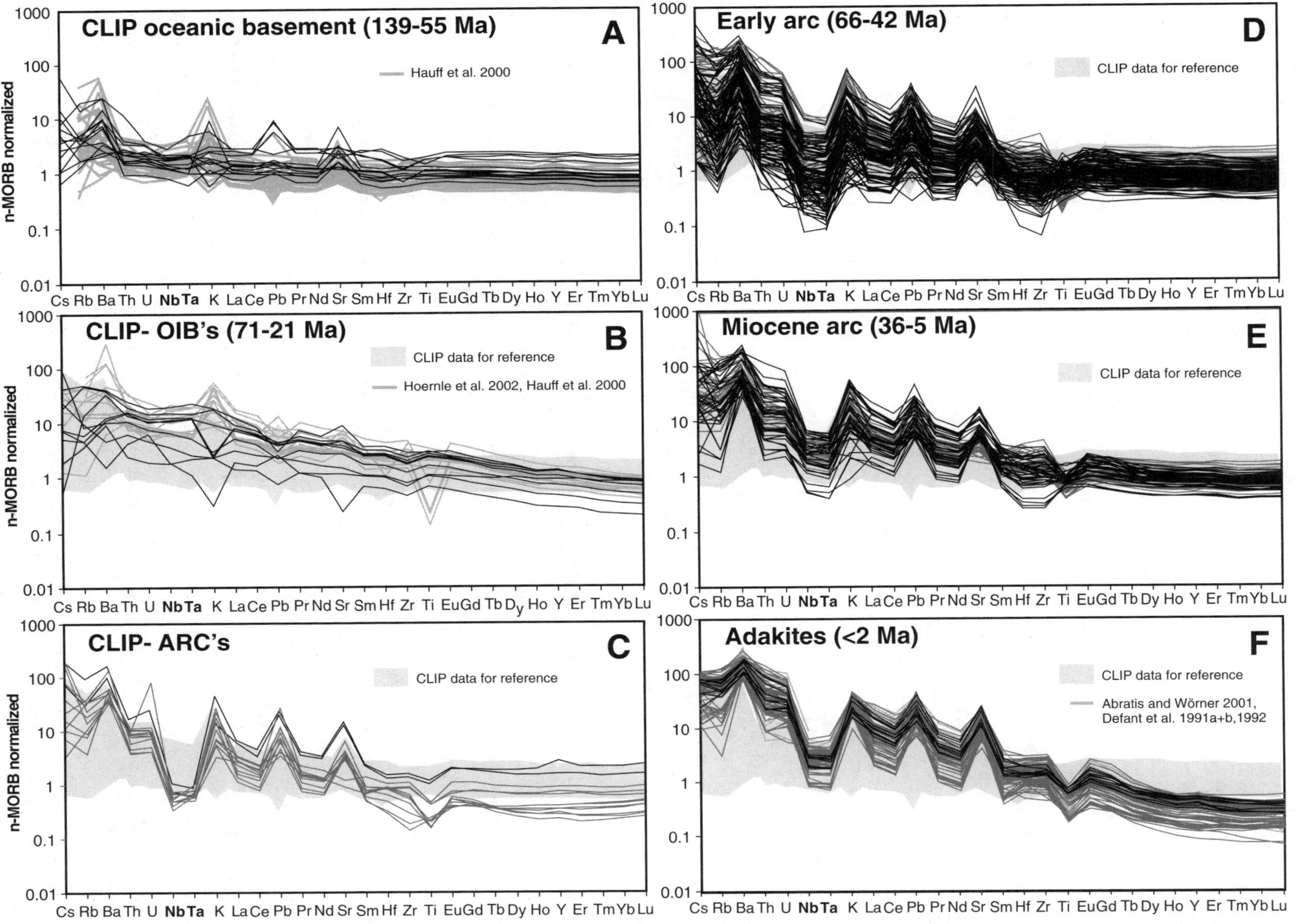

Figure 4. Chondrite-normalized (following Sun and McDonough, 1989) trace-element patterns for volcanic and intrusive rocks from central and western Panama. The data set has been filtered for compositions: <57% SiO_2 is in black and >57% SiO_2 is in dark gray to avoid effects of fractional crystallization and variable degrees of melting on the trace-element patterns. Light-gray patterns are literature data from Hoernle et al. (2002, 2004). (A) Caribbean large igneous province (CLIP) oceanic basement from Azuero and Soná. The CLIP field is also given for reference in the other diagrams to aid comparison. (B) Accreted ocean-island basalt (OIB) terranes derived from the Galapagos hotspot. Plots A and B include published analyses from Hoernle et al. (2002) for reference. (C) Rare arc rocks from Azuero and Soná. (D) "Early" arc rocks (66–42 Ma) that are dominated by a large set of samples from the Chagres igneous complex. (E) Miocene arc centers of the Cordillera de Panama of western Panama. (F) Adakites (including data from Abratis and Wörner, 2001; Defant et al., 1991a, 1991b). ARC—arc rocks of variable age; n-MORB—normal mid-oceanic-ridge basalt. See text for further discussion.

K_2O contents of many rocks and the post-emplacement, low-temperature chemical alteration. The majority of rocks (~60% of the samples analyzed) have SiO_2 contents that range between 48% and 57%. Rocks of intermediate composition, such as amphibole-bearing andesites, are present but rare. In essence, the rocks of the Chagres igneous complex define an assemblage that is bimodal in composition between basalts, basaltic andesites, and their intrusive counterparts on one hand, and granodiorites, tonalites, and granites, on the other.

Normalized trace-element patterns (Fig. 4D) display typical arc signatures, i.e., depletions in Nb and Ta and enrichment in fluid-mobile elements such as Sr, Pb, K, and Ba. Note, however, that the less incompatible trace elements on the right side of the diagram have a flat pattern and are similar in elemental abundances to CLIP rocks. The large range in Nb and Ta abundances is quite distinct, however, and trace-element patterns show Nb-Ta troughs of highly variable magnitude reflected by variable Nb and Ta values and a relatively large range in Ba/Nb ratios. Both of these observations argue for a mantle wedge with variable extents of enrichment prior to the influence of a (large ion lithophile element [LILE]–enriched) slab fluid. In comparison, the early arc mantle wedge was initially highly variable, i.e., partly more depleted and partly more enriched relative to the earlier CLIP mantle source. This compositionally variable mantle was then subsequently overprinted by slab components from a subducted oceanic plate, which resulted in enrichment of the fluid-mobile elements. It is interesting to note, that the (as yet undated) arc rocks from the CLIP terranes (see previous discussion; Buchs et al., 2007) are more restricted in composition with respect to normalized Nb and Ta vales compared to these (younger?) early arc rocks.

Age dating of the arc rocks has been initiated using Ar-Ar techniques, and initial results indicate ages between 66 and 42 Ma (Wörner et al., 2006).

Further east in Panama, "early" arc rocks consisting of low-K gabbros, tonalites, granites, and andesites with similar ages of 61–55 Ma have been reported by Maury et al. (1995) from areas between the Panama Canal and the Darien region. Thus, the first occurrence of arc magmatism in Panama is dated at 66 Ma (Maury et al., 1995; Buchs et al., 2007). Although the compiled data for this group of "early" arc rocks are dominated by the large set of samples from the Chagres igneous complex (Fig. 4), there is no systematic difference in composition between these "early" arc rocks and other arc rocks in western Panama of similar age.

A primary conclusion drawn from these observations and our initial age data (Wörner et al., 2006; Buchs et al., 2007) is that subduction and arc magmatism commenced around 66 Ma, lasted until at least 42 Ma, and involved large volumes of low-K tholeiitic basaltic magmas and their derivatives. Even though most CLIP ages are significantly older than the onset of arc magmatism, there is in fact some overlap in ages between 50 and 66 Ma (see compilation by Hoernle and Hauff, 2007) between CLIP basement and arc magmatism. It is interesting to note, that Hoernle et al. (2004) concluded that OIB seamounts began to be accreted onto the CLIP basement at around 66 Ma. We, therefore, conclude that there should have been a significant rearrangement in the plate-tectonic setting in Panama at the Cretaceous-Tertiary boundary, which is expressed in the onset of subduction and arc magmatism at that time.

The highly variable range of trace-element patterns represents both tholeiitic and calc-alkaline magmas and this implies that the mantle source either changed or remained variable during this period (66 Ma to 42 Ma), and thus it is geochemically distinct compared to that from which the older CLIP magmas were derived more than 70 m.y. ago. As sufficient age data on this series of rocks are lacking, it is not possible to determine at what time the change from tholeiitic to calc-alkaline magmas occurred within the range from 66 to 42 Ma, and whether or not it was synchronous across western and central Panama. A similar change has been documented for arc rocks in Costa Rica, however, at a much more recent time (17 Ma; Alvarado et al., 1992; Abratis and Wörner, 2001).

The Miocene Arc (36–5 Ma)

Younger sections of volcanic arc rocks occur along the entire Cordillera de Panama in west-central Panama (Drummond et al., 1995; de Boer et al., 1988, 1991). The oldest K-Ar ages for this group of rocks fall in the range of 36–29 Ma (Kesler et al., 1977). These rocks are calc-alkaline andesites and dacites located within the Cordillera de Panama, where morphological expression of these volcanic centers is poor due to deep erosion and cover by younger deposits. This may be the reason why these Miocene arc centers can easily be overlooked, as emphasized by de Boer et al. (1991) and Maury et al. (1995). However, series of younger stratovolcanoes and isolated centers in the Cordilleran and forearc regions have been identified on the basis of morphology and have yielded K-Ar and Ar-Ar ages from 21 to 5 Ma (de Boer et al., 1988, 1991; Defant et al., 1992; Drummond et al., 1995; this study).

Trace-element patterns of all "younger arc" rocks (younger than 35 Ma) have a less-pronounced Nb-Ta depletion relative to LILE compared to older arc rocks in this region, and, in contrast to the volcanic rocks produced during the earlier period of magmatism, very few of the Miocene arc rocks have depleted trace-element patterns (i.e., Nb and Ta normalized concentrations below 1; Fig. 4E). This indicates that the change from (depleted) tholeiitic trace-element patterns to more mature (enriched) arc magmatism must have been completed prior to 36 Ma, which is significantly earlier than in Costa Rica (17 Ma; Alvarado et al., 1992; Drummond et al., 1995; Abratis and Wörner, 2001).

Magmatic Gap and HREE-Depleted Andesites and Dacites (<2 Ma)

"Normal" arc magmatism in Central Panama ceased at around 6 Ma. Lonsdale and Klitgord (1978) first identified a gap in magmatism between ca. 6 and 3 Ma in west-central Panama and southeastern Costa Rica. Following this magmatic gap, younger volcanic products (1–3 Ma) are invariably HREE-depleted rocks (Abratis and Wörner, 2001; Fig. 4F). This observation is explained by the cessation of active spreading in the Panama Basin during late Miocene time. More recently, the Pleistocene uplift of the Cordillera de Talamanca in the region where young arc volcanoes are absent was proposed to have been caused by the collision of the Cocos Ridge with the subduction zone in southern Costa Rica. As discussed by Abratis and Wörner (2001), the timing of this collision was determined from sedimentologic and tectonic evidence as having occurred between 3 and 4 Ma (von Eynatten et al., 1993; Krawinkel et al., 2000). However, Silver et al. (2004) presented evidence that a first "disturbance" in the deep-marine environment occurred at 8–10 Ma, which may have been related to the initiation of collision. However, the cessation of "normal" arc magmatism throughout Panama at this early time probably had a different cause, since the full ridge collision and associated Cordilleran uplift occurred much later, and uplift is only located in the present Cordillera de Talamanca. Rather, as indicated by Londsdale and Klitgord (1978), the convergence between the Cocos plate and Panama, and thus also active subduction, stopped at this time as plate motions were progressively accommodated inside the Caribbean plate by the Panama deformed belt (Fig. 1). In essence, there are two unrelated causes for the magmatic gap in southern Costa Rica and Panama: (1) a plate-tectonic rearrangement after cessation of subduction of the Cocos plate throughout Panama at 6 Ma, and (2) the collision of the Cocos Ridge in Costa Rica beginning at about 4 to 5 m.y. ago.

Following this evolution, scattered volcanic centers in southern Costa Rica, as well as a few large stratovolcanoes in western Panama, erupted andesites and dacites with an unusually depleted heavy REE character (Fig. 4F; Defant et al., 1991a, 1991b; Johnston and Thorkelson, 1997). This particular trace-element pattern in arc rocks has been attributed to the melting of basalts at high pressure under conditions where garnet is a stable residual phase (Defant et al., 1991a, 1991b; Drummond et al., 1995; Kay et al., 1993; Johnston and Thorkelson, 1997; Abratis and Wörner, 2001; Goss and Kay, 2007). A slab window and melting of Cocos plate MORB following subduction of the Panama spreading center have been proposed as an explanation (Johnston and Thorkelson, 1997; Thorkelson and Breitsprecher, 2005). An alternative view was presented by Abratis and Wörner (2001), who argued that the Pb-isotope composition of these rocks is not consistent with the melting of MOR basalts, and, therefore, they favored melting of the leading edge of the Cocos Ridge after its collision with the Central American land bridge as an alternative explanation. Figure 4F includes data for El Baru volcano in westernmost Panama from this study and data from other centers in the Cordillera de Panama and in southern Costa Rica (Defant et al., 1991a, 1991b; Drummond et al., 1995; Abratis and Wörner, 2001). All HREE-depleted arc rocks in this region have very similar trace-element patterns. In addition to the fact that Cocos plate MORB is inconsistent with the Pb-isotope composition of these rocks, preliminary trace-element modeling requires a basaltic precursor more enriched in incompatible trace elements compared to MORB. From both of these lines of evidence, we conclude that likely sources for these HREE-depleted andesites and dacites include CLIP basement or Galapagos ridge basalts (e.g., from subducted portions of the Cocos or Coiba Ridge).

DISCUSSION: CHANGING MANTLE SOURCES THROUGH TIME

This study has documented systematic variations in magma compositions for Tertiary volcanic and intrusive rocks (66–6 Ma) from west-central Panama based on a compilation of published and new data. These compositional variations reflect changing magma sources and mantle source compositions over the last 66 m.y. On a wider scope, the igneous basement of the Central American land bridge has been constructed since ca. 139 Ma from basaltic rocks derived from a large oceanic igneous province (CLIP) produced by the Galapagos plume (Hauff et al., 1997, 2000; Hoernle et al., 2002, 2004; Sinton et al., 1997, 1998; Thompson et al., 2004). At this time, the thickened CLIP oceanic crust must have been underlain by a plume-mantle, which had been variably mixed with asthenospheric mantle material during plume ascent and spreading below the lithosphere (Hoernle et al., 2004; Sinton et al., 1997, 1998; Thompson et al., 2004). Magmas, which subsequently were derived from this mixed plume mantle until 55 Ma (Hoernle et al., 2004), range from 1 to 30 in chondrite-normalized compositions for the most incompatible high field strength elements (HFSEs) such as Nb and Ta, whereas Zr contents are much more restricted (Figs. 4A and 4B). It is important to note that arc magmatism ("early arc"; Fig. 4D) followed directly after the main phase of CLIP magmatism from 66 Ma onward. Immobile HFSEs, which should be unaffected by enrichments from the subducting slab, exhibit a similar range of chondrite-normalized trace-element contents as the CLIP rocks, but they extend to much lower values (0.01 for chondrite-normalized Nb and Ta). The mantle wedge of the "early" arc system consisted of mantle material that was only in part similar to that of the previously underlying the Caribbean plate: it contained, in addition, a much more depleted component. Either a depleted asthenospheric mantle, which was underlying the plume source, was introduced into the melting region of the arc mantle wedge, or an entirely new, depleted mantle source was introduced into the melting region of the arc rocks laterally. In any case, arc magmatism initiated with large melt volumes and eruption rates in an almost entirely marine setting at 66 Ma, as documented by the Chagres igneous complex and other age data (Maury et al., 1995; Buchs et al., 2007). The enigmatic arc rocks in the CLIP terranes all have relatively low Nb-Ta contents compared to CLIP rocks, and they were entirely derived from a depleted mantle source. However, the "early arc" rocks, while showing an overall large range in Nb and Ta values, include compositions that were derived from an even more depleted mantle. The large variation in early arc rocks (Fig. 4D) could represent a transitional phase of magmatism when subduction was initially established. We do not know the ages of the enigmatic arc rocks found on the CLIP terranes (Fig. 4C), but if an age around 65 Ma from the Osa Peninsula (Buchs et al., 2007) is confirmed for other occurrences on Soná and Azuero, then these CLIP-arc rocks could hold a clue to better understanding of arc initiation at the Cretaceous-Tertiary boundary in Central America. High-precision age dating of these rocks is necessary to better constrain changing plate arrangements at this time.

Based on existing evidence, we propose a major plate-tectonic rearrangement at this time (55–65 Ma), which involved: (1) initiation of voluminous submarine arc magmatism in Panama, (2) a sudden change in magma sources toward variably depleted mantle and strong slab fluid signatures, (3) onset of seamount accretion onto the active southern Caribbean plate margin, and (4) cessation of CLIP magmatism.

With time, and increasing maturation of the arc system, the mantle source beneath the Cordillera de Panama became more homogeneous. Chondrite-normalized Nb-Ta values for volcanic rocks that erupted between 66 and 36 Ma cluster in a restricted compositional range (Fig. 4E). Two interpretations, or possibly some combination thereof, can explain this observation: either the mantle wedge became thoroughly mixed and homogenized and/or the mantle wedge was replaced by a more homogeneous mantle. In either case, two conclusions can be drawn: (1) strongly depleted mantle material, which gave rise to magmas with chondrite-normalized Nb-Ta values near 0.01, ceased to exist in central Panama, and (2) the mantle wedge and the subduction system must have changed significantly between the "early" arc and the Miocene arc system. One possibility that can explain this change in mantle sources is the tectonic reorganization of the Pacific plate after the breakup of the Farallon plate at around this time (Lonsdale, 2005). Arc magmatism appears to have been derived from the same, relatively homogeneous mantle wedge between 36 and 5 Ma, since large compositional heterogeneities or changes in trace-element signatures are not observed during this time.

As indicated by both morphological evidence and age-dating, "normal" arc magmatism ceased in southern Costa Rica and west-central Panama at ca. 5 Ma as a consequence of a major change in tectonic setting and dynamics of the subduction system. Renewed, but highly localized magmatic activity in western Panama since ca. 2 Ma is characterized by HREE-depleted andesites and dacites (adakites) that have similar radiogenic isotope signatures to Galapagos-derived CLIP and OIB basalts (Hoernle et al., 2004; Abratis and Wörner, 2001).

We consider that (1) cessation of "normal" arc magmatism around 5 Ma, (2) uplift of the Cordillera de Talamanca in southeastern Costa Rica and western Panama, (3) collision of the Cocos and Coiba Ridges, and (4) development of the Panama fracture zone and Panama deformed belt (Fig. 1) are all related events that resulted from little to no plate convergence and thus the end of active subduction below Panama as documented by Mann and Kolarsky (1995). Apparently, another major plate rearrangement also occurred at this time. The "Occam's razor" principle (i.e., that the simplest explanation tends to be the correct one), demands that all of these events should be in some way related to one another and also to the origin of the HREE-depleted andesite and dacite magmatism in this region. Since the geochemical data strongly indicate that Galapagos Ridge crust, rather than Cocos plate MORB, was the precursor to these magmas, and because the crust in Panama is not sufficiently thick to stabilize garnet in the residue of partial melting, one is left with two options to explain the observed HREE-depleted magmatic geochemistry. (1) Westernmost Panama is underlain by a slab window, and melting occurs at the leading edge of subducted Cocos and Coiba Ridge (Abratis and Wörner, 2001), or (2) CLIP- and ridge-type material is being tectonically eroded and involved in the magma source region (Goss and Kay, 2007, and discussion therein). In the course of such tectonic erosion (von Huene and Scholl, 1991), material from the upper plate may be removed (tectonically eroded), subducted below western Panama and southeastern Costa Rica and then partially melted in the region of magma generation of the subduction zone (Goss and Kay, 2007). More "normal" arc magmatism is absent in the region, and thus subduction, slab dehydration, and mantle-wedge melting do not occur here. At the same time, arc magmatism is observed in abundance to the northwest and southeast, respectively (Carr et al., 1990), of the arc magmatic gap in northwestern Costa Rica and northern Colombia. Therefore, we find it difficult to envision melting of tectonically eroded material only occurring where normal arc magmatism is absent, even though CLIP and Galapagos Ridge basalts occur (and may become tectonically eroded almost everywhere along the leading upper plate). Thus, we favor an interpretation involving slab window formation and melting of the exposed leading edges of subducted oceanic ridges that were originally derived from the Galapagos hotspot. De Boer et al. (1995) suggested that mantle-wedge material was introduced to the northwest below central Costa Rica and to the southeast (westernmost Panama) by the subducted mass of the Cocos Ridge. A similar idea was postulated by Hoernle et al. (2008), who argued, based mainly on Pb-isotopic composition of arc rocks, that a geochemical component from the subducted seamount chain and ridges can be identified in the arc magma source below Costa Rica. We would like to emphasize that at present, the plate-tectonic situation is highly complex, that geochemical signatures in arc magmas can be highly varied, and that mantle sources can move laterally. In this context, it is very difficult to interpret the compositional characteristics observed for rocks from Panama, which are highly variable spatially, cover a time span of many millions of years in a plate-tectonic setting that may have changed rapidly and that may have been equally complicated in the past as it is today.

SUMMARY AND CONCLUSIONS

The geologic development of Panama can be divided into four stages: (1) development of a genetically complex basement prior to ca. 70 Ma, (2) development of an "early" volcanic arc between ca. 65 and 40 Ma, (3) development of a "younger" volcanic arc between ca. 40 and 5 Ma, and (4) localized HREE-depleted magmas since ca. 2 Ma. Three distinct trace-element signatures are recognized in the oldest (younger than 95 Ma to older than 66 Ma) basement rocks: (1) flat trace-element patterns in oceanic basement of the Caribbean large igneous province (CLIP), (2) enriched OIB signatures in CLIP terranes, and (3) an arc signature in CLIP rocks. The mantle source involved in CLIP magmatism is related to the Galapagos plume. The change from intraplate CLIP magmatism occurred at ca. 66 Ma. The arc magmas generated between 66 and 42 Ma were derived from a highly compositionally variable mantle source that ranged from highly depleted to somewhat less depleted in character, onto which a slab signature was imprinted. The "younger" arc magmatism that began at ca. 40 Ma originated from a relatively homogeneous, and slightly enriched, mantle wedge. The changes observed in mantle sources for these Panamanian igneous rocks may be related to replacement of plume-mantle in the mantle wedge and/or homogenization by corner flow, and involvement of less depleted mantle material with time. Cessation of "normal" arc magmatism occurred around 5 Ma as a result of the collision of the Cocos Ridge in southeastern Costa Rica and cessation of convergence in Panama. The occurrence of HREE-depleted magmas in part of the arc-magmatic gap during the past 2 m.y. was the result of melting of the leading edge of the subducting Cocos and Coiba Ridges at the margins of a slab window.

REFERENCES CITED

Abratis, M., and Wörner, G., 2001, Ridge collision, slab-window formation, and the flux of Pacific asthenosphere into the Caribbean realm: Geology, v. 29, p. 127–130, doi: 10.1130/0091-7613(2001)029<0127:RCSWFA>2.0.CO;2.

Alvarado, G.E., Kussmaul, S., Chiesa, S., Gillot, P.Y., Appel, H., Wörner, G., and Rundle, C., 1992, Resumen cronoestratigrafico de las rocas igneas de Costa Rica basado en dataciones radiometricas (Chronostratigraphic review of igneous rocks of Costa Rica based on radiometric dates): Journal of South American Earth Sciences, v. 6, p. 151–168, doi: 10.1016/0895-9811(92)90005-J.

Buchs, D.M., Baumgartner, P.O., and Arculus, R., 2007, Late Cretaceous arc initiation on the edge of an oceanic plateau (southern Central America): Eos (Transactions, American Geophysical Union), v. 88, no. 52, Fall Meeting supplement, abstract T13C-1468.

Carr, M.J., Feigenson, M.D., and Bennett, E.A., 1990, Incompatible element and isotopic evidence for tectonic control of source mixing and melt extraction along the Central American arc: Contributions to Mineralogy and Petrology, v. 105, no. 4, p. 369-380.

Coates, A.G., Aubry, M.P., Berggren, W.A., and Collins, L.S., 2000, New evidence for the earliest stages in the rise of the Isthmus of Panama from Bocas del Toro, Panama: Geological Society of America Abstracts with Programs, v. 32, no. 7, p. 146.

Collins, L.S., Coates, A.G., Berggren, W.A., Aubry, M.-P., and Zhang, J., 1996, The late Miocene Panama isthmian strait: Geology, v. 24, p. 687–690, doi: 10.1130/0091-7613(1996)024<0687:TLMPIS>2.3.CO;2.

de Boer, J.Z., Defant, M.J., Stewart, R.H., Restrepo, J.F., Clark, L.F., and Ramirez, A., 1988, Quaternary calc-alkaline volcanism in western Panama: Regional variation and implication for tectonic framework: Journal of South American Earth Sciences, v. 1, p. 275–293, doi: 10.1016/0895-9811(88)90006-5.

de Boer, J.Z., Defant, M.J., Stewart, R.H., and Bellon, H., 1991, Evidence for active subduction below western Panama: Geology, v. 19, p. 649–652, doi: 10.1130/0091-7613(1991)019<0649:EFASBW>2.3.CO;2.

de Boer, J.Z., Drummond, M.S., Bordelon, M.J., Defant, M.J., Bellon, H., and Maury, R.C., 1995, Cenozoic magmatic phases of the Costa Rican island arc (Cordillera de Talamanca), *in* Mann, P., ed., Geologic and Tectonic Development of the Caribbean Plate Boundary in Southern Central America: Geological Society of America Special Paper 295, p. 35–55.

Defant, M.J., Clark, L.F., Stewart, R.H., Drummond, M.S., de Boer, J.Z., Maury, R.C., Bellon, H., Jackson, T.E., and Restrepo, J.F., 1991a, Andesite and dacite genesis via contrasting processes: The geology and geochemistry of El Valle Volcano, Panama: Contributions to Mineralogy and Petrology, v. 106, p. 309–324, doi: 10.1007/BF00324560.

Defant, M.J., Richerson, P.M., de Boer, J.Z., Stewart, R.H., Maury, R.C., Bellon, H., Drummond, M.S., Feigenson, M.D., and Jackson, T.E., 1991b, Dacite genesis via both slab melting and differentiation: Petrogenesis of La Yeguada volcanic complex, Panama: Journal of Petrology, v. 32, p. 1101–1142.

Defant, M.J., Jackson, T.E., Drummond, M.S., de Boer, J.Z., Bellon, H., Feigenson, M.D., Maury, R.C., and Stewart, R.H., 1992, The geochemistry of young volcanism throughout western Panama and southeastern Costa Rica: An overview: Journal of the Geological Society of London, v. 14, p. 569–579.

Drummond, M.S., Bordelon, M., de Boer, J.Z., Defant, M.J., Bellon, H., and Feigenson, M.D., 1995, Igneous petrogenesis and tectonic setting of plutonic and volcanic rocks of the Cordillera de Talamanca, Costa Rica–Panama, Central American arc: American Journal of Science, v. 295, p. 875–919.

Escalante, G., 1990, The geology of southern Central America and western Colombia, *in* Dengo, G., and Case, J.E., eds., The Caribbean Region: Boulder, Colorado, Geological Society of America, Geology of North America, v. H, p. 201–230.

Goossens, P.J., Rose, W.I., and Flores, D., 1977, Geochemistry of tholeiites of the Basic Igneous Complex of northwestern South America: Geological Society of America Bulletin, v. 88, p. 1711–1720, doi: 10.1130/0016-7606(1977)88<1711:GOTOTB>2.0.CO;2.

Goss, A., and Kay, S.M., 2007, Steep REE patterns and enriched Pb isotopes in southern Central American arc magmas: Evidence for forearc subduction erosion?: Geochemistry, Geophysics, Geosystems, v. 7, Q05016, doi: 10.1029/2005GC001163.

Harmon, R.S., ed., 2004, The Chagres Basin: Multidisciplinary Profile of a Tropical Watershed: Dordrecht, the Netherlands, Springer, Water Science and Technology Library, v. 52, p. 1–355.

Hauff, F., Hoernle, K., Schmincke, H.-U., and Werner, R., 1997, A mid-Cretaceous origin for the Galapagos hotspot; volcanological, petrological and geochemical evidence from Costa Rican oceanic crustal segments: Geologische Rundschau, v. 86, p. 141–155.

Hauff, F., Hoernle, K., van der Bogaard, P., Alvarado, G., and Garbe-Schoenberg, D., 2000, Age and geochemistry of basaltic complexes in western Costa Rica; contributions to the geotectonic evolution of Central America: Geochemistry, Geophysics, Geosystems, v. 1, 1009, doi: 10.1029/1999GC000020.

Haug, G.H., and Tiedemann, R., 1998, Effect of the formation of the Isthmus of Panama on Atlantic Ocean thermohaline circulation: Nature, v. 393, p. 673–676, doi: 10.1038/31447.

Hoernle, K., and Hauff, V., 2007, Oceanic igneous complexes, *in* Bundschuh, J., and Alvarado, G., eds., Geology of Central America: Leiden, the Netherlands, Taylor & Francis, p. 523–547.

Hoernle, K., van der Bogaard, P., Werner, R., Lissinna, B., Hauff, V., Allvarado, G., and Garbe-Schönberg, D., 2002, Missing history (16–71 Ma) of the Galapagos hotspot: Implications for the tectonic and biological evolution of the Americas: Geology, v. 30, p. 795–798, doi: 10.1130/0091-7613(2002)030<0795:MHMOTG>2.0.CO;2.

Hoernle, K., Hauff, V., and van der Bogaard, P., 2004, 70 m.y. history (139–69 Ma) for the Caribbean large igneous province: Geology, v. 32, p. 697–700, doi: 10.1130/G20574.1.

Hoernle, K., Abt, D.L., Fischer, K.M., Nichols, H., Hauff, V., Abers, G.A., van der Bogaard, P., Heydolph, K., Alvarado, G., Protti, M., and Strauch, W., 2008, Arc-parallel flow in the mantle wedge beneath Costa Rica and Nicaragua: Nature, v. 451, p. 1094–1097, doi: 10.1038/nature06550.

Johnston, S.T., and Thorkelson, D.J., 1997, Cocos-Nazca slab window beneath Central America: Earth and Planetary Science Letters, v. 146, p. 465–474, doi: 10.1016/S0012-821X(96)00242-7.

Kay, S.M., Ramos, V.A., and Marquez, M., 1993, Evidence in Cerro Pampa volcanic rocks for slab-melting prior to ridge-trench collision in southern South America: The Journal of Geology, v. 101, p. 703–714.

Kesler, S.E., Sutter, J.F., Issigonis, M.J., Jones, L.M., and Walker, R.L., 1977, Evolution of porphyry copper mineralization in an oceanic island arc; Panama: Economic Geology and the Bulletin of the Society of Economic Geologists, v. 72, no. 6, p. 1142–1153.

Krawinkel, H., Seyfried, H., Calvo, C., and Astorga, A., 2000, Origin and inversion of sedimentary basins in Southern Central America: Zeitschrift für Angewandte Geologie: Sonderheft, v. 1, p. 71–77.

Lonsdale, P., 2005, Creation of the Cocos and Nazca plates by fission of the Farallon plate: Tectonophysics, v. 404, no. 3–4, p. 237–264.

Lonsdale, P., and Klitgord, K.D., 1978, Structure and tectonic history of the eastern Panama Basin: Geological Society of America Bulletin, v. 89, no. 7, p. 981–999.

Mann, P., and Kolarsky, R.A., 1995, East Panama deformed belt; structure, age, and neotectonic significance, *in* Mann, P., ed., Geologic and Tectonic Development of the Caribbean Plate Boundary in Southern Central America: Geological Society of America Special Paper 295, p. 111–130.

Maury, R.C., Defant, M.J., Bellon, H., de Boer, J.Z., Stewart, R.W., and Cotten, J., 1995, Early Tertiary arc volcanics from eastern Panama, *in* Mann, P., ed., Geologic and Tectonic Development of the Caribbean Plate Boundary in Southern Central America: Geological Society of America Special Paper 295, p. 29–34.

Meschede, M., and Barckhausen, U., 2001, The relationship of the Cocos and Carnegie Ridges: Age constraints from paleogeographic reconstructions: International Journal of Earth Science, v. 90, p. 386–392.

Pindell, J.L., and Barret, S.F., 1990, Geological evolution of the Caribbean region: A plate tectonic perspective, *in* Dengo, G., and Case, J.E., eds., The Caribbean Region: Boulder, Colorado, Geological Society of America, Geology of North America, v. H, p. 405–432.

Pindell, J.L., Kennan, L., Maresch, W.V., Stanek, K.P., Draper, G., and Higgs, R., 2005, Plate-kinematics and crustal dynamics of circum-Caribbean arc-continent interactions: Tectonic controls on basin development in proto-Caribbean margins, *in* Avé Lallemant, H.G., and Sisson, V.B., eds., Caribbean-South American Plate Interactions, Venezuela: Geological Society of America Special Paper 394, p. 7–52.

Pindell, J., Kennan, L., Stanek, K.P., Maresch, W.V., and Draper, G., 2006, Foundations of Gulf of Mexico and Caribbean evolution: Eight controversies resolved: Geological Acta, v. 4, p. 303–341.

Reynaud, C., Jaillard, E., Lapierre, H., Mamberti, M., and Mascle, G.H., 1999, Oceanic plateau and island arcs of southwestern Ecuador; their place in the geodynamic evolution of northwestern South America: Tectonophysics, v. 307, no. 3–4, p. 235–254.

Silver, E., Costa-Pisani, P., Hutnak, M., Fisher, A., DeShon, H., and Taylor, B., 2004, An 8–10 Ma tectonic event on the Cocos plate offshore Costa Rica; results of Cocos Ridge collision?: Geophysical Research Letters, v. 31, p. L18601, doi: 10.1029/2004GL020272.

Sinton, C.W., Duncan, R.A., and Denyer, P., 1997, Nicoya Peninsula, Costa Rica: A single suite of Caribbean oceanic plateau magmas: Journal of Geophysical Research, v. 102, p. 15,507–15,520, doi: 10.1029/97JB00681.

Sinton, C.W., Duncan, R.A., Storey, M., Lewis, J., and Estrada, J.J., 1998, An oceanic flood basalt province within the Caribbean plate: Earth and Planetary Science Letters, v. 155, p. 221–235, doi: 10.1016/S0012-821X(97)00214-8.

Sun, S.S., and McDonough, W.F., 1989, Chemical and isotopic systematics of oceanic basalts: Implications for mantle composition and processes, *in* Saunders, A.D., and Norry, M.J., eds., Magmatism in the Ocean Basins: Geological Society of London Special Publication 42, p. 313–345.

Thompson, P.M.E., Kempton, P.D., White, R.V., Kerr, A.C., Tarney, J., Saunders, A.D., Fitton, J.G., and McBirney, A., 2004, Hf-Nd isotope constraints on the origin of the Cretaceous Caribbean plateau and its relationship to the Galapagos plume: Earth and Planetary Science Letters, v. 217, p. 59–75, doi: 10.1016/S0012-821X(03)00542-9.

Thorkelson, D.J., and Breitsprecher, K., 2005, Partial melting of slab window margins: Genesis of adakitic and non-adakitic magmas: Lithos, v. 79, p. 25–41.

von Eynatten, H., Krawinkel, H., and Winsemann, J., 1993, Plio-Pleistocene outer arc basins in southern Central America, *in* Frostick, L.E., and Steel, R.J., eds., Tectonic Controls and Signatures in Sedimentary Successions: International Association of Sedimentologists Special Publication 20, p. 399–414.

von Huene, R., and Scholl, D.W., 1991, Observations at convergent margins concerning sediment subduction, subduction erosion, and the growth of continental crust: Reviews of Geophysics, v. 29, p. 279–316, doi: 10.1029/91RG00969.

Wörner, G., Harmon, R.S., Wegner, W., and Singer, B., 2006, Linking America's backbone: Geological development and basement rocks of central Panama, *in* Abstracts with Programs, Geological Society of America Conference "Backbone of the Americas," Mendoza, Argentina, 3–7 April 2006: Boulder, Colorado, Geological Society of America, p. 60.

Manuscript Accepted by the Society 5 December 2008

Printed in the USA

The Geological Society of America
Memoir 204
2009

Mode and timing of terrane accretion in the forearc of the Andes in Ecuador

Cristian Vallejo*
Wilfried Winkler[†]
Geological Institute, Eldgenössische Technische Hochschule Zürich, CH-8092 Zürich, Switzerland

Richard A. Spikings[†]
Department of Mineralogy, University of Geneva, CH-1205 Geneva, Switzerland

Leonard Luzieux[§]
Geological Institute, Eldgenössische Technische Hochschule Zürich, CH-8092 Zürich, Switzerland

Friedrich Heller[†]
Geological Institute, Eldgenössische Technische Hochschule Zürich, CH-8093 Zürich, Switzerland

François Bussy[†]
Institute of Mineralogy and Geochemistry, University of Lausanne, CH-1015 Lausanne, Switzerland

ABSTRACT

The volcanic basement of the Ecuadorian Western Cordillera (Pallatanga Formation and San Juan unit) is made up of mafic and ultramafic rocks that once formed an oceanic plateau. Radiometric ages from these rocks overlap with a hornblende $^{40}Ar/^{39}Ar$ plateau age of 88 ± 1.6 Ma obtained for oceanic plateau basement rocks of the Piñon Formation in coastal Ecuador, and with ca. 92–88 Ma ages reported for oceanic plateau sequences in the Caribbean and western Colombia. These results suggest that the oceanic plateau rocks of the Western Cordillera and flat forearc in Ecuador are derived from the Late Cretaceous Caribbean-Colombia oceanic plateau.

Intraoceanic island-arc sequences (Rio Cala Group) overlie the plateau in the Western Cordillera and yield crystallization ages that range between ca. 85 and 72 Ma. The geochemistry and radiometric ages of island-arc lavas from the Rio Cala Group, combined with the age range and geochemistry of their turbiditic, volcaniclastic products, indicate that the arc was initiated by westward subduction beneath the Caribbean Plateau. They are coeval with island-arc rocks of coastal Ecuador (Las Orquideas, San Lorenzo, and Cayo Formations) and Colombia (Ricaurte Arc). These island-arc

*Corresponding author; present address: Geoconsult Ltda, Pasaje Jardín 168 y 6 de Diciembre. Edificio Century Plaza, Quito, Ecuador; e-mail: cristian.vallejo@alumni.ethz.ch.
[†]E-mails: Winkler—wilfried.winkler@erdw.ethz.ch; Spikings—spikings@terre.unige.ch; Heller—heller@mag.ig.erdw.ethz.ch; Bussy—Francois.Bussy@unil.ch.
[§]Present address: Holcim Group Support Ltd., CH-5113 Holderbank, Aargau, Switzerland; e-mail: leonard.luzieux@holcim.com.

Vallejo, C., Winkler, W., Spikings, R.A., Luzieux, L., Heller, F., and Bussy, F., 2009, Mode and timing of terrane accretion in the forearc of the Andes in Ecuador, *in* Kay, S.M., Ramos, V.A., and Dickinson, W.R., eds., Backbone of the Americas: Shallow Subduction, Plateau Uplift, and Ridge and Terrane Collision: Geological Society of America Memoir 204, p. 197–216, doi: 10.1130/2009.1204(09). For permission to copy, contact editing@geosociety.org.

units may be related to the Late Cretaceous Great Arc of the Caribbean. Paleomagnetic analyses of volcanic rocks of the Piñon and San Lorenzo Formations of the southern external forearc show that they erupted at equatorial or low southern latitudes.

The initial collision between South America and the Caribbean-Colombia oceanic plateau caused rock uplift and exhumation (>1 km/m.y.) within the continental margin during the Late Cretaceous (ca. 75–65 Ma). Magmatism associated with the Campanian–early Maastrichtian Rio Cala Arc ceased during the Maastrichtian because the collision event blocked the subduction zone below the oceanic plateau. Paleomagnetic data from basement and sedimentary cover rocks in the coastal forearc reveal 20°–50° of clockwise rotation during the Campanian, which was synchronous with the collision of the oceanic plateau and arc sequence with South America.

East-dipping subduction beneath the accreted oceanic plateau formed the latest Maastrichtian to early Paleogene (ca. 60 Ma) Silante volcanic arc, which was deposited in a terrestrial environment. Subsequently, Paleocene to Eocene volcanic rocks of the Macuchi unit were deposited, and these probably represent a continuation of the Silante arc. This submarine volcanism was coeval with the deposition of siliciclastic rocks of the Angamarca Group, which were mainly derived from the emerging Eastern Cordillera.

INTRODUCTION

Subduction of ocean floor at a convergent plate boundary can be regarded as a steady-state process that can be interrupted by the arrival of a buoyant object. Arcs and continents are familiar buoyant objects involved in collisions. In this contribution, we analyze a less familiar kind of collision, that of an oceanic plateau against a continental margin. We study the southernmost allochthonous fragment of oceanic plateau affinity in the Northern Andes, which now forms the basement of the Western Cordillera of Ecuador.

Western Ecuador and Colombia are characterized by a large positive Bouguer anomaly that spans the Western Cordillera and external forearc region (Case et al., 1971; Feininger and Seguin, 1983), and that results from a thick (>10 km), mafic, crystalline basement. The mafic rocks define allochthonous structural blocks that accreted in the Cretaceous, resulting in orogenesis, surface uplift, and exhumation (Feininger and Bristow, 1980; McCourt et al., 1984; Spikings et al., 2001, 2005; Toro and Jaillard, 2005). Consequently, the northwestern margin of the South American plate grew rapidly via accretion during the Cretaceous, and it is hence unique along the Andean active margin.

Previous work in the Western Cordillera of Ecuador (e.g., Wolf, 1892; Sauer, 1965; Egüez, 1986; Kerr et al., 2002a; Jaillard et al., 2004; Spikings et al., 2005) has partly constrained the stratigraphic and tectonic evolution of this ocean-continent accretionary complex. Nevertheless, various geological problems remain, mainly because of poor rock exposure and a lack of accurate radiometric age data. Key questions include: (1) the age and origin of the allochthonous volcanic basement of the Western Cordillera and coastal plain, (2) the time of accretion of the allochthonous blocks, and (3) the ways in which the rocks of the Western Cordillera evolved during the Paleogene. This study addresses these questions and presents new qualitative and quantitative data from the Western Cordillera of Ecuador.

REGIONAL GEOLOGY

Ecuador is divided into five morphotectonic regions (Fig. 1): (1) the coastal lowlands, where a mafic crystalline basement sequence (Feininger and Bristow, 1980; Jaillard et al., 1995; Reynaud et al., 1999) is covered by Paleogene to Neogene forearc deposits, (2) the Western Cordillera, which is composed of mafic and intermediate volcanic and intrusive rocks tectonically juxtaposed with mostly turbiditic deposits of Late Cretaceous to Oligocene age, and (3) the Interandean Depression, which lies east of the Western Cordillera and hosts thick Pliocene to Pleistocene volcanic deposits (e.g., Winkler et al., 2005). Small basement inliers of metamorphic and mafic crystalline rocks (e.g., Bruet, 1949) sporadically occur throughout the depression. The Interandean Depression extends north into Colombia, and it is bound against the Western Cordillera (Fig. 2) by the Calacalí-Pujilí fault (in Ecuador) and the Cauca Patia fault (in Colombia). These faults define at least part of the Late Cretaceous suture between the South American plate and mafic allochthonous blocks. (4) The Eastern Cordillera (Cordillera Real in Ecuador) is composed of Paleozoic metamorphic rocks and Mesozoic granitoids (Aspden and Litherland, 1992; Litherland et al., 1994; Pratt et al., 2005). It is separated from the Interandean Depression by the Peltetec fault, which is a continuation of the Romeral fault of Colombia. (5) Last, the Oriente Basin, including the Subandean zone, is a Late Cretaceous–Holocene foreland basin that developed on the South American plate in response to growth of the Eastern Cordillera (e.g., Martin-Gombojav and Winkler, 2008).

Western Cordillera of Ecuador

The Western Cordillera of Ecuador (Fig. 2) consists of allochthonous oceanic blocks that accreted against the South American plate margin during the Late Cretaceous to Eocene

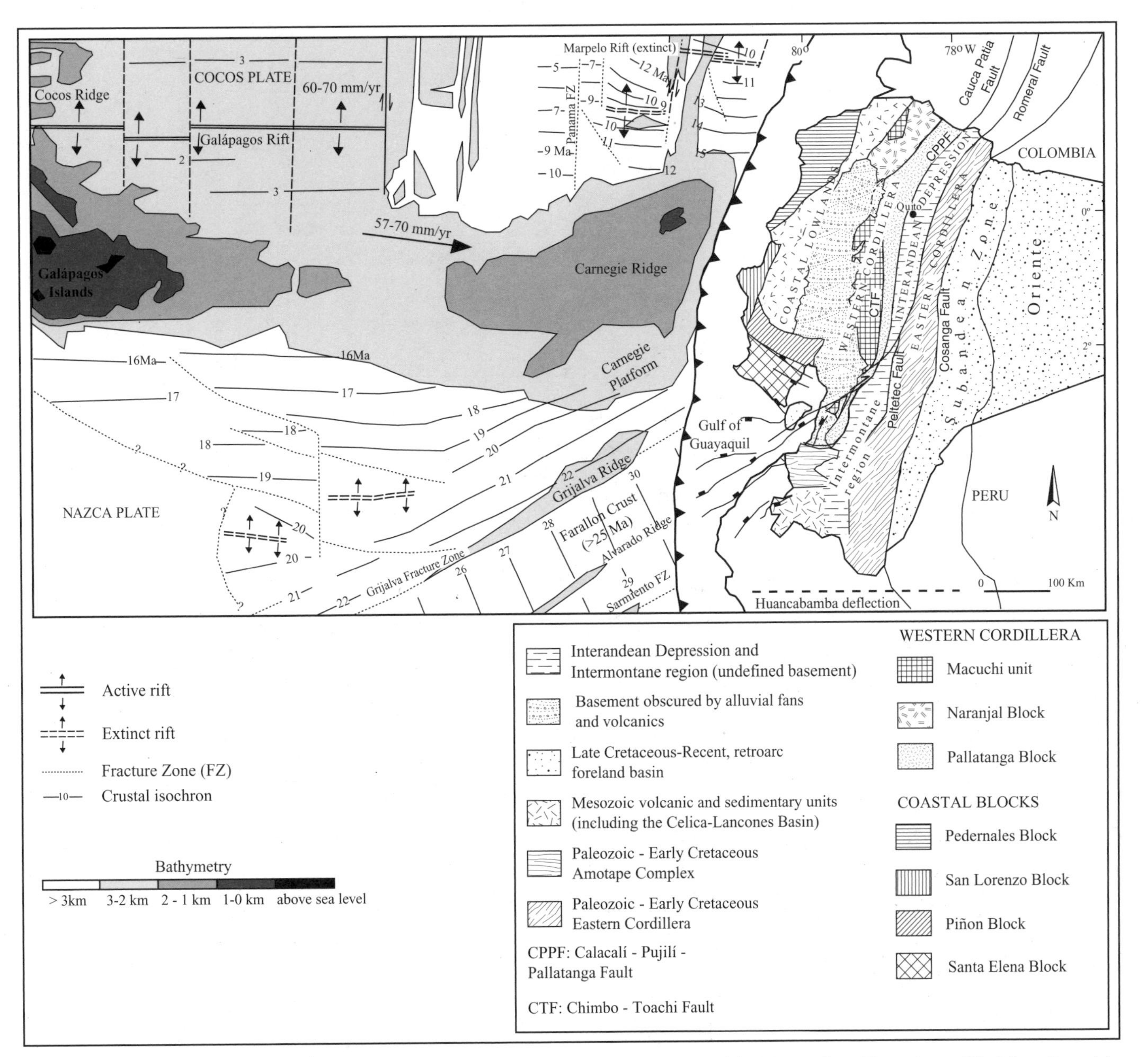

Figure 1. Geological setting of Ecuador, simplified bathymetry, and magnetic anomalies of the Nazca plate. Figure is modified from Lonsdale (1978) and Spikings et al. (2001).

(Goosens and Rose, 1973; Feininger and Bristow, 1980; Hughes and Pilatasig, 2002; Kerr et al., 2002a; Jaillard et al., 2004; Pratt et al., 2005; Spikings et al., 2005; Vallejo et al., 2006). Transcurrent fault displacement along approximately N-S–trending faults has structurally juxtaposed volcano-sedimentary successions of similar lithologies, but different ages, within the Western Cordillera (Fig. 2).

The Pallatanga block is exposed along the eastern border of the Western Cordillera, and it is separated from the continental margin by a suture zone (the Calacalí-Pujilí-Pallatanga fault) that represents part of the Late Cretaceous ocean-continent suture. Basement basalts and gabbros of the Pallatanga block (Fig. 2; Pallatanga Formation) yield enriched mid-ocean-ridge basalt (E-MORB) and oceanic plateau geochemical affinities (Lebrat et al., 1985; Lapierre et al., 2000; Hughes and Pilatasig, 2002; Kerr et al., 2002a; Mamberti et al., 2003), suggesting that the Pallatanga Formation erupted from a mantle plume. Furthermore, it also has been suggested that the rocks of the Pallatanga Formation are genetically related to the Caribbean-Colombia oceanic plateau (Lapierre et al., 2000; Spikings et al., 2001; Kerr et al., 2002a). The Pallatanga Formation is overlain by pelagic sediments of late Campanian to Maastrichtian age (Toro and Jaillard, 2005; Vallejo,

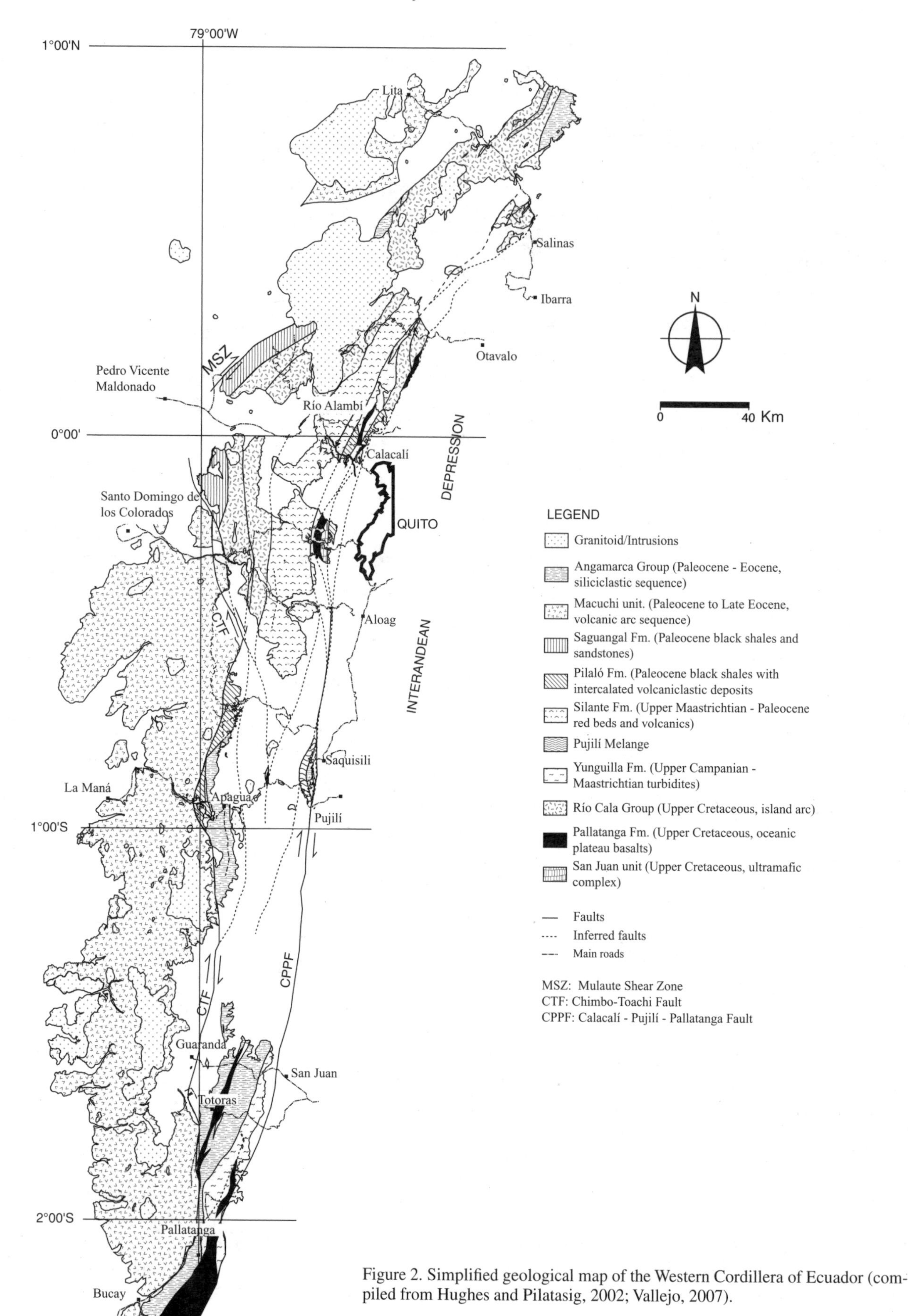

Figure 2. Simplified geological map of the Western Cordillera of Ecuador (compiled from Hughes and Pilatasig, 2002; Vallejo, 2007).

2007), and it is faulted against siliciclastic rocks of the late Campanian–early Maastrichtian Yunguilla Formation (Jaillard et al., 2004). The Late Cretaceous–Paleocene rocks of the Rio Cala Group, the Maastrichtian Silante Formation, and Eocene basin fill sequences of the Angamarca Group are interpreted to rest unconformably on top of the Pallatanga Formation (Fig. 3).

The Macuchi unit (McCourt et al., 1997) is located along the western border of the Western Cordillera (Fig. 2), and its eastern boundary coincides with the Chimbo-Toachi fault (Hughes and Bermúdez, 1997; Hughes and Pilatasig, 2002). The Macuchi unit contains basaltic pillow lavas, lithic tuffs of basaltic and andesitic composition, basaltic breccias, high-level andesitic intrusions, reworked volcanic material in turbiditic beds, and cherts. Volcaniclastic beds comprise ~80% of the sequence. The Macuchi unit was erupted through either MORB (Boland et al., 2000) or oceanic plateau (Chiaradia and Fontboté, 2001) rocks. Previous radiometric and biostratigraphic data suggest that the Macuchi unit spans the Paleocene to late Eocene (e.g., Egüez, 1986; Hughes and Pilatasig, 2002; Spikings et al., 2005). In this study, we present $^{40}Ar/^{39}Ar$ ages (plagioclase and groundmass) that support an Eocene age for the Macuchi unit (see following). Earlier interpretations of the Macuchi unit maintain that it was deposited in an intraoceanic island arc (e.g., Egüez, 1986; Aguirre and Atherton, 1987), and hence has an allochthonous origin, and that it was accreted during the Eocene (Hughes and Bermúdez, 1997; Hughes et al., 1998).

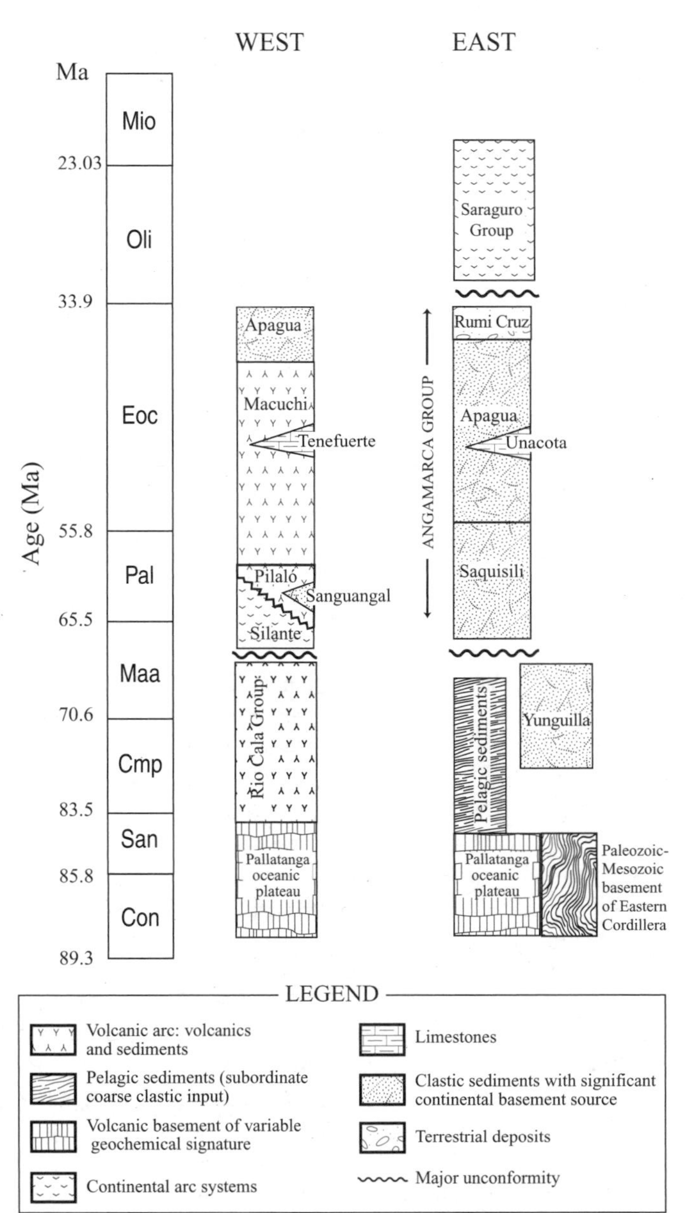

Figure 3. Stratigraphic columns of the volcanic and sedimentary formations of the Western Cordillera. Western (left side) and eastern (right side) stratigraphies are distinguished. On both sides, a late Maastrichtian unconformity separates the composite oceanic allochthonous basement from the postcollisional volcanic arc and sedimentary rocks.

Caribbean Plateau

The basement of the Caribbean region is composed of anomalously thick oceanic crust (Mauffret and Leroy, 1997; Sinton et al., 1998; Mauffret et al., 2001), which is interpreted as an oceanic plateau (e.g., Lapierre et al., 2000; Kerr et al., 2003; Mamberti et al., 2003), and it is commonly referred to as the Caribbean-Colombia oceanic plateau. The lithological, geochemical, and chronological similarities between rocks with oceanic plateau affinities in Ecuador and Colombia (Fig. 4) and the Caribbean suggest that these formations belong to the same large igneous province.

Fragments of the Caribbean-Colombia oceanic plateau are reported in Panama, Costa Rica, northern Venezuela, Aruba, Curacao, the Dominican Republic, Colombia, and Ecuador (Revillon et al., 2000; Lapierre et al., 2000; Spikings et al., 2001; Kerr et al., 2003). Oceanic plateaus form in deep-ocean basins as broad, flat-topped features, and their surfaces are 2000 m or more above the seafloor (e.g., Mann and Taira, 2004). Their crustal thicknesses are typically >20 km (Sinton et al., 1998; Revillon et al., 2000). Most authors agree that the Caribbean Plateau originated in the Pacific Ocean (e.g., Duncan and Hargraves, 1984; Pindell et al., 1988; Burke, 1988; Pindell and Barrett, 1990; Montgomery et al., 1994; Kerr et al., 2003). The Galápagos mantle plume, located at the junction between the Nazca and Cocos plates, may have formed the Caribbean-Colombia oceanic plateau (e.g., Duncan and Hargraves, 1984; Roperch et al., 1987; Acton et al., 2000; Spikings et al., 2001; Hoernle et al., 2002; Thompson et al., 2004). The South and North American continental plates rifted apart during the Triassic to Late Cretaceous, resulting in the formation of a continental gap. This permitted the Caribbean-Colombia oceanic plateau to migrate toward the ENE from the Pacific Ocean into its present position. Subduction of proto-Caribbean oceanic crust below the leading edge of the oceanic plateau resulted in the formation of an island arc (Burke, 1988; White et al., 1999; Pindell et al., 2005).

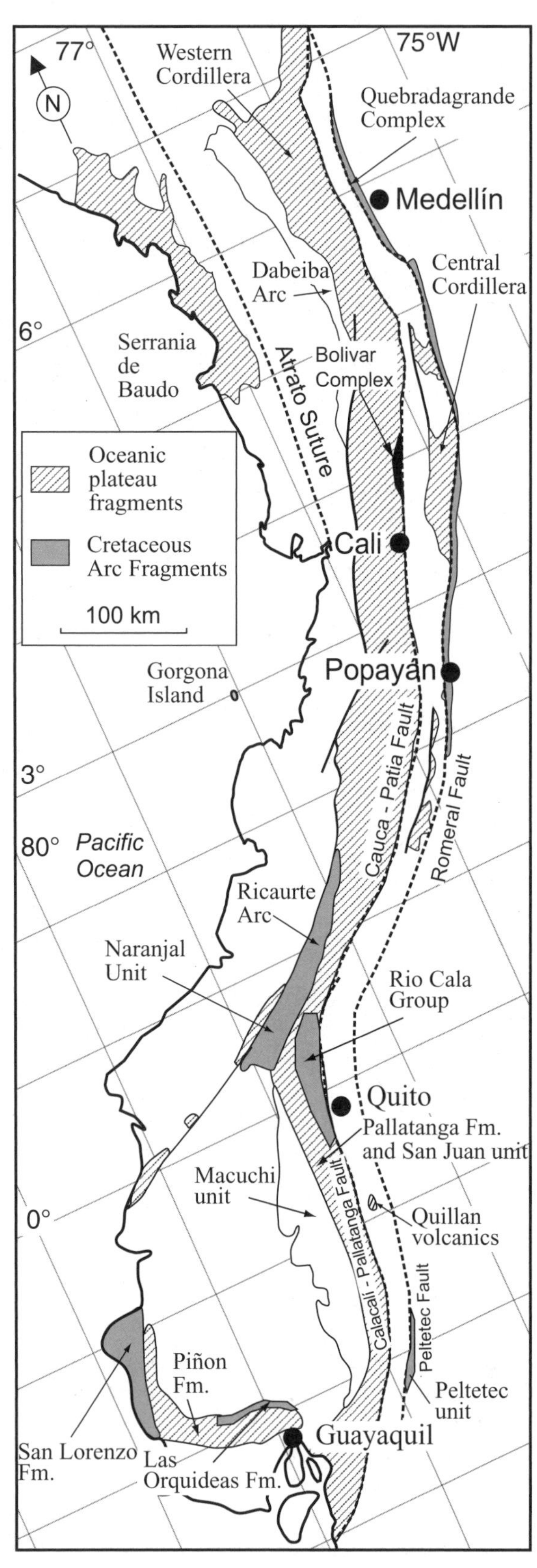

Figure 4. Exposure of Cretaceous oceanic plateau and island-arc sequences in western Colombia and Ecuador (modified from Kerr et al., 2003).

The Caribbean-Colombia oceanic plateau was probably wider than the intercontinental gap between the North and South American plates, resulting in collisions between those plates and the Caribbean-Colombia oceanic plateau during the Late Cretaceous (e.g., Ross and Scotese, 1988; Burke, 1988; Spikings et al., 2001; Pindell et al., 2005). Faulted units of transitional (T)-MORB and E-MORB picrites, basalts, dolerites, and ankaramites in the forearc and arc regions of Ecuador and Colombia (McCourt et al., 1984; Kerr et al., 2002a; Luzieux et al., 2006) record early fragmentation of the plateau, possibly during collision with the continental plate.

The majority of radiometric and biostratigraphic ages for the Caribbean Plateau in the Caribbean and in Colombia range between 99 and 88 Ma (Wadge et al., 1982; Donnelly et al., 1990; Walker et al., 1991, 1999; Kerr et al., 1997, 1999, 2002b; Sinton et al., 1997, 1998; Hauff et al., 2000; Mauffret et al., 2001), although a younger, minor phase of volcanism may have occurred at 76–72 Ma (e.g., Kerr et al., 2003).

Fragments of the Caribbean Plateau have been recognized in Colombia to the west of the Romeral fault, on the western side of the Central Cordillera, and Gorgona Island (Fig. 4; Restrepo and Toussaint, 1974; McCourt et al., 1984; Millward et al., 1984; Aspden and McCourt, 1986; Aspden et al., 1987; Megard, 1987; Spadea et al., 1989; Nivia, 1996; Kerr et al., 1998, 2003; Kerr and Tarney, 2005). The rocks with oceanic plateau affinity of Colombia range between 91.7 ± 2.7 and 70.0 ± 3.5 Ma (Bourgois et al., 1982; Kerr et al., 1997; Sinton et al., 1998; Walker et al., 1999).

Age of the Oceanic Plateau Basement of Western Ecuador

The San Juan unit (Hughes and Bermúdez, 1997) is an ultramafic sequence that includes serpentinized peridotites, layered cumulate fine-grained peridotites, dunites, layered cumulate olivine gabbros, fine- to coarse-grained amphibole-bearing gabbros, norites, locally anorthosites, and dolerites. Hughes and Pilatasig (2002) defined the San Juan unit as the ultramafic root of the Pallatanga block. Mamberti et al. (2003) utilized geochemical data to suggest that the San Juan unit represents a magmatic chamber that existed within an oceanic plateau.

Zircons extracted from a layered gabbro mapped as the San Juan unit (Hughes et al., 1998) yield a weighted mean U/Pb (sensitive high-resolution ion microprobe [SHRIMP]) age of 87.10 ± 1.66 Ma (Vallejo et al., 2006; Vallejo, 2007). However, Lapierre et al. (2000) obtained an Early Cretaceous Sm/Nd age from an isotropic gabbro that is also mapped as the San Juan unit. This clear conflict may be due to erroneous mapping. The gabbro dated by Lapierre et al. (2000) may form part of an older Early Cretaceous ultramafic basement sequence in the easternmost Western Cordillera (e.g., Litherland et al., 1994; Jaillard et al., 2004), whereas the Late Cretaceous U/Pb age matches the time of crystallization of mafic components of the Pallatanga Formation. The Early Cretaceous rock sequence reported by Lapierre et al. (2000) probably belongs to the ultramafic-mafic Peltetec unit (Fig. 4), which is mainly exposed along the western border of the Eastern Cordillera, and which yields both enriched MORB and subduction-related signatures. The Peltetec unit may have accreted to the South American continent during the Aptian (Litherland et al., 1994). Notably, the Late Cretaceous U/Pb age obtained for the San Juan unit is similar to a plateau hornblende $^{40}Ar/^{39}Ar$ age of 88 ± 1.6 Ma (2σ) obtained from mantle plume–derived basalts of the Piñon Formation in the coastal forearc of Ecuador (Luzieux et al., 2006; Luzieux, 2007).

Amphibolite inliers (Totoras amphibolite unit) occur in the central part of the Western Cordillera (Fig. 2; Lugo, 2003). Beaudon et al. (2005) presented a detailed geochemical investigation of the amphibolites and associated granulites. They concluded that these oceanic plateau–derived amphibolites experienced peak pressure-temperature (*P-T*) conditions of 6–9 kbar and 800–850 °C. The amphibolites produced a hornblende plateau $^{40}Ar/^{39}Ar$ age of 84.69 ± 2.23 Ma (Vallejo et al., 2006; Vallejo, 2007).

The radiometric ages obtained in the oceanic plateau basement of western Ecuador overlap elsewhere with the peak of ages (92–88 Ma) obtained from basalts of the present-day Caribbean Plateau, using the $^{40}Ar/^{39}Ar$ method (Sinton et al., 1998; Kerr et al., 2003).

Rio Cala Island Arc

The Rio Cala Group contains several Late Cretaceous volcaniclastic formations that were deposited on top of the Pallatanga Formation (Vallejo, 2007; Fig. 3). It includes andesites and basalts of the Rio Cala Formation (Boland et al., 2000), turbidite beds of the Natividad Formation (Boland et al., 2000), turbidites and intercalated volcanic rocks of the Pilatón and Mulaute Formations (Kehrer and van der Kaaden, 1979; Egüez, 1986; Hughes and Bermúdez, 1997), and high-Mg andesites (boninites) of the La Portada Formation (Van Thournout, 1991; Kerr et al., 2002a). Biostratigraphic ages from the associated turbidite beds range from Santonian to early Maastrichtian (Egüez, 1986; Hughes and Bermúdez, 1997; Boland et al., 2000).

Geochemical and isotopic results from volcanic rocks of this group depict an intraoceanic island-arc signature (Cosma et al., 1998; Mamberti, 2001; Vallejo et al., 2006). Allibon et al. (2005) showed that the lavas of the Rio Cala Formation were formed by subduction beneath thickened oceanic crust in an intraoceanic setting; Nd and Pb isotopic data show that the volcanic rocks of the Rio Cala Formation result from mixing of a Pacific MORB mantle, subducted pelagic sediments, and an oceanic plateau component. Whole-rock Nd isotopic data and pyroxene compositions are typical for a primitive arc source (Cosma et al., 1998; Mamberti, 2001) that extruded on top of an oceanic plateau (Mamberti, 2001; Allibon et al., 2005; Vallejo et al., 2006).

The presence of pumpellyite, epidote, and chlorite is also consistent with an intraoceanic island-arc setting. Such low-grade metamorphism occurs by the interaction between volcanic rocks and seawater under a moderate to high thermal gradient (Aguirre and Atherton, 1987).

Provenance analyses (heavy mineral data) of sedimentary rocks of the Rio Cala Group (Pilatón, Mulaute, and Natividad Formations) show that they originated from a volcanic arc (Fig. 5) and did not receive typical continental material (e.g., detrital zircon, rutile, tourmaline), suggesting that they were deposited far from the continent (Vallejo, 2007). This corroborates evidence presented by Hughes et al. (1998), who indicated that the source of the sediments of the Pilatón Formation was exclusively andesitic to basaltic in composition.

We obtained an imprecise plateau $^{40}Ar/^{39}Ar$ age (groundmass) of 66.74 ± 7.16 Ma (2σ) from volcanic rocks of the Rio Cala Formation. Despite the large error, this age is consistent with Campanian-Maastrichtian biostratigraphic ages obtained from sedimentary rocks of the Natividad Formation. Similarly, Boland et al. (2000) and Kerr et al. (2002a) obtained Santonian to early Campanian biostratigraphic ages for the La Portada Formation, which is considered to be the oldest formation within the Rio Cala Group.

Basalts and gabbros of the San Lorenzo and Las Orquideas Formations in coastal Ecuador (Fig. 4) have island-arc geochemical affinities and overlie plateau-derived rocks of the Piñon Formation (Reynaud et al., 1999; Kerr et al., 2002a; Luzieux et al., 2006). Lebras et al. (1987) reported a $^{40}Ar/^{39}Ar$ age of 72.7 ± 1.4 Ma for an island-arc basalt in the San Lorenzo Formation. According to Luzieux (2007), magnetostratigraphic results document volcanic activity during the Campanian in the Las Orquideas island-arc sequence, which is exposed northwest of Guayaquil (Fig. 4).

The association of Santonian to Maastrichtian intraoceanic island-arc rocks overlying oceanic plateau rocks is documented throughout the circum-Caribbean region (e.g., Frost and Snoke, 1989; Donnelly et al., 1990). Additional evidence for Late Cretaceous island arcs occurs in southern Colombia, where Spadea and Espinosa (1996) documented Campanian radiolaria from rocks intercalated with island-arc lavas (Ricaurte Arc; Fig. 4). The coeval island-arc rocks in the external forearc and Western Cordillera of Ecuador, together with equivalents in Colombia and the Caribbean region, presumably formed within the same intraoceanic subduction system that was active between the Santonian and early Maastrichtian.

An ophiolite-bearing tectonic mélange (Fig. 2) located along the eastern margin of the Western Cordillera of Ecuador is interpreted to be part of the suture between allochthonous oceanic blocks and the continental margin, and it is interpreted to have formed in the Late Cretaceous. The Pujilí granite, which occurs as tectonic blocks within the Pujilí mélange (Hughes et al., 1998; Vallejo et al., 2006) yields high concentrations of large ion lithophile elements (LILEs), a negative Nb anomaly, strong light rare earth element (LREE) enrichment ($[La/Yb]_N$ = 33.7), and depletion of heavy (H) REEs. Additional important parameters are MgO = 0.21 wt%, SiO_2 = 56.92wt%, Al_2O_3 = 10.04 wt%, Sr = 303 ppm, and Nb = 1.03 ppm. Some of these characteristics are similar to Archean TTGs (tonalite-trondhjemite-granodiorite complexes) and adakites. ε_{Nd} values of +6 suggest

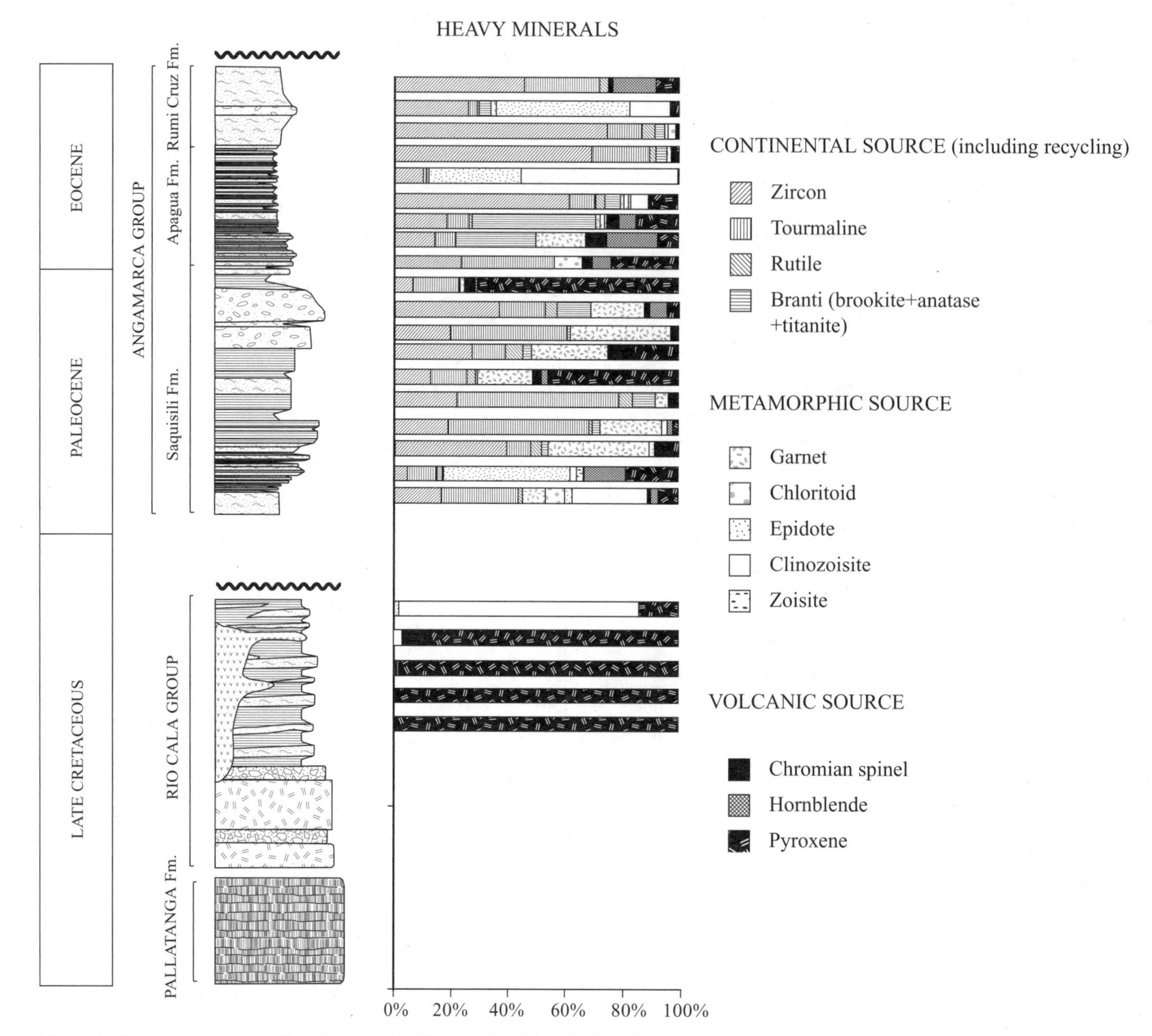

Figure 5. Composite stratigraphic column of the Western Cordillera (left) and heavy mineral frequencies (right) in pre- and postaccretionary sedimentary formations. A clear change is observed in the composition of the Late Cretaceous Rio Cala Group, which was derived from intraoceanic volcanic sources, whereas Paleocene sediments of the Angamarca Group were shed from granitic and metamorphic sources that formed part of a continental plate, and now constitute the Eastern Cordillera.

that the magma was not contaminated with old continental crust, and these values lie within the range of values obtained from the Caribbean-Colombia oceanic plateau (Thompson et al., 2004). Negative Nb and Ti anomalies suggest that the Pujilí granite formed in a subduction zone.

The Pujilí granite yields a zircon U/Pb (SHRIMP) crystallization age of 85.5 ± 1.4 Ma (Vallejo et al., 2006), which is indistinguishable from plateau $^{40}Ar/^{39}Ar$ ages of 86 ± 1 (muscovite) and 82 ± 1 (biotite) Ma (Spikings et al., 2005). Late Cretaceous granitic rocks with island-arc signatures and Late Cretaceous mafic oceanic plateau rocks are also reported on Aruba (White et al., 1999). The Aruba batholith and the Pujilí granite yield similar isotopic ratios (ε_{Nd} ~7–6) and geochemical signatures. Tonalites of the Aruba batholith yield a zircon ion-microprobe (secondary ionization mass spectrometry) age of 86 ± 1 Ma (R. Spikings, 2006, personal commun.), which is indistinguishable from the zircon U/Pb age yielded by the Pujili granite. Collectively, these data suggest that they may have crystallized within the same subduction zone. White et al. (1999) assumed that the Aruba batholith was produced by partial melting of the Caribbean Plateau above a subduction zone with an anomalously hot mantle wedge. We suggest a similar scenario for the Pujilí granite. However, the HREE depletions of the Pujilí granite indicate that it was formed below the garnet stability field (~60–70 km depth), suggesting that it may also have a component derived from melting of a subducted slab, accounting for the negative Nb and Ti anomalies.

EVIDENCE FOR THE INITIAL COLLISION OF THE CARIBBEAN PLATEAU IN ECUADOR

Subduction-related magmatism below the leading edge of the oceanic plateau is recorded in the Western Cordillera and forearc region of Ecuador. It ended during the Maastrichtian, implying that west-dipping subduction ceased. The termination of arc magmatism coincided with rapid changes in paleomagnetic declination (20°–50° vertical-axis clockwise rotation; Luzieux et al., 2006; Luzieux, 2007) between 73 and 70 Ma in both the Piñon and the San Lorenzo blocks. These dramatic events, which occurred simultaneously over large distances (e.g., 1000 km), were produced by the collision between South America and the Caribbean-Colombia oceanic plateau (Vallejo et al., 2006).

The earliest significant cooling and exhumation event detected along the continental margin by $^{40}Ar/^{39}Ar$ (white mica, biotite) and fission-track (zircon, apatite) thermochronology occurred during 75–65 Ma in the Amotape Complex (Spikings et al., 2001, 2005). This is confirmed by plateau $^{40}Ar/^{39}Ar$ ages from Triassic migmatites (U/Pb zircon age of 227 ± 2 Ma; Litherland et al., 1994) of 68.5 ± 0.4 (white mica) and 68.6 ± 0.5 (biotite) Ma from the southern Eastern Cordillera (Vallejo et al., 2006), which are indicative of rapid cooling through 380–330 °C during the Maastrichtian.

In the Oriente Basin, east of the Eastern Cordillera (Fig. 1), the Tena Formation was deposited in an overfilled, retroarc foreland basin, which received increasing proportions of metamorphic minerals due to episodic uplift and erosion of the Eastern Cordillera (e.g., Ruiz et al., 2004; Martin-Gombojav and Winkler, 2008). Similarly, Paleocene to Eocene turbidites of the Angamarca Group, which directly overlie the mafic basement of the Western Cordillera, contain increasing amounts of heavy minerals derived from continental and metamorphic basement (Fig. 5). This distinguishes them clearly from sediments of the Rio Cala Group, which were exclusively derived from volcanic source rocks.

The evidence from both the study region (proximal to the collision zone) and elsewhere in the Northern Andes strongly favors the hypothesis that the leading edge of the Caribbean Plateau, and an overlying arc, collided with the Ecuadorian sector of South America during the late Campanian–Maastrichtian (ca. 75–65 Ma). Shear sense indicators in the Calacalí-Pujilí-Pallatanga fault (Hughes and Pilatasig, 2002) and paleomagnetically constrained block rotations (Luzieux, 2007) prove a dextral sense of movement associated with east-northeastward–oriented collision of the Caribbean Plateau.

Postaccretionary Volcanic Arc

The Silante Formation, exposed northwest of Quito (Fig. 2), provides important constraints on the Cenozoic evolution of the Western Cordillera. The Silante Formation consists of porphyritic andesites and basalts (Tandapi volcanic facies), red beds, conglomerates, and tuffs deposited in a terrestrial environment. Sedimentary clasts lithologically similar to the Pallatanga or Rio Cala Group are found in the conglomerates, indicating reworking of older volcanic and sedimentary formations. It unconformably overlies the Pilatón Formation (Rio Cala Group). The calc-alkaline composition of the Silante lavas (Kehrer and van der Kaaden, 1979; Hughes and Bermúdez, 1997; Cosma et al., 1998) contrasts with the mostly tholeiitic signatures of the Rio Cala Group and the Pallatanga Formation. The environmental and geochemical change between the Rio Cala Group and Silante Formation suggests that an important geological event separated them.

Zircon fission-track data from tuffs intercalated within red beds of the Silante Formation yield an age of 16.8 ± 0.8 Ma (Hughes and Bermúdez, 1997). However, zircon fission-track ages are at least partially reset at temperatures higher than ~220 °C (Tagami et al., 1998), and hence the age only represents a minimum crystallization age. Wallrabe-Adams (1990) presented a K/Ar (whole-rock) age of 52.7 ± 2.9 Ma from rocks previously mapped as Silante Formation. However, detailed field mapping shows that the rocks dated by this author are intercalated within marine sediments of the younger Pilaló Formation (Fig. 3).

The Pilaló Formation overlies the Silante Formation along the Alambi River section (Fig. 2). This formation includes coarse-grained turbiditic sandstones, black shales, matrix-supported breccias with volcanic clasts of andesitic composition, andesites, siltstones, wood fragments, and reworked tuffs. The volcanic breccias contain fragments of red, oxidized andesites derived from erosion of the underlying Silante Formation. Savoyat et al. (1970) reported Maastrichtian to Danian biostratigraphic ages for sedimentary rocks of the Pilaló Formation.

The Saguangal Formation (Fig. 3) is defined here as a shallow-marine sedimentary succession of possibly Paleocene age (Vallejo, 2007) that overlies volcaniclastic rocks of the Rio Cala Group. Heavy minerals extracted from the Saguangal Formation (Vallejo, 2007) show that it was derived from mixed continental, metamorphic, and volcanic source rocks. Single-grain geochemistry data from pyroxenes in these sandstones indicate a volcanic source with a calc-alkaline affinity (Vallejo, 2007). These data suggest that the Saguangal Formation was partially derived from erosion of volcanic rocks of the Silante Formation.

Age Dating of the Postaccretionary Series

Four whole-rock samples (andesites and basalts) of the Tandapi volcanic facies (Silante Formation) collected northwest of Quito and two samples of the Macuchi unit were analyzed using the $^{40}Ar/^{39}Ar$ method. Whole-rock sample preparation

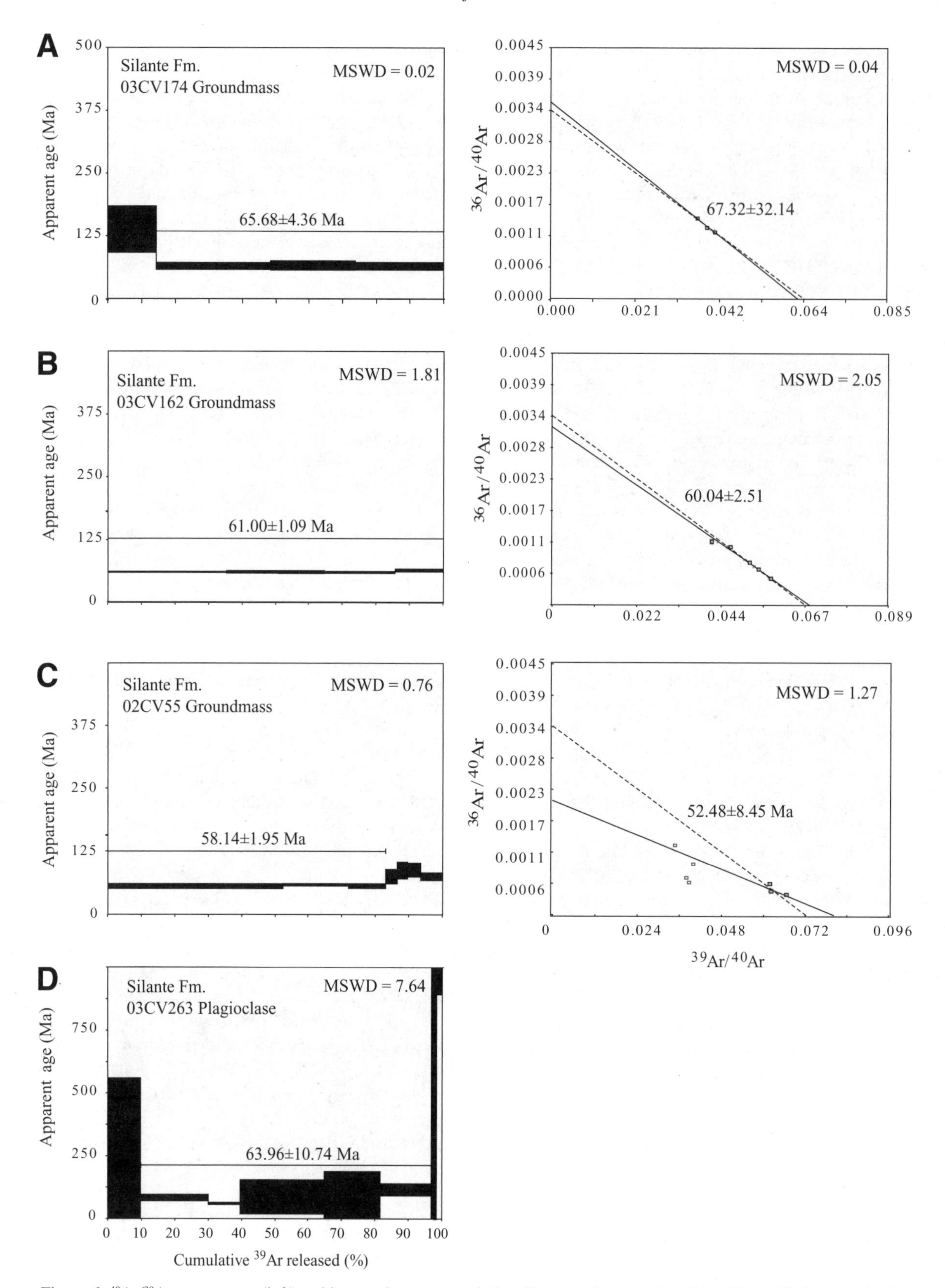

Figure 6. $^{40}Ar/^{39}Ar$ age spectra (left) and inverse-isotope correlation diagrams for samples of the Silante Formation. A plateau age (indicated) has been determined from the weighted mean of contiguous, concordant step-ages. Boxes for each step extend vertically to ±2σ. MSWD—mean square of weighted deviations.

followed that described in Koppers et al. (2000), and $^{40}Ar/^{39}Ar$ analyses were performed following the procedure described in Marshik et al. (2007). Plateaus are herein defined as the release of ≥50% of the total ^{39}Ar gas released in three or more successive steps, the ages of which are concordant within 2σ error (Dalrymple and Lanphere, 1971).

Sample 02CV174 (UTM: 763379/9995871) is a fine- to medium-grained basalt that has an ophitic to granular texture. The $^{40}Ar/^{39}Ar$ step-heating of a groundmass concentrate yielded a plateau age of 65.68 ± 4.36 Ma (2σ; Fig. 6A). The plateau age is indistinguishable from the imprecise inverse isochron age of 67.32 ± 32.14 Ma (2σ; mean square of weighted deviations [MSWD] = 1.04). The plateau age is considered to approximate the crystallization age of the volcanic rock and is consistent with Maastrichtian–Paleocene biostratigraphic ages of overlying sedimentary rocks of the Pilaló Formation.

Sample 03CV162 (UTM: 768285/10001600) is an andesite that has a porphyritic texture with phenocrysts of plagioclase and clinopyroxene. A phenocryst-free, groundmass separate yielded a plateau $^{40}Ar/^{39}Ar$ age of 61 ± 1.09 Ma (2σ; Fig. 6B), which is indistinguishable from its inverse isochron age of 60.04 ± 2.51 Ma (MSWD = 2.05).

Sample 02CV55 (UTM: 766935/10002688) is a porphyritic andesite, with plagioclase and clinopyroxene phenocrysts and a matrix composed of glass and microliths of plagioclase and minor pyroxene. A groundmass separate yielded a $^{40}Ar/^{39}Ar$ age of 58.14 ± 1.95 Ma (2σ; Fig. 6C), which is indistinguishable from its inverse isochron age of 52.48 ± 8.45 Ma (2σ).

Diorite 03CV263 (UTM: 769015/10001259) forms part of the Silante Formation close to a faulted contact with turbidites of the Yunguilla Formation. The rock contains plagioclase, clinopyroxene, and quartz. A plagioclase separate yielded an imprecise weighted mean age of 63.96 ± 10.74 (2σ) and an MSWD of 7.64 (Fig. 6D), indicating that the data do not strictly define a plateau. However, the weighted mean age is consistent with similar ages obtained for the Silante Formation in the same area.

Sample 02CV126 (UTM: 725129/9965028) forms part of the Macuchi unit (Fig. 7A). The sample is a fine-grained, moderately clinopyroxene-plagioclase–phyric dolerite with a subophitic to slightly intersertal texture. The $^{40}Ar/^{39}Ar$ step-heating

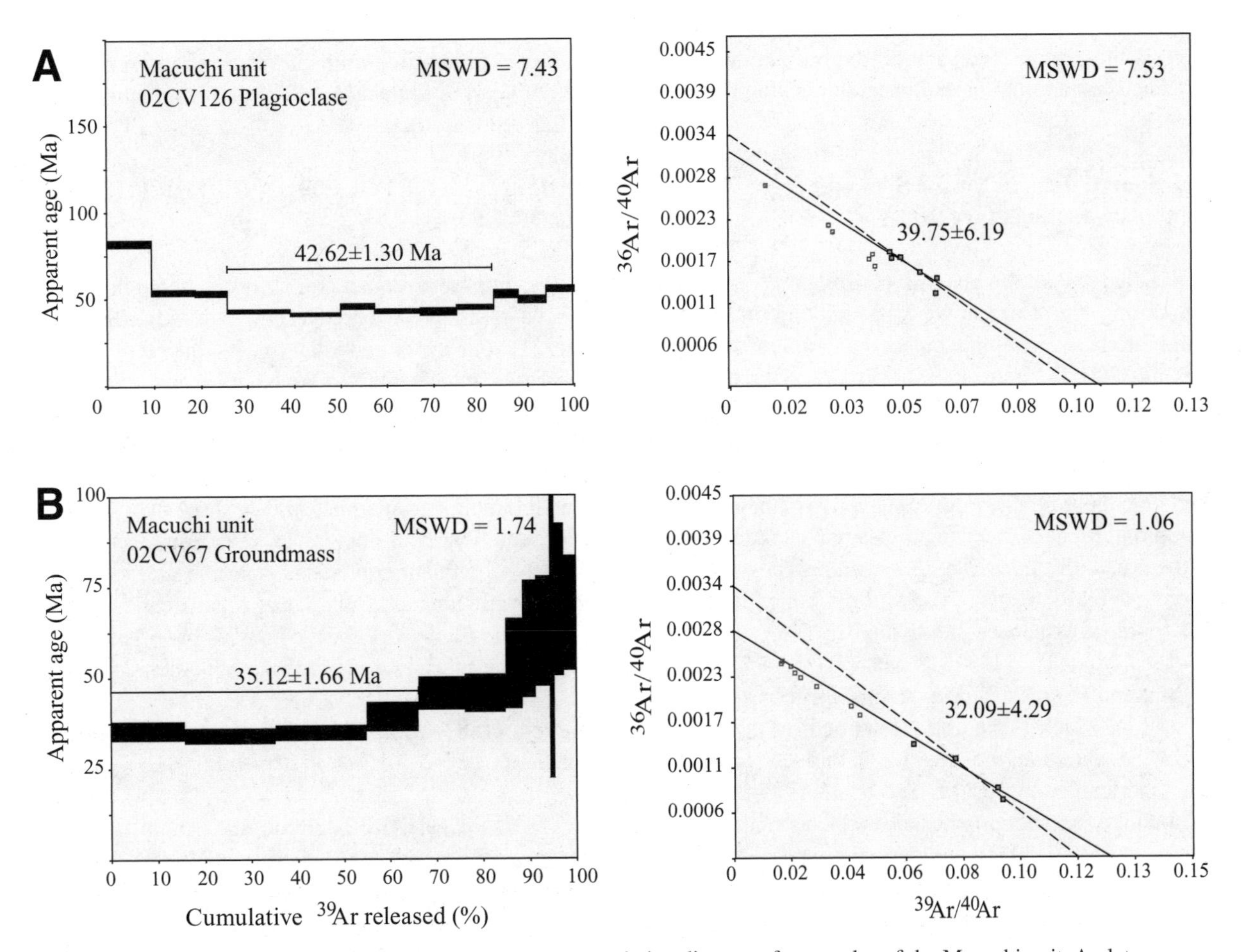

Figure 7. $^{40}Ar/^{39}Ar$ age spectra (left) and inverse-isotope correlation diagrams for samples of the Macuchi unit. A plateau age (indicated) has been determined from the weighted mean of contiguous, concordant step-ages. Boxes for each step extend vertically to ±2σ. MSWD—mean square of weighted deviations.

yielded a saddle-shaped age spectrum, suggesting that excess ^{40}Ar is present within the sample, although this could not be categorically identified due to the low precision of the inverse isochron (initial ^{40}Ar/^{36}Ar ratio of 317.18 ± 23.93; Fig. 7A). The weighted mean age for the six steps, which define the flattest part of the age spectrum, and span ~54% of the ^{39}Ar released, is 42.62 ± 1.3 Ma (2σ; MSWD = 7.43). The weighted mean age is concordant with the inverse isochron age of 39.75 ± 6.19 (2σ; MSWD = 7.53) and is considered to represent a maximum ^{40}Ar/^{39}Ar plagioclase age for the sample.

Sample 02CV67 (UTM: 784106/10095939) is part of an andesite lava flow collected in the northern part of the Western Cordillera. The andesite has a porphyritic texture, with phenocrysts of plagioclase and clinopyroxene, and a matrix with a trachytic texture composed of subparallel-oriented acicular microliths of plagioclase. The ^{40}Ar/^{39}Ar step-heating analysis of a groundmass separate yields an overall discordant age spectra, where the oldest ages occur late within the step-wise, gas release experiment (Fig. 7B). However, heating steps at lower and intermediate temperatures yielded a concordant, plateau age of 35.12 ± 1.66 Ma (2σ), with an acceptable MSWD of 1.74, which is concordant within the inverse isochron age of 32.09 ± 4.29 Ma. The imprecise, initial ^{40}Ar/^{36}Ar ratio of 354.05 ± 79 suggests that excess ^{40}Ar may not be present, and therefore the plateau age approximates the crystallization age of the volcanic rock.

Age of the Source Regions for the Syn- and Postaccretionary Sedimentary Series

Siliciclastic rocks of the Yunguilla Formation (Fig. 3) were deposited during the Campanian-Maastrichtian in a NNE-trending forearc basin along the paleocontinental margin (e.g., Jaillard et al., 2005). Figure 8A shows U/Pb detrital zircon ages (laser ablation–inductively coupled plasma–mass spectrometry [ICP-MS]) for a sandstone sample (00RS33, UTM: 771490/10001497) within the Yunguilla Formation. The ages are plotted on concordia diagrams, together with the age population histogram. Most of the data points are concordant or plot close to the concordia line. The ages obtained provide information on the age of the source rocks of the Yunguilla Formation. The zircon ages can be subdivided in five populations: (1) 1789–1867 Ma, (2) 1112–1318 Ma, (3) 755–978 Ma, (4) 384–639 Ma, and (5) 69–100 Ma. A single zircon yielded an age of 2642 Ma, which is the oldest recorded for this sample. The ages of the source regions overlap with ages reported for the South American craton (Litherland et al., 1985), U/Pb zircon ages obtained for metamorphosed sediments of the Eastern Cordillera (Chew et al., 2007), and detrital zircons deposited in the proximal retroarc foreland basin (Subandean zone and Oriente) as shown by Martin-Gombojav and Winkler (2008). This detrital grain age correlation clearly shows that the Yunguilla Formation was sourced from rocks located in the Eastern Cordillera. However, the three youngest zircons with a weighted mean age of 72.4 ± 6.4 Ma (2σ) match with the biostratigraphic age of the Yunguilla Formation (Jaillard et al., 2004) and therefore suggest the minor presence of a coeval volcanic source.

The LA-ICP-MS detrital zircon ages from a sandstone sample (sample 03CV137, UTM: 747194/10023517) of the Saguangal Formation (Fig. 8B) also reveal five distinctive age populations (Vallejo, 2007): (1) 1563–1914 Ma, (2) 1017–1151 Ma, (3) 567–881 Ma, (4) 229–478 Ma, and (5) 56–86 Ma. Proterozoic ages overlap the ages reported for the South American craton (Litherland et al., 1985) and the Eastern Cordillera (Chew et al., 2007; Martin-Gombojav and Winkler, 2008). A weighted mean age of 58.8 ± 8.9 Ma was calculated for the youngest zircons. Most of these grains are euhedral due to little transport, which may indicate that they were derived from a coeval volcanic source.

In addition, we performed U/Pb (SHRIMP) dating on detrital zircons recovered from clastic rocks of the Macuchi unit (Fig. 8C). The zircons are rounded crystals, which indicate multiple recycling. The obtained crystallization ages range between 538.9 ± 5.8 and 589.9 ± 7.4 Ma (1σ). These ages are different than the biostratigraphic and radiometric ages obtained for the Macuchi unit elsewhere, which indicate a Paleocene to Eocene age (e.g., Egüez, 1986; Hughes and Pilatasig, 2002). Therefore the Precambrian–Cambrian zircon ages represent detrital input of older sources to the Macuchi unit, presumably derived from the Eastern Cordillera.

GEODYNAMIC EVOLUTION OF THE WESTERN CORDILLERA

We propose a refined model for the geological evolution of the Ecuadorian forearc based on new data and earlier studies (e.g., Egüez, 1986; Hughes and Pilatasig, 2002; Kerr et al., 2002a). We consider the mantle plume–derived mafic and ultramafic rocks of the basement of the Western Cordillera and the coastal region to have been originally part of the same oceanic plateau (Vallejo et al., 2006; Luzieux et al., 2006), which was the ca. 88 Ma Caribbean-Colombia oceanic plateau. We show that the geology can be accounted for by a single Late Cretaceous accretionary event, and that subsequent continental volcanic arc activity has been almost continuous since the latest Maastrichtian. An improved stratigraphic framework for the igneous basement and sedimentary and volcanic cover supports our hypothesis.

Coniacian (88 Ma): Origin of the Mafic Crystalline Basement

The U/Pb (SHRIMP) zircon age obtained from the San Juan unit, 87.1 ± 1.66 Ma (Vallejo et al., 2006), which we interpret to represent the basement of the Western Cordillera, is consistent with a ^{40}Ar/^{39}Ar age (hornblende) of 88.8 ± 1.6 Ma from the coastal basement Piñon Formation (Luzieux et al., 2006). These rocks crystallized at the same time as the Caribbean Plateau (ca. 91–88 Ma; Sinton et al., 1998). Paleomagnetic

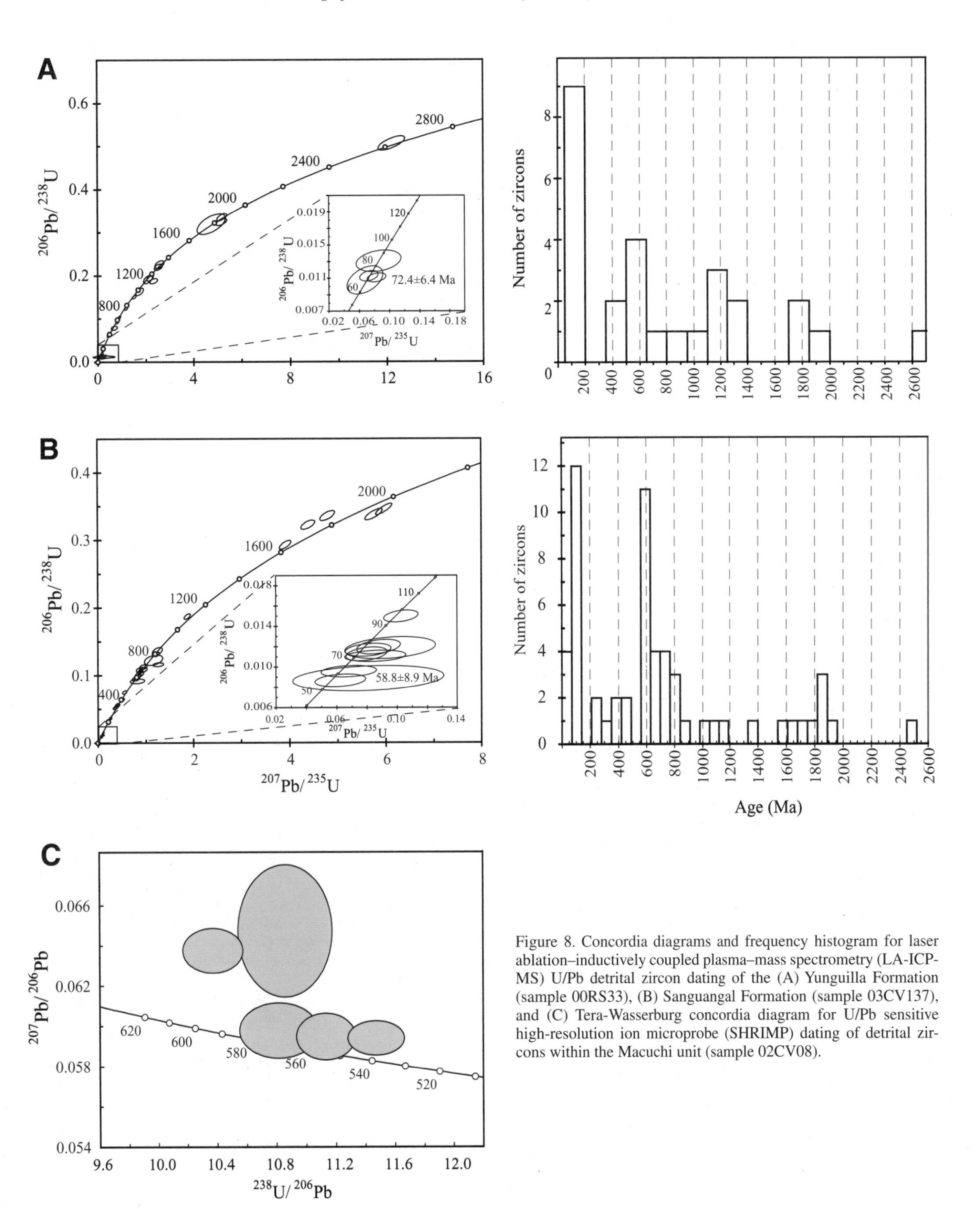

Figure 8. Concordia diagrams and frequency histogram for laser ablation–inductively coupled plasma–mass spectrometry (LA-ICP-MS) U/Pb detrital zircon dating of the (A) Yunguilla Formation (sample 00RS33), (B) Sanguangal Formation (sample 03CV137), and (C) Tera-Wasserburg concordia diagram for U/Pb sensitive high-resolution ion microprobe (SHRIMP) dating of detrital zircons within the Macuchi unit (sample 02CV08).

analyses of the Piñon Formation (Luzieux et al., 2006; Luzieux, 2007) show that it crystallized at equatorial or low southern latitudes. However, we are unable to constrain their original longitudinal separation from South America. The estimated ~20 km thickness of the Caribbean-Colombia oceanic plateau (Sinton et al., 1998; Revillon et al., 2000), combined with an approximate geothermal gradient of 40 °C/km, gave rise to the required pressure and temperature conditions required to generate the Totoras amphibolite at the base of the plateau, which retrogressed through ~500 °C at 84.69 ± 2.22 Ma (Vallejo et al., 2006; Vallejo, 2007).

Late Cretaceous paleotectonic reconstructions of the Caribbean-Colombia oceanic plateau and the Farallon plate relative to South America are poorly constrained. Plate reconstructions of the Caribbean plate are inaccurate for Ecuador because of a lack of quantitative data. Duncan and Hargraves (1984) proposed a model in which the ca. 90 Ma Caribbean Plateau formed above the Galápagos hotspot (Fig. 9A), and its eastern edge was located ~2200 km west of South America. Consequently, the plateau would have had to drift eastward to collide with northwestern South America. However, this is not consistent with the northeast drift of the Farallon plate (Pilger, 1981; Engebretson et al., 1985). Furthermore, Pindell et al. (2006) suggested that the Galápagos hotspot was located ~1000 km west of the site where the Caribbean Plateau was extruded (Fig. 9B).

Such contrasting reconstructions reflect the paucity of data, and as suggested by Pindell et al. (2006), the Caribbean Plateau may have originated close to the South American margin. However, by considering the northeastward movement of the Farallon plate during the Late Cretaceous, it is likely that the leading edge of the Caribbean Plateau collided obliquely with South America.

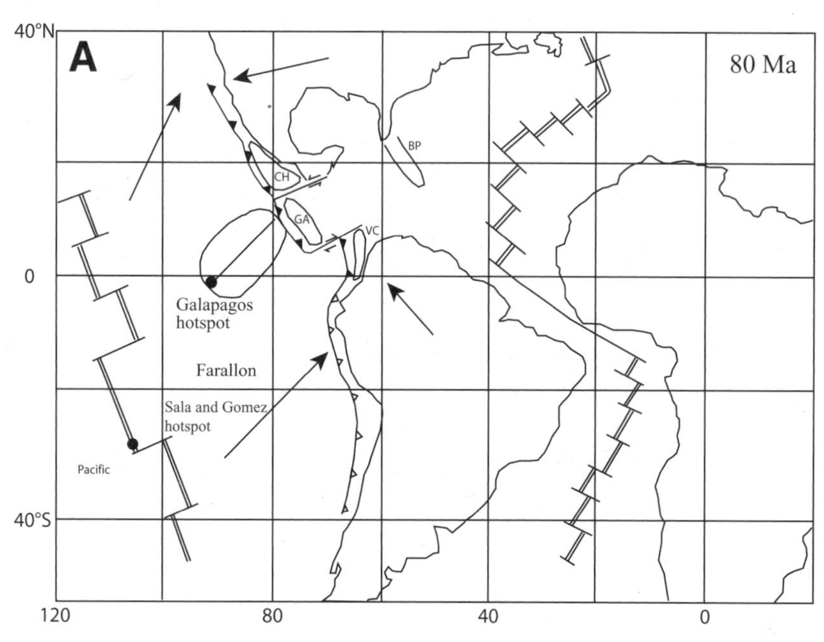

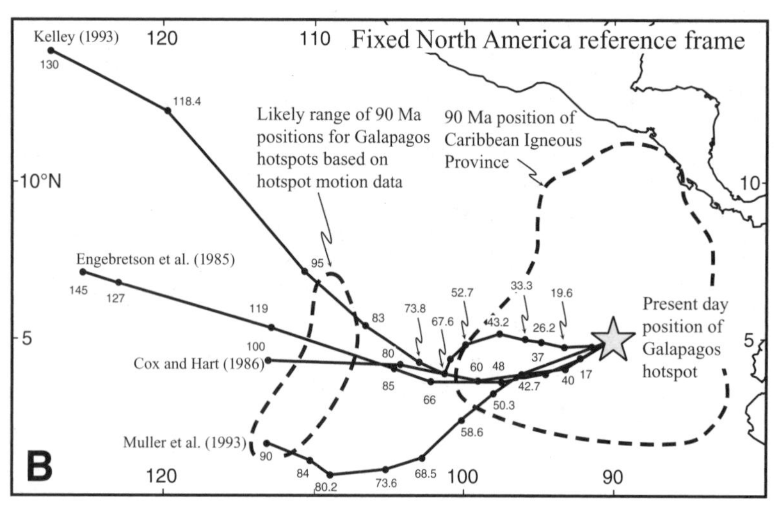

Figure 9. Contrasting interpretations for the positions of the Galápagos hotspot during the Late Cretaceous and its relationship with the Caribbean Plateau. (A) Paleotectonic reconstruction of the Caribbean region at 80 Ma. Plate positions and motions are in the hotspot reference frame (Duncan and Hargraves, 1984). In this model, the Caribbean Plateau originated at the Galápagos hotspot. (B) Position of the Caribbean plate at ca. 90 Ma (Pindell et al., 2006). These authors suggest that the Caribbean Plateau was located at least 1000 km east of the Galápagos hotspot at that time, implying that the plateau did not extrude from the Galápagos. BP—Bahamas Platform; CH—Chortis Block; GA—Great Arc of the Caribbean; VC—Villa de Cura.

Santonian to Early Campanian (85–83 Ma): Initiation of Subduction below the Oceanic Plateau

The time elapsed between the eruption of the oceanic plateau and the initiation of the island-arc sequence was only ~3 my. Initiation of a subduction zone is a consequence of lateral compositional buoyancy contrast within the lithosphere (Niu et al., 2003). This suggests that the initiation of subduction below an oceanic plateau is produced by the difference in density between the young and buoyant oceanic plateau and the surrounding and older MORB oceanic crust, allowing the denser MORB to subduct beneath the buoyant overriding oceanic plateau, if frictional forces are overcome. The ca. 85 Ma rocks of the Pujilí granite and boninites of the La Portada Formation were probably formed as a result of west-dipping subduction along the leading edge of the plateau (Fig. 10A). Furthermore, rocks with adakitic affinities (Pujilí granite) and boninites are broadly considered to indicate initiation of subduction beneath thickened oceanic crust (e.g., Sajona et al., 1993; Niu et al., 2003; Escuder Viruete et al., 2006).

Campanian to Maastrichtian (85–75 Ma): Rio Cala Island Arc

The Rio Cala arc sequence is a series of volcaniclastic turbidites with intercalated basalts. The turbiditic sedimentary rocks were deposited in submarine fans derived from contemporaneous intraoceanic island-arc volcanoes. Trace- and major-element geochemistry of the sedimentary and volcanic facies is typical of rocks formed in an island arc. The island-arc rocks of the Western Cordillera temporally and geochemically correlate with island-arc rocks of the coast (e.g., San Lorenzo Formation). In addition, geochemical and isotopic characteristics of the Rio Cala Group overlap with those of basalts of the Late Cretaceous Great Arc in the present-day Caribbean (e.g., Thompson et al., 2004; Vallejo, 2007). This chronostratigraphic and geochemical evidence suggests that these igneous rocks represent parts of the same island-arc system. Trace-element and isotopic (Pb) data support a model whereby the island-arc rocks of the Rio Cala Formation erupted through the oceanic plateau (Allibon et al., 2005). Palinspastic constraints suggest that the Rio Cala island arc developed on the plateau by west-dipping subduction of the Farallon plate (Fig. 10B). The absence of regional-scale, subduction-related igneous activity along the Ecuadorian continental margin from 85 to 65 Ma corroborates the absence of eastward-directed subduction beneath the continental margin. As a consequence, closure of the oceanic basin between the Caribbean plateau-arc complex and the South America continental margin was solely accommodated by westward subduction beneath the Caribbean Plateau.

Further support for this model comes from the Yunguilla Formation, which was deposited coeval with island-arc volcanism, and was tectonically squeezed between the South American continental margin and the plateau-arc sequences. Detrital grains were eroded from rocks of the present-day Eastern Cordillera (Fig. 10B), which defined the contemporaneous continental plate margin (Vallejo, 2007). The Yunguilla Formation was deposited in a north-south–oriented passive margin–type basin along the South American continental margin, predating the final collision of the Rio Cala island arc.

Latest Campanian to Maastrichtian (Ca. 73–70 Ma): Collision of the Pallatanga Block

During the Maastrichtian several events occurred: (1) subduction-related Rio Cala arc magmatism stopped, (2) there was a rapid change in the paleomagnetic declination of the Piñon and the San Lorenzo blocks (Luzieux et al., 2006), (3) there was strong deformation of the Yunguilla Formation, and (4) the Eastern Cordillera and the Amotape Complex underwent rapid cooling and exhumation (>1 km/m.y.) (Spikings et al., 2000, 2001, 2005). All these arguments suggest that the collision between South America and the Caribbean-Colombia oceanic plateau started at ca. 73 Ma (Fig. 10C).

Latest Maastrichtian: Silante Arc

After collision, the calc-alkaline Silante arc was established on the amalgamated margin due to a polarity change of subduction zone. The change from earlier island-arc tholeiites to the high-K, calc-alkaline volcanic rocks is coincident with east-dipping subduction during the Maastrichtian (Fig. 10D). Furthermore, any invoked event would have to account for a transition from submarine island-arc volcanism of the Rio Cala Group to the continental volcanic arc of the Silante Formation, which was most likely driven by the collision of the Pallatanga block, and subduction-zone polarity change.

The latest Maastrichtian to Paleocene radiometric eruption ages of the Silante Formation also corroborates Maastrichtian to Danian biostratigraphic ages (e.g., Savoyat et al., 1970) from sedimentary rocks that overlie the Silante Formation, northwest of Quito (Pilaló Formation). The radiometric ages from the Silante Formation indicate that a sudden change from marine to terrestrial environment, and the initiation of continental arc volcanism, occurred before 60 Ma. Emersion of the landmass during this period was perhaps caused by isostatic rebound following rapid exhumation (e.g., Spikings et al., 2005), combined with prolonged compressive stress. Rock uplift and erosion was marked by the deposition of red beds of the Tena in the Oriente Basin, indicating that a terrestrial environment was widespread at the end of the Maastrichtian.

Latest Maastrichtian to Paleocene $^{40}Ar/^{39}Ar$ ages obtained from the Silante Formation suggest that it correlates with the Sacapalca Formation of southern Ecuador. The Sacapalca Formation yielded a zircon fission-track age of 66.9 ± 5.8 Ma (Hungerbühler et al., 2002), and it is intruded by the El Tingo pluton, which yields a K/Ar (biotite) age of 50 ± 3 Ma (Kennerly, 1980).

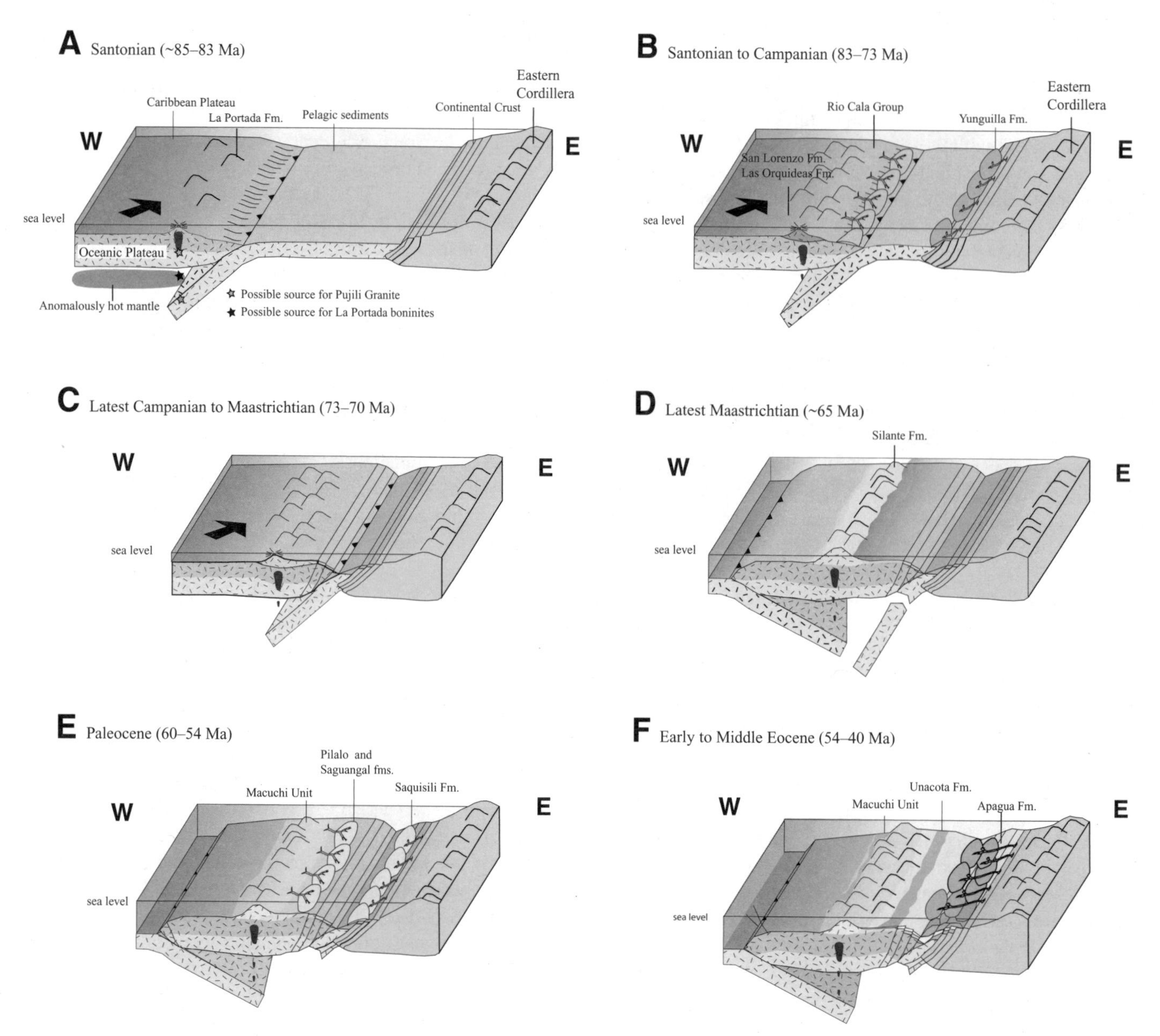

Figure 10. Schematic paleotectonic model for the development of the Western Cordillera and surrounding areas from Late Cretaceous to Eocene (not to scale).

Paleocene to Eocene: Macuchi Arc

During the early Paleocene, marine conditions dominated in the forearc (Fig. 10E). Mixed volcaniclastic and siliciclastic sediments were deposited at northern latitudes (Pilaló and Saguangal Formations; Fig. 4). The volcanic detritus of these formations probably originated from the Silante and Macuchi arc complexes. However, the Saguangal Formation also hosts metamorphic and granitic detritus derived from the Eastern Cordillera, which represents further evidence for the complete welding of the present-day Ecuadorian forearc.

The Macuchi arc was established and may represent the Paleocene–Eocene continuation of the Silante arc. The presence of detrital zircons of Cambrian age within the Macuchi unit strongly suggests that the Macuchi Arc was closer to an older source region like the Eastern Cordillera. This evidence indicates that the Macuchi unit formed close to the South American continent. In addition, in a section in central Ecuador (UTM: 721613/9766370), turbidites of the Eocene Apagua Formation are observed to conformably overlie undeformed rocks of the Macuchi unit, supporting the hypothesis that there was not an accretion event during the Eocene. Furthermore, it is geometrically challenging because the adjacent Piñon block to the west (Fig. 1) had already accreted to the South American continent during the Late Cretaceous (Luzieux et al., 2006).

Siliciclastic rocks of the Angamarca Group (Saquisilí and later the Apagua Formation) were deposited to the east of the Macuchi arc, and detritus was supplied from the emerging Eastern Cordillera. The heavy mineral assemblages record a progressive increase in the proportion of minerals in the Angamarca Group, which eroded from metamorphosed sialic crust (Fig. 5), reflecting uplift and erosion of the Eastern Cordillera. Accelerating exhumation of metamorphic rocks is also recorded by reducing lag-times of fission-track ages of detrital zircons in the Amazon foreland Basin (Subandean zone; e.g., Ruiz et al., 2007). Multiple reworking events of zircons derived from the Amazon craton is observed on both sides of the Eastern Cordillera (Vallejo, 2007; Martin-Gombojav and Winkler, 2008). In summary, exhumation of the source regions was caused by accretion of the Pallatanga and Piñon blocks, and subsequent deformation of the continental margin.

CONCLUSIONS

1. The ca. 88 Ma radiometric ages from oceanic plateau sequences in western Ecuador, when combined with their geochemical affinities, support a genetic relationship with the coeval Caribbean-Colombia oceanic plateau.

2. The initiation of subduction at the leading edge of the Caribbean-Colombia oceanic plateau was responsible for the generation of the Santonian–early Campanian (ca. 85–75 Ma) boninites of the La Portada unit (Rio Cala Group), and the ocean-island-arc–related Pujilí granite at ca. 85 Ma. These rocks were produced during the early stages of arc magmatism via west-dipping subduction beneath the Caribbean-Colombia oceanic plateau. The radiometric ages and chemical compositions of the Late Cretaceous island-arc rocks of the Western Cordillera correlate with the island-arc rocks of coastal Ecuador, Colombia, and the Caribbean region. Therefore, it is reasonable to suggest that they represent the southward extension of the Great Arc of the Caribbean.

3. The oceanic plateau and overlying island arc drifted east and collided with South America during the Campanian (ca. 73 Ma). The termination of arc magmatism in the early Maastrichtian (ca. 70 Ma) was caused by clogging of the subduction zone during collision between the buoyant Caribbean Plateau and South America. Paleomagnetic data from basement and sedimentary cover rocks in the coastal region indicate a contemporaneous 20°–50° clockwise rotation during 73–70 Ma (Luzieux et al., 2006).

4. The collision between South America and the Caribbean-Colombia oceanic plateau resulted in rapid exhumation (>1 km/m.y.) and surface uplift in the Eastern Cordillera as shown by Spikings et al. (2001, 2005). The occurrence of enhanced erosion throughout the continental margin from ca. 75 to 65 Ma coincided with clastic sedimentation in the forearc and backarc of the South American continental arc during the late Campanian and Maastrichtian.

5. Collision of the Caribbean Plateau and South America initiated eastward subduction beneath the accreted oceanic-plateau fragments. The new active margin gave rise to the latest Maastrichtian-Paleocene (ca. 65 Ma) Silante volcanic arc, which developed in a terrestrial environment.

6. During the Paleocene to Eocene, volcanic rocks of the Macuchi arc formed as the temporal continuation of the Silante arc. Submarine volcanism was coeval with the deposition of siliciclastic sedimentary rocks of the Angamarca Group and Saguangal Formation, which were mainly shed from the emerging Eastern Cordillera.

7. We find no evidence for the accretion of the Macuchi arc during the late Eocene as suggested by previous authors, who also have claimed that such accretion was responsible for structural inversion of the Angamarca Basin. Furthermore, it is geometrically and stratigraphically difficult to reconcile late Eocene accretion of the Macuchi unit, given that it is currently located between the Piñon block and the Pallatanga block, which amalgamated during the Late Cretaceous.

ACKNOWLEDGMENTS

We thank Arturo Egüez, John Aspden, Kevin Burke, Andrew Kerr, and Bernardo Beate for fruitful discussions about the geology of Ecuador and the Caribbean region. Fieldwork benefited from the knowledge of Peter Hochuli, Efrain Montenegro, and Janeth Gaibor. E. Jaillard and W. Pratt are acknowledged for their constructive and thorough reviews. This work was supported by the Swiss National Science Foundation, projects 2-77193-02 and 2-77504-04.

REFERENCES CITED

Acton, G.D., Galbrun, B., and King, J.W., 2000, Paleolatitude of the Caribbean plate since the Late Cretaceous, *in* Leckie, R.M., Sigurdsson, H.R., Acton, G.D., and Draper, G., et al., Proceedings of the Ocean Drilling Program, Scientific Results, Volume 165: College Station, Texas, Ocean Drilling Program, p. 149–173.

Aguirre, L., and Atherton, M.P., 1987, Low-grade metamorphism and geotectonic setting of the Macuchi Formation, Western Cordillera of Ecuador: Journal of Metamorphic Geology, v. 5, p. 473–494, doi: 10.1111/j.1525-1314.1987.tb00397.x.

Allibon, J., Monjoie, P., Lapierre, H., Jaillard, E., Bussy, F., and Bosch, D., 2005, High Mg-basalts in the Western Cordillera of Ecuador: Evidence of plateau root melting during Late Cretaceous arc magmatism, *in* Sempéré, T., ed., Proceedings of the Sixth International Symposium on Andean Geodynamics, Program and Abstracts: Barcelona, Spain, Universitat de Barcelona, Instituto Geológico y Minero de España, IRD Éditions, p. 3335.

Aspden, J.A., and Litherland, M., 1992, The geology and Mesozoic collisional history of the Cordillera Real, Ecuador, *in* Oliver, R.A., et al., eds., Andean Geodynamics: Tectonophysics, v. 205, p. 187–204.

Aspden, J.A., and McCourt, W.J., 1986, Mesozoic oceanic terranes in the central Andes of Colombia: Geology, v. 14, p. 415–418, doi: 10.1130/0091-7613(1986)14<415:MOTITC>2.0.CO;2.

Aspden, J.A., McCourt, W.J., and Brook, M., 1987, Geometrical control of subduction-related magmatism: The Mesozoic and Cenozoic plutonic history of western Colombia: Journal of the Geological Society of London, v. 144, p. 893–905, doi: 10.1144/gsjgs.144.6.0893.

Beaudon, E., Martelat, J.E., Amortegui, A., Lapierre, H., and Jaillard, E., 2005, Métabasites de la cordillère occidentale d'Équateur, témoins du soubassement océanique des Andes d'Équateur: Comptes Rendus Geoscience, v. 337, p. 625–634, doi: 10.1016/j.crte.2005.01.002.

Boland, M.P., Pilatasig, L.F., Ibandango, C.E., McCourt, W.J., Aspden, J.A., Hughes, R.A., and Beate, B., 2000, Geology of the Western Cordillera between 0°–1°N: Quito, Ecuador, Proyecto de Desarrollo Minero y Control Ambiental, Programa de Informacion Cartografica y Geológica Informe 10, Corporación de Desarrollo e Investigación Geológica, Minera y Metalúrgica–Ministerio de Energia Ecuador–British Geological Survey, 72 p.

Bourgois, J., Calle, B., Tournon, J., and Toussaint, J.F., 1982, The Andean ophiolitic megastructures on the Buga-Buenaventura transverse (Western Cordillera–Valle Colombia): Tectonophysics, v. 82, p. 207–229, doi: 10.1016/0040-1951(82)90046-4.

British Geological Survey and Corporación de Desarrollo e Investigación Geologico, Minero y Metalúrgico, 1997a, Geological Map of the Western Cordillera, Ecuador between 0° and 1° N: Keyworth, UK, British Geological Survey, scale 1:200,000.

British Geological Survey and Corporación de Desarrollo e Investigación Geologico, Minero y Metalúrgico, 1997b, Geological Map of the Western Cordillera, Ecuador between 0° and 1° S: Keyworth, UK, British Geological Survey, scale 1:200,000.

British Geological Survey and Corporación de Desarrollo e Investigación Geologico, Minero y Metalúrgico, 1997c, Geological Map of the Western Cordillera, Ecuador between 1° and 2° S: Keyworth, UK, British Geological Survey, scale 1:200,000.

British Geological Survey and Corporación de Desarrollo e Investigación Geologico, Minero y Metalúrgico, 1997d, Geological Map of the Western Cordillera, Ecuador between 2° and 3° S: Keyworth, UK, British Geological Survey, scale 1:200,000.

British Geological Survey and Corporación de Desarrollo e Investigación Geologico, Minero y Metalúrgico, 1997e, Geological Map of the Western Cordillera, Ecuador between 3° and 4° S: Keyworth, UK, British Geological Survey, scale 1:200,000.

Bruet, F., 1949, Les enclaves des laves des volcans de Quito, République de l'Équateur: Bulletin de la Société Géologique de France, v. 19, p. 477–491.

Burke, K., 1988, Tectonic evolution of the Caribbean: Annual Review of Earth and Planetary Sciences, v. 16, p. 201–230, doi: 10.1146/annurev.ea.16.050188.001221.

Case, J.E., Duran, A.R., and Moore, W.R., 1971, Tectonic investigations in western Colombia and eastern Panama: Geological Society of America Bulletin, v. 82, p. 2685–2712, doi: 10.1130/0016-7606(1971)82[2685:TIIWCA]2.0.CO;2.

Chew, D., Schaltegger, U., Kosler, J., Whitehouse, M.J., Gutjahr, M., Spikings, R.A., and Miskovic, A., 2007, U-Pb geochronologic evidence for the evolution of the Gondwanan margin of the north-central Andes: Geological Society of America Bulletin, v. 119, p. 697–711, doi: 10.1130/B26080.1.

Chiaradia, M., and Fontboté, L., 2001, Radiogenic lead signatures in Au-rich volcanic-hosted massive sulfide ores and associated volcanic rocks of the early Tertiary Macuchi Island Arc (Western Cordillera of Ecuador): Economic Geology and the Bulletin of the Society of Economic Geologists, v. 96, p. 1361–1378.

Cosma, L., Lapierre, H., Jaillard, E., Laubacher, G., Bosch, D., Desmet, A., Mamberti, M., and Gabriele, P., 1998, Petrographie et geochimie des unites magmatiques de la Cordillère Occidentale d'Equateur (0°30′S): Implications tectoniques: Bulletin de la Société Géologique de France, v. 169, p. 739–751.

Cox, A., and Hart, R.B., 1986, Plate Tectonics—How It Works: Oxford, Blackwell Scientific Publications, 392 p.

Dalrymple, G.B., and Lanphere, M.A., 1971, $^{40}Ar/^{39}Ar$ technique of K-Ar dating: A comparison with the conventional technique: Earth and Planetary Science Letters, v. 12, p. 300–308, doi: 10.1016/0012-821X(71)90214-7.

Donnelly, T., Beets, D., Carr, M., Jackson, T., Klaver, G., Lewis, J., Maury, R., Schellekens, H., Smith, A., Wadge, G., and Westercamp, D., 1990, History and tectonic setting of the Caribbean magmatism, *in* Dengo, G., and Case, J., eds., The Caribbean Region: Boulder, Colorado, Geological Society of America, Geology of North America, v. H, p. 339–374.

Duncan, R.A., and Hargraves, R.B., 1984, Plate tectonic evolution of the Caribbean region in the mantle reference frame, *in* Bonini, W.E., Hargraves, R.B., and Shagam, R., eds., The Caribbean–South America Plate Boundary and Regional Tectonics: Geological Society of America Memoir 162, p. 81–93.

Egüez, A., 1986, Evolution Cénozoïque de la Cordillère Occidentale Septentrionale d'Equateur (0°15′S–01°10′S), les Mineralisations Associées [Ph.D. thesis]: Paris, Université Pierre et Marie Curie, 116 p.

Engebretson, D.C., Cox, A., and Gordon, R.G., 1985, Relative Motions between Oceanic and Continental Plates in the Pacific Basin: Geological Society of America Special Paper 206, 59 p.

Escuder Viruete, J., Díaz de Neira, A., Hernáiz Huerta, P.P., Monthel, J., García Senz, J., Joubert, M., Lopera, E., Ullrich, T., Friedman, R., Mortensen, J., and Pérez-Estaún, A., 2006, Magmatic relationships and ages of Caribbean Island arc tholeiites, boninites and related felsic rocks: Lithos, v. 90, p. 161–186, doi: 10.1016/j.lithos.2006.02.001.

Feininger, T., and Bristow, C.R., 1980, Cretaceous and Paleogene geologic history of Coastal Ecuador: Geologische Rundschau, v. 69, p. 849–874, doi: 10.1007/BF02104650.

Feininger, T., and Seguin, M.K., 1983, Simple Bouguer gravity anomaly field and the inferred crustal structure of continental Ecuador: Geology, v. 11, p. 40–44, doi: 10.1130/0091-7613(1983)11<40:SBGAFA>2.0.CO;2.

Frost, C., and Snoke, A., 1989, Tobago, West Indies, a fragment of a Mesozoic oceanic island arc: Petrochemical evidence: Journal of the Geological Society of London, v. 146, p. 953–964, doi: 10.1144/gsjgs.146.6.0953.

Goossens, P.J., and Rose, W.I., 1973, Chemical composition and age determination of tholeiitic rocks in the Basic Igneous Complex, Ecuador: Geological Society of America Bulletin, v. 84, p. 1043–1052, doi: 10.1130/0016-7606(1973)84<1043:CCAADO>2.0.CO;2.

Hauff, F., Hoernle, K., Bogaard, P.V.D., Alvarado, G.E., and Garbe-Schonberg, D., 2000, Age and geochemistry of basaltic complexes in western Costa Rica: Contributions to the geotectonic evolution of Central America: Geochemistry, Geophysics, Geosystems, v. 1, no. 5, doi: 10.1029/1999GC000020.

Hoernle, K., Van den Bogaard, P., Werner, R., Lissina, B., Hauff, F., Alvarado, G., and Garbe-Schonberg, D., 2002, Missing history (16–71 Ma) of the Galápagos hotspot: Implications for the tectonic and biological evolution of the Americas: Geology, v. 30, p. 795–798, doi: 10.1130/0091-7613(2002)030<0795:MHMOTG>2.0.CO;2.

Hughes, R., and Bermúdez, R., 1997, Geology of the Cordillera Occidental of Ecuador between 0° 00′and 1° 00′S: Quito, Ecuador, Proyecto de Desarollo Minero y Control Ambiental, Programa de Información Cartográfica y Geológica Report 4 (CODIGEM–British Geological Survey), 75 p.

Hughes, R.A., and Pilatasig, L.F., 2002, Cretaceous and Tertiary terrane accretion in the Cordillera Occidental of the Andes of Ecuador: Tectonophysics, v. 345, p. 29–48, doi: 10.1016/S0040-1951(01)00205-0.

Hughes, R., Bermúdez, R., and Espinel, G., 1998, Mapa Geologico de la Cordillera Occidental del Ecuador entre 0°–1° S: Quito, Ecuador, Corporación de Desarrollo e Investigación Geológica, Minera y Metalúrgica–Ministerio de Energia Ecuador–British Geological Survey, scale 1:200,000.

Hungerbühler, D., Steinmann, M., Winkler, W., Seward, D., Egüez, A., Peterson, D.E., Helg, U., and Hammer, C., 2002, Neogene Andean geodynamics of southern Ecuador: Earth-Science Reviews, v. 57, p. 75–124, doi: 10.1016/S0012-8252(01)00071-X.

Jaillard, E., Ordoñez, M., Benitez, S., Berrones, G., Jimenez, N., Montenegro, G., and Zambrano, I., 1995, Basin development in an accretionary, oceanic-floored fore-arc setting: Southern coastal Ecuador during Late Cretaceous–late Eocene time, *in* Tankard, A.J., Suárez Soruco, R., and Welsink, H.J., eds., Petroleum Basins of South America: American Association of Petroleum Geologists (AAPG) Memoir 62, p. 615–631.

Jaillard, E., Ordoñez, M., Suarez, J., Toro, J., Iza, D., and Lugo, W., 2004, Stratigraphy of the Late Cretaceous Paleogene deposits of the Cordillera Occidental of central Ecuador: Geodynamic implications: Journal of South American Earth Sciences, v. 17, p. 49–58, doi: 10.1016/j.jsames.2004.05.003.

Jaillard, E., Bengtson, P., and Dhondt, A., 2005, Late Cretaceous marine transgressions in Ecuador and northern Peru: A refined stratigraphic framework: Journal of South American Earth Sciences, v. 19, p. 307–323, doi: 10.1016/j.jsames.2005.01.006.

Kehrer, W., and van der Kaaden, G., 1979, Notes on the geology of Ecuador, with special reference to the Western Cordillera: Geologisches Jahrbuch, v. 35, p. 5–57.

Kelley, K., 1993, Relative motions between North America and oceanic plates of the Pacific Basin during the past 130 Ma [M.Sc. thesis]: Western Washington University, 189 p.

Kennerly, J.B., 1980, Outline of the geology of Ecuador: British Geological Survey Overseas Geologic and Mineral Resources, no. 55: London, 20 p.

Kerr, A.C., and Tarney, J., 2005, Tectonic evolution of the Caribbean and northwestern South America: The case for accretion of two Late Cretaceous oceanic plateaus: Geology, v. 33, p. 269–272, doi: 10.1130/G21109.1.

Kerr, A.C., Marriner, G.F., Tarney, J., Nivia, A., Saunders, A.D., Thirlwall, M.F., and Sinton, C.W., 1997, Cretaceous basaltic terranes in western Colombia: Elemental, chronological and Sr-Nd isotopic constraints on petrogenesis: Journal of Petrology, v. 38, p. 677–702, doi: 10.1093/petrology/38.6.677.

Kerr, A.C., Tarney, J., Nivia, A., Marriner, G.F., and Saunders, A.D., 1998, The internal structure of oceanic plateaus: Inferences from obducted Cretaceous blocks in western Colombia and the Caribbean: Tectonophysics, v. 292, p. 173–188, doi: 10.1016/S0040-1951(98)00067-5.

Kerr, A.C., Iturralde-Vinent, M.A., Saunders, A.D., Babbs, T.L., and Tarney, J., 1999, A new plate tectonic model of the Caribbean: Implications from a geochemical reconnaissance of Cuban Mesozoic volcanic rocks: Geological Society of America Bulletin, v. 111, p. 1581–1599, doi: 10.1130/0016-7606(1999)111<1581:ANPTMO>2.3.CO;2.

Kerr, A.C., Aspden, J.A., Tarney, J., and Pilatasig, L.F., 2002a, The nature and provenance of accreted oceanic blocks in western Ecuador: Geochemical and tectonic constraints: Journal of the Geological Society of London, v. 159, p. 577–594, doi: 10.1144/0016-764901-151.

Kerr, A.C., Tarney, J., Kempton, P.D., Spadea, P., Nivia, A., Marriner, G.F., and Duncan, R.A., 2002b, Pervasive mantle plume head heterogeneity: Evidence from the Late Cretaceous Caribbean-Colombian oceanic plateau: Journal of Geophysical Research, v. 107, no. B7, 2140, doi: 10.1029/2001JB000790.

Kerr, A.C., White, R.V., Thompson, P.M., Tarnez, J., and Saunders, A.D., 2003, No oceanic plateau–no Caribbean plate? The seminal role of an oceanic plateau in Caribbean plate evolution, *in* Bartolini, C., Buffler, R.T., and Blickwede, J., eds., The Circum-Gulf of Mexico and the Caribbean: Hydrocarbon Habitats, Basin Formation, and Plate Tectonics: American Association of Petroleum Geologists (AAPG) Memoir 79, p. 126–168.

Koppers, A.A.P., Staudigel, H., and Wijbrans, J.R., 2000, Dating crystalline groundmass separates of altered Cretaceous seamount basalts by the ^{40}Ar/^{39}Ar incremental heating technique: Chemical Geology, v. 166, p. 139–158, doi: 10.1016/S0009-2541(99)00188-6.

Lapierre, H., Bosch, D., Dupuis, V., Polve, M., Maury, R., Hernandez, J., Monie, P., Yeghicheyan, D., Jaillard, E., Tardy, M., Mercier de Lepinay, B., Mamberti, M., Desmet, A., Keller, F., and Senebier, F., 2000, Multiple plume events in the genesis of the peri-Caribbean Cretaceous oceanic plateau province: Journal of Geophysical Research, v. 105, p. 8403–8421, doi: 10.1029/1998JB900091.

Lebras, M., Megard, F., Dupuy, C., and Dostal, J., 1987, Geochemistry and tectonic setting of pre-collision Cretaceous and Paleogene volcanic rocks of Ecuador: Geological Society of America Bulletin, v. 99, p. 569–578, doi: 10.1130/0016-7606(1987)99<569:GATSOP>2.0.CO;2.

Lebrat, M., Mégard, F., Juteau, T., and Calle, J., 1985, Pre-orogenic volcanic assemblage and structure in the Western Cordillera of Ecuador between 1°40′S and 2°20′S: Geologische Rundschau, v. 74, p. 343–351, doi: 10.1007/BF01824901.

Litherland, M., Klinck, B.A., O'Connor, E.A., and Pitfield, P.E., 1985, Andean-trending mobile belts in the Brazilian Shield: Nature, v. 314, p. 345–348, doi: 10.1038/314345a0.

Litherland, M., Aspden, J., and Jemielita, R.A., 1994, The Metamorphic Belts of Ecuador: British Geological Survey Overseas Memoir 11, 147 p.

Lonsdale, P., 1978, Ecuadorian subduction system: American Association of Petroleum Geologists (AAPG) Bulletin, v. 62, p. 2454–2477.

Lugo, W., 2003, Geología y petrología de la serie detrítica del Cretácico superior–Eoceno de la Cordillera Occidental (entre San Juan y Guaranda) [Tesis]: Quito, Escuela Politécnica Nacional, 96 p.

Luzieux, L.D.A., 2007, Origin and Late Cretaceous–Tertiary Evolution of the Ecuadorian Forearc [Ph.D. thesis]: Zürich, Switzerland, Institute of Geology, ETH Zürich, 197 p., http://e-collection.ethbib.ethz.ch/show?type=diss&nr=16983, doi: 10.1016/j.epsl.2006.07.008.

Luzieux, L.D.A., Heller, F., Spikings, F., Vallejo, C.F., and Winkler, W., 2006, Origin and Cretaceous tectonic history of the coastal Ecuadorian forearc between 1°N and 3°S: Paleomagnetic, radiometric and fossil evidence: Earth and Planetary Science Letters, v. 249, p. 400–414.

Mamberti, M., 2001, Origin and Evolution of Two Distinct Cretaceous Oceanic Plateaus Accreted in Western Ecuador (South America): Petrological, Geochemical and Isotopic Evidence [Ph.D. thesis]: Lausanne, Université de Lausanne, 241 p.

Mamberti, M., Lapierre, H., Bosch, D., Ethien, R., Jaillard, E., Hernandez, J., and Polve, M., 2003, Accreted fragments of the Late Cretaceous Caribbean-Colombian plateau in Ecuador: Lithos, v. 66, p. 173–199, doi: 10.1016/S0024-4937(02)00218-9.

Mann, P., and Taira, A., 2004, Global tectonic significance of the Solomon Islands and Ontong Java Plateau convergent zone: Tectonophysics, v. 389, p. 137–190, doi: 10.1016/j.tecto.2003.10.024.

Martin-Gombojav, N., and Winkler, W., 2008, Recycling of Proterozoic crust in the Andean Amazon foreland of Ecuador: Implications for orogenic development of the Northern Andes: Terra Nova, v. 20, p. 22–31.

Marshik, R., Spikings, R., and Kusku, I., 2007, Geochronology and stable isotope signature of alteration related to hydrothermal iron oxide ores in Central Anatolia: Mineralium Deposita, v. 43, no. 1, doi: 10.1007/s00126-007-0160-4.

Mauffret, A., and Leroy, S., 1997, Seismic stratigraphy and structure of the Caribbean igneous province: Tectonophysics, v. 283, p. 61–104, doi: 10.1016/S0040-1951(97)00103-0.

Mauffret, A., Leroy, S., Vila, J.-M., Hallot, E., Mercier de Lèpinay, B., and Duncan, R., 2001, Prolonged magmatic and tectonic development of the Caribbean Igneous Province revealed by a diving submersible survey: Marine Geophysical Researches, v. 22, p. 17–45, doi: 10.1023/A:1004873905885.

McCourt, W.J., Aspden, J.A., and Brooks, M., 1984, New geological and geochronological data from the Colombian Andes: Continental growth by multiple accretion: Journal of the Geological Society of London, v. 141, p. 831–845.

McCourt, W.J., Duque, P., and Pilatasig, L.F., 1997, Geology of the Western Cordillera of Ecuador between 1–2°S: Quito, Ecuador, Proyecto de Desarrollo Minero y Control Ambiental, Programa de Informacion Cartografica y Geologica Informe 3 (Corporación de Desarrollo e Investigación Geológica, Minera y Metalúrgica–British Geological Survey); scale 1:200,000.

Megard, F., 1987, Cordilleran Andes and marginal Andes: A review of Andean geology north of the Arica elbow (18°), *in* Manger, J.W.H., and Francheteau, J., eds., Circum-Pacific Orogenic Belts and Evolution of the Pacific Ocean Basin: American Geophysical Union Geodynamic Monograph 18, p. 71–95.

Millward, D., Marriner, G.F., and Saunders, A.D., 1984, Cretaceous tholeiitic volcanic rocks from the Western Cordillera of Colombia: Journal of the Geological Society of London, v. 141, p. 847–860, doi: 10.1144/gsjgs.141.5.0847.

Montgomery, H., Pessagno, E., Lewis, J., and Schellekens, J., 1994, Paleogeography of Jurassic fragments in the Caribbean: Tectonics, v. 13, p. 725–732, doi: 10.1029/94TC00455.

Müller, R.D., Royer, J.-Y., and Lawver, L.A., 1993, Revised plate motions relative to the hotspots from combined Atlantic and Indian Ocean hotspot tracks: Geology, v. 21, p. 275–278, doi: 10.1130/0091-7613(1993)021<0275:RPMRTT>2.3.CO;2.

Niu, Y., O'Hara, M.J., and Pearce, J., 2003, Initiation of subduction zones as a consequence of lateral compositional buoyancy contrast within the lithosphere: A petrological perspective: Journal of Petrology, v. 44, p. 851–866, doi: 10.1093/petrology/44.5.851.

Nivia, A., 1996, The Bolivar mafic-ultramafic complex, SW Colombia: The base of an obducted oceanic plateau: Journal of South American Earth Sciences, v. 9, p. 59–68, doi: 10.1016/0895-9811(96)00027-2.

Pilger, R., 1981, Plate reconstructions, aseismic ridges and low-angle subduction beneath the Andes: Geological Society of America Bulletin, v. 92, p. 448–456, doi: 10.1130/0016-7606(1981)92<448:PRARAL>2.0.CO;2.

Pindell, J.L., and Barrett, S.F., 1990, Geological evolution of the Caribbean region: A plate tectonic perspective, *in* Dengo, G., and Case, J.E., eds., The Caribbean Region: Boulder, Colorado, Geological Society of America, Geology of North America, v. H, p. 405–432.

Pindell, J.L., Cande, S.C., and Pitam, W.C., 1988, A plate-kinematic framework for models of Caribbean evolution: Tectonophysics, v. 155, p. 121–138, doi: 10.1016/0040-1951(88)90262-4.

Pindell, J., Kennan, L., Maresch, W., Stanek, K., Draper, G., and Higgs, R., 2005, Plate kinematics and crustal dynamics of circum-Caribbean arc-continent interactions; tectonic controls on basin development in proto-Caribbean margins, *in* Avé-Lallemant, H.G., and Sisson, V.B., eds., Caribbean–South American Plate Interactions, Venezuela: Geological Society of America Special Paper 394, p. 7–52.

Pindell, J., Kennan, L., Stanek, K., Maresch, W., and Draper, G., 2006, Foundations of Gulf of Mexico and Caribbean evolution: Eight controversies resolved: Geologica Acta, v. 4, p. 303–341.

Pratt, W.T., Duque, P., and Ponce, M., 2005, An autochthonous geological model for the eastern Andes of Ecuador: Tectonophysics, v. 399, p. 251–278, doi: 10.1016/j.tecto.2004.12.025.

Restrepo, J.A., and Toussaint, J.F., 1974, Obducción cretácea en el Occidente Colombiano: Anales Fac Minas, Medellín, v. 58, p. 73–105.

Revillon, S., Hallot, E., Arndt, N.T., Chauvel, C., and Duncan, R.A., 2000, A complex history for the Caribbean Plateau: Petrology, geochemistry, and geochronology of the Beata Ridge, South Hispaniola: The Journal of Geology, v. 108, p. 641–661, doi: 10.1086/317953.

Reynaud, C., Jaillard, E., Lapierre, H., Mamberti, M., and Mascle, G., 1999, Oceanic plateau island arcs of southwestern Ecuador: Their place in the geodynamic evolution of northwestern South America: Tectonophysics, v. 307, p. 235–254, doi: 10.1016/S0040-1951(99)00099-2.

Roperch, P., Megard, F., Laj, C., Mourier, T., Clube, T., and Noblet, C., 1987, Rotated oceanic blocks in western Ecuador: Geophysical Research Letters, v. 14, p. 558–561, doi: 10.1029/GL014i005p00558.

Ross, M.I., and Scotese, C.R., 1988, A hierarchical tectonic model of the Gulf of Mexico and Caribbean region: Tectonophysics, v. 155, p. 139–168, doi: 10.1016/0040-1951(88)90263-6.

Ruiz, G.M.H., Seward, D., and Winkler, W., 2004, Detrital thermochronology—A new perspective on hinterland tectonics; an example from the Andean Amazon Basin, Ecuador: Basin Research, v. 16, p. 413–430, doi: 10.1111/j.1365-2117.2004.00239.x.

Ruiz, G.M.H., Seward, D., and Winkler, W., 2007, Heavy minerals combined with detrital grain-age studies furthers understanding of source region thermotectonic histories: An example from the Andean Amazon Basin in Ecuador, *in* Mange, M.A., and Wright, D.T., eds., Heavy Minerals in Use: Developments in Sedimentology, v. 58: Amsterdam, Elsevier, p. 907–934.

Sajona, F.G., Maury, R.C., Bellon, H., Cotten, J., Defant, M.J., Pubellier, M., and Rangin, C., 1993, Initiation of subduction and the generation of slab melts in western and eastern Mindanao, Philippines: Geology, v. 21, p. 1007–1010, doi: 10.1130/0091-7613(1993)021<1007:IOSATG>2.3.CO;2.

Sauer, W., 1965, Geología del Ecuador: Quito, Ministerio de Educación, 583 p.

Savoyat, F., Vernet, R., Sigal, J., Mosquera, C., Granja, B., and Guevara, R., 1970, Estudio General de la Cuenca de Esmeraldas. Estudio Micropantologico de las Formaciones de la Sierra: Servicio Nacional de Geología y Minas–Institut Français du Pétrole: Quito, Ecuador, Institut Français du Pétrole, 87 p.

Sinton, C.W., Duncan, R.A., and Denyer, P., 1997, Nicoya Peninsula, Costa Rica: A single suite of Caribbean oceanic plateau magmas: Journal of Geophysical Research, v. 102, p. 15,507–15,520, doi: 10.1029/97JB00681.

Sinton, C.W., Duncan, R.A., Storey, M., Lewis, J., and Estrada, J.J., 1998, An oceanic flood basalts province within the Caribbean plate: Earth and Planetary Science Letters, v. 155, p. 221–235, doi: 10.1016/S0012-821X(97)00214-8.

Spadea, P., and Espinosa, A., 1996, Petrology and chemistry of Late Cretaceous volcanic rocks from the southernmost segment of the Western Cordillera of Colombia (South America): Journal of South American Earth Sciences, v. 9, p. 79–90, doi: 10.1016/0895-9811(96)00029-6.

Spadea, P., Espinosa, E., and Orrego, A., 1989, High-Mg extrusive rocks from the Romeral zone ophiolites in the southwestern Colombian Andes: Chemical Geology, v. 77, p. 303–321, doi: 10.1016/0009-2541(89)90080-6.

Spikings, R.A., Seward, D., Winkler, W., and Ruiz, G.M., 2000, Low temperature thermochronology of the northern Cordillera Real, Ecuador: Tectonic insights from zircon and apatite fission-track analysis: Tectonics, v. 19, p. 649–668, doi: 10.1029/2000TC900010.

Spikings, R.A., Winkler, W., Seward, D., and Handler, R., 2001, Along-strike variations in the thermal and tectonic response of the continental Ecuadorian Andes to the collision with heterogeneous oceanic crust: Earth and Planetary Science Letters, v. 186, p. 57–73, doi: 10.1016/S0012-821X(01)00225-4.

Spikings, R.A., Winkler, W., Hughes, R.A., and Handler, R., 2005, Thermochronology of allochthonous blocks in Ecuador: Unraveling the accretionary and post-accretionary history of the Northern Andes: Tectonophysics, v. 399, p. 195–220, doi: 10.1016/j.tecto.2004.12.023.

Tagami, T., Galbraith, R.F., Yamada, R., and Laslett, G.M., 1998, Revised annealing kinetics of fission tracks in zircon and geological implications, *in* Van den Haute, P., and de Corte, F., eds., Advances in Fission-Track Geochronology: Dordrecht, Kluwer Academic Publishers, p. 99–112.

Thompson, P.M., Kempton, P.D., White, R.V., Kerr, A.C., Tarney, J., Saunders, A.D., Fitton, J.G., and McBirney, A., 2004, Hf-Nd isotope constraints on the origin of the Cretaceous Caribbean Plateau and its relationship to the Galápagos plume: Earth and Planetary Science Letters, v. 217, p. 59–75, doi: 10.1016/S0012-821X(03)00542-9.

Toro, J.A., and Jaillard, E., 2005, Provenance of the Upper Cretaceous to Upper Eocene clastic sediments of the Western Cordillera of Ecuador: Geodynamic implications: Tectonophysics, v. 399, p. 279–292, doi: 10.1016/j.tecto.2004.12.026.

Vallejo, C., 2007, Evolution of the Western Cordillera in the Andes of Ecuador (Late Cretaceous–Paleogene) [Ph.D. thesis]: Zürich, Switzerland, Institute of Geology, ETH Zürich, 208 p., http://e-collection.ethbib.ethz.ch/show?type=diss&nr=17023.

Vallejo, C., Spikings, R.A., Winkler, W., Luzieux, L., Chew, D., and Page, L., 2006, The early interaction between the Caribbean Plateau and the NW South American plate: Terra Nova, v. 18, p. 264–269.

Van Thournout, F., 1991, Stratigraphy, Magmatism and Tectonism in the Ecuadorian Northwestern Cordillera: Metallogenic and Geodynamic Implications [Ph.D. thesis]: Leuven, Katholieke Universiteit Leuven, 150 p.

Wadge, G., Jackson, T.A., Issacs, M.C., and Smith, T.E., 1982, The ophiolitic Bath-Dunrobin Formation, Jamaica: Significance for Cretaceous plate margin evolution in the NW Caribbean: Journal of the Geological Society of London, v. 139, p. 321–333, doi: 10.1144/gsjgs.139.3.0321.

Walker, R.J., Echeverria, I.M., Shirey, S.B., and Horan, M.F., 1991, Re-Os isotopic constraints on the origin of volcanic rocks, Gorgona Island, Colombia: Os isotopic evidence for ancient heterogeneities in the mantle: Contributions to Mineralogy and Petrology, v. 107, p. 150–162, doi: 10.1007/BF00310704.

Walker, R.J., Storey, M.J., Kerr, A.C., Tarney, J., and Arndt, N.T., 1999, Implications of ^{187}Os isotopic heterogeneities in a mantle plume: Evidence from Gorgona Island and Curacao: Geochimica et Cosmochimica Acta, v. 63, p. 713–728, doi: 10.1016/S0016-7037(99)00041-1.

Wallrabe-Adams, H.J., 1990, Petrology and geotectonic development of the western Ecuadorian Andes: The Basic Igneous Complex: Tectonophysics, v. 185, p. 163–182, doi: 10.1016/0040-1951(90)90411-Z.

White, R., Tarney, J., Kerr, A., Saunders, A., Kempton, P., Pringle, M., and Klaver, G., 1999, Modification of an oceanic plateau, Aruba, Dutch Caribbean: Implications for the generation of continental crust: Lithos, v. 46, p. 43–68, doi: 10.1016/S0024-4937(98)00061-9.

Winkler, W., Villagómez, D., Spikings, R., Abegglen, P., Tobler, St., and Egüez, A., 2005, The Chota basin and its significance for the inception and tectonic setting of the Inter-Andean Depression in Ecuador: Journal of South American Earth Sciences, v. 19, p. 5–19, doi: 10.1016/j.jsames.2004.06.006.

Wolf, T., 1892, Geografía y Geología del Ecuador: Leipzig, Brockhaus, 671 p. (Reprinted, 1933, Geography and Geology of Ecuador, Flanagan, J.W., transl., Toronto, Grand & Toy, 684 p.)

Manuscript Accepted by the Society 5 December 2008

Printed in the USA

The Geological Society of America
Memoir 204
2009

Influence of the subduction of the Carnegie volcanic ridge on Ecuadorian geology: Reality and fiction

François Michaud*
Cesar Witt[†]
Université Pierre et Marie Curie, Géoazur (IRD-CNRS-UNS), la Darse, BP48, 06235, Villefranche/Mer, France, and
Departamento de Geologia, Escuela Politécnica Nacional, Andalucia n/s, C.P. 17-01-2755, Quito, Ecuador

Jean-Yves Royer*
Université Européenne de Bretagne and Université de Brest, Centre National de la Recherche Scientifique, Institut Universitaire Européen de la Mer, Laboratoire Domaines Océaniques, Place Copernic, 29280 Plouzané, France

ABSTRACT

The proposed ages for the collision of the Carnegie Ridge with the South America trench, offshore Ecuador, range from 1 to 15 Ma. In this time frame, many geological features of Ecuador are commonly linked to the subduction of the Carnegie Ridge. (1) After the ridge collided with the trench at ca. 15 Ma, the subsequent interplate coupling produced high exhumation rates of volcanic materials at ca. 9 Ma. (2) The oblique convergence of the Carnegie Ridge would have caused the northward drift of the North Andean block and the opening of the Gulf of Guayaquil. (3) During the late Miocene, the subduction of the Carnegie Ridge would have triggered a regional tectonic inversion along the forearc. (4) Along the collision front of the ridge with the trench, subduction-related erosion is occurring, and the Ecuadorian continental margin is being uplifted in the present day. (5) The chemistry of the active volcanic arc is explained as resulting from the arrival of the Carnegie Ridge into the trench. For instance, the adakitic signal, which appears at 1.5 Ma, is thought to be ridge-induced. (6) The buoyancy of the subducted Carnegie Ridge would explain the flatness of the slab beneath Ecuador. In this paper, we review the geological evolution of the Northern Andes in order to establish which of these geological events may be related to the subduction of the Carnegie Ridge. This review suggests that there is no clear deformation linked with the subduction of the Carnegie Ridge or with its landward prolongation postulated at depth.

INTRODUCTION

When prominent bathymetric features subduct at convergent plate margins, they influence interplate coupling and forearc tectonic deformation roughly in proportion to their size, although other factors such as sediment thickness may affect the process (Cloos, 1993; Geist et al., 1993; Scholz and Small, 1997; Dominguez et al., 1998; Fisher et al., 1991; Gardner et al., 1992, 2001; Corrigan et al., 1990; Meffre and Crawford, 2001; Bilek et al., 2003). Several processes, such as basal and frontal

*E-mails: Michaud—micho@geoazur.obs-vlfr.fr; Royer—jyroyer@univ-brest.fr.
[†]Current address: Fault Dynamics Research Group, Department of Geology, Royal Holloway, University of London, Egham, Surrey, TW20 0EX, UK; c.witt@es.rhul.ac.uk.

Michaud, F., Witt, C., and Royer, J.-Y., 2009, Influence of the subduction of the Carnegie volcanic ridge on Ecuadorian geology: Reality and fiction, *in* Kay, S.M., Ramos, V.A., and Dickinson, W.R., eds., Backbone of the Americas: Shallow Subduction, Plateau Uplift, and Ridge and Terrane Collision: Geological Society of America Memoir 204, p. 217–228, doi: 10.1130/2009.1204(10). For permission to copy, contact editing@geosociety.org.

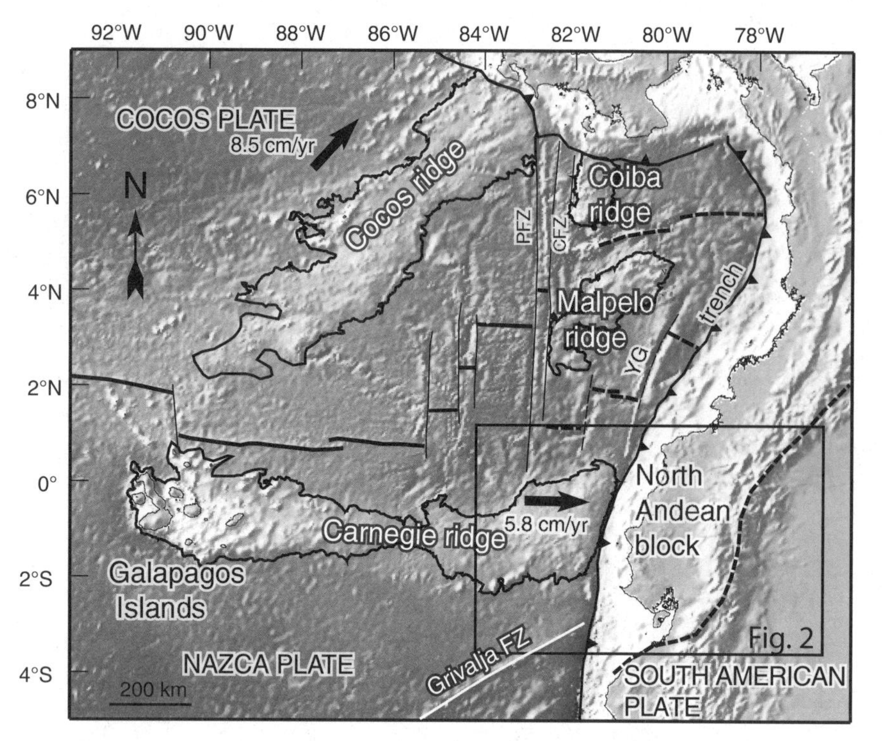

Figure 1. Topographic (onshore) and bathymetric (offshore) map of the Northern Andean region showing the major features mentioned in text. Data are from etopo2 grid of the National Geophysical Data Center–National Oceanic and Atmosphere Administration. CFZ—Coiba Fracture Zone; PFZ—Panama Fracture Zone; YG—Yaquina graben. Dotted black lines—North Andean block boundary.

subduction-erosion, underplating, interplate coupling increase, and forearc uplift and subsidence, have been attributed to ridge subduction. Furthermore, subducted ridges have been recently considered as barriers inhibiting seismogenic rupture (Kodaira et al., 2000). The impingement of bathymetric features results in an increase of the interplate coupling between the subducting and overriding plates and hence in tectonic deformation mostly along the upper plate (i.e., Scholz and Small, 1997; Taylor et al., 2005). Although the upper plate response is geometrically simple, it combines various mechanisms that are difficult to define. Furthermore, several aspects, such as sediment thickness, subduction regime, convergence obliquity, and strength of the overriding plate, can modify the type and amount of ridge-induced deformation in the upper plate.

The Carnegie Ridge is an ~200-km-wide oceanic plateau abutting the Ecuadorian trench (Figs. 1 and 2). Off the Ecuadorian margin (3°S–1°N) close to the trench, the Carnegie Ridge presents a relief up to 1000 m in depth and a crustal thickness ranging from 14 km (Graindorge et al., 2004) to 19 km (Sallarès and Charvis, 2003). The crust of the Carnegie Ridge is thus ~10 km thicker than the crust of the surrounding oceanic basins.

Whether the Carnegie Ridge is just entering the trench or has been subducting for several million years is a matter of debate. The proposed ages for the Carnegie Ridge–trench collision, offshore Ecuador, vary over a wide range from 1 to 3 Ma (Lonsdale, 1978; Lonsdale and Klitgord, 1978; Cantalamessa and Di Celma, 2004) to 8 Ma (Daly, 1989; Gutscher et al., 1999) up to 15 Ma (Pilger, 1984; Spikings et al., 2001, 2005). Many geological features of Ecuador are commonly ascribed to the Carnegie Ridge subduction (Hall and Wood, 1985), not only in the forearc area, but also as far from the trench as the backarc area (Fig. 3). Some of these features include: (1) subduction-erosion (Calahorrano, 2005; Sage et al., 2006) and current coastal uplift (Cantalamessa and Di Celma, 2004; Pedoja et al., 2006a, 2006b); (2) northward drift of the North Andean block and opening of the Gulf of Guayaquil (Lonsdale, 1978; Witt et al., 2006); (3) high exhumation rates of volcanic materials in the Andes (Steinmann et al., 1999; Spikings et al., 2001); (4) the presence of a flat slab corresponding to a landward prolongation of the Carnegie Ridge beneath the overriding plate (Gutscher et al., 1999); and (5) changes in the chemistry of the active volcanic arc related to the flat-slab geometry (Gutscher et al., 2000; Bourdon et al., 2002; Samaniego et al., 2002).

In this paper, we review the proposed hypotheses for Carnegie Ridge subduction and its imprint on Ecuadorian geology.

EVIDENCE FOR CARNEGIE RIDGE SUBDUCTION

Morphology of the Trench Walls

The trench area off Ecuador has now been fully covered with swath-bathymetry data (PUGU cruise of R/V *Atalante*; SO-158, SO-159, and SO-162 cruises of the R/V *Sonne*). The depth of the trench off Ecuador (at the Carnegie subduction zone) is less than 2800 m, whereas, offshore Peru and Colombia, it is up to 3700 m depth, indicating that the Carnegie Ridge is currently subducting at the trench. On the trench outer wall, the swath-bathymetry data display closely spaced trench-parallel normal faults. These clear-cut normal faults formed in response to the bending of the oceanic plate during subduction (Collot et al., 2006a, 2006b). This indicates that the Carnegie Ridge is being dragged down by the

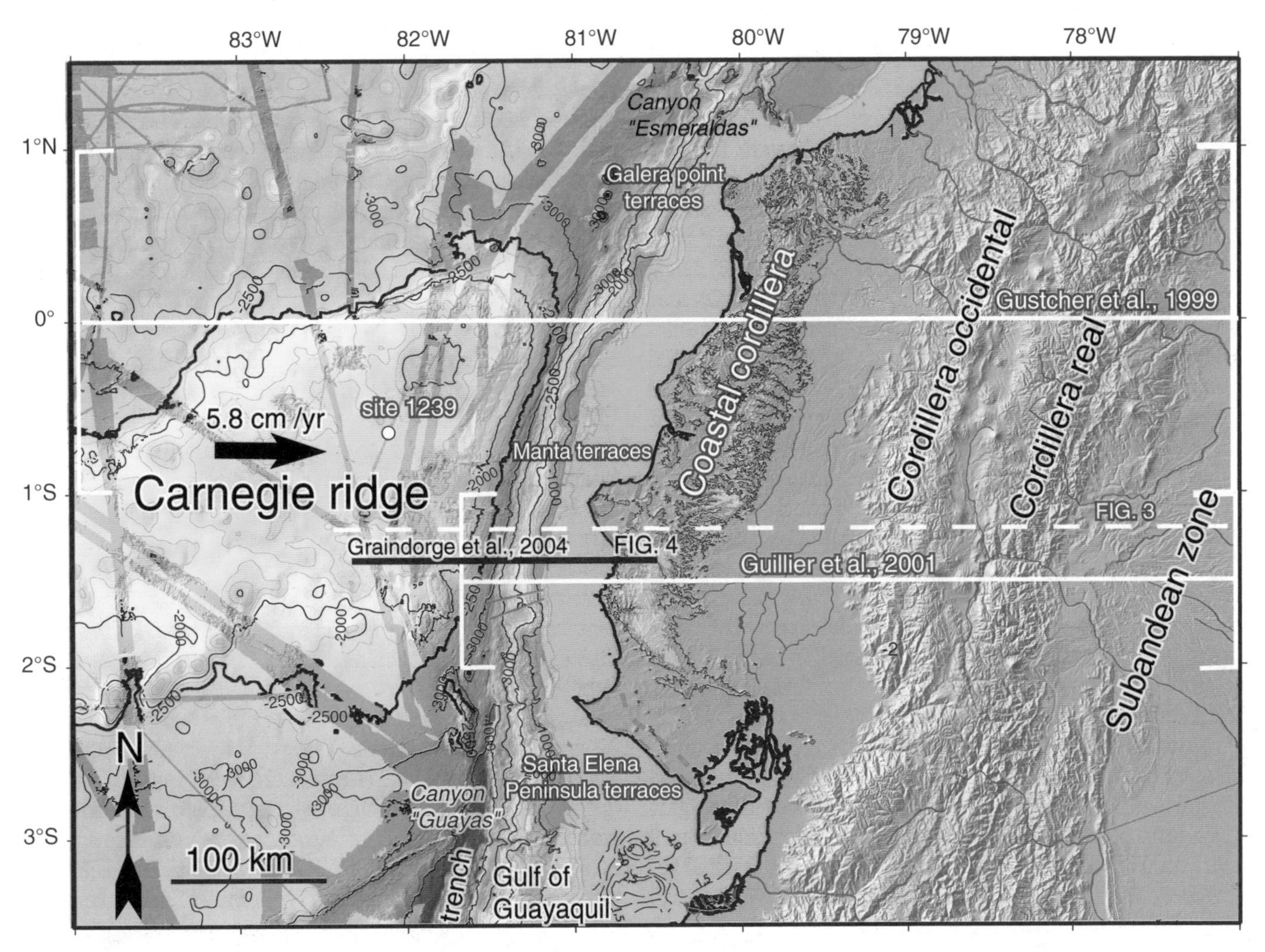

Figure 2. Elevation map of the Carnegie Ridge–Ecuador margin collision area. Bathymetry is from compilation of multibeam and Smith and Sandwell (1997) data (areas not covered by swath bathymetry data are in pale colors); compilation is based on the Pugu and Amadeus cruises (Collot et al., 2006a, 2006b), Salieri and NOAA-NGDC (National Oceanic and Atmosphere Administration–National Geophysical Data Center) databank. Land topography is from Shuttle Radar Topography Mission data from National Aeronautics and Space Administration (NASA); size of the grid cell = 150 m. Offshore, the 2500 m isobath in black demarcates the Carnegie Ridge. Inland, the isocontour in black demarcates the topography up to 200 m, and the isocontour in red demarcates the topography up to 400 m. In the Gulf of Guayaquil area, the dotted black lines correspond to the two-way traveltime (twtt) depth of the top of the Pliocene (from Deniaud et al., 1999), showing the important stage of sediment storage during the Pleistocene. The regions between white brackets correspond to the areas used by Gutscher et al. (1999) and Guillier et al. (2001) to plot earthquakes beneath the continent along a profile. White dashed-line is the cross section of Figure 3; solid black line locates the cross section of Figure 4. Red lines are seismic profiles published by Sage et al. (2006). White circle locates Site 1239 of Ocean Drilling Program (ODP) Leg 202 (Mix et al., 2003). Thick red dashed lines locate the marine terraces from Pedoja et al. (2006a).

subduction. South of 1°30′S, several indentations on the trench inner wall are attributed to multiple seamount impacts (Sage et al., 2006). These seamount impacts are in line with a seaward seamount chain (Fig. 2), which corresponds to the southward limit of the Carnegie Ridge (Michaud et al., 2005, 2006). The shallower depth of the trench along the Carnegie Ridge collision zone and the indentation of the continental margin by incoming bathymetric highs thus provide other evidence for the subduction of the ridge beneath the margin. The short-term tectonic effects of relatively small impinging seamounts should occur as distinct events from the broad interaction of the Carnegie Ridge with the forearc.

Prolongation of the Carnegie Ridge at Depth

The sedimentary cover of the Carnegie Ridge is mainly carbonated and 500 m thick. Seismic units related to this sedimentary cover can be followed for up to 10 km beneath the margin (Sage et al., 2006). This observation suggests that the Carnegie Ridge is currently subducting. Moreover, wide-angle refraction crustal models, orthogonal to the trench, indicate a crustal thickness between 19 km (Sallarès et al., 2005) and 14 km (Graindorge et al., 2004) at the zone of Carnegie Ridge subduction (Figs. 2 and 4). These models indicate that the Carnegie Ridge

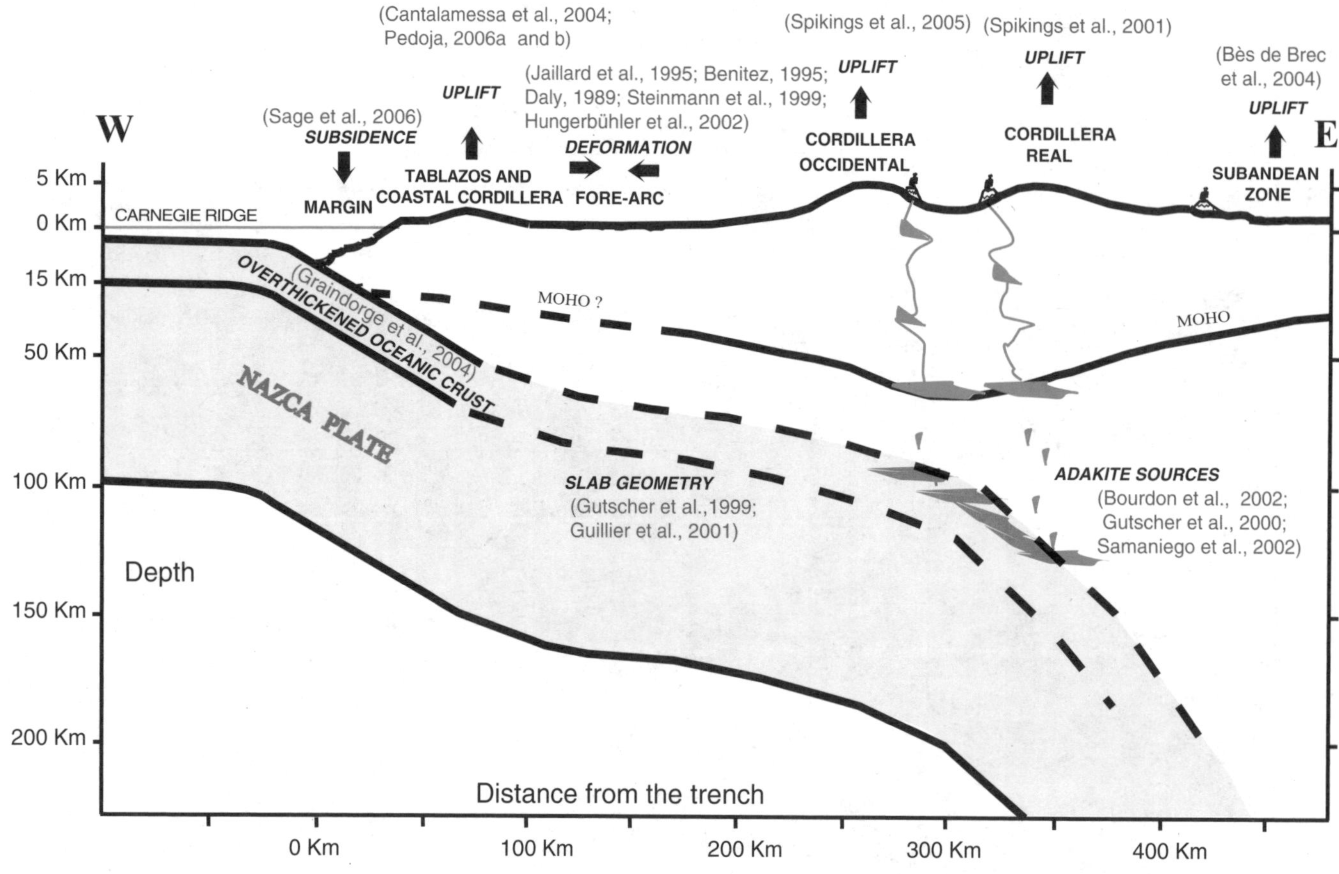

Figure 3. Cross section (location on Fig. 2) showing the geological events attributed to the Carnegie Ridge subduction.

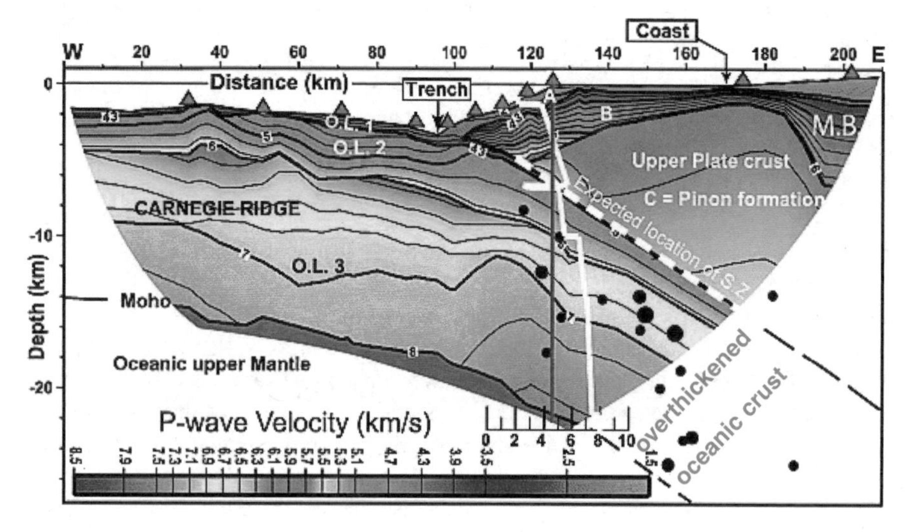

Figure 4. Velocity model across the Ecuadorian margin (from Graindorge et al., 2004). This cross section (location on Fig. 2) shows that the thick Carnegie Ridge crust extends at least 60 km beneath the upper plate. Black circles indicate position of earthquakes from Engdahl's catalogue projected over a distance of 50 km on each side of the profile. M.B.—Manabi Basin, S.Z.—inferred seismogenic zone, O.L.—oceanic layer. One-dimensional (1-D) velocity model at 125.40 km (OBS 10) is plotted on the two-dimensional (2-D) model.

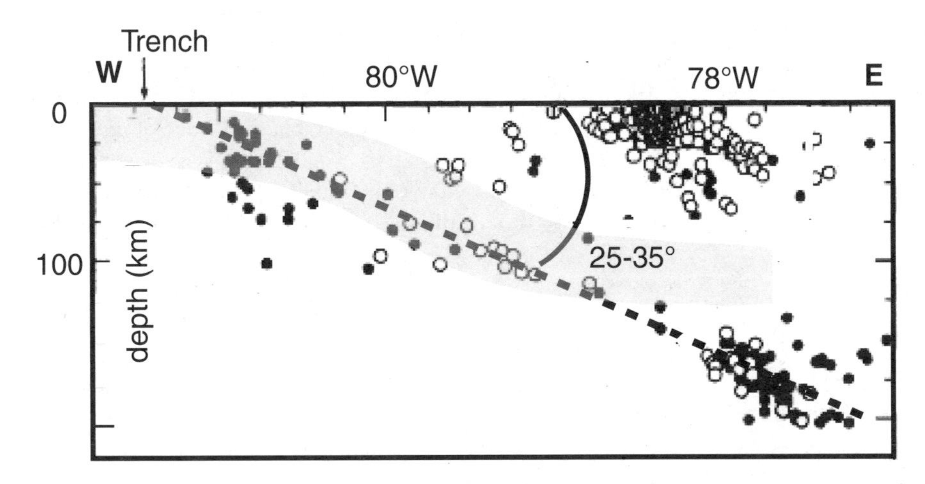

Figure 5. Comparison between the slab geometry from Gutscher et al. (1999; postulated flat slab in gray) and from Guillier et al. (2001; dashed line). The cross section corresponds to the latitude 1°S–2°S (see Fig. 2). Circles are from Guillier et al. (2001): Open circles represent events located using temporary network data; filled circles represent earthquake location from Engdahl et al. (1998).

extends at least 60 km beneath the upper plate (Graindorge et al., 2004). Taking into account the convergence rate at this latitude (5.8 ± 0.2 cm; Trenkamp et al., 2002), the prolongation of the thick Carnegie Ridge crust beneath the margin suggests that the Carnegie Ridge has been subducting since at least 1.4 Ma (Graindorge et al., 2004).

Interplate Coupling along the Forearc

Interplate coupling along a ridge subduction zone is a response of the physical barrier and the buoyancy effects of the ridge itself (Cloos, 1993). Global positioning system (GPS) data along the Ecuadorian forearc show that the eastward-directed velocity vectors in front of the Carnegie Ridge are larger than in its surrounding zones (Trenkamp et al., 2002; White et al., 2003). Although the GPS constraints are evidence of the last seismic cycle, we consider the along-strike variation in the eastward movement of the forearc as evidence for a high long-term degree of coupling between the subducting and overriding plates at the site where the Carnegie Ridge enters subduction. Elastic modeling of GPS-derived data predicts at least 50% locking of the convergence movement at site (Trenkamp et al., 2002; White et al., 2003).

Slab Geometry

Based on the depth distribution of earthquakes from seismological catalogues (Mb > 4), Gutscher et al. (1999) suggested a flat-slab geometry resulting from the buoyancy of subducted portions of the Carnegie Ridge, although a lack of seismic events has prevented the slab from being followed between 80 and 150 km. An additional collection of earthquakes (2.2 < Mb < 5.1) recorded in northern Ecuador by a dense network of 54 stations (Guillier et al., 2001) shows instead a continuous plunging slab down to a depth of ~200 km and an angle varying from 35° in the north to 25° in the south. Thus, the hypocenters do not reveal any major change in the slab dip angle at depths associated with the Carnegie Ridge (Figs. 1 and 5), and subduction of the Carnegie Ridge does not significantly affect the geometry of the slab.

UPPER-PLATE DEFORMATION RELATED TO CARNEGIE RIDGE SUBDUCTION

Continental Slope Deformation

In many active margins, severe deformation is observed at the front of the overriding plate where seamounts or aseismic ridges subduct (Cloos, 1993; Scholz and Small, 1997; Dominguez et al., 1998). Numerous bathymetric and seismic records show that continental slope deformation depends mainly on parameters such as the nature of the asperity, the structure and dip of the oceanic plate, and also the rheology and tectonic regime of the overriding plate (Clift and Vannucchi, 2004). During the SISTEUR cruise (Collot et al., 2002), several seismic profiles were shot across the Ecuadorian margin. Seismic units dipping seaward are affected by extensional faults (Sage et al., 2006). The units are truncated at the interplate boundary. Such geometry is considered to be the signature of a subduction-erosion regime working at depth. South of the Carnegie Ridge, the Ecuadorian margin is affected by pervasive N-S–directed, seaward-dipping normal faulting resulting from subduction-erosion processes at depth (Collot et al., 2002; Calahorrano, 2005). At the latitude of the Gulf of Guayaquil, N-S–trending, seaward-dipping normal faults have been active at least since middle Miocene times (Witt et al., 2006). This fault system connects southward with the extensive structures of the northern Peruvian margin (Bourgois et al., 2007), where subduction-erosion processes have been active since Miocene times (i.e., Bourgois et al., 1988; von Huene et al., 1996; Sosson et al., 1994). Therefore, the subduction-erosion process beneath the continental margin seems to extend from northern Peru to the northern flank of the Carnegie Ridge and seems to have been active at least since Miocene times (von Huene et al., 1996). Consequently, it is most likely that the subduction of the Carnegie Ridge is not the only factor controlling subduction-erosion along the Ecuadorian continental margin. However, the arrival of the Carnegie Ridge at the trench surely increased this process. This scenario is probably similar to that of the zones affected by the Nazca ridge subduction, where the arrival of the ridge at the trench at ca. 11 Ma increased a pre-emplaced subduction-erosion regime (Hampel, 2002; Hampel and Kukowski, 2004).

Coastal Uplift

Subduction of large buoyant ridges or plateaus may produce important uplift of wide areas due mainly to isostatic adjustments. Two mechanisms may account for a rapid vertical motion along ridge-subduction zones: a displacement mechanism, in which the forearc is displaced upward by the volume of the object passing beneath the forearc, and a crustal shortening mechanism, in which the forearc thickens and uplifts because of horizontal shortening (Taylor et al., 2005). Other mechanisms such as underplating have also been proposed for forearc uplift in response to ridge subduction.

Coastal uplift opposite the Carnegie Ridge is indicated by marine terraces exposed at 200–300 m. Several authors relate the uplift of the Ecuadorian coast to the subduction of the Carnegie Ridge (Cantalamessa and Di Celma; 2004; Pedoja et al., 2006a, 2006b). The initial uplift along the Ecuadorian coast has been dated by these authors as late Pliocene–early Pleistocene. From this period to the present, the rate of coastal uplift in front of the Carnegie Ridge would be constant and equal to ~0.4 mm/yr (Cantalamessa and Di Celma, 2004) or would range between 0.42–0.51 mm/yr (Pedoja et al., 2006a). The latter authors interpret the slight increase in uplift rates (0.1–0.27 mm/yr) with respect to those observed south of the Carnegie Ridge subduction area (north Peruvian coast) as a direct effect of the Carnegie Ridge subduction (combined with the effects of a concave subduction zone and erosion of the coastal margin). Maximum coastal uplift at rates of up to 6 mm/yr have been measured in front of the Cocos Ridge subduction zone (Gardner et al., 1992), whereas maximum rates of 0.5 mm/yr have been measured in front of the Nazca Ridge collision zone (Hsu, 1992). These uplift rates suggest that the vertical uplift imprinted by Carnegie Ridge subduction is not significant. It seems that the transmission of vertical strain is not efficient at the Carnegie Ridge subduction zone compared with the transmission of the horizontal strain as evidenced from GPS data (Trenkamp et al., 2002; White et al., 2003). Consequently, it is probable that migration of the North Andean block reduces the ridge-induced vertical strain.

The Coastal Cordillera is an ~N-S–trending chain located ~120 km east of the trench. To the east, it limits the marine terraces observed along the Ecuadorian coast. The highest points of the Coastal Cordillera of Ecuador are ~600–700 m above sea level (Fig. 2). The age of uplift of the Coastal Cordillera is poorly known. Benítez (1995) related a detritism paroxysm along the southern Ecuadorian forearc basins with significant uplift of the Coastal Cordillera during Oligocene times. However, Benítez (1995) proposed that Carnegie Ridge subduction produced a major period of uplift and tectonic inversion along the Ecuadorian forearc (and along the Coastal Cordillera) during Quaternary times. If uplift rates were more or less constant during the Quaternary and considering that the maximum altitude of the oldest marine terrace is up to 360 ± 10 m and the highest marine terrace (Pedoja et al., 2006a, 2006b) is dated by extrapolation as 700 ka to 1 Ma, the small difference in elevation could indicate that uplift of the Coastal Cordillera is recent (1–1.5 Ma). The uplift of the Coastal Cordillera probably altered the drainage patterns and associated sediment transfer, which were diverted toward the north and the south through the Esmeraldas and Guayas River systems, respectively (Fig. 5). In the Gulf of Guayaquil area, the Quaternary subsiding zone and corresponding sedimentary infill are linked to the sedimentary contribution of the Guayas River system. The combination of a very high subsidence rate and high sediment input produced an accumulation of at least 3500 m of Quaternary sediments (Deniaud et al., 1999; Witt et al., 2006). Therefore, the evolution of the drainage system through time probably controlled the sediment input reaching the continental platform at the Gulf of Guayaquil area. This suggests that the early Pleistocene main sediment storage stage was probably coeval with the alteration of the Guayas River system and uplift of the Coastal Cordillera.

Forearc Deformation

One of the major aspects used in defining ridge-induced forearc deformation is the migration of subducted parts of the Carnegie Ridge beneath the forearc in response to obliquity between the ridge axis and the convergence movement and/or the trench zone. However, it is difficult to assess the migration rate of the Carnegie Ridge beneath the margin and the coastal forearc zone. The E-W–trending Carnegie Ridge parallels the convergence direction and subducts almost orthogonally to the trench. This geometry predicts a very limited migration of the subducted parts of the Carnegie Ridge beneath the forearc. However, maximum elevation zones seaward from the trench are disposed in a NE-directed hinge (Fig. 2), which could have migrated to the south after subduction. Such a southward migration of this bathymetric high has not been considered in previous publications.

Basin evolution in the Ecuadorian forearc is controlled by a N-S–directed strain produced by the northward migration of an upper-plate sliver since early Miocene times (Daly, 1989; Benítez, 1995). Four major forearc basins developed in this context, from north to south they are: the Borbón, the Manabí, the Progreso, and the Gulf of Guayaquil Basins.

Daly (1989) explored the possible relationship between Carnegie Ridge subduction and the evolution of the forearc area. He proposed that, during the late Miocene, the Carnegie Ridge subduction triggered a regional tectonic inversion along the forearc. However, he noted that Carnegie Ridge subduction occurred too late to be the only factor responsible for the inversion. Benítez (1995) proposed that the uplift of the Manabí and Progreso Basins originated from a Quaternary tectonic inversion caused by Carnegie Ridge subduction. Subsequently, Deniaud et al. (1999) proposed a Pliocene age for Carnegie Ridge subduction. Similarly, an important post-Pliocene change in the extensional tectonic regime that characterizes the Borbón Basin was associated with the subduction of the ridge by Aalto and Miller (1999); they proposed a Quaternary age for Carnegie Ridge subduction. Nonetheless, the major event in the Ecuadorian forearc for at least the last 10 m.y. is the main subsidence step of the Gulf of Guayaquil area. The Gulf of Guayaquil is located in the forearc zone near to the landward prolongation of the southernmost edge of the Carnegie Ridge. Here, ~3500 m of Quaternary sediments have resulted from the northward migration of the North Andean block (Deniaud et al., 1999; Dumont et al., 2005; Witt et al., 2006). The major deformation period in the Gulf of Guayaquil area occurred during the Quaternary. Taking into account the strong dependence of the subsidence in the Gulf of Guayaquil area with respect to the northward drift of the North Andean block, Witt et al. (2006) suggested that major depocenter formation along the Gulf of Guayaquil is associated with an acceleration of the North Andean block drift rate during the Quaternary. No major kinematic reorganization has occurred along the Nazca–South America plate boundary since the Pliocene (5.2 Ma) that can explain the higher rate of Quaternary northward drift of the North Andean block (Pardo-Casas and Molnar, 1987; Somoza, 1998). Therefore, Witt et al. (2006) postulated that the increase in interplate coupling associated with collision of the Carnegie Ridge or collision of an along-strike positive relief feature must have played a major role in controlling the North Andean block northward drift, and hence its acceleration period since the Pliocene–early Pleistocene boundary.

Andean Uplift

Fission track-derived uplift rates predict high exhumation periods at 15 and 9 Ma during the landward prolongation of the proposed Carnegie Ridge flat-slab segment (Spikings et al., 2001), and they are considered to have been ridge induced. Beyond the uncertainty of the flat slab at depth, the along-strike segmentation of the Andean chain is questionable. Similar 9 Ma high exhumation rates along the southern segments of the Ecuadorian Andes (i.e., up to 150 km southward from the Carnegie Ridge subduction zone) are also related to the subduction of the ridge (Steinmann et al., 1999; Hungerbühler et al., 2002). Furthermore, in northern Peru (~10°S) the same 9–10 Ma cooling period has been documented using fission-track data (Garver et al., 2005). Similarly, along the Eastern Cordillera of Colombia, a comparable cooling event of late Miocene age has also been inferred from fission-track analyses (Gómez et al., 2003, 2005). Widespread cooling events have been observed in the Andes for at least the past 25 m.y. (Gregory-Wodzicki, 2000), possibly representing an ongoing process. It is most likely that there is not one specific cooling period linked with Carnegie Ridge subduction. Moreover (U-Th)/He thermal history models identify specific cooling periods along the Ecuadorian Andes that commenced at 5.5–3.3 Ma (Spikings and Crowhurst, 2004), some of which represent a redefinition of older fission-track–derived ages.

CHEMISTRY OF THE VOLCANIC ARC

Important geochemical changes from calc-alkaline to adakitic magmas have been observed in different volcanic complexes along the Ecuadorian arc. The proposed ages of this major geochemical change range from 1.6 Ma to 0.1 Ma (Bourdon et al., 2003; Samaniego et al., 2002, 2005; Hidalgo et al., 2005). It is postulated that subduction of the Carnegie Ridge produced a high geothermal gradient that favored slab melting and adakitic rock generation (Bourdon et al., 2003; Samaniego et al., 2005; Hidalgo et al., 2007). The main mechanism for increasing the geothermal gradient may have been the flattening of the slab produced by the subduction of the Carnegie Ridge (Gutscher et al., 1999; Bourdon et al., 2003). Under such conditions, the subducted slab would have remained under pressure-temperature (*P-T*) conditions that allowed partial melting to occur for a long period of time (Gutscher et. al., 2000). Gutscher et al. (1999) suggested that the slab dip is shallow along the trajectory of the Carnegie Ridge and folded or torn on either side, such that the slab dips steeply to the north and to the south of the ridge. In a slightly different model, Samaniego et al. (2005) proposed that the arrival of the Carnegie Ridge at the subduction zone (but not necessary the slab geometry) modified the geothermal gradient along the Wadati-Benioff zone, favoring partial melting of the slab. In fact, slab melting needs a high geothermal gradient, which can be attained in different ways, such as a young slab, a slab tear, or a high strain stress.

Garrison and Davidson (2003) noted that on the basis of observed seismological data (mostly the Gutscher et al. [2000] data), the large-scale effect of slab flattening owing to subduction of the Carnegie Ridge was questionable. Moreover seismological data from a local network (Guillier et al., 2001) unambiguously confirmed show that the slab is not flat beneath Ecuador, but, instead, it regularly dips to the east with an angle of 25°–35° (Fig. 5). Finally, as pointed out by Andrade et al. (2005), the time window related to Carnegie Ridge subduction largely overpasses the time window related to geochemical observations, and thus the latter could probably represent only a small consequence of a larger and more complex process (Andrade et al., 2005). Furthermore, if Carnegie Ridge subduction is changing the Ecuadorian arc magmas petrogenesis, this process has not yet affected or arrived in the easternmost parts of the arc.

Geochemical evidence along the Western Cordillera of Ecuador points out that adakitic signals could be as old as late Pliocene (Chiaradia et al., 2004; Somers et al., 2005). Instead of being related to Carnegie Ridge subduction and/or to a flat-slab setting, these authors proposed that the late Pliocene adakitic magmas could be derived from partial melting of the deep mafic root of the Western Cordillera. Although slab melting cannot be rule out, both the geochemistry and distribution of the volcanic zone can be explained by alternative, relatively simple models of mantle-wedge melting followed by assimilation and garnet fractionation on ascent through the thick Andean crust (Garrison and Davidson, 2003; Bryant et al., 2006).

Furthermore, abundant evidence for adakites in the Andes suggests that slab tearing or a flat slab is not necessarily required for generation of adakites (Kay, et al., 2005). Where the upper crust is >40 km thick, in situ crust is a potential factor in producing an adakitic signature. Where adakitic magmas erupt in unstable and migrating arc configurations, upper-plate components introduced into the mantle by forearc subduction-erosion must be considered. Although slab melting cannot be ruled in the Ecuadorian volcanic arc, adakitic magmas may be attributed to a combination of melting along the base of thickened lower crust and crust entering the mantle through subduction-erosion. The thickness of the crust in the Ecuadorian volcanic arc zone is at least 50 km (Feininger and Seguin, 1983; Prévot et al., 1996), and the subduction erosion has been working since at least late Miocene time (Witt et al., 2006; Sage et al., 2006). As for the Juan Fernandez Ridge (Kay et al., 2005), the scale of these correlations shows that the Carnegie Ridge subduction may be only a minor perturbation in a much larger geodynamic scheme driven by plate interaction or mantle flow. Recent studies along the Cocos Ridge subduction zone (Hoernle et al., 2008) show that ridge subduction could even be partially responsible for the origin of trench-parallel convergence movement in the mantle wedge.

KINEMATIC DISCUSSION

Several ages for the Carnegie Ridge collision with the trench axis have been inferred from offshore studies (e.g., Malfait and Dinkelman, 1972; Rea and Malfait, 1974; Hey, 1977; Lonsdale, 1978, 2005; Lonsdale and Klitgord, 1978; Pilger, 1984; Wilson and Hey, 1995; Meschede et al., 1998; Sallarès and Charvis, 2003). The resulting picture is a complex history of plate-tectonic interactions between the Cocos-Nazca spreading center and the Galapagos hotspot, after the breakup of the Farallon plate ca. 25 Ma. Some models suggest that the ridge-trench collision occurred at the Pliocene–early Pleistocene boundary (Rea and Malfait, 1974; Hey, 1977; Lonsdale, 1978; Lonsdale and Klitgord, 1978; Wilson and Hey, 1995). However, older ages for ridge collision also have been proposed, including the late Miocene (Pilger, 1984; Spikings et al., 2001, 2005) or the early Miocene (Malfait and Dinkelman, 1972).

Most of the previous geologic arguments provide indirect and nondefinitive evidence on how far east and how deep the Carnegie Ridge extends on the Nazca slab. This discussion raises two questions: (1) How long is the trace of the Galapagos hotspot? (2) How long would the Carnegie Ridge be if it entered subduction at 15, 8, or 1.5 Ma? The first question can be addressed by looking at the conjugate trace of the Carnegie Ridge on the Cocos plate. From its conjugate location relative to the Carnegie Ridge across an extinct spreading center and a series of conjugate magnetic anomalies, the Malpelo Ridge seems to be the best candidate (Fig. 1; Lonsdale and Klitgord, 1978; MacMillan et al., 2004; Lonsdale, 2005). The eastern extremity of the Malpelo Ridge does not abut the Yaquina fracture zone, which to the north dies south of the Panama Basin, whereas to the south, the trough marking the fracture zone can be traced to the flank of the Carnegie Ridge (Figs. 1 and 2). Although the Malpelo Ridge does not reach the Yaquina fracture zone, Sallarès and Charvis (2003) speculated on the possibility of a further eastern extension of the Galapagos hotspot trace east of the fracture zone. This extension would now be subducted beneath Colombia, and its conjugate would represent the subducted part of the Carnegie Ridge. Nevertheless, this hypothesis, suggesting a continuation of the Malpelo Ridge east of the Yaquina graben between 2°N and 4°N, is not supported by wide-angle and multichannel seismic data (Marcaillou et al., 2006), which do not show an eastward prolongation of the Malpelo Ridge. Moreover, the data set of Gailler et al. (2007) shows that the velocity structure of the subducting oceanic crust of the corresponding area is consistent with a "normal" oceanic crust and shows no evidence of hotspot influence. The oldest age of 17 Ma reported for Malpelo Island (Hoernle et al., 2002; Werner and Hoernle, 2003) provides an estimate of the age of the conjugate Carnegie Ridge (Lonsdale, 2005), which is very close to the age of the first sediments cored on the Carnegie Ridge at Site 1239 of Ocean Drilling Program (ODP) 202 (Mix et al., 2003). Based on the crustal thickness, Sallarès and Charvis (2003) inferred an age of 20 Ma for the Malpelo Ridge. These are the best age estimates available for the oldest part of the Galapagos track that is still present on the Nazca plate.

The second question can be addressed by reconstructing the location of the Carnegie Ridge relative to South America at 1.5, 8, and 15 Ma (Fig. 6). This reconstruction is based on a revised plate circuit for South America–Africa (Müller et al., 1999), Africa–East Antarctica (Royer et al., 2006; J.-Y. Royer, 2008, personal commun.), East Antarctica–West Antarctica (Cande et al., 2000), West Antarctica–Pacific (Cande et al., 1995; J. Stock, 2008, personal commun.), and Pacific-Nazca (Tebbens and Cande, 1997). Figure 6 also shows the motion of the Galapagos hotspot relative to South America since 15 Ma and the predicted track of the Galapagos hotspot on the Nazca plate; both tracks assume that the Galapagos hotspot is fixed relative to the Pacific hotspot reference frame (HS3-NUVEL1A from Gripp and Gordon, 2002; the rate for the last 4–7 m.y. was extrapolated to 20 Ma). The small departure of the predicted trail of the Galapagos hotspot from the present-day Carnegie Ridge (Fig. 2) means either that the Galapagos hotspot is not fixed relative to the Pacific hotspot or that Nazca-Pacific motion is incorrect, or both. It shows that the eastern end of the Carnegie Ridge could be 20–25 m.y. old, older than, but compatible with, the age inferred from the Malpelo Island (17 Ma) and similar to the age inferred for the Malpelo Ridge by Sallarès and Charvis (2003). This predicted age coincides with the estimated age of onset of the Galapagos hotspot and breakup of the Farallon plate (Hey, 1977; Lonsdale and Klitgord, 1978; Lonsdale 2005). Based on this argument, the Carnegie Ridge could only have entered subduction at 6–8 Ma. In addition, one can see that

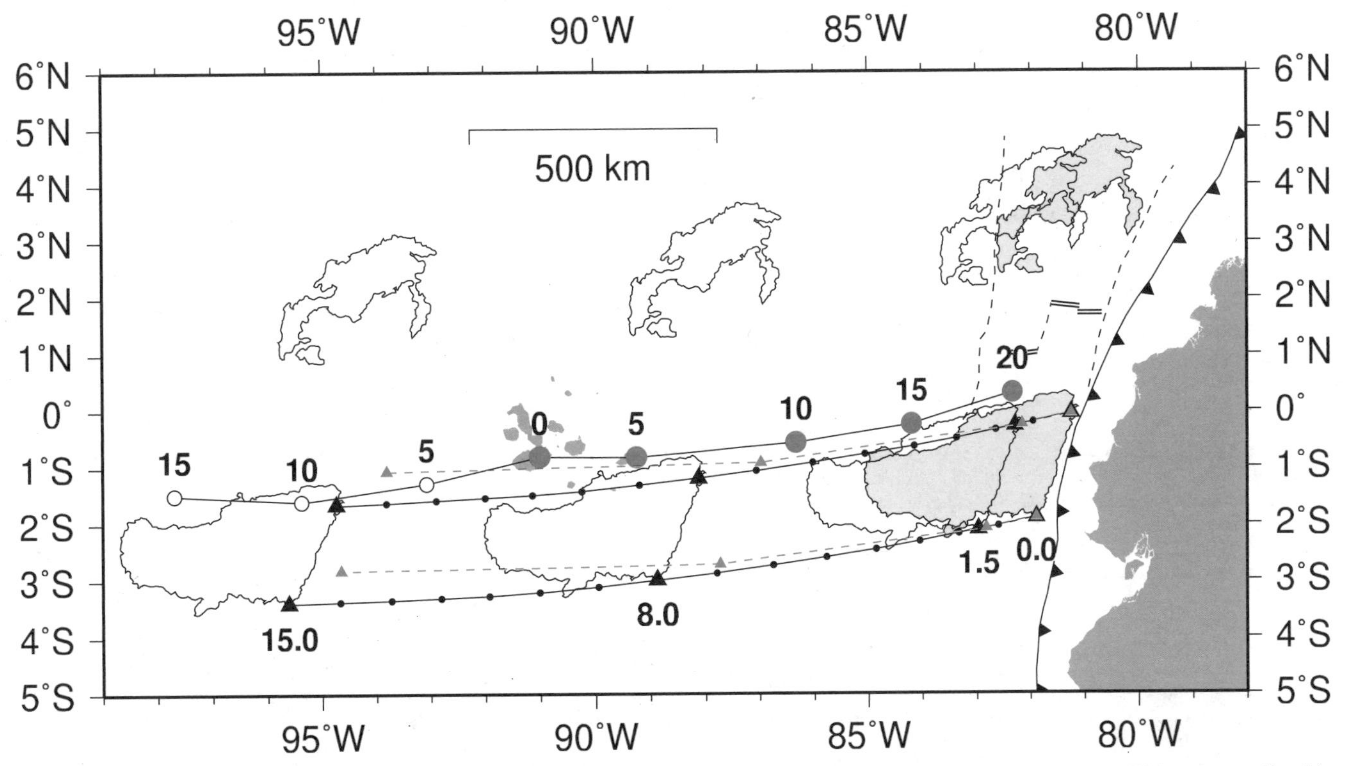

Figure 6. Kinematic reconstruction of the Carnegie Ridge relative to South America (fixed). The Carnegie and Malpelo Ridges are outlined by the 2200 m isobath. Gray shading shows their present-day location. Tracks are plotted every million year (full dots). Triangles show the location of the Carnegie Ridge eastern edge at 1.5, 8, and 15 Ma. Red tracks and triangles are based on Somoza's rotation model (2002). Note that the Malpelo Ridge has not been reconstructed. The open circles represent the relative motion of the Galapagos hotspot relative to South America (assuming the Galapagos hotspot belongs to the Pacific hotspot reference frame). The green symbols and track show the motion of the Nazca plate relative to the Galapagos hotspot (i.e., Galapagos hotspot track on the Nazca plate); the departure of this track from the present-day Carnegie Ridge means either that the Galapagos hotspot is not fixed relative to the Pacific hotspot or that the Nazca-Pacific motion is incorrect, or both.

if the Carnegie Ridge subducted 15, 8, or 1.5 m.y. ago, the subducted part of the plateau would have had an west-east extension of 1500, 780, and 120 km, respectively. Estimates derived from Somoza's rotation model (2002) would be very similar (Fig. 6). Geological expression of subduction of the Carnegie Ridge only requires a 300 km penetration of the Carnegie Ridge beneath Ecuador, which equates to initiation of subduction only 3–4 m.y. ago (80 km/m.y.). A 1500 or 780 km extension of the Carnegie Ridge beneath Ecuador would probably be seen in the earthquake distribution at depth, which does not seem to be the case. However, the Galapagos hotspot could have been active since the Cretaceous, producing the Caribbean igneous province at plume head arrival (e.g., Duncan and Hargraves, 1984) and Paleocene volcanic remnants along Central America (Hoernle et al., 2002). The track of the Galapagos hotspot relative to South America shows that, in the last 15 m.y., the hotspot has moved ~750 km eastward relative to the South American Trench (or conversely) and is not fixed relative to the trench, as it is often assumed.

Since we do not know the extent of the Galapagos hotspot trail beyond 20–25 Ma, plate kinematics cannot provide a definitive answer as to when the Carnegie Ridge entered subduction.

CONCLUSIONS

The deformation of the margin toe and uplift of the coast are not necessarily related to Carnegie Ridge subduction. There appears to be no distinct cooling period linked with the subduction of the Carnegie Ridge in the Andes. In the absence of a flat slab (Guillier et al., 2001), the geometry of the slab cannot be used as an argument for the continuation at depth of the Carnegie Ridge (Gutscher et al., 1999). The adakitic signal can be explained by alternative models (e.g., Kay et al., 2005) that do not require the subduction of the Carnegie Ridge.

These observations suggest that the vertical uplift imprinted by the Carnegie Ridge subduction is not significant. Considering that the transmission of vertical strain is not efficient at the Carnegie Ridge subduction zone when compared with the transmission of the horizontal strain as evidenced from GPS data (Trenkamp et al., 2002; White et al., 2003), it is probable that the migration of the North Andean block is inhibiting ridge-induced vertical strain.

From this review, we conclude that there is no clear segmentation of the deformation linked with Carnegie Ridge subduction or with its postulated landward prolongation at depth.

ACKNOWLEDGMENTS

We wish to thank J. Bustillos and L. Peñafiel, who participated at the beginning of this study. Thanks are extended for the helpful reviews of P. Samaniego and an anonymous reviewer. Thanks are also due to the Departamento de Geologia de la Escuela Politecnica Nacional (Ecuador, Quito).

REFERENCES CITED

Aalto, K.R., and Miller, W., III, 1999, Sedimentology of the Pliocene Upper Onzole Formation, an inner-trench slope succession in northwestern Ecuador: Journal of South American Earth Sciences, v. 12, p. 69–85, doi: 10.1016/S0895-9811(99)00005-X.

Andrade, D., Monzier, M., Martin, H., and Cotton, J., 2005, Systematic time-controlled geochemical changes in the Ecuadorian volcanic arc, *in* Institut de Recherche pour le Développment, ed., 6th International Symposium on Andean Geodynamics (ISAG 2005, Barcelona), Extended Abstracts: Barcelona, Spain, p. 46–49.

Benítez, S., 1995, Évolution géodynamique de la province côtière sud-équatorienne au Crétacé supérieur Tertiaire: Géologie Alpine, v. 71, p. 5–163.

Bès de Berc, S., Baby, P., Soula, J.C., Rosero, J., Souris, M., Cristophoul, F., and Vega, J., 2004, La superficie Mera-Upano: Marcador geomorfológico de la incisión fluviatil y del levantamineto tectónico de la zona Subandina Ecuatoriana, *in* Baby, P., Rivadeneira, M., and Barragán, R., eds., La Cuenca Oriente: Geología y Petróleo, p. 153–167.

Bilek, S.L., Schwartz, S.Y., and DeShon, H.R., 2003, Control of seafloor roughness on earthquake rupture behavior: Geology, v. 31, p. 455–458, doi: 10.1130/0091-7613(2003)031<0455:COSROE>2.0.CO;2.

Bourdon, E., Eissen, J.-P., Gutscher, M.-A., Monzier, M., Samaniego, P., Robin, C., Bollinger, C., and Cotton, J., 2002, Slab melting and slab melt metasomatism in the Northern Andean volcanic zone: Adakites and high-Mg andesites from Pichincha volcano (Ecuador): Bulletin de la Société Géologique de France, v. 173, no. 3, p. 195–206, doi: 10.2113/173.3.195.

Bourdon, E., Eissen, J.-P., Gutscher, M.-A., Monzier, M., Hall, M., and Cotten, J., 2003, Magmatic response to early aseismic ridge subduction: The Ecuadorian margin case (South America): Earth and Planetary Science Letters, v. 205, p. 123–138, doi: 10.1016/S0012-821X(02)01024-5.

Bourgois, J., von Huene, R., Pautot, G., and Huchon, P., 1988, Jean Charcot Seabeam survey along ODP Leg 112 northern transect, *in* Suess, E., von Huene, R., et al., Proceedings of the Ocean Drilling Program (ODP), Initial Reports, Volume 112: College Station, Texas, Ocean Drilling Program, p. 131–137.

Bourgois, J., Bigot-Cormier, F., Bourles, D., Braucher, R., Dauteuil, O., Witt, C., and Michaud, F., 2007, Tectonic records of strain buildup and abrupt co-seismic stress release across the northwestern Peru coastal plain, shelf, and continental slope during the past 200 kyr: Journal of Geophysical Research, v. 112, p. B04104, doi: 10.1029/2006JB004491.

Bryant, J.A., Yogodzinski, G.M., Hall, M.L., Lewicki, J.L., and Bailey, D.G., 2006, Geochemical constraints in the origin of volcanic rocks from the Andean Northern volcanic zone, Ecuador: Journal of Petrology, v. 47, p. 1147–1175, doi: 10.1093/petrology/egl006.

Calahorrano, A., 2005, Structure de la marge du Golfe de Guayaquil (Equateur) et propriété physique du chenal de subduction, à partir de données de sismique marine réflexion et refraction [Ph.D. thesis]: Paris VI, Université Pierre et Marie Curie, 227 p.

Cande, S.C., Raymond, C.A., Stock, J., and Haxby, W.F., 1995, Geophysics of the Pitman fracture zone and Pacific-Antarctic plate motions during the Cenozoic: Science, v. 270, p. 947–953, doi: 10.1126/science.270.5238.947.

Cande, S.C., Stock, J.M., Müller, R.D., and Ishihara, T., 2000, Cenozoic motion between East and West Antarctica: Nature, v. 404, p. 145–150, doi: 10.1038/35004501.

Cantalamessa, G., and Di Celma, C., 2004, Origin and chronology of Pleistocene marine terraces of Isla de la Plata and of flat, gently dipping surfaces of the southern coast of Cabo San Lorenzo (Manabí, Ecuador): Journal of South American Earth Sciences, v. 16, p. 633–648, doi: 10.1016/j.jsames.2003.12.007.

Chiaradia, M., Fontboté, L., and Paladines, A., 2004, Metal sources in mineral deposits and crustal rocks of Ecuador (1°N–4°S): A lead isotope synthesis: Mineralium Deposita, v. 39, p. 204–222, doi: 10.1007/s00126-003-0397-5.

Clift, P., and Vannucchi, P., 2004, Controls on tectonic accretion versus erosion in subduction zones: Implications for the origin and recycling of the continental crust: Geophysics: Geophysical Review, v. 42, RG2001, doi: 10.1029/2003RG000127.

Cloos, M., 1993, Lithospheric buoyancy and collisional orogenesis: Subduction of oceanic plateaus, continental margins, island arcs, spreading ridges and seamounts: Geological Society of America Bulletin, v. 105, p. 715–737, doi: 10.1130/0016-7606(1993)105<0715:LBACOS>2.3.CO;2.

Collot, J.-Y., Charvis, P., Gutscher, M.-A., and Operto, S., 2002, Exploring the Ecuador-Colombia active margin and inter-plate seismogenic zone: Eos (Transactions, American Geophysical Union), v. 83, no. 17, p. 189–190, doi: 10.1029/2002EO000120.

Collot, J.-Y., Charvis, P., Gutscher, M.-A., and Operto, S., 2002, Exploring the Ecuador-Colombia active margin and inter-plate seismogenic zone: Eos (Transactions, American Geophysical Union), v. 83, no. 17, p. 189–190.

Collot, J.-Y., Legonidec, Y., Michaud, F., Marcaillou, B., Alvarado, A., Ratzov, G., Sosson, M., Lopez, E., Silva, P., and El Personal Científico y Técnico del INOCAR, 2006a, *in* Instituto Oceanográfico de la Armada, ed., Mapas del Margen Continental del Norte de Ecuador y del Suroeste de Colombia: Batimetría, Relieve, Reflectividad Acústica e Interpretación Geológica: Publicación IOA-CVM-03-Post, Guayaquil, scale 1:800,000.

Collot, J.-Y., Michaud, F., Legonidec, Y., Calahorrano, A., Sage, F., Alvarado, A., and El Personal Científicoy Técnico del INOCAR, 2006b, *in* Instituto Oceanográfico de la Armada, ed., Mapas del Margen Continental Centro y Sur de Ecuador: Batimetría, Relieve, Reflectividad Acústica e Interpretación Geológica: Publicación IOA-CVM-04-Post, Guayaquil, scale 1:800,000.

Corrigan, J., Mann, P., and Ingle, J., 1990, Forearc response to subduction of the Cocos Ridge, Panama–Costa Rica: Geological Society of America Bulletin, v. 102, p. 628–652, doi: 10.1130/0016-7606(1990)102<0628:FRTSOT>2.3.CO;2.

Daly, M., 1989, Correlations between Nazca/Farallon plate kinematics and forearc basin evolution in Ecuador: Tectonics, v. 8, p. 769–790, doi: 10.1029/TC008i004p00769.

Deniaud, Y., Baby, P., Basile, C., Ordoñez, M., Montenegro, G., and Mascle, G., 1999, Opening and tectonic and sedimentary evolution of the Gulf of Guayaquil: Neogene and Quaternary fore-arc basin of the south Ecuadorian Andes: Comptes Rendus à l'Académie des Sciences, Paris, v. 328, p. 181–187.

Dominguez, S., Lallemand, S., Malavieille, J., and Schnuerle, P., 1998, Oblique subduction of the Gagua Ridge beneath the Ryukyu accretionary wedge system: Insights from marine observations and sandbox experiments: Marine Geophysical Researches, v. 20, p. 383–402, doi: 10.1023/A:1004614506345.

Dumont, J.-F., Santana, E., Vilema, W., Pedoja, K., Ordonez, M., Cruz, M., Jimenez, N., and Zambrano, I., 2005, Morphological and microtectonic analysis of Quaternary deformation from Puna and Santa Clara Islands, Gulf of Guayaquil, Ecuador (South America): Tectonophysics, v. 399, no. 1–4, p. 331–350.

Duncan, R.A., and Hargraves, R.B., 1984, Plate tectonics evolution of the Caribbean region in the mantle reference frame, *in* Bonni, W.E., Hargraves, R.B., and Shagam, R., eds., The Caribbean–South American Plate Boundary and Regional Tectonics: Geological Society of America Memoir 162, p. 81–93.

Engdahl, E.R., van der Hilst, R., and Buland, R., 1998, Global teleseismic earthquake relocation with improved travel times and procedures for depth determination: Bulletin Seismological Society of America, v. 88, p. 722–743.

Feininger, T., and Seguin, M.K., 1983, Simple Bouguer gravity anomaly field and the inferred crustal structure of the continental Ecuador: Geology, v. 11, p. 40–44, doi: 10.1130/0091-7613(1983)11<40:SBGAFA>2.0.CO;2.

Fisher, M.A., Collot, J.-Y., and Geist, E.L., 1991, The collision zone between the North d'Entrecasteaux Ridge and the New Hebrides Island arc: 2. Structure from multichannel seismic data: Journal of Geophysical Research, v. 96, p. 4479–4495, doi: 10.1029/90JB00715.

Gailler, A., Charvis P., and Flueh E., 2007, Segmentation and South American plates along the Ecuador subduction zone from wide-angle seismic profiles: Earth and Planetary Science Letters, v. 260, no. 3–4, p. 444–464.

Gardner, T.W., Verdonck, D., Pinter, N.M., Slingerland, R., Furlong, K.P., Bullard, T.F., and Wells, S.G., 1992, Quaternary uplift astride the aseismic Cocos Ridge, Pacific coast, Costa Rica: Geological Society of America Bulletin, v. 104, p. 219–232, doi: 10.1130/0016-7606(1992)104<0219:QUATAC>2.3.CO;2.

Gardner, T., Marshall, J., Merritts, D., Bee, B., Burgette, R., Burton, E., Cooke, J., Kehrwald, N., Protti, M., Fisher, D., and Sak, P.B., 2001, Holocene forearc block rotation in response to seamount subduction, southeastern Peninsula de Nicoya, Costa Rica: Geology, v. 29, p. 151–154, doi: 10.1130/0091-7613(2001)029<0151:HFBRIR>2.0.CO;2.

Garrison, J., and Davidson, J., 2003, Dubious case for slab melting in the Northern volcanic zone of the Andes: Geology, v. 31, no. 6, p. 565–568, doi: 10.1130/0091-7613(2003)031<0565:DCFSMI>2.0.CO;2.

Garver, J.I., Reiners, P.W., Walker, L.J., Ramage, J.M., and Perry, S.E., 2005, Implications for timing of Andean uplift from thermal resetting of radiation-damaged zircon in the Cordillera Huaybusash, northern Peru: The Journal of Geology, v. 113, p. 117–138, doi: 10.1086/427664.

Geist, E.L., Fisher, M.A., and Scholl, D.W., 1993, Large-scale deformation associated with ridge subduction: Geophysical Journal International, v. 115, p. 344–366, doi: 10.1111/j.1365-246X.1993.tb01191.x.

Gómez, E., Jordan, T., and Allmendinger, R., 2003, Controls on architecture of the Late Cretaceous to southern Middle Magdalena Valley Basin, Colombia: Geological Society of America Bulletin, v. 115, p. 131–147, doi: 10.1130/0016-7606(2003)115<0131:COAOTL>2.0.CO;2.

Gómez, E., Jordan, T., and Allmendinger, R., 2005, Syntectonic Cenozoic sedimentation in the northern Middle Magdalena Valley Basin of Colombia and implications for exhumation of the Northern Andes: Geological Society of America Bulletin, v. 117, no. 5/6, p. 547–569, doi: 10.1130/B25454.1.

Graindorge, D., Calahorrano, A., Charvis, P., Collot, J.-Y., and Bethoux, N., 2004, Deep structures of the margin and the Carnegie Ridge, possible consequence on great earthquake recurrence interval: Geophysical Research Letters, v. 31, doi: 10.1029/2003GL018803.

Gregory-Wodzicki, K.M., 2000, Uplift history of the Central and Northern Andes: A review: Geological Society of America Bulletin, v. 112, p. 1091–1105, doi: 10.1130/0016-7606(2000)112<1091:UHOTCA>2.3.CO;2.

Gripp, A.E., and Gordon, R.G., 2002, Young tracks of hotspots and current plate velocities: Geophysical Journal International, v. 150, p. 321–361, doi: 10.1046/j.1365-246X.2002.01627.x.

Guillier, B., Chatelain, J.-L., Yepes, H., Poupinet, G., and Fels, J.-F., 2001, Seismological evidence on the geometry of the orogenic system in central–northern Ecuador (South America): Geophysical Research Letters, v. 28, p. 3749–3752, doi: 10.1029/2001GL013257.

Gutscher, M.A., Malavieille, J., Lallemand, S., and Collot, J.-Y., 1999, Tectonic segmentation of the North Andean margin: Impact of the Carnegie Ridge collision: Earth and Planetary Science Letters, v. 168, p. 255–270, doi: 10.1016/S0012-821X(99)00060-6.

Gutscher, M.-A., Maury, R., Eissen, J.-P., and Bourdon, E., 2000, Can slab melting be caused by flat subduction?: Geology, v. 28, p. 535–538, doi: 10.1130/0091-7613(2000)28<535:CSMBCB>2.0.CO;2.

Hall, M.L., and Wood, C.A., 1985, Volcano-tectonic segmentation of the northern Andes: Geology, v. 13, p. 203–207, doi: 10.1130/0091-7613(1985)13<203:VSOTNA>2.0.CO;2.

Hampel, A., 2002, The migration history of the Nazca Ridge along the Peruvian active margin: A re-evaluation: Earth and Planetary Science Letters, v. 203, p. 665–679, doi: 10.1016/S0012-821X(02)00859-2.

Hampel, A., and Kukowski, N., 2004, Ridge subduction at an erosive margin: The collision zone of the Nazca Ridge in southern Peru: Journal of Geophysical Research, v. 109, p. B02101, doi: 10.1029/2003JB002593.

Hey, R., 1977, Tectonic evolution of the Cocos-Nazca spreading center: Geological Society of America Bulletin, v. 88, p. 1404–1420, doi: 10.1130/0016-7606(1977)88<1404:TEOTCS>2.0.CO;2.

Hidalgo, S., Monzier, M., Martin, H., Cotten, J., Fornari, M., and Eissen, J.-P., 2005, New geochemical and geochronological data for the Atacazo-Ninahuilca volcanic complex (Ecuador), *in* Institut de Recherche pour le Développement, ed., 6th International Symposium on Andean Geodynamics (ISAG 2005, Barcelona), Extended Abstracts: Barcelona, Spain, p. 379–382.

Hidalgo, S., Monzier, M., Martin, H., Chazot, G., Eissen, J.-P., and Cotten, J., 2007, Adakitic magmas in the Ecuadorian volcanic front: Petrogenesis of the Iliniza volcanic complex (Ecuador): Journal of Volcanology and Geothermal Research, v. 159, p. 366–392, doi: 10.1016/j.jvolgeores.2006.07.007.

Hoernle, K.A., van den Bogaard, P., Werner, R., Hauff, F., Lissina, B., Alvarado, G.E., and Garbe-Schönberg, D., 2002, Missing history (16–71 Ma) of the Galapagos hotspot: Implications for the tectonic and biological evolution of the Americas: Geology, v. 30, p. 795–798.

Hoernle, K.A., Abt, D.L., Fischer, K.M., Nochols, H., Hauff, F., Abers, G.A., van den Bogaard, P., Heydolph, K., Alvarado, G., Protti, M., and Strauch, W., 2008, Arc-parallel flow in the mantle wedge beneath Costa Rica and Nicaragua: Nature, v. 451, p. 1094–1098, doi: 10.1038/nature06550.

Hsu, J.T., 1992, Quaternary uplift of the Peruvian coast related to the subduction of the Nazca Ridge: 13.5 to 15.6° south latitude: Quaternary International, v. 15/16, p. 87–97, doi: 10.1016/1040-6182(92)90038-4.

Hungerbühler, D., Steinmann, M., Winkler, W., Seward, D., Egüez, A., Peterson, D.E., Helg, U., and Hammer, C., 2002, Neogene stratigraphy and Andean geodynamics of southern Ecuador: Earth-Science Reviews, v. 57, p. 75–124, doi: 10.1016/S0012-8252(01)00071-X.

Jaillard, E., Ordonez, M., Benitez, S., Berrones, G., Jiménez, N., Montenegro, G., and Zambrano, I., 1995, Basin development in an accretionary, oceanic-floored fore-arc setting: Southern coastal Ecuador during Late Cretaceous–late Eocene time, *in* Tankard, A.J., Suarez, S.R., and Welsink, H.J., eds., Petroleum Basins of South America: American Association of Petroleum Geologists Memoir 62, p. 615–631.

Kay, S.M., Godoy, E., and Kurtz, A., 2005, Episodic arc migration, crustal thickening, subduction erosion, and magmatism in the south-central Andes: Geological Society of America Bulletin, v. 117, p. 67–88, doi: 10.1130/B25431.1.

Kodaira, S., Takahashi, N., Nakanishi, A., Miura, S., and Kaneda, Y., 2000, Subducted seamount imaged in the rupture zone of the 1946 Nankaido earthquake: Science, v. 289, p. 104–106, doi: 10.1126/science.289.5476.104.

Lonsdale, P., 1978, Ecuadorian subduction system: American Association of Petroleum Geologists Bulletin, v. 62, p. 2454–2477.

Lonsdale, P., 2005, Creation of the Cocos and Nazca plates by fission of the Farallon plate: Tectonophysics, v. 404, p. 237–264, doi: 10.1016/j.tecto.2005.05.011.

Lonsdale, P., and Klitgord, K., 1978, Structure and tectonic history of the eastern Panama Basin: Geological Society of America Bulletin, v. 89, p. 981–999, doi: 10.1130/0016-7606(1978)89<981:SATHOT>2.0.CO;2.

MacMillan, I., Gans, P.B., and Alvarado, G., 2004, Middle Miocene to present plate tectonic history of the southern Central American volcanic arc: Tectonophysics, v. 392, p. 325–348, doi: 10.1016/j.tecto.2004.04.014.

Malfait, B.T., and Dinkelman, M.G., 1972, Circum-Caribbean tectonics and igneous activity and the evolution of the Caribbean plate: Geological Society of America Bulletin, v. 83, p. 251–272, doi: 10.1130/0016-7606(1972)83[251:CTAIAA]2.0.CO;2.

Marcaillou, B., Charvis, P., and Collot, J.-Y., 2006, Structure of the Malpelo Ridge (Colombia) from seismic and gravity modeling: Marine Geophysical Researches, v. 27, p. 289–300, doi: 10.1007/s11001-006-9009-y.

Meffre, S., and Crawford, A.J., 2001, Collision tectonics in the New Hebrides arc (Vanuatu): The Island Arc, v. 10, p. 33–50, doi: 10.1046/j.1440-1738.2001.00292.x.

Meschede, M., Barckhausen, U., and Worm, H.-U., 1998, Extinct spreading on the Cocos Ridge: Terra Nova, v. 10, p. 211–216, doi: 10.1046/j.1365-3121.1998.00195.x.

Michaud, F., Chabert, A., Collot, J.-Y., Sallarès, V., Flueh, E.R., Charvis, P., Graindorge, D., Gutscher, M.-A., and Bialas, G., 2005, Fields of multikilometer-scale sub-circular depressions in the Carnegie Ridge sedimentary blanket: Effect of underwater carbonate dissolution?: Marine Geology, v. 216, p. 205–219, doi: 10.1016/j.margeo.2005.01.003.

Michaud, F., Collot, J.-Y., Alvarado, A., López, E., and El Personal Científico y Técnico del INOCAR, 2006, *in* Instituto Oceanografico de la Armada, ed., Batimetría y Relieve Continental e Insular: Publicación IOA-CVM-01-Post, Guayaquil, scale 1:15,000,000.

Mix, A.C., Tiedemann, R., Blum, P., and Leg 202 Shipboard Scientific Party, 2003, Proceedings of the Ocean Drilling Program (ODP) Initial Reports, Volume 202: College Station, Texas, Ocean Drilling Program, p. 1–145.

Müller, R.D., Royer, J.-Y., Cande, S.C., Roest, W.R., and Maschenkov, S., 1999, New constraints on the Late Cretaceous/Tertiary plate tectonic evolution of the Caribbean, *in* Mann, P., ed., Caribbean Basins, Volume 4: Sedimentary Basins of the World: Amsterdam, Elsevier Science, p. 33–59.

Pardo-Casas, F., and Molnar, P., 1987, Relative motion of the Nazca (Farallon) and South American plates since Late Cretaceous time: Tectonics, v. 6, p. 233–248, doi: 10.1029/TC006i003p00233.

Pedoja, K., Orlieb, L., Dumont, J.F., Lamothe, M., Ghaleb, B., Auclair, M., and Labrousse, B., 2006a, Quaternary coastal uplift along the Talara Arc (Ecuador, northern Peru) from new marine terrace data: Marine Geology, v. 228, p. 73–91, doi: 10.1016/j.margeo.2006.01.004.

Pedoja, K., Dumont, J.F., Lamothe, M., Ortlieb, L., Collot, J.Y., Ghaleb, B., Auclair, M., Alvarez, V., and Labrousse, B., 2006b, Plio-Quaternary uplift of the Manta Peninsula and La Plata Island and the subduction of the Carnegie Ridge, central coast of Ecuador: Journal of South American Earth Sciences, v. 22, p. 1–21, doi: 10.1016/j.jsames.2006.08.003.

Pilger, R.H., 1984, Cenozoic plate kinematics, subduction and magmatism: South American Andes: Journal of the Geological Society of London, v. 141, p. 793–802, doi: 10.1144/gsjgs.141.5.0793.

Prévot, R., Chatelain, J.-L., Guillier, B., and Yepes, H., 1996, Tomographie des Andes Equatoriennes: Évidence d'une continuité des Andes Centrales: Comptes Rendus de l'Académie des Sciences, Série II, Sciences de la Terre et des Planètes, v. 323, p. 833–840.

Rea, D.K., and Malfait, B.T., 1974, Geologic evolution of the northern Nazca plate: Geology, v. 2, p. 317–320, doi: 10.1130/0091-7613(1974)2<317:GEOTNN>2.0.CO;2.

Royer, J.-Y., Gordon, R.G., and Horner-Johnson, B.C., 2006, Motion of Nubia relative to Antarctica since 11 Ma: Implications for Nubia-Somalia, Pacific–North America, and India-Eurasia motion: Geology, v. 34, p. 501–504, doi: 10.1130/G22463.1.

Sage, F., Collot, J.-Y., and Ranero, C.R., 2006, Interplate patchiness and subduction-erosion mechanisms: Evidence from depth migrated seismic images at the Central Ecuador convergent margin: Geology, v. 34, p. 997–1000, doi: 10.1130/G22790A.1.

Sallarès, V., and Charvis, P., 2003, Crustal thickness constraints on the geodynamic evolution of the Galapagos volcanic province: Earth and Planetary Science Letters, v. 214, p. 545–559, doi: 10.1016/S0012-821X(03)00373-X.

Sallarès, V., Charvis P., Flueh, E.R., Bialas J., and the Salieri Scientific Team, 2005, Seismic structure of the Carnegie Ridge and the nature of the Galapagos melt anomaly: Geophysical Journal International, v. 161, p. 763–788, doi: 10.1111/j.1365-246X.2005.02592.x.

Samaniego, P., Martin, H., Robin, C., and Monzier, M., 2002, Transition from calc-alkalic to adakitic magmatism at Cayambe volcano, Ecuador: Insights into slab melts and mantle wedge interactions: Geology, v. 30, p. 967–970, doi: 10.1130/0091-7613(2002)030<0967:TFCATA>2.0.CO;2.

Samaniego, P., Martin, H., Monzier, M., Robin, C., Fornari, M., Eissen, J.P., and Cotton, J., 2005, Temporal evolution of magmatism in the Northern volcanic zone of the Andes: The geology and petrology of Cayambe volcanic complex (Ecuador): Journal of Petrology, v. 46, p. 2225–2252, doi: 10.1093/petrology/egi053.

Scholz, C.H., and Small, C., 1997, The effect of seamount subduction on seismic coupling: Geology, v. 25, p. 487–490, doi: 10.1130/0091-7613(1997)025<0487:TEOSSO>2.3.CO;2.

Smith, W.H.F., and Sandwell, D.T., 1997, Global seafloor topography from satellite altimetry and ship depth soundings: Science, v. 277, p. 1957–1962.

Somers, C., Amórtegui, A., Lapierre, H., Jaillard, E., Bussy, F., and Brunet, P., 2005, Miocene adakitic intrusions in the Western Cordillera of Ecuador, *in* Institut de Recherche pour le Développement, ed., 6th International Symposium on Andean Geodynamics (ISAG 2005, Barcelona), Extended Abstracts: Barcelona, Spain, p. 679–680.

Somoza, R., 1998, Updated Nazca (Farallon)–South America relative motions during the last 40 My: Implications for mountain building in the central Andean region: Journal of South American Earth Sciences, v. 11, p. 211–215, doi: 10.1016/S0895-9811(98)00012-1.

Sosson, M., Bourgois, J., and Mercier de Lépinay, B., 1994, SeaBeam and deep-sea submersible *Nautile* surveys in the Chiclayo canyon off Peru (7°S): Subsidence and subduction-erosion of an Andean-type convergent margin since Pliocene time: Marine Geology, v. 118, p. 237–256, doi: 10.1016/0025-3227(94)90086-8.

Spikings, R., and Crowhurst, P., 2004, (U-Th)/He thermochronometric constraints on the late Miocene–Pliocene tectonic development of the northern Cordillera Real and the Interandean Depression, Ecuador: Journal of South American Earth Sciences, v. 17, p. 239–251, doi: 10.1016/j.jsames.2004.07.001.

Spikings, R., Winkler, W., Seward, D., and Handler, R., 2001, Along-strike variations in the thermal and tectonic response of the continental Ecuadorian Andes to the collision with heterogeneous oceanic crust: Earth and Planetary Science Letters, v. 186, p. 57–73, doi: 10.1016/S0012-821X(01)00225-4.

Spikings, R., Winkler, W., Hughes, R., and Handler, R., 2005, Thermochronology of allochthonous terranes in Ecuador: Unravelling the accretionary and post-accretionary history of the Northern Andes: Tectonophysics, v. 399, p. 195–220, doi: 10.1016/j.tecto.2004.12.023.

Steinmann, M., Hungerbuhler, D., Seward, D., and Winkler, W., 1999, Neogene tectonic evolution and exhumation of the southern Ecuadorian Andes: A combined stratigraphy and fission-track approach: Tectonophysics, v. 307, p. 255–276, doi: 10.1016/S0040-1951(99)00100-6.

Taylor, F.W., Mann, P., Bevis, M.G., Edwards, R.L., Hai Cheng, Cutler, K.B., Gray, S.C., Burr, G.S., Beck, J.W., Phillips, D.A, Cabioch, G., and Recy, J., 2005, Rapid forearc uplift and subsidence caused by impinging bathymetric features: Examples from the New Hebrides and Solomon arcs: Tectonics, v. 24, TC6005, doi: 10.1029/2004TC001650.

Tebbens, S.F., and Cande, S.C., 1997, Southeast Pacific tectonic evolution from early Oligocene to present: Journal of Geophysical Research, v. 102, p. 12,061–12,084.

Trenkamp, R., Kellogg, J., Freymueller, J., and Mora, H., 2002, Wide plate margin deformation, southern Central America and northwestern South America; CASA GPS observations: Journal of South American Earth Sciences, v. 15, p. 157–171, doi: 10.1016/S0895-9811(02)00018-4.

Von Huene, R., Pecher, I.A., and Gutscher, M.A., 1996, Development of the accretionary prism along Peru and material flux after subduction of Nazca Ridge: Tectonics, v. 15, p. 19–33, doi: 10.1029/95TC02618.

Werner, R., and Hoernle, K., 2003, New volcanological and volatile data provide strong support for the continuous existence of the Galapagos Islands over the past 17 million years: International Journal of Earth Sciences, v. 92, p. 904–911, doi: 10.1007/s00531-003-0362-7.

White, C., Trenkamp, R., and Kellogg, J., 2003, Recent crustal deformation and the earthquake cycle along the Ecuador-Colombia subduction zone: Earth and Planetary Science Letters, v. 216, p. 231–242, doi: 10.1016/S0012-821X(03)00535-1.

Wilson, D.S., and Hey, R., 1995, History of rift propagation and magnetization intensity for the Cocos-Nazca spreading center: Journal of Geophysical Research, v. 100, no. B7, p. 10,041–10,056, doi: 10.1029/95JB00762.

Witt, C., Bourgois, J., Michaud, F., Ordoñez, M., Jiménez, N., and Sosson, M., 2006, Development of the Gulf of Guayaquil (Ecuador) as an effect of the North Andean block tectonic escape since the lower Pleistocene: Tectonics, v. 25, p. TC3017, doi: 10.1029/2004TC001723.

Manuscript Accepted by the Society 5 December 2008

Printed in the USA

The Geological Society of America
Memoir 204
2009

Shallowing and steepening subduction zones, continental lithospheric loss, magmatism, and crustal flow under the Central Andean Altiplano-Puna Plateau

Suzanne Mahlburg Kay*
Department of Earth and Atmospheric Sciences, Snee Hall, Cornell University, Ithaca, New York 14853, USA

Beatriz L. Coira*
Consejo Nacional de Investigaciones Científicas y Técnicas (CONICET), Casilla de Correo 258, (4600) San Salvador de Jujuy, Argentina

ABSTRACT

Integrated magmatic, structural, and geophysical data provide a basis for modeling the Neogene lithospheric evolution of the high Central Andean Puna-Altiplano Plateau. Reconstruction of three transects south of the Bolivian orocline in the Altiplano and Puna Plateau shows processes in common, including subduction characterized by relatively shallow and changing slab dips, crustal shortening, delamination of thickened lower crust and lithosphere, crustal melting, eruption of giant ignimbrites, and deep crustal flow. Temporal similarities in events in the three transects can be correlated with changes in the rate of westward drift of South America and slab rollback. Temporal differences between the three transects can be attributed to variations in Nazca plate geometry in response to southward subduction of the aseismic Juan Fernandez Ridge. Subduction of the north-south arm of the ridge can explain an Oligocene flat slab under the Altiplano, and subduction of a northeast arm of the ridge can explain a long period of relatively shallow subduction characterized by local steepening and shallowing. Major episodes of ignimbrite eruption and delamination have occurred over steepening subduction zones as the ridge has passed to the south. Late Miocene to Holocene delamination of dense lithosphere is corroborated by published seismic images.

The southern Altiplano transect (17°S–21°S) is notable for high, structurally complex Western and Eastern Cordilleras flanking the Altiplano Basin, the eastern border of which is marked by late Miocene ignimbrites. The broad Subandean fold-and-thrust belt lies to the east. The Neogene evolution can be modeled by steepening of a shallowly subducting plate, leading to mantle and crustal melting that produced widespread volcanism including large ignimbrites. Major uplift of the plateau at 10–6.7 Ma was dominantly a response to crustal thickening related to Subandean

*E-mails: Kay—smk16@cornell.edu; Coira—bcoira2400@yahoo.com.ar.

Kay, S.M., and Coira, B.L., 2009, Shallowing and steepening subduction zones, continental lithospheric loss, magmatism, and crustal flow under the Central Andean Altiplano-Puna Plateau, *in* Kay, S.M., Ramos, V.A., and Dickinson, W.R., eds., Backbone of the Americas: Shallow Subduction, Plateau Uplift, and Ridge and Terrane Collision: Geological Society of America Memoir 204, p. 229–259, doi: 10.1130/2009.1204(11). For permission to copy, contact editing@geosociety.org.

shortening and peak lower-crustal flow into the Altiplano from the bordering cordilleras as the ignimbrites erupted, and partly a response to delamination along the eastern Altiplano border. A smaller ignimbrite volume than in the northern Puna suggests the Altiplano lithosphere never reached as high a degree of melting as to the south. An Oligocene flat-slab stage can explain extensive Oligocene deformation of the high plateau region.

The northern Puna transect at ~21°S–24°S is notable for voluminous ignimbrites (>8000 km^3) and a narrower Subandean fold-and-thrust belt that gives way southward to a thick-skinned thrust belt. The evolution can be modeled by an early Miocene amagmatic flat slab that underwent steepening after 16 Ma, which led to mantle melting that culminated in widespread ignimbrite eruptions beginning at 10 Ma, peaking in the backarc at ca. 8.5–6 Ma, restricted to the near arc by 4.5 Ma, and ending by 3 Ma. The formation of eclogitic residual crust caused periodic lower-crustal and lithospheric mantle delamination. Late Miocene uplift was largely due to crustal thickening in response to crustal shortening, magmatic addition, and delamination. Crustal flow played only a minor role. The high degree of mantle and crustal melting can be explained as a response to steepening of the early Miocene flat slab.

The southern Puna transect at ~24°S–~28°S is notable for eastward frontal arc migration at 8–3 Ma, intraplateau basins bounded by high ranges, long-lived Miocene stratovolcanic-dome complexes, voluminous 6–2 Ma ignimbrites, 7–0 Ma backarc mafic flows, and the latest Miocene uplift of the reverse-faulted Sierras Pampeanas ranges to the east. Its evolution can be modeled by a moderately shallow slab producing widespread volcanism with subsequent steepening by 6 Ma, leading to delamination of dense lithosphere culminating in the eruption of the voluminous Cerro Galan ignimbrite at 2 Ma.

INTRODUCTION

The Central Andes are often used as the example for defining processes associated with subduction of oceanic crust beneath a continental margin. The most prominent feature in the Central Andes is the extensive Puna-Altiplano Plateau (Fig. 1), which, after the Tibetan Plateau, is the world's highest (average elevation 3700 m) and largest (700 × 200 km) continental plateau. Unlike the Tibetan Plateau, the uplift of the Puna-Altiplano Plateau is unrelated to continental collision; it is instead a response to a compressional tectonic regime associated with collision between a continental plate and a relatively shallow subducting oceanic slab (e.g., Isacks, 1988). As under the Tibetan Plateau, the crust is exceptionally thick—seismic data indicate thicknesses up to 80 km (e.g., Beck et al., 1996; Yuan et al., 2002; McGlashan et al., 2008). Unlike the Tibetan Plateau, the Central Andean plateau is covered by diffuse chains of Neogene volcanic centers, the generation of which is critical in understanding the evolution of the plateau.

Most recent overview papers have emphasized the structural and geophysical characteristics of the Puna-Altiplano Plateau (e.g., Beck and Zandt, 2002; McQuarrie et al., 2005; Oncken et al., 2006; Sobolev et al., 2006). The principal shortening is Eocene to middle Miocene in age under the plateau and latest Miocene to Holocene in age in the eastern foreland. The principal uplift is considered to be late Miocene in age (e.g., Oncken et al., 2006; McQuarrie et al., 2008). The uplift has been related to some combination of crustal thickening associated with crustal shortening together with limited magmatic addition (e.g., Isacks, 1988; Allmendinger et al., 1997; Oncken et al., 2006; Babeyko et al., 2006), delamination of the lower continental crust and lithosphere (e.g., Kay and Kay, 1993; Kay et al., 1994a; Sobolev and Babeyko, 2005; Garzione et al., 2006; Molnar and Garzione, 2007; Sobolev et al., 2006), and intracrustal flow (Husson and Sempere, 2003; Gerbault et al., 2005). Recent reviews have also emphasized the importance of the westward drift of South America (e.g., Oncken et al., 2006) as proposed by Silver et al. (1998).

The objective of this paper is to integrate the Neogene and Quaternary magmatic history of the plateau with its structural and geophysical characteristics as a basis for an overview of the Neogene tectonic evolution. The generation of the magmas reflects the thermo-mechanical evolution of the crustal and mantle lithosphere and the underlying geometry of the subducting plate. The magmatic history presented expands on that in Coira et al. (1993), Kay et al. (1999), and Trumbull et al. (2006). Emphasis is placed on the general spatial and temporal eruption pattern and composition of the erupted magmas, the giant dacitic ignimbrite complexes, the distinctive mafic flows, and the relationship of magmatism to the structural evolution and modern geophysical characteristics. The similarities and differences of the southern Altiplano, northern Puna, and southern Puna regions of the plateau south of the Bolivian orocline (Figs. 1 and 2) are examined. In each region, the magmatic and structural histories and geophysical framework are briefly outlined, and an integrated model of the Neogene tectonic evolution is presented. Finally, the internal and external controls

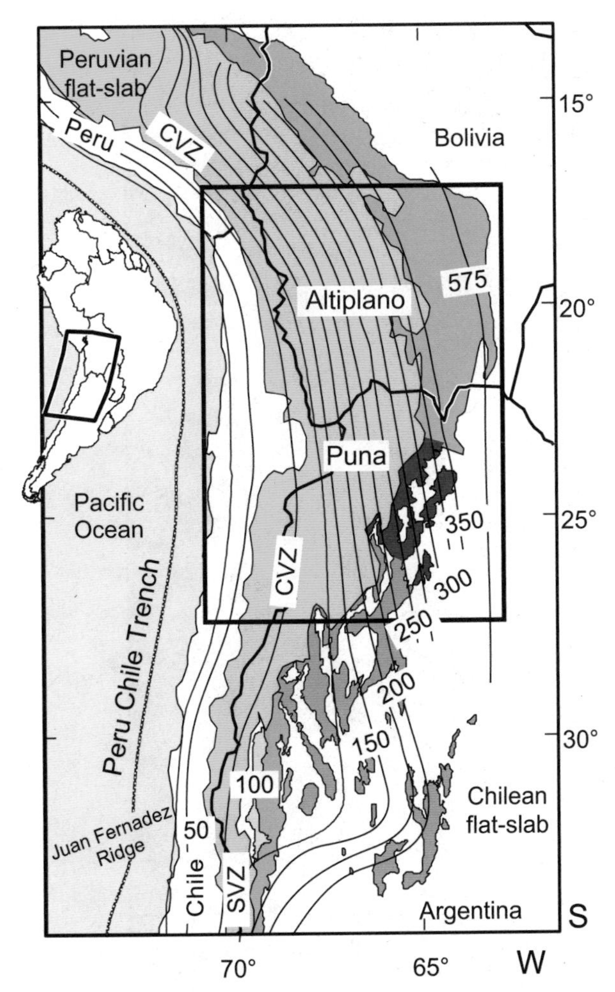

Figure 1. Map of the Puna-Altiplano region of the Central Andes modified from Allmendinger et al. (1997). Areas above 3700 m in elevation are shown in pink, major thin-skinned fold-and-thrust belts are shown in green, thick-skinned belts are shown in dark gray, and block uplifts of the Sierras Pampeanas are shown in gray. The ends of the active frontal volcanic arc in the Central volcanic zone and Southern volcanic zone are shown by CVZ and SVZ labels. Contours to the Wadati-Benioff zone from Cahill and Isacks (1992) are labeled at 50 km intervals. The region in the box is that discussed in this paper and shown in Figure 2.

on the upper and lower plates, including the influence of the subduction of the aseismic Juan Fernandez Ridge (oceanic plateau) on the subducted Nazca plate, are examined.

OVERALL PLATE-TECTONIC AND GEOLOGIC FRAMEWORK

A special feature of the Oligocene- to Eocene-age Nazca plate subducting under the plateau is its relatively shallow subduction angle (<30°) compared to other regions around the circum-Pacific, where subduction angles are rarely less than 45°. To the north and south of the plateau, even shallower segments underlie the volcanically quiescent arc over the Peruvian and Chilean-Pampean flat-slab segments (Fig. 1; see Isacks, 1988; Cahill and Isacks, 1992). The geometry of the subducting plate under the plateau also varies with a central segment that dips at ~30°, transitioning abruptly into the Peruvian flat slab and more gradually into the Chilean-Pampean flat slab (Fig. 1). The main section of the slab affected by flattening is at ~75–135 km depth. The origin of the shallower segments has been variously attributed to subduction of the Juan Fernandez and Nazca Ridges on the downgoing Nazca plate (e.g., Pilger, 1981, 1984; Yañéz et al., 2001) and complex interactions between the Nazca and overriding South American plates (e.g., Isacks, 1988).

The Altiplano-Puna Plateau segment is the widest part of the high Andean Cordillera, and it encompasses the active Central Volcanic Zone arc. The plateau is largely built on a late Precambrian to Paleozoic deformed basement (see Ramos, this volume). A generalized west to east section south of the Bolivian orocline crosses the Coastal Cordillera, forearc Central Depression, Chilean Precordillera, Cordillera Occidental (western), Altiplano-Puna Plateau, Cordillera Occidental (eastern), Subandean belt, and composite foreland basin (e.g., Allmendinger et al., 1997; Fig. 2).

West of the plateau, the Coastal Cordillera is largely composed of Jurassic to Middle Cretaceous arc sequences cut by the intra-arc Jurassic–Early Cretaceous Atacama strike-slip fault system. Part of the Jurassic–Early Cretaceous arc runs offshore at the oroclinal bend in northernmost Chile, where its western margin appears to have been removed by forearc subduction-erosion. To the east, the Neogene sediment-filled Central Depression is flanked by the Paleozoic basement and Mesozoic-Tertiary volcanic and sedimentary cover of the Chilean Precordillera, which are cut by the Paleogene Domeyko strike-slip fault system. The Precordillera hosts the Eocene–early Oligocene giant porphyry copper deposits of northern Chile. The Cordillera Occidental, which contains the modern Central volcanic zone arc front, where peaks reach over 6600 m, marks the western limit of the high Andes.

The Cordillera Occidental and the modern arc front are both displaced eastward around the Atacama Basin near 23°S to 24°S. The internally drained Altiplano-Puna Plateau to the east has an extensive cover of late Cenozoic sedimentary-evaporite filled basins called salars and transverse chains of Cenozoic volcanic centers. The highest regions (>6300 m) are largely formed by post–middle Miocene andesitic to dacitic stratovolcanoes and giant ignimbrite complexes that erupted in a major flare-up to create one of the largest ignimbrite provinces on Earth. North of ~21.5°S, the southern Altiplano is dominantly a sediment-filled basin over a largely Paleozoic sedimentary basement. To the south, the Puna is broken into ranges with high peaks and basins, and it has a larger component of pre–late Paleozoic magmatic rocks in the basement.

The eastern border of the Altiplano-Puna Plateau is bounded by contractional deformational belts that vary along strike. The Cordillera Oriental and Subandean fold-and-thrust belts border the Altiplano and northernmost Puna. The Cordillera Oriental, where peaks reach up to 6000 m, is mainly composed of deformed Paleozoic sedimentary and Tertiary magmatic rocks

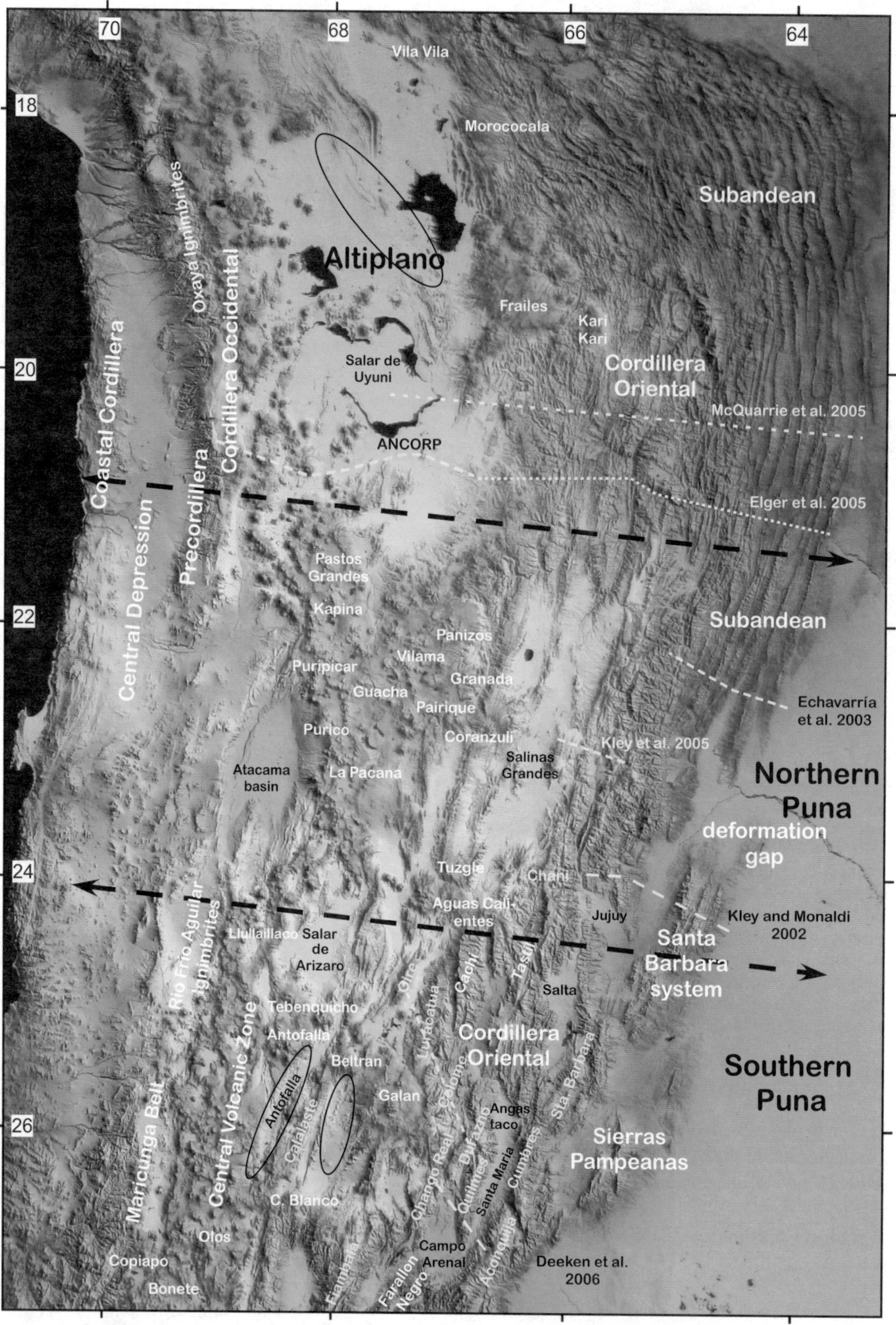

Figure 2. Satellite Radar Topography Mission (SRTM) image showing the southern Altiplano, northern Puna, and southern Puna regions and major geologic provinces discussed in this paper. Ranges (yellow text), basins (black text), and magmatic centers (white text) referred to in the text are labeled. The structural cross sections shown in Figures 4, 6, and 10 are indicated by dashed yellow lines. Monogenetic and simple polygenetic mafic volcanic centers are concentrated in the regions outlined by black ellipses. Highest regions are in red (darkest red is near 4500 m) and blue tones (above ~5500 m), intermediate regions are in gray (darkest gray is near 2500 m) and white (clearest white is near 3800 m) tones, and lower regions are in green (yellow green is near 1500 m) and brown (darkest brown is near 600 m) tones. The large magmatic centers are largely in the red tones with the highest peaks in blue. Used with permission from Jay Hart.

affected by Eocene to late Miocene folding and thrusting. The much lower-elevation Subandean belt to the east is largely composed of Paleozoic sedimentary rocks cut by late Miocene to Holocene thin-skinned thrusts. East of the central Puna region, the Cordillera Oriental consists of deformed Late Precambrian to Triassic sedimentary and magmatic sequences, and the Subandean belt is replaced by the thick-skinned Santa Barbara system, which formed by late Neogene inversion of a complex Cretaceous to Paleogene rift system. Finally, east of the southernmost Puna region, both belts are replaced by the high-angle, reverse-faulted basement uplifts of the Sierras Pampeanas, which continue southward into the Chilean-Pampean flat-slab region.

MAGMATIC, STRUCTURAL, AND GEOPHYSICAL FRAMEWORK OF THE SOUTHERN ALTIPLANO PLATEAU

The first of the three regions considered is the southern Altiplano Plateau between 17°S and 21°S (Fig. 2). In this area, the plateau is largely a sediment-filled basin bounded to the west by the Cordillera Occidental and the Neogene arc front and to the east by the western Cordillera Oriental and the Miocene Morococala and Los Frailes ignimbrites (Fig. 3). Further east, the Cordillera Oriental is commonly divided into the Interandean zone, bounded by the Interandean thrust, and the Subandean belt, bounded on the east by the Subandean thrust (Fig. 4). Much has been published on the structural evolution (McQuarrie, 2002; McQuarrie et al., 2005), surface uplift (Garzione et al., 2006; Molnar and Garzione, 2007; Ege et al., 2007), and basin development (DeCelles and Horton, 2003) in this region. The overall geologic evolution is reviewed in Lamb and Hoke (1997), Lamb et al. (1997), Kennan (2000), Barke and Lamb (2006), and Barke et al. (2007). Geophysical images are presented by Myers et al. (1998), Swenson et al. (2000), Beck and Zandt (2002), Leidig and Zandt (2003), and Zandt et al. (2003), among others. An interpreted seismic transect near 20°S from Beck and Zandt (2002) is shown in Figure 5. The history and chemical compositions of the magmatic rocks are far less known than to the south.

Two competing hypothesis for the uplift of the Altiplano have been presented. On one side, McQuarrie et al. (2005), Barke and Lamb (2006), and Hoke and Lamb (2007), among others, argue that uplift is consistent with crustal shortening and periodic loss of upper-plate lithosphere. The distribution of crustal thickening relative to shortening across the high region has led to complementary hypotheses of deep crustal flow into the Altiplano (Husson and Sempere, 2003; Gerbault et al., 2005). In the competing view, Garzione et al. (2006) and Molnar and Garzione (2007) have argued that catastrophic delamination of the eclogitic lower crust and lithosphere during the late Miocene largely explains the timing and scale of uplift of the Altiplano.

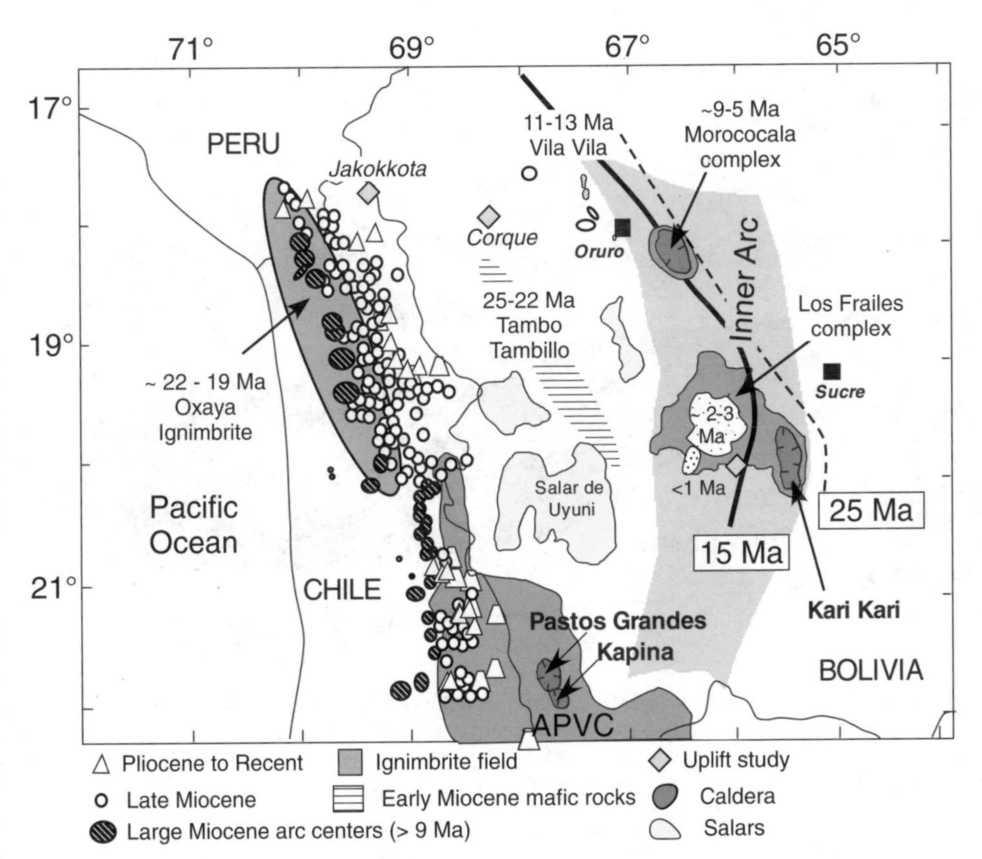

Figure 3. Map showing the Altiplano magmatic centers considered in the text. Large ignimbrite fields are shown in blue; salars are shown in yellow. Solid and dashed lines, respectively, indicate the eastern extent of volcanism at 25 and 15 Ma according to Lamb and Hoke (2007). Yellow diamonds are locations of leaf and oxygen studies used in determining paleoelevations (Corque region in Ghosh et al., 2006; others in Gregory-Wodzicki, 2000). Squares are cities. Map is largely based on information compiled and presented by Coira et al. (1993), Wörner et al. (2000), Barke et al. (2007), and Lamb and Hoke (2007). APVC—Altiplano-Puna volcanic complex.

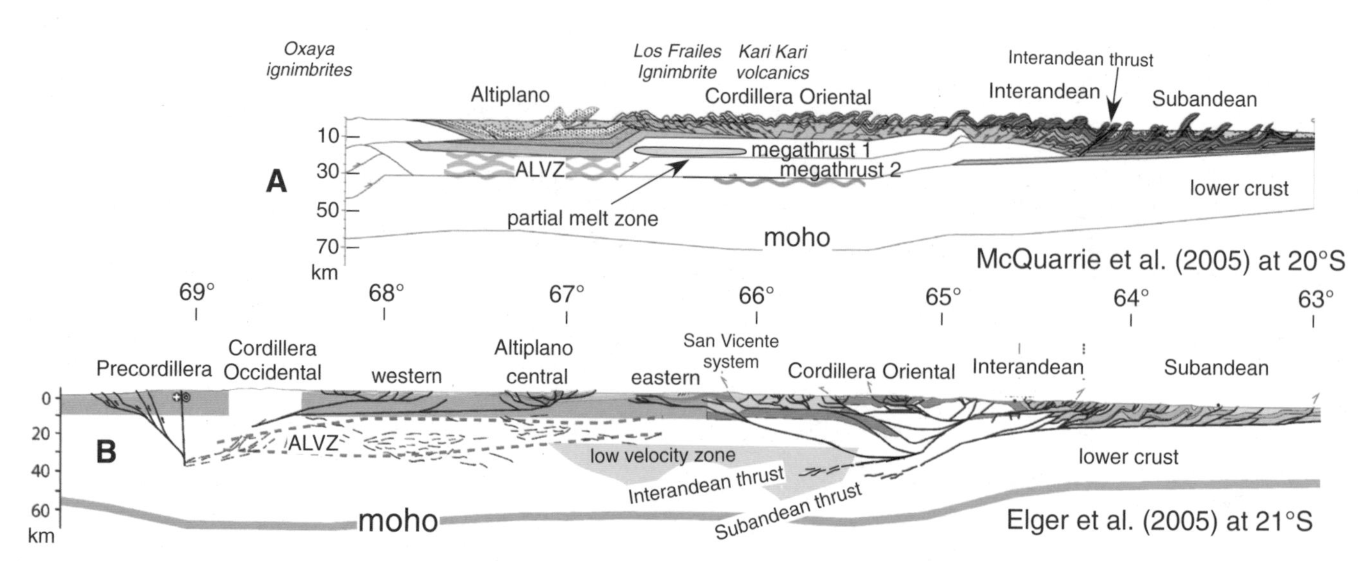

Figure 4. Crustal-scale structural sections at the same scale across the southern Altiplano showing the variations in upper-crustal deformational style from west to east and two interpretations on propagating the structures to depth. Profiles locations are shown in Figure 2. (A) Profile near 20°S from McQuarrie (2002) shows two megathrusts at midcrustal level; megathrust 1 is argued to have been active from ca. 40 to 20 Ma, and megathrust 2 is argued to have been active from ca. 20 to 15 Ma. The Altiplano low-velocity zone (ALVZ), the low-velocity zone below the Los Frailes complex (labeled "partial melt zone"), and the low-velocity zone near 30 km under the Cordillera Oriental (wavy line) were added by McQuarrie et al. (2005) based on the summary figure of Beck and Zandt (2002; see Fig. 5). (B) Section near 21°S from Elger et al. (2005) along the ANCORP Working Group (2003) and Re-Fu-Ca geophysical profiles (Heit et al., 2008a) showing major structures merging into the ductile lower crust (see text and Fig. 9A). Region labeled "low velocity zone" has the lowest crustal velocity in P-wave tomographic image of Heit et al. (2008a).

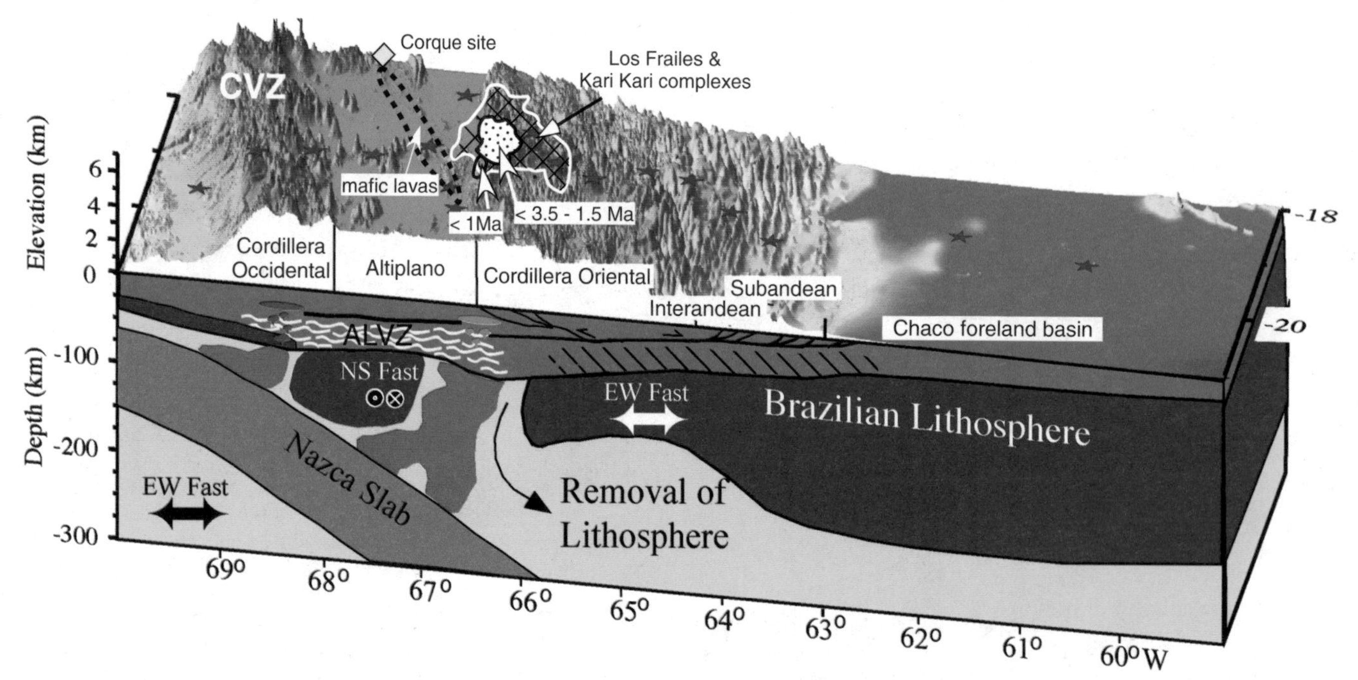

Figure 5. Interpreted seismic profile for the southern Altiplano from Beck and Zandt (2002) with the Los Frailes complex and zone of Neogene mafic lavas (dashed region) superimposed from Figure 3 and Barke and Lamb (2006). The Corque site is the area of the oxygen studies of Ghosh et al. (2006) and Garzione et al. (2006). ALVZ—Altiplano low-velocity zone. Zones of partial melt in the mid- to upper crust are inferred under the Central Volcanic Zone (CVZ) arc (red) and Los Frailes ignimbrite (pink) and are underlain by low-velocity mantle (red). NS fast and EW fast indicate directions of SKS shear-wave splitting (Polet et al., 2000). Blue stars show location of seismic instruments in the University of Arizona experiment. See text for discussion.

Neogene Magmatic History of the Southern Altiplano

The post–26 Ma magmatic history of the Altiplano from 17°S to 21°S has been summarized by Coira et al. (1993), Wörner et al. (2000), and Hoke and Lamb (2007), and references therein. Figure 3 shows a generalized map of the main volcanic features discussed below. The magmatism in this region postdates a time of virtual volcanic quiescence and uplift from ca. 38 to 27 Ma, which James and Sacks (1999) have proposed corresponds to a time of flat-slab subduction.

As reviewed by Hoke and Lamb (2007), the first magmas to erupt after the magmatic quiescence were mafic to felsic volcanic rocks reaching up to 300 km into the backarc (Fig. 3). The latest Oligocene to early Miocene Tambo-Tambillo basaltic and shoshonitic lava flows and sills that intrude and cover Oligocene to early Miocene backarc red beds over a region of 1000 km^2 are shown in the lined field in Figure. 3. Based on rare earth element (REE) modeling, these intraplate-like magmas are suggested to have largely formed in a spinel lherzolite mantle at depths of less than 90 km (Hoke and Lamb, 2007). The widespread rhyodacitic Oxaya ignimbrite group in the arc region, which erupted between 22.7 and 19.4 Ma, extends from 17°S to 20°S and has an estimated volume of over 3000 km^3 (see García et al., 2000; Wörner et al., 2000). The ignimbrite eruptions were followed by large andesitic shield volcanoes with $^{40}Ar/^{39}Ar$ ages ranging from 20 to 19 to 9 Ma on the western slope of the Cordillera Occidental (see Wörner et al., 2000; García et al., 2004), showing that andesitic volcanism was active in the Central Volcanic Zone arc region until at least 9 Ma.

Miocene to Pliocene andesitic to dacitic volcanic centers in the eastern Altiplano and western Cordillera Oriental include the Los Frailes, Morococala, and Vila Vila region volcanics in the "Inner arc" region (Fig. 3). The Los Frailes complex covers an area of 7500 km^2 and shows a broad span of ages (see Schneider and Halls, 1985; Barke et al., 2007). The oldest centers are small ca. 25–20 Ma dacitic domes and stocks in the northwest, and ca. 16–9 Ma domes and ignimbrites along the western and southeast margins. The most voluminous deposits are 9–5 Ma ash flows that erupted from multiple centers and are exposed in the northern part of the complex. The youngest flows are 3.5–1.5 and younger than 1 Ma dacites in the central and southern region. The total erupted volume is estimated at 2000 km^3 (Schneider and Halls, 1985). The eruptions from the Morococala center further north cover an area of 1800 km^2 (Barke et al., 2007). The Morococala ignimbrites consist of three major ash-flow units erupted from distinct centers between 8.4 and 6.5 Ma. The oldest are andalusite-bearing two-mica rhyolitic tuffs; the younger are biotite-quartz latitic tuffs and domes (Koeppen et al., 1987). The 13–11 Ma Vila Vila shoshonites further north erupted north of the inactive Eucalyptus fault, and the 8–5 Ma intermediate to acidic Eucalyptus lavas and dikes erupted along the trace of the Eucalyptus fault (see Hoke and Lamb, 2007).

Pliocene to Pleistocene mafic volcanic flows with ages younger than 5 Ma erupted up to 200 km east of the modern arc front. They principally form basaltic to high-K basaltic andesitic monogenetic centers, cinder cones, and maars (Davidson and de Silva, 1992). Hoke and Lamb (2007) argued that 80% of the melting that produced the near-arc magmas was at less than 100 km, and more than 90% of that for the backarc magmas was at less than 80 km. Helium isotopic data show that mantle melting is ongoing under the Altiplano and western Cordillera Oriental, including under the Los Frailes region.

After 3 Ma, andesitic to dacitic centers were concentrated in a relatively narrow east-west zone near the modern volcanic front. The active centers are predominately large dacitic ignimbrite-dome complexes (see Stern et al., 2004). A conspicuous gap in activity east of the Salar de Uyuni (Fig. 3), called the Pica Gap, correlates with subduction of the Iquique aseismic ridge (see Wörner et al., 2000).

Deformation and Uplift of the Southern Altiplano

The deformational history of the Altiplano is reviewed in Lamb and Hoke (1997), Lamb et al. (1997), Kennan et al. (1997), McQuarrie et al. (2005), and Oncken et al. (2006), among others. Deformation began in the Chilean Precordillera at around 85 Ma (see Mpodozis et al., 2005), in the Cordillera Occidental by 70 Ma, and then moved eastward. DeCelles and Horton (2003) use the history of the sedimentary basins to argue that deformation propagated eastward during the Tertiary. The deformational history from the Altiplano to the Subandean belt is discussed relative to the cross sections at 20°S (McQuarrie, 2002; McQuarrie et al., 2005) and 21°S (Elger et al., 2005) shown in Figure 4 (located in Fig. 2). A third section at 15°S–17°S can be found in McQuarrie et al. (2008). The upper-crustal differences between the sections largely reflect along-strike changes in the structures. The lower-crustal differences reflect different ways to accommodate the structures at depth. McQuarrie (2002) viewed these systems as being controlled by an older (ca. 40–20 Ma), upper megathrust and a younger (<20 Ma), lower megathrust, whereas Elger et al. (2005) considered the west-central Altiplano and Cordillera Oriental thrust systems to be largely independent systems that merge into the ductile lower crust. In both models, deformation ceased in the Cordillera Oriental and the Altiplano as the Subandean fold-and-thrust belt became active after 10 Ma (e.g., Gubbels et al., 1993).

In the section at 20°S, the model presented by McQuarrie (2002) and McQuarrie et al. (2005) considers that 250–200 km of shortening occurred during the Late Cretaceous to Paleocene on a narrow thrust belt west of the Altiplano and that 330–300 km of additional shortening has occurred since 40 Ma in the Cordillera Oriental and Subandean zone. They suggest a fairly constant shortening rate of ~8–10 mm/yr over the last 40 m.y., and a dramatic deceleration between 25 Ma and ca. 15–8 Ma (see McQuarrie et al., 2008). The slowdown coincides with the deposition of the 24–21 Ma Salla Formation over back thrusts in the Cordillera Oriental, and the eruption of the Oxaya ignimbrites and mafic flows (Fig. 3). In the McQuarrie et al. (2005) model, earlier deformation is attributed to slip on a 200–400-km-long upper megathrust that initiated at 40 Ma, had slowed by 20 Ma, and was inactive by 10 Ma. Structures argued to have formed above this megathrust include west-vergent Oligocene and early Miocene back thrusts in the western Cordillera Oriental, east-vergent Oligocene thrusts in the eastern Cordillera Oriental, east-vergent early Miocene thrusts in the Interandean zone, and east-vergent, ca. 18–13 Ma out-of-sequence thrusts in the eastern Cordillera Oriental. Slip on a younger and deeper megathrust is used to explain the deeper regional detachment under the Cordillera Oriental that Kley (1996) showed was necessary to accommodate the geometry of Subandean thrusting. The transfer of slip from the upper to the lower megathrust is suggested to have begun by 20 Ma, with most slip on the lower megathrust by ca. 15 Ma. McQuarrie et al. (2005) speculated that the crust under the Cordillera Oriental could have been as thick as 60 km by 20 Ma.

The kinematic model of Elger et al. (2005) inferred for the deep crust in the section near 21°S (Fig. 4B) is very different and essentially incompatible with that in Figure 4A, as the west-central Altiplano and Cordillera Oriental thrust systems are viewed as two largely independent systems in which the main deformation peaks occurred out of phase. Starting in the west, Tertiary deformation is argued to have begun in the Chilean Precordillera at 48–38 Ma with shortening rates up to ~1.2 mm/yr. In the Altiplano to the east, at least 65 km of shortening is suggested to have been accumulated from 33 to 27 Ma and from ca. 19 to 8 Ma, where the intervening slowdown coincided with a time of a high deformation rate in the Cordillera Oriental (see following). The thrusts in the Altiplano generally occur on an east-verging western, a doubling-verging central, and a west-verging eastern system. The western and central systems are argued to be kinematically coupled above a joint detachment system, over which the highest deformation rate was achieved after 20 Ma (highest individual rates are >2 and 4 mm/yr). The eastern Altiplano system reached a much lower highest average rate (<0.25 mm/yr) before ca. ~25 Ma. Thrusting occurred on all three systems at a maximum combined rate of 4.7 mm/yr between ca. 11 and 8 Ma as the Los Frailes and Morococala ignimbrites erupted. Further east, Ege et al. (2007) argued for exhumation and deformation by 40–36 Ma in the central Cordillera Oriental, where thrusting began on both the west-vergent San Vincent system on the west and the east-vergent Interandean system on the east by the early Oligocene. The San Vincente system was active in irregular order until ca. 20 Ma; the Interandean system propagated eastward after ca. 30 Ma. The deformation rate peaked at 9–6 mm/yr between 30 and 17 Ma and had slowed by 12–8 Ma as deformation ended and moved into the Subandean belt. Ege et al. (2007) suggested that ~15% of the 140–150 km of Subandean shortening calculated by Kley (1996) occurred from 30 to 18 Ma.

Taking into account the entire region, Oncken et al. (2006) suggested that the overall deformation rate reached a maximum at 8 mm/yr by the late Oligocene and then fluctuated near 8 mm/yr until thrusting shifted into the Subandean belt at 8–7 Ma, where it propagated eastward at a variable, but high, rate of ~16–8 mm/yr. The current shortening rate from global positioning system (GPS) studies is 9 ± 1.5 mm/yr (Bevis et al., 2001). Oncken et al. (2006) argued that the general pattern and relative rates of deformation in the transect near 21°S generally apply from 15° to 23°S, and along-strike changes are due to minor variations in deformation rates.

The most precise surface uplift data for the Altiplano come from Ghosh et al. (2006) and Garzione et al. (2006), who used oxygen isotopic compositions on carbonates from a sedimentary sequence in the Corque syncline region north of 18°S (Fig. 3) to argue for rapid late Miocene uplift. The age of the studied sequence is constrained by magnetostratigraphy and $^{40}Ar/^{39}Ar$ ages on the interbedded 10.4 Ma Ulloma, 9.0 Ma Callapa, and 5.4 Ma Toba 76 tuffs. Garzione et al. (2006) combined their results with those from a leaf physiognomy study of the 10.6 Ma Jakokkota flora northwest of Corque (Gregory-Wodzicki, 2000) to argue that the absolute surface elevation of the Altiplano was ~500 m at 12–10.3 Ma, after which it increased to near 2000 m at 7.6 Ma, and then rose to 4000 m by 6.8 Ma. The plateau surface is suggested to have risen ~2500–3500 m between 10.3 and 6.8 Ma, giving an average rate of 1.03 ± 0.12 mm/yr (Ghosh et al., 2006). To the east in the Cordillera Oriental, Barke and Lamb (2006) used a combination of methods to argue for an uplift of 1705 ± 695 m since 12–9 Ma.

Geophysical Framework of the Southern Altiplano

The interpreted seismic transect near 20°S of Beck and Zandt (2002) is shown in Figure 5. They used this profile to argue for a weak, locally compensated lower crust, melt zones over a variably strong lithosphere west of the Cordillera Oriental, and a stronger crust and lithosphere to the east. Crustal thicknesses are variable across the region. Under the Cordillera Occidental, the crustal thickness at 20°S is ~70 km, assuming an average crustal velocity of 6.0 km/s (see Beck and Zandt, 2002; McGlashan et al., 2008). Beneath the Altiplano, the thickness is inferred to vary from 59 to 64 km (Fig. 5), and the composition is inferred to be siliceous based on an average P-wave velocity of 5.8–6.25 km/s. Yuan et al. (2002) argued that a thin mafic base is present. McGlashan et al. (2008) indicated crustal thicknesses of 75–73 km under the Altiplano just north of 21°S. Further east, the crust under the Cordillera Oriental is shown to thin from ~74 km in the west to ~60 km in the east, and it is considered to be silicic based on a P-wave velocities of 5.75–6.0 km/s (Wigger et al., 1994; Beck and Zandt, 2002). Yuan et al. (2002) and McGlashan et al. (2008) indicated crustal thicknesses up to 80–82 km just north of 21°S. Beck and Zandt (2002) suggested thicknesses of ~40 km under the Subandean zone and near 32 km under the Chaco foreland basin.

Low-velocity zones within the crust are considered to be zones of partial melt. Anomalies near 15 km depth are interpreted as melt zones under the active Central Volcanic Zone arc and the Los Frailes volcanic complex (see Beck and Zandt, 2002). The one under the Los Frailes region occurs below volcanic rocks dated at younger than 3.5 Ma (Barke and Lamb, 2006; Fig. 3). A more extensive low-velocity zone under the Cordillera Oriental and Altiplano, which is interpreted as a partial melt zone, was called the Altiplano low-velocity zone by Yuan et al. (2002). Further east, a low-velocity zone occurs at ~30 km (Beck and Zandt, 2002). The best interpretation for both the Altiplano low-velocity zone and this low-velocity zone appears to be a migmatite zone.

As shown in Figure 4, McQuarrie al. (2005) correlated portions of the boundaries of their postulated megathrusts with crustal velocity anomalies. Under the Altiplano, the upper and lower boundaries of megathrust 1, which they argued had substantially slowed by 20 Ma and was inactive by 15 Ma, are at the boundaries of the Altiplano low-velocity zone. Under the Cordillera Oriental, the base of megathrust 1 falls just below the interpreted partial melt zone under the Los Frailes volcanic field, and megathrust 2 falls is at the base of the low-velocity zone near 30 km, which they interpreted as the brittle-ductile transition. The chronology allows the major melting that produced the modern low-velocity zones to have largely occurred after slip on the megathrusts had ceased, facilitating partial melting zones at the old thrust boundaries. However, a problem remains if the crust was already silicic at 20–15 Ma, since a brittle-ductile transition at 30 km indicates a geothermal gradient no hotter than 10 °C/km. Such a low gradient seems inconsistent with volcanic activity in the Los Frailes volcanic system before 15 Ma.

The thickness of the upper-mantle lithosphere is also variable. The base was placed at 125–150 km under a high Vp (P-wave velocity) and Vs (shear-wave velocity) and moderate Q (quality factor) region beneath the western and central Altiplano by Myers et al. (1998), although it is important to note that there are still few direct measurements of lithospheric thickness in this region. Zones of thin or missing mantle lithosphere are shown west and east of this region, and the low-velocity anomaly under the Los Frailes volcanic region is interpreted as an area where lithosphere has been removed or severely altered. The concentration of Miocene lavas above this anomaly led Hoke and Lamb (2007) to suggest that this is a long-lived zone of lithosphere removal. High P-wave upper-mantle velocities under the Cordillera Oriental further east indicate a strong mantle lithosphere that thickens under the Subandean belt (see Beck and Zandt, 2002). The area of the profile showing E-W fast shear-wave splitting (see Fig. 5) has been suggested to be part of the Brazilian Shield (Polet et al., 2000; Beck and Zandt, 2002).

Lithospheric-Scale Model for the Evolution of the Southern Altiplano

Based on the magmatic, structural, and geophysical data discussed here, a model for the Neogene evolution of the Altiplano near 19°S to 20°S is shown in a series of sections in Figure 6A. These sections are to scale where possible. The slab geometries in the sections are based on modern-day slab profiles made using Figure 1 (Cahill and Isacks, 1992) at the latitude shown on Figure 6A. Previous models for the Altiplano by James and Sacks (1999) and Hoke and Lamb (2007) have shown a flat-slab segment before 25 Ma and a moderately shallow slab at 25–15 or 25–21 Ma, followed by cartoons with modern dip.

Period 1: 26–17 Ma

The 26–17 Ma section is shown with the modern form of the slab at 28°S in accord with similarities between the 26–17 Ma bimodal arc-backarc volcanic rocks and less than 3 Ma volcanic rocks near 27°S–28°S today (Kay et al., 1994b, 2008). This is the geometry of the shallowest slab segment supporting volcanism in the modern Central Andes. This geometry is also consistent with the model of James and Sacks (1999), which calls for a relatively shallow slab at 25–15 Ma, and the backarc deformation in the Cordillera Oriental at this time (McQuarrie et al., 2005; Oncken et al., 2006). The backarc lithosphere is shown with a thickness of less than 100 km in accord with melting depths for backarc mafic flows from Hoke and Lamb (2007).

James and Sacks (1999) argued for a flat-slab stage preceding this stage to explain a 38–27 Ma magmatic hiatus and widespread Oligocene deformation. The end of the flat-slab stage is correlated with initiation of widespread arc-backarc volcanism, and Hoke and Lamb (2007) inferred crustal delamination at this time. Hydration of the continental lithosphere during the flat-slab stage was argued by James and Sacks (1999) to weaken the lithosphere and facilitate backarc deformation. There are some questions regarding whether a flat-slab stage is required. The first is that Eocene to early Miocene deformation, which was thought to be largely concentrated in the Altiplano, is now known to extend and be widespread further south (e.g., Coutand et al., 2001; Deeken et al., 2006; see following). The second is that mantle hydration under the modern Chilean-Pampean flat slab is not supported by seismic studies (Wagner et al., 2006). A third is that a relative magmatic lull at this time can be at least partially attributed to oblique subduction (e.g., Pardo Casas and Molnar, 1987; Somoza, 1998).

Period 2: 16–11 Ma

The slab geometry at 16–11 Ma is shown as being slightly steeper than that at 25–17 Ma and the shape being that of the modern slab near 27°S. The continuation of volcanism across the backarc and stronger deformation in the Altiplano are consistent with mantle melting and lithospheric weakening above a more moderately dipping slab. By ca. 18 Ma, the deformation rate was slowing in the Cordillera Oriental as it was increasing under the Altiplano (Elger et al., 2005). In the alternative model of McQuarrie et al. (2005), all of the slip had been transferred from megathrust 1 to megathrust 2 by 15 Ma. The crustal thickening pattern in Figure 6A reflects the growing influence of the ductile behavior of the middle to lower crust under the Altiplano. Evidence that enriched mantle lithosphere was beginning to be removed and incorporated into the magma source region above the shallowly dipping slab fits with an analogy between the 13–11 Ma Vila Vila shoshonites and the Pleistocene Puna shoshonites near 24°S (Coira and Kay, 1993; see following).

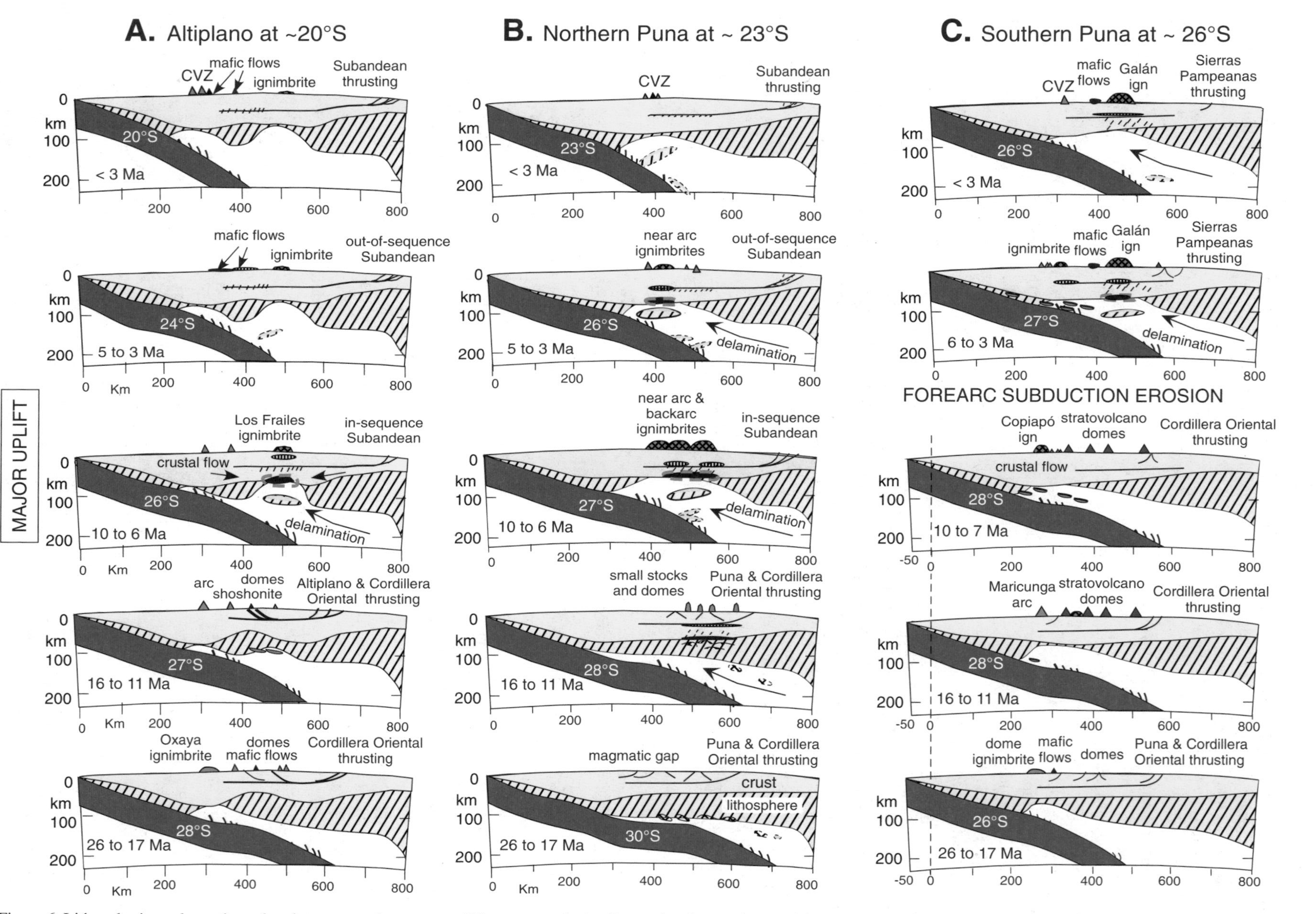

Figure 6. Lithospheric-scale sections showing proposed sequence of Neogene events leading to the observed magmatic and deformation features of the (A) southern Altiplano, (B) northern Puna, and (C) southern Puna regions. The geometry of the subducting plate is based on profiles of the modern subducting plate in Figure 1 at the latitude indicated on the downgoing slab. The latitude chosen for the slab profile is largely based on the distribution and chemistry of the overlying volcanic centers as discussed in the text. Arrows indicate regions of decompression melting above steepening slabs. Wavy lines above the slabs represent zones of dehydration. Vertical dashed line in C shows the area added to the east of the modern trench to account for the material removed during a peak in forearc subduction-erosion between 7 and 3 Ma. See discussion in text. CVZ—Central Volcanic Zone; ign—ignimbrite.

Period 3: 10–6 Ma

The average slab dip and geometry of the subduction system at 10–6 Ma are considered to be like that near 26°S today. The rationale is that the large-volume, ca. 9–5 Ma Los Frailes and Morococala ignimbrites, which are now located 275–300 km above the subducting slab, erupted when the slab was at a depth of less than 150–160 km, as was the case for the eruption of the large-volume young Puna ignimbrites (Figs. 1, 2, and 7). Significant slab steepening ca. 9–8 Ma facilitated crustal and lithosphere delamination and increased melting to produce the ignimbrites. The major uplift of the Altiplano between 10.7 and 6 Ma (Garzione et al., 2006) corresponds with ongoing heating and crustal melting that produced the Los Frailes and Morococala ignimbrites, delamination of lower crust and lithosphere, and crustal flow. A delaminated block is shown being removed from under the Los Frailes volcanic complex in accord with seismic evidence for a thin lithosphere (Myers et al., 1998). The size is based on the proposed delaminated block in the P-wave tomographic image under the northern Puna (Schurr et al., 2006), which is about the size shown by Kay et al. (1994a) under the southern Puna. This was also the time that deformation ceased in the Altiplano and Cordillera Oriental as thrusting began in the thin-skinned Subandean belt at a high rate (~8–16 mm/yr; Elger et al., 2005). Compensation for upper-crustal shortening in the Subandean belt allows lower-crustal thickening to play a major role in uplift to the west (e.g., Isacks, 1988). Barke and Lamb (2006) showed that ~1700 m of uplift in the eastern Cordillera Oriental at this time is easily balanced by 62–100 km of Subandean belt shortening. Myers et al. (1998) concluded that the Brazilian Shield was underthrusting by this time, thickening the lithosphere in the west (Fig. 6).

Many authors have called for removal of continental lithosphere under the Altiplano and the Cordillera Oriental at this time (e.g., Myers et al., 1998; Beck and Zandt, 2002; Sobolev and Babeyko, 2005; Barke and Lamb, 2006). The question is one of scale. Molnar and Garzione (2007) took the extreme view that uplift was almost entirely driven by massive lower-crustal and lithospheric removal. Using an Airy isostasy model, they argued that only wholesale rapid removal of dense eclogitic lower crust and mantle lithosphere could cause ~2500–3500 m of surface uplift between 10.3 and 6.8 Ma. However, different combinations of input parameters than the ones they use, which are within the limits of uncertainty, including a shortening rate of 13–15 mm/yr rather than 10 mm/yr, an initial thickness of 51–53 km rather than 50 km, an initial plateau width of 280–290 km rather than 300 km, a mantle density of ~3350 kg/m^3 rather than 3300 kg/m^3, and a crustal density of ~2675 kg/m^3 rather than 2800 kg/m^3, can accommodate some 2500–2600 m of uplift in 4 m.y. A faster deformation rate is in accord with the range from 8 to 16 mm/yr given by Elger et al. (2005) for the Subandean belt. A lower-crustal density fits with bulk crustal seismic velocities and the view of Lucassen et al. (2000), who state that the crust was relatively silicic by the end of the Paleozoic. Further, Hoke and Lamb (2007) pointed to a problem in the way in which to form and maintain a more than 40-km-thick mafic eclogitic root at the base of an ~100-km-thick crust during periodic and widespread Miocene mafic and silicic volcanism. Another problem is that the removal of either an 80-km-thick by 390-km-wide block from beneath the whole plateau or a combination of a 45-km-thick by 100-km-wide block under the western Altiplano and a 140-km-thick by 180-km-wide block under the western border of the Altiplano, as proposed by Garzione et al. (2008), is incompatible with the slab geometry for that time shown in Figure 6A.

Another process for thickening the Altiplano crust is lower-crustal flow. The generally flat topography and smaller ignimbrite volume in the Altiplano compared to the northern Puna (Fig. 2) support the model of Husson and Sempere (2003), in which uplift can be largely explained by crustal flow from both the Cordillera Occidental and Cordillera Oriental. They used two observations to support crustal flow into the Altiplano. The first is the observation from Beck et al. (1996) that the modern crust under the Altiplano is too thick and that under the Cordillera Occidental and Cordillera Oriental is too thin to balance measured crustal shortening. The second is a plot of cumulative uplift from Kennan (2000), which shows that major uplift of the Cordilleras Occidental and Oriental preceded that of the Altiplano and that the Cordillera Occidental was over 3000 m, the Cordillera Oriental at nearly 2000 m, and the Altiplano at ~1000 m when rapid uplift of the Altiplano began at 10 Ma. Using these constraints, Husson and Sempere (2003) proposed that temperature and viscosity conditions for lower-crustal flow became optimum near 12–10 Ma and that the crust flowed from under the 40–50-km-thick marginal cordilleras into the Altiplano. Their model is in accord with both uplift and modern crustal thicknesses. Flow of silicic crust from the edges of the plateau helps to explain the low average crustal seismic velocities in the Altiplano (Swenson et al., 2000). Lower-crustal flow also facilitates delamination as mid- to lower crust injected from the Cordillera Oriental and Oriental depressed the lower Altiplano lower crust, forcing it into a deeper and hotter region where eclogite formation is accelerated. In another model, Gerbault et al. (2005) proposed that a thicker crust and denser mantle lithosphere under the Altiplano than the northern Puna caused northerly lower-crustal channel flow into the Altiplano, producing uplift with little surface deformation.

The steepening subduction zone below the Altiplano could have been a key factor in the progressive heating and melting that produced the Los Frailes and Morococala ignimbrite eruptions between 9 and 5 Ma. Mantle-derived mafic magmas produced by decompression melting above the steepening, but still moderately dipping, slab provided the heat source. Dehydration of the slab facilitated eclogitization of the thickened lower crust. Mafic magmas ponded in the basal crust produced hybridized partial melts that segregated and migrated upward, leaving a mafic residue that was denser than the underlying mantle (Dufek and Bergantz, 2005). Delamination of this dense lower crust along with lithospheric mantle left the lower crust subject to further melting, producing the large volumes of magma expelled in the giant Frailes and Morococala ignimbrite eruptions. The failure of the weak-

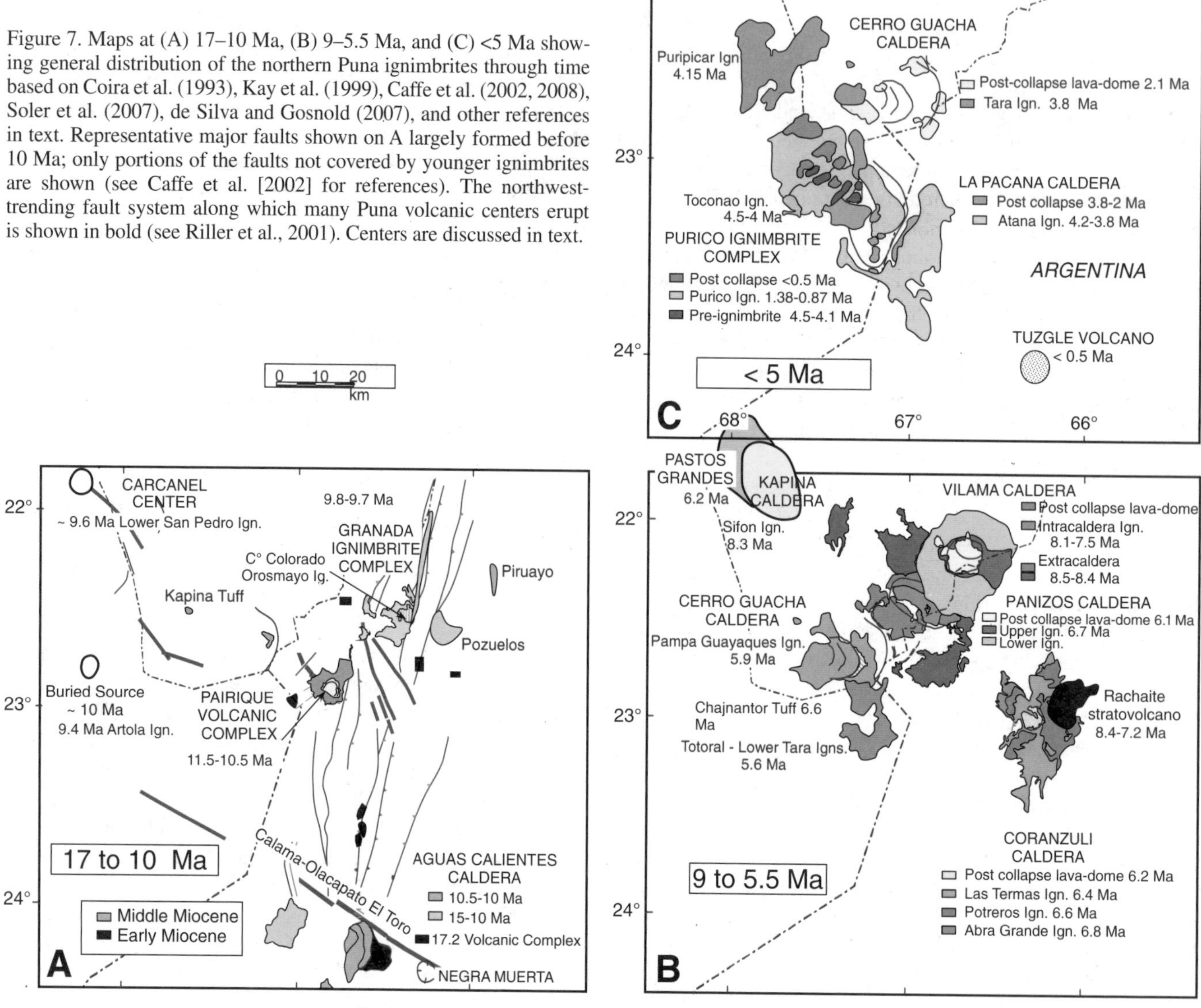

Figure 7. Maps at (A) 17–10 Ma, (B) 9–5.5 Ma, and (C) <5 Ma showing general distribution of the northern Puna ignimbrites through time based on Coira et al. (1993), Kay et al. (1999), Caffe et al. (2002, 2008), Soler et al. (2007), de Silva and Gosnold (2007), and other references in text. Representative major faults shown on A largely formed before 10 Ma; only portions of the faults not covered by younger ignimbrites are shown (see Caffe et al. [2002] for references). The northwest-trending fault system along which many Puna volcanic centers erupt is shown in bold (see Riller et al., 2001). Centers are discussed in text.

ened Altiplano–eastern Cordillera Oriental lower crust and melt release associated with the large ignimbrite eruptions could have triggered the expansion of the thrust belt into the foreland. The ability of the crust to yield to the east could have contributed to less delamination beneath the Altiplano. The largest ignimbrites erupted at the limit between the Altiplano and Eastern Cordillera, where delamination has been argued to be responsible for the apparently missing underlying lithosphere (Myers et al., 1998; Beck and Zandt, 2002; Hoke and Lamb, 2007). In contrast, the presence of mantle lithosphere to the west beneath the Altiplano suggests that this region was not subject to delamination.

Period 4: 5–0 Ma

In the last stage, the subducting plate steepened to its modern geometry. Andesitic and dacitic arc magmatism was confined to dome/stratovolcanic complexes in the frontal arc region by the steeper subduction zone. In accord with a cooler and more refractory crust, large backarc ignimbrite eruptions ended. Small intraplate-like mafic lavas with calculated melting depths of <100 km (Hoke and Lamb, 2007), which are in accord with lithospheric thicknesses from seismic data (Fig. 5), imply a more brittle lower crust, which allows mafic magmas to reach the surface.

MAGMATIC, STRUCTURAL, AND GEOPHYSICAL FRAMEWORK OF THE NORTHERN PUNA

The next region to be considered is the northern Puna region between 21°S and 24°S (Fig. 2). Important differences with the southern Altiplano are summarized by Whitman et al. (1996), Allmendinger et al. (1997), and Yuan et al. (2002), among others. The most distinctive features are the widespread late Miocene to Pliocene ignimbrite complexes that are flanked by the Altiplano Basin to the north and discontinuous basins to the south (Fig. 2). These ignimbrite complexes, which are shown in the maps in Figure 7, are generally referred to as the Altiplano-Puna volcanic complex or APVC (de Silva, 1989; de Silva and Gosnold, 2007). Where not covered by volcanic rocks, the Puna surface is characterized by a series of narrow ridges and broad intervening basins created by east- and west-dipping, high-angle reverse faults that largely put Ordovician sedimentary and magmatic units over Tertiary sequences. The Cordillera Oriental to the east is a high-standing, imbricated stack of thrust slivers and thrust-bounded folds that expose Proterozoic to Cambrian sedimentary and magmatic basement (Mon et al., 2005). Contractional deformation under the plateau ranges from at least Eocene to middle Miocene in age (e.g., Coutand et al., 2001; Hongn et al., 2007). A representative structural section is shown in Figure 8A. Further east, the foreland consists of the thin-skinned, Subandean fold-and-thrust belt, where the total shortening reaches up to 150 km north of 23°S, and the Santa Barbara system to the south, where late Neogene, thick-skinned thrusts invert Cretaceous to Paleocene rift basins (see Kley and Monaldi, 2002). Representative sections across the Subandean ranges and the Santa Barbara system are shown in Figures 8B and 8C.

The same tectonic elements as in the Altiplano have been argued to explain the uplift and evolution of the northern Puna. Following the uplift model of Isacks (1988) and utilizing the ignimbrite history, a role for a thickening crust and thermal uplift over a steepening subduction zone was advocated by Coira et al. (1993), Kay et al. (1999), and Kay and Mpodozis (2001). More recent models show a significant role for lower-crust and mantle delamination (e.g., Sobolev and Babeyko, 2005; Kay et al., 2008) that is supported by seismic interpretations (e.g., Schurr et al., 2003, 2006; see Fig. 9). A key question is: why is there far more ignimbrite volcanism in the northern Puna than to the north or the south? One answer is more mafic magma input into the crust over a steepening slab and a delaminating lithosphere leading to higher temperatures that optimize crustal melt.

Neogene Magmatic History of the Northern Puna

The Neogene magmatic history of the northern Puna region has been reviewed by Coira et al. (1993), Kay et al. (1999), and de Silva and Gosnold (2007). The region near 24°S was discussed by Petrinovic et al. (1999) and Mazzuoli et al. (2008). The major centers and ignimbrites between ~22° and 24.5°S are under study; a working summary is shown in Figure 7. The general history can be divided into four stages: (1) small widespread centers emplaced between 17 and 10 Ma, (2) intermediate to giant arc to backarc calderas erupted from ca. 9 to 5.5 Ma, (3) intermediate to giant calderas erupted in the near arc region from 5.6 to 3 Ma, and (4) large dacitic ignimbrite-lava dome complexes erupted in the arc region after 3 Ma.

The eruption of middle Miocene magmatic centers in the Puna was preceded by a general magmatic hiatus across the region that lasted from the Oligocene until ca. 17–16 Ma. Neogene magmatism began with the emplacement of small porphyritic stocks, domes, and ignimbrites in the far backarc from ca. 17 to 11 Ma (see Caffe et al., 2002). They were succeeded by small to mid-size caldera complexes as the major ignimbrite activity began at 11–10 Ma (Fig. 7A). The early complexes include the 11–10 Ma, <10 km^3 Pairique complex (all volumes in dense rock equivalent [DRE]) (Caffe et al., 2008), the ca. 10 Ma, ~60 km^3 Granada complex (Caffe et al., 2008), and the 9.4 Ma, >150 km^3 Artola ignimbrite from an unknown source (de Silva and Gosnold, 2007). The ca. 10 Ma Aguas Calientes caldera near 24°S erupted ~270 km^3 (Petrinovic et al., 1999). The peak of the giant backarc ignimbrite eruptions occurred between ca. 8.7 and 6.3 Ma (Fig. 7B). The backarc centers include the ca. 8.5 Ma, ~1400 km^3 Vilama caldera ignimbrites (Soler et al., 2007), the ca. 6.7 Ma, ~650 km^3 Panizos caldera ignimbrites (Ort, 1993; Ort et al., 1996), and the ca. 6.8–6.4 Ma, ~650 km^3 Coranzulí caldera ignimbrites (Seggiaro and Aniel, 1989; Seggiaro, 1994). Other ignimbrites closer to the arc region in the list of de Silva and Gosnold (2007) include the 8.3 Ma, >1000 km^3 Sifón ignimbrite, which was likely derived from the Kapina or Pastos Grandes caldera, the 6.2 Ma, >1000 km^3 Pastos Grandes caldera ignimbrites, and the 5.6 Ma, >500 km^3 Totoral–Lower Tara ignimbrite from

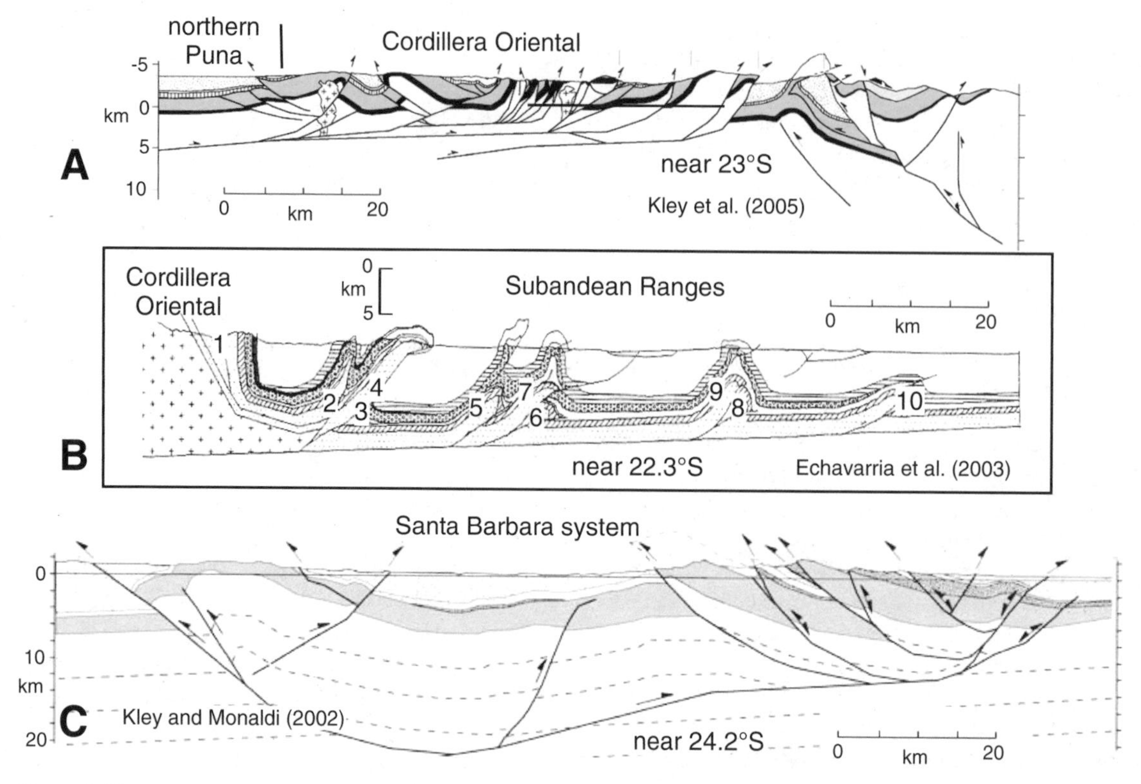

Figure 8. Upper-crustal cross sections at the same scale showing differences in structural style in the northern Puna, Cordillera Oriental, Subandean ranges, and Santa Barbara system. Profiles are located in Figure 2. (A) Section from Kley et al. (2005) near 23°S showing the shallow crustal imbricate style of thrusting and minor folds of the Cordillera Oriental and the mixed vergence of the northern Puna and Cordillera Oriental structures. (B) Section from Echavarría et al. (2003) south of 22°S illustrating the thin-skinned in-sequence and out-of-sequence thrusts of the Subandean belt; numbers on thrusts show order in which they became active. (C) Section near 24.2°S from Kley and Monaldi (2002) showing the thick-skinned and normal fault inversion style typical of the Santa Barbara system.

the Cerro Guacha caldera. Two other units that could be from the Cerro Guacha caldera are the 6.6 Ma Chajnantor Tuff and the 5.9 Ma Pampa Guayaques ignimbrite (Barquero-Molino, 2003).

In the latest Miocene and Pliocene, ignimbritic eruptions were largely concentrated in the arc region (Fig. 7C), and backarc activity was mainly limited to late-stage silicic andesites (see Coira et al., 1993). Initially, large ignimbrite eruptions continued with the La Pacana caldera, erupting the 5.6 Ma, >500 km^3 Pujsa ignimbrite, the 4.5–4 Ma, >100 km^3 Toconao ignimbrite, and the giant 3.8–4.2 Ma, >1600 km^3 Atana ignimbrite (Gardeweg and Ramírez, 1987; de Silva, 1989; Lindsay et al., 2001; de Silva and Gosnold, 2007). Other large ignimbrites with uncertain sources include the giant ca. 4.2 Ma, ~1200 km^3 Puripicar ignimbrite (possibly from the Guacha caldera; de Silva and Gosnold, 2007) and the smaller ca. 3.8 Ma, >300 km^3 Tara ignimbrite (Lindsay et al., 2001). Seven other ignimbrites in Chile and Bolivia with ages from ca. 5 to 1 Ma and volumes less than 150 km^3 are listed by de Silva and Gosnold (2007). They include the ca. 1 Ma, ~100 km^3 Purico ignimbrite (Schmitt et al., 2001). The ignimbrites with ages younger than 3 Ma have volumes of 150 km^3 or less. By 3 Ma, volcanism was concentrated in the modern Central Volcanic Zone (see review in Stern, 2004), and activity in the backarc was very minor (see Coira et al., 1993).

Deformation and Uplift of the Northern Puna

The structural setting of the northern Puna and the foreland fold-and-thrust belts to the east is variable along strike (Fig. 2). The cross sections in Figure 8 illustrate the structural styles of the northern Puna and Cordillera Oriental (Kley et al., 2005), the Subandean Ranges (Echavarría et al., 2003), and the Santa Barbara system (Kley and Monaldi, 2002).

The northern Puna, which widens to the south as the Cordillera Oriental narrows, is characterized by east- and west-vergent thrusts, whereas the Cordillera Oriental is primarily an imbricated stack of basement-involved high-angle thrusts and folds (e.g., Marrett et al., 1994; Mon et al., 2005; Allmendinger et al., 1997; Kley et al., 2005; Fig. 8A). The deformation age and structural style of the northern Puna and Cordillera Oriental have been analyzed by Cladouhos et al. (1994), Coutand

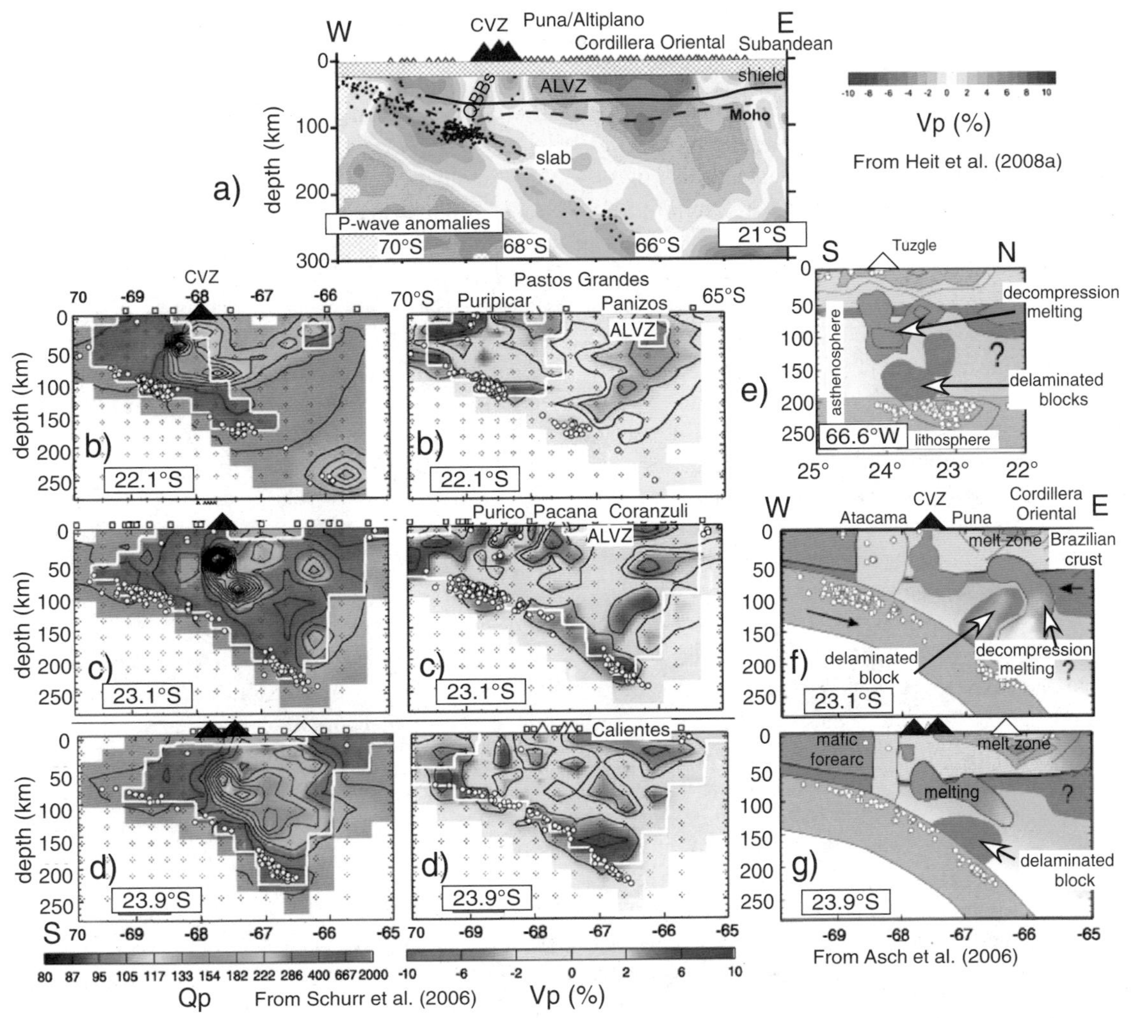

Figure 9. Seismic P-wave tomographic images of the northern Puna region from north to south at the same scale along with selected Qp (attenuation generally correlates with temperature; low values are highest temperatures) and interpreted images. Dots are earthquake locations. (A) P-wave tomographic image of Heit et al. (2008a). The solid line is the Moho on the Elger et al. (2005) section in Figure 4B; the dashed line is the receiver function Moho of Heit et al. (2008a). (B–D) Qp and Vp tomographic sections at 22.1°S, 23.1°S, and 23.9°S from Schurr et al. (2006) with the position of some of the major ignimbrite calderas from Figure 7 superimposed. The white lines outline the regions where the images are most robust. (E–G) Interpreted east-west and north-south sections through the northern Puna region modified from Asch et al. (2006). Features on the sections are discussed in the text. ALVZ—Altiplano low-velocity zone; CVZ—Central Volcanic Zone; QBBs—Quebrada Blanca Bright spot.

et al. (2001), Riller and Oncken (2003), and Monaldi et al. (2008), among others. Coutand et al. (2001) argued for shortening by the late Eocene in the Puna and by the late Eocene to early Oligocene in the Cordillera Oriental. Hongn et al. (2007) provided clear paleontological evidence for middle Eocene deformation on the Puna–Cordillera Oriental transition along the Cretaceous-Paleogene rift basin border. Important middle to late Miocene shortening in the plateau, which is considered to have coincided with the structural isolation of intermontane basins, seems to have terminated by ca. 10–9 Ma (e.g., Gubbels et al., 1993). As in southern Bolivia, early to middle Miocene strata can be up to 2000 m thick, whereas late Miocene to Pliocene-Pleistocene sequences are thin or absent. Balanced cross sections suggest a minimum of 40–50 km of total shortening across the region (Cladouhos et al., 1994; Coutand et al., 2001). The main deformation is associated with NW-SE to E-W shortening and subvertical stretching. Irregular eastward propagation and mixed fault vergence reflect control by preexisting Mesozoic structures. Faulting in some parts of the Cordillera Oriental is young, as shown by seismic studies (Cahill et al., 1992) and thrusts that cut Quaternary strata near 24°S (Marrett et al., 1994).

Riller and Oncken (2003) described three types of faults associated with Neogene deformation in the Puna and argued that their geometries are strongly controlled by the Paleozoic structural grain. The first are northwest-trending faults that can show sinistral displacement and have Neogene volcanic rocks localized along them. The second are north-striking faults that mostly acted as reverse faults in the Neogene. The third are northeast-trending faults with dextral motion that can have newly formed in the Neogene. These faults accommodated the maximum Neogene shortening. Riller and Oncken (2003) suggested that combined motion on north- and northeast-striking faults produced rhomb-shaped closed basins surrounded by transpressive ranges.

Further east, the structural style of the foreland belt is more variable than to the north, as the Subandean fold-and-thrust belt ends abruptly at 23°S where the Silurian shales that form the detachment levels disappear (see Kley and Monaldi, 2002). Echavarría et al. (2003) recognized two stages of thin-skinned thrusting in a transect through the Subandean belt near 22.3°S. The first is marked by in-sequence eastward propagating thrusting that began at 9–8.5 Ma (thrusts 1–3 in Fig. 8B), and the second initiated with out-of-sequence thrusting at ca. 4.5 Ma and then propagated eastward (thrusts 4–10 in Fig. 8B). A total shortening of 60 km or 36% has been estimated for the last million years at a deformation rate of 13 mm/yr or higher from 9 to 7 Ma and a slower rate from 7 to 2 Ma. An apparent deformation gap occurs south of where the Subandean belt ends at 23.5°S and where the northern Puna ignimbrite field also essentially ends (Fig. 2). The 400-km-long Santa Barbara system begins at 23.5°S and is predominantly characterized by west-verging, high-angle thrust faults, many of which are inverted normal faults from Cretaceous to Paleogene rifting (see Kley et al., 2005). Kley and Monaldi (2002) calculated 30–25 km of shortening at 24°S to 25°S on preexisting north- and northeast-striking dextral strike-slip faults that flatten into a detachment at ~10 km. The shortening at 24.2°S, where Mesozoic extension was less than 10 km, is 26–21 km.

Acocella et al. (2007) summarized field measurements to argue for strain partitioning in a transect near 24°S. They argued that generalized contractional regimes in the Puna and the Santa Barbara system with 15% and 25%–30% shortening, respectively, bound a horizontal shear regime in the Cordillera Oriental, where 24%–30% of two-dimensional (2-D) shortening needs to be augmented to account for ~62% dextral and 29% sinistral shear. Their general conclusion is in accord with three-dimensional (3-D) modeling by Hindle et al. (2005), who argued that shortening percentages based on 2-D cross sections underestimate total shortening in the northern Puna region. Barke et al. (2007) used paleomagnetic data to show that rotation of the Cordillera Oriental relative to stable South America since 15 Ma reaches a maximum (~13.5° clockwise) at 22°S to 23°S and decreases to zero by ~18°S and 23°S. As large-scale Cordillera Oriental faults capable of absorbing this rotation became inactive at 9 Ma, rotation either ended or was absorbed on small-scale faults. A correlation between the relative magnitude of rotation and Subandean shortening, which is 86 km near 18°S, ~47 km at 22°S, and ~33 km at 23°S, is consistent with continuing rotation.

Geophysical Framework of the Northern Puna

Summaries of the geophysical characteristics of the southernmost Altiplano and northern Puna from 21°S to 24°S based on reflection seismology, receiver functions, and traveltime tomography are given by Yuan et al. (2002), ANCORP Working Group (2003), Asch et al. (2006), Schurr et al. (2006), and Heit et al. (2008a). An interpreted P-wave tomographic image at 21°S from Heit et al. (2008a) is presented in Figure 9A, P-wave tomographic and P-wave attenuation (Qp) images from Schurr et al. (2006) at 22.1°S, 23.1°S, and 23.9°S are shown in Figures 9B to 9D, and interpreted images from Asch et al. (2006) are given in Figures 9E to 9G. Lower seismic velocities and higher attenuation in the crust and mantle under the northern Puna than under the Altiplano are interpreted to indicate a hotter crust and a thinner mantle lithosphere under the Puna (e.g., Whitman et al., 1992).

Starting in the west, the Chilean Coastal Cordillera and Longitudinal Valley are characterized by high seismic velocities to a depth of 60 km (Fig. 9). The high velocities are attributed to dense cumulates and mafic rocks associated with Jurassic magmas that were emplaced in Precambrian to Paleozoic crust all along the margin. A change to a lower-velocity crust and lithosphere under the active Central Volcanic Zone arc is attributed to melts, fluids, and intrusives above the subducting slab. The transition to the forearc is sharpest near 22°S to 23°S, where the lowest Qp values (Figs. 9B and 9C) indicate the highest temperature under the arc, and where the largest Pliocene and younger ignimbrites occur (Puripicar and Tara at 22°S; Purico and La Pacana at 23°S; Fig. 7C). A crustal high-velocity zone east of the Central Volcanic Zone at 21°S (Fig. 9A) was attributed to mafic intrusives and pre-Mesozoic crust by Heit et al. (2008a).

The distinctive reflective zone in the crust on the ANCORP Working Group (2003) profile between the West Fissure fault and the Central Volcanic Zone arc at ~21°S, known as the Quebrada Blanca Bright spot, appears as the low-velocity anomaly labeled QBBs on Figure 9A. Koulakov et al. (2006) attributed the earthquake nest in the slab below the bright spot to mantle hydration and the passage and breakup of a lithospheric block removed by forearc subduction-erosion. They attributed the bright spot to fluids and crustal melts associated with the removed block. Yoon et al. (2008) associated the bright spot with a complex network of ascending fluids or partial melts initiated by ascending fluids released from the subducted plate, but this does not explain why bright spots are not seen everywhere along the margin, unless there is a limit to the seismic resolution at the other latitudes.

Continuing east, the northern Puna and western Cordillera Oriental crust is characterized by a low-velocity anomaly with an irregular lower boundary at ~20 kms that weakens downwards. This is the Altiplano low-velocity zone (ALVZ) identified by Yuan et al. (2000) with receiver functions. On the ANCORP Working Group (2003) profile near 21°S, this zone lies just above a package of reflectors near 20 km depth that stretches from the arc front into the eastern Altiplano. The seismic character of the crust has led to the interpretation that much of the region under the northern Altiplano-Puna volcanic complex (Fig. 2) is underlain by a midcrustal magma chamber that extends from at least 21.5°S to 23.5°S and 68.5°W to 66°W (Chmielowski et al., 1999; Zandt et al., 2003). Leidig and Zandt (2003) interpreted the Altiplano low-velocity zone either as a magma body at a depth of 18–17 km overlain by anisotropic crust or as a magma body at a depth of 16–14 km overlain by a pluton or partial melt zone that reaches up to 10 km below the surface.

In reality, the term magma chamber is a misnomer because the low-velocity region is most likely a zone of partial melt. The receiver functions defining the top and bottom of the Altiplano low-velocity zone were interpreted by Yuan et al. (2000) as the upper limit of fluids and melts associated with the brittle-ductile transition, and the upper boundary of dry, hot refractory crust that has lost melts and volatiles. The presence of a distributed melt phase with possibly a thin layer of saline fluid at the top is supported by high heat flow (>>100 mW/m^2) and magnetotelluric and magnetic data in a profile near 22°S (Schilling et al., 2006). As argued by these authors, a melt fraction near 20% matches that expected in migmatites, conserves the interconnectivity needed to produce the conductivity anomalies, and leaves a sufficiently solid rock framework for strongly attenuated S-waves to propagate.

The low-velocity anomalies in the crust are not uniform in intensity. Two low-velocity anomalies are seen under the Cordillera Oriental at 21°S (Fig. 9A). The western one near 67°W is along strike with the anomaly under the Los Frailes ignimbrite near 20°S (Fig. 5) and is interpreted as a zone of partial melt (see Heit et al., 2008a). A more intense anomaly to the east extends into the lower crust. Superimposing the structural cross section from Figure 4B onto Figure 9A shows that the generally undeformed western Altiplano is located over the western anomaly and that the dashed portions of the Interandean and Subandean thrusts, which according to Elger et al. (2005) fade into the lower crust, project into the eastern anomaly. The eastern anomaly is best interpreted as generally felsic crust containing some partial melt. Further east, the thrusts are over a higher-velocity region as expected for colder, denser crust overridden by the Interandean and Subandean thrust in the uplift of the eastern Cordillera Oriental and Subandean belt. The solid line above the dashed S-wave receiver function Moho from Heit et al. (2008a) is the Moho from Figure 4B, which is more in line with that from other studies (ANCORP Working Group, 2003; Yuan et al., 2002; McGlashan et al., 2008).

The strongest low-velocity anomalies in the crust between 21°S and 23.5°S (Figs. 9B–9D) are generally below the large, late Miocene ignimbrite calderas (Fig. 7). The correlation is most obvious on the profile at 23.1°S, where the strong anomalies coincide with the 4.2–3.8 Ma La Pacana and 6.8–6.2 Ma Coranzulí calderas, as well as the smaller 0.5–1.4 Ma Purico complex. The anomaly on the 23.9°S section is in the general area of the Miocene Aguas Calientes caldera system and the young Tuzgle Volcano (Fig. 7C).

The character of the mantle under the southernmost Altiplano and northern Puna is variable. On the profile at 21°S (Fig. 9A), a prominent low-velocity anomaly east of 66°S was attributed by Heit et al. (2008a) to magma accumulating at or below the Moho. The low-velocity anomaly at 60–70 km at 23.1°S (Fig. 9C) could have a similar explanation. A less well-constrained anomaly at ~100–190 km depth near 65°W in Figure 9A generally occurs where receiver functions in Heit et al. (2007) show the lithosphere thinning abruptly. Heit et al. (2008a) associated this thinning with asthenospheric mantle replacing delaminated lithosphere. A striking feature on the section at 23.1°S is a prominent high-velocity-Qp anomaly at 66°W at a depth of 100–150 km that is ~100 km across (Fig. 9C). This region, the seismic characteristics of which are consistent with it being relatively dense, cold, and dry, has been interpreted as detached lithosphere (Fig. 9F) by Schurr et al. (2006). Interestingly, the size is again similar to the delaminated block shown under the southern Puna by Kay et al. (1994a). The block is along strike with the low-velocity anomalies at 21°S (Fig. 9A) and 20°S (Fig. 5) that are interpreted as asthenosphere replacing delaminated lithosphere. Schurr et al. (2006) pointed out that the block has a similar seismic signature to the underlying slab, consistent with eclogite in both. The westward dip fits with the block possibly being delaminated from beneath the La Pacana caldera region (Figs. 7 and 9C). The low-velocity region around the proposed delaminated block is interpreted to indicate decompression melting in the surrounding mantle wedge (Figs. 9C and 9F). The section at 23.9°S (Fig. 9D) shows a similar high-Vp and high-Qp anomaly at 130–200 km depth above the high-velocity anomaly and earthquake hypocenters marking the descending slab. Schurr et al. (2006) interpreted this anomaly as a delaminated block sitting above the slab. The low-velocity, Qp anomaly from ~50–120 km above the block is interpreted as a zone of decompression melting that extends into the crust under the arc region. The north-south profile in Figure 9E shows the region south of 24°S as underlain by a complex low-velocity, Qp mantle anomaly. This anomaly, which also has high Vp/Vs, shows two branches in the east-west sections in Schurr et al. (2003) that are interpreted as zones of partial melt. The western branch dips eastward and extends from beneath the arc crust to ~130–150 km, where it joins with a more vertical eastern branch extending into the crust under Tuzgle volcano (Fig. 9E). Like Coira and Kay (1993) and Kay et al. (1994a), Schurr et al. (2006) associate volcanism at Tuzgle with a region of delaminated lithosphere and argued that nearby Quaternary shoshonites contain melts of disrupted mantle lithosphere.

Lithospheric-Scale Model for the Evolution of the Northern Puna

A proposed summary of the Neogene evolution of the northern Puna near 22°S–23°S based on the magmatic, structural, and geophysical data presented above is shown in Figure 6B. As in Figure 6A, the sections are drawn to scale where possible, and the slab profile is based on the modern slab geometry in Figure 1 at the latitude indicated. The model, which is modified from Kay et al. (1999) and Kay and Mpodozis (2001), shows a steepening subduction zone.

Period 1: 26–17 Ma

The model of Kay et al. (1999) showed a flat-slab segment under the northern Puna in this period to explain a virtual lull in volcanism at a time of contractional deformation and basin formation across the northern Puna backarc. The slab geometry in Figure 6B is equivalent to that above the amagmatic modern Chilean-Pampean flat-slab region near 30°S today. Schematic faults represent deformation occurring in the Cordillera Oriental and Puna. The base of the lithosphere is shown as being actively removed as proposed by Kay and Abbruzzi (1996) for the Chilean-Pampean flat slab. The crust is shown as thicker to the west in accord with Eocene-Oligocene deformation in that region (e.g., Mpodozis et al., 1995, 2005). The proposal of a thinner crust to the east is based on geochemical data indicating that garnet was not an important residual phase associated with dacitic magmas in that region until after 14 Ma (Kay et al., 1999; Caffe et al., 2002).

Period 2: 16–11 Ma

As in Kay et al. (1999), the slab is shown to be steepening by 16–14 Ma to account for the eruption of small domes, porphyritic stocks, and ignimbrites (see Caffe et al., 2002; Fig. 7C). The geometry of the slab is shown as that near 28°S today, since this is the shallowest slab supporting volcanism. The crust is thickened in the east in accord with geochemical evidence for garnet appearing as a residual phase to dacitic magmas after 14 Ma (Caffe et al., 2002; Petrinovic et al., 1999). Mafic melts formed by decompression melting in an enlarging mantle wedge are shown accumulating at the base of the crust and producing hybridized crustal melts that then segregate and accumulate at the brittle-ductile transition (Kay et al., 1999). The accumulation of melts near 20 km is supported by geophysical images of melt zones at this level (ANCORP Working Group, 2003; Yuan et al., 2002; Zandt et al., 2003). Initial steepening of the slab followed by small eruptions at 17–16 Ma and the first giant ignimbrites by 8.5 Ma is consistent with the model of Babeyko et al. (2002), in which peak lower- to midcrustal temperatures were achieved ~20–10 m.y. after the onset of increased basal heat heating. Clockwise rotation of the Cordillera Oriental after 15 Ma (Prezzi and Alonso, 2002; Barke et al. 2007) can explain dilation of northwest-trending faults facilitating eruptions along them (see Riller and Oncken, 2003).

Period 3: 10–6 Ma

Further steepening of the slab from 10 to 6 Ma is consistent with widespread and voluminous 10–6.2 Ma ignimbrites erupting across the plateau. The slab geometry shown in Figure 6B is that near 27°S today, in keeping with a depth of less than 150–160 km below the eastern ignimbrites. Rapid uplift of the Puna Plateau at the same time as the Altiplano Plateau (Garzione et al., 2006) is supported by the onset of extensive ignimbrite eruptions. Decompression mantle melting above a steepening slab provides a heat source to produce intermediate volume ignimbrites by 10 Ma and giant volume ignimbrites by 8.5 Ma (Soler et al., 2007), and to sustain large backarc eruptions until ca. 6 Ma (Fig. 7B). As discussed for the southern Altiplano, the erupted ignimbrites are considered to be hybridized crustal and mantle melts formed in the lower crust, which accumulated at midcrustal levels before erupting. A larger magnitude of slab steepening under the northern Puna region than under the Altiplano (cf. Figs. 6A and 6B) led to a larger volume of both mantle and crustal melts. Dense eclogitic residual crust generated in the melting process and by eclogitization of the lower crust above the former shallowly dipping dehydrating slab caused the delamination of basal eclogitic crust and underlying lithosphere. The size and shape of the delaminated piece are based on the seismic images of Schurr et al. (2006) (Fig. 9). Removal of the delaminated block would further enhance crustal melting by exposing silicic crust to partially molten mantle. The intensity of crustal melting can explain the scarcity of erupted mafic lavas and extensional and strike-slip faults used by Kay et al. (1994a) as evidence for delamination in the southern Puna.

As in the Altiplano, the large ignimbrite eruptions postdate contractional deformation on the plateau and coincide with the transfer of deformation into the thin-skinned Subandean zone. Kay et al. (1999) speculated that crustal melts accumulating at the brittle-ductile transition under the Puna could lead to periodic failure of the ductile crust and facilitate both extraction of ignimbrite melts and thrusting in the Subandean belt. A comparison between the times of major ignimbrite eruptions and motion on Subandean thrusts (Echavarría et al., 2003; see Fig. 8B) shows that the initiation of thrusting (1 in Fig. 8B) at 9–8.5 Ma coincided with time of the first large eruptions (Vilama at 8.5 Ma, Sifón at 8.3 Ma; Fig. 7B). Thrusts 2 and 3 in Figure 6B then propagated eastward in an in-sequence pattern with activity at 7.6 and 6.9 Ma, overlapping in time with the eruptions of the other backarc ignimbrites, which lasted until 6 Ma (Fig. 7C). The change to an out-of-sequence thrusting episode near 4.4 Ma coincided with the restriction of large ignimbrite eruptions to the arc region (Figs. 7C). The combination of contraction in the ductile lower backarc crust and lithospheric delamination can explain a westward retreat of Subandean thrusting as melting and ignimbrite eruption accelerated in the arc region.

A much discussed issue in the Puna region is an insufficient amount of observed crustal shortening in two dimensions to account for crustal thicknesses measured in seismic and gravimetric studies (Kley and Monaldi, 1998). Hindle et al. (2005) divided the Central Andes into structural blocks and rotated them based on plate convergence and GPS vectors to analyze displacements in three dimensions. They explained the resulting thickness excesses and deficits with a midcrustal flow model. Their model, which started with a uniform 40-km-thick crust at 25 Ma, requires 7–10 km of thickening by crustal flow under the northern Puna. However, Kay and Coira (2008) showed that Puna ignimbrite magmas can be explained as near 50:50 mixtures of crust and new mantle–derived magmas, resulting in an additional 5 km of crust averaged across the region given a 4 or

5:1 intrusive:extrusive ratio and an arc magma production rate of 72–90 km^3 km^{-1} yr^{-1} (see Jicha et al., 2006). Although delamination removes mafic crust and reduces the net crustal addition, the crustal flow required is less, and volume uncertainties allow net northward crustal flow as modeled by Gerbault et al. (2005). The concentration of the northern Puna ignimbrites at the latitude of maximum paleomagnetic rotation in the last 15 m.y. in the Cordillera Oriental (Barke et al., 2007) and the minimum in Subandean thrusting indicate that block rotation could facilitate Miocene magma accumulation at depth and its escape at the surface.

Periods 4 and 5: 5 Ma to Holocene

Further steepening of the slab is shown in the cartoon at 5–3 Ma with the resulting slab geometry similar to that at 26°S today. This geometry puts the La Pacana caldera ~130 km above the slab at the eruption time. Delamination under the arc region is depicted by a block like that in the interpreted seismic image in Figure 9C. Older blocks further east are shown above the slab as in the image in Figure 9D. Mantle-derived magmas accumulating in the lower crust produced hybrid melts containing crustal melts that reflect generation in a region with a long arc magmatic history. This melting culminated in the giant ignimbrite episode at 4.2–3.8 Ma (Puripicar, La Pacana, and Tara; Fig. 7C). The second stage of Subandean thrusting, which initiated the out-of-sequence thrusting at 4.5 Ma, overlaps the eruption of these ignimbrites. The period 5 cartoon shows the modern slab geometry, shortening in the Subandean belt, and andesitic and dacitic volcanism concentrated in the arc front. The lithospheric and crustal thicknesses and delaminated blocks are based on the interpreted seismic image in Figure 9C.

MAGMATIC, STRUCTURAL, AND GEOPHYSICAL FRAMEWORK OF THE SOUTHERN PUNA

The last of the three regions considered is the southern Puna region between 24°S to 28°S, which is separated from the northern Puna region by the NW-trending Olacapato–El Toro lineament (Figs. 2 and 7A). The southern Puna differs from the northern Puna in that the pattern of gradual shallowing of the subducting slab leading to the Chilean-Pampean flat slab begins south of 24°S, a gap in intermediate depth Wadati-Benioff seismicity occurs at 25°S to 27°S, normal and strike-slip faults (Marrett et al., 1994) are associated with younger than 7 Ma mafic volcanic rocks (Kay et al., 1994a, 1999; Risse et al., 2008), and smaller more discontinuous and diachronous basins occur (Vandervoort et al., 1995). Other southern Puna characteristics are a higher average topography and a thinner lithosphere, which is inferred from uppermost mantle shear wave attenuation, and a low effective elastic thickness of less than 10 km (Whitman et al., 1992, 1996; Tassara et al., 2006). Evidence for a thinner lithosphere near the Olacapato–El Toro lineament near 24°S comes from seismic tomography (Schurr et al., 2003, 2006; Fig. 9E) and shoshonitic lavas along faults (Coira and Kay, 1993; Kay et al., 1994a).

Magmatic differences between the northern and southern Puna were initially pointed out by Alonso et al. (1984) and later discussed by Coira et al. (1993) and Kay et al. (1999, 2008a). As in the northern Puna, the large southern Puna volcanic centers are concentrated on northwest-trending shear zones that mark zones of lithospheric weakness (e.g., Riller and Oncken, 2003). From northwest to southeast, the most prominent chain includes the modern Llullaillaco Volcano in the Central Volcanic Zone, the largely late Miocene to Pliocene backarc Antofalla, Beltran, and Tebenquicho Volcanoes, and the 6–4 Ma Cerro Galán caldera complex (Fig. 2). Post–7 Ma glassy andesitic to dacitic flows, which are characteristic of anhydrous and hot magmas that rise rapidly to the surface, are also a distinctive feature of the southern Puna (e.g., Kay et al., 1994a, 1999, 2008a).

The Cordillera Oriental to the east is bordered by the Santa Barbara system, which together give way to the thick-skinned foreland fragmented by the reverse-fault–bounded basement uplifts of the Sierras Pampeanas and intervening basins near 26°S (see Ramos et al., 2002; Figs. 1 and 2). The boundary between the Santa Barbara system and the Sierras Pampeanas coincides with the southern termination of the Cretaceous Salta rift system (see Kley and Monaldi, 2002).

Magmatic History of the Southern Puna

The main late Oligocene to Holocene magmatic features of the southern Puna region are shown in Figures 2 and 10. Summaries of the magmatic history in the southern Puna are given in Coira et al. (1993), Kay et al. (1994a, 1999), Trumbull et al. (2006), and Schnurr et al. (2007), and for the arc region in Chile, summaries are provided in Mpodozis et al. (1995, 1996), Kay et al. (1994b, 2008a), and Kay and Mpodozis (2002).

From 26 to 18 Ma, large rhyodacitic ignimbrite and dome complexes, which overlap the age of the Oxaya ignimbrites north of 22°S, were widespread in the frontal arc region in the Maricunga Belt in Chile (Fig. 10A; Mpodozis et al. 1995). In the backarc, the ca. 24 Ma Segerstrom mafic flows just east of the Maricunga arc (Kay et al., 1999) and small dacitic to rhyodacitic centers farther east (Coira and Pezzutti, 1976; Petrinovic et al., 1999) have ages that overlap those of backarc mafic flows and silicic centers in the Altiplano (see Hoke and Lamb, 2007). This picture changed at ca. 18–16 Ma as frontal arc volcanism slowed, and more widespread small stratovolcanoes and ignimbrites spread into the backarc (Fig. 10A).

In the middle Miocene, further changes occurred as large andesitic stratovolcano complexes erupted at the arc front, and volcanism initiated at long-lived mafic andesitic to dacitic backarc stratovolcanic/dome complexes like Beltran, Antofalla, and Tebenquicho (e.g., Coira et al., 1993; Kraemer et al., 1999; Richards et al., 2006; Fig. 10B). Volcanism was also occurring at the northern Puna transition near 24°S (Petrinovic et al., 1999; Fig. 7A). By the late Miocene, activity in the Maricunga Belt was largely reduced to the eruption of the Copiapó rhyodacitic ignimbrite-dome complex (Mpodozis et al., 1995; Fig. 10B). In the

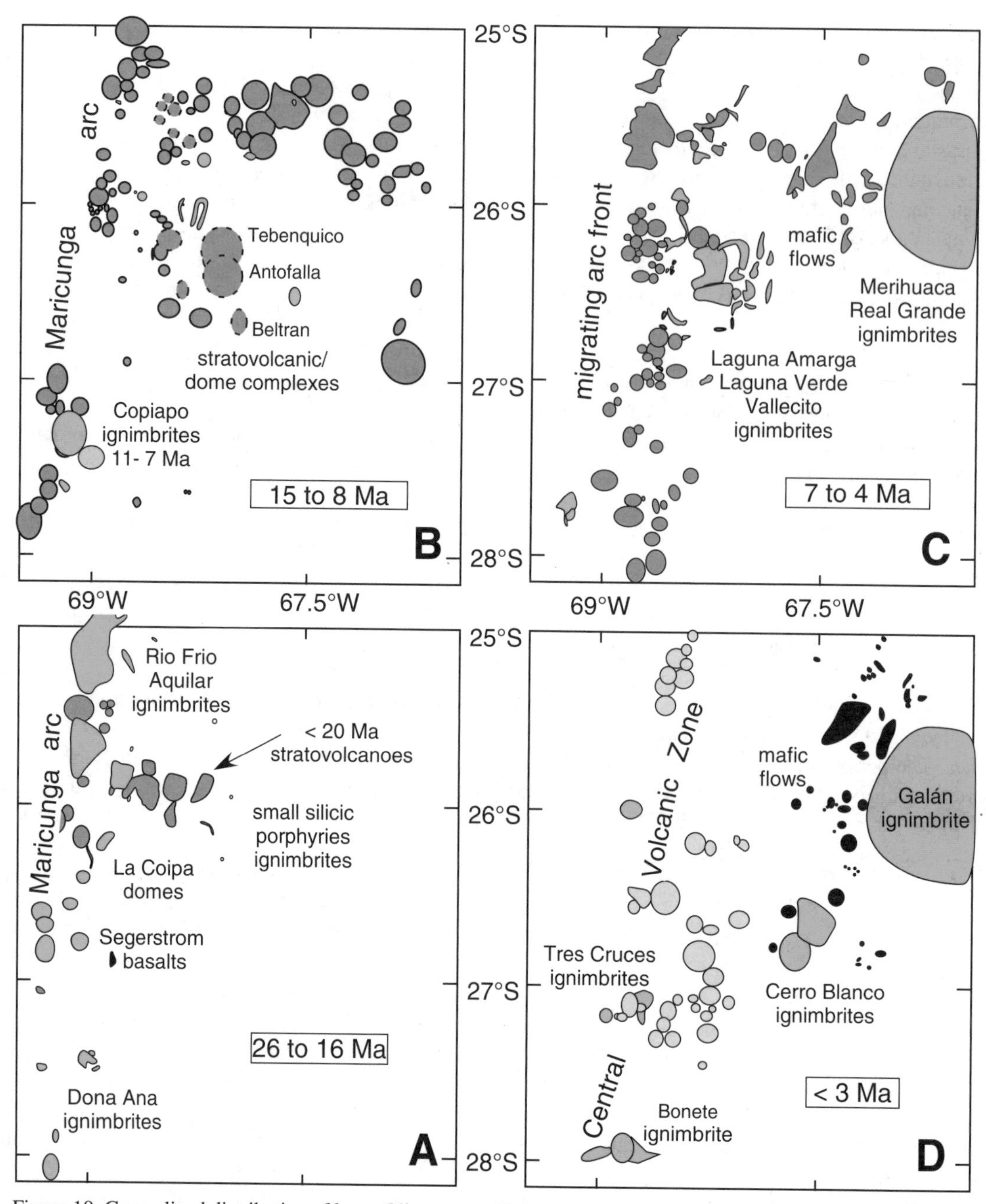

Figure 10. Generalized distribution of latest Oligocene to Holocene magmatic centers in the southern Puna region. Silicic domes and ignimbrites are shown in light red, the youngest mafic lavas are shown in black, and all other centers are in gray. Compare with Figure 2. Figure is modified from Kay et al. (2008a).

backarc, widespread 11–10 Ma ignimbrite eruptions preceded further andesitic to dacitic activity at the long-lived stratovolcanic complexes, and eruptions occurred in the eastern Cordillera Oriental and Sierras Pampeanas (Sasso and Clark, 1998). Middle to late Miocene backarc andesitic to dacitic volcanic rocks all have subduction-related chemical signatures.

In the latest Miocene to Pliocene (Fig. 10C), volcanism terminated at the Copiapó complex at 7–6 Ma as ca. 8–5 Ma andesites and dacites were erupting from the dying Maricunga arc to the future Central Volcanic Zone arc front (Mpodozis et al., 1996). Major changes in backarc volcanism also occurred after ca. 6.7 Ma as mafic lavas erupted from monogenetic and simple polygenetic centers emplaced along normal and strike-slip faults (Kay et al., 1994a; Marrett et al., 1994), the ca. 6 Ma Merihuaca and ca. 4 Ma Real Grande ignimbrites erupted from the Cerro Galán complex (both ~500 km^3; Sparks et al., 1985), and glassy andesites and dacites with strong arc-like chemical signatures that equilibrated with residual eclogitic mafic crust at depth erupted along faults (Kay et al., 1994a). To the west, voluminous ca. 4.8–3.8 Ma ignimbrites (Laguna Verde, Laguna Amarga, Vallecito) erupted in the Central Volcanic Zone region (Mpodozis et al., 1996; Siebel et al., 2001). In the easternmost Cordillera Oriental and Sierras Pampeanas (Fig. 2), K-rich andesitic to dacitic eruptions reached a peak at 8–7 Ma and ended at 5 Ma (Sasso and Clark, 1998).

By the latest Pliocene (Fig. 10D), the volcanic arc front was stabilizing in the modern Central Volcanic Zone (Mpodozis et al., 1996), and the giant 1000 km^3 Cerro Galán ignimbrite erupted at ca. 2.0 Ma (Sparks et al., 1985). Elsewhere, intraplate-like mafic lavas erupted near Cerro Galán, arc-like mafic magmas erupted to the north and south (Kay et al., 1994a; Risse et al., 2008), and the Pleistocene silicic ignimbrites/domes of the Tres Cruces complex and Cerro Blanco and Incapillo calderas erupted in the south (Siebel et al., 2001; Schnurr et al., 2007; Kay et al., 2008a). In contrast to the strong arc tendencies of the late Miocene and early Pliocene lavas, the younger than 3 Ma mafic lavas have both intraplate and arc signatures and, as in the Altiplano, reflect major melting at mantle depths of less than 100 km.

Deformation and Uplift of the Southern Puna

The history of deformation and uplift of the southern Puna region has been discussed by Vandervoort et al. (1995), Kraemer et al. (1999), Carrapa et al. (2005, 2006), Deeken et al. (2006), Coutand et al. (2006), and Strecker et al. (2007). East- to southeast-trending cross sections in Figure 11 from the Complejo Oire in the Puna to Cerro Durazno in the Cordillera Oriental (location in Fig. 2) modified from Deeken et al. (2006) summarize the history of uplift. Deformation began in the southern Puna and Cordillera Oriental in the Eocene to Oligocene (Coutand et al., 2001, 2006; Carrapa et al., 2005; Hongn et al., 2007), and the most recent exhumation of the region began in the early Miocene and reached the easternmost Cordillera Oriental by 14–7 Ma. Exhumation was largely over on the plateau by the middle Miocene, and major uplift in the late Miocene led to drainage reorganization east of the plateau margin (Carrapa et al., 2006). The major uplift of the Sierras Pampeanas to the east had begun by 6 Ma (see Ramos et al., 2002; Mortimer et al., 2007).

In detail, pre–late Miocene deformation in the Puna region progressed from west to east, leading to the diachronous formation of internally drained basins depending on local uplift. Kraemer et al. (1999) showed that foreland basin sedimentation at 27°S to 25°S started with the Quiñoas Formation, which contains 37.6 ± 0.3 Ma tuffs. Carrapa et al. (2005) argued that a Quiñoas source component could have come from local highlands to the east (Coutand et al., 2001). Local high-relief and a 37–35 Ma magmatic source are also needed to explain the accumulation of the Geste Formation near 25.5°S (DeCelles et al., 2007). Depositional similarities of Quiñoas and Altiplano sediments (Horton et al., 2002) led Carrapa et al. (2005) to suggest that a semicontinuous

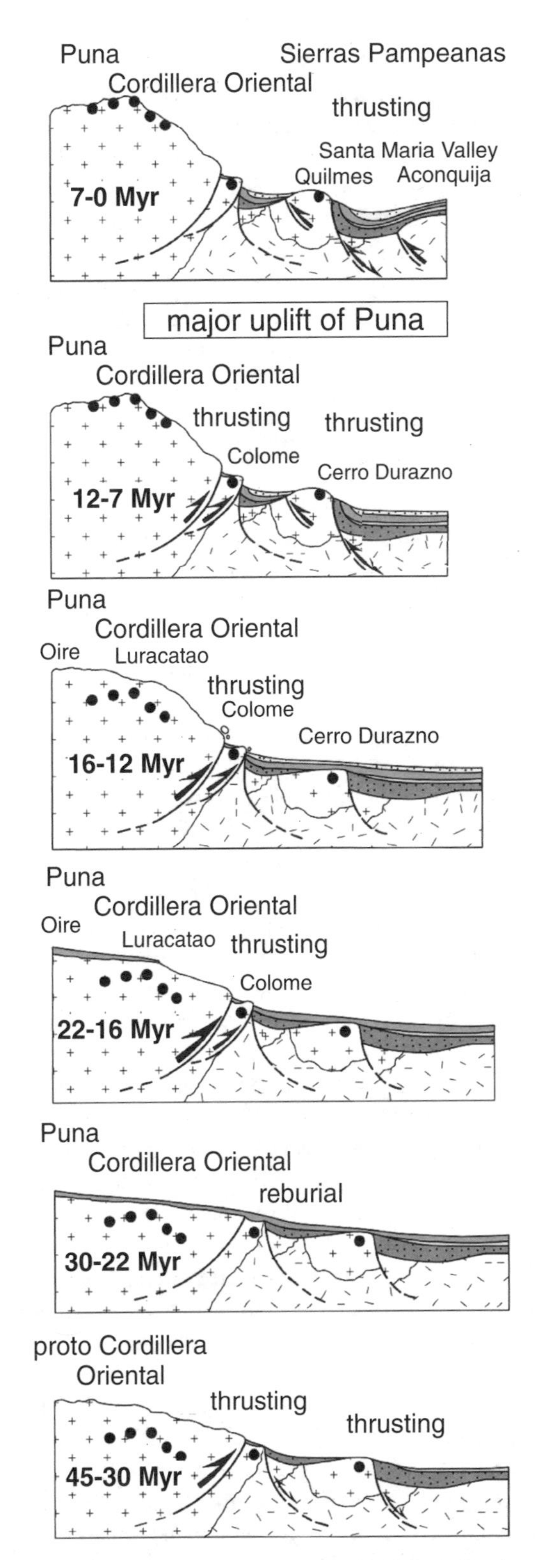

Figure 11. Generalized cross sections modified from Deeken et al. (2006) summarizing the structural and uplift history of the easternmost Puna, Cordillera Oriental, and Sierras Pampeanas in a northwest-southeast transect from ~25°S to 26°S. See Figure 2 for location of transect. Dark circles indicate change in elevation of samples from fission-track study of Deeken et al. (2006). Arrows indicate active faults. Dark layers are post–30 Ma sedimentary deposits. Thrusts are those shown in profiles in Figure 6C for the southern Puna.

Eocene foreland basin extended all along the plateau. Kraemer et al. (1999) showed that reverse faults were active in the southern Puna by the late Oligocene, and Carrapa et al. (2005) demonstrated that the Calalaste Range was a source for Puna sedimentary sequences containing tuffs with ages of 28.9 ± 0.8 Ma and 24.2 ± 0.9 Ma. Carrapa et al. (2006) argued for Oligocene exhumation of the southernmost Puna near 27°S. Vandervoort et al. (1995) and Coutand et al. (2001) proposed that internal drainage developed in the Puna at 24°S to 26°S between 24.2 and 15 Ma.

Further east, Deeken et al. (2006) suggested accelerated exhumation at ca. 40–35 Ma in the Cordillera Oriental related to shortening on new west-dipping reverse faults in the west and reactivated east-dipping faults in the east (Fig. 11). Evidence for late Eocene to Oligocene exhumation and deformation also comes from the Chango Real at 38–29 Ma (Coutand et al., 2001), Santa Rosa de Tastil at ca. 30 Ma (Andriessen and Reutter, 1994), and Angastaco Basin sediments near 26°S (Coutand et al., 2006). Deeken et al. (2006) argued that varying degrees of reburial of Eocene to Oligocene structures across the Cordillera Oriental in a generally eastward-propagating foreland basin system began at 30–25 Ma (Fig. 11).

According to Deeken et al. (2006), the modern exhumation of the southern Cordillera Oriental had begun by 22.5–21 Ma as the plateau margin propagated eastward. Exhumation began on west-dipping thrusts at 22.5–21 Ma at Luracatao and at 23–19 Ma at Colome and was uplifting Cerro Durazno on a reactivated east-dipping fault by 14–7 Ma (Fig. 11). By 16–15 Ma, sediments derived from the Luracatao and Cachi Ranges were reaching foreland basins to the east, whereas sediments from the Puna were absent (Coutand et al., 2006). Range crests were at 2000 m by 14 Ma (Deeken et al., 2006), and the deformation front had reached the Angastaco Basin by 13–10 Ma (Coutand et al., 2006). Between 9 and 5 Ma, the Cordillera Oriental became a barrier to moisture in the Puna.

The Sierras Pampeanas began to uplift on both east- and west-dipping, high-angle reverse faults in the latest Miocene. Sobel and Strecker (2003) suggested that sediments covering the Cumbres Calchaquíes Range, which began to accumulate in the middle Miocene, were being actively exhumed after 6 Ma. Mortimer et al. (2007) argued that the ranges surrounding the El Cajón–Campo del Arenal Basin were uplifted in an out-of-sequence pattern starting at 6–5 Ma. First, the Chango Real in the west was uplifted on a west-dipping fault, then the Aconquija Range was uplifted in the east on an east-dipping fault, and, finally, the Sierra de Quilmes was uplifted on a west-dipping fault with deformation progressing to the south until 3.4 Ma. Uplifts further east like the Cerro de los Colorados created a moisture barrier into the Angastaco Basin by 3.4–2.4 Ma (see Coutand et al., 2006).

Geophysical Framework of the Southern Puna Region

Geophysical data are still limited and preliminary in the southern Puna region. Gravity data suggest that the thinnest crust under the plateau occurs in this region (Tassara et al., 2006), in accord with crustal thicknesses of 42–49 km near 24°S (Yuan et al., 2002; McGlashan et al., 2008). Gravity data also support a thin lithosphere near 25°S (Tassara et al., 2006), consistent with high shear wave (Sn) attenuation (Whitman et al., 1992). A passive seismic profile from 69°W to 66°W near 25.5°S in Heit (2005) indicates crustal thicknesses up to 60 km; no data are yet available from an extensive passive seismic array installed in 2008 between 25°S and 28°S (Kay et al., 2008b).

A preliminary view of the Puna near 25.5°S comes from a combined tomographic and receiver function image by Heit (2005). The most striking feature on the P-wave tomographic image is a strong low-velocity anomaly above the slab that originates at a depth greater than 200 km and extends upward, connecting with two low-velocity crustal anomalies just below the Moho. The nature of this anomaly is consistent with a hot mantle and thin lithosphere above a recently delaminated portion of southern Puna lithosphere. The upper- to midcrustal low-velocity anomalies, which are strongest near 20–25 km depth, are interpreted as zones of partial melt. The weaker anomaly near 68°S is under the Antofalla Volcano; the stronger at 67°W to 66°W is north of the Cerro Galán caldera (Fig. 2). The intervening higher-velocity crust beneath the Salar de Antofalla coincides with the north-south–trending Central Andean gravity high isostatic residual (Götze and Krause, 2002; Götze and Kirchner, 1997). Receiver functions that image boundaries that dip west on the eastern sides of the Antofalla anomaly and the Salar de Antofalla and east on the western side of the Cerro Galan anomaly could be related to major north-south–trending, range-bounding, high-angle reverse faults (see map in Riller and Oncken, 2003). A sharp change to a higher-velocity crust east of Cerro Galan near 66°W marks the western limit of the Cordillera Oriental.

Lithospheric-Scale Model for the Evolution of the Southern Puna Region

A summary of the Neogene evolution of the southern Puna region near 26°S based on the magmatic, structural, and geophysical data presented above is shown in Figure 6C. Again, the sections are drawn to scale where possible, and the slab geometry is based on modern slab profiles from Figure 1. The model, which expands on that of Kay et al. (1994a, 2008a), shows moderate shallowing of the slab followed by resteepening. The oldest sections have 50 km added to the west of the modern trench to compensate for forearc subduction-erosion inferred from frontal arc migration at ca. 7–4 Ma (Kay and Mpodozis, 2002).

Period 1: 26–17 Ma

The concentration of andesitic/dacitic volcanism in the arc region, small-volume mafic lavas just behind the arc, and small dacitic domes further east are consistent with a slab geometry similar to that today. The lithospheric thickness under the Puna Plateau is put near 120 km based on the S-wave receiver function analyses of Heit et al. (2008b) for the eastern Sierras Pampeanas at 30°S. The arc and near-arc crust are put at 40–45 km based on lack of evidence for residual garnet in dacitic lavas (Kay et al., 1994b). The voluminous 26–18 Ma dacitic-rhyolitic domes and ignimbrites in the arc and ca. 24 Ma Segerstrom mafic lavas to the east indicate a deformational environment with local extension. A contractional setting is needed by 17–16 Ma in the arc region to explain deformation east of the Maricunga Belt (e.g., Mpodozis and Clavero, 2002) and by ca. 22 Ma in the backarc to explain thrusting in the Puna (Kraemer et al., 1999) and the Cordillera Oriental (Deeken et al., 2006; Fig. 11).

Period 2: 16–11 Ma

Shallowing of the subducting slab in this period explains the eastward broadening of arc magmatism exemplified by the initiation of the large, long-lived Puna stratovolcanoes such as Antofalla, Beltrán, and Tebenquicho at ca. 14 Ma (Coira and Pezzutti, 1976; Kraemer et al., 1999; Richards et al., 2006) and volcanism in the Farallón Negro region at ca. 12 Ma (Sasso and Clark, 1998). The arc-like chemical signature in the backarc lavas strongly supports a role for an underlying slab. The slab geometry is drawn as that near 28°S today, consistent with the Farallón Negro volcanic center being 200 km above the slab at ~600 km east of the trench (see Kay and Mpodozis, 2002). The lithosphere is shown as thinning beneath the expanding arc under the Puna.

Period 3: 10–7 Ma

The averaged slab dip is shown as the same as that at 11 Ma, since volcanism did not propagate further eastward, and a magmatic gap did not develop. A peak in the intensity of volcanism in the Sierras Pampeanas at 8–7 Ma (Sasso and Clark, 1998) is consistent with a shallow dip. The biggest magmatic change is at the Maricunga arc front, where the Copiapó ignimbrites erupted before the arc shut off at 7–6 Ma. A thick crust beneath the arc is indicated by evidence for residual garnet in equilibrium with the dacitic magmas (Kay et al., 1994b). In line with modern estimates (McGlashan et al., 2008), the crust is shown as thickening to 65 km. Crustal flow can explain why crustal shortening estimates are too low to accommodate this thickening (Hindle et al., 2005). The model of Yang et al. (2003) shows southward crustal flow into the region from ca. 20 to 10 Ma and northward flow after 10 Ma.

Period 4: 6–3 Ma

Major changes in magmatic and deformational style occurred across the arc and backarc after 7 Ma. The averaged slab geometry in Figure 6C is shown as that at 27°S today, consistent with moderate steepening. A 50 km eastward translation of the arc front from the Maricunga Belt to the modern Central Volcanic Zone region between 7 and 3 Ma is shown in line with a major pulse of forearc subduction-erosion (Kay and Mpodozis, 2002). The numerous volcanic centers and ca. 4 Ma ignimbrites, which occur between the Maricunga and the modern Central Volcanic Zone arc fronts, reflect this frontal arc instability. The material lost by subduction-erosion is shown being incorporated into the mantle wedge and the 50 km of forearc added to the east of the modern trench in the older sections is removed. Major changes in the Puna at this time are consistent with delamination of the lower crust and mantle lithosphere as suggested by Kay et al. (1994a, 1999). The block is shown with the dimensions in the seismic image of Schurr et al. (2006) in the northern Puna, which are close to those shown in Kay et al. (1994a). Evidence supporting delamination comes from a mixed stress regime with normal, strike-slip, and contractional faults (Marrett et al., 1994), eruption of mafic lavas along faults after 6.7 Ma, eruption of the ignimbrites from Cerro Galán (Sparks et al., 1985), major surface uplift in the latest Miocene (Carrapa et al., 2006), and geophysical evidence for a thin lithosphere and crust (Whitman et al., 1992, 1996; Tassara et al., 2006). As to the north, mantle magmas produced above a steepening slab ascended into the crust to generate hybrid melts that segregated and accumulated, producing the low-velocity crustal anomalies on seismic images (Heit, 2005). Delamination subjected the former midcrust to melting. As to the north, large ignimbrite eruptions occurred at times of changes in the foreland as the Cerro Galán ignimbrites started to erupt as the Sierras Pampeanas began to uplift at 6–5 Ma. Faulting in the easternmost Cordillera Oriental, the Sierras Pampeanas, and the Santa Barbara system coincided with the peak of forearc subduction-erosion.

Period 5: 3–0 Ma

By 3 Ma, the slab had steepened to acquire its modern geometry. The eruption of the ~1000 km^3 Cerro Galán ignimbrite at ca. 2.0 Ma resulted from the voluminous hybrid dacitic magmas that accumulated in the crust above the steepening slab. The combination of delamination and a thicker mantle wedge led to heating of the slab surface, producing the observed reduction in intermediate slab seismicity (Cahill and Isacks, 1992). As argued by Kay et al. (1994a), the thermal condition of the wedge restricted the hydrothermally induced arc signature where the lithosphere delaminated, resulting in intraplate-like mafic magmas near Cerro Galán. Seismic data in Whitman et al. (1992) and Heit (2005) suggest that the southern Puna is currently underlain by a thin lithosphere and a low-velocity mantle wedge.

CONTROLS ON MAGMATIC AND TECTONIC PROCESSES

Processes in common for all three transects in Figure 6 include relatively shallow and changing subduction zone dips, crustal shortening, delamination of thickened lower crust and lithosphere, crustal melting and eruption of giant ignimbrites, and lower-crustal flow. Control on where and when these processes occur comes from internal and external forces superimposed on the overlying South American and underlying lower Nazca plate. An understanding of the crustal and mantle processes requires a trench to forearc perspective.

Role of the Upper Plate

Babeyko et al. (2006) used numerical modeling to evaluate major locally influenced factors that control Andean deformation rates and concluded that the most important is the mechanical heterogeneity imposed by the distribution of the large sedimentary packages that control upper-crustal slip. Other locally influential factors evaluated in order of importance are eclogitization and removal of thickened lower crust, crustal melting and internal convection in felsic crust, and finally fluvial erosion. The presence of shale detachments in the thin-skinned Subandean belt and their absence in the thick-thinned foreland belts to the south have long been viewed as playing a potential role in differences between the Altiplano and Puna regions (e.g., Allmendinger and Gubbels, 1996). The concentration of giant Puna ignimbrites west of where shortening decreases to a minimum in the Subandean belt suggests a potential correlation between melting and lithosphere behavior as thin-skinned thrusting releases stress more efficiently than thick-skinned thrusting. An argument against this being a major control in overall along-strike differences comes from comparisons between the Altiplano and southern Puna, which have similar ignimbrite volumes but are flanked, respectively, by the most and the least shortened areas in the foreland belt.

Sobolev et al. (2006) used numerical modeling to evaluate external factors controlling Andean deformation rates and concluded that the velocity of the westward drift of South America is the most important factor, in accord with Silver et al. (1998). Marrett and Strecker (2000), Hindle et al. (2002), and Oncken et al. (2006) demonstrated the poor correlation of Puna-Altiplano deformation rates with Nazca–South American convergence rates (Pardo Casas and Molnar, 1987; Somoza 1998) and argued that the westward drift of South America is critical. Kay and Copeland (2006) made the same case for Oligocene to Miocene evolution of the Neuquén region of the south-central Andes. The graph in Figure 12 shows the rate of westward hinge motion (slab rollback) that Oncken et al. (2006) calculated from subtracting the estimated shortening and subduction-erosion rates at 21°S from the South American plate velocity using a sliding 5 m.y. average. The periods of acceleration, deceleration, and nearly constant rates of westward hinge motion emerge in the time frames used in constructing the lithospheric cross sections in Figure 6, as well as in the general periods long used to characterize events from Peru to south-central Chile (see Coira et al., 1993; Kay and Mpodozis, 2002). Specifically, a low rate of ~1.35 cm/yr at 26 Ma that accelerates to ~1.65 cm/yr by 18 Ma correlates with Altiplano and southern Puna bimodal arc and near-arc volcanism in a region with little contraction (Oxaya and Rio Frio–Dona Ana ignimbrite groups and mafic lavas). A higher relatively constant velocity of 1.65–1.7 cm/yr from ca. 17 to 11 Ma correlates with andesitic-dacitic volcanism in a contractional regime all along the margin and generally shallow subduction. A deceleration to 1.4 cm/yr by 7 Ma correlates with volcanism across the backarc and slab steepening by 7 Ma. A reacceleration to 1.6 cm/yr from ca. 8–7 to 3 Ma correlates with deformation east of the plateau and arc-front migration in a peak of subduction-erosion in the south. A nearly constant rate of 1.6 cm/yr from 3 to 0 Ma coincides with stabilization of the Central Volcanic Zone arc and less contraction along the margin. The correlation strongly supports a relation with slab rollback and western drift of South America, which may not be independent (see discussion in Oncken et al., 2006).

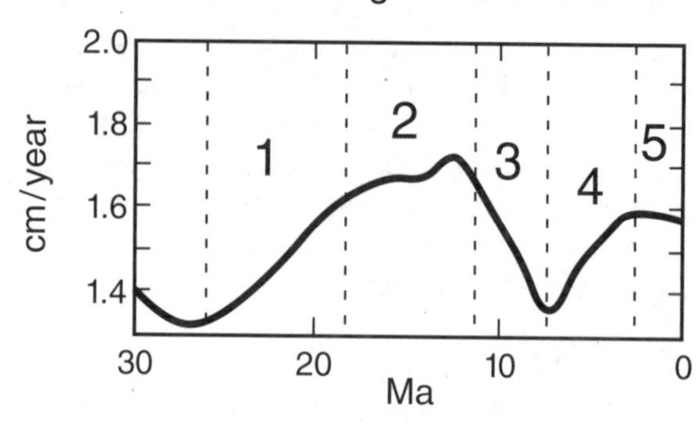

Figure 12. Rate of westward hinge motion or rollback at 21°S latitude in cm/yr as calculated by Oncken et al. (2006) by subtracting the deformation rate in the backarc plus the forearc subduction-erosion rate from the westward drift rate for South America on a sliding 5 m.y. scale compared to time periods shown on the sections. The deformation rate fluctuates from ~1.3 to 1.7 cm/yr (Oncken et al., 2006), the forearc subduction rate is not well known but is has little effect on the shape of the curve, and the westward drift rate generally increases from ~2.2 cm/yr at 25 Ma to 2.9 cm/yr by 3 Ma (Silver et al., 1998). See Oncken et al. (2006) for a discussion of the rates used in the calculation. The changes in slope of the curve generally correspond to the times of tectonic change in the Central Andes chosen independently in constructing the lithospheric sections in Figure 6. See discussion in text.

A second factor considered important by Sobolev et al. (2006) is shear coupling at the plate interface, which, from their numerical models, they suggested is controlled by delaminated material entering the mantle wedge. In their model, a large amount would impede corner-flow, decreasing arc magma productivity and increasing coupling and upper-plate deformation. In contrast, when corner-flow is not impeded, arc magma productivity would be high and deformation low. In contrast to this model as shown in Figure 6, delamination is accompanied by increased rates of both magmatism and deformation (e.g., northern Puna section at 10–6 Ma).

Role of the Lower Plate and the Juan Fernandez Ridge

Preexisting lithospheric differences, the westward drift of South America, and delamination-driven plate coupling are all upper-plate factors. Together, they do not explain many features in Figure 6, such as the magmatic gap in the northern Puna, differences in volcanic volumes in different sectors, and the pattern of slab steepening and shallowing. A missing variable is the lower-plate control on slab dip. The models of Sobolev et al. (2006) and Babeyko et al. (2006) used slab dips no shallower than that at 20°S–21°S today, yet Figure 6 shows this as the steepest Neogene slab dip at any time or place in the southern Altiplano and Puna. A long-standing suggestion has been that subducting ridges cause Andean flab slabs and that subduction of the aseismic Juan Fernandez Ridge on the Nazca plate is a factor in controlling the slab geometry under the Puna and Chilean-Pampean flat slab (Pilger, 1981, 1984). The influence of the Juan Fernandez Ridge on the evolution of the southern Altiplano and Puna regions can be evaluated by comparing the sections in Figure 6 with the model of Yañéz et al. (2001) for southward subduction of the ridge (Fig. 13). The latitudes next to the west to east profiles in each time step in Figure 13 show the modern latitude of the subducting slab used to make the sections in Figure 6. As the Nazca plate shallows southward from 20°S to 30°S (Fig. 1), higher numbers indicate shallower dips. The vertical line marks the arc-front region, circles indicate the major ignimbrite centers based on size, and the patterned line shows the younger than 10 Ma foreland contractional belts east of the plateau. As in the comparison of Neogene Chilean-Pampean flat-slab volcanic rocks with the position of the Juan Fernandez Ridge in Kay and Mpodozis (2002), the ridge width is shown as ~200 km. The real width would be variable and influence the extent to which the ridge controls the geometry of the subducting plate (see Yañéz et al., 2001).

The reconstructed Juan Fernandez Ridge segment in Figure 13 that subducted under the Altiplano-Puna had a northeast-trending arm that linked a north-south segment to the north with the nearly east-west segment that began to subduct under the Chilean-Pampean flat-slab region at ca. 10 Ma. A comparison with Figure 6 shows the northeast-trending arm arriving under the southern Altiplano arc at 23 Ma. The middle part of the arm was then under the Altiplano and northern Puna arc by 14 Ma, under the entire arc region by 12 Ma, leaving the Altiplano and northern Puna arc by 10–8 Ma, and had gone from the southern Puna arc by ca. 5 Ma. The sweep of the northeast-trending arm under the region is in accord with slab dips similar to those near 27°S–28°S today under the entire region from 14 to 10 Ma, and a broad magmatic arc with long-lasting activity that began earlier in the north and ended later in the south.

The pattern of ridge subduction is distinctive in the Altiplano where the north-south ridge arm was subducting until the northeast-trending arm began passing near 25 Ma (Fig. 13). The amagmatic flat-slab period of James and Sacks (1999) and a high deformation rate in the Altiplano (see Oncken et al., 2006) correspond with subduction of the north-south arm. The flat-slab stage passed with the arrival of the northeast-trending arm, but the slab dip remained shallow until the arm passed to the south at 12–10 Ma. The subduction of the north-south arm can account for a long period of flat-slab subduction, and the change at 25 Ma contributed to volcanism resuming across the region. A major ignimbrite flare-up did not immediately follow the flat-slab stage as the slab remained relatively flat. Using the deformation rates in Elger et al. (2005) and Oncken et al. (2006), the intensity of thrusting was highest in the Cordillera Oriental as the bend in the ridge was arriving and passing (ca. 28–18 Ma), and then it increased in the western and central Altiplano as the northeast arm passed under the region at ca. 18–10 Ma. The northeast arm was arriving at ca. 25 Ma to 15–8 Ma, when McQuarrie et al. (2005) argued for a dramatic slowdown in deformation rate. Thrusting was transferred into the Subandean belt as the ridge crest was leaving the backarc at ca. 10 Ma. After the passage of the northeast-trending arm, the slab under the southern Altiplano steepened as major uplift and the eruption of the giant Los Frailes ignimbrite occurred between 10 and 7 Ma.

Events in the Puna correspond with the passage of the northeast-trending arm (see Fig. 13). The amagmatic period in the northern Puna region shown with a flat-slab configuration in Figure 6B occurs as the ridge is still arriving from the north and ends as the ridge crust passes to the south at 14 Ma. The intermediate-volume ignimbrites erupt after the ridge crest passes, and the giant ignimbrite erupt after the ridge passes. A slightly later initiation of deformation in the southern Subandean belt is consistent with the latter passage of the ridge. The steepening and shallowing of the subduction zone under the southern Puna (Fig. 6C) generally correspond with the approach and passage of the ridge. The large Cerro Galán ignimbrites erupted after the passage of the ridge. The major uplift of the Sierras Pampeanas at 6–5 Ma corresponds with the passage of the crest and the arriving east-west arm.

CONCLUSIONS

An appraisal of the latest Oligocene to Neogene magmatic and deformational patterns and modern geophysical characteristics of the Altiplano-Puna Plateau south of the Bolivian orocline suggests that the dominant factors controlling the Neogene evolution are the rates of westward drift of the South American plate and of the western hinge (Fig. 12) as discussed by Oncken et al. (2006), and a relatively shallow, but changing geometry of the subducting Nazca plate. In the last 26 m.y., the slab under the southern Altiplano has shallowed from a geometry like that near 28°S today, and the slab under the northern Puna has shallowed from a flat-slab like that at 30°S today. In the same period, the slab under the southern Puna region had a geometry similar to today at 26–17 Ma, shallowed moderately from 16 to 7 Ma and then resteepened after 7 Ma. Across the region, the transfer of deformation from the plateau to the foreland belts occurred when the slab geometry was like that at 27°S today and the large to giant ignimbrites erupted on the plateau at times of steepening of the subduction zone. A comparison of the changes in slab geometries with the model of Yañéz et al. (2001) for the southward subduction of the aseismic Juan Fernandez Ridge on the Nazca plate suggests that a dominant factor in controlling slab dip, mantle wedge shape, magmatism, and transfer of deformation from the plateau to the foreland was the subduction of the Juan Fernandez Ridge. The Oligocene to middle Miocene deformation pattern in the Altiplano region can be reconciled with subduction of the north-south–trending arm followed by the northeast-trending arm of the Juan Fernandez Ridge. The pre–middle Miocene deformation cannot be related to the Juan Fernandez Ridge and requires another cause related to the rate of westward drift of South America or other factors.

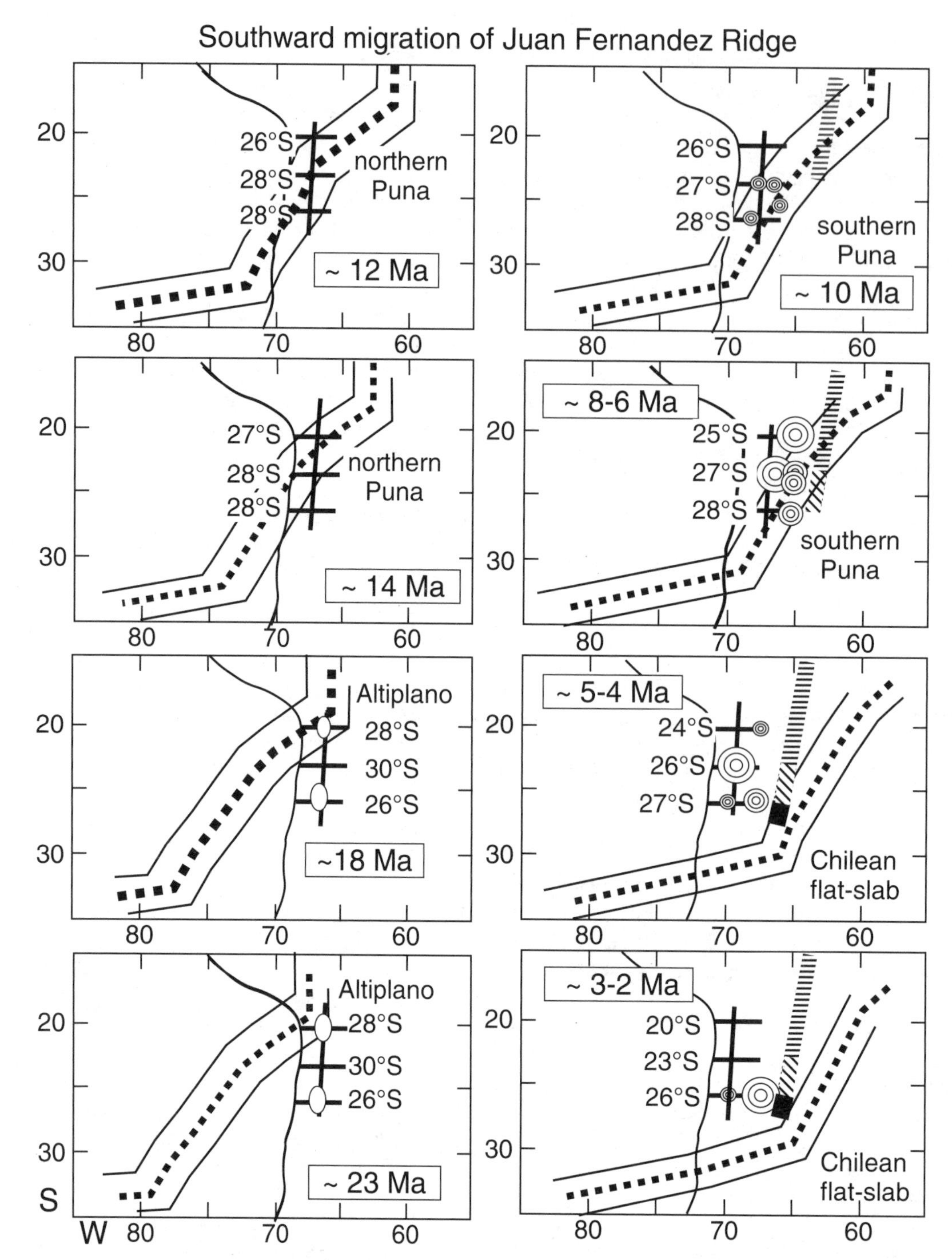

Figure 13. Southward migration of the Juan Fernandez Ridge from the model of Yañéz et al. (2001) compared to slab dips from Figure 6 for the Altiplano at 20°S, the northern Puna at 23°S, and the southern Puna at 26°S. The trace of the Juan Fernandez Ridge is the heavy dotted line bracketed by solid lines at ~100 km on each side. The other solid line is the continental margin used for reference to adjust for cumulative convergence by Yañéz et al. (2001). The coarser lines from the reference margin extend to the Los Frailes Ignimbrite in the Altiplano, the Coranzulí ignimbrite in the northern Puna, and the Cerro Galán ignimbrite in the southern Puna, respectively. The vertical crosscutting line shows the approximate position of the frontal arc. The numbers in degrees next to the line are the modern latitudinal equivalents of the slab profiles shown in Figure 6. The white ellipses show regions of important early Miocene ignimbrites. The concentric circles represent the intermediate to giant ignimbrites, which are scaled for size. The broad line shows the Subandean belt (horizontal striped part), the Santa Barbara zone (slanted line part), and the Sierras Pampeanas (black part). As discussed in the text, the slab dips, magmatic pattern, and uplift of the ranges to the east show a broad correspondence to the southward passage of the ridge.

A general observation is that andesitic volcanism generally expanded eastward over the shallowing slab as the ridge arrived and narrowed westward over the steepening slab after the ridge passed. Smaller ignimbrite eruptions can occur above the subducting ridge, but the largest eruptions occurred as the slab steepened after the ridge passed. The proposed temporal and spatial pattern of delamination of thick eclogitic crust and lithosphere as the slab steepened is consistent with eclogitization associated with dehydration of a relatively shallowly dipping slab and formation of hybrid melts as mafic magmas intrude the lower crust. The potential for delamination of eclogitic mafic crust is enhanced by slab steepening as decompression melting in the expanding mantle wedge lowers the mantle viscosity. Furthermore, a less viscous mantle wedge enhances slab steepening, according to Manea and Gurnis (2007), who showed that lower-mantle wedge viscosity causes slab dip to increase.

Late Miocene uplift of the southern Altiplano region can be attributed to lower-crustal flow, thickening in response to shortening, and a moderate amount of delamination, consistent with a broad flat topography, a moderate ignimbrite volume, and extensive shortening to the east. The Neogene evolution can be modeled by steepening of a shallowly subducting plate, leading to mantle and crustal melting that produces widespread volcanism and large local ignimbrites. Late Miocene mass redistribution by crustal flow explains the distribution of pre–late Miocene shortening and is consistent with less crustal and mantle melting over a more gradual steepening of the slab than in the northern Puna region. Late Miocene delamination under the Los Frailes ignimbrite complex is consistent with modern geophysical evidence that shows a very thin lithosphere under the region. The massive delamination event argued for by Garzione et al. (2006) and Molnar and Garzione (2007) to explain plateau uplift is difficult to reconcile with the deformational and magmatic history of the southern Altiplano region.

The presence of the huge ignimbrite volumes in the northern Puna region correlates with the greatest extent of slab steepening, since the largest amount of mafic magma is produced by decompression melting in the mantle, triggering the most crustal melting. The evolution of the region can be modeled by an early Miocene amagmatic flat-slab segment that underwent steepening after 16 Ma, leading to mantle melting that culminated in widespread ignimbrite eruptions beginning at 10 Ma, peaking at ca. 8.5–4 Ma, and ending by 3 Ma. Following the 3-D modeling of Hindle et al. (2005), uplift can be largely attributed to crustal thickening. Little crustal flow is needed as the local deficit in thickening can be largely balanced by magmatic addition. Dense eclogitic crust formed in the melting process and by eclogitization above a shallowly dipping dehydrating slab causes the most delamination to occur in this region.

More recent delamination has led to a hotter modern mantle wedge under the southern Puna region. The evolution of this region can be modeled by a moderately shallowly dipping slab that produced widespread backarc volcanism, followed by delamination of dense lithosphere above a steeper slab at 7–6 Ma, culminating in the voluminous Cerro Galán ignimbrite eruption at 2 Ma. Seismic images provide evidence for delamination near ~24°S and for very low mantle velocities above the slab near 25°S. A currently installed passive seismic network will soon provide a more detailed picture.

A remaining question is the temporal distribution of deformation throughout the plateau. Even though deformation began in the Eocene to Oligocene, the major deformation could be Miocene in the Puna, in contrast to the Altiplano, where the principal period of deformation is older.

ACKNOWLEDGMENTS

This paper results from many years of working in the Central Andes and the contributions of many people. We particularly thank Bryan Isacks, Robert Kay, Constantino Mpodozis, Víctor Ramos, Pablo Caffe, Richard Allmendinger, Teresa Jordan, Dean Whitman, Susan Beck, Jose Viramonte, Neil McGlashan, and Gregory Hoke for fruitful discussions on the uplift and magmatic and deformational processes in the Central Andes. This work has been most recently supported by grants from the Agencia Nacional de Promoción Científica y Tecnológica PICT 7-12420 and CONICET-PIP 2638 to Beatriz Coira and U.S. National Science Foundation grants 0087515 and 0538245. The Shuttle Radar Topography Mission image in Figure 2 was processed by Jay Hart of Ithaca, New York, for the Cornell Andes project. We thank C. Mpodozis, R. Trumbull, E. Sandvol, C. Andronicos, and A. Folguera for their reviews and constructive comments and suggestions on improving the paper.

REFERENCES CITED

Acocella, V., Vezzoli, L., Omarini, R., Matteini, M., and Mazzuoli, R., 2007, Kinematic variations across Eastern Cordillera at 24°S (Central Andes): Tectonic and magmatic implications: Tectonophysics, v. 434, p. 81–92, doi: 10.1016/j.tecto.2007.02.001.

Allmendinger, R.W., and Gubbels, T., 1996, Pure and simple shear plateau uplift, Altiplano-Puna, Argentina and Bolivia: Tectonophysics, v. 259, p. 1–13, doi: 10.1016/0040-1951(96)00024-8.

Allmendinger, R.W., Jordan, T.E., Kay, S.M., and Isacks, B.L., 1997, The evolution of the Altiplano-Puna Plateau of the Central Andes: Annual Review of Earth and Planetary Sciences, v. 25, p. 139–174, doi: 10.1146/annurev.earth.25.1.139.

Alonso, R., Viramonte, J., and Gutierrez, R., 1984, Puna Austral—Bases para el subprovincialismo geológico de la Puna Argentina, *in* IX Congreso Geológico Argentino: Actas, v. 1, p. 43–63.

ANCORP Working Group, 2003, Seismic imaging of a convergent continental margin and plateau in the central Andes (Andean Continental Research Project 1996 ANCORP'96): Journal of Geophysical Research, v. 108, p. 2328, doi: 10.1029/2002JB001771.

Andriessen, P., and Reutter, K., 1994, K-Ar and fission-track mineral age determination of igneous rocks related to multiple magmatic arc systems along the 23°S latitude of Chile and NW Argentina, *in* Reutter, K., Scheuber, E., and Wigger, P., eds., Tectonics of the Southern Central Andes: New York, Springer, p. 141–153.

Asch, G., Schurr, B., Bohm, M., Yuan, X., Haberland, C., Heit, B., Kind, R., Woelbern, I., Bataille, K., Comte, D., Pardo, M., Viramonte, J., Rietbrock, A., and Geisse, P., 2006, Seismological studies of the Central and Southern Andes, *in* Oncken, O., et al., eds., The Andes—Active Subduction Orogeny: Berlin, Springer-Verlag, Frontiers in Earth Sciences, v. 1, p. 439–451.

Babeyko, A.Y., Sobolev, S.V., Trumbull, R.B., Oncken, O., and Lavier, L.L., 2002, Numerical models of crustal scale convection and partial melting beneath the Altiplano-Puna Plateau: Earth and Planetary Science Letters, v. 199, p. 373–388, doi: 10.1016/S0012-821X(02)00597-6.

Babeyko, A.Y., Sobolev, S.V., Vietor, T., Oncken, O., and Trumbull, R.B., 2006, Weakening of the upper plate during tectonic shortening: Thermo-mechanical causes and consequences, *in* Oncken, O., et al., eds., The Andes—Active Subduction Orogeny: Berlin, Springer-Verlag, Frontiers in Earth Sciences, v. 1, p. 495–512.

Barke, R., and Lamb, S., 2006, Late Cenozoic uplift of the Eastern Cordillera, Bolivian Andes: Earth and Planetary Science Letters, v. 249, p. 350–367, doi: 10.1016/j.epsl.2006.07.012.

Barke, R., Lamb, S., and MacNiocaill, C., 2007, Late Cenozoic bending of the Bolivian Andes: New paleomagnetic and kinematic constraints: Journal of Geophysical Research, v. 112, p. B01101, doi: 10.1029/2006JB004372.

Barquero-Molino, M., 2003, ^{40}Ar/^{39}Ar Chronology and Paleomagnetism of Ignimbrites and Lavas from the Central Volcanic Zone, Northern Chile, and ^{40}Ar/^{39}Ar Chronology of Silicic Ignimbrites from Honduras and Nicaragua [M.S. thesis]: Madison, University of Wisconsin, 70 p.

Beck, S., and Zandt, G., 2002, The nature of orogenic crust in the central Andes: Journal of Geophysical Research, v. 107, no. B10, p. 2230, doi: 10.1029/2000JB000124.

Beck, S., Zandt, G., Myers, S., Wallace, T., Silver, P., and Drake, L., 1996, Crustal thickness variations in the central Andes: Geology, v. 24, p. 407–410.

Bevis, M., Kendrick, E., Smalley, R., Brooks, B., Allmendinger, R. and Isacks, B., 2001, On the strength of interplate coupling and the rate of back arc convergence in the central Andes: An analysis of the interseismic velocity field: Geochemistry, Geophysics, Geosystems, v. 2, 2001GC000198.

Caffe, P.J., Trumbull, R.B., Coira, B.L., and Romer, R.L., 2002, Petrogenesis of early Neogene magmatism in the northern Puna; implications for magma genesis and crustal processes in the Central Andean Plateau: Journal of Petrology, v. 43, p. 907–942, doi: 10.1093/petrology/43.5.907.

Caffe, P.J., Soler, M.M., Coira, B.L., Onoe, A.T., and Cordani, U.G., 2008, The Granada ignimbrite: A compound pyroclastic unit and its relationship with Upper Miocene caldera volcanism in the northern Puna: Journal of South American Earth Sciences, v. 25, p. 464–484, doi: 10.1016/j.jsames.2007.10.004.

Cahill, T.A., and Isacks, B.L., 1992, Seismicity and shape of the subducted Nazca plate: Journal of Geophysical Research, v. 97, p. 17,503–17,529, doi: 10.1029/92JB00493.

Cahill, T.A., Isacks, B.L., Whitman, D., Chatelain, J.-L., Perez, A., and Ming Chiu, J., 1992, Seismicity and tectonics in Jujuy province, northwestern Argentina: Tectonics, v. 11, p. 944–959, doi: 10.1029/92TC00215.

Carrapa, B., Adelmann, D., Hilley, G.E., Mortimer, C., Sobel, E.R., and Strecker, M.R., 2005, Oligocene range uplift and development of plateau morphology in the southern central Andes: Tectonics, v. 24, p. TC4011, doi: 10.1029/2004TC001762.

Carrapa, B., Strecker, M.R., and Sobel, E., 2006, Cenozoic orogenic growth in the Central Andes: Evidence from sedimentary rock provenance and apatite fission track thermochronology in the Fiambalá Basin, southernmost Puna Plateau margin (NW Argentina): Earth and Planetary Science Letters, v. 247, p. 82–100, doi: 10.1016/j.epsl.2006.04.010.

Chmielowski, J., Zandt, G., and Haberland, C., 1999, The Central Andean Altiplano-Puna magma body: Geophysical Research Letters, v. 26, p. 783–786, doi: 10.1029/1999GL900078.

Cladouhos, T.T., Allmendinger, R.W., Coira, B., and Farrar, E., 1994, Late Cenozoic deformation in the Central Andes: Fault kinematics from the northern Puna, northwestern Argentina and southwestern Bolivia: Journal of South American Earth Sciences, v. 7, p. 209–228, doi: 10.1016/0895-9811(94)90008-6.

Coira, B., and Kay, S.M., 1993, Implications of Quaternary volcanism at Cerro Tuzgle for crustal and mantle evolution of the Puna Plateau, Central Andes, Argentina: Contributions to Mineralogy and Petrology, v. 113, p. 40–58, doi: 10.1007/BF00320830.

Coira, B., and Pezzutti, N., 1976, Vulcanismo cenozoico en el ámbito de La Puna catamarqueña (25°30′–25°50′ Lat. S. y 68°–68°30′ Long. O.): Asociación Geológica Argentina Revista, v. 31, p. 33–52.

Coira, B., Kay, S.M., and Viramonte, J., 1993, Upper Cenozoic magmatic evolution of the Argentine Puna—A model for changing subduction geometry: International Geology Review, v. 35, p. 677–720.

Coutand, I., Cobbold, P.R., de Urreizieta, M., Gautier, P., Chauvin, A., Gapais, D., Rosello, E.A., and López-Gamundi, O., 2001, Style and history of Andean deformation, Puna plateau, northwestern Argentina: Tectonics, v. 20, p. 210–234, doi: 10.1029/2000TC900031.

Coutand, I., Carrapa, B., Deeken, A., Schmitt, A., Sobel, E., and Strecker, M., 2006, Propagation of orographic barriers along an active range front: Insights from sandstone petrography and detrital apatite fission track thermochronology in the intramontane Angastaco Basin, NW Argentina: Basin Research, v. 18, p. 1–26, doi: 10.1111/j.1365-2117.2006.00283.x.

Davidson, J.P., and de Silva, S.L., 1992, Volcanic rocks from the Bolivian Altiplano: Insights into crustal structure, contamination, and magma genesis in the Central Andes: Geology, v. 20, p. 1127–1130, doi: 10.1130/0091-7613(1992)020<1127:VRFTBA>2.3.CO;2.

DeCelles, P.G., and Horton, B.K., 2003, Early to middle Tertiary foreland basin development and the history of Andean crustal shortening in Bolivia: Geological Society of America Bulletin, v. 115, p. 58–77, doi: 10.1130/0016-7606(2003)115<0058:ETMTFB>2.0.CO;2.

DeCelles, P.G., Carrapa, B., and Gehrels, G., 2007, Detrital zircon U-Pb ages provide provenance and chronostratigraphic information from Eocene synorogenic deposits in northwestern Argentina: Geology, v. 35, p. 323–326, doi: 10.1130/G23322A.1.

Deeken, A., Sobel, E.R., Coutand, I., Haschke, M., Riller, U., and Strecker, M.R., 2006, Development of the southern Eastern Cordillera, NW Argentina, constrained by apatite fission track thermochronology: From Early Cretaceous extension to middle Miocene shortening: Tectonics, v. 25, p. TC6003, doi: 10.1029/2005TC001894.

de Silva, S.L., 1989, Altiplano-Puna volcanic complex of the central Andes: Geology, v. 17, p. 1102–1106, doi: 10.1130/0091-7613(1989)017<1102:APVCOT>2.3.CO;2.

de Silva, S.L., and Gosnold, W.D., 2007, Episodic construction of batholiths: Insights from the spatiotemporal development of an ignimbrite flare-up: Journal of Volcanology and Geothermal Research, v. 167, p. 320–325, doi: 10.1016/j.jvolgeores.2007.07.015.

Dufek, J., and Bergantz, G., 2005, Lower crustal magma genesis and preservation: A stochastic framework for the evaluation of basalt-crust interaction: Journal of Petrology, v. 46, p. 2167–2195, doi: 10.1093/petrology/egi049.

Echavarría, L., Hernandez, R., Allmendinger, R., and Reynolds, J., 2003, Subandean thrust and fold belt of northwestern Argentina: Geometry and timing of the Andean evolution: American Association of Petroleum Geologists Bulletin, v. 87, p. 965–985.

Ege, H., Sobel, E.R., Scheuber, E., and Jacobshagen, V., 2007, Exhumation history of the southern Altiplano Plateau (southern Bolivia) constrained by apatite fission track thermochronology: Tectonics, v. 26, p. TC1004, doi: 10.1029/2005TC001869.

Elger, K., Oncken, O., and Glodny, J., 2005, Plateau-style accumulation of deformation: Southern Altiplano: Tectonics, v. 24, p. TC4020, doi: 10.1029/2004TC001675.

García, M., Gardeweg, M., Hérail, G., and Pérez de Arce, C., 2000, La ignimbrita Oxaya y la Caldera Lauca: Un evento explosivo de gran volumen del Mioceno Inferior en la Región de Arica (Andes Centrales, 18–19°S), *in* IX Congreso Geologico Chileno: Actas, v. 2: Puerto Varas, Chile, p. 286–290.

García, M., Gardeweg, M., Clavero, J., and Hérail, G., 2004, Hoja Arica. Región de Tarapacá: Santiago, Servicio Nacional de Geología y Minería, Carta Geológica de Chile 84, scale 1:250,000.

Gardeweg, M., and Ramírez, C.F., 1987, La Pacana caldera and the Atana ignimbrite—A major ash-flow and resurgent caldera complex in the Andes of northern Chile: Bulletin of Volcanology, v. 49, p. 547–566, doi: 10.1007/BF01080449.

Garzione, C.N., Molnar, P., Libarkin, J.C., and MacFadden, B.C., 2006, Rapid late Miocene rise of the Bolivian Altiplano: Evidence for removal of mantle lithosphere: Earth and Planetary Science Letters, v. 241, p. 543–556, doi: 10.1016/j.epsl.2005.11.026.

Garzione, C.N., Hoke, G.D., Libarkin, J.C., Withers, S., MacFadden, B., Eiler, J., Ghosh, P., and Mulch, A., 2008, Rise of the Andes: Science, v. 320, p. 1304–1307, doi: 10.1126/science.1148615.

Gerbault, M., Martinod, J., and Herail, G., 2005, Possible orogeny-parallel lower crustal flow and thickening in the Central Andes: Tectonophysics, v. 399, p. 59–72, doi: 10.1016/j.tecto.2004.12.015.

Ghosh, P., Garzione, C.N., and Eiler, J.M., 2006, Rapid uplift of the Altiplano revealed in ^{13}C–^{18}O bonds in paleosol carbonates: Science, v. 311, p. 511–515, doi: 10.1126/science.1119365.

Götze, H.-J., and Kirchner, A., 1997, Interpretation of gravity and geoid in the Central Andes between 20° and 29°S: Journal of South American Earth Sciences, v. 10, p. 179–188, doi: 10.1016/S0895-9811(97)00014-X.

Götze, H.-J., and Krause, S., 2002, The Central Andean gravity high, a relic of an old subduction complex?: Journal of South American Earth Sciences, v. 14, p. 799–811, doi: 10.1016/S0895-9811(01)00077-3.

Gregory-Wodzicki, K.M., 2000, Uplift history of the Central and Northern Andes: A review: Geological Society of America Bulletin, v. 112, p. 1091–1105, doi: 10.1130/0016-7606(2000)112<1091:UHOTCA>2.3.CO;2.

Gubbels, T.L., Isacks, B.L., and Farrar, E., 1993, High level surfaces, plateau uplift, and foreland development, Central Bolivian Andes: Geology, v. 21, p. 695–698, doi: 10.1130/0091-7613(1993)021<0695:HLSPUA>2.3.CO;2.

Heit, B.S., 2005, Teleseismic Tomographic Images of the Central Andes at 21°S and 25.5°S: An Inside Look at the Altiplano and Puna Plateaus [Ph.D. thesis]: Berlin, Germany, Freien Universität, http://www.diss.fu-Berlin.de/2005/319/indexe.html.

Heit, B., Sodoudi, F., Yuan, X., Bianchi, M., and Kind, R., 2007, An S receiver function analysis of the lithospheric structure in South America: Geophysical Research Letters, v. 34, p. L14307, doi: 10.1029/2007GL030317.

Heit, B.S., Koulakov, I., Asch, G., Yuan, X., Kind, R., Alcocer-Rodriguez, I., Tawackoli, S., and Wilke, W., 2008a, More constraints to determine the seismic structure beneath the Central Andes at 21°S using teleseismic tomography analysis: Journal of South American Earth Sciences, v. 25, p. 22–36.

Heit, B., Yuan, X., Bianchi, M., Sodoudi, F., and Kind, R., 2008b, Crustal thickness estimation beneath the southern central Andes at 30°S and 36°S from S wave receiver function analysis: Geophysical Journal International, v. 174, p. 249–254, doi: 10.1111/j.1365-246X.2008.03780.x.

Hindle, D., Kley, J., Klosko, E., Stein, S., Dixon, T., and Norabuena, E., 2002, Consistency of geologic and geodetic displacements during Andean orogenesis: Geophysical Research Letters, v. 29, doi: 10.1029/2001GL013757.

Hindle, D., Kley, J., Oncken, O., and Sobolev, S.V., 2005, Crustal flux and crustal balance from shortening in the Central Andes: Earth and Planetary Science Letters, v. 230, p. 113–124, doi: 10.1016/j.epsl.2004.11.004.

Hoke, L., and Lamb, S., 2007, Cenozoic behind-arc volcanism in the Bolivian Andes, South America: Implications for mantle melt generation and lithospheric structure: Journal of the Geological Society of London, v. 164, p. 795–814, doi: 10.1144/0016-76492006-092.

Hongn, F., del Papa, C., Powell, J., Petrinovic, I., Mon, R., and Deraco, V., 2007, Middle Eocene deformation and sedimentation in the Puna–Eastern Cordillera transition (23°–26°S): Control by preexisting heterogeneities on the pattern of initial Andean shortening: Geology, v. 35, p. 271–274, doi: 10.1130/G23189A.1.

Horton, B.K., Hampton, B.A., LaReau, B.N., and Baldellón, E., 2002, Tertiary provenance history of the northern and central Altiplano (central Andes, Bolivia): A detrital record of plateau-margin tectonics: Journal of Sedimentary Research, v. 72, p. 711–726, doi: 10.1306/020702720711.

Husson, L., and Sempere, T., 2003, Thickening the Altiplano crust by gravity-driven crustal channel flow: Geophysical Research Letters, v. 30, doi: 10.1029/2002GL016877.

Isacks, B.L., 1988, Uplift of the Central Andean Plateau and bending of the Bolivian orocline: Journal of Geophysical Research, v. 93, p. 3211–3231, doi: 10.1029/JB093iB04p03211.

James, D.E., and Sacks, I.S., 1999, Cenozoic formation of the central Andes: A geophysical perspective, *in* Skinner, B., ed., Geology and Ore Deposits of the Central Andes: Society of Economic Geology Special Publication 7, p. 1–25.

Jicha, B.R., Scholl, D.W., Singer, B.S., Yogodzinski, G.M., and Kay, S.M., 2006, Revised age of Aleutian Island arc formation implies high rate of magma production: Geology, v. 34, p. 661–664, doi: 10.1130/G22433.1.

Kay, R.W., and Kay, S.M., 1993, Delamination and delamination magmatism: Tectonophysics, v. 219, p. 177–189, doi: 10.1016/0040-1951(93)90295-U.

Kay, S.M., and Abbruzzi, J.M., 1996, Magmatic evidence for Neogene lithospheric evolution of the Central Andean flat-slab between 30° and 32°S: Tectonophysics, v. 259, p. 15–28, doi: 10.1016/0040-1951(96)00032-7.

Kay, S.M., and Coira, B.L., 2008, Implications of chemical and isotopic variations in Neogene Puna Plateau ignimbrites for Central Andean crustal evolution: Geochimica et Cosmochimica Acta, v. 72, p. A456.

Kay, S.M., and Copeland, P., 2006, Early to middle Miocene backarc magmas of the Neuquén Basin: Geochemical consequences of slab shallowing and the westward drift of South America, *in* Kay, S.M., and Ramos, V.A., eds., Evolution of an Andean Margin: A Tectonic and Magmatic View from the Andes to the Neuquén Basin (35°–39°S Lat.): Geological Society of America Special Paper 407, p. 185–213, doi: 10.1130/2006.2407(9).

Kay, S.M., and Mpodozis, C., 2001, Central Andean ore deposits linked to evolving shallow subduction systems and thickening crust: GSA Today, v. 11, no. 3, p. 4–9, doi: 10.1130/1052-5173(2001)011<0004:CAODLT>2.0.CO;2.

Kay, S.M., and Mpodozis, C., 2002, Magmatism as a probe to the Neogene shallowing of the Nazca plate beneath the modern Chilean flat-slab: Journal of South American Earth Sciences, v. 15, p. 39–59, doi: 10.1016/S0895-9811(02)00005-6.

Kay, S.M., Coira, B., and Viramonte, J., 1994a, Young mafic back-arc volcanic rocks as indicators of continental lithospheric delamination beneath the Argentine Puna Plateau, Central Andes: Journal of Geophysical Research, v. 99, p. 24,323–24,339, doi: 10.1029/94JB00896.

Kay, S.M., Mpodozis, C., Tittler, A., and Cornejo, P., 1994b, Tertiary magmatic evolution of the Maricunga mineral belt in Chile: International Geology Review, v. 36, p. 1079–1112.

Kay, S.M., Mpodozis, C., and Coira, B., 1999, Magmatism, tectonism and mineral deposits of the Central Andes (22°–33°S latitude), *in* Skinner, B.J., ed., Geology and Ore Deposits of the Central Andes: Society of Economic Geology Special Publication 7, p. 27–59.

Kay, S.M., Coira, B.L., and Mpodozis, C., 2008a, Field trip guide to the Neogene to Recent evolution of the Puna Plateau and the southern Central Volcanic Zone, *in* Kay, S.M., and Ramos, V.A., eds., Field Trip Guides to the Backbone of the Americas in the Southern and Central Andes: Boulder, Colorado, Geological Society of America Field Guide 13, p. 117–181, doi: 10.1130/2008.0013(05).

Kay, S.M., Beck, S.L., Heit, B., and McGlashan, N., 2008b, Características geofísicas de la Puna de Jujuy y regiones adyacentes en la plateau de los Andes Centrales, *in* Coira, B., and Zappettini, E.O., eds., Geología y Recursos Naturales de Jujuy, Relatorío del XVII Congreso Geológico Argentino, Asociación Geológica Argentina, p. 385–396.

Kennan, L., 2000, Large-scale geomorphology of the Andes: Interrelationships of tectonics, magmatism and climate, *in* Summerfield, M.A., ed., Geomorphology and Global Tectonics: New York, Wiley, p. 165–199.

Kennan, L., Lamb, S.H., and Hoke, L., 1997, High-altitude palaeosurfaces in the Bolivian Andes: Evidence for late Cenozoic surface uplift, *in* Widdowson, M., ed., Palaeosurfaces: Recognition, Reconstruction and Palaeoenvironmental Interpretation: Geological Society of London Special Publication 120, p. 307–323.

Kley, J., 1996, Transition from basement-involved to thin-skinned thrusting in the Cordillera Oriental of southern Bolivia: Tectonics, v. 15, p. 763–775, doi: 10.1029/95TC03868.

Kley, J., and Monaldi, C.R., 1998, Tectonic shortening and crustal thickness in the Central Andes: How good is the correlation?: Geology, v. 26, p. 723–726, doi: 10.1130/0091-7613(1998)026<0723:TSACTI>2.3.CO;2.

Kley, J., and Monaldi, C.R., 2002, Tectonic inversion in the Santa Barbara system of the Central Andean foreland thrust belt, northwestern Argentina: Tectonics, v. 21, p. 1–18, 1061, doi: 10.1029/2002TC902003.

Kley, J., Rosello, E.A., Monaldi, C.R., and Habighorst, B., 2005, Seismic and field evidence for selective inversion of Cretaceous normal faults, Salta rift, northwest Argentina: Tectonophysics, v. 399, p. 155–172, doi: 10.1016/j.tecto.2004.12.020.

Koeppen, R.P., Smith, R.L., Kunk, M.L., Flores, A.M., Luedke, R.G., and Sutter, J.F., 1987, The Morococala volcanics: Highly peraluminous rhyolite ash flow magmatism in the Cordillera Oriental, Bolivia: Geological Society of America Abstracts with Programs, Annual Program, v. 19, p. 731.

Koulakov, I., Sobolev, S., and Asch, G., 2006, P- and S-velocity images of the lithosphere–asthenosphere system in the Central Andes from local source tomographic inversion: Geophysical Journal International, v. 167, p. 106–126, doi: 10.1111/j.1365-246X.2006.02949.x.

Kraemer, B., Adelmann, D., Alten, M., Schnurr, W., Erpenstein, K., Kiefer, E., van den Bogaard, P., and Gorler, K., 1999, Incorporation of the Paleogene foreland into the Neogene Puna Plateau: The Salar de Antofalla area, NW Argentina: Journal of South American Earth Sciences, v. 12, p. 157–182, doi: 10.1016/S0895-9811(99)00012-7.

Lamb, S., and Hoke, L., 1997, Origin of the high plateau in the Central Andes, Bolivia, South America: Tectonics, v. 16, p. 623–649, doi: 10.1029/97TC00495.

Lamb, S., Hoke, L., Kennan, L., and Dewey, J., 1997, Cenozoic evolution of the Central Andes in Bolivia and northern Chile, *in* Burg, J.P., and Ford, M., eds., Orogeny through Time: Geological Society of London Special Publication 121, p. 237–264.

Leidig, M., and Zandt, G., 2003, Highly anisotropic crust in the Altiplano-Puna volcanic complex of the central Andes: Journal of Geophysical Research, v. 108, doi: 10.1029/2001JB000649.

Lindsay, J.M., de Silva, S., Trumbull, R., Emmermann, R., and Wemmer, K., 2001, La Pacana caldera, N. Chile: A re-evaluation of the stratigraphy and volcanology of one of the world's largest resurgent calderas: Journal of Volcanology and Geothermal Research, v. 106, p. 145–173, doi: 10.1016/S0377-0273(00)00270-5.

Lucassen, F., Becchio, R., Wilke, H.G., Franz, G., Thirlwall, M.F., Viramonte, J., and Wemmer, K., 2000, Proterozoic-Paleozoic development of the basement of the Central Andes (18°–26°)—A mobile belt of the South American craton: Journal of South American Earth Sciences, v. 13, p. 697–715, doi: 10.1016/S0895-9811(00)00057-2.

Manea, V., and Gurnis, M., 2007, Subduction zone evolution and low viscosity wedges and channels: Earth and Planetary Science Letters, v. 264, p. 22–45, doi: 10.1016/j.epsl.2007.08.030.

Marrett, R.A., and Strecker, M.R., 2000, Response of intracontinental deformation in the Central Andes to late Cenozoic reorganization of South American plate motions: Tectonics, v. 19, p. 452–467, doi: 10.1029/1999TC001102.

Marrett, R.A., Allmendinger, R.W., Alonso, R.N., and Drake, R.E., 1994, Late Cenozoic tectonic evolution of the Puna Plateau and adjacent foreland, northwestern Argentine Andes: Journal of South American Earth Sciences, v. 7, p. 179–208, doi: 10.1016/0895-9811(94)90007-8.

Mazzuoli, R., Vezzoli, L., Omarini, R., Acocella, V., Gioncada, A., Matteini, M., Dini, A., Guillou, H., and Hauser, N., Uttini, A., and Scaillet, S., 2008, Miocene magmatism and tectonics in the easternmost sector of the Calama–Olocapato–El Toro fault system in Central Andes at ~24°S: Insights into the Eastern Cordillera evolution: Geological Society of America Bulletin, v. 120, p. 1493–1517, doi: 10.1130/B26109.1.

McGlashan, N., Brown, L.D., and Kay, S.M., 2008, Crustal thicknesses in the Central Andes from teleseismically recorded depth phase precursors: Geophysical Journal International, v. 175, p. 1013–1022.

McQuarrie, N., 2002, The kinematic history of the Central Andean fold-thrust belt, Bolivia: Implications for building a high plateau: Geological Society of America Bulletin, v. 114, p. 950–963, doi: 10.1130/0016-7606(2002)114<0950:TKHOTC>2.0.CO;2.

McQuarrie, N., Horton, B.K., Zandt, G., Beck, S., and DeCelles, P.G., 2005, Lithospheric evolution of the Andean fold-thrust belt, Bolivia, and the origin of the Central Andean plateau: Tectonophysics, v. 399, p. 15–37, doi: 10.1016/j.tecto.2004.12.013.

McQuarrie, N., Barnes, B.B., and Ehlers, T.A., 2008, Geometric, kinematic, and erosional history of the Central Andean Plateau, Bolivia (15–17°S): Tectonics, v. 27, doi: 10.1029/2006TC002054.

Molnar, P., and Garzione, C., 2007, Bounds on the viscosity coefficient of continental lithosphere from removal of mantle lithosphere beneath the Altiplano and Eastern Cordillera: Tectonics, v. 26, p. TC2013, doi: 10.1029/2006TC001964.

Mon, R., Monaldi, C.R., and Salfity, J.A., 2005, Curved structures and inference fold patterns associated with lateral ramps in the Eastern Cordillera, Central Andes: Tectonophysics, v. 399, p. 173–179, doi: 10.1016/j.tecto.2004.12.021.

Monaldi, C.R., Salfity, J.A., and Kley, J., 2008, Preserved extensional structures in an inverted Cretaceous rift basin, northwestern Argentina: Outcrop examples and implications for fault reactivation: Tectonics, v. 27, p. TC1011, doi: 10.1029/2006TC001993.

Mortimer, E., Carrapa, B., Coutand, I., Schoenbohm, L., Sobel, E., Sosa Gomez, J., and Strecker, M.R., 2007, Fragmentation of a foreland basin in response to out-of-sequence basement uplifts and structural reactivation: El Cajón–Campo del Arenal Basin, NW Argentina: Geological Society of America Bulletin, v. 119, p. 637–653, doi: 10.1130/B25884.1.

Mpodozis, C., and Clavero, J., 2002, Tertiary tectonic evolution of the southwestern edge of the Puna Plateau: Cordillera Claudio Gay (26°–27°S), *in* Proceedings of the 5th International Symposium on Andean Geodynamics: Toulouse, p. 445–448.

Mpodozis, C., Cornejo, P., Kay, S.M., and Tittler, A., 1995, La Franja de Maricunga: Síntesis de la evolución del frente volcánico Oligoceno-Mioceno de la zona sur de los Andes Centrales: Revista Geológica de Chile, v. 22, p. 273–314.

Mpodozis, C., Kay, S.M., Gardeweg, M., and Coira, B., 1996, Geología de la región de Ojos del Salado (Andes centrales, 27°S): Implicancias de la migración hacia el este del frente volcánico Cenozoico Superior, *in* XIII Congreso Geológico Argentino: Actas, v. 3, p. 539–54.

Mpodozis, C., Arriagada, C., Basso, M., Roperch, P., Cobbold, P., and Reich, M., 2005, Mesozoic to Paleogene stratigraphy of the Atacama (Purilactis) Basin, Antofagasta region, northern Chile: Insights into the earlier stages of Central Andean tectonic evolution: Tectonophysics, v. 399, p. 125–154, doi: 10.1016/j.tecto.2004.12.019.

Myers, S.C., Beck, S., Zandt, G., and Wallace, T., 1998, Lithospheric-scale structure across the Bolivian Andes from tomographic images of velocity and attenuation for P and S waves: Journal of Geophysical Research, v. 103, p. 21,233–21,252, doi: 10.1029/98JB00956.

Oncken, O., Hindle, D., Kley, J., Elger, K., Victor, P., and Schemmann, K., 2006, Deformation of the Central Andean upper plate system—Facts, fiction, and constraints for plateau models, *in* Oncken, O., et al., eds., The Andes—Active Subduction Orogeny: Berlin, Springer-Verlag, Frontiers in Earth Sciences, v. 1, p. 3–28.

Ort, M., 1993, Eruptive processes and caldera formation in a nested downsag-collapse caldera: Cerro Panizos, Central Andes mountains: Journal of Volcanology and Geothermal Research, v. 56, p. 221–252, doi: 10.1016/0377-0273(93)90018-M.

Ort, M., Coira, B., and Mazzoni, M., 1996, Generation of a crust-mantle magma mixture: Magma sources and contamination at Cerro Panizos, Central Andes: Contributions to Mineralogy and Petrology, v. 123, p. 308–322.

Pardo-Casas, F., and Molnar, P., 1987, Relative motion of the Nazca (Farallon) and South American plates since Late Cretaceous time: Tectonics, v. 6, p. 233–248, doi: 10.1029/TC006i003p00233.

Petrinovic, I.A., Mitjavilla, J., Viramonte, J.G., Marti, J., Becchio, R., Arnosio, M., and Colombo, F., 1999, Geoquímica y geocronología de las secuencias neógenas de trasarco, en el extremo oriental de la cadena volcánica transversal del Quevar, noroeste de Argentina: Acta Geológica Hispana, v. 34, p. 255–273.

Pilger, R.H., Jr., 1981, Plate reconstructions, aseismic ridges, and low-angle subduction beneath the Andes: Geological Society of America Bulletin, v. 92, p. 448–456, doi: 10.1130/0016-7606(1981)92<448:PRARAL>2.0.CO;2.

Pilger, R.H., Jr., 1984, Cenozoic plate kinematics, subduction and magmatism: South American Andes: Journal of the Geological Society of London, v. 141, p. 793–802, doi: 10.1144/gsjgs.141.5.0793.

Polet, J., Silver, P., Zandt, G., Ruppert, S., Bock, G., Kind, R., Reudloff, A., Asch, G., Beck, S., and Wallace, T., 2000, Shear wave anisotropy beneath the Andes from the BANJO, SEDA, and PISCO experiments: Journal of Geophysical Research, v. 105, p. 6287–6304, doi: 10.1029/1999JB900326.

Prezzi, C.B., and Alonso, R.N., 2002, New paleomagnetic data from the northern Argentine Puna: Central Andes rotation pattern reanalyzed: Journal of Geophysical Research, v. 107, p. 2041, doi: 10.1029/2001JB000225.

Ramos, V.A., Cristallini, E.O., and Perez, D.J., 2002, The Pampean flat-slab of the Central Andes: Journal of South American Earth Sciences, v. 15, p. 59–78, doi: 10.1016/S0895-9811(02)00006-8.

Richards, J.P., Ulrich, T., and Kerrick, R., 2006, The late Miocene Quaternary Antofalla volcanic complex, southern Puna, NW Argentina: Protracted history, diverse petrology and economic potential: Journal of Volcanology and Geothermal Research, v. 152, p. 197–239, doi: 10.1016/j.jvolgeores.2005.10.006.

Riller, U., and Oncken, O., 2003, Growth of the Central Andean plateau by tectonic segmentation is controlled by the gradient in crustal shortening: The Journal of Geology, v. 111, p. 367–384, doi: 10.1086/373974.

Riller, U., Petrinovic, I., Ramelow, J., Strecker, M., and Oncken, O., 2001, Late Cenozoic tectonism, collapse caldera and plateau formation in the central Andes: Earth and Planetary Science Letters, v. 188, p. 299–311, doi: 10.1016/S0012-821X(01)00333-8.

Risse, A., Trumbull, R.B., Coira, B., Kay, S.M., and van den Bogaard, P., 2008, $^{40}Ar/^{39}Ar$ geochronology of basaltic volcanism in the back-arc region of the southern Puna Plateau, Argentina: Journal of South American Geology, v. 26, p. 1–15.

Sasso, A., and Clark, A.H., 1998, The Farallon Negro Group, northwest Argentina: Magmatic, hydrothermal and tectonic evolution and implications for Cu-Au metallogeny in the Andean back-arc: Society of Economic Geology Newsletter, v. 34, p. 1–18.

Schilling, F.R., Trumbull, R.B., Brasse, H., Haberland, C., Asch, G., Bruhn, D., Mai, K., Haak, V., Giese, P., Muñoz, M., Ramelow, J., Rietbrock, A., Ricaldi, E., and Vietor, T., 2006, Partial melting in the Central Andean crust: A review of geophysical, petrophysical, and petrologic evidence, *in* Oncken, O., et al., eds., The Andes—Active Subduction Orogeny: Berlin, Springer-Verlag, Frontiers in Earth Sciences, v. 1, p. 459–474.

Schmitt, A.K., deSilva, S.L., Trumbull, R.B., and Emmermann, R., 2001, Magma evolution in the Purico ignimbrite complex, northern Chile: Evidence for zoning of a dacitic magma by injection of rhyolitic melts following mafic recharge: Contributions to Mineralogy and Petrology, v. 140, p. 680–700.

Schneider, A., and Halls, C., 1985, Chronology of eruptive processes and mineralization of the Frailes volcanic field, Eastern Cordillera, Bolivia: Comunicaciones, Departamento de Geología, Universidad de Chile, v. 35, p. 214–224.

Schnurr, W.B.W., Trumbull, R.B., Clavero, J., Hahne, K., Siebel, W., and Gardeweg, M., 2007, Twenty-five million years of silicic volcanism in the southern Central Volcanic Zone of the Andes: Geochemistry and magma genesis of ignimbrites from 25° to 27°S, 67° to 72°W: Journal of Volcanology and Geothermal Research, v. 266, p. 27–46.

Schurr, B., Asch, G., Rietbrock, A., Trumbull, R., and Haberland, C., 2003, Complex patterns of fluid and melt transport in the Central Andean subduction zone revealed by attenuation tomography: Earth and Planetary Science Letters, v. 215, p. 105–119, doi: 10.1016/S0012-821X(03)00441-2.

Schurr, B., Rietbrock, A., Asch, G., Kind, R., and Oncken, O., 2006, Evidence for lithospheric detachment in the central Andes from local earthquake tomography: Tectonophysics, v. 415, p. 203–223, doi: 10.1016/j.tecto.2005.12.007.

Seggiaro, R.E., 1994, Petrología, Geoquímica y Mecanismos de Erupción del Complejo Volcánico Coranzulí [Ph.D. thesis]: Salta, Argentina, Universidad Nacional de Salta, 137 p.

Seggiaro, R.E., and Aniel, B., 1989, Los ciclos piroclásticos del área Tiomayo-Coranzulí, Provincia de Jujuy: Asociación Geológica Argentina Revista, v. 44, p. 394–401.

Siebel, W., Schnurr, W., Hahne, K., Kraemer, B., Trumbull, R.B., van den Bogaard, P., and Emmermann, R., 2001, Geochemistry and isotope systematics of small- to medium-volume Neogene-Quaternary ignimbrites in the southern Central Andes: Evidence for derivation from andesitic magma sources: Chemical Geology, v. 171, p. 213–237, doi: 10.1016/S0009-2541(00)00249-7.

Silver, P.G., Russo, R.M., and Lithgow-Bertelloni, C., 1998, Coupling of South American and African plate motion and plate deformation: Science, v. 279, p. 60–63, doi: 10.1126/science.279.5347.60.

Sobolev, S.V., and Babeyko, A.Y., 2005, What drives orogeny in the Andes?: Geology, v. 33, p. 617–620, doi: 10.1130/G21557.1.

Sobel, E., and Strecker, M.R., 2003, Uplift, exhumation and precipitation: Tectonic and climatic control of late Cenozoic landscape evolution in the northern Sierras Pampeanas, Argentina: Basin Research, v. 15, p. 431–451, doi: 10.1046/j.1365-2117.2003.00214.x.

Sobolev, S.V., Babeyko, A.Y., Koulakov, I., Oncken, O., and Vietor, T., 2006, Mechanism of the Andean orogeny: Insight from the numerical modeling, *in* Oncken, O., et al., eds., The Andes—Active Subduction Orogeny: Berlin, Springer-Verlag, Frontiers in Earth Sciences, v. 1, p. 509–531.

Soler, M.M., Caffe, P., Coira, C., Onoe, A.T., and Kay, S.M., 2007, Geology of the Vilama caldera: Correlations and a new interpretation of a large scale explosive event in the Central Andean Plateau during the Upper Miocene: Journal of Volcanology and Geothermal Research, v. 164, p. 27–53, doi: 10.1016/j.jvolgeores.2007.04.002.

Somoza, R., 1998, Updated Nazca (Farallón)–South America relative motions during the last 49 m.y.; implications for mountain building in the Central Andean region: Journal of South American Earth Sciences, v. 11, p. 211–215, doi: 10.1016/S0895-9811(98)00012-1.

Sparks, R.S.J., Francis, P.W., Hamer, R.D., Pankhurst, R.J., O'Callaghan, L.L., Thorpe, R.S., and Page, R.S., 1985, Ignimbrites of the Cerro Galán caldera, NW Argentina: Journal of Volcanology and Geothermal Research, v. 24, p. 205–248, doi: 10.1016/0377-0273(85)90071-X.

Stern, C.R., 2004, Active Andean volcanism: Its geologic and tectonic setting: Revista Geológica de Chile, v. 31, p. 161–206.

Strecker, M.R., Alonso, R.N., Bookhagen, B., Carrapa, B., Hilley, G.E., Sobel, E.R., and Trauth, M.H., 2007, Tectonics and climate of the southern Central Andes: Annual Review of Earth and Planetary Sciences, v. 35, doi: 10.1146/annurev.earth.35.031306.140158.

Swenson, J.L., Beck, S.L., and Zandt, G., 2000, Crustal structure of the Altiplano from broadband regional waveform modeling; implications for the composition of thick continental crust: Journal of Geophysical Research, v. 105, p. 607–621, doi: 10.1029/1999JB900327.

Tassara, A., Götze, H.-J., Schmidt, S., and Hackney, R., 2006, Three-dimensional density model of the Nazca plate and the Andean continental margin: Journal of Geophysical Research, v. 111, p. B09404, doi: 10.1029/2005JB003976.

Trumbull, R.B., Riller, U., Oncken, O., Schueber, E., Munier, K., and Hongn, F., 2006, The time-space distribution of Cenozoic arc volcanism in the Central Andes: A new data compilation and some tectonic considerations, *in* Oncken, O., et al., eds., The Andes—Active Subduction Orogeny: Berlin, Springer-Verlag, Frontiers in Earth Sciences, v. 1, p. 29–44.

Vandervoort, D.S., Jordan, T.E., Zeitler, P., and Alonso, R.N., 1995, Chronology of internal drainage development and uplift, southern Puna Plateau, Argentine Central Andes: Geology, v. 23, p. 145–148, doi: 10.1130/0091-7613(1995)023<0145:COIDDA>2.3.CO;2.

Wagner, L.S., Beck, S., Zandt, G., and Ducea, M.N., 2006, Depleted lithosphere, cold, trapped asthenosphere, and frozen melt puddles above the flat slab in central Chile and Argentina: Earth and Planetary Science Letters, v. 245, p. 289–301, doi: 10.1016/j.epsl.2006.02.014.

Whitman, D., Isacks, B.L., Chalelain, J.L., Chiu, J.M., and Perez, A., 1992, Attenuation of high-frequency seismic waves beneath the Central Andean Plateau: Journal of Geophysical Research, v. 97, p. 19,929–19,947, doi: 10.1029/92JB01748.

Whitman, D., Isacks, B.L., and Kay, S.M., 1996, Lithospheric structure and along-strike segmentation of the Central Andean Plateau: Tectonophysics, v. 259, p. 29–40, doi: 10.1016/0040-1951(95)00130-1.

Wigger, P.J., Schmitz, M., Araneda, M., Asch, G., Baldzuhn, S., Giese, P., Heinsohn, W.D., Martinez, E., Ricaldi, E., Röwer, P., and Viramonte, J., 1994, Variation in the crustal structure of the southern Central Andes deduced from seismic refraction investigations, *in* Reutter, K.J., Scheuber, E., and Wigger, P.J., eds., Tectonics of Southern Central Andes: Structure and Evolution of an Active Continental Margin: New York, Springer-Verlag, p. 23–48.

Wörner, G., Hammerschmidt, K., Henjes-Kunst, F., Lezaun, J., and Wilke, H., 2000, Geochronology ($^{40}Ar/^{39}Ar$, K-Ar and He-exposure ages) of Cenozoic magmatic rocks from northern Chile (18–22°S): Implications for magmatism and tectonic evolution of the Central Andes: Revista Geológica de Chile, v. 27, p. 205–240.

Yañéz, G.A., Ramiro, C.R., von Huene, R., and Diaz, J., 2001, Magnetic anomaly interpretation across the southern Central Andes (32°–34°S): The role of the Juan Fernández Ridge in the late Tertiary evolution of the margin: Journal of Geophysical Research, v. 106, p. 6325–6345, doi: 10.1029/2000JB900337.

Yang, Y., Liu, M., and Stein, S., 2003, A 3-D geodynamic model of lateral crustal flow during Andean mountain building: Geophysical Research Letters, v. 30, 2093, doi: 10.1029/2003GL018308.

Yoon, M., Buske, S., Shapiro, S.A., and Wigger, P., 2008, Reflection image spectroscopy across the Andean subduction zone: Tectonophysics, doi: 10.1016/j.tecto.2008.03.014.

Yuan, X., Sobolev, S.V., Kind, R., Oncken, O., Bock, G., Asch, G., Schurr, B., Graeber, F., Rudloff, A., Hanka, W., Wylegalla, K., Tibi, R., Haberland, C., Rietbrock, A., Glese, P., Wigger, P., Rower, P., Zandt, G., Beck, S., Wallace, T., Pardo, M., and Comte, D., 2000, Subduction and collision processes in the Central Andes constrained by converted seismic phases: Nature, v. 408, p. 958–961, doi: 10.1038/35050073.

Yuan, X., Sobolev, S.V., and Kind, R., 2002, Moho topography in the Central Andes and its geodynamic implications: Earth and Planetary Science Letters, v. 199, p. 389–402, doi: 10.1016/S0012-821X(02)00589-7.

Zandt, G.M., Leidig, J., Chmielowski, J., Baumont, D., and Yuan, X., 2003, Seismic detection and characterization of the Altiplano-Puna magma body, Central Andes: Pure and Applied Geophysics, v. 160, p. 789–807, doi: 10.1007/PL00012557.

Manuscript Accepted by the Society 5 December 2008

Printed in the USA

The Geological Society of America
Memoir 204
2009

Flat-slab subduction and crustal models for the seismically active Sierras Pampeanas region of Argentina

Patricia Alvarado*
CONICET (Consejo Nacional de Investigaciones Científicas y Técnicas)–Departamento de Geofísica y Astronomía, Facultad de Ciencias Exactas, Físicas y Naturales, Universidad Nacional de San Juan, Meglioli 1160 S (5407) Rivadavia, San Juan, Argentina

Mario Pardo
Departamento de Geofísica, Facultad de Ciencias Físicas y Matemáticas, Universidad de Chile, Blanco Encalada 2002, Santiago, Chile

Hersh Gilbert
Department of Earth and Atmospheric Sciences, 550 Stadium Mall Dr., Purdue University, West Lafayette, Indiana 47907, USA

Silvia Miranda
Departamento de Geofísica y Astronomía, Facultad de Ciencias Exactas, Físicas y Naturales, Universidad Nacional de San Juan, Meglioli 1160 S (5407) Rivadavia, San Juan, Argentina

Megan Anderson
Department of Geology, Colorado College, 14 E. Cache La Poudre, Colorado Springs, Colorado 80903, USA

Mauro Saez
Departamento de Geofísica y Astronomía, Facultad de Ciencias Exactas, Físicas y Naturales, Universidad Nacional de San Juan, Meglioli 1160 S (5407) Rivadavia, San Juan, Argentina

Susan Beck
Department of Geosciences, University of Arizona Gould-Simpson Building #77, 1040 E. 4th St., Tucson, Arizona 85721, USA

ABSTRACT

The Sierras Pampeanas in the west-central part of Argentina are a modern analog for Laramide uplifts in the western United States. In this region, the Nazca plate is subducting beneath South America almost horizontally at about ~100 km depth before descending into the mantle. The flat-slab geometry correlates with the inland prolongation of the subducted oceanic Juan Fernández Ridge. This region of Argentina is characterized by the termination of the volcanic arc and uplift of the active basement-cored Sierras Pampeanas. The upper plate shows marked differences in seismic properties that are interpreted as variations in crustal composition in agreement with the

*alvarado@unsj.edu.ar

Alvarado, P., Pardo, M., Gilbert, H., Miranda, S., Anderson, M., Saez, M., and Beck, S., 2009, Flat-slab subduction and crustal models for the seismically active Sierras Pampeanas region of Argentina, *in* Kay, S.M., Ramos, V.A., and Dickinson, W.R., eds., Backbone of the Americas: Shallow Subduction, Plateau Uplift, and Ridge and Terrane Collision: Geological Society of America Memoir 204, p. 261–278, doi: 10.1130/2009.1204(12). For permission to copy, contact editing@geosociety.org.

presence of several Neoproterozoic to Paleozoic accreted terranes. In this paper, we combine the results from the CHile-ARgentina Geophysical Experiment (CHARGE) and the CHile-ARgentina Seismology Measurement Experiment (CHARSME) passive broadband arrays to better characterize the flat-slab subduction and the lithospheric structure. Stress tensor orientations indicate that the horizontal slab is in extension, whereas the upper plate backarc crust is under compression.

The Cuyania terrane crust exhibits high P-wave seismic velocities (Vp ~6.4 km/s), high P- to S-wave seismic velocity ratios (Vp/Vs = 1.80–1.85), and 55–60 km crustal thickness. In addition, the Cuyania terrane has a high-density and high-seismic-velocity lower crust. In contrast, the Pampia terrane crust has a lower Vp value of 6.0 km/s, a lower Vp/Vs ratio of 1.73, and a thinner crust of ~35 km thickness. We integrate seismic and gravity studies to evaluate crustal models that can explain the unusually low elevations of the western Sierras Pampeanas.

Flat-slab subduction models based on CHARGE and CHARSME seismic data and gravity observations show a good correlation with the predicted Juan Fernández Ridge path beneath South America, the deep Moho depths in the Andean backarc, and the high-density and high-seismic-velocity lower crust of the Cuyania terrane. The Cuyania terrane is also the region characterized by more frequent and larger-magnitude crustal earthquakes.

INTRODUCTION

What causes flat-slab subduction and its associated seismicity is still enigmatic (Gutscher et al., 2000; Hacker et al., 2003). Several authors have suggested that the nearly horizontal position of the subducted plate is temporary (a few million years) and induces compression far inland from the trench in the upper plate (e.g., Allmendinger et al., 1990; Jordan et al., 1983). It has also been suggested that flat-slab subduction took place in western North America during the Laramide orogeny (e.g., Bird, 1988; Saleeby, 2003; Wells and Hoisch, 2008; Humphreys, this volume) and possibly in the central Altiplano-Puna region, providing fundamental controls on formation of the crust and major ore deposits (e.g., Isacks et al., 1986; Kay et al., 1999; James and Sacks, 1999; Kay and Mpodozis, 2001). There are not many places on Earth to explore the occurrence of present-day flat-slab subduction in detail, since only 10% of the total length of modern convergent margins exhibits a nearly horizontally positioned subducting plate (Gutscher et al., 2000). In the last decade, the increasing use of regional passive-source broadband seismic experiments in the study of the crust and the lithosphere has provided us with refined observations and images (Owens et al., 1993; IRIS, 2004), which give new insights into the understanding of flat-slab subduction processes.

One of the most seismically active regions in South America, west-central Argentina near 31°S, was the site of several recent temporary broadband seismic network deployments (Fig. 1) (Pardo et al., 2004; Beck et al., 2005; Alvarado et al., 2005). At this latitude, the oceanic Nazca plate is subducting horizontally beneath the Andean Cordillera and the Sierras Pampeanas at a convergence rate of 6.7 cm/yr along an azimuth of ~78° (Vigny et al., 2009). Seismicity in modern and historical times indicates that the Andean backarc region overriding the flat-slab subduction zone is active at both crustal (<35 km) and mantle (~100 km) depths (Smalley et al., 1993; Pardo et al., 2002; Alvarado et al., 2007). The region located immediately at the southern end in the transition between flat to "normal" subduction also experiences earthquakes within the (normally) subducted slab and within the upper plate. This shallow seismicity is located along the active arc of the Andean Cordillera (Barrientos et al., 2004) and in its backarc region (Fig. 1) (Salazar, 2005; Alvarado et al., 2007).

A distinctive feature of this active margin is that the continental upper plate varies in composition and changes in character from west to east (Alvarado et al., 2005). These differences record a series of successive accretionary events, which mark the crustal evolution of this region in the last 600 m.y. as reviewed by Ramos (2000).

In this paper, we address the question of how terrane inheritance affects the present distribution of earthquakes in the Andean backarc crust over the flat-slab subduction zone in Argentina. We also discuss the correlation between crustal thickness and elevation. We make use of gravity observations and recent determinations from the CHile-ARgentina Geophysical Experiment (CHARGE) and the CHile-ARgentina Seismology Measurement Experiment (CHARSME), which provide constraints on the crust and upper-mantle structure, as well as the characterization of crustal and intermediate-depth seismicity.

REGIONAL SEISMIC NETWORKS

The CHARGE project deployed 22 broadband IRIS-PASSCAL seismic stations in the south-central Andes from the Chilean coast to the central part of Argentina between late 2000 and mid-2002. This seismic network was designed to map regional differences in slab geometry and the corresponding structure of the overriding plate along two east-west–oriented transects: (1) the

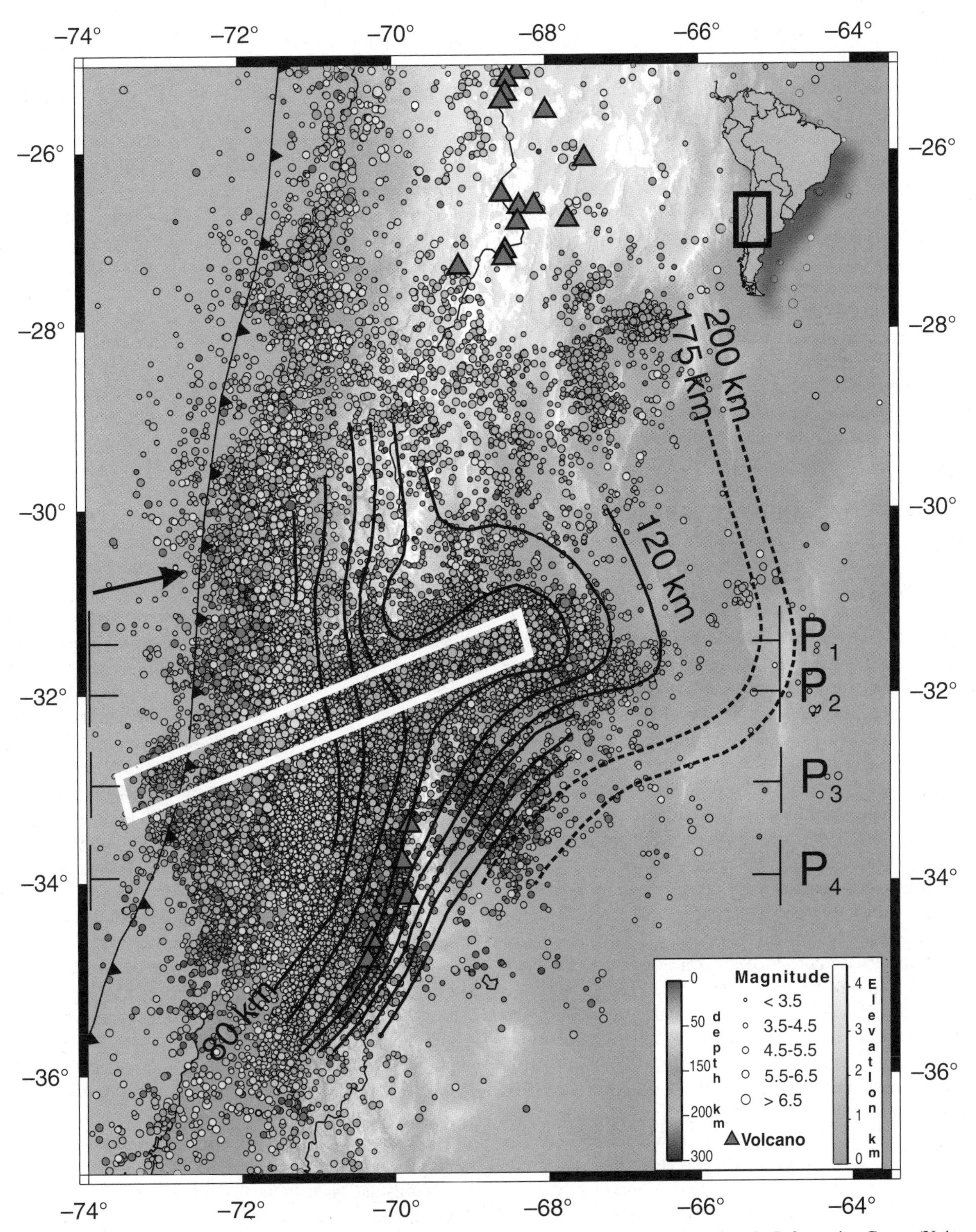

Figure 1. Points showing distribution of Preliminary Determination of Earthquakes–National Earthquake Information Center (United States Geological Service) seismicity during the past 30 yr. The convergence between the Nazca plate and the South American plate at a rate of 6.7 cm/yr along an azimuth of 78° is represented by the black arrow (Vigny et al., 2009). The solid black lines are contours of the top of the subducted Nazca plate from Anderson et al. (2007). Dashed line contours mark slab depths from Cahill and Isacks (1992). Triangles are active volcanoes from Stern (2004). Note the correlation among the flat-slab subduction, the inland prolongation of the Juan Fernández Ridge (large white rectangle), and the absence of active volcanism. P_1–P_4 indicate the location of the profiles shown in Figure 3.

northern line, located at ~30°S, is above the horizontal part of the subducted Nazca plate and (2) the southern line, located at ~36°S, is over a "normally" subducted slab. The CHARGE network collected more than 12 terabytes of continuous broadband seismic data during 18 months and involved research efforts from the University of Arizona (USA), the National University of San Juan (Argentina), the National Institute for Seismic Disaster Mitigation-INPRES (Instituto Nacional de Prevención Sísmica) (Argentina), and the University of Chile in Santiago, Chile (Fig. 2).

The subsequent CHARSME seismic network consisted of 29 broadband seismic stations deployed between 31°S and 34°S. The network extended from the coast of Chile to 67°W longitude. This array continuously recorded seismicity from November 2002 to March 2003 as part of an international collaboration between the University of Chile (Chile), the Institut de Recherche pour le Développement (France), the National Institute for Seismic Disaster Mitigation-INPRES (Argentina), and the National University of San Juan (Argentina) (Fig. 2).

THE CONTINENTAL SOUTH AMERICAN UPPER PLATE: A MOSAIC OF ACCRETED TERRANES

The continental crust of the western margin of South America between 30°S and 36°S is composed of several accreted terranes in the southwestern part of Gondwana (Fig. 2) (e.g., Ramos et al., 1986; Ramos, 2000). These terranes and the boundaries between them have strongly influenced younger tectonic events. The tectonic evolution of this region has been described in terms of a "parautochthonous" hypothesis, which involves blocks that were rifted from Gondwana and then reamalgamated through the subduction process, or "allochthonous" models related to the collision of exotic blocks to Gondwana and accretion by subduction.

By 560–530 Ma, the Rio de La Plata craton and other minor basement blocks that make up the crust of Brazil, Uruguay, and the eastern part of Argentina were amalgamated to Gondwana (Ramos, 1988; Rapela et al., 1998; Basei et al., 2001; Trindade et al., 2006), where the western side was the active margin. The exhumed Sierras de Córdoba in central Argentina is the expression of the Pampean magmatic arc that developed in an approximate north-south orientation above a west-dipping subduction zone (Rapela et al., 1998). Immediately to the east, there is a metasedimentary sequence, which has been interpreted as an accretionary prism related to the Pampean orogeny (Kraemer et al., 1995). This configuration continued until the oblique collision of the Pampia terrane with Gondwana at 530–515 Ma, now bounded to the east by the Rio de Plata craton (Rapela et al., 2007).

Another major orogenic (Famatinian) system, which existed from 490 to 460 Ma, is associated with a convergence regime that ran along more than 1700 km of the Pampia western margin (Astini et al., 1995; Pankhurst et al., 1998, 2000). A simultaneous deformation front to the west has been related to the collision of a basement terrane of Grenville age and Laurentian affinity, which included the Precordillera (e.g., Ramos et al., 1986; Dalla Salda et al., 1992; Thomas and Astini, 2003; Castro de Machuca et al., 2008). Recent studies have interpreted this terrane as a composite block (Cuyania) made up of two terranes (Pie de Palo and Precordillera) that were earlier assembled before being accreted to western Gondwana (Ramos, 2004). This is based on the following observations: (1) the lack of an active subduction margin between the Pie de Palo basement terrane and the Precordillera terrane in Ordovician times; i.e., there is no evidence for an Ordovician magmatic arc within the Cuyania terrane (Vujovich et al., 2004); and (2) the existence of a suture between the Pie de Palo and Precordillera defined by ophiolite rocks of Grenville age (Vujovich and Kay, 1998).

The prevailing model, that the Cuyania terrane separated from Laurentia and was transferred to Gondwana as the Iapetus Ocean was closing between the Cuyania and Pampia terranes, is debated by some authors (e.g., Vaughan and Pankhurst, 2008). Alternative interpretations point to major horizontal motion parallel to the Gondwana margin that might have controlled the accretion of the Grenville Pie de Palo–Precordillera basement block (e.g., Baldis et al., 1989; Aceñolaza et al., 2002; Finney et al., 2005; Rapela et al., 2007; Casquet et al., 2008).

The existence of an ophiolite belt along the western margin of the Precordillera at the boundary with the Frontal Cordillera (Caminos, 1979) has led to proposals for a late Paleozoic accretionary event. An allochthonous accreted terrane called Chilenia is suggested to have collided with Cuyania at ca. 360 Ma (Ramos et al., 1986; Davis et al., 2000). A question remains as to whether Chilenia was a piece that rifted from Cuyania and was then reattached, or if, like Cuyania, it is a far-traveled exotic terrane from Laurentia (Ramos et al., 1986; Dalla Salda et al., 1992; Rapela et al., 1998).

Exposures of pre-Andean basement rocks in the suggested accreted terranes show marked differences in their composition. The eastern Sierras Pampeanas are mainly composed of low-pressure, high-temperature schists and gneisses with sedimentary protoliths, granitoids that can reach batholithic dimensions, and subordinate basic and ultrabasic rocks (e.g., Martino et al., 1995; Baldo et al., 1996; Varela et al., 2000; Rapela et al., 2002; Otamendi et al., 2006). In contrast, the western Sierras Pampeanas and Precordillera terranes are characterized by basic-ultrabasic rocks, metacarbonate and calc-silicate metamorphic sequences, and fewer granitoids (e.g., McDonough et al., 1993; Quenardelle and Ramos, 1999; Ramos et al., 2000; Vujovich et al., 2004).

Superimposed on the terrane accretion history, there are Mesozoic extensional events related to the opening of the South Atlantic Ocean. This Mesozoic extension largely occurred along the former Paleozoic terrane sutures associated with continental rift deposits like those in the Cuyo Basin on the western margin of the Cuyania terrane near Mendoza (Fig. 2), one of the oldest (ca. 240 Ma) synrift systems (Ramos and Kay, 1991; Franzese et al., 2003). Rifting, intraplate volcanism, strike-slip faulting, half-graben extension, and synrift basin development in the hanging wall of the sutures lasted until ca. 165 Ma (Ramos, 1999; Rossello and Mozetic, 1999). At ca. 130 Ma, the combination of South Atlantic spreading and subduction along the western margin of

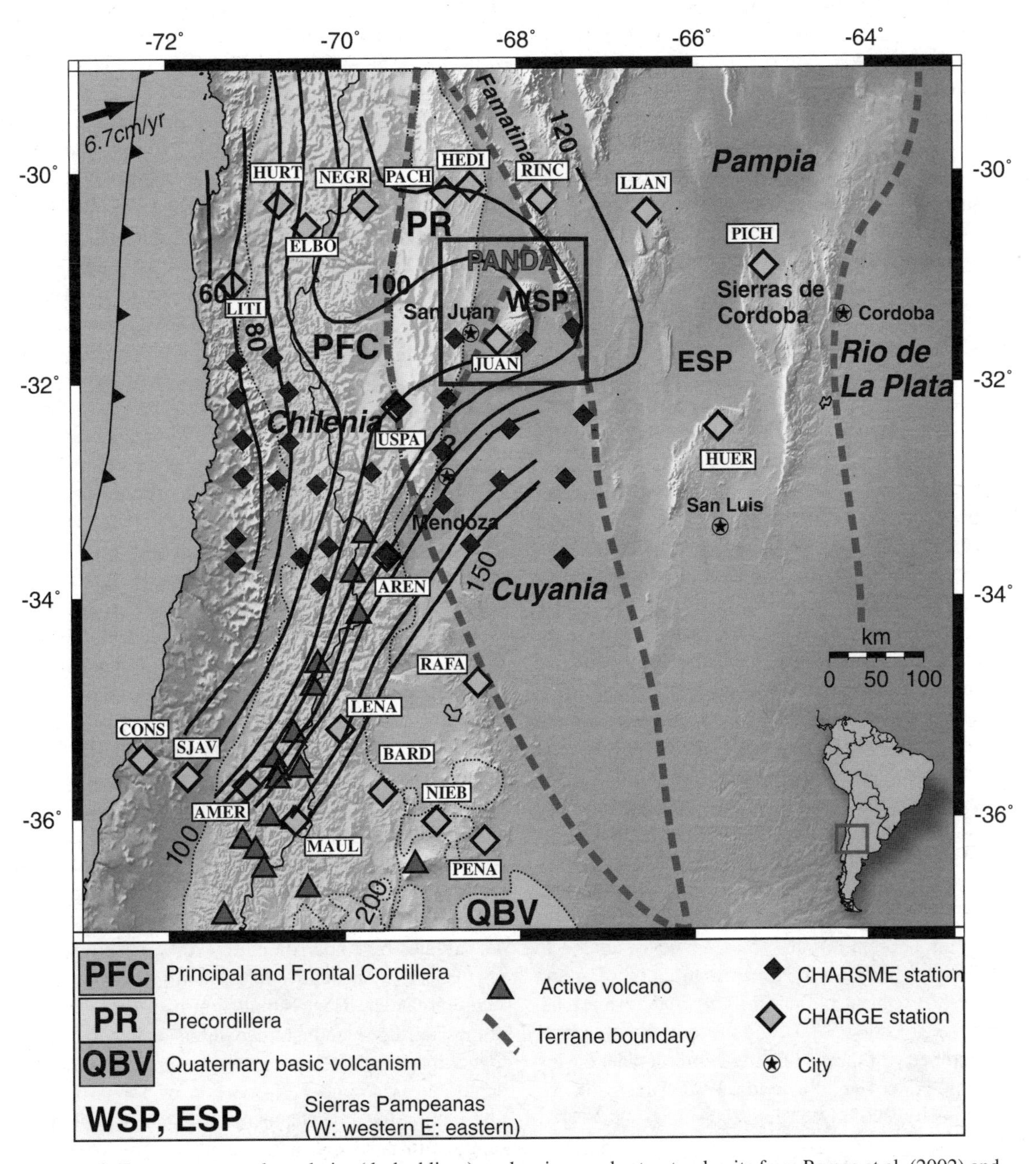

Figure 2. Terranes, terrane boundaries (dashed lines), and main morphostructural units from Ramos et al. (2002) and Rapela et al. (2007) shown relative to temporary seismic networks of the short-period Portable Array for Numerical Data Acquisition (PANDA) deployment (small box) (Regnier et al., 1994) and the broadband CHARGE stations (large diamonds) (Beck et al., 2005) and CHARSME stations (small diamonds) (Deshayes et al., 2008). Contours to the top of the Benioff zone from Anderson et al. (2007) are at 10 km intervals. CHARGE—Chile-Argentina Geophysical Experiment; CHARSME—Chile-Argentina Seismology Measurement Experiment.

Gondwana produced compressional reactivation and tectonic inversion of older structures associated with the proto–Sierras Pampeanas uplifts (Ramos et al., 2002), as well as the development of the Andean magmatic arc (Mpodozis and Ramos, 1989). Thus, the accretionary and extensional pre-Andean processes created anisotropy and fracture (weakness) zones that later controlled thin- and thick-skinned Andean deformation by inverting ancient normal faults.

This paper discusses the terrane inheritance, accretionary and extensional history of the continental overriding plate, subduction of the Juan Fernández Ridge beneath South America, and the Neogene evolution of the flat-slab subduction and its associated upper-plate deformation.

SEISMICITY AND SHAPE OF THE SUBDUCTED NAZCA PLATE

The most prominent feature of the modern subduction zone at ~31.5°S is the flattening of the oceanic subducting Nazca slab at ~100 km depth beneath the Andean backarc region (Fig. 1; e.g., Barazangi and Isacks, 1976; Cahill and Isacks, 1992). The modern flat-slab geometry is argued to have been developed by ca. 7–5 Ma from a previous steeper subduction angle position based on cessation of the volcanic arc in both the arc and retroarc (Kay et al., 1991). The effects of the change in slab geometry on the overriding continental plate are recorded in the distribution, petrology, geochemistry, and isotopic signatures of the volcanic rocks (e.g., Kay and Mpodozis, 2002). The linked pattern of volcanism has been interpreted to result from a gradual flattening of the slab over time that caused a decrease in rate and cessation of arc volcanism and an eastward migration and broadening away from the trench. Simultaneously, the retroarc region developed a thin-skinned fold-and-thrust belt in the Precordillera and a series of discontinuous basement-involved thrusts that uplifted the Sierras Pampeanas (Figs. 1 and 2; e.g., Allmendinger et al., 1990; Kay and Abbruzzi, 1996). This region of Argentina is often referred to as a modern analog to the Laramide orogeny in western North America (Jordan and Allmendinger, 1986) because of the similarity in deformational style of the crust (e.g., Erslev, 1993), and the models for a flat or subhorizontal geometry of the subducting Farallon plate during the early Cenozoic (Coney and Reynolds, 1977; Dickinson and Snyder, 1978; Saleeby, 2003; DeCelles, 2004).

The first images of the geometry of the subducting Nazca slab beneath South America were obtained by mapping the Wadati-Benioff zone using global seismicity (e.g., Barazangi and Isacks, 1976; Cahill and Isacks, 1992). The early studies outlined the change in the Nazca slab angle of inclination in western South America, indicating a gradual transition to a more horizontal position between 28°S and 33°S. More recent studies have refined the geometry of the slab by relocating multiple intermediate-depth seismicity recorded at local-to-regional distances in the PANDA, CHARGE, and CHARSME networks (Figs. 2 and 3; Reta 1992; Smalley et al., 1993; Pardo et al., 2002; Anderson et al., 2007). The new, more detailed contours in Figure 2 inferred from relocated slab hypocenters show the uppermost portion of the Nazca plate flattening at ~100 km depth between 31°S and 32°S and having an elongated shape that extends to the northeast for ~200 km before resuming descent into the mantle (Fig. 1; Anderson et al., 2007).

The flat-slab shape and the pattern of strains derived from slab earthquake focal mechanisms support flattening due to subduction of a buoyant ridge related to the inland prolongation of the subducting Juan Fernández Ridge (JFR in Figs. 1 and 3) (Smalley and Isacks, 1987; Pardo et al., 2002; Anderson et al., 2007). Most workers agree that the Juan Fernández Ridge continues on the subducting plate beneath the South American continent (e.g., Yañéz et al., 2001). Studies by Pilger and Handschumacher (1981) suggested that the Juan Fernández Ridge formed above a fixed hotspot or a propagating intraplate fracture. Stuessy et al. (1984) provided potassium-argon ages consistent with an island-seamount chain origin from a Pacific hotspot. Estimations of the contribution of the subducted oceanic Nazca plate to marine and land magnetic observations by Yañéz et al. (2001) provide a paleoreconstruction kinematic model of the Juan Fernández hotspot chain subduction back into the Miocene. The magnetic model predicts a rapid southward migration of the ridge-continent collision point until ca. 11 Ma. Subsequently, the margin of South America moved westward at a slower rate of 3.5 cm/yr (Chase, 1978), in agreement with the shallowing of the Nazca plate, eastward volcanic arc migration, and broadening of the volcanic arc in the last ~10 m.y. (Kay et al., 1991; Kay and Mpodozis, 2002). A direct association of flat-slab subduction with the passage of the Juan Fernández Ridge, however, is not accepted by all. Specifically, some three-dimensional (3-D) laboratory experiments indicate that a ridge of modest dimensions subducted perpendicular to the trench, as is the case for the present-day Juan Fernández Ridge, would not be sufficiently buoyant to produce the flat-slab geometry observed in western Argentina (Martinod et al., 2005).

Studies by Anderson et al. (2007) have pointed to modern seismic activity at depths of less than 100 km along the predicted Juan Fernández Ridge subduction path (Fig. 1). The region of proposed ridge subduction exhibits a larger concentration of higher-magnitude seismic events compared to the intermediate-depth events around it. Inspection of T-axis orientations from first-motion focal mechanisms during the CHARGE experiment suggests that the slab is in extension. This extension is argued to result from a combination of buoyant forces of the subducted Juan Fernández Ridge and the slab pulling away from the ridge to the north and south.

Constraints from the CHARSME network are consistent with these observations and interpretations (Figs. 1 and 3). Table 1 summarizes the stress tensor information in the flat-slab subduction zone calculated by Salazar (2005) using the relocated CHARSME seismicity. The events used in this analysis were recorded by at least eight stations, had root mean squared location errors of <0.2, hypocenter errors of <10 km, and first-motion

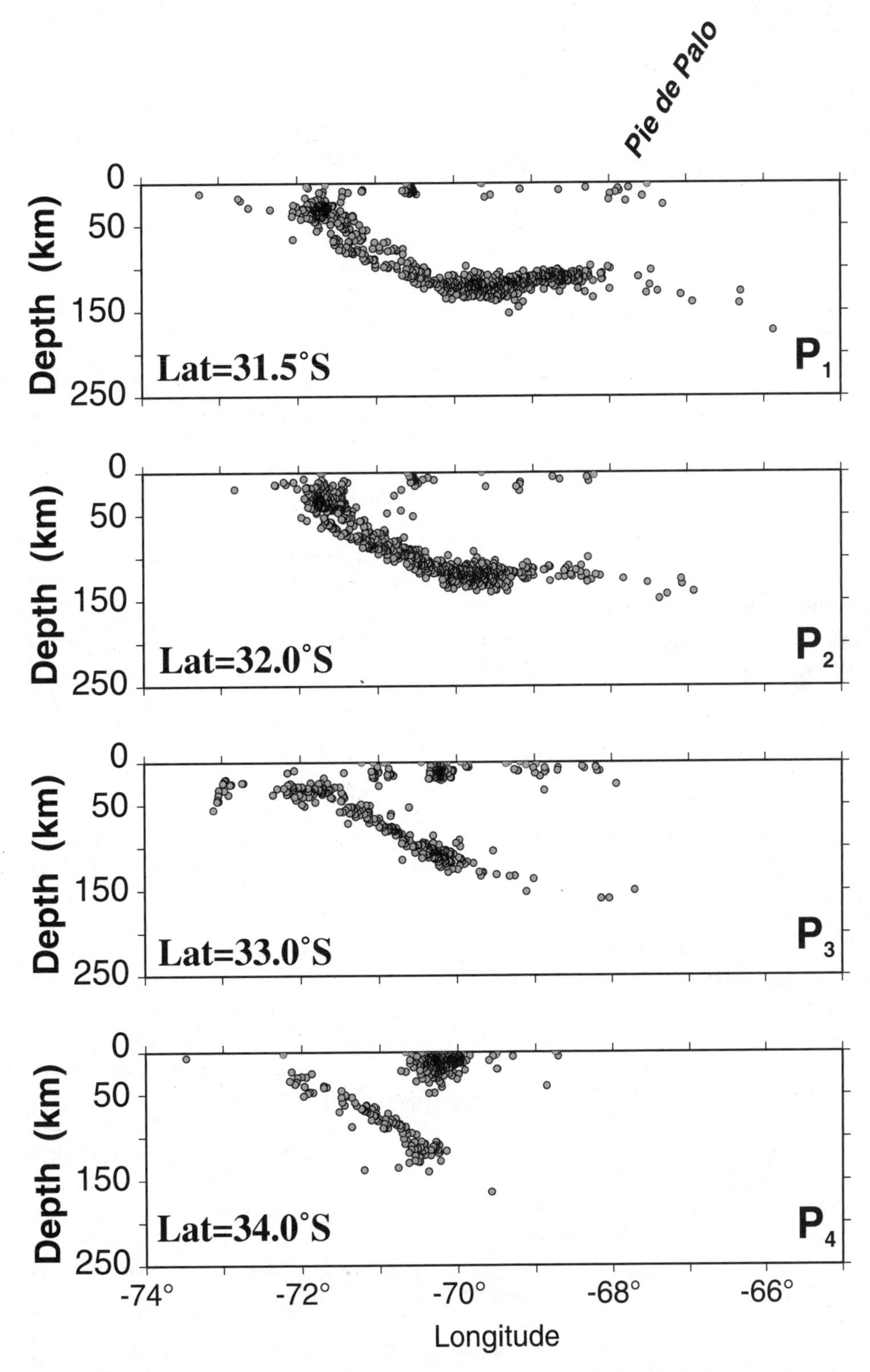

Figure 3. East-west cross sections at different latitudes (see Fig. 1 for location) showing seismicity that was mainly recorded by the CHARSME experiment (November 2002–March 2003) and relocated by body-wave tomography of Pardo et al. (2004).

TABLE 1. STRESS TENSOR ORIENTATIONS FROM SALAZAR (2005) USING RIVERA AND CISTERNAS (1990) METHODOLOGY AND THE CHARSME DATA

Seismogenic zone	σ_1	σ_2	σ_3
CHARSME results (Salazar, 2005)			
Coupled zone (0–70 km)	257	166	54
	20	7	62
Intermediate depth (100–200 km)	82	321	230
	83	3	5
Backarc crust (0–45 km)	98	188	329
	4	5	81
Results from Pardo et al. (2002)			
Shallow depth (0–70 km)	275	185	81
	8	2	78
Intermediate depth (100–200 km)	74	201	292
	77	6	8
Backarc (0–45 km)	254	75	344
	28	55	0

Note: Each major component orientation is given by its azimuth and plunge for different seismogenic zones in the flat-slab subduction segment. Results for the backarc crust are based on Harvard Centroid Moment Tensor (CMT) solutions. Previous results from Pardo et al. (2002) are also shown for comparison.

focal mechanisms calculated from a minimum of 10 polarities. Some 250 earthquakes were used to estimate the stress tensor orientations for three distinct seismogenic zones using the method of Rivera and Cisternas (1990), which calculates the stress tensor orientation from the Euler angles based on fault-plane focal mechanisms. The stress tensor selected is that which best represents the observed slip vectors for all of the focal mechanisms considered.

According to Salazar (2005), the results in Table 1 correspond to an interplate zone (focal depths < 70 km), an intermediate-depth zone (focal depths between 100 km and 200 km), and a crustal upper-plate zone (focal depths < 45 km). The results for the crustal upper-plate zone include focal mechanisms reported by the Harvard Centroid Moment Tensor (CMT) catalog due to the limited amount of data from this zone in this study. The resulting stress tensor orientations at the interplate zone indicate compression: σ_1 is mainly oriented in an east-west direction, and σ_3 is almost vertical. In contrast, the intermediate depth region exhibits σ_3 along the slab plunging ~8° and σ_1 almost vertical, providing clear evidences for a slab in extension. Previous results for the same regions from Pardo et al. (2002) are in agreement (see Table 1).

P- and S-wave receiver function studies show a weak signal associated with the top of the subducted slab beneath broadband seismic stations along a west-east profile at ~30°S (see Figs. 2 and 4 for location of the northern CHARGE transect) (Gilbert et al., 2006; Heit et al., 2008). Based on the Wadati-Benioff contours of Anderson et al. (2007) in Figure 1, the inclination of the slab to the north at that latitude might be responsible for broadened waveforms in the receiver functions, making the slab difficult to image using this technique (Figs. 1 and 4). The absence of a sharp discontinuity in the receiver functions from the slab in this region could also be due to a small impedance contrast at this depth resulting from compositional effects.

Near 31.5°S, the nearly continuous horizontal Wadati-Benioff zone as defined by earthquakes is shown to bend upward, so that the Nazca plate shows a slight dip to the west (instead of the east) (see cross-section P_1 in Figs. 1 and 3). In earlier studies, this effect was ascribed to an artifact caused by earthquake location programs that used seismic velocity structures consisting of horizontal layers in a flat Earth. To avoid possible depth-shift errors in slab earthquake locations, crustal and upper-mantle models need to account for 3-D seismic velocity structures and the presence of the Andean root (Smalley and Isacks, 1987; Pujol et al., 1991). We note that the tomographic analyses from Pardo et al. (2004) and Wagner et al. (2005), which used more realistic seismic velocity structures for the crust and upper mantle, retain a tendency for a slight westward dip of the Wadati-Benioff zone beneath the backarc. To the east of this region, below the Sierra Pie de Palo near 68°W, the intense intermediate-depth seismic activity diminishes, and the slab again turns downward and begins to descend into the mantle (Fig. 3). We argue that the cause of the apparent bulge in the subducted slab and the change in slab seismic activity could be due to relief on the Nazca plate, such as subduction of the Juan Fernández Ridge seamount chain. Recent studies suggest that subducted seamounts can influence the seismic behavior in subduction zones by changing the physical relief or the physical properties of faults materials (von Huene, 2008). Importantly, bathymetric studies to the west in Chile have confirmed the presence of at least one prominent magnetic offshore anomaly, which is interpreted as a subducted seamount (e.g., O'Higgins, Papudo seamounts) on the Juan Fernández Ridge (von Huene et al., 1997).

Although we do not have direct observational evidence for active within-slab structures to account for the distribution of the earthquakes at ~100 km depth, outer-rise earthquakes and seismic-reflection images provide possible evidence for faulting that could rupture a significant part of the oceanic lithosphere and generate seismicity (Fromm et al., 2006). However, at the same time, we cannot rule out a brittle rheologic control associated with bending of the slab in producing this seismicity. To the east of ~68°W, the reduction in earthquake density beneath the Sierra Pie de Palo (Fig. 3) could be related to the local release of volatiles, which inhibits the nucleation of slab seismicity. If so, this is likely a local effect, since tomographic images do not show evidence for the presence of fluids or for a hydrated mantle above the slab (Wagner et al., 2005). Magnetotelluric observations, which should be sensitive to fluid conductivity, are not available under the Sierra Pie de Palo (Fig. 3) to compare with those that have been collected and related to fluid release under the Sierras de Córdoba to the east (Booker et al., 2004).

Phase changes like the basalt to eclogite transition, which have been related with the release of fluids (e.g., Kirby et al., 1996), could be delayed in the eastern region of the flat-slab subduction segment. The point where the slab descends into the mantle could coincide with the basalt to eclogite phase change (increase in the density of the slab), causing slab pull force and the pattern of observed earthquakes. If this is the case, the

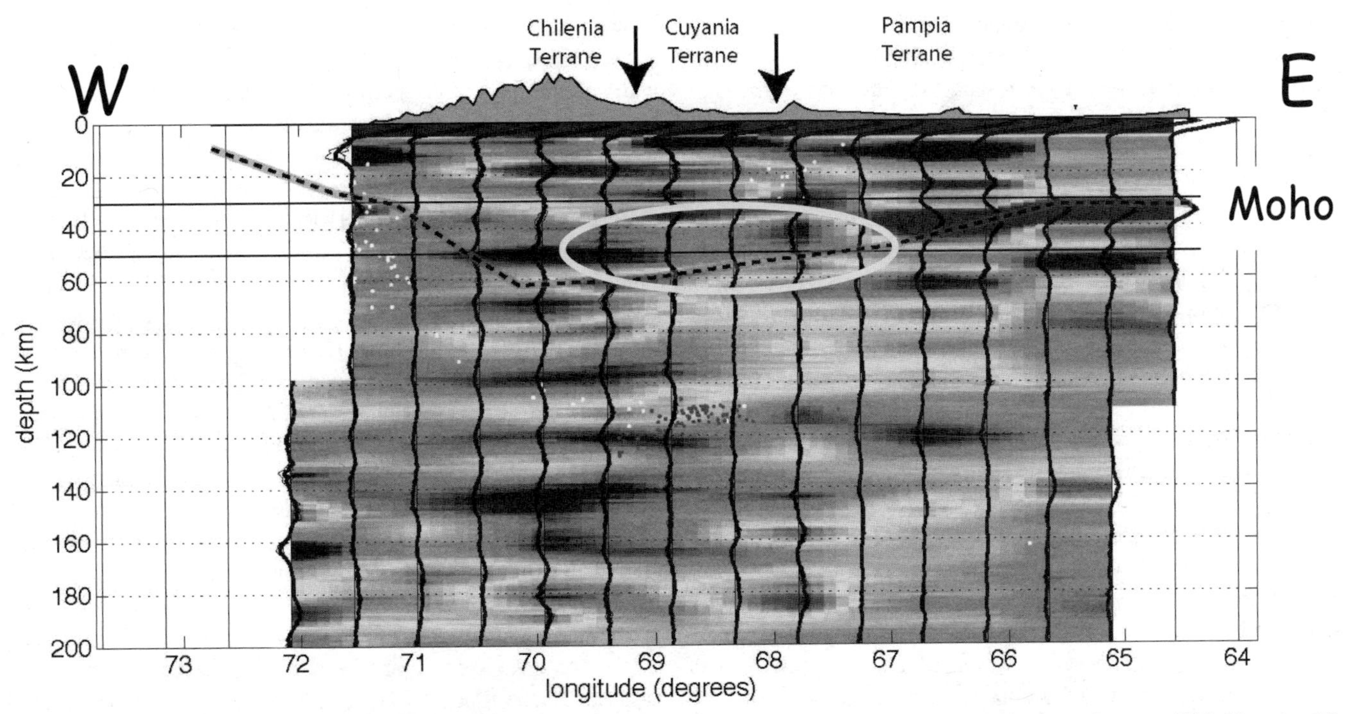

Figure 4. P-wave receiver function results from Gilbert et al. (2006) obtained using data from the transect of seismic stations at ~30°S. Note the differences in crustal structure. The eastern Pampia terrane shows a strong Moho signal and a 30–35-km-thick crust. In contrast, the western Cuyania and Chilenia terranes exhibit more complexity in their crustal structure and a thicker crust. The seismic Moho determinations using Pn phases and regional broadband waveform modeling by Fromm et al. (2004) and Alvarado et al. (2005) are shown by a dashed line. The oval indicates the region with a small-amplitude Moho signal interpreted as a weak impedance contrast probably related to partial eclogitization of the lower crust.

eclogite transformation should be accompanied by a release of water and some earthquake activity in the slab (Dobson et al., 2002; Hacker et al., 2003) to the east of Sierra Pie de Palo. More detailed seismic studies will help to constrain the structure of the slab and the pattern of seismicity associated with it.

LITHOSPHERIC STRUCTURE ABOVE THE FLAT SLAB

Nearly 200 km of crustal shortening along thrust belts over the flat-slab region has occurred since the Oligocene and produced the present-day tectonic structures across the region (e.g., Allmendinger et al., 1990; Vietor and Echtler, 2006). The crust in the high Cordillera and Precordillera above the flat slab thickened from near 40 km at 20 Ma to its present thickness of over 60 km by 6 Ma (Kay et al., 1991; Kay and Abbruzzi, 1996; Gilbert et al., 2006). This thickening was probably mostly related to shortening associated with thrusting along thin-skinned thrust faults above a décollement in the Precordillera, and to a lesser extent to thick-skinned block faults in the Sierras Pampeanas (Fig. 2). Changes in convergence parameters between the South American and Nazca plates due to the breakup of the Farallon plate into the Nazca and Cocos plates ushered in the present period of orthogonal convergence (e.g., Cande and Leslie, 1986). The latest early Miocene (20–16 Ma) can be characterized as a transitional period in the flat-slab region from an extensional to a compressional tectonic regime. Initial shallowing of the slab occurred during this period, as indicated by eastward broadening of the arc, high-angle thrusting in the Principal Cordillera, the termination of mafic backarc volcanism, and the inception of deformation in the Precordillera (Kay et al., 1991; Kay and Abruzzi, 1996).

A spatial correlation exists between the flat portion of the subducting Nazca slab and areas of foreland deformation in the western Sierras Pampeanas basement-cored uplifts (Fig. 2; see also Allmendinger et al., 1990; Ramos et al., 2002). Such a correlation can also be found within the flat-slab region in Peru (James and Snoke, 1994). Likewise, the shallowly subducting Farallon slab has been inferred to be responsible for producing a period of regional compression during the early Cenozoic in western North America (Coney and Reynolds, 1977; Dickinson and Snyder, 1978; Saleeby, 2003; DeCelles, 2004). However, the physical connection between deformation in the upper plate and the shallowly subducting slab remains enigmatic. Many ideas have been proposed for such a spatial correlation including: (1) coupling of the slab with the overriding lithosphere (Bird, 1988), (2) cooling and strengthening of the lithosphere caused by the pinching out of the asthenosphere above it, allowing for transmission of compressive stresses at the trench far inland (Gutscher, 2002), (3) a greater area of compressional coupling across the subduction interface, and thus more extensive shortening of the continental crust (Saleeby, 2003), (4) a simple correlation of crustal composition and structures available for compressional deformation (DeCelles, 2004), and (5) weakening of the crust due to eastward migration of volcanism with the flattening of the slab (Ramos et al., 2002). Many of these models give different predictions for what should be seen in the crustal and lithospheric structure under the Sierras Pampeanas.

The hypotheses that have been put forward to explain Laramide crustal deformation include block-tilting along through-going high-angle faults to low-angle thrusts that sole in the midcrust or extend well into the lower crust. Different suites of models propose shear thickening or lithospheric buckling of the crust to produce block uplifts (e.g., Fletcher, 1984; Verrall, 1989; Oldow et al., 1989). Seismic data indicate a relatively flat Moho under some Laramide arches (Prodehl and Lipman, 1989), making block-tilting models, which predict thrust faults cutting the Moho, unlikely. The evidence for rootless Laramide arches is also inconsistent with models that predict pure shear thickening of the lower crust under individual arches. Erslev (1993) noted that Laramide structures like the Beartooth, Bighorn, and Laramide uplifts define anastomosing arches, which span the entire Laramide orogeny, rather than distinct individual uplifts. Detachments with fault-propagating folding appear to fit the overall geometry of Laramide arches (Lowell, 1983; Oldow et al., 1989; Erslev, 1993). The continuity of Laramide arches leads to suggestions for the existence of a single master thrust that would connect the Laramide uplifts (Erslev, 1993).

Crustal structures within the Sierras Pampeanas illuminated by the CHARGE array could indicate the presence of midcrustal discontinuities at around 20 and 40 km depth that mark midcrustal detachment surfaces (Gilbert et al., 2006; Calkins et al., 2006), which would be in agreement with the studies of Cominguez and Ramos (1991) and Zapata and Allmendinger (1996) (Fig. 4). The accreted terranes that make up the Sierras Pampeanas and the boundaries between these terranes have influenced present-day deformation. Studies have shown that Cenozoic faulting reactivated shear zones formed during earlier collisional events (e.g., Ramos, 1994; Schmidt et al., 1995). More seismic sampling within the Sierras Pampeanas, spanning from the Sierras de San Juan to Sierras de Córdoba (Fig. 2), is needed to determine if a master detachment like that predicted for Laramide deformation in North America (Erslev, 1993) connects individual uplifts within the Sierras Pampeanas. In this case, a continuous pattern of faults that bound the uplifted blocks would connect to a master thrust fault at depth. In contrast, the crustal structure within the Sierras Pampeanas could be related to a discontinuous pattern of separated high-angle faults.

SEISMIC EVIDENCE FOR AN OVERTHICKENED CONTINENTAL CRUST WITH PARTIAL ECLOGITIZATION IN THE LOWER CRUST

CHARGE seismic studies have determined Moho depths along transects at ~30°S and 36°S using head-wave refraction analyses (Fromm et al., 2004), seismic moment tensor inversion of moderate crustal earthquakes (Alvarado et al., 2005), teleseismic receiver functions (Gilbert et al., 2006), and forward modeling of long-period (10–100 s) seismic waveforms (Alvarado et al., 2007). The details of these studies can be found in the cited references. Taken together, these studies indicate crustal thicknesses of more than 60 km under the Principal Cordillera in the northern CHARGE transect (Fig. 2). These studies also show that thick crust persists to the east under the Precordillera and western Sierras Pampeanas (Figs. 2 and 4), where the crust is 55–60 km thick, consistent with observations from McGlashan et al. (2008) using precursor seismic phases and Heit et al. (2008) using S-wave receiver functions. The crustal thickness estimates for the Precordillera and western Sierras Pampeanas from these studies are high compared to average global crustal thicknesses of ~41 km (Christensen and Mooney, 1995) and need to be explained.

Looking in detail, studies on a local scale by Regnier et al. (1994) used seismicity from the Wadati-Benioff zone to estimate crustal thickness beneath the PANDA seismic stations by determining S-to-P converted phases at the Moho (Fig. 2). They argued for a Moho depth of 60 km under the Precordillera in the eastern Andean foothills. The depth to the Moho in the western Sierras Pampeanas, which is a region dominated by thrust faults that uplift crystalline basement blocks was put at ~52 km.

Data from the CHARGE array have been used to identify differences in seismic velocities and structures across the various accreted terranes in the Sierras Pampeanas region (see Figs. 4–9) (Alvarado et al., 2005; Gilbert et al., 2006; Alvarado et al., 2007). The Cuyania terrane in the west under the western Sierras Pampeanas is characterized by a 50-km-thick crust, high P-wave velocities (Vp ~6.4 km/s), and a high P- to S-wave seismic velocity ratio (Vp/Vs) of ~1.8–1.85. High Vp/Vs ratios like these generally correspond to mafic crust or reflect the presence of fractures and faulting. To the east, the crust under the Pampia terrane is thinner; thicknesses are between 30 and 40 km, Vp is near 6.0 km/s, and Vp/Vs is less than 1.75. These values generally correspond to a more felsic crust. The Famatina belt between the Cuyania and Pampia terranes has an intermediate Vp/Vs ratio and a transitional crustal thickness.

Figure 5 shows results from Alvarado et al. (2007) for the Cuyania terrane, which principally consists of the Precordillera and the western Sierras Pampeanas region (Fig. 2). Alvarado et al. (2007) used single earthquake-station paths to do forward modeling of three-component, 10–80-s-period broadband waveforms assuming well-constrained focal mechanisms. These waveforms are sensitive to absolute P- and S-seismic waves that travel in the crust. The best correlation values between synthetic and observed seismograms were obtained for a 55-km-thick crust with seismic velocities that were highest at the deepest (bottom 15–20 km) levels. P-wave teleseismic and local receiver function analyses from Gilbert et al. (2006) and Calkins et al. (2006) for a CHARGE station located in the western flank of Sierra Pie de Palo (station JUAN in Fig. 2) are more sensitive to impedance contrasts across discontinuities.

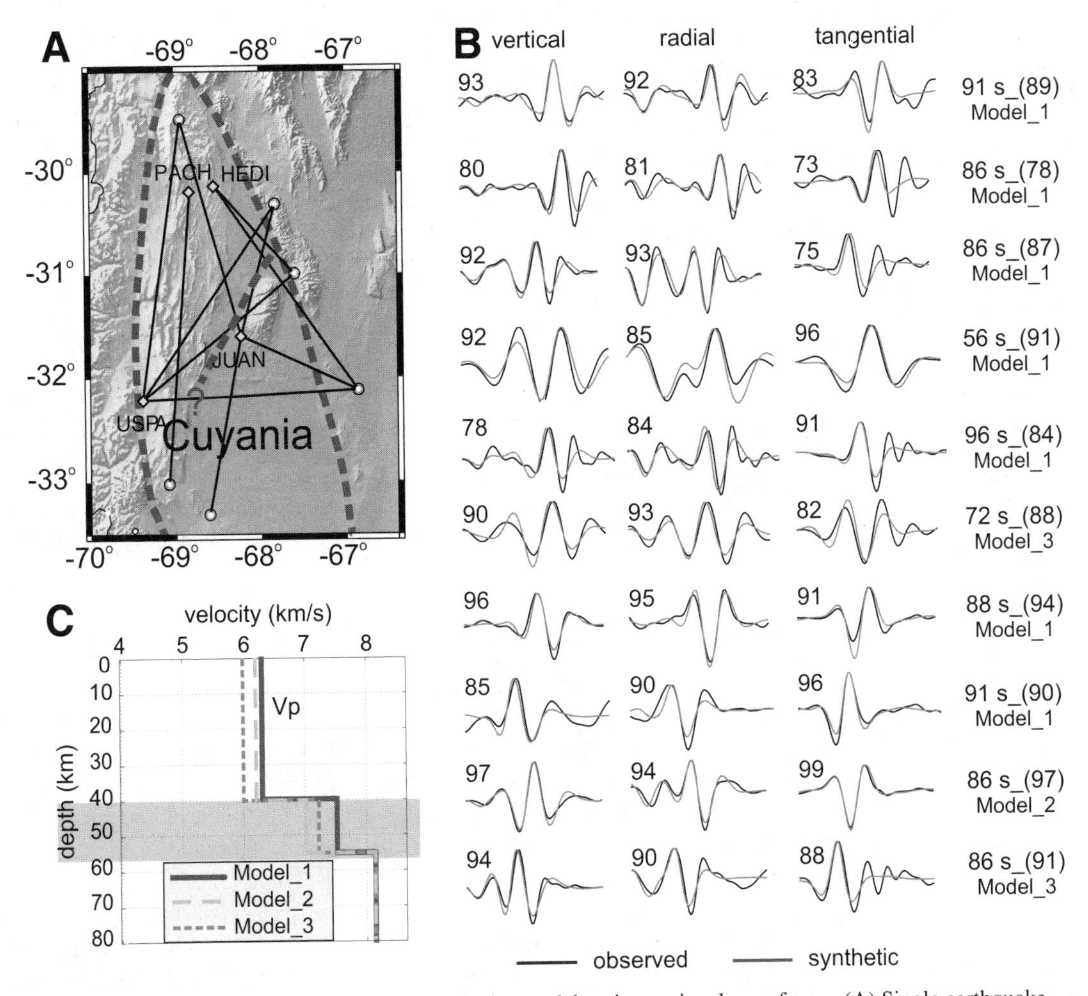

Figure 5. Test of different crustal seismic velocity models using regional waveforms. (A) Single earthquake-CHARGE station paths investigated. (B) Broadband synthetic seismograms calculated for one crustal seismic velocity structure (model 1, 2, or 3) compared with observed data. The percentage of cross correlation is shown on top of each seismic waveform component, and the maximum cross correlation is shown in the last column in parentheses, together with the record length and crustal model used in C. (C) Crustal models that produce the maximum correlation in the forward modeling shown in A and B. Note that the models have a high-velocity lower crust under the Cuyania terrane (modified from Alvarado et al., 2007).

The results from these studies are consistent with an ~55-km-thick crust and a gradual increase in seismic velocities at the lower crust between depths of 35 km to 50 km (Fig. 6).

A map of the differences in crustal composition using seismic velocity structures across the region from west to east is in agreement with patterns of crustal seismicity in the Andean back-arc crust. The western terranes, which have a more mafic average composition and a thicker crust, nucleate seismicity at deeper levels (focal depths <35 km) (Smalley and Isacks, 1987; Regnier et al., 1992; Smalley et al., 1993; Alvarado et al., 2005, 2007). In contrast, to the east, in the eastern Sierras Pampeanas, refined focal hypocentral depths estimated by broadband waveform modeling show depths of <20 km (Alvarado et al., 2005). The shallower earthquake nucleation depth in this region is consistent with a more fragile upper-crustal regime consistent with the exposed rocks in the region, and a more quartz-rich rheologic control of the brittle-ductile transition (Alvarado et al., 2005, 2007).

The highest concentration and largest size historic and modern crustal earthquakes have occurred in the western terranes (Chinn and Isacks, 1983; Langer and Hartzell, 1996; Alvarado and Beck, 2006; INPRES, 2009). Thrust-fault focal mechanism solutions dominate the region around Sierra Pie de Palo in the western Sierras Pampeanas and its transition to the Precordillera to the west (Regnier et al., 1992; Alvarado et al., 2005; Salazar, 2005; Siame and Bellier, 2005). Based on these seismic observations, the high Vp/Vs values, and the mafic composition of the exposed rocks in this region, the brittle-ductile transition in the crust under this area is considered to be at a greater depth, consistent with a deeper nucleation of the crustal earthquakes. In addition, we note the spatial correlation of: (1) the high crustal

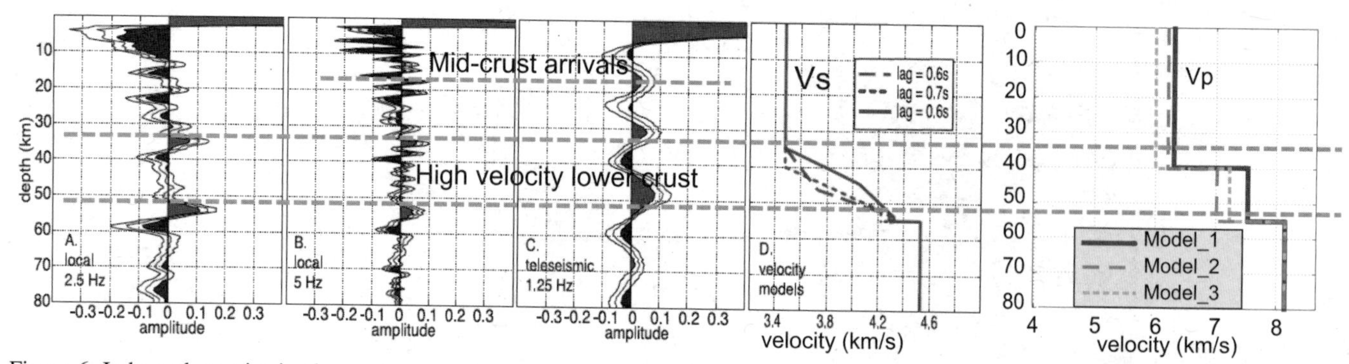

Figure 6. Independent seismic observations used to constrain a preferred crustal model for the Cuyania terrane. The first two diagrams show observations of P-wave receiver functions (2.5 Hz and 5 Hz) from Calkins et al. (2006) beneath station JUAN (see Fig. 2). The third diagram shows an analysis for teleseismic data from Gilbert et al. (2006). The receiver functions are consistent with a crustal model in which the seismic velocity gradually increases in the lower crust (fourth diagram). The far-right diagram shows the crustal models in Figure 5 obtained from seismic waveform modeling along a single path by Alvarado et al. (2007). Note that the high-velocity lower crust and the midcrustal phases could be linked to décollement faults.

seismicity right above the flat-slab subduction segment along the trend of the suggested inland prolongation of the subducting Juan Fernández Ridge, (2) the larger magnitudes and the higher number of intermediate-depth earthquakes with hypocenters along the subducted Juan Fernández Ridge than in the surrounding slab (Anderson et al., 2007), (3) the dry, cold, magnesium-rich nature of the upper-mantle wedge between the flat slab and the continental Moho (Wagner et al., 2005), and (4) the presence of a high-density partially eclogitized lower continental crust in this region. Such a combination could potentially contribute to the crustal seismicity by producing an effective guide in transferring stresses from the slab to the upper plate.

We speculate that the partially eclogitized lower crust in the flat-slab region could extend into the nearby provinces to the south and/or to the north above where the slab is subducting with a more "normal" angle (Fig. 2). If so, such a lower crust might be important in transferring stresses north and south into these regions where some large crustal earthquakes have occurred. Finally, there is an apparent concentration of crustal earthquakes along some of the terrane boundaries (Alvarado et al., 2005) as has also been observed in other regions in South America. For example, crustal seismicity in northeast Brazil seems to occur along boundaries between basement blocks, which have been imaged using upper-mantle seismic tomography (Peres-Rocha, 2008).

CRUSTAL MODELS BASED ON SEISMIC AND GRAVITY OBSERVATIONS

Inspection of the seismically determined Moho and topography in the flat-slab region shows that the average elevations for the Precordillera (~2000 m) and western Sierras Pampeanas (~1000 m) are lower than would be expected from isostasy (Fig. 7A). This can be shown using a methodology like that used by Introcaso et al. (1992), in which a 36-km-thick crustal block is considered to be isostatically compensated using an Airy isostasy model. A simple relationship ($M = 36 + 6.5 \times H$) between crustal thickness (M) and elevation (H) describes the predicted crustal thickness at a particular elevation, where the factor of 6.5 accounts for the differences in density between the crust (e.g., 2.9 g/cm^3) and upper mantle (e.g., 3.3 g/cm^3). Figure 7A shows the seismic crustal thicknesses constrained for the CHARGE station NEGR (Fig. 2), located at ~4400 m elevation, to be ~64 km. This thickness is consistent with the crustal thickness of 62 km obtained from receiver functions (Gilbert et al., 2006). Overall, the stations in the high Cordillera appear to be isostatically compensated with the exception of the station USPA (Fig. 2). In contrast, most of the stations in the Andean backarc, east of 69.5°W, are at elevations lower than would be expected.

Figure 7B shows the relationship between the seismically determined crustal thicknesses for the region plotted versus Bouguer anomalies from the database of the Instituto Geográfico Militar from Argentina. The plot shows that variations in crustal thickness of more than 20 km are not reflected in the Bouguer anomalies. For instance, a Bouguer anomaly increase of ~100 mGal (between –100 and –200 mGal) is observed for an increase from 50 to 60 km in the seismically observed crustal thickness. Studies by Yuan et al. (2002) in the Altiplano-Puna region show larger Bouguer anomalies between –200 and –400 mGal for similar crustal thickness variation from 50 to 60 km. Hence, we cannot explain the low elevations in the western Sierras Pampeanas and Precordillera by crustal thickness alone and require some excess density in the crust or upper mantle.

In the absence of seismic data, studies by Introcaso et al. (1992) used a one-layer simple crustal model to evaluate the Airy isostasy along east-west profiles at about ~30°S and to the south. These gravity models (Fig. 8A) were consistent with a maximum crustal thickness of 70 km beneath the high Cordillera, which is in

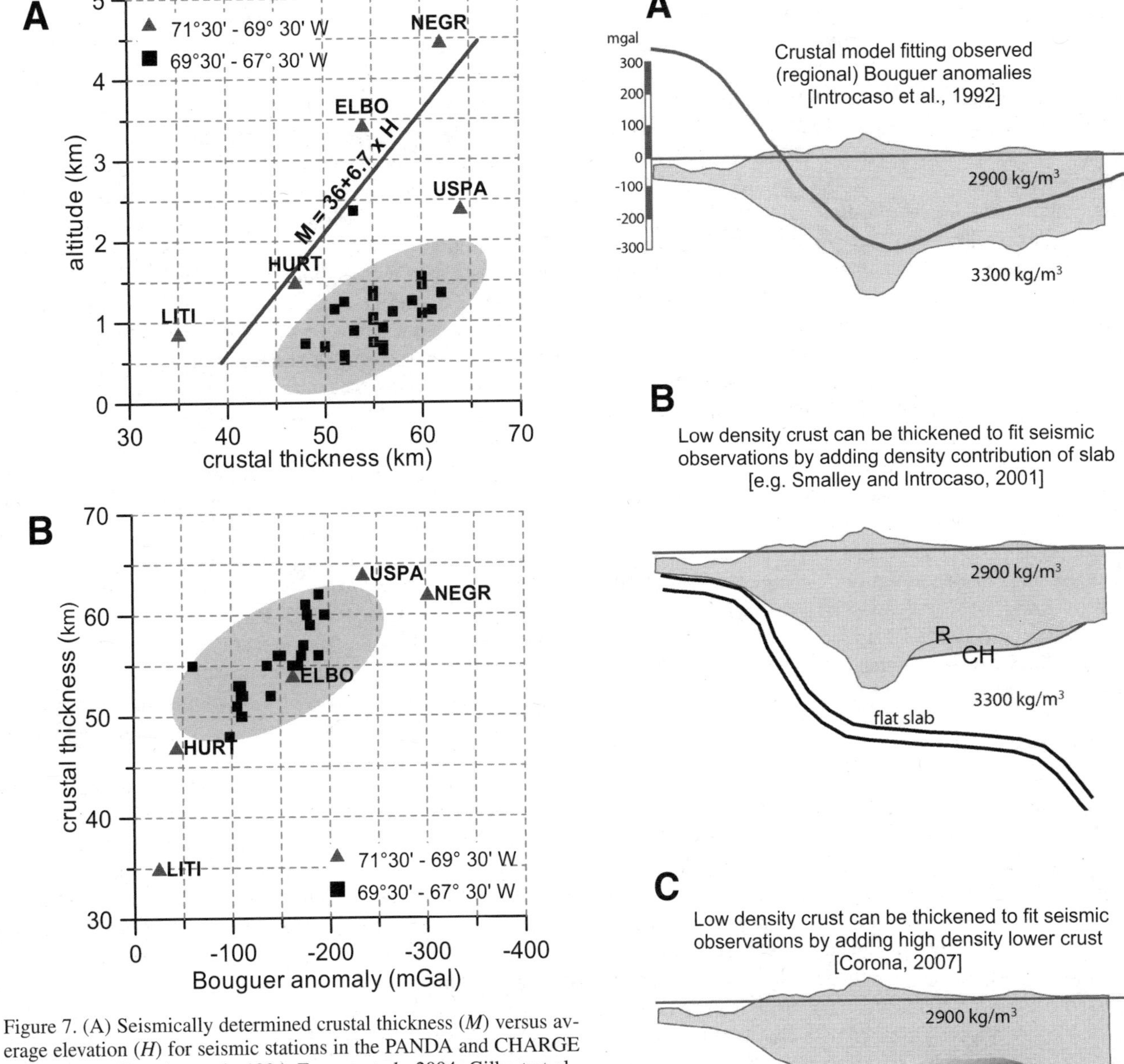

Figure 7. (A) Seismically determined crustal thickness (M) versus average elevation (H) for seismic stations in the PANDA and CHARGE deployments (Regnier et al., 1994; Fromm et al., 2004; Gilbert et al., 2006; Alvarado et al., 2007). The data for the stations in the high Cordillera between ~69.5°W and 71.5°W are shown as triangles; those for stations in the Precordillera and western Sierras Pampeanas between 67.5°W and 69.5°W are shown by squares. The solid line, which is defined by $M = 36 + 6.5 \times H$, shows where isostatic compensation should occur using a reference crustal thickness of 36 km at sea level. Note that, except for USPA, the stations at high elevation are generally isostatically compensated at the Moho. In contrast, the Precordillera and western Sierras Pampeanas (Cuyania terrane) stations lie to the east plot below the line, indicating that their elevations are too low to be isostatically compensated under these assumptions. (B) Observed Bouguer gravity anomalies versus seismic crustal thickness for the same stations. Note that crustal thickness generally correlates with Bouguer anomalies, especially for the stations in the dotted oval. Thus, the amplitudes of the gravity anomalies in the Precordillera–western Sierras Pampeanas (squares) are relatively low compared to those generally measured for similar crustal thicknesses elsewhere, consistent with a high-density crust under the Cuyania stations.

Figure 8. Seismic, gravity, and topographic crustal models near 30°S along the northern CHARGE transect (see Fig. 2). (A) Crustal model with Moho constraints obtained from observed regional Bouguer anomalies. This model predicts a thick continental crust (Introcaso et al., 1992). (B) Alternative models that fit the Bouguer anomalies and seismic Moho determinations by Regnier et al. (1994) (R) and the CHARGE studies (CH) require a low-density thick crust and the contribution of the subducted Nazca plate (e.g., Smalley and Introcaso, 2001). The slab is not at the same scale as the crust. (C) A thick crust with an increasing density or high-density lower crust under the Cuyania terrane can fit the seismically observed Moho, the observed Bouguer gravity anomalies, and the observed topography (Corona, 2007). This model is consistent with partial eclogitization of the lower crust in the western terranes.

good agreement with recent seismic determinations of 60–65 km (e.g., Alvarado et al., 2007). In contrast, the same types of one-layer simple crustal models were more difficult to apply further to the east due to the narrow width of the Sierras Pampeanas and their low elevations. Introcaso et al. (1992) tried to use the effect of lithospheric rigidity to account for the apparent excess in density needed for isostatic balance in the middle- to upper-crustal levels in this region.

Based on the Moho determinations from the PANDA network in the Precordillera and the western Sierras Pampeanas (Regnier et al., 1994), several authors subsequently postulated that the contribution of the horizontally subducted Nazca plate could play an important role in explaining isostatic compensation at slab depths in simple one-density "thick" crust models (Gimenez et al., 2000; Smalley and Introcaso, 2001) (see Fig. 8B). However, such a model is difficult to reconcile with the expected buoyancy of the subducting plate that has been called upon to account for the slab flattening (Yañéz et al., 2001).

A more recent study by Corona (2007) combined seismic, gravity, and topographic observations along a cross section at ~30°S latitude. This study shows that all of the observations can be accounted for a model with a thick continental crust and a high-density lower crust (e.g., +0.224 g/cm^3 density contrast) (Fig. 8C), which is consistent with the idea of partial eclogitization of the lower crust suggested by Gilbert et al. (2006) and Alvarado et al. (2007) (see Fig. 5).

CONCLUSIONS

Modern seismic observations from the CHARGE and CHARSME experiments in the region of flat-slab subduction of the oceanic Nazca plate beneath Chile and Argentina refine the seismicity and shape of the Wadati-Benioff zone. The hypocenters define a "finger" of seismicity at about ~100 km depth that extends along the subducting plate between 31°S and 32°S located in western Argentina. This "finger" shows a strong correlation with the suggested inland prolongation of the Juan Fernández Ridge. Physical processes related with buoyancy forces from the ridge and slab pull could be responsible for the shape and associated seismicity of the subducted plate. Stress tensor orientations show that the horizontal slab is in extension, whereas stress tensor orientations in the interplate (coupled) region beneath central Chile exhibit a typical pattern of compression. Stress tensors indicate that the backarc crust of the upper plate is also in compression.

Variations in the seismic properties of the Andean backarc crust can be correlated with the regions of the various pre-Mesozoic accreted terranes in the flat-slab region. The Cuyania terrane, which corresponds with the central region including the Precordillera and the western Sierras Pampeanas, has a high Vp value (~6.4 km/s), a high Vp/Vs ratio (~1.8), a thick crust (~55 km), and a complex crustal structure. In contrast, the Pampia terrane, which corresponds with the eastern Sierras Pampeanas, is characterized by a lower Vp value (~6.0 km/s), a lower Vp/Vs ratio (~1.73), a thinner crust (~35 km), and a simpler crustal structure. These observations are consistent with a more mafic composition or more fractures or faults in the western terranes versus a more quartz-rich composition in the eastern Pampia terrane.

Seismic and gravity studies are consistent with the model for a high-density lower crust in the Precordillera and western Sierras Pampeanas (Cuyania terrane) region. A partially eclogitized, high-density lower crust can explain the overall lower elevations of this region. This situation contrasts with the more common crustal situation in other parts of the Andes, where thickened crust is associated with high topography, and crustal thickening is a response to Andean age, rather than pre-Andean shortening. We note that the spatial correlation of high crustal seismicity, high-density lower crust, flat-slab geometry, and high slab seismic activity at ~100 km may result in stresses being transmitted from the slab to the crust. In view of intense historical and modern crustal seismicity, the history of deformation, and midcrustal receiver function interfaces that could reflect structural décollements, we argue that the relatively low elevations in the western Sierras Pampeanas reflect a combination of the crustal deformation history, partial eclogitization of the lower crust, and, perhaps, a compensational effect from the underlying flat-slab segment.

The presence of the flat slab, the dry, cold lithosphere above the flat slab, and a high-density lower crust can potentially enhance the nucleation of large crustal earthquakes by providing an efficient mechanism to transfer stresses from the slab into the upper plate (Fig. 9). The probable extension of partially eclogitized lower crust to the north and to the south of the flat-slab region may play an important role as a stress guide in nucleating strong crustal seismicity, even though the slab geometry is more normal.

Ongoing broadband seismic experiments in Argentina (e.g., Sierras pampeanas Experiment using a Multicomponent Broadband Array [SIEMBRA], Lithospheric Dynamics in the Southernmost Andean Plateau [PUNA], Eastern Sierras Pampeanas [ESP]) will continue to provide insights into the evolution of flat-slab subduction processes.

ACKNOWLEDGMENTS

Financial support for this study came from the U.S. National Science Foundation (grants EAR-9811870 and EAR-0510966), Argentina (projects PICT-0122, CICITCA E814), Chile (FONDECYT 1990355, 1020972, 1050758), and the French Géosciences Azur-L'Institut de recherché pour le développement. We acknowledge Instituto Nacional de Prevención Sísmica, the National University of San Juan (Argentina), and the Servicio Sismológico from the University of Chile for their logistical support and assistance in the field and the entire CHARGE and CHARSME Working Groups. The instruments used in the CHARGE and CHARSME field experiments were provided by the PASSCAL facility of the Incorporated Research Institutions for Seismology IRIS (USA) and Lithoscope (France). The CHARGE data are available through the IRIS Data Management Center. The National Science Foundation under Cooperative Agreement EAR-0004370 supports

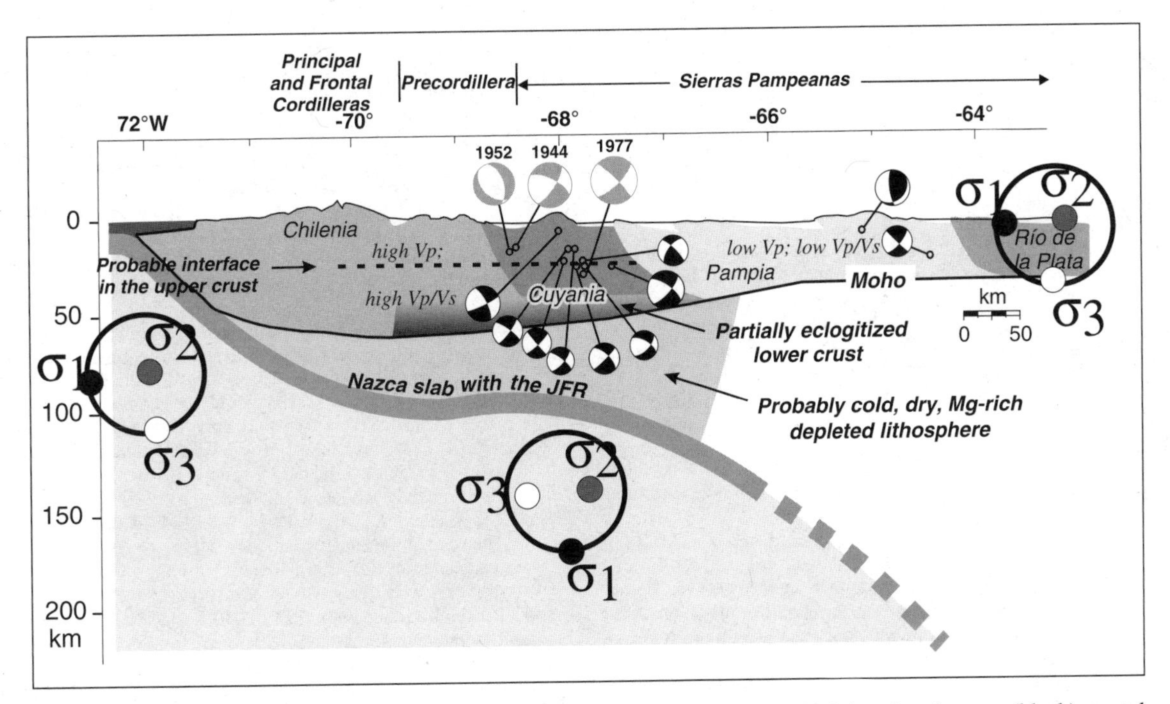

Figure 9. Schematic cross section over the flat-slab subduction showing the historical (gray) and recent (black) crustal earthquake focal mechanisms obtained from seismic moment tensor inversions (Langer and Hartzell, 1996; Alvarado et al., 2005; Alvarado and Beck, 2006). The mechanisms are in vertical projections on the back hemisphere, with dark compressional quadrants (modified from Alvarado et al., 2007). Also shown are the stress tensor orientations from Salazar (2005) for the continental upper plate, interplate (coupled), and intermediate-depth seismogenic zones (see Table 1). The regions of the Chilenia, Cuyania, Pampia, and Rio de la Plata terranes are shown relative to the changing geophysical characteristics across the transect. JFR—Juan Fernández Ridge.

the facilities of the IRIS Consortium. Maps were generated using GMT software (Wessel and Smith, 1998). We thank Suzanne Kay, Víctor Ramos, Augusto Rapalini, Tony Monfret, Tonya Richardson, and Larry Brown for their comments, suggestions, and corrections, which improved this paper.

REFERENCES CITED

Aceñolaza, E.G., Miller, H., and Toselli, A.J., 2002, Proterozoic–early Paleozoic evolution in western South America: A discussion: Tectonophysics, v. 354, p. 121–137, doi: 10.1016/S0040-1951(02)00295-0.

Allmendinger, R.W., Figueroa, D., Snyder, D., Beer, J., Mpodozis, C., and Isacks, B.L., 1990, Foreland shortening and crustal balancing in the Andes at 30°S latitude: Tectonics, v. 9, p. 789–809, doi: 10.1029/TC009i004p00789.

Alvarado, P., and Beck, S., 2006, Source characterization of the San Juan (Argentina) crustal earthquakes of 15 January 1944 (Mw 7.0) and 11 June 1952 (Mw 6.8): Earth and Planetary Science Letters, v. 243, p. 615–631, doi: 10.1016/j.epsl.2006.01.015.

Alvarado, P., Beck, S., Zandt, G., Araujo, M., and Triep, E., 2005, Crustal deformation in the south-central Andes backarc terranes as viewed from regional broad-band seismic waveform modeling: Geophysical Journal International, v. 163, no. 2, p. 580–598, doi: 10.1111/j.1365-246X.2005.02759.x.

Alvarado, P., Beck, S., and Zandt, G., 2007, Crustal structure of the south-central Andes Cordillera and backarc region from regional waveform modeling: Geophysical Journal International, v. 170, no. 2, p. 858–875, doi: 10.1111/j.1365-246X.2007.03452.x.

Anderson, M., Alvarado, P., Zandt, G., and Beck, S., 2007, Geometry and brittle deformation of the subducting Nazca plate, central Chile and Argentina: Geophysical Journal International, v. 171, no. 1, p. 419–434, doi: 10.1111/j.1365-246X.2007.03483.x.

Astini, R.A., Benedetto, J.L., and Vaccari, N.E., 1995, The early Paleozoic evolution of the Argentine Precordillera as a Laurentian rifted, drifted, and collided terrane: A geodynamic model: Geological Society of America Bulletin, v. 107, no. 3, p. 253–273, doi: 10.1130/0016-7606(1995)107<0253:TEPEOT>2.3.CO;2.

Baldis, B.A., Peralta, S., and Villegas, R., 1989, Esquematizaciones de una posible transcurrencia del terreno de Precordillera como fragmento continental procedente de áreas pampeanobonaerenses: Facultad de Ciencias Naturales, Universidad Nacional de Tucumán, Serie Correlación Geológica, v. 5, p. 81–100.

Baldo, E.G.A., Demange, M., and Martino, R.D., 1996, Evolution of the Sierras de Cordoba, Argentina: Tectonophysics, v. 267, no. 1–4, p. 121–142, doi: 10.1016/S0040-1951(96)00092-3.

Barazangi, M., and Isacks, B.L., 1976, Spatial distribution of earthquakes and subduction of the Nazca plate beneath South America: Geology, v. 4, no. 11, p. 686–692, doi: 10.1130/0091-7613(1976)4<686:SDOEAS>2.0.CO;2.

Barrientos, S., Vera, E., Alvarado, P., and Monfret, T., 2004, Crustal seismicity in Central Chile: Journal of South American Earth Sciences, v. 16, no. 8, p. 759–768, doi: 10.1016/j.jsames.2003.12.001.

Basei, M.A.S., Siga, O., Jr., Harara, O.M., Preciozzi, F., Sato, K., and Kaufuss, G., 2001, Precambrian terranes of African affinities in the southeastern part of Brazil and Uruguay, *in* Proceedings of the 3rd South American Symposium on Isotope Geology: Pucón, Chile, Extended Abstracts, p. 98–101.

Beck, S.L., Gilbert, H., Wagner, L.S., Alvarado, P., Anderson, M., Zandt, G., Araujo, M., and Triep, E., 2005, The lithospheric structure of the Sierras Pampeanas region of Argentina: Geological Society of America Abstracts with Programs, v. 37, no. 7, p. 553.

Bird, P., 1988, Formation of the Rocky Mountains, western United States: A continuum computer model: Science, v. 239, p. 1501–1507, doi: 10.1126/science.239.4847.1501.

Booker, J.R., Favetto, A., and Pomposiello, M.C., 2004, Low electrical resistivity associated with plunging of the Nazca flat-slab beneath Argentina: Nature, v. 429, no. 6990, p. 399–403, doi: 10.1038/nature02565.

Cahill, T., and Isacks, B.L., 1992, Seismicity and shape of the subducted Nazca plate: Journal of Geophysical Research, v. 97, no. B12, p. 17,503–17,529, doi: 10.1029/92JB00493.

Calkins, J.A., Zandt, G., Gilbert, H.J., and Beck, S.L., 2006, Crustal images from San Juan, Argentina, obtained using high frequency local event receiver functions: Geophysical Research Letters, v. 33, no. 7, L07309, p. 1–4, doi: 10.1029/2005GL025516.

Caminos, R., 1979, Cordillera Frontal: Córdoba, *in* Castellanos, T.G., Sersic, J.L., Amuchastegui, S., Caputto, R., Cocucci, A.E., Fuchs, G.L., Gordillo, C.E., and Melo, C.R., eds., 2nd Simposio de Geología Regional Argentina: Córdoba, Argentina, Academia Nacional de Ciencias, v. 1, p. 394–453.

Cande, S.C., and Leslie, R.B., 1986, Late Cenozoic tectonics of the southern Chile trench: Journal of Geophysical Research, v. 91, p. 471–496, doi: 10.1029/JB091iB01p00471.

Casquet, C., Pankhurst, R.J., Rapela, C.W., Galindo, C., Fanning, C.M., Chiaradia, M., Baldo, E., González-Casado, J.M., and Dalquist, J.A., 2008, The Maz terrane: A Mesoproterozoic domain in the western Sierras Pampeanas (Argentina) equivalent to the Arequipa-Antofalla block of southern Peru? Implications for Western Gondwana margin evolution, *in* Casquet, C., Pankhurst, R.J., and Vaughan, P.M., eds., The Western Gondwana Margin: Proterozoic to Mesozoic: Gondwana Research, v. 13, no. 2, p. 163–175, doi: 10.1016/j.gr.2007.04.005.

Castro de Machuca, B., Arancibia, G., Morata, D., Belmar, M., Previley, L., and Pontoriero, S., 2008, *P-T-t* evolution of an early Silurian medium-grade shear zone on the west side of the Famatinian magmatic arc, Argentina: Implications for the assembly of the western Gondwana margin, *in* Casquet, C., Pankhurst, R.J., and Vaughan, P.M., eds., The Western Gondwana Margin: Proterozoic to Mesozoic: Gondwana Research, v. 13, no. 2, p. 216–226.

Chase, C.G., 1978, Plate kinematics: Americas, East-Africa, and the rest of the world: Earth and Planetary Science Letters, v. 37, no. 3, p. 355–368, doi: 10.1016/0012-821X(78)90051-1.

Chinn, D., and Isacks, B., 1983, Accurate source depths and focal mechanisms of shallow earthquakes in western South America and in the New Hebrides Island arc: Tectonics, v. 2, no. 6, p. 529–563, doi: 10.1029/TC002i006p00529.

Christensen, N.I., and Mooney, W.D., 1995, Seismic velocity structure and composition of the continental crust: A global view: Journal of Geophysical Research, v. 100, p. 9761–9788, doi: 10.1029/95JB00259.

Cominguez, A.H., and Ramos, V.A., 1991, La estructura profunda entre Precordillera y Sierras Pampeanas de la Argentina: Evidencias de la sísmica de reflexión profunda: Revista Geológica de Chile, v. 18, no. 1, p. 3–14.

Coney, P.J., and Reynolds, S.J., 1977, Cordilleran Benioff zones: Nature, v. 270, no. 5636, p. 403–406, doi: 10.1038/270403a0.

Corona, G., 2007, Estructura Litosférica del Sistema Andes-Sierras Pampeanas en la Banda 30°S–31°S a Partir de Datos de Gravedad y Sísmicos [Undergraduate final work thesis]: San Juan, Argentina, Facultad de Ciencias Exactas, Físicas y Naturales, Universidad Nacional de San Juan, 50 p.

Dalla Salda, L., Cingolani, C., and Varela, R., 1992, Early Paleozoic orogenic belt of the Andes in southwestern South America: Result of Laurentia-Gondwana collision?: Geology, v. 20, p. 617–620, doi: 10.1130/0091-7613(1992)020<0617:EPOBOT>2.3.CO;2.

Davis, J.S., Roeske, S.M., McClelland, W.C., and Kay, S.M., 2000, Mafic and ultramafic crustal fragments of the southwestern Precordillera terrane and their bearing on tectonic models of the early Paleozoic in western Argentina: Geology, v. 28, no. 2, p. 171–174, doi: 10.1130/0091-7613(2000)28<171:MAUCFO>2.0.CO;2.

DeCelles, P.G., 2004, Late Jurassic to Eocene evolution of the Cordilleran thrust belt and foreland basin system, western USA: American Journal of Science, v. 304, no. 2, p. 105–168, doi: 10.2475/ajs.304.2.105.

Deshayes, P., Monfret, T., Pardo, M., and Vera, E., 2008, Three-dimensional P- and S-wave seismic attenuation models in central Chile–western Argentina (30°–34°S) from local recorded earthquakes: Nice, France, Proceedings of the 7th International Symposium on Andean Geodynamics Extended Abstracts, p. 184–187.

Dickinson, W.R., and Snyder, W.S., 1978, Plate tectonics of the Laramide orogeny, *in* Matthews, V.I., ed., Laramide Folding Associated with Basement Block Faulting in the Western U.S.: Geological Society of America Memoir 151, p. 355–366.

Dobson, D.P., Meredith, P.G., and Boon, S.A., 2002, Simulation of subduction zone seismicity by dehydration of serpentine: Science, v. 298, no. 5597, p. 1407–1410, doi: 10.1126/science.1075390.

Erslev, E.A., 1993, Thrusts, back-thrusts, and detachment of Rocky Mountain foreland arches, *in* Schmidt, C.J., Chase, R.B., and Erslev, E.A., eds., Laramide Basement Deformation in the Rocky Mountain Foreland of the Western United States: Geological Society of America Special Paper 280, p. 339–358.

Finney, S.C., Peralta, S.P., Gehrels, G.E., and Marsaglia, K.M., 2005, The early Paleozoic history of the Cuyania (greater Precordillera) terrane of western Argentina: Evidence from geochronology of detrital zircons from Middle Cambrian sandstones, *in* Aceñolaza, G.F., ed., Ordovician Revisited: Reconstructing a Unique Period in Earth History: Geologica Acta, v. 3, p. 330–354.

Fletcher, R.C., 1984, Instability of lithosphere undergoing shortening: A model for Laramide foreland structures: Geological Society of America Abstracts with Programs, v. 16, no. 2, p. 83.

Franzese, J., Spalletti, L., Gómez Pérez, I., and Macdonald, D., 2003, Tectonic and paleoenvironmental evolution of Mesozoic sedimentary basins along the Andean foothills of Argentina (32°–54°S): Journal of South American Earth Sciences, v. 16, p. 81–90, doi: 10.1016/S0895-9811(03)00020-8.

Fromm, R., Zandt, G., and Beck, S.L., 2004, Crustal thickness beneath the Andes and Sierras Pampeanas at 30°S inferred from Pn apparent phase velocities: Geophysical Research Letters, v. 31, doi: 10.1029/2003GL019231.

Fromm, R., Alvarado, P., Beck, S., and Zandt, G., 2006, The April 9, 2001, Juan Fernandez Ridge (Mw 6.7) tensional outer-rise earthquake and its aftershock sequence: Journal of Seismology, v. 10, p. 163–170, doi: 10.1007/s10950-006-9013-3.

Gilbert, H., Beck, S., and Zandt, G., 2006, Lithospheric and upper mantle structure of central Chile and Argentina: Geophysical Journal International, v. 165, no. 1, p. 383–398, doi: 10.1111/j.1365-246X.2006.02867.x.

Gimenez, M., Martinez, M.P., and Introcaso, A., 2000, A crustal model based mainly on gravity data in the area between the Bermejo Basin and the Sierras de Valle Fértil, Argentina: Journal of South American Earth Sciences, v. 13, no. 3, p. 275–286, doi: 10.1016/S0895-9811(00)00012-2.

Gutscher, M.A., 2002, Andean subduction styles and their effect on thermal structure and interplate coupling: Journal of South American Earth Sciences, v. 15, no. 1, p. 3–10, doi: 10.1016/S0895-9811(02)00002-0.

Gutscher, M.A., Spakman, W., Bijwaard, H., and Engdahl, E.R., 2000, Geodynamics of flat subduction: Seismicity and tomographic constraints from the Andean margin: Tectonics, v. 19, p. 814–833, doi: 10.1029/1999TC001152.

Hacker, B.R., Peacock, S.M., Abers, G.A., and Holloway, S.D., 2003, Subduction factory: 2. Are intermediate-depth earthquakes in subducting slabs linked to metamorphic dehydration reactions?: Journal of Geophysical Research, v. 108, doi: 10.1029/2001JB001129.

Heit, B., Yuan, X., Bianchi, M., Sodoudi, F., and Kind, R., 2008, Crustal thickness estimation beneath the southern central Andes at 30°S and 36°S from S-wave receiver function analysis: Geophysical Journal International, v. 174, no. 1, p. 249–254, doi: 10.1111/j.1365-246X.2008.03780.x.

Humphreys, E., 2009, this volume, Relation of flat subduction to magmatism and deformation in the western United States, *in* Kay, S.M., Ramos, V.A., and Dickinson, W.R., eds., Backbone of the Americas: Shallow Subduction, Plateau Uplift, and Ridge and Terrane Collision: Geological Society of America Memoir 204, doi: 10.1130/2009.1204(04).

INPRES (Instituto Nacional de Prevención Sísmica), 2009, Catálogo de Terremotos Históricos: http://www.inpres.gov.ar (February 2009).

Introcaso, A., Pacino, M.C., and Fraga, H., 1992, Gravity, isostasy and Andean crustal shortening between latitudes 30°S and 35°S: Tectonophysics, v. 205, p. 31–48, doi: 10.1016/0040-1951(92)90416-4.

IRIS (Incorporated Research Institutions for Seismology), 2004, The IRIS Consortium, twenty years of support for seismological research (1984–2004), *in* Reviews: Washington, D.C., The Incorporated Research Institutions for Seismology, p. 1–72.

Isacks, B.L., Kay, S.M., Fielding, E.J., and Jordan, T., 1986, Andean volcanism; icing on the cake: Eos (Transactions, American Geophysical Union), v. 67, no. 44, p. 1073.

James, D.E., and Sacks, S., 1999, Cenozoic deformation in the Central Andes: A geophysical perspective, *in* Skinner, B.J., ed., Geology and Ore Deposits of the Central Andes: Society of Economic Geologists Special Paper 7, p. 1–25.

James, D.E., and Snoke, J.A., 1994, Structure and tectonics in the region of flat subduction beneath central Peru: Crust and uppermost mantle: Journal of Geophysical Research, v. 99, no. B4, p. 6899–6912, doi: 10.1029/93JB03112.

Jordan, T.E., and Allmendinger, R.W., 1986, The Sierras Pampeanas of Argentina: A modern analogue of Rocky Mountain foreland deformation: American Journal of Science, v. 286, p. 737–764.

Jordan, T.E., Allmendinger, R.W., Brewer, J.A., Ramos, V.A., and Ando, C.J., 1983, Andean tectonics related to geometry of subducted Nazca plate: Geological Society of America Bulletin, v. 94, p. 341–361, doi: 10.1130/0016-7606(1983)94<341:ATRTGO>2.0.CO;2.

Kay, S.M., and Abbruzzi, J.M., 1996, Magmatic evidence for Neogene lithospheric evolution of the central Andean flat-slab between 30°S and 32°S: Tectonophysics, v. 259, p. 15–29, doi: 10.1016/0040-1951(96)00032-7.

Kay, S.M., and Mpodozis, C., 2001, Central Andean ore deposits linked to evolving shallow subduction systems and thickening crust: GSA Today, v. 11, no. 3, p. 4–9, doi: 10.1130/1052-5173(2001)011<0004:CAODLT>2.0.CO;2.

Kay, S.M., and Mpodozis, C., 2002, Magmatism as a probe to the Neogene shallowing of the Nazca plate beneath the modern Chilean flat-slab: Journal of South American Earth Sciences, v. 15, p. 39–57, doi: 10.1016/S0895-9811(02)00005-6.

Kay, S.M., Mpodozis, C., Ramos, V.A., and Munizaga, F., 1991, Magma source variations for mid-late Tertiary magmatic rocks associated with a shallowing subduction zone and a thickening crust in the central Andes (28° to 33°S) Argentina, *in* Harmon, R.S., and Rapela, C.W., eds., Andean Magmatism and Its Tectonic Setting: Geological Society of America Special Paper 265, p. 113–137.

Kay, S.M., Mpodozis, C., and Coira, B., 1999, Magmatism, tectonism, and mineral deposits of the central Andes (22°–33°S), *in* Skinner, B., ed., Geology and Ore Deposits of the Central Andes: Society of Economic Geology Special Publication 7, p. 27–59.

Kirby, S., Engdahl, E.R., and Denlinger, R., 1996, Intermediate-depth intraslab earthquakes and arc volcanism as physical expressions of crustal and uppermost mantle metamorphism in subducting slabs (Overview), *in* Bebout, G.E., Scholl, D.W., Kirby, S.H., and Platt, J.P., eds., Subduction from Top to Bottom: American Geophysical Union Geophysical Monograph 96, p. 195–215.

Kraemer, P.E., Escayola, M.P., and Martino, R.D., 1995, Hipótesis sobre la evolución tectónica Neoproterozoica de las Sierras Pampeanas de Córdoba (30°40′–32°40′): Revista de la Asociación Geológica Argentina, v. 50, p. 47–59.

Langer, C.J., and Hartzell, S., 1996, Rupture distribution of the 1977 western Argentina earthquake: Physics of the Earth and Planetary Interiors, v. 94, p. 121–132, doi: 10.1016/0031-9201(95)03080-8.

Lowell, J.D., 1983, Foreland deformation, *in* Lowell, J.D., ed., Rocky Mountain Foreland Basins and Uplifts: Denver, Colorado, Rocky Mountain Association of Geologists, 1–8 p.

Martino, R., Kraemer, P., Escayola, M., Giambastiani, M., and Arnosio, M., 1995, Transecta de las Sierras de Córdoba a los 33°S: Revista de la Asociación Geológica Argentina, v. 50, p. 60–77.

Martinod, J., Funiciello, F., Faccenna, C., Labanieh, S., and Regard, V., 2005, Dynamical effects of subducting ridges: Insights from 3-D laboratory models: Geophysical Journal International, v. 163, p. 1137–1150, doi: 10.1111/j.1365-246X.2005.02797.x.

McDonough, M.R., Ramos, V.A., Isachsen, C.E., Bowring, S.A., and Vujovich, G.I., 1993, Edades preliminares de circones del basamento de la Sierra de Pie de Palo, Sierras Pampeanas Occidentales de San Juan: Sus implicancias para el supercontinente Proterozoico de Rodinia, *in* Proceedings of the 12th Congreso Geológico Argentino and 2nd Congreso de Exploración de Hidrocarburos: Mendoza, Argentina, v. 3, p. 340–342.

McGlashan, N.A., Brown, L.D., and Kay, S., 2008, Crustal thicknesses in the Central Andes from teleseismically recorded depth phase precursors: Geophysical Journal International, v. 175, p. 1013–1022, doi: 10.1111/j.1365-246X.2008.03897.x.

Mpodozis, C., and Ramos, V.A., 1989, The Andes of Chile and Argentina: Earth Science Series, v. 11, p. 59–90.

Oldow, J.S., Bally, A.W., Avé Lallemant, H.G., and Leeman, W.P., 1989, Phanerozoic evolution of the North American Cordillera: United States and Canada, *in* Bally, A.W., and Palmer, A.R., eds., An Overview: Boulder, Colorado, Geological Society of America, Geology of North America, v. A, p. 139–232.

Otamendi, J.E., Demichelis, A.H., Tibaldi, A.M., and De la Rosa, J.D., 2006, Genesis of aluminous and intermediate granulites; a case study in the eastern Sierras Pampeanas, Argentina: Lithos, v. 89, no. 1–2, p. 66–88, doi: 10.1016/j.lithos.2005.09.007.

Owens, T.J., Randall, G.E., Wu, F.T., and Zeng, R., 1993, Passcal instrument performance during the Tibetan Plateau Passive Seismic Experiment: Bulletin of the Seismological Society of America, v. 83, no. 6, p. 1959–1970.

Pankhurst, R.J., Rapela, C.W., Saavedra, J., Baldo, E., Dahlquist, J., Pascua, I., and Fanning, C.M., 1998, The Famatinian magmatic arc in the central Sierras Pampeanas, *in* Pankhurst, R.J., and Rapela, C.W., eds., The Proto-Andean Margin of South America: Geological Society of London Special Publication 142, p. 343–367.

Pankhurst, R.J., Rapela, C.W., and Fanning, C.M., 2000, Age and origin of coeval TTG, I- and S-type granites in the Famatinian Belt of the NW Argentina, *in* Barbarin, B., et al., eds., Fourth Hutton Symposium on the Origin of Granites and Related Rocks: Transactions of the Royal Society of Edinburgh, Earth Sciences, v. 91, parts 1–2, p. 151–168.

Pardo, M., Comte, D., and Monfret, T., 2002, Seismotectonic and stress distribution in the central Chile subduction zone: Journal of South American Earth Sciences, v. 15, p. 11–22, doi: 10.1016/S0895-9811(02)00003-2.

Pardo, M., Monfret, T., Vera, E., Yañéz, G., and Eisenberg, A., 2004, Flat-slab to steep subduction transition zone in central Chile–western Argentina: Body wave tomography and state of stress: Eos (Transactions, American Geophysical Union), series 85, v. 47, Fall Meeting supplement, abstract B164.

Peres-Rocha, M., 2008, Tomografia Sísmica com Ondas P e S para o Estudo do Manto Superior no Brasil [Ph.D. thesis]: São Paulo, Brazil, Universidade de São Paulo, 86 p. (in Portuguese).

Pilger, R.H., Jr., and Handschumacher, D.W., 1981, The fixed-hotspot hypothesis and origin of the Easter-Sala y Gomez-Nazca trace: Geological Society of America Bulletin, v. 92, no. 7, p. I437–I446.

Prodehl, C., and Lipman, P.W., 1989, Crustal structure of the Rocky Mountain region, *in* Pakiser, L.C., and Mooney, W.D., eds., Geophysical Framework of the Continental United States: Geological Society of America Memoir 172, p. 249–284.

Pujol, J., Chiu, J.M., Smalley, R., Jr., Regnier, M., Isacks, B.L., Chatelain, J.L., Vlasity, J., Vlasity, D., Castano, J.C., and Puebla, N., 1991, Lateral velocity variations in the Andean foreland in Argentina determined with the JHD method: Bulletin of the Seismological Society of America, v. 81, p. 2441–2457.

Quenardelle, S.M., and Ramos, V.A., 1999, Ordovician western Sierras Pampeanas magmatic belt; record of Precordillera accretion in Argentina, *in* Ramos, V.A., and Keppie, D., eds., Laurentia-Gondwana Connections before Pangea: Geological Society of America Special Paper 336, p. 63–86.

Ramos, V.A., 1988, Tectonics of the Late Proterozoic–early Paleozoic: A collisional history of southern South America: Episodes, v. 11, no. 3, p. 168–174.

Ramos, V.A., 1994, Terranes of southern Gondwanaland and their control in the Andean structure (30°–33°S latitude), *in* Reutter, K.J., Scheuber, E., and Wigger, P.J., eds., Tectonics of the Southern Central Andes: New York, Springer-Verlag, p. 249–261.

Ramos, V.A., 1999, Evolución tectónica de la Argentina: Servicio Geológico Minero Argentino, Buenos Aires, Argentina, Anales, v. 29, p. 715–785.

Ramos, V.A., 2000, The southern Central Andes, *in* Cordani, U.G., Thomaz Filho, A., and Campos, D.A., eds., Tectonic Evolution of South America: Rio de Janeiro, Brazil, In-Folo Producao Editorial, Grafica e Programacao Visual, p. 561–604.

Ramos, V.A., 2004, Cuyania, an exotic block to Gondwana: Review of a historical success and the present problems: Gondwana Research, v. 7, no. 4, p. 1009–1026, doi: 10.1016/S1342-937X(05)71081-9.

Ramos, V.A., and Kay, S.M., 1991, Triassic rifting and associated basalts in the Cuyo Basin, central Argentina, *in* Harmon, R.S., and Rapela, C.W., eds., Andean Magmatism and Its Tectonic Setting: Geological Society of America Special Paper 265, p. 79–91.

Ramos, V.A., Jordan, T.E., Allmendinger, R.W., Mpodozis, C., Kay, S.M., Cortés, J.M., and Palma, M.A., 1986, Paleozoic terranes of the central Argentine–Chilean Andes: Tectonics, v. 5, no. 6, p. 855–880, doi: 10.1029/TC005i006p00855.

Ramos, V.A., Escayola, M., Mutti, D.I., and Vujovich, G.I., 2000, Proterozoic–early Paleozoic ophiolites of the Andean basement of southern South America, *in* Dilek, Y., Moores, E.M., Elthon, D., and Nicolas, A., eds., Ophiolites and Oceanic Crust; New Insights from Field Studies and the Ocean Drilling Program: Geological Society of America Special Paper 349, p. 331–349.

Ramos, V.A., Cristallini, E.O., and Pérez, D.J., 2002, The Pampean flat-slab of the central Andes: Journal of South American Earth Sciences, v. 15, p. 59–78, doi: 10.1016/S0895-9811(02)00006-8.

Rapela, C.W., Pankhurst, R.J., Casquet, C., Baldo, E., Saavedra, J., Galindo, C., and Fanning, C.M., 1998, The Pampean orogeny of the southern proto-Andes: Cambrian continental collision in the Sierras de Córdoba, *in* Pankhurst, R.J., and Rapela, C.W., eds., The Proto-Andean Margin of Gondwana: Geological Society of London Special Publication 142, p. 181–218.

Rapela, C.W., Baldo, E.G., Pankhurst, R.J., and Saavedra, J., 2002, Cordieritite and leucogranite formation during emplacement of highly peraluminous magma: The El Pilón granite complex, Sierras Pampeanas, Argentina: Journal of Petrology, v. 43, no. 6, p. 1003–1028, doi: 10.1093/petrology/43.6.1003.

Rapela, C.W., Pankhurst, R.J., Casquet, C., Fanning, C.M., Baldo, E.G., González-Casado, J.M., Galindo, C., and Dahlquist, J., 2007, The Río de la Plata craton and the assembly of SW Gondwana: Earth-Science Reviews, v. 83, p. 49–82, doi: 10.1016/j.earscirev.2007.03.004.

Regnier, M., Chatelain, J.L., Smalley, R., Jr., Chiu, J.M., Isacks, B., and Araujo, M., 1992, Seismotectonics of Sierra Pie de Palo, a basement block uplift in the Andean foreland of Argentina: Bulletin of the Seismological Society of America, v. 82, p. 2549–2571.

Regnier, M., Chiu, J.M., Smalley, R., Jr., Isacks, B.L., and Araujo, M., 1994, Crustal thickness variation in the Andean foreland, Argentina, from converted waves: Bulletin of the Seismological Society of America, v. 84, p. 1097–1111.

Reta, M.C., 1992, High-Resolution View of the Wadati-Benioff Zone and Determination of the Moho Depth in San Juan, Argentina [M.Sc. thesis]: Memphis, Memphis State University, 98 p.

Rivera, L., and Cisternas, A., 1990, Stress tensor and fault plane solutions for a population of earthquakes: Bulletin of the Seismological Society of America, v. 80, p. 600–614.

Rossello, E.A., and Mozetic, M.E., 1999, Caracterización estructural y significado geotectónico de los depocentros cretácicos continentales del centro-oeste argentino: Boletim do 5th Simposio do Cretáceo do Brasil y 1st Simposio sobre el Cretácico de América del Sur: Sierra Negra, p. 107–113.

Salazar, P., 2005, Análisis del Campo de Esfuerzos en la Zona de Subducción Bajo Chile Central (30°–34°S) [M.Sc. thesis]: Santiago, Chile, University of Chile, 196 p. (in Spanish).

Saleeby, J., 2003, Segmentation of the Laramide slab—Evidence from the southern Sierra Nevada region: Geological Society of America Bulletin, v. 115, no. 6, p. 655–668, doi: 10.1130/0016-7606(2003)115<0655:SOTLSF>2.0.CO;2.

Schmidt, C.J., Astini, R.A., Costa, C.H., Gardini, C.E., and Kraemer, P.E., 1995, Cretaceous rifting, alluvial fan sedimentation, and Neogene inversion, southern Sierras Pampeanas, Argentina, *in* Tankard, A., Suarez, R., and Welsink, H.J., eds., Petroleum Basins of South America: American Association of Petroleum Geologists Memoir 62, p. 341–358.

Siame, L.L., and Bellier, O., 2005, Deformation partitioning in flat subduction setting: Case of the Andean foreland of western Argentina (28°S–33°S): Tectonics, v. 24, p. TC5003, doi: 10.1029/2005TC001787.

Smalley, R.F., Jr., and Introcaso, A., 2001, Estructura de la corteza y del manto superior en el antepaís andino de San Juan (Argentina): Anales de la Academia Nacional de Ciencias Exactas, Físicas y Naturales, v. 53, p. 203–215.

Smalley, R.F., Jr., and Isacks, B.L., 1987, A high-resolution local network study of the Nazca plate Wadati-Benioff zone under western Argentina: Journal of Geophysical Research, v. 92, no. B13, p. 13,903–13,912, doi: 10.1029/JB092iB13p13903.

Smalley, R.F., Jr., Pujol, J., Regnier, M., Chiu, J.M., Chatelain, J.L., Isacks, B.L., Araujo, M., and Puebla, N., 1993, Basement seismicity beneath the Andean Precordillera thin-skinned thrust belt and implications for crustal and lithospheric behavior: Tectonics, v. 12, p. 63–76, doi: 10.1029/92TC01108.

Stern, C.R., 2004, Active Andean volcanism: Its geologic and tectonic setting: Revista Geológica de Chile, v. 31, no. 2, p. 161–206.

Stuessy, T.F., Foland, K.A., Sutter, J.F., Sanders, R.W., and Silva, M., 1984, Botanical and geological significance of potassium-argon dates from the Juan Fernández Islands: Science, v. 225, no. 4657, p. 49–51.

Thomas, W.A., and Astini, R.A., 2003, Ordovician accretion of the Argentine Precordillera terrane to Gondwana: A review: Journal of South American Earth Sciences, v. 16, p. 67–79, doi: 10.1016/S0895-9811(03)00019-1.

Trindade, R.I.F., D'Agrella-Filho, M.S., Epof, I., and Brito Neves, B.B., 2006, Paleomagnetism of Early Cambrian Itabaiana mafia dikes (NE Brazil) and the final assembly of Gondwana: Earth and Planetary Science Letters, v. 244, p. 361–377, doi: 10.1016/j.epsl.2005.12.039.

Varela, R., Roverano, D., and Sato, A.M., 2000, Granito El Peñón, Sierra de Umango: Descripción, edad Rb/Sr e implicancias geotectónicas: Revista de la Asociación Geológica Argentina, v. 55, no. 4, p. 407–413.

Vaughan, A.P.M., and Pankhurst, R.J., 2008, Tectonic overview of the west Gondwana margin, *in* Casquet, C., Pankhurst, R.J., and Vaughan, P.M., eds., The Western Gondwana Margin: Proterozoic to Mesozoic: Gondwana Research, v. 13, no. 2, p. 150–162.

Verrall, P., 1989, Speculations on the Mesozoic-Cenozoic tectonic history of the western United States, *in* Tankard, A.J., and Balkwill, H.R., eds., Extensional Tectonics and Stratigraphy of the North Atlantic Margins: American Association of Petroleum Geologists Memoir 46, p. 615–631.

Vietor, T., and Echtler, H., 2006, Episodic Neogene southward growth of the Andean subduction orogen between 30°S and 40°S—Plate motions, mantle flow, climate, and upper-plate structure, *in* Oncken, O., et al., eds., The Andes—Active Subduction Orogeny: Berlin, Springer, p. 375–400.

Vigny, C., Rudloff, A., Ruegg, J.C., Madariaga, R., Campos, J., and Alvarez, M., 2009, Upper plate deformation measured by GPS in the Coquimbo gap, Chile: Physics of the Earth and Planetary Interiors.

von Huene, R., 2008, When seamounts subduct: Science, v. 321, p. 1165–1166, doi: 10.1126/science.1162868.

von Huene, R., Corvalán, J., Flueh, E.R., Hinz, K., Korstgard, J., Ranero, C.R., Weinrebe, W., and C.O.N.D.O.R. Scientists, 1997, Tectonic control of the subducting Juan Fernández Ridge on the Andean margin near Valparaiso, Chile: Tectonics, v. 16, no. 3, p. 474–488, doi: 10.1029/96TC03703.

Vujovich, G., and Kay, S.M., 1998, A Laurentian? Grenville-age oceanic arc/back-arc terrane in the Sierra de Pie de Palo, western Sierras Pampeanas, Argentina, *in* Pankhurst, R.J., and Rapela, C.W., eds., The Proto-Andean Margin of Gondwana: Geological Society of London Special Publication 142, p. 159–180.

Vujovich, G.I., van Staal, C.R., and Davis, W., 2004, Age constraints on the tectonic evolution and provenance of the Pie de Palo Complex, Cuyania composite terrane, and the Famatinian orogeny in the Sierra de Pie de Palo, San Juan, Argentina, *in* Vujovich, G.I., Fernandes, L.A.D., and Ramos, V.A., eds., Cuyania, an Exotic Block to Gondwana: Gondwana Research, v. 7, no. 4, p. 1041–1056.

Wagner, L.S., Beck, S., and Zandt, G., 2005, Upper mantle structure in the south central Chilean subduction zone (30 degrees to 36 degrees S): Journal of Geophysical Research, v. 110, doi: 10.1029/2004JB003238.

Wells, M.L., and Hoisch, T.D., 2008, The role of mantle delamination in widespread Late Cretaceous extension and magmatism in the Cordillera orogen, western United States: Geological Society of America Bulletin, v. 120, p. 515–530, doi: 10.1130/B26006.1.

Wessel, P., and Smith, W.H.F., 1998, New, improved version of the Generic Mapping Tool released: Eos (Transactions, American Geophysical Union), v. 79, p. 579, doi: 10.1029/98EO00426.

Yañéz, G.A., Ranero, C.R., von Huene, R., and Díaz, J., 2001, Magnetic anomaly interpretation across the southern central Andes (32°–34°S): The role of the Juan Fernández Ridge in the late Tertiary evolution of the margin: Journal of Geophysical Research, v. 106, no. B4, p. 6325–6345, doi: 10.1029/2000JB900337.

Yuan, X., Sobolev, S., and Kind, R., 2002, Moho topography in the central Andes and its geodynamic implications: Earth and Planetary Science Letters, v. 199, p. 389–402, doi: 10.1016/S0012-821X(02)00589-7.

Zapata, T.R., and Allmendinger, R.W., 1996, Thrust front zone of the Precordillera, Argentina: A thick-skinned triangle zone: American Association of Petroleum Geologists Bulletin, v. 80, p. 359–381.

Manuscript Accepted by the Society 5 December 2008

Printed in the USA